Geospatial Information to Support the UN's Sustainable Development Goals

This unique book investigates the 2030 Agenda for Sustainable Development through geospatial information lenses, making the case that geospatial information is central to the SDGs and is imperative to understand the place, location, communities, and the people to identify where the root of the problem lies. Combined with other layers of information, it supports advanced analysis and visualizations and explains how SDG indicators can be interpreted and measured either through direct use of geospatial data itself or through integration with other statistical data. International case studies and coverage of all the SDGs help users evaluate impacts and monitor progress to achieve the SDGs.

Features

- Demonstrates how geospatial information can be used to assess SDG targets and indicators.
- Provides 16 applications and real-world case studies of SDGs and their related indicators where geospatial information contributes significantly to the implementation and monitoring.
- Addresses practical and technological issues related to geospatial information, data acquisition, and integration approaches.
- Illustrates the integration of geospatial information with statistical and other development information
- Discusses the impact that applications have on policy and society to make informed decisions and implement better policies.

This book is for professionals, researchers, senior undergraduates and graduate students, working on or studying projects focused on specific GIS applications in support of sustainable development. It is also an excellent resource for policymakers and NGOs involved with UN SDGs.

GIS in Action
Series Editor

GIS and Housing: Principles and Practices
Laxmi Ramasubramanian, Jochen Albrecht, and
Deborah Rojas De Leon

Geospatial Information to Support the UN's Sustainable Development Goals
Edited by Etta D. Jackson and Amor Laaribi

For more information about the series, please visit: https://www.routledge.com/GIS-in-Action/book-series/CRCGISINACTI

Geospatial Information to Support the UN's Sustainable Development Goals

Edited by
Etta D. Jackson and Amor Laaribi

CRC Press
Taylor & Francis Group
Boca Raton London New York

CRC Press is an imprint of the
Taylor & Francis Group, an **informa** business

Designed cover image: © Shutterstock

First edition published 2026
by CRC Press
2385 NW Executive Center Drive, Suite 320, Boca Raton FL 33431

and by CRC Press
4 Park Square, Milton Park, Abingdon, Oxon, OX14 4RN

CRC Press is an imprint of Taylor & Francis Group, LLC

ISBN: 9781032957449 (hbk)
ISBN: 9781032958712 (pbk)
ISBN: 9781003586319 (ebk)

DOI: 10.1201/9781003586319

For Product Safety Concerns and Information please contact our EU representative: GPSR@taylorandfrancis.com
Taylor & Francis Verlag GmbH,
Kaufingerstraße 24,
80331 München, Germany.

Contents

Preface

I have worked for over 16 years at the UN Statistics Division in charge of geospatial information, and was the coordinator of the inter-governmental United Nations Committee of Experts on Global Geospatial Information Management (UN-GGIM) since its inception, and promoted the use of geospatial information and its integration with statistical information in support of sustainable development. I have also delivered many consultancies on the use of geospatial information and the integration of statistical and geospatial data in support of sustainable development. These experiences allowed me to acquire rare insights into the various challenges countries are experiencing in implementing the SDG indicators and to monitor and achieve the SDGs.

The "data-gap" challenge, either in terms of data collection, availability, access and use, or data quality has been recognized of paramount importance by the 2030 Agenda for Sustainable Development which recommends that Member States undertake new data acquisition and integration approaches to supply reliable, timely, accessible and usable, and disaggregated data that is essential to make informed, evidence-based decisions. In this regard, beside the traditional data sources (surveys, censuses, administrative data), new sources of data such as geospatial data, including earth observation, have emerged as crucial to cope with the increasing complexity of national and global development challenges and issues, and to understand the interrelationships across the economic, social and environmental dimensions (the three pillars of sustainable development), and implement better policies for sustainable development, to ultimately "leave no one behind"[1].

The geographic location ("Where") dimension is being recognized as central to the Sustainable Development Goals (SDGs) as geospatial data is absolutely imperative to understand the place, the location, the communities we are working with, the people we are serving and the spatial context about that place, allowing us to interpret these data more comprehensively and identify where the root of the problem lies. Moreover, geospatial technology is being increasingly recognized as an enabler for facilitating data accessibility and sharing, and improving the quality, availability and timeliness of a huge volume of information, leading geospatial information to emerge as an indexing reference, linking the three pillars, social, economic and environmental, of sustainable development.

For example, to address the growing demand for information on small geographic areas in order to monitor the development goals and indicators at local and community levels, the 2030 Agenda "calls out geospatial information and earth observations as key methods for tracking progress and informing people about these global development policies"[2]. In addition, to improve the quality of census data, the United Nations recommended the need for countries to keep

[1] See "Transforming our world: the 2030 Agenda for Sustainable Development", at https://sustainabledevelopment.un.org/post2015/transformingourworld.

[2] See more details in the book: "GIS and the 2020 Census – Modernizing Official Statistics", Esri Press, 2019. Available as an e-book and in print.

abreast of the technological advances made since the previous round of censuses, especially in the area of Geographic Information Systems (GIS) and Global Navigation Satellite Systems (GNSS)/Global Positioning System (GPS), and that the adoption of GIS for the future Round of Censuses should be a major strategic decision[3]. By combining diverse layers of information and supporting advanced analysis and visualizations through customized maps or 3D models, GIS tools have become a significant way to unlock new insights and evaluate impacts that would never have been possible by looking at socio-economic or geospatial data in isolation. In this regard, policymakers and managers of SDGs are in need of guidelines providing them help to choose, appropriate with their national conditions and needs, the best advanced geospatial information technology that can contribute to implement the SDG indicators and monitor the progress of the SDGs.

While this book is centered on the use of GIS and other related geospatial technology in support of the SDGs, it offers a set of illustrative case studies to show how geospatial information contributes directly or significantly to calculate the SDG indicators or targets. What distinguishes this reference book, one of the rare books in the field of geospatial information in support of SDGs, is the fact that it offers concrete examples of using geospatial information to measure and visualize the SDG indicators through GIS maps, and reconciles the challenging specific needs of the countries to improve the quality of their data with the sate-of-the-art geospatial technologies and related methodologies, advocating for the strategic use of geospatial information and other innovative technologies in support of SDGs.

Amor LAARIBI, PhD.
Former Coordinator of the UN Committee of Experts on Global Geospatial Information Management (UN-GGIM); former Substantive Secretary of the UN Experts Group on the Integration of Statistical and Geospatial Information and the UN Group of Experts on Geographical Names (UNGEGN).

[3] See UN Principles and Recommendations – Revision 3, 2017, United Nations Publications, New York. Available at: https://unstats.un.org/unsd/publication/seriesM/Series_M67Rev3en.pdf.

Editor Bios

Etta D. Jackson has worked in the field of education for 25 years as a teacher, guidance counselor, and district administrator in both New York and Wisconsin. Her passion for wanting to make a difference led her to the founding of The Institute for Conscious Global Change (ICGC), an international non-profit NGO organization in Special Consultative Status with the Economic and Social Council (ECOSOC) of the United Nations. She now serves as the CEO of the Institute. This New York-based organization was founded in 2007 to provide visual but tangible development solutions to support the unfinished mandate of the United Nations Millennium Development Goals (MDGs) and now SDGs of the 2030 Agenda for Sustainable Development. With the aid of Geographic Information Systems (GIS), Earth Observation and GeoDesign technologies, ICGC engages governments and citizens to develop integrated, holistic and comprehensive development models for implementation in each country. Dr. Jackson is also the author of four books and many articles.

Amor Laaribi has over 30 years of extensive technical, managerial, and policy experience in the development, implementation and evaluation of geospatial information strategies, programs and projects in support of decision-making and sustainable development. He joined the United Nations Statistics Division in 2000 to define and plan the priorities of the Division in geospatial information and to promote its use and integration with census/statistical information in developing countries. Dr. Laaribi has written/co-written two other books with reputable publishers.

Contributors

Nashon J. Adero
Taita Taveta University
Kenya

Salsebil Bel Hadj Ali
Research Unit Sustainable Management of Water and Soil Resources
Higher School of Engineers of Medjez El Bab (ESIM)
University of Jendouba Tunisia

Saudi Aramco
Dharan, Eastern Province, Saudi Arabia

Natalia da Silveira Arruda
Arizona State University

Dara Carney-Nedelman
Appalachian State University

Rosario Casanova
Universidad la República
Uruguay

Jeanne-Marie Col
City University of New York (CUNY)
and
Institute for Conscious Global Change

Williams S. Ebhota
Institute for Systems Science
Durban University of Technology
Durban, South Africa

Phoenix Eyre
Terran Development Operations
Los Angeles, California

Hishmi Jamil Husain
Institute of Artificial Intelligence
Shaoxing University
Shaoxing, Zhejiang province, China
and
Lead Strategy Market Analyst

Etta D. Jackson
Founder and CEO
The Institute for Conscious Global Change (ICGC)

Lotfi Kaabi
ESG & SDG Expert
Metapolicy
Toulouse, France

Lamia Khemiri
Regional Center for Remote Sensing of North Africa States (CRTEAN)
Tunis, Tunisia

Nasim Khonsari
Westcliff University
College of Business
Westcliff University
Irvine, CA, USA

Josphat Makanga
Regional Centre for Mapping of Resources for Development (RCMRD)

Mustapha Mimouni
Sahara and Sahel Observatory
Remote Sensing/GIS
Tunis, Tunisia

Mirriam Omala-Gauvin
Platform for African Women in Smart Climate Agriculture (PAWISCA)

Shirley Orr
Association of Public Health Nurses
Executive Director

Saied Pirasteh
Institute of Artificial Intelligence
Shaoxing University
Shaoxing, Zhejiang province, China

Tammineni Rajitha
Institute of Rural Management Anand
India

Lincoln T. Satkunarajah
Engineering and Technology Expert, Inventor and Investor
Founder and CEO Linkay Technologies Inc.
Co-Founder Linkay Ventures Inc.
New York, USA

Mahdieh Shirmohammadi
Institute of Artificial Intelligence
Shaoxing University
Shaoxing, Zhejiang province, China

Patricia Solís
Arizona State University

Pavel Y. Tabakov
Institute for Systems Science
Durban University of Technology
Durban, South Africa

Fatma Trabelsi
Research Unit Sustainable Management of Water and Soil Resources
Higher School of Engineers of Medjez El Bab (ESIM)
University of Jendouba Tunisia

Cecilia A. Wandiga
Executive Director
Centre for Science & Technology Innovations (CSTI)

Shem O. Wandiga
Managing Trustee
Centre for Science & Technology Innovations (CSTI)

Luis Evence Zoungrana
Sahara and Sahel Observatory
Remote Sensing/GIS
Tunis, Tunisia

Introduction

In 2015, the international community adopted the 2030 Agenda for Sustainable Development as a universal approach that provides a new framework for development policies for the next 15 years. Unlike the Millennium Development Goals (MDGs) (2000–2015), which were limited in scope (8 goals and 21 targets) and coverage (dedicated to developing countries) and defined by a small UN team; the SDGs (20016- 2030) was the result of a global consultation, covering all social, economic and environmental challenges, 17 goals their 169 targets and 234 Indicators, and involves all the stakeholders including all 193 Member States, Civil Society, Business, Academia and Philanthropic Organizations. The SDGs were built on the progress made under the MDGs and considered the lessons learned from their implementation, including the shortages – chief among them the "data gap". The SDGs Indicator Framework encompasses some 247 indicators, where a set of indicators for each target has been established to track progress toward the goals.

Monitoring needs to be carried out for each indicator to measure progress until 2030, when the final evaluation will inform about the level of success of the SDGs implementation. Most of the indicators include elements of geography, place, and location. It has been estimated that approximately 20% of the SDG indicators can be interpreted and measured either through direct use of geospatial data itself or through integration with other statistical data. The geospatial community knows well that GIS is a data-driven system and data constitutes its most costly component GIS-based proof of concepts would be helpful to showcase to the policy makers the support geospatial information can bring across all sectors.

This book examines the 2030 Agenda through the geospatial information lenses, making the case that geospatial information is fundamental to supporting sustainable development, highlighting the opportunities to raise awareness of their importance, and particularly providing academic authors and leading professionals the opportunity to combining their research, consultancy and relevant project work focusing on specific GIS applications in support of sustainable development. It is composed of a maximum of thirteen (13) chapters, each corresponding to a goal or a set of goals, and grouped in four parts, namely:

1. Eradication of Extreme Poverty, Food Security, Healthy Lives
2. Innovations, Safe Communities and Life of Dignity
3. Climate, Food Consumption, and Environmental Sustainability
4. Strong Institutions and Sustainable Development

Providing insights on the indicators and written by leading researchers and practitioners in their field. Each chapter is approximately 20–30 pages maximum, including color illustrations. The chapters are structured to focus on the pertinent theoretical concepts from Geoscience and the application discipline(s), address the relevant practical and technological issues, provide case studies and discuss the impact that applications can have on policy and society.

The 2030 Agenda for Sustainable Development recommends that countries undertake new data acquisition and integration approaches to supply reliable, timely, accessible, usable, and disaggregated data essential to set priorities, making informed decisions and implement better policies.

Geospatial information is being recognized as central to the SDGs and is absolutely imperative to understand the place, the location, the communities we are working with, the people that we are serving, allowing us to identify where the root of the problem lies.

Combined with other layers of information, it supports advanced analysis and visualizations through customized maps or 3D models, helping users in evaluating impacts, monitoring the progress to achieve the SDGs. This book is unique for examining the role of geospatial information in supporting sustainable development, including how the SDG indicators can be interpreted and measured either through direct use of geospatial data itself or through integration with other statistical data. This book provides illustrative examples of SDGs and their related indicators where geospatial information contributes significantly to the implementation and monitoring of the 2030 Agenda for Sustainable Development.

The lead editors and chapter authors would like to acknowledge that Chapters 1, 4, 6, 8, and 9 used Chapter GPT 4 to assist with language improvement, refinement, grammar, and coherence. Chapter 2 utilized Chat GPT 3.5 for the same purposes, and Chapter 5 used Chat GPT 3.0.

Part 1

Eradication of Extreme Poverty, Food Security, Healthy Lives

1 SDG #1—Leveraging Geographic Information Systems (GIS) to Address Multidimensional Poverty

Phoenix Eyre and Lotfi Kaabi

INTRODUCTION

Key Considerations for Governments, Operational Communities, and NGOs

In translating Geographic Information Systems (GIS) insights into effective SDG#1 interventions, governments, non-goverment organizations (NGOs), and local operators must navigate data, technology, and stakeholder dynamics. The following considerations outline the guiding principles for collecting and integrating spatial data, diagnosing poverty hotspots, optimizing resource delivery, and building resilient, community-owned systems. By attending to these foundational elements, practitioners can ensure that GIS-driven programs are inclusive, scalable, and sustainable.

Framing Note on SDG 17 and Funders' Role

While governments are the primary actors responsible for implementing SDG 1, effective outcomes depend on coordinated partnerships. Funders—whether public, philanthropic, or corporate—play a critical enabling role across all tiers of action. Their support is most impactful when aligned with government-led strategies, citizen needs, and inclusive stakeholder frameworks, as emphasized by SDG 17. The following considerations, shown in Tables 1.1 to 1.18, reflect this multi-actor approach, showing how public leadership, stakeholder expertise, and community ownership can jointly deliver sustainable results.

DOI: 10.1201/9781003586319-2

TABLE 1.1
Data Collection and Integration

General Best Practices	• Develop comprehensive, real-time databases that integrate demographic, economic, environmental, and infrastructure data. • Ensure inclusivity by capturing data from marginalized and underserved communities. • Utilize platforms such as Esri (Esri, 2024), Google Earth Engine (Gorelick et al., 2017), and OpenStreetMap for mapping and data collection (OpenStreetMap Contributors, 2024).		
Scale	**Government**	**NGO/IGO/Private Sector**	**Community-Level/Personal**
Practical Tools	Use enterprise GIS platforms like Esri ArcGIS Enterprise to manage centralized geodatabases. Leverage cloud-based geospatial platforms like Google Earth Engine and Amazon Web Services (AWS) Earth Data for large-scale imagery and AI analytics. Integrate national spatial data infrastructures (NSDIs) and public mapping portals from sources like data.gov for open government data.	Adopt cost-effective GIS tools: Quantum GIS (QGIS) or Geographic Resources Analysis Support System (GRASS GIS) for desktop analysis, ArcGIS Online or Mapbox for cloud maps, and mobile data apps (KoBoToolbox, Open Data Kit) for field surveys. Use free satellite data services (Sentinel Hub, NASA Earthdata) and open-source AI tools like TensorFlow for analytics. Employ participatory platforms like Survey123 or OpenStreetMap (OSM) for crowdsourcing local data.	Use free participatory mapping apps and open data: OpenStreetMap editors (iD, JOSM), mobile apps like MapSwipe or Ushahidi for local mapping. Employ low-cost GPS units or smartphone GPS and engage in "mapping parties." Utilize low-tech sensors like DIY air-quality kits and paper maps digitized by community volunteers. Provide simple open-source GIS (QGIS Desktop) or smartphone GIS viewers for community members to visualize data.
Recommended Analysis and Policy	Integrate AI-powered analytics tools like IBM Watson, Microsoft AI for Earth, TensorFlow for pattern recognition and predictive modeling (Abadi et al., 2015).	Promote open data-sharing initiatives while respecting privacy and ethical considerations. Streamline public participation by creating open-source GIS tools that allow individuals to contribute real-time data.	
Financing Pathways	Fund GIS infrastructure through national budgets and development bank loans from World Bank, Asian Development Bank (ADB), and issue geospatial or "smart city" bonds.	Leverage donor grants such as United States Agency for International Development (USAID), United Nations Development Program (UNDP), and EU, foundations that fund spatial projects.	Rely on micro-grants or community loan programs such as GIS loan program offering up to ~$25K to rural communities.

(Continued)

TABLE 1.1 *(Continued)*
Data Collection and Integration

Scale	Government	NGO/IGO/Private Sector	Community-Level/Personal
	Explore Public Private Partnerships (PPPs) for data networks from telecom companies contributing remote sensing data. Tap international SDG/ climate funds or trust funds for data innovation such as World Bank's Data Innovation Fund.	Participate in multi-donor or UN Trust Funds like the World Bank Data Innovation Fund, which funded 38 projects with \$25K–\$250K each. Seek private philanthropy and corporate social responsibility (CSR) partnerships, and tech firms who often sponsor open-mapping initiatives. Consider cooperative models or consortium funding among NGOs for shared GIS services.	Encourage community crowdfunding or local cooperative funds for mapping projects. Partner with local NGOs or cooperatives to access small grants. Leverage volunteer networks such as HOT/ OpenStreetMap communities which operate at low cost and use in-kind contributions like volunteer time and donated hardware.

TABLE 1.2
Mapping Poverty Hotspots and Vulnerability Assessments

General Best Practices	• Use GIS to identify areas with the highest poverty and vulnerability. Integrate multiple data layers (health, education, infrastructure) to pinpoint overlapping deprivations. • Prioritize basic needs first, as in *Maslow's Hierarchy of Needs,* by mapping access to food, water, and shelter before higher-order services. • Incorporate high-frequency data sources (Gorelick et al., 2017). Combine daily or weekly satellite imagery using Planet Labs and Sentinel and community-generated data crowdsourced from OSM edits and participatory surveys to update hotspot maps regularly and capture sudden shocks. This real-time approach fills micro-level data gaps and empowers communities to geo-locate local assets and risks. • Validate and ground-truth hotspots. Hold community workshops and mobile-PGIS surveys with, for example, Ushahidi or MapSwipe so residents can mark vulnerable locations. Local knowledge confirms or refines modeled hotspots, ensuring maps reflect on-the-ground realities (Alvarez et al., 2021).		
Scale	**Government**	**NGO/IGO/Private Sector**	**Community-Level/Personal**
Practical Tools	Use advanced spatial analysis tools (ArcGIS Spatial Analyst, Enterprise GeoAnalytics) to overlay poverty, health, and education layers.	Utilize open-source spatial statistics tools like Hot Spot Analysis in QGIS or R to identify clusters of poverty.	Conduct community-led mapping workshops with analog and digital tools.

(Continued)

TABLE 1.2 *(Continued)*
Mapping Poverty Hotspots and Vulnerability Assessments

Scale	Government	NGO/IGO/Private Sector	Community-Level/Personal
	Incorporate real-time satellite imagery with Planet Labs daily imagery, ESA Sentinel data and remote sensing night lights and NDVI for socio-economic indicators. Apply AI-driven geospatial services like IBM PAIRS and Microsoft Planetary Computer for change detection.	Employ lightweight GIS apps such as ESRI Survey123 combined with Dashboard for visualizing hotspots. Use cloud platforms like Google Earth Engine with open poverty datasets such as WorldPop for dynamic mapping. Engage citizen-science platforms like Maptionnaire for local vulnerability inputs.	Use Participatory GIS(PGIS) apps like Ushahidi or Harava, where community members mark vulnerable locations. Deploy low-cost drones or balloon mapping for local-area imagery. Collect local data such as water points, clinics, etc., with smartphone apps or GPS trackers, then combine them in OSM.
Recommended Analysis and Policy	Overlay multi-dimensional poverty indicators such as access to healthcare, education, and clean water. Use AI-driven satellite imagery from Planet Labs and Maxar Technologies to monitor changes in socio-economic conditions. Embed hotspot analysis into planning. For example, use mapped clusters to target mobile clinics, food aid, or cash transfers to areas with overlapping deprivations. Integrate hotspot maps into national development plans and budgets, focusing resources on identified priority zones. Tie infrastructure investments such as roads and schools to these maps so the poorest, most isolated areas receive service upgrades first.	Collaborate on shared hotspot datasets. Pool data and tools so all agencies work from the same maps, avoiding duplication. Use AI-driven satellite products like nightlights and high-res images to monitor socioeconomic trends in hotspots. Project proposals and donor reports should cite hotspot maps to justify targeting like a village falling into a poverty cluster.	Engage directly with the process. Local groups should lead map-validation events, annotate maps with forgotten assets or risks, and use validated maps in advocacy. Participatory mapping empowers residents to share their knowledge and influence planning decisions. Community organizations can lobby officials using hotspot maps as evidence of need, ensuring interventions align with local priorities.
Financing Pathways	Allocate a public budget for national poverty mapping initiatives. Leverage PPPs, such as energy companies who are funding data collection in service areas.	Leverage public-private partnerships to co-finance mapping projects with energy and telecommunications firms operating in impacted areas.	Community groups can tap small NGO micro-grants for participatory mapping.

(Continued)

TABLE 1.2 ***(Continued)***

Mapping Poverty Hotspots and Vulnerability Assessments

Scale	Government	NGO/IGO/Private Sector	Community-Level/Personal
	Tap bilateral/multilateral aid from the World Bank, UN-Habitat, for example, targeted at poverty and vulnerability mapping. Issue "development bonds" or green bonds that include GIS components. Use sustainability funds such as Green Climate Fund to support climate-related vulnerability maps.	Secure grants from development agencies such as USAID and EU's Global Challenges Fund for hotspot mapping. Partner with global health and development funds like the Bill & Melinda Gates Foundation and Global Fund that integrate spatial targeting. Utilize corporate partnerships: Telecom or energy companies may co-fund mapping in areas they serve. Explore impact-investment or blended finance for resilience projects that include GIS mapping.	Use crowdfunding for social-impact mapping projects. Engage local businesses or diaspora for sponsorship. Local cooperatives such as farmer co-ops that can pool funds to map agricultural vulnerability. In some cases, Community Development Finance Institutions (CDFIs) or microfinance institutions may offer loans for resilience projects, as illustrated by GIS loan funds in rural areas.

TABLE 1.3

Resource Allocation and Service Delivery Optimization

General Best Practices	• Use spatial analysis to optimize where and how to deliver resources (food aid, healthcare, education). For example, employ network analysis to find the shortest routes and reduce delivery time. • Leverage advanced GIS tools and AI. Implement enterprise GIS (ArcGIS Network Analyst, GeoAI tools) to model supply chains and forecast demand across regions. Continuously update plans with real-time data (e.g. mobile phone usage, utility usage) to adapt to shifting needs. • **Engage local input.** Integrate community feedback into allocation models. Use open-source routing engines like GraphHopper, Open-Source Routing Machine (OSRM) in QGIS) for transparent planning, and collect field data via mobile apps like Survey123 or QField to monitor distributions. Encourage participatory mapping of local services by villagers mapping health centers and markets, etc., so that models reflect on-the-ground conditions.

(Continued)

TABLE 1.3 *(Continued)*
Resource Allocation and Service Delivery Optimization

Scale	Government	NGO/IGO/ Private Sector	Community-Level/Personal
Practical Tools	Implement enterprise GIS with network analysis like Esri Network Analyst, Bentley OpenCities to optimize routes for aid delivery and utilities. Integrate GIS with national logistics systems (GPS tracking of trucks, Geo-fencing). Use big data (mobile Call Detail Records (CDRs), utility usage in spatial platforms to forecast demand. Deploy AI-based planning (e.g. ArcGIS GeoAI, Palantir) to model resource needs across regions.	Apply open-source routing engines (GraphHopper, OSRM) and cost-distance tools in QGIS for planning. Use free dashboard solutions like Google Data Studio and Tableau Public to link to mobile-collected data for real-time tracking. Utilize field GIS on tablets (ArcGIS Survey123, QField) for monitoring distributions. Integrate with platforms like District Health Information Software (DHIS2) for health service allocation.	Encourage community mapping of local services such as health centers, markets using simple tools like OSM, paper-to-digital workflows, etc. Use mobile apps (CommCare, ODK) for villagers to report service gaps. Organize participatory planning sessions using printed maps to plan optimal village resource distribution. Leverage local knowledge like fishermen's networks marking fish stock locations integrated via community GIS.
Analysis and Policy Recommendations	Implement AI-based decision-making tools like SAS Analytics to forecast resource demands. Institutionalize GIS-driven logistics. Require that infrastructure and aid plans use spatial analysis in budgeting—for example, mandate GIS route optimization for national health or food programs.	Develop mobile GIS dashboards like Tableau or ArcGIS Dashboards to provide real-time data for logistics and service delivery. Work with governments to align GIS insights with national development strategies to ensure long-term sustainability.	Participate in planning. Local groups should map service gaps such as broken pumps, no schools, etc., using simple tools like paper maps and mobile apps to inform models. Volunteer networks can update GIS data through Missing Maps events to ensure accuracy. Communities may form cooperatives to co-finance and co-manage small infrastructure issues like boreholes and use collective GIS mapping to decide locations.

(Continued)

TABLE 1.3 *(Continued)*
Resource Allocation and Service Delivery Optimization

Scale	Government	NGO/IGO/ Private Sector	Community-Level/Personal
	Issue infrastructure bonds to fund smart delivery systems, demonstrating data-driven planning to investors. Align GIS insights with national development strategies to ensure sustainability.	Adopt open tools to coordinate service delivery. Use open-source GIS and dashboard platforms such as QGIS, DHIS2, and Tableau Public to visualize resource gaps and track distributions. Share geodatabases with partners so NGOs can cross-check coverage such as one NGO's water points vs. another's clinics. Apply AI analytics like Palantir Foundry and SAS to collect data to forecast needs in remote areas and develop mobile dashboards for real-time monitoring by field teams.	
Financing Pathways	Use public infrastructure budgets and targeted grants for service mapping (e.g. water and health facility geodata). Leverage PPPs with logistics and telecom firms, as suggested in 1.2, to co-fund delivery optimization. Issue infrastructure bonds for smart logistics, given that many governments use municipal bonds for utilities. Access multilateral project loans (World Bank, ADB) with components for GIS-driven service delivery.	Fund through program grants such as WFP and UNICEF logistic grants. Partner in consortium proposals to donors like USAID Logistics Cluster and EU civil protection funds. Explore CSR partnerships with private logistics companies for co-financed delivery projects. Pursue social impact bonds or performance-based contracts like pay-for-success for efficiency improvements.	Rely on communal savings or rotating funds to pay for simple mapping exercises. Seek small grants from NGOs or local governments for community resource mapping. Form cooperatives to co-finance infrastructure as suggested in 1.6 and include mapping as a shared asset. At times, community asset-mapping can be incorporated into larger NGO projects as an in-kind contribution.

TABLE 1.4
Monitoring and Evaluation of Development Programs

General Best Practices	• Utilize remote sensing and GIS for impact assessment. For example, use night-time lights to proxy electrification, or landcover change to measure reforestation. Establish geospatial Monitoring and Evaluation (M&E) indicators (satellite vegetation indices, change in infrastructure) to objectively track outcomes. • Deploy geospatial dashboards. Maintain up-to-date GIS databases of project sites and outputs. Use platforms such as ArcGIS Dashboards, and Power BI to integrate spatial layers to show facilities and assets built so stakeholders can instantly see progress. Automate ingestion of new survey rounds and imagery so deprivation layers refresh biannually. • **Engage beneficiaries.** Collect geo-referenced feedback from communities. Implement participatory M&E methods to produce SMS reports and community scorecards with GPS tags so residents report on project status, for example, "3 pumps out of 5 wells are operational." Publicize results on open portals like Ushahidi maps to close the feedback loop and build trust.		
Scale	**Government**	**NGO/IGO/ Private Sector**	**Community-Level/Personal**
Practical Tools	Deploy national M&E dashboards such as Power BI, ArcGIS dashboards that integrate GIS layers like program sites and outcome indicators. Use remote sensing proxies that show changes in night-time lights for electrification as evaluation metrics. Implement geospatial databases that track project inputs and outputs geographically. Employ satellite monitoring with Sentinel land cover to measure environmental outcomes of programs.	Use mobile data-collection apps for real-time monitoring (Akvo FLOW, Survey123). Implement open-source M&E platforms with GIS DevResults or Umoja with GIS modules. Empower beneficiaries with reporting tools like Ushahidi SMS platforms for feedback. Provide data portals where communities and donors can view project maps.	Encourage citizen monitoring: SMS or app-based reporting of project progress like community water point status via FrontlineSMS. Use simple participatory M&E methods (community scorecards) geo-tagged on maps. Provide public portals (like Ushahidi maps) where community outcomes are posted. Leverage open data from government projects for grassroots analysis.
Analysis and Policy Recommendations	Implement dynamic dashboards (Microsoft Power BI, Google Data Studio, Tableau) to visualize progress.	Train AI agents to collect data to provide real-time project design feedback, ensuring interventions remain adaptive.	Empower citizen monitoring. Train local volunteers to input geotagged data using, for example, Akvo FLOW or FrontlineSMS, on project implementation.

(Continued)

TABLE 1.4 *(Continued)*
Monitoring and Evaluation of Development Programs

Scale	Government	NGO/IGO/ Private Sector	Community-Level/Personal
	Integrate GIS into national M&E frameworks. Tie funding disbursements to spatial targets like "X" new clinics geocoded and use satellite checks for accountability. Establish independent audit protocols that use geospatial evidence like third-party satellite imagery reviews to verify reported results. Support open data portals where project maps and progress are published for public scrutiny.	Establish an independent monitoring framework to ensure transparency and accountability among stakeholders. Build real-time dashboards for stakeholders. Use tools like Google Data Studio or Tableau to share updated maps with donors and communities. Train local M&E staff in data analysis so they can adapt programs on the fly, for instance, redirect aid if an area's situation worsens. Collaborate with research institutions for spatial evaluations of programs.	Communities can map realized versus planned activities and publish scorecards on public dashboards. Participatory budgeting can allocate small grants for community-led audits, making residents co-responsible for GIS-based project tracking.
Financing Pathways	Budget for M&E tools within project funding. Tie government grant/ funding disbursements to geospatial indicators to encourage data collection. Leverage results-based financing: Donors provide funds only if GIS-monitored targets are met. Consider allocating part of project costs to independent geospatial audit firms (some projects are integrating spatial monitoring in procurement).	Seek dedicated M&E funding from donors such as CDC and DFAT, who have budgets for impact tracking. Use impact bonds or matching funds that require transparent GIS-based reporting. Engage third-party evaluators funded by oversight agencies who use GIS in their evaluations. Partner with research institutions who often have their own funding to monitor projects.	Allocate small budgets for community audit teams in project design that are often built into participatory development grants. Use social accountability funds to grant that empower communities to monitor projects. Encourage participatory budgeting so communities vote on allocating funds for local monitoring, which itself is a financing mechanism for Monitoring and Evaluation (M&E).

TABLE 1.5
Climate Resilience and Sustainable Land Management

General Best Practices	• Map environmental risks. Identify and prioritize areas prone to desertification, flood, or deforestation that exacerbate poverty. Overlay climate hazards like floodplains and drought indices with poverty maps to highlight compounding risks. • **Plan multi-benefit land use.** Use GIS to design land-use strategies that balance economic activity like farming with conservation such as watersheds and forests. Employ decision-support tools like landcover for change detection, climate projection layers to evaluate tradeoffs (Gorelick et al., 2017). • **Incorporate local knowledge.** Engage communities in climate mapping by setting up rain-gauge networks and participatory flood-mapping. Map traditional land uses such as sacred groves and customary fields through PGIS, so cultural resources are protected. This bottom-up data enriches resilience planning and ensures interventions that respect local practices (Alvarez et al., 2021).		
Scale	**Government**	**NGO/IGO/ Private Sector**	**Community-Level/ Personal**
Practical Tools	Use GIS for climate risk assessment: Integrate climate projections using, for example, WorldClim and NASA POWER in national planning GIS. Employ software like IBM Environmental Intelligence Suite or ArcGIS Climate Resilience tools for scenario modeling. Map land-cover change with satellite data with Sentinel-2, Landsat to detect deforestation or degradation. Utilize geospatial decision-support systems for land-use planning like computer-aided design (CAD)-based GIS extensions for agriculture.	Leverage free climate data platforms such as Google Earth Engine climate datasets, TerrSet, and InVEST for risk mapping. Use open-source land management tools like Open Foris and QGIS forestry plugins to map tree cover and soil erosion. Integrate mobile climate adaptation apps like apps for farmers' weather data reporting.	Promote community-led climate monitoring: Rain-gauge networks, community weather stations, and participatory flood-mapping exercises. Use low-cost tools like GPS trackers to map migration corridors of livestock and participatory apps like MBON and Epicollect to collect local climate impacts. Map traditional land-use such as sacred groves and ancestral fields with PGIS.
Analysis and Policy Recommendations	Incorporate AI models such as IBM Environmental Intelligence Suite to predict climate vulnerabilities.	Build predictive models. Deploy AI-driven climate vulnerability tools (like IBM's Environmental Intelligence Suite) to forecast future hotspots of food insecurity or disaster risk.	Develop community-owned land management co-ops that give locals partial ownership in conservation efforts.

(Continued)

TABLE 1.5 *(Continued)*
Climate Resilience and Sustainable Land Management

Scale	Government	NGO/IGO/ Private Sector	Community-Level/ Personal
	Use GIS to enforce climate-smart policies. For example, designate protected zones or agroforestry areas based on poverty and climate risk layers. Establish Payments for Ecosystem Services (PES) schemes using GIS to verify outcomes such as mapped carbon stocks for agroforestry projects. Integrate these priorities into climate and development budgets like Green Climate Fund projects.	Use these forecasts to pre-position aid such as food reserves in areas predicted to suffer drought and design adaptation programs. Support community co-ops by helping set up landscape management associations; GIS can give members partial ownership of mapped resources and track compliance with conservation goals.	Take active roles. Community groups should participate in climate monitoring like tracking rainfall and soil moisture, and then feed data into GIS. They can use resilience maps to plan local adaptations like drought-tolerant crops where GIS shows high future stress. Cooperatives can manage shared resources – for instance, farmers planting and mapping trees together under PES schemes – turning mapping efforts into income opportunities.
Financing Pathways	Access international climate finance (Green Climate Fund, Adaptation Fund) that is earmarked for resilience mapping. Issue green or sustainability bonds to fund climate-proof infrastructure (with GIS for site selection). Partner with conservation finance for carbon credits and PES schemes where mapping supports verification. Incorporate GIS funding into national climate budgets or disaster risk reduction programs.	Obtain grants from environmental donors like UNDP, UNEP and GEF for resilience mapping. Collaborate with international research programs like Consultative Group for International Agricultural Research (CGIAR) climate projects. Engage corporate environmental CSR funds for mining or agri-companies funding of watershed mapping as part of CSR. Seek blended finance solutions such as grants and concessional loans for sustainable land-use projects.	Enable community Payment for Eco-System (PES) schemes where farmers get paid for planting trees that they map and manage. Use cooperative models for sustainable agriculture funding. Apply for small grants from climate adaptation windows of major funds. Local NGOs can facilitate micro-loans or rotating funds to adopt resilient practices often linked to geospatial planning.

TABLE 1.6
Urban Planning and Infrastructure Development

General Best Practices	• Support equitable urban development by mapping informal settlements and prioritizing basic service expansion. • Map informal settlements and service gaps. Continuously survey and digitize slum areas using open tools like OpenStreetMap editors or UAV/drone mapping to guide upgrades. Always update slum boundaries and asset inventories so that interventions like sanitation and electricity reflect current conditions. • Integrate planning and safety layers. Overlay infrastructure plans with hazard maps (flood zones, earthquake fault lines) and socioeconomic data (income, land tenure) to avoid building on risky or inequitable sites. Use 3D GIS and BIM integrations such as Autodesk InfraWorks or OpenCities for holistic city models. • Inclusive community engagement. Conduct participatory urban mapping à la Missing Maps so residents can document streets, buildings, and problems such as no streetlights. Provide residents with simple mapping apps and encourage photomapping infrastructure needs. This ensures marginalized voices can shape planning.		
Scale	**Government**	**NGO/IGO/ Private Sector**	**Community-Level/ Personal**
Practical Tools	Integrate GIS with urban planning suites: Use Autodesk InfraWorks or Bentley OpenCities Planner for 3D urban modeling (as noted in Best Practices). Use city-level GIS dashboards for utilities such as water network GIS and power grid GIS. Deploy mobile GIS for site surveys of informal settlements to collect data on slum amenities. Utilize BIM-GIS integration for new infrastructure and digital twins of city assets.	Employ open-source urban GIS (OSM) for mapping informal areas. Use planning toolkits like QGIS Urban plugin and OpenTripPlanner for transit. Develop community-accessible maps of zoning or hazard areas. Use participatory data collection (Map Kibera model) for slum mapping.	Mobilize communities to map local infrastructure needs using free tools like Missing Maps project methods such as GPS mapping and smartphone apps. Use mobile apps to report service deficits and photomapping potholes or broken streetlights. Participate in local "open data hackathons" to build apps for community use.

(Continued)

TABLE 1.6 *(Continued)*
Urban Planning and Infrastructure Development

Scale	Government	NGO/IGO/ Private Sector	Community-Level/ Personal
Analysis and Policy Recommendations	Implement AI-driven simulations to optimize road networks, water supply systems, and energy grids. Invest in digital city models. Use AI-driven simulations to optimize utilities and transport networks—for example, test water distribution under growth scenarios using city “digital twins.” Leverage municipal bonds to finance smart infrastructure like roads and mass transit that are planned with GIS guidance. Prioritize slum upgrading programs by citing mapped data, like issue development loans for neighborhoods where poverty and informal settlement overlays coincide.	Use storytelling and advocacy. Create GIS story maps to visualize living conditions in informal areas, persuading authorities and donors to fund inclusive policies. Support open-data initiatives so developers and researchers can build free apps like OpenTripPlanner for transit that improves access. Guide community organizations in zoning workshops, ensuring plans reflect local spatial data.	Organize around shared maps. Host mapping hackathons to collect neighborhood data or join global efforts like the Missing Maps project to build skills. Use mobile apps to report infrastructure faults by photographing potholes and broken pipes. Through participatory budgeting or citizen surveys mapped in GIS, neighborhoods can vote on small infrastructure projects and guide municipal spending.
Financing Pathways	Issue municipal or infrastructure bonds, often used for roads and transit and integrate GIS in the project planning to secure financing, as investors favor data-driven planning. Tap urban development loans from World Bank Urban Fund and ADB Urban projects. Set up PPPs with developers, especially those developers who invest in mapping city services in exchange for development rights. Use impact investors and sovereign wealth funds interested in sustainable city projects, as suggested.	Leverage grants from urban-focused agencies like UN-Habitat and Cities Alliance. Engage international funds for low-income city upgrades from Global Fund for Cities. Collaborate with infrastructure funds (ADB, AfDB) on slum-mapping components. Encourage CSR from local construction or telecom companies for community mapping that contributes to community broadband mapping.	Promote participatory budgeting where citizens vote on funding local infrastructure projects, which can fund both mapping and interventions. Support local savings groups to contribute to small-scale neighborhood improvements that require mapping. Organize cooperatives that partially own infrastructure like community solar projects and fund mapping of resources.

TABLE 1.7
Public Engagement and Community-Led Development

Scale	Government	NGO/IGO/ Private Sector	Community-Level/ Personal
General Best Practices	• Apply PGIS. Engage residents directly in mapping exercises: Use simple drawing maps, community storyboards, or mobile-PGIS apps so locals set mapping priorities. This builds ownership and trust. • Empower marginalized groups. Provide accessible mapping tools such as multilingual interfaces, icon-based apps and printed maps so all can participate. Target trainings to underrepresented groups like women's cooperatives, youth clubs, and incorporate local knowledge through such methods as oral histories and indigenous landmarks into maps. • Create feedback loops. Ensure mapping outcomes are shared back to the community via such instruments as story maps or public dashboards so people see how their input shaped decisions. This transparency fosters civic engagement and makes GIS a tool for social inclusion.		
Practical Tools	Launch public GIS portals and story-maps using Esri StoryMaps and ArcGIS Hub for citizen engagement. Use social media mapping using Twitter/ YouTube geotag analysis for public feedback. Provide open APIs for community developers to build on government data.	Leverage GIS-based storytelling platforms like Esri StoryMaps to communicate challenges and successes. Use community storytelling GIS like AudioMap and Story Maps to amplify local voices. Implement PGIS projects with NGOs, facilitating workshops. Employ citizen science platforms like iNaturalist for environmental engagement.	Drive PGIS with simple tools: Participatory maps drawn on tablets or paper, community photomapping, and local GIS workshops. Use Ushahidi or SMS polls to let citizens report needs or vote on priorities. Promote bottom-up data collection on citizen health mapping, crime mapping, for example.
Analysis and Policy Recommendations	Institutionalize participation. Adopt open data and open government policies so citizens can access and use GIS data. Implement participatory budgeting that uses GIS evidence: Allow communities to nominate projects on maps and vote on funds allocation, ensuring government spending aligns with mapped needs.	Invest in civic mapping initiatives. Fund training programs and microgrants for community tech hubs where residents learn GIS skills. Partner with local NGOs to facilitate GIS workshops (as capacity building and trust-building), enabling residents to lead mapping projects.	Develop a co-op ownership model where communities partially own infrastructure and services, ensuring long-term engagement. Take ownership of Data. Form local GIS cooperatives or clubs that manage shared resources like GPS units and community servers.

(Continued)

TABLE 1.7 *(Continued)*

Public Engagement and Community-Led Development

Scale	Government	NGO/IGO/ Private Sector	Community-Level/ Personal
	Recognize co-operative ownership models, where communities partially own services such as water and energy and maintain associated GIS data.	Support storytelling platforms (e.g. Ushahidi, StoryMaps) to amplify local voices on issues like land rights or service delivery.	Use mapping to advocate: For example, produce community-led story maps highlighting needs and present them to officials. Encourage tech-savvy youth to develop local apps or SMS polls, so residents can report problems in real time. By embedding GIS into civic action, communities ensure that development is demand driven.
Financing Pathways	Allocate civic tech funds or digital government grants that are often available from donor programs like E-Government funds. Support public participatory platforms like geoviz tools via grants or budget. Encourage PPPs for community broadband and mapping centers.	Use crowdfunding for community story-map initiatives. Secure community development grants from agencies like USAID community-driven development programs that include GIS engagement. Explore social impact investors who fund civic engagement like Agence Française de Développement's lab for cities.	Form local co-ops that partly "own" services so communities pay into a fund—such as cooperative ownership of water or energy—and then map those assets. Use self-help group savings or rotating funds to finance local GIS initiatives. Leverage small community grants from Rotary Clubs or local NGOs for civic mapping projects.

TABLE 1.8

Capacity Building and Institutional Strengthening

General Best Practices	• Establish GIS training programs. Create national or regional GIS centers and online courses like World Bank Open Learning Campus, ESRI Academy for public officials, and NGO staff. Mandate basic GIS skills in planning and statistics agencies. Use AI chatbots or tutorial apps for on-demand learning support. • Promote peer learning. Organize regular workshops (online/offline) on free GIS tools (QGIS, ODK), and set up "train-the-trainer" programs so skills multiply within communities. Encourage participation in conferences and certification schemes (like GISCert) for knowledge exchange.

(Continued)

TABLE 1.8 *(Continued)*
Capacity Building and Institutional Strengthening

- Tailor materials for all learners. Provide training resources in local languages and low-literacy formats with icon-based apps and illustrated guides. Support community GIS clubs and peer tutoring such as young volunteers teaching elders to use GPS to broaden adoption at the grassroots level.

Scale	Government	NGO/IGO/Private Sector	Community-Level/Personal
Practical Tools	Establish national GIS training centers and e-learning platforms such as World Bank Open Learning Campus and, ESRI Academy. Mandate GIS training for planning and statistics agencies. Deploy AI-powered training assistants like chatbots and tutorials for on-demand learning.	Organize training workshops online/offline on free GIS tools (QGIS, ODK). Develop toolkits and "train-the-trainer" programs. Use platforms like GISCert or international conferences for knowledge exchange.	Encourage grassroots learning through community GIS clubs, peer-to-peer training with youth volunteers teaching elders to use GPS. Provide multi-lingual and low-literacy GIS resources using map "story" templates and icon-based apps.
Analysis and Policy Recommendations	Budget for ongoing GIS capacity. Allocate funds for training and hiring GIS specialists, and partner with development banks or UNDP for technical assistance grants given that many programs support governance and tech training. Build public–private partnerships with universities and tech firms to co-develop GIS curricula and labs, for example, sponsored GIS research labs.	Provide training for government officials, local organizations, and community leaders in GIS applications. Foster collaboration between academic institutions, research organizations, and development agencies.	Leverage local resources by forming community-driven GIS groups that pool small fees or fundraise to buy shared equipment such as GPS devices and computers. Encourage hackathons or mapping events, often supported by grants, where locals build tools for community needs. Cooperatives such as farmers or youth co-ops can allocate part of their income towards paying for GIS training or software licenses, turning capacity building into a communal investment.

(Continued)

TABLE 1.8 ***(Continued)***

Capacity Building and Institutional Strengthening

Scale	Government	NGO/IGO/Private Sector	Community-Level/Personal
	Embed GIS training into formal education such as high schools and universities to sustain a local talent pipeline. Establish AI-powered chatbots and training assistants to support ongoing education and skills development. Strengthen governance by ensuring transparency through publicly accessible GIS portals.	Strengthen institutional links. Collaborate with academic and research institutions to co-fund GIS training programs for local organizations. Use grant funding such as USAID and Sida capacity-building programs to subsidize courses on spatial data for low-income country practitioners. Offer mentorship and data-sharing networks so NGOs can learn from each other's GIS experiences.	
Financing Pathways	Budget for ongoing GIS training and hiring in the public sector. Seek technical assistance grants from development partners like UNDP governance programs, World Bank and TA funds. Build PPPs with universities or private sector for GIS curriculum development (e.g. sponsored labs).	Apply for capacity-building grants from USAID Global Programs and Sida that fund local tech skills. Leverage multilateral funds from the African Development Bank ICT training funds. Partner with academia for subsidized courses (often funded by research grants).	Use nominal membership fees in community groups to buy shared equipment such as GPS units and computers. Organize local hackathons or crowdsourcing events sometimes funded by grants or local businesses. Promote cooperative education models that provide profits from a community service fund for GIS training workshops.

TABLE 1.9
Summary of GIS Methodologies and Applications for MPI Indicators

GIS Methodology	Application	MPI Indicator (Component)	SDG Indicator Code
Real-time Data Integration	APIs for ArcGIS, Google Earth Engine, and OpenStreetMap	Standard of Living: Cooking Fuel, Sanitation	1.4.1
PGIS Surveys	Mobile apps for crowd-sourced mapping	Education: School Attendance	4.1.2
Remote Sensing and Satellite Analytics	Night-time lights, land cover change detection	Health: Child Mortality	3.2.1
Hotspot & Cluster Analysis	Getis-Ord Gi* and Local Moran's I for poverty cluster identification	Standard of Living: Electricity, Water Access	1.4.1
Composite Index Mapping	Rasterization of weighted MPI indicators	Health: Nutrition, Child Mortality	2.2.1; 3.2.1
Spatial Accessibility and Isochrone	Travel-time analysis to essential services	Standard of Living: Access to Healthcare	3.8.1
Predictive ML Modeling	Random Forest, Gradient Boosting for poverty score estimation	Education: Years of Schooling	4.1.2
Scenario Simulation Dashboards	What-if tools for infrastructure and service planning	Education: School Attendance	4.1.2
Edge and Cloud AI Inference Pipelines	On-site GPU servers and cloud microservices	Standard of Living: Cooking Fuel, Sanitation	1.4.1
Ethical AI and Explainability	SHAP, LIME for feature importance and bias auditing	Cross-cutting: All MPI Indicators	1.2.1

TABLE 1.10
Data Integration and Composite Index Mapping

General Best Practices	• Indicator Layering: Assemble spatial layers for each MPI indicator (e.g. school enrollment, child mortality, access to cooking fuel) using census data, household surveys (DHS, MICS), and high-resolution remote sensing proxies. • Weighted Rasterization: Convert each indicator into a standardized raster (0–1) reflecting deprivation intensity. Apply MPI weights to generate a composite poverty surface, visualized in GIS as a continuous heat map. • Spatial Smoothing: Use kernel density or Bayesian hierarchical smoothing to reduce noise in thinly sampled areas, ensuring stable estimates at micro-levels.

(Continued)

TABLE 1.10 ***(Continued)***
Data Integration and Composite Index Mapping

Scale	Government	NGO/IGO/Private Sector	Community-Level/ Personal
Practical Tools			
Analysis and Policy Recommendations	Incorporate composite index maps into policy. For example, use these maps to direct multi-sectoral programs in areas with high combined deprivation such as health, education, and utilities. Embed map results in national poverty reduction plans, so allocations of funds and services reflect the full context of need.	Use composite maps for targeting and evaluation. Design programs where weighted poverty is greatest; cite index maps in proposals to justify intervention choices. Regularly update indices and involve local surveys, since combining official data with community input yields richer insights. Share index methodologies openly with partners to harmonize targeting across agencies.	Participate in integration. Contribute local survey data such as condition of schools or clinics, to refine layers and weightings. Use the visual index maps to advocate for resources; community leaders can show maps at meetings to illustrate local needs. Likewise, local feedback can guide which indicators to emphasize in national MPI calculations.
Financing Pathways	Public Budgets: Fund MPI data updates and GIS mapping. Multilateral Loans: Access funds from international development banks. Innovation Bonds: Issue bonds for GIS-based infrastructure. Partnerships: Co-finance with private data providers.	Grants: Seek funding from foundations or UN for MPI mapping. Public–Private Partnerships: Collaborate with telecom or satellite firms for shared data. CSR Investment: Engage corporates in mapping for service expansion.	Small Grants: Use local development funds for mapping projects. Citizen Science: Offer stipends or in-kind support for data collection. Use crowdsourcing to fund mapping initiatives via local donations or international grants.

TABLE 1.11
Hotspot Analysis and Spatial Statistics

General Best Practices	• Hotspot Detection: Apply spatial statistics such as Getis–Ord *Gi* or Local Moran's *I* to identify significant clusters of high poverty—these "poverty hotspots" flag priority intervention zones. Perform significance tests and validate clusters against local knowledge to avoid spurious results. • Cluster Typologies: Classify areas by combining composite poverty with contextual variables such as market access, flood risk and urban density. This yields typologies like "remote extreme poverty" vs. "periurban vulnerability" that help to tailor interventions. • Continuous Monitoring: Re-run hotspot analysis annually or after major interventions. Track how clusters shrink or shift over time so programs can adapt.

Scale	**Government**	**NGO/IGO/Private Sector**	**Community-Level/ Personal**
Practical Tools			
Analysis and Policy Recommendations	Target hotspot regions with integrated programs. Once clusters are identified, focus cross-sector projects there; for example, build schools, clinics and roads in a hotspot. Use hotspot maps to streamline resource allocation, for example, concentrate subsidies or social services in the highest clusters to maximize impact.	Coordinate via shared hotspot data. Use the cluster maps to avoid service overlaps. NGOs can agree that Organization A covers health in Cluster X while Organization B focuses on education. Hotspot maps also help justify funding: Donors are more willing to invest when interventions clearly target statistically proven poverty centers.	Leverage cluster maps for advocacy. Residents in identified hotspots can use official maps to demand services. Local leaders should lobby policymakers by showing spatial evidence of multi-dimensional deprivation. When communities understand these hot spots, they can also self-organize local solutions like forming cooperatives in areas where geography or infrastructure make them vulnerable.
Financing Pathways	Public Budget Lines: Allocate funds for statistical mapping and GIS staff. Development Loans: Include spatial cluster analysis in funded social protection or infrastructure projects. Smart Subsidy Planning: Use hotspot maps to justify efficient resource targeting.	Donor Funding: Include hotspot mapping in grant proposals for poverty reduction. Data Partnerships: Collaborate with academic institutions for analysis in exchange for co-authorship or access to findings. Impact Reporting: Use hotspot targeting to boost donor confidence.	Small Grants: Fund participatory mapping of hotspot areas. Advocacy Tools: Provide funds or training to local leaders to use hotspot maps for lobbying. Integrate into existing NGO field projects. Community-Led Mapping: Integrate into existing NGO field projects.

TABLE 1.12
Predictive Modeling and Scenario Planning

General Best Practices	• Integrate satellite proxies into models. Train machine-learning models with random forests and gradient boosting using remote sensing indices such as nightlights, vegetation, and land use alongside socioeconomic grids to predict poverty where surveys are scarce. Ensure model training data that are unbiased and representative of all groups. • Build "What-If" tools. Develop interactive GIS dashboards that let planners adjust inputs such as number of schools and road lengths to immediately see simulated effects on MPI scores. These simulations help prioritize investments by showing which combinations yield the greatest poverty reduction (Esri, 2024). • Maintain model transparency and updates. Use interpretable algorithms and update models regularly with new data so predictions stay current. Complement model results with stakeholder input to check plausibility.		
Scale	**Government**	**NGO/IGO/Private Sector**	**Community-Level/ Personal**
Practical Tools			
Analysis and Policy Recommendations	Machine Learning Integration: Train spatially explicit models such as random forests and gradient boosting to predict composite poverty scores in data-sparse regions using remote sensing indices such as night-time lights, vegetation cover, and gridded socioeconomic datasets (Gorelick et al., 2017). Use forecasting to drive policy. For example, if models predict that a district's poverty will worsen without intervention, preemptively allocate resources there. Incorporate scenario results into budgeting to include plan for building additional clinics in a modeled "high impact" scenario. Integrate predictive maps with early warning systems for disasters such as droughts and floods to trigger timely aid.	What-If Simulations: Build interactive GIS dashboards that let planners adjust inputs such as expanding school networks or improving road access to simulate impacts on MPI scores, optimizing investment portfolios. Build technical capacity. Train program staff to understand model outputs and use them in project design. When writing proposals, use model-driven evidence such as predicting which new road could lift X people out of poverty to strengthen the case. Align monitoring frameworks with scenarios: track the actual impacts versus predicted to learn and refine models.	Engage with Predictions. Simplify and share modeled scenarios using maps of projected poverty under different assumptions in local forums. Educated residents can use these to plan adaptive strategies like diversifying crops in areas forecasting as future "cold spots" of food security. Transparency about projections also helps build trust in data-driven planning.

(Continued)

TABLE 1.12 *(Continued)*
Predictive Modeling and Scenario Planning

Scale	Government	NGO/IGO/Private Sector	Community-Level/ Personal
Financing Pathways	Innovation Funds: Allocate R&D budgets for predictive modeling tools. Smart Planning Budgets: Fund GIS simulations as part of infrastructure or social planning. Public–Academic Partnerships: Co-fund model development with universities.	Tech Grants: Apply for grants focused on AI for good, forecasting, or planning innovation. CSR and Analytics Partnerships: Collaborate with tech companies to co-develop modeling tools. Proposal Strengthening: Use model outputs to justify funding.	Community Grants: Use small grants to localize model inputs and validate outputs. Awareness Campaigns: Fund simplified outputs like scenario posters/ maps) for village planning meetings. Digital Literacy Funding: Train local users.

TABLE 1.13
Monitoring Change Over Time

General Best Practices

- Temporal GIS: Automate updates by routinely ingesting new survey data and satellite imagery so deprivation layers refresh yearly or biannually. Maintain a time-series archive of poverty maps to analyze trends.
- Change Detection: Compare classified maps like land cover and nighttime lights over time to spot areas that are improving or worsening. Highlight zones of rapid change that are positive or negative so stakeholders can react quickly to deploy aid if new famine indicators emerge.
- Accessibility Over Time: Periodically recompute service-access isochrones with updated road or settlement data to see if connectivity improves or gaps persist. Use these dynamics to assess whether infrastructure investments have had the intended effect.

Scale	Government	NGO/IGO/Private Sector	Community-Level/ Personal
Practical Tools			
Analysis and Policy Recommendations	Make GIS change analysis part of adaptive management. If maps show little improvement after a program's first year, re-evaluate strategies. Tie midterm reviews to spatial metrics such as target "X" new schools in 2 years—check GIS to confirm. Use changes to meet SDG reporting by demonstrating spatial progress or lack thereof in poverty reduction.	Commit to longitudinal data sharing. Release project GPS logs like locations of new wells or schools so the community and other groups can see change. Employ open dashboards where maps evolve, inviting external analysts or donors to verify progress.	Participate in monitoring. Organize periodic local surveys like annual income or asset surveys, and feed the results into maps. Community members can also use simple mobile apps to report newly fixed or broken services. By contributing to continuous mapping, residents ensure that changes such as increased electrification or new roads that are visible and credited, while emerging problems get noticed early.

(Continued)

TABLE 1.13 *(Continued)*
Monitoring Change Over Time

Scale	Government	NGO/IGO/Private Sector	Community-Level/Personal
Financing Pathways	Performance-Based Budgets: Link funding to verified spatial progress. Public Monitoring Systems: Invest in national GIS change-tracking platforms. Infrastructure Audits: Allocate funds for evaluating ROI via spatial data.	M&E Grants: Secure donor funding for longitudinal impact tracking. Tech Partnerships: Co-finance real-time dashboards with private firms. Open Data Incentives: Use spatial evidence to meet transparency and compliance conditions.	Local Monitoring Funds: Use micro-grants for community-led change tracking. Mobile Tools Support: Provide devices, apps, or airtime to report local updates. Citizen Audit Budgets: Finance grassroots verification efforts.

TABLE 1.14
Equity-Focused Service Targeting

General Best Practices

- Accessibility Analysis: Combine poverty maps with service layers such as health centers, schools, and water points to compute travel-time or network isochrones. This reveals underserved pockets where services are too distant despite overall coverage.
- Resource Allocation Models: Integrate poverty intensity and accessibility scores into location–allocation models to site new facilities. For instance, prioritize placing a clinic where poverty is high *and* travel time is longest, maximizing impact.
- Incorporate social equity factors. Overlay non-spatial data (like gender-disaggregated poverty, disability prevalence) to ensure targeting benefits the most excluded. Ensure facility siting accounts for cultural or linguistic minority communities often missed by geography alone.

Scale	Government	NGO/IGO/Private Sector	Community-Level/Personal
Practical Tools			
Analysis and Policy Recommendations	Use GIS targeting in planning. For example, mandate that rural schools or clinics be built in villages identified as both poor and hard-to-reach. Allocate extra budget to such "equity zones." Monitor service coverage maps to reduce gaps: If analysis shows a town's poorest quarter is far from markets, add a local distribution point or transport subsidy.	Advocate with equity maps. Present target-area analyses to policymakers—like "Village X has no school within 10 km and high child poverty" to push for investments. Use these maps to direct NGO interventions for food deliveries, mobile clinics, etc., specifically into identified service deserts. Also, tailor program design like school-feeding to the specific access barriers revealed.	Use targeted maps for local action. Community groups can review accessibility maps to organize initiatives, for instance, running a community bus service if the nearest hospital is far. They can petition local governments by showing on maps which neighborhoods lack lighting or clean water. When the community has mapped evidence, it's easier to secure municipal or NGO attention.

(Continued)

TABLE 1.14 *(Continued)*
Equity-Focused Service Targeting

Scale	Government	NGO/IGO/Private Sector	Community-Level/Personal
Financing Pathways	Equity Budget Lines: Dedicate funds to serve "equity zones" with poor access. Conditional Grants: Link funding to spatial equity metrics like service gaps in remote and poor areas. Social Inclusion Funds: Finance culturally targeted infrastructure.	Targeted Aid Grants: Apply for funding tied to reaching hard-to-reach groups. Data-Driven CSR: Partner with companies to expand access in underserved zones. Impact Justification: Use equity maps to strengthen proposals and funding cases.	Local Advocacy Funds: Support community lobbying using equity maps. Micrograntss: Finance small-scale solutions like community transport to health posts. Collaborative Planning Budgets: Fund shared service access reviews.

TABLE 1.15
Participatory Validation

General Best Practices	• Mobile PGIS surveys: Deploy community-driven mapping apps like Ushahidi and StreetComplete so residents can verify, refine, and annotate poverty and service maps in real time. This ensures local knowledge complements quantitative analysis. • Visual Storytelling: Use narrative GIS such as StoryMaps and AudioMap to present multidimensional poverty data as engaging stories. When stakeholders see their own communities' names and stories on a map, buy-in increases, and they support tailored interventions. • Workshop mapping: Hold public mapping sessions where residents review draft maps. Let them mark errors such as a new well not on the map or name unlabeled features. Use printed maps or interactive displays in local languages to make this accessible. These sessions validate the data and build consensus on map accuracy.		
Scale	**Government**	**NGO/IGO/Private Sector**	**Community-Level/Personal**
Practical Tools			
Analysis and Policy Recommendations	Institutionalize community validation. Require that major projects include a "community sign-off" GIS step: Plans are not finalized until affected residents have reviewed and approved the mapped data. This could be formalized in planning regulations or funding criteria. Treat participatory maps as supplemental to official records, giving them legal weight.	Close the feedback loop. After collecting and analyzing spatial data, always return to communities with maps and ask for corrections. Use these validated maps in funding proposals and reports to donors: A project that shows "these maps were co-validated by the village" demonstrates credibility. Provide training so community representatives can interpret and present GIS findings such as in town halls.	Leverage validated maps for advocacy. Use them in meetings with officials, media, or donors to illustrate needs. For example, a community can create its own story map highlighting infrastructure gaps and share it on social networks or local radio. Validation increases trust, so when community members quote the maps, policymakers take them more seriously. In essence, participatory maps give residents a data-backed voice to negotiate development.

(Continued)

TABLE 1.15 ***(Continued)***
Participatory Validation

Scale	Government	NGO/IGO/Private Sector	Community-Level/Personal
Financing Pathways	Inclusive Planning Budgets: Fund community validation as part of infrastructure or social planning. Regulatory Requirements: Tie planning approval to local sign-off via participatory maps. Decentralized Funding: Enable local units to contract validation activities	Participatory Grants: Apply for donor funds that prioritize community engagement. Accountability Mechanisms: Use validated maps in reporting and audits. Partnerships: Co-finance validation with local civil society or media.	Microgrants: Support local mapping workshops such as printing and facilitation. Tech Kits: Provide phones, tablets, or data to community mappers. Storytelling Funds: Support creation of local story maps for advocacy.

TABLE 1.16
Spatial Implementation Matrix

Step	Heatmap (Density)	Water Points	Waste Sites	Energy Nodes	Coordination Hub	Monitoring Refresh
1. Map and Prioritize Needs	Layer population density; flag red hotspots.	Overlay existing points to see coverage gaps	Overlay existing sites to see coverage gaps	Overlay existing nodes to see coverage gaps.	Register all layers in shared GeoAI Streamer.	—
2. Plan Clean-Water Interventions	Identify high-need villages via hotspots.	Model 1-mile service buffers around gaps	—	—	Enter planned borehole projects.	Refresh to show new interventions.
3. Design Waste-Management Systems	Target dense clusters for waste routing.	—	Cluster dg zones; pilot co-ops in those areas.	—	Enter waste collection schedules.	Refresh to show route performance.
4. Deploy Sustainable Energy Infrastructure	Pinpoint dense, underserved red zones.	—	—	Prioritize gaps for PV/ microgrids.	Enter energy project plans.	Refresh to show energy uptime.

(Continued)

TABLE 1.16 *(Continued)*
Spatial Implementation Matrix

Step	Heatmap (Density)	Water Points	Waste Sites	Energy Nodes	Coordination Hub	Monitoring Refresh
5. Forge Partnerships and Governance	Validate selected hotspots against policy goals.	Confirm borehole plans with Ministry of Health	Confirm collection plans with Ministry of Environment.	Confirm energy plans with Ministry of Energy.	All ministries enter and update policy entries.	Reflects policy changes on the map.
6. Monitor, Evaluate, and Optimize	Track shifts in density vs. service coverage.	IoT sensor data mapped around water points.	Sensor/ feedback overlays around waste sites.	Sensor/ usage overlays around energy nodes.	All stakeholders review via coordination hub.	Auto-refresh triggers alerts for new gaps.

TABLE 1.17
Implementation Roadmap

Phase	Activities
Scoping and Engagement	• Define MPI indicators • Secure data-sharing agreements • Stakeholder workshops
Prototype Development	• Deploy core GIS workflows • Integrate initial AI pipelines on pilot geography • Launch beta dashboard
Validation and Feedback	• Conduct PGIS field workshops • Retrain models via active learning • Produce interim report
Scale-Up and Capacity Building	• Train "power users" in GeoAI workflows • Integrate outputs into national SDG reporting • Convene governance committee
Sustainability and Innovation	• Publish open-data GIS portal • Establish public–private co-investments • Pilot edge-AI nodes and drone mapping

TABLE 1.18
Cost Estimates and Funding

Methodology/System	Free/Open-Source	Low-Cost (< $1 000)	Mid-Tier ($1-10K)	High-Tier (> $10 000)	Grant and Funding Programs
Data Collection and Integration	QGIS; OpenStreetMap; GEE (academic tier)	Crowd-sourcing tool setup (~$500)	ArcGIS Online Single-User (~$1 500/yr)	Enterprise-scale GIS deployment (>$10 000/yr)	Esri Nonprofit Program; Google.org; MapGive
Mapping Hotspots and Vulnerability	QGIS hotspot and clustering plugins	PlanetScope imagery subs (~$600/yr)	Maxar basemap access (~$2 000/yr)	Custom AI-segmentation services (>$15 000)	World Bank Space for Smarter Development; Planet Impact
Resource Allocation Dashboards	Tableau Public; Power BI Free	Pantheon or simple hosting (~$500/yr)	Tableau Creator or Power BI Pro (~$840/yr)	Palantir Foundry enterprise (>$20 000)	Tableau Foundation; Power BI for Nonprofits
Remote Sensing and Analytics	Google Earth Engine free tier	Sentinel Hub starter (~$400/yr)	PlanetScope monthly (~$1 200/mo)	Maxar Vivid contract (>$25 000)	ESA Earth Innovation; NASA ARSET
PGIS	uMap; OpenStreetMap	KoBoToolbox survey platform (~$200/yr)	Mergin Maps mobile syncing (~$500/yr)	Custom mobile-app development (>$10 000)	Humanitarian OpenStreetMap Team; USAID GeoCenter
IoT and Sensors (Water)	Arduino/RPi sensor kits (~$50/unit)	LoRaWAN sensor kits (~$300/unit)	Xylem Smart Water station (~$2 000/station)	Veolia smart-water network (>$20 000)	Climate Justice Fund; Global Innovation Fund
Waste-Management Mapping	OSM + WasteAtlas free tools	Community GPS kits (~$200)	RecycleBank pilot program (~$2 500)	Covanta/Suez circular-economy systems (>$50 000)	UN-Habitat Circularity; TerraCycle grants
Energy-Access Analysis	Google Sunroof API; PVGIS	Solar irradiance meters (~$150)	HOMER Pro license (~$1 200/yr)	Siemens/Tesla smart-grid integration (>$30 000)	SolarAid; Schneider Electric Foundation
AI/ML Pipelines	TensorFlow; PyTorch	Cloud GPU credits (~$400)	AWS SageMaker or GCP AI Platform (~$1 000/mo)	Palantir/Microsoft custom AI solutions (>$25 000)	Google AI Impact Challenge; Microsoft AI for Good
Edge and Cloud Infrastructure	AWS/GCP Free Tier; Docker on shared host	DigitalOcean droplet (~$5/mo)	AWS EC2 small instances (~$70/mo)	On-prem GPU cluster (>$15 000)	AWS Imagine Grant; Google Cloud for Startups

Table 1.9 Explained

- GIS Methodology: Core spatial and analytic techniques deployed in poverty assessments.
- Application: Specific technological implementations and platforms used.
- SDG 1 Indicator (MPI Component): Corresponding multidimensional poverty indicators as defined by the UN MPI framework, illustrating how each GIS method maps to real-world deprivation measures.

GIS FOR MULTI-DIMENSIONAL POVERTY ASSESSMENT

Multi-dimensional poverty moves beyond income-based metrics to capture deprivations in health, education, living standards, and more. The UN's Multi-Dimensional Poverty Index (MPI) offers a structured framework of ten indicators across three dimensions (Alkire & Santos, 2010). GIS enables the spatial disaggregation of these indicators at fine scales—down to villages, urban neighborhoods, or even street blocks—revealing localized patterns of deprivation that national averages obscure.

POPULATION DENSITY MAPPING

GIS-based poverty mapping gains critical nuance when combined with population density surfaces (Tatem, 2017; Henebry et al., 2018). While the MPI composite highlights areas of severe deprivation, density layers reveal where the largest numbers of deprived individuals reside. This dual lens ensures interventions are prioritized not only by depth of poverty but also by scale of impact for the "maximum people helped per unit resource" approach.

Data Sources

- Use gridded population products: WorldPop, LandScan, GPW v4 (Tatem, 2017; WorldPop, 2024).
- Validate against recent census micro-data or mobile-phone call-detail record (CDR) proxies where available.

Methods

- **Kernel-Density Estimation:** Smooth census points into a continuous surface reflecting true settlement patterns.
- **Dasymetric Refinement:** Use land-cover/built-area masks (e.g. from Sentinel) to refine rural vs. urban density (Henebry et al., 2018).
- **Raster Integration:** Resample your population raster to match your MPI composite grid (e.g. 100 m or 1 km resolution).

Integration with MPI

- **Weighted Composite:** Multiply your deprivation heat map by normalized density to get a "population-weighted poverty surface."
- **Hotspot Prioritization:** Identify cells where both density and MPI exceed critical thresholds—these become dual-priority zones.

Visualization

- **Bivariate Choropleths:** Show deprivation on one axis, density on the other, so readers instantly see where high-poverty, high-density overlaps occur.
- **Dasymetric Overlays:** Display population-density contours beneath your poverty map to contextualize resource-allocation decisions.

AI AND MACHINE LEARNING FOR REAL-TIME POVERTY ASSESSMENT VIA GIS

In rapidly changing contexts, static maps and periodic surveys cannot capture the full dynamics of poverty shocks and recovery. By embedding AI and machine-learning pipelines into GIS platforms, practitioners can ingest live data streams—from satellites, sensors, and mobile networks—and translate them into continuously updated poverty indicators. The following section outlines how to architect real-time data ingestion, extract socio-economic features, deploy adaptive models at the edge and in the cloud, and visualize insights through decision-support dashboards. This approach empowers stakeholders to anticipate crises, fine-tune interventions on the fly, and foster resilient communities under SDG 1.

Real-Time Data Ingestion

- **Satellite and Aerial Imagery Streams:** Leverage high-frequency imagery from constellations like PlanetScope and ICEYE to capture daily or weekly changes in built environments, agricultural fields, and infrastructure. Integrate these feeds directly into GIS platforms using such platforms as ArcGIS Pro, Google Earth Engine via APIs for continuous monitoring.
- **Internet-of-Things (IoT) Sensors**: Deploy low-cost environmental sensors (e.g. for air quality, water levels, soil moisture) alongside community-managed weather stations. Ingest sensor telemetry into spatial databases (PostGIS, GeoJSON streams) to detect deprivations linked to environmental stressors such as flooding, drought, and pollution that exacerbate poverty.
- **Mobile and Social Media Proxies:** Harness anonymized CDRs like mobile money transactions, and geotagged social media posts. Use streaming Extract, Transform, Load (ETL) tools like Kafka and NiFi to funnel these streams into GIS for near-instantaneous mapping of population movements, market access, and livelihood shocks.

AI-Driven Feature Extraction

- **Deep Learning on Imagery:** Train convolutional neural networks like UNet, SegNet to segment features like informal shelters, unpaved roads, and water points. Embed model inference pipelines into GIS via ArcGIS Image Server or custom Docker containers so that every new image layer is automatically annotated with socio-economic proxies (Paszke et al., 2019).
- **Natural Language Processing (NLP)**: Apply transformer-based models such as **Bidirectional Encoder Representations from Transformers**

(BERT) or RoBERTa to parse text from community feedback, SMS surveys, or local news feeds. Geo-reference extracted insights such as mentions of illness outbreaks, school closures, market price spikes—to enrich poverty indicator layers in real time (Abadi et al., 2015).

- **Anomaly Detection**: Use unsupervised models such as autoencoders and isolation forests on time series of satellite-derived night-time light intensity or vegetation indices to flag unexpected drops in economic activity or food security, prompting rapid field verification.

Online Learning and Model Updating

- **Incremental Training**: Implement online algorithms like incremental random forest, streaming K-means that update model parameters as new labeled data like ground surveys, PGIS validations, arrive, ensuring models adapt to evolving local contexts without full retraining.
- **Active Learning Loops**: Integrate model uncertainty measures to drive targeted data collection campaigns. When the model flags high-uncertainty pixels or features, use mobile apps to dispatch field teams or crowdsource verifications, closing the loop between predictions and ground truth.

Edge-to-Cloud Architectural Patterns

- **Edge Computing Nodes:** Install GPU-enabled edge servers in district offices or aboard drones to perform initial inference on imagery and sensor data, reducing latency for local decision-making.
- **Federated Learning**: For sensitive data like mobile records, health surveys, etc., train local models on-device and aggregate only model weights, not raw data, centrally, preserving privacy while enhancing model generalization across regions (Paszke et al., 2019).

Decision Support and Visualization

- **AI-Powered Dashboards**: Build dynamic dashboards like Dash, Power BI, with Python SDK that fuse AI outputs with GIS layers. Features include real-time alerts when poverty indicators cross critical thresholds, scenario-comparison tools for "what-if" planning, and automated reporting modules.
- **Digital Twin Environments:** Create virtual replicas of target regions that simulate infrastructure investments or policy interventions. Link GIS-based twins with reinforcement-learning agents to explore optimal resource allocation strategies under varying budget and climate scenarios.

Ethical AI and Governance

- **Bias Auditing**: Routinely evaluate model performance across demographic groups and geographies to detect and correct biases—ensuring that AI predictions do not perpetuate existing inequalities.
- **Transparent Model Explainability:** Integrate explainable-AI tools like SHAP and LIME within GIS so that planners can understand which features like road proximity and land cover change, drive poverty risk scores.

- **Community Oversight:** Establish participatory AI councils, where local representatives review AI-derived poverty maps and co-design model refinements, reinforcing trust and accountability.

Future Directions

- **Multi-Modal Fusion:** Advance toward unified architectures that jointly process imagery, text, sensor streams, and tabular data, delivering richer, more nuanced poverty assessments.
- **Predictive Early Warning Systems:** Combine climate-risk forecasts, market-price models, and migration projections in GIS-AI hybrids to anticipate spikes in multi-dimensional poverty and trigger pre-emptive interventions (Paszke et al., 2019).
- **Autonomous Field Robotics:** Pilot ground and aerial robots equipped with computer-vision models to autonomously survey remote areas, generating high-resolution poverty proxies, and accelerating data refresh cycles.

CASE STUDIES

Below are two illustrative applications of the chapter's methodologies, one leveraging real CDR data for dynamic poverty mapping in Côte d'Ivoire, the other a hands-on GIS workflow simulation in rural Malawi.

Case Study One: Côte d'Ivoire—Mobile Phone-Driven Poverty Mapping

Context: Traditional household surveys in Côte d'Ivoire occur only every few years, leaving large gaps in up-to-date poverty data.

Step 1: Anonymized CDR Extraction

- Partnered with Orange to access six months of anonymized Call Detail Records; voice/SMS events are then aggregated at cell-tower level (Orange Telecom, 2015).
- Ensured GDPR-style Safeguards: User IDs hashed; spatial resolution coarsened to 1 km^2 grid cells.

Step 2: Feature Engineering in GIS

- Computed tower-catchment flows and mobility metrics with radius of gyration and location entropy.
- Generated raster proxies—average daily unique users, in-out flux, diurnal entropy—using QGIS.

Step 3: ML-Driven Poverty Prediction

- Trained a Random Forest on grids where the recent DHS survey clusters provided wealth-index scores.
- Achieved $R^2 \approx 0.72$ out-of-sample, demonstrating strong correspondence between CDR features and poverty (Smith & Mashadi, 2015).

Step 4: Continuous Poverty Surface

- Applied the trained model across the national grid daily, producing up-to-date poverty heat maps that captured transient economic stressors like market closures.

Step 5: Operational Impact

- Deployed the heat maps in an ArcGIS Online dashboard for the Ministry of Social Affairs (Esri, 2024).
- Enabled rapid targeting of emergency cash transfers during a cocoa-price slump—reaching 120 000 extra beneficiaries within 48 hours.

Lessons Learned

- Timeliness: CDRs reveal economic shocks weeks before surveys.
- Equity Risks: Tower-coverage bias can under-represent remote areas.
- Ethics and Trust: Public data-use agreements and transparency are essential.

Case Study Two: Simulated Use Case—Government-Led Sustainable Development in Rural Malawi

This simulated scenario illustrates how a government-led approach—supported by stakeholders such as NGOs, funders, and technology partners could implement a GIS-powered poverty reduction strategy in rural Malawi. While fictional, this use case reflects the architecture of an SDG 17-aligned project, where national authorities lead planning and coordination, and external stakeholders contribute tools, funding, and technical expertise to ensure inclusive and sustainable outcomes (Kaabi, 2021).

GEO AI-BASED PROCESS

Step 1: Identifying the Needs

Using GIS technology, we first map out the region's existing infrastructure, water sources, waste disposal sites, and energy access points (QGIS Development Team, 2024). AI-assisted data analysis cross-references satellite imagery, local surveys, and economic activity indicators to establish key gaps as seen in Figure 1.1.

Why it Matters: A structured approach using GIS helps stakeholders avoid redundant efforts and ensures that interventions target real needs, reducing costs and increasing efficiency.

Step 2: Clean Water Accessibility

Technology and Process:

- Low-cost Option: Use Open-source GIS like QGIS and Google Earth to analyze water source locations and contamination levels.
- Mid-tier Investment: IoT-enabled water sensors from Xylem or IBM's Weather Company to track water quality.
- High-end Infrastructure: Comprehensive smart water management systems like those from Veolia or GE Digital.

Human Impact: A mother no longer must walk five miles for potable water, freeing time for education and economic activities.

Step 3: Waste Management Strategy

Technology and Process:

- Low-cost Option: Community-led mapping with tools like OpenStreetMap and WasteAtlas to identify informal waste sites.

- Mid-tier Investment: Partnerships with organizations like TerraCycle or RecycleBank to implement structured waste collection and processing.
- High-end Infrastructure: Full-scale circular economy waste processing systems from Covanta or Suez.

Human Impact: Reducing waste-borne diseases and pollution improves quality of life and economic potential for small businesses.

Step 4: Sustainable Energy Systems

Technology and Process:

- Low-cost option: Solar mapping via Google Sunroof and basic photovoltaic panel deployment.
- Mid-tier Investment: Decentralized microgrid solutions from companies like Schneider Electric or Tesla.
- High-end Infrastructure: Nationwide smart grid integration using Siemens or ABB technology.

Human Impact: Reliable power access enables local schools, healthcare facilities, and businesses to operate efficiently.

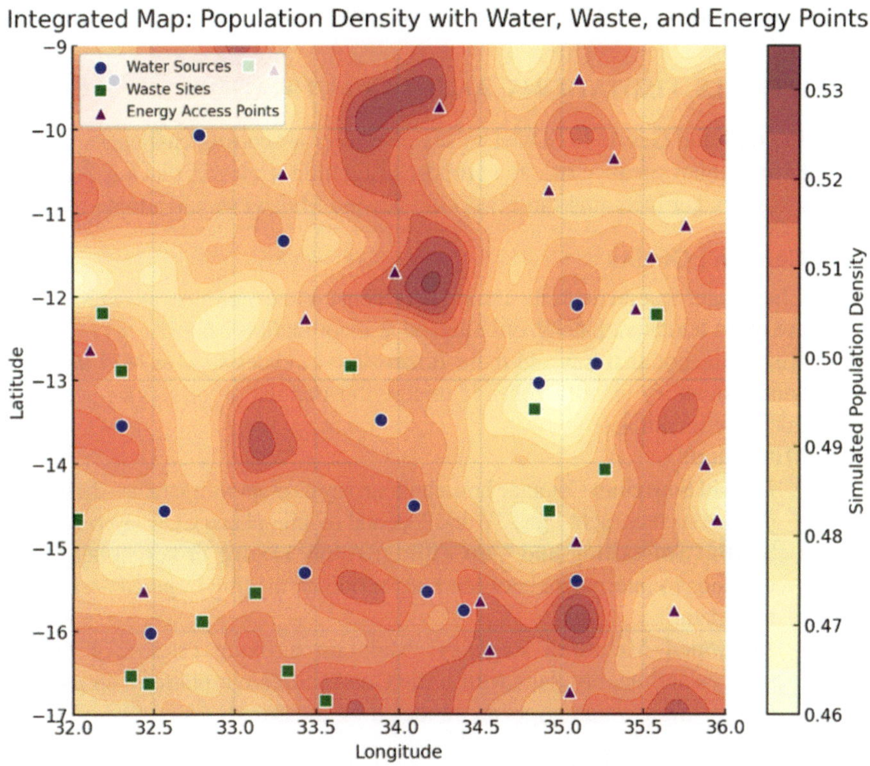

FIGURE 1.1 This map illustrates how GIS layers and population data converge, helping prioritize interventions in higher-density areas.

Step 5: Implementation and Public–Private Partnerships

Strategic Approach:

- Government Collaboration: Working with Malawi's Ministry of Energy and Water Development to ensure policy alignment.
- Corporate Investment: Attracting CSR-driven funding from energy and tech companies.
- Community Co-Ownership: Establishing a local cooperative model where residents have a vested interest in maintaining systems.

  Human Impact: Ownership builds trust and sustainability, empowering the community to drive its own development.

Step 6: AI-Driven Real-Time Feedback and Monitoring

- Free option: AI models trained on public data sets to predict project impact.
- Mid-tier investment: AI-driven dashboards from companies like Esri's ArcGIS for monitoring sustainability goals.
- High-end infrastructure: AI platforms such as IBM Environmental Intelligence Suite or Microsoft's Planetary Computer provide advanced analytics for predictive maintenance and impact tracking, under government leadership and aligned with national priorities.

  Human Impact: Transparent tracking reassures stakeholders that development efforts are achieving tangible results.

CHALLENGES, LIMITATIONS, AND RISK MITIGATION

Applying GIS and GeoAI to SDG 1 raises several cross-cutting challenges. Below, we outline four key areas of risk and propose mitigation strategies that any practitioner or policymaker including those working beyond Malawi should adopt.

Data Gaps and Quality

- **Challenge:** Spatial datasets like surveys, census, and administrative records are often outdated, patchy, or collected at incompatible scales.
- **Risk:** Hotspot analyses and multi-dimensional indices such as MPI layers can misidentify priority zones if underlying population or service-access maps are incomplete.
- **Mitigation:**
  - **Hybrid Sourcing:** Combine household survey clusters with satellite-derived proxies such as night-lights, WorldPop, and mobile-phone CDR layers.
  - **Spatial Smoothing and Gap Filling:** Apply kriging or kernel-density methods to interpolate between known points, tagging imputed areas in metadata.
  - **Versioned Metadata Registry:** Maintain a living catalog of data sources and quality flags within the GeoAI Streamer platform so users always know which layers are modeled vs. measured.

Technical Capacity and Sustainability

- **Challenge:** Many local governments and NGOs lack skills in advanced GIS workflows, machine-learning integration, or dashboard development.

- **Risk:** Complex, one-off pilots fail to transition into routine operations, leading to "orphaned" systems.
- **Mitigation:**
  - ***Modular Training Packages*:** Develop short, self-contained "GIS-as-a-Service" modules—covering core tasks like hotspot detection, buffer analysis, and federated model updates.
  - ***AI-Powered Assistants*:** Embed chat-bot guides and code snippets directly in the GeoAI Streamer UI to scaffold common analyses.
  - ***Train-the-Trainer Networks:*** Certify local "power users" who can then cascade workshops across districts, ensuring knowledge remains in-country.

Ethical, Privacy, and Governance Concerns

- ***Challenge:*** Integration of mobile metadata, household surveys, and IoT sensor feeds can risk re-identification or inadvertent disclosure of sensitive socio-economic conditions.
- ***Risk:*** Breaches of trust, legal challenges under data-protection laws, or community backlash.
- ***Mitigation:***
  - **Federated Architecture:** Keep raw data within custodial agencies—only aggregate parameters or model weights leave each node.
  - **Spatial Coarsening and Anonymization:** Impose minimum mapping units (e.g. 1 km^2 grids) and hashed IDs for any individual-level inputs.
  - **Governance Frameworks:** Establish multi-stakeholder oversight committees, including data-ethics boards and community representatives, to vet use cases and policies.

Financial and Infrastructure Constraints

- ***Challenge*:** Real-time GeoAI workflows can strain limited IT budgets, connectivity, and hardware in remote settings.
- ***Risk*:** Data sync delays, dashboard outages, or unsustainable recurring costs.
- ***Mitigation:***
  - **Hybrid Deployment Models:** Combine edge-compute nodes for local processing and caching with cloud bursts only during off-peak hours.
  - **Open-Source First:** Leverage QGIS, GRASS, R, Python, and serverless functions to minimize licensing fees.
  - **Cloud Credit Partnerships:** Secure grants or credits from major providers such as AWS, Google Cloud, or Azure to underwrite GPU-accelerated training and large-scale storage.

Key Takeaway: By proactively embedding these mitigation strategies into project design—documenting trade-offs in a transparent metadata registry, upskilling local teams, and enforcing robust privacy safeguards for GIS-AI interventions for SDG 1 can move beyond proof-of-concept into enduring, scalable solutions across diverse contexts.

Policy Implications and Governance Recommendations: To translate GeoAI-enabled poverty monitoring from pilot projects into sustained, country-wide practice, and ultimately across all SDGs, governments, and stakeholders should adopt the following policy and governance measures:

Embed GIS-AI Monitoring into National SDG Frameworks

- ***Rationale:*** Spatially explicit, real-time poverty data closes the feedback loop between policy targets and on-the-ground outcomes.
- ***Action Steps:***
  - Incorporate GeoAI outputs (poverty-heat maps, service-coverage indicators) into national SDG dashboard portals.
  - Mandate regular, such as quarterly, updates of GIS-AI metrics alongside traditional survey reports.
- ***Example:*** Country X's Ministry of Planning publishes an integrated GIS-AI poverty indicator alongside its annual SDG report, enabling mid-year course corrections.

Standardize Open-Data Policies with Licensing and Privacy Safeguards

- ***Rationale*:** Harmonized data governance promotes interoperability, re-use, and public trust, while protecting sensitive information.
- ***Action Steps:***
  - Define clear licensing tiers such as open or restricted for each dataset type like survey, CDR, or satellite.
  - Enforce minimum anonymization and spatial-coarsening standards in official data-release protocols.
  - Provide "data passports" within GeoAI Streamer documenting provenance, quality, and usage rights.
- ***Example:*** A regional data-sharing platform requires all contributors to certify compliance with coarsening and metadata guidelines before publishing any layer.

Establish Multi-Stakeholder Steering Committees

- ***Rationale:*** Inclusive governance ensures that technical insights inform policy, and that community needs shape analytical priorities.
- ***Action Steps:***
  - Convene representatives from line ministries such as Health, Education and Social Affairs, civil society, academia, and private sector such as telecom and energy.
  - Define clear terms of reference: Data-governance, budget oversight, and joint funding mechanisms for GeoAI initiatives.
  - Schedule periodic review meetings aligned with SDG reporting cycles.
- **Example:** Augmented Governance in Action.

Metapolicy's augmented governance system provides a modular, AI-driven infrastructure that interlinks finance, public institutions, and civil society in a single, interoperable platform, thereby operationalizing vertical and horizontal SDG coordination without fragmenting across silos

Incentivize Cross-Sector Partnerships

- ***Rationale:*** Private–public collaborations accelerate technology deployment, share risk, and mobilize additional financing.
- ***Action Steps:***
  - Offer tax credits or matching-grant schemes for companies that co-invest in GIS-AI infrastructure such as cell-tower data access and sensor networks.
  - Launch innovation challenges or hackathons that prize solutions integrating GeoAI Streamer outputs with service-delivery platforms.
  - Embed partnership requirements into CSR guidelines for development-sector firms.
- **Example:** A telecom operator receives a regulatory fee reduction in exchange for anonymized CDR provisioning and joint funding of rural connectivity hubs.

Key Takeaway: By codifying these policies—embedding GeoAI metrics in national SDG frameworks, standardizing data governance, convening multi-stakeholder bodies, and incentivizing partnerships, countries can ensure that spatial-AI innovations become a permanent, scalable component of their sustainable development toolkit.

TOOLKIT AND RESOURCES

Open-Source Software and Platforms

- **GIS Desktop and Libraries:** QGIS (QGIS Development Team, 2024), GRASS GIS, OpenLayers.
- **Web and Cloud APIs:** Google Earth Engine (Gorelick et al., 2017), OpenStreetMap Overpass (OpenStreetMap Contributors, 2024), Mapbox.
- **Metapolicy:** A comprehensive AI-driven platform (Kaabi, 2021) that allows developing, assessing, and streamlining public policy horizontal and vertical coherence across multiple policymaking levels. Features include federated ML pipelines to safeguard data, dynamic dashboards integrating satellite, sensor, and mobile proxies, and automated ESG/SDG compliance outputs like SPI scorecards and XBRL exports.

Data Sources and APIs

- Household and Demographics: DHS Program, WorldPop (Tatem, 2017; WorldPop, 2024).
- Humanitarian and Economic: Global Humanitarian Data Exchange.

Learning Materials

- **Online Courses and MOOCs:** UN-Habitat GIS Academy; Esri MOOC "AI for Good." (Esri, 2024)
- **Documentation:** Official guides for QGIS (QGIS Development Team, 2024), GRASS GIS, TensorFlow (Abadi et al., 2015), PyTorch (Paszke et al., 2019).

GLOSSARY

- Federated Learning: A technique where models train locally and share only parameters, preserving data privacy (Paszke et al., 2019).
- Isochrone: A map showing areas accessible within equal travel times to a given point.
- MPI: A composite index measuring deprivations across health, education, and living standards (Alkire & Santos, 2010).
- PGIS : GIS workflows that actively involve community members in data collection and analysis (Alvarez et al., 2021).

CONCLUSION

This chapter has outlined a forward-thinking approach to integrating GIS, AI, and machine learning for comprehensive, real-time poverty assessment and intervention design. By combining methodological rigor (Alkire & Santos, 2010; Gorelick et al., 2017), ethical governance, and participatory engagement (Alvarez et al., 2021), stakeholders can drive targeted, sustainable progress toward SDG 1. Continued innovation, capacity building, and robust policy frameworks will be critical to ensure these tools translate into inclusive, lasting change.

REFERENCES

Abadi, M. et al. (2015). TensorFlow: Large-scale machine learning on heterogeneous distributed systems (arXiv:1603.04467) [Preprint]. Retrieved from https://arxiv.org/abs/1603.04467.

Alkire, S., & Santos, M. E. (2010). Acute Multidimensional Poverty: A New Index for Developing Countries (OPHI Working Paper No. 38; UNDP HDRO Background Paper 2010/11). Oxford Poverty & Human Development Initiative, University of Oxford. Available at: https://ssrn.com/abstract=1815243

Esri. (2024). ArcGIS Online. Retrieved from https://www.arcgis.com

Gorelick, N. et al. (2017). Google Earth Engine: Planetary-scale geospatial analysis for everyone. *Remote Sensing of Environment*, *202*, 18–27. https://doi.org/10.1016/j.rse.2017.06.031

Kaabi, L. (2021). Multidimensional Poverty Assessment Mechanism, Unesco UNESCO-funded project, Implementation Partner - Commission of Social Solidarity and Poverty Reduction, Sudan, 2021, https://www.metapolicy.net/mpam

Mennis, J., & Hultgren, T. (2006). Generating surface models of population using dasymetric mapping. *Cartography and Geographic Information Science*, *33*(3), 179–194..

Malakar, K.D., & Roy, S. (2024). Community Cartography and Participatory GIS. In *Mapping Geospatial Citizenship* (pp. 63–75). SpringerBriefs in GIS. Springer, Cham. https://doi.org/10.1007/978-3-031-63107-8_5

OpenStreetMap Contributors. (2024). OpenStreetMap. Retrieved from https://www.openstreetmap.org

Orange Telecom. (2015). Data for Development (D4D) Challenge: Côte d'Ivoire. Retrieved from https://www.d4d.orange.com

Paszke, A. et al. (2019). PyTorch: An imperative style, high-performance deep learning library. *Advances in Neural Information Processing Systems*, *32*, 8024–8035.

QGIS Development Team. (2024). QGIS geographic information system. Open-Source Geospatial Foundation Project. Retrieved from https://qgis.org

Smith-Clarke, C., Mashhadi, A., & Capra, L. (2014). *Poverty on the Cheap: Estimating Poverty Maps Using Aggregated Mobile Communication Networks.* In *Proceedings of the 18th ACM Conference on Computer Supported Cooperative Work & Social Computing (CSCW 2014).* ACM Digital Library.

Tatem, A. J. (2017). WorldPop, open data for spatial demography. *Scientific Data*, 4, Article 170004. https://doi.org/10.1038/sdata.2017.4

WorldPop. (2024). WorldPop Population Data. Retrieved from https://www.worldpop.org

2 SDG #2—Technology for Regenerative and Sustainable Agriculture for Food Security

Mirriam Omala-Gauvin and Josphat Makanga

INTRODUCTION

The 2030 Agenda for Sustainable Development Goals (SDGs), launched in 2015, is a global agenda to end extreme poverty, eradicate hunger, protect the planet, and promote peace and prosperity for everyone everywhere by 2030. In Africa, the implementation of the 17 SDGs with the 169 targets and 234 indicators is meant to be guided by national policies. In Africa, these goals are aligned with Agenda 2063 and supported by institutional frameworks, with the African Union playing a key role in coordination and advocacy. The main question remains: How can the achievement of SDG#2 be accelerated? The clock is ticking, and more than ever, the world is plagued with conflicts, food crisis, climate instigated emergencies, and pandemics. The GIS and its applications can accelerate the achievement of the SDGs. In the area of agriculture and to ensure food security, and sustainable and regenerative agriculture across the African continent, the African Union, established the Comprehensive Africa Agriculture Development Programme (CAADP), launched in 2003 under the New Partnership for Africa (NEPAD) and endorsed by the African Union. This has emerged as a flagship initiative aimed at revitalizing Africa's long-neglected agricultural sector. Its overarching goal is to promote sustainable agricultural growth and poverty reduction by addressing food insecurity, building agricultural value chains, and managing natural resources holistically. CAADP represents a significant shift in continental agricultural policy, prioritizing evidence-based planning, inclusive stakeholder engagement, regional harmonization, and increased investment. CAADP, being a continental framework, enhances Africa-wide food productivity, that provides food security and ensures the regenerative practices through the implementation of policies at National level. Key achievements include increased political commitment, the creation of investment plans aligned with national strategies, the strengthening of regional economic communities (RECs) in guiding implementation, and enhanced donor coordination through mechanisms like the Multi-Donor Trust Fund. Although challenges remain, such as limited post-compact follow-up and uneven national ownership, the CAADP has laid a strong foundation for agricultural transformation and continues to serve as a vital platform for mobilizing resources and partnerships to meet Africa's agricultural and food security goals.

DOI: 10.1201/9781003586319-3

Efforts are further strengthened through funding, capacity building, and strong regional and international partnerships across key sectors like agriculture, health, and education. SDG 2 – Zero Hunger seeks to end hunger, achieve food security, improve nutrition, and promote sustainable agriculture, with a focus on ensuring access to safe and nutritious food for all, particularly the poor and vulnerable. Progress is measured through key indicators such as the prevalence of undernourishment, child stunting and wasting, agricultural productivity and incomes of small-scale farmers, the sustainability of food production systems, and the genetic diversity of crops and livestock.

Since its inception, SDG 2 has driven significant global progress in combating food and nutritional security with notable success stories in Africa. The growing adoption of geospatial technology has further accelerated its impact through the introduction of numerous geospatially enabled tools. Several geospatial tools are supporting climate-resilient agriculture in Africa. Some of the success stories include-*Africa Agriculture Adaptation Atlas (AAAA),* which maps climate-smart agricultural zones, crop suitability, and gender-sensitive enterprise options to guide adaptive decision-making; the *AgroClimate,* which is a platform that provides agro-climatic advisories that help farmers manage risks and optimize practices under changing climate conditions. Kenya's *Crop Monitor,* which uses satellite data to monitor crop conditions, offers timely insights to support farm-level interventions. Others include Africa *Soil Information Service (AfSIS)* that delivers high-resolution soil and agronomy maps, enabling better-informed land use and input application decisions across the region and the Risk Management System for Plant Pests and Diseases, which is an Earth Observation (EO) platform promoting data-driven, evidence-based decision-making through monitoring of plant pests and disease occurrences. Other solutions that do not necessary harness the power of Geospatial technologies but have proved impactful toward the attainment of SDG 2, include an innovative solution in rural Nigeria that is helping farmers significantly reduce post-harvest losses caused by pests and environmental factors. And the Kenya Integrated Agriculture Management Information System (KIAMIS), which is an Open Data Platform for the Ministry of Agriculture and Livestock Development that provides farmers with agriculture information and government services country-wide.

Sustainable Development Goal 2 or SDG#2 States

End hunger, achieve food security, improved nutrition, and promote sustainable agriculture.

The targets under SDG2 are the following:

1. By 2030, end hunger and ensure access by all people, in particular the poor and people in vulnerable situations, including infants, to safe, nutritious, and sufficient food all year round
2. By 2030, end all forms of malnutrition, including achieving, by 2025, the internationally agreed targets on stunting and wasting in children under five years of age, and address the nutritional needs of adolescent girls, pregnant and lactating women and older persons

3. By 2030, double the agricultural productivity and incomes of small-scale food producers, in particular women, indigenous peoples, family farmers, pastoralists and fishers, including through secure and equal access to land, other productive resources and inputs, knowledge, financial services, markets and opportunities for value addition and non-farm employment
4. By 2030, ensure sustainable food production systems and implement resilient agricultural practices that increase productivity and production, that help maintain ecosystems, that strengthen capacity for adaptation to climate change, extreme weather, drought, flooding and other disasters and that progressively improve land and soil quality
5. By 2020, maintain the genetic diversity of seeds, cultivated plants and farmed, and domesticated animals and their related wild species, including through soundly managed and diversified seed and plant banks at the national, regional and international levels and promote access to and fair and equitable sharing of benefits arising from the utilization of genetic resources and associated traditional knowledge, as internationally agreed
6. Increase investment, including through enhanced international cooperation, in rural infrastructure, agricultural research and extension services, technology development and plant and livestock gene banks in order to enhance agricultural productive capacity in developing countries, in particular, least developed countries
7. Correct and prevent trade restrictions and distortions in world agricultural markets, including through the parallel elimination of all forms of agricultural export subsidies and all export measures with equivalent effect, in accordance with the mandate of the Doha Development Round
8. Adopt measures to ensure the proper functioning of food commodity markets and their derivatives and facilitate timely access to market information, including food reserves, in order to help limit extreme food price volatility.

GEOSPATIAL AND EARTH OBSERVATION TOOLS USED IN ENVIRONMENTAL/AGRICULTURE MONITORING

Geospatial technologies encompass a range of fields used to collect, analyze, and visualize location-based data for better decision-making. It constitutes Global Positioning System/Global Navigation Satellite Systems (GPS/GNSS) that provide accurate positioning and navigation tools essential for mapping and environmental data collection. Remote Sensing, which involves acquiring EO data from satellites and airborne sensors, monitors land cover, vegetation condition, climate hazards and trends in land use change. Drone technologies enable rapid, high-resolution and flexible data collection at local scales, supporting applications such as crop monitoring, infrastructure inspection, and disaster risk assessment. Other components include geographic information systems (GIS), which integrate, analyze and visualize spatial data, forming the backbone of spatial decision support systems. These outputs are often presented through Cartography, the art and science of map-making that communicates geospatial insights clearly. *Geo-design* combines spatial analysis with design thinking to plan and evaluate scenarios for land use, urban planning, and environmental management in a participatory and data-driven way. Increasingly, artificial intelligence (AI) and machine learning (ML), neural network programming provide predictive insights of large-scale environmental data, and are being used to automate feature extraction, detect spatial patterns and enhance predictive modeling in large and complex geospatial datasets. Ground-penetrating radar (GPR) and light detection and ranging (LiDAR) are emerging EO technologies that enhance geospatial analysis. GPR is used for subsurface mapping, while LiDAR, uses laser pulses to generate highly accurate 3D models of terrain, vegetation, and infrastructure, even under dense canopy cover. These tools complement traditional remote sensing methods by offering enhanced resolution and depth insights, expanding the possibilities for precise and detailed geospatial analysis.

Geospatial technologies and EO play crucial and complementary roles in planning, implementing, and managing a sustainable agriculture sector in any country by supporting data-driven decision-making across the entire agricultural value chain. EO satellites routinely capture high-resolution data as they orbit the Earth along predetermined paths, providing consistent, timely, and accurate information on land cover, crop health, soil conditions, and more. Additionally, these satellites can be dynamically tasked to observe specific areas outside their usual path, enabling rapid response in urgent situations such as natural disasters or emergency agricultural assessments. When integrated with geospatial tools like GIS, GPS, unmanned aerial vehicles (UAVs)/drones, GPR, and LiDAR, Geospatial technology becomes even more powerful by enhancing precision agriculture.

Optimizing resource use and supporting long-term agricultural resilience and sustainability have become urgent imperatives in the face of climate change, erratic rainfall patterns, prolonged droughts, and increasing pest pressures. In many developing countries, such as Kenya, population growth, coupled with limited youth participation in agriculture, places the burden of feeding nations on an aging farming population. This makes transforming the agriculture sector not a matter of choice, but of survival. Geospatial technologies, including advanced tools like geo-artificial intelligence (GeoAI), are essential in this transformation. By integrating vast datasets from remote sensing, mobile platforms, and other sources, GeoAI enables accurate insights into crop health, soil moisture, biomass, and yield potential. In a climate-constrained world, food production must be faster, more efficient, nutritionally sound, and environmentally sustainable if we are to end hunger and ensure food access to all people, the poor in vulnerable situations and infants, all year round. GIS-embedded tools such as crop monitors and suitability analytics play a vital role in matching the most appropriate crops and value chains to optimal agroecological zones. Geospatial technologies like GIS and GPS further enhance yield estimation and early action, in case of low yield estimates, by mapping, analyzing, and interpreting spatial patterns in agricultural activities, thereby driving precision farming, efficient resource allocation, early risk detection, and reliable yield forecasting. Through collaboration, Agri-players across all levels of the value chain, from policymakers to smallholder farmers, can adopt sustainable strategies that enhance productivity and respond proactively to environmental and market dynamics.

GEOSPATIAL TECHNOLOGIES IN EMPOWERING GRASSROOT FARMERS

Many nations across Africa are increasingly leveraging geospatial technologies to empower citizens and optimize farm management practices. In Kenya, this transformation is exemplified by the Ministry of Agriculture and Livestock Development, with support from the Regional Centre for Mapping of Resources for Development (RCMRD), which has digitized the traditional farmers' handbook and converted outdated agro-ecological maps into dynamic, digital agro-ecological zone (AEZ) maps with a mobile version that can easily be accessed in a farmers mobile phone for real time decision on one's farming preference. This digitization effort forms a core part of the government's strategy to shift from subsidizing food prices to investing in the means of production, a move fully aligned with the bottom-up economic transformation agenda (BETA) (BETA, 2023). By harnessing the power of GIS, remote sensing, and other geospatial tools, the government can now deliver more targeted support to small-scale producers, particularly women, indigenous communities, and livestock farmers, through interventions like AEZ (UNFAO), specific seeds and fodder varieties, blended fertilizers, geo-tagged tractors, and localized soil testing services. The government is using geospatial technologies to empower farmers to contribute to the SDGs with the aim of doubling agricultural productivity, enhancing resilience, and improving livelihoods across Kenya. See Figure 2.1 for Digital Kenya AEZ.

The presence of water towers in a country plays a critical role in maintaining the water balance within ecosystems. Given their vastness and often remote locations, water towers are best monitored using satellite remote sensing. These natural reservoirs are crucial for regulating the availability of water across sectors such as agriculture, industry, and domestic use. However, their ability to function sustainably depends on preserving the surrounding forest cover within ecological thresholds. To support monitoring and conservation, geospatial technologies offer powerful capabilities for tracking and analysing these changes by identifying any adverse situations in these ecosystems. The technologies enable detailed monitoring of water level fluctuations upstream and downstream, evaluation of consumption trends, and assessment of the environmental impacts of overuse. This data-driven approach is essential for mitigating ecological degradation and preventing potential environmental disasters. In addition, spatial relationships between features such as points, lines, and polygons, also known as topology, enable agricultural stakeholders to understand how key elements interact across the landscape. This may include analysing the proximity of water sources to irrigation schemes, markets, labor hubs, and varying climate conditions to ensure sustainable food production, sound ecosystem health and resilient agricultural practices that increase productivity and conservation opportunities. These spatial insights are essential for planning sustainable food production systems as they support the implementation of climate-resilient agricultural practices that enhance productivity, protect ecosystems, and build adaptive capacity to climate change, extreme weather, drought, flooding, and other environmental challenges. Figure 2.2 illustrates the land use classification to determine productivity and identify conservation areas.

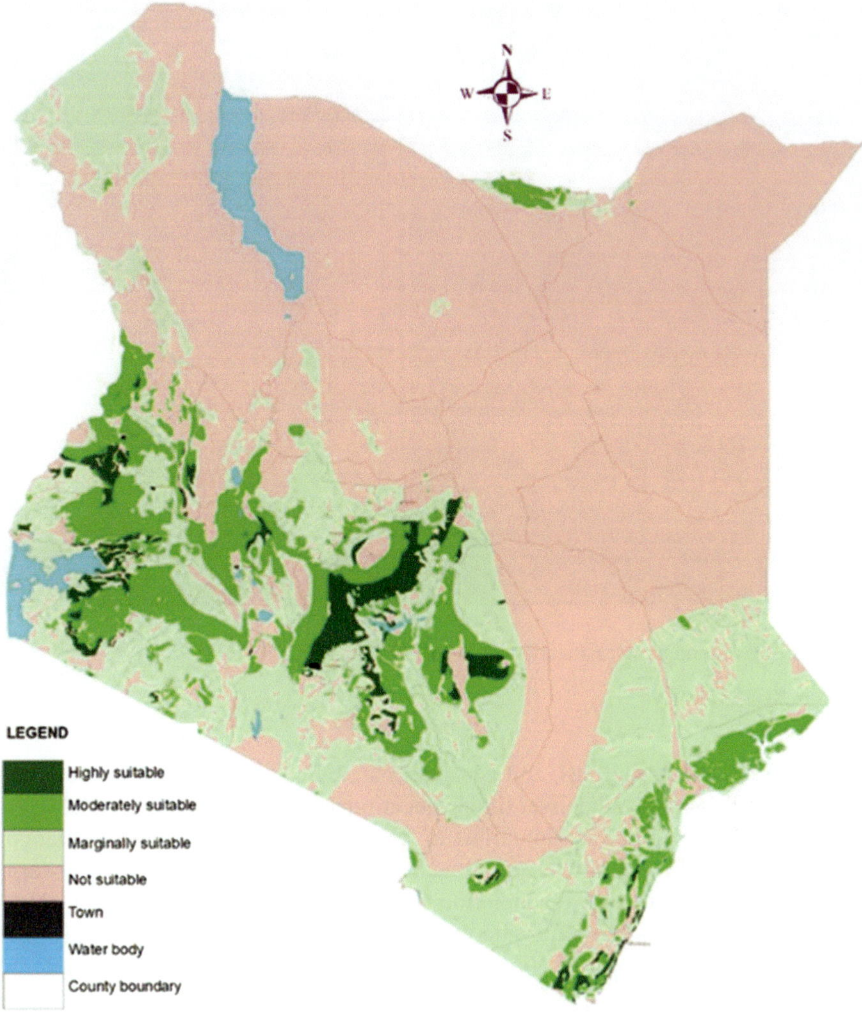

FIGURE 2.1 Digital maize suitability map of Kenya: https://statistics.kilimo.go.ke/en/aez-dataset/ https://statistics.kilimo.go.ke/en/aezdataset/

TACKLING CLIMATE CHALLENGES IN THE AGRICULTURE SECTOR USING EARTH OBSERVATION DATA AND INNOVATIVE TOOLS

In Sub-Saharan Africa, particularly the Horn of Africa, encompassing Kenya, Somalia, Djibouti, Ethiopia, and Eritrea—climate change is dramatically reshaping the geographic landscape. Shifting rainfall patterns, accelerated desertification, fluctuating surface and groundwater levels, and changing agricultural zones are increasingly disrupting ecosystems and human livelihoods. These transformations pose serious threats to food security, biodiversity, and sustainable development across the region. EO technologies including satellite data, archived

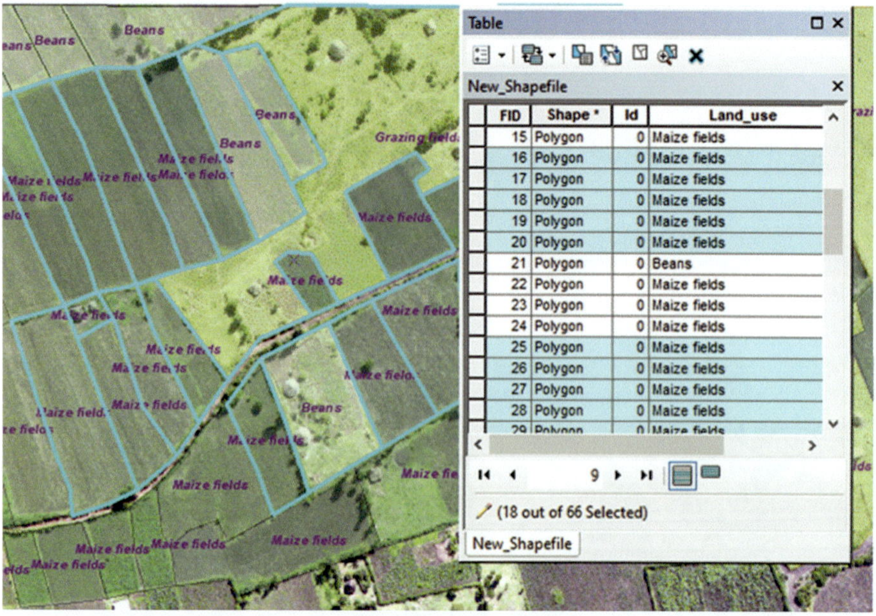

FIGURE 2.2 Land use classification from aerial photograph.

Source: RCMRD.

agriculture research findings, and geospatial analytics offer authoritative tools to visualize the changes and deliver timely, location-specific insights that enable governments, scientists, and communities to track related environmental challenges such as water body fluctuations and drought severity. This in return assists governments in designing proactive adaptation strategies. By visualizing climate stressors on maps, critical intersections between vulnerable land uses, communities, and ecosystems can be identified, allowing for targeted, practical, and locally tailored interventions. Additionally, EO enables the detection of recurring crop failures, the degradation of indigenous vegetation cover, and the identification of strategic opportunities for land restoration and enhanced farm management-key adaptations in strengthening the resilience of the agricultural sector amid growing environmental uncertainty.

CASE STUDY—AFRICA AGRICULTURE ADAPTATION ATLAS

With support from the Sustainable Agriculture Foundation-Africa (SAF-A) and the Gates Foundation, RCMRD is spearheading the implementation of the AAAA use case across Kenya's 38 counties. The initiative aims to equip the agricultural sector with data-driven, research-backed practices that enhance climate resilience and drive increased productivity at the county level. The AAAA is a data-rich platform developed by the International Centre for Tropical Agriculture (CIAT) that provides a guide to tackle climate challenges in the agriculture sector. Its focus is on climate-smart solutions including crop diversification, improved water management, and livestock resilience. The platform has been compiled from research and observation data of over 80 years of research from 1938 to 2018. It contributes guidance on adaptation investments and policies by providing interactive notebooks that are developed to tackle different topics related to climate hazards adaptations for policymakers, researchers, and investors. The insights delve into climate risks, impacts, vulnerability, and solutions to effectively mitigate these challenges, while stressing the importance of adapting tested farming methods such as crop diversification, improved water management, and resilient livestock breeding.

RCMRD is implementing the Kenyan use case and leveraging on its strength in geospatial capacity to build knowledge products that can be visualized as insightful maps that can easily be understood by the Kenya county agriculture personnel, including the departmental heads, crops and livestock officers, data officers, and gender specialists. The AAAA is a uniquely valuable platform designed to advance the targets under SDG 2—Zero Hunger by providing a data-rich, interactive environment for agricultural planning and climate resilience. What sets the AAAA apart is its structured approach to addressing each specific indicator under SDG targets 2.3 and 2.4, such as improving agricultural productivity, ensuring sustainable food systems, and enhancing resilience to climate-related shocks.

Through the integration of EO data, geospatial analysis, and field-based insights, the platform enables users, ranging from policymakers to farmers, to visualize trends, analyze regional disparities, and make evidence-based decisions. By transforming complex data into accessible maps and dashboards, the Atlas supports the monitoring of progress, identification of vulnerabilities, and the design of localized adaptation strategies, making it a practical tool for achieving food and nutrition security across Africa against the adverse impacts changing climate, as seen in Figure 2.3.

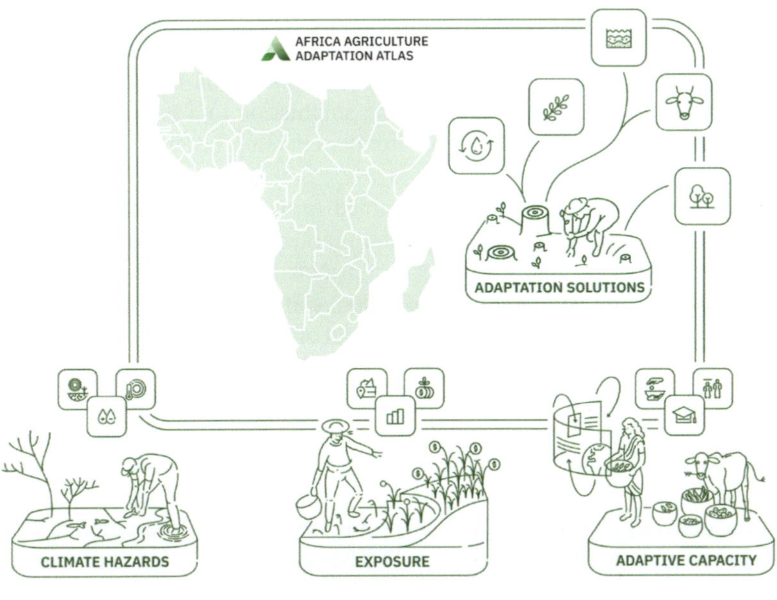

FIGURE 2.3 Africa Agriculture adaptation atlas landing page: https://adaptationatlas.cgiar.org/

LEVERAGING GEOSPATIAL MOBILE APPLICATIONS FOR PEST AND DISEASE MANAGEMENT

Pests and diseases remain a major threat to food and nutritional security in sub-Saharan Africa, causing significant yield losses in key staple and cash crops such as maize, cassava, sorghum, and vegetables. These losses not only compromise household food availability but also limit income generation and access to diverse, nutrient-rich diets causing worsening malnutrition, especially among vulnerable populations. The increasing frequency and intensity of destructive pests like the fall armyworm De Groote et al (2020) and desert locusts have been closely linked to changing climate patterns, such as rising temperatures and erratic rainfall, which create favorable breeding and migration conditions.

In response to these growing challenges, GIS and EO technologies have emerged as vital tools in the real-time monitoring, early warning, and spatial analysis of pest and disease outbreaks. By integrating satellite imagery, weather data, and ground-level observations, these technologies help model pest risk zones, track movement patterns, and support timely interventions such as pesticide application, use of biological control agents, or adoption of pest-resistant crop varieties. This data-driven approach strengthens surveillance systems and enhances decision-making, reducing both pre- and post-harvest losses in a region already struggling with food insecurity.

A practical application of this approach is exemplified by the Integrated Multi-Pest Surveillance Tool, co-developed by Kenya's Ministry of Agriculture and Livestock Development and the RCMRD. The tool was designed to address the challenge of migratory pests and invasive weeds by enabling geospatially referenced reporting through a mobile application. Initial piloting and stakeholder validation were conducted in three counties: Nyeri, Kakamega, and Kilifi, where feedback from County Governments and Ward Agricultural Officers (WAOs) was incorporated into the tool's refinement. Traditionally, WAOs relied on text-based reports to inform county offices of pest incidents, leading to delays in response and limited data usability. The mobile tool now allows real-time data submission to a national server at the Plant Protection and Food Safety Directorate, enabling faster analysis, diagnosis, and intervention.

In Kakamega County, WAOs were trained to use the tool for continuous pest surveillance. The initial 30-day test period marked the start of a sustained effort to integrate technology into local pest management practices, improving responsiveness and enabling structured feedback loops after control measures are implemented. This initiative underscores the potential of GIS and EO in transforming pest and disease management, especially in the face of climate variability and increasing agricultural vulnerabilities. The implementation of risk management strategies for plant pests and diseases has brought together a wide range of institutions under the Multi-Institutional Technical Team (MITT), including the Desert Locust Control Organization (DLCO). Through strengthened international cooperation, the team has collaboratively developed a pest surveillance infrastructure tailored for rural farming communities. In addition, the MITT has guided the design of user interfaces

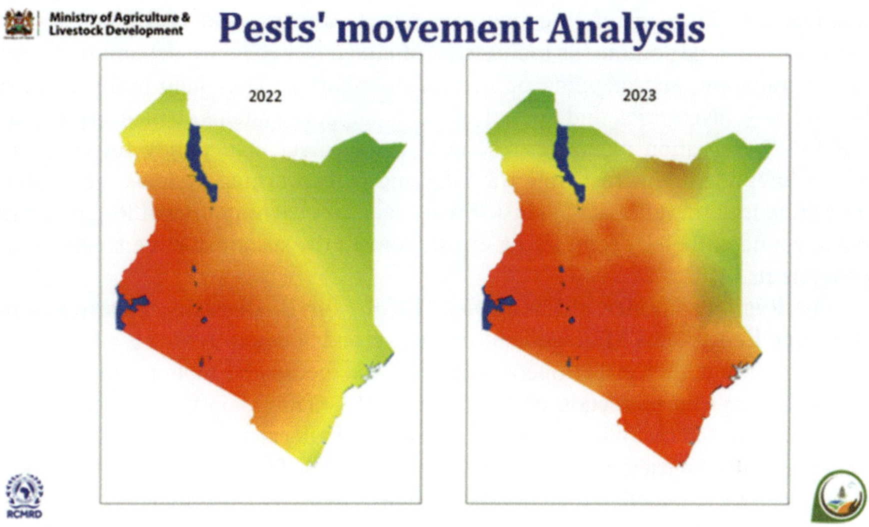

FIGURE 2.4 Pests suitability maps developed from three reported counties in Kenya

Source: Pest Protection and Food Safety Directorate

for surveillance tools and identified priority pests and diseases, thereby driving targeted investments in rural surveillance systems and promoting agricultural research and extension services.

EARTH OBSERVATION IN INVENTORYING AND PROTECTING BIODIVERSITY

Protected areas, along with plant and animal species, are central to biodiversity conservation and significantly support local livelihoods, as seen in Figure 2.4 above. They underpin vital ecosystem services such as food production, clean water supply, medicinal resources, and natural safeguards against disasters and climate change. Effectively managed protected area networks are key to fulfilling global environmental goals, notably the Convention on Biological Diversity and the SDGs. Specifically, they contribute to SDG Target 2.5 and Indicators 2.5.1 and 2.5.2, which emphasize the conservation of genetic diversity in seeds, cultivated plants, and domesticated or farmed animals, including their wild relatives. This includes supporting well-managed, diversified seed and plant banks across national, regional, and international levels, and ensuring equitable access and benefit-sharing of genetic resources and associated traditional knowledge, as internationally agreed. Geospatial technologies play a transformative role in the identification, mapping, inventory, and management of protected areas and biodiversity. Through satellite imagery, remote sensing, and GIS applications, these tools enable continuous monitoring of land cover changes, detection of habitat degradation, and mapping of species distribution. They help build detailed biodiversity inventories and spatial decision-support systems, which are essential for conservation planning, resource management, and policy formulation. When integrated with real-time tracking and early warning systems, geospatial data enhances surveillance against threats such as poaching, illegal logging, and encroachment, thereby reinforcing efforts toward the sustainable management of ecosystems.

The Regional Centre of Excellence (RCoE) for Biodiversity, Forestry, and Seascape Ecosystems Management in Eastern and Southern Africa is a European Union-funded initiative that operationalizes these geospatial tools to improve the understanding and protection of critical natural systems. Jointly implemented by the Centre for International Forestry Research (CIFOR) and the RCMRD, the project spans 24 countries: Angola, Botswana, Comoros, Djibouti, Eritrea, Eswatini, Ethiopia, Kenya, Lesotho, Madagascar, Malawi, Mauritius, Mozambique, Namibia, Rwanda, Seychelles, Somalia, South Africa, South Sudan, Sudan, Tanzania, Uganda, Zambia, and Zimbabwe.

Its core objective is to ensure that forests, seascapes, and biodiversity within these countries are sustainably managed through the provision of high-quality, timely spatial data and information. This supports ecosystem resilience and continued access to essential services such as food, water, energy, and shelter, while enabling sustainable economic development and human well-being.

In support of this mission, the program has also produced a comprehensive Map Book series. These publications highlight spatial data products, biodiversity trends, and decision-support tools focused on sustainable land use, conservation planning, and biodiversity protection across the region. The series is designed to guide policy and program implementation aligned with the Kunming-Montreal Global Biodiversity Framework targets, Figure 2.5.

FIGURE 2.5 Regional center of excellence for biodiversity, forestry, and seascape management: https://rcoe-geoportal.rcmrd.org/

Geospatial tools provide powerful aids in assessing and monitoring a region's land use, land cover, and soil characteristics. By visualizing this data through maps and charts, trends in environmental degradation, such as changes in forest cover, wetland pollution, cropland expansion trends, and grassland depletion, can be effectively analyzed. In densely forested equatorial regions, GIS can be used to estimate tree cover and forest carbon sequestration, which are vital for climate regulation and ecosystem services. Such insights are critical for promoting sustainable agroforestry and informed settlement planning that ensures that commercial crops like coffee, tea, oil palm, soybeans, cocoa, and tobacco do not drive deforestation. Since soil quality is closely tied to crop yields, addressing soil and land degradation is essential for food and nutritional security. These forms of degradation are often triggered by excessive erosion from heavy rains or prolonged drought that destabilise the protective land cover and threaten both environmental health and agricultural productivity.

The Global Monitoring for Environment and Security and Africa (GMES) Initiative

The Global Monitoring for Environment and Security and Africa (GMES) program is in the forefront both providing EO services geared to addressing the land degradation menace as well as empowering communities in identifying restorative opportunities through its programs. The GMES and Africa program represents landmark efforts in positioning Africa as a key player in harnessing space and geospatial technologies for sustainable management and development. Rooted in the African Union's Agenda 2063, these initiatives aim to strengthen the continent's capacity in space science and EO, while promoting African ownership of data and services critical for managing its natural resources and addressing climate-related challenges. The program also seeks to advance the objectives of the United Nations Forum on Forests by highlighting the vital role forests play in sustaining planetary health, stabilizing the climate, and supporting human well-being. Occupying 31% of the Earth's land surface, forests are essential for nurturing biodiversity, regulating water cycles, and sustaining the livelihoods of over 1.6 billion people worldwide.

The GMES and Africa program, jointly funded by the African Union Commission and the European Union, focuses on empowering African institutions to develop and operationalize services based on EO and geospatial technologies. It supports African regional centers of excellence, such as the RCMRD, to deliver customized services in areas like wetlands monitoring and assessment, land degradation assessment, coastal monitoring, and natural disaster preparedness. These services help governments, researchers, and communities make informed decisions on how to sustainably manage Africa's rich but vulnerable natural ecosystems.

By fostering local expertise, infrastructure, and innovation, the Africa Space Initiative and GMES and Africa are helping to build a continental ecosystem that moves Africa from being a passive consumer of space-derived data to an active contributor and innovator in the global space and geospatial arena. The GMES & Africa initiative has been providing information to key stakeholders including policy makers, scientists, businesses, academia, and end users, through consortia of institutions implementing EO services and natural resource management projects across the five regions of Africa. One of the key services of GMES and Africa Program is the land degradation Index Maps that portray actual and vulnerable land degradation hotspots in geographic extent, severity, linking the level of degradation to loss productivity as a threat to food security. Land degradation service products are actively utilized by key government ministries to support environmental restoration and sustainable agriculture. The Ministry of Environment, for instance, leverages these products to identify priority areas for ecosystem restoration, particularly in

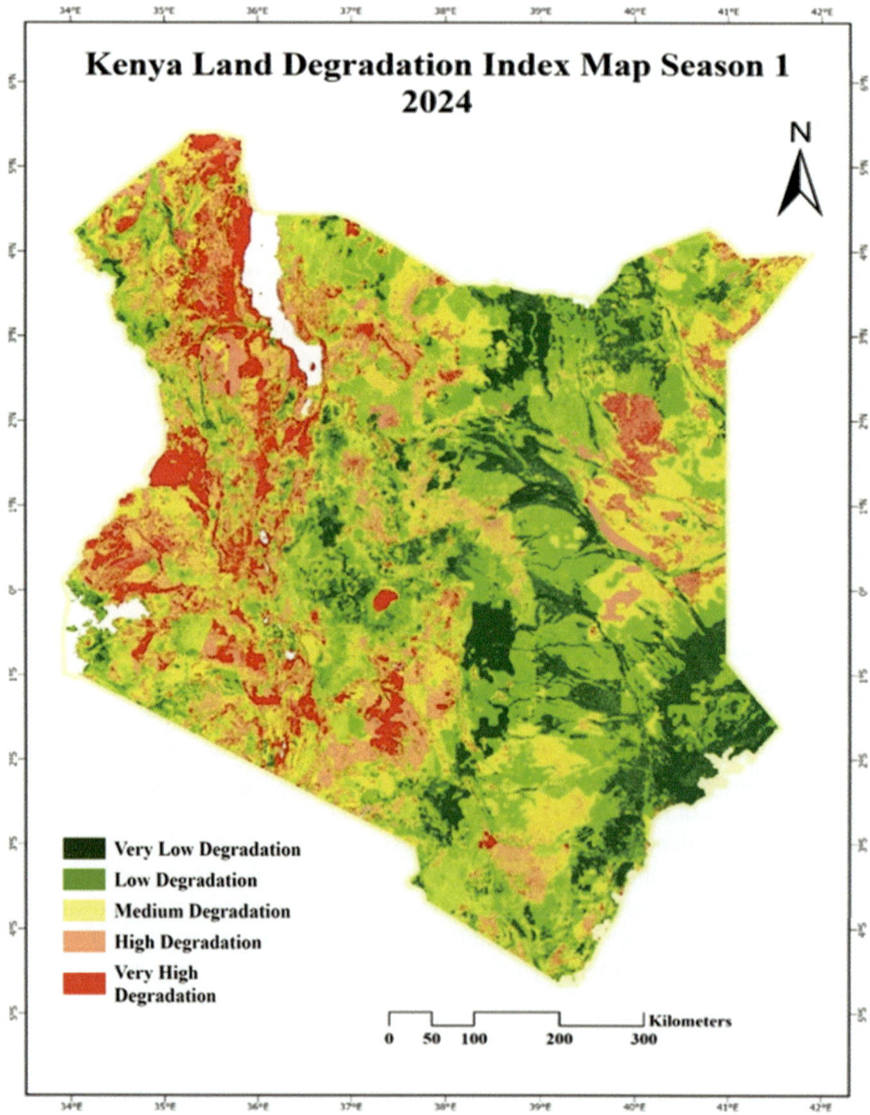

FIGURE 2.6 Land degradation index map of Kenya 2024: https://gmesgeoportal.rcmrd.org

support of Kenya's ambitious target of planting 15 billion trees by 2032. Similarly, the Ministry of Agriculture and Livestock Development uses the data to promote sustainable food production systems and implement climate-resilient agricultural practices. These practices aim to boost productivity, preserve healthy ecosystems, enhance resilience to climate change impacts – including extreme weather, droughts, and floods – and gradually improve soil and land quality across the country as illustrated in Figure 2.6. Water surfaces and bodies, as captured through EO and GIS technologies, are essential assets for agriculture, especially in supporting irrigation during dry seasons.

CASE STUDY FROM NORTH HORR, Kenya

A powerful case comes from North Horr, Kenya, located in the arid Horn of Africa, a region that has endured four consecutive failed rainy seasons. With crops failing, boreholes drying up, and livestock dying en masse, the situation has escalated into a humanitarian crisis. Malnourished children can no longer attend school, and access to potable water is nearly non-existent, resulting in a rise in diseases like typhoid and bilharzia, also known as schistosomiasis, even among children. Despite enormous personal sacrifices and enlisting local leaders using their own resources to fund deep borehole drilling with technical teams from Nairobi, the efforts yielded no results. Had GIS and GPR been employed, precise groundwater mapping could have guided successful borehole placement, providing vital water for drinking, livestock, and small-scale irrigation. This would have transformed

FIGURE 2.7 Prolonged severe arid and drought conditions in North Horr, Kenya. Courtesy of Hon. Ashe Wario, Member of Parliament, North Horr Constituency, Kenya

lives, and revived ecosystems, restoring dignity and building food and water resilience and has the potential to do so in the future with predictable funding. Ultimately, GIS technologies hold the power to turn crises into opportunities. By guiding decisions with data, they support regenerative, eco-centric livelihoods that balance people and planet. Whether enabling borehole drilling in arid zones or helping farmers transition to sustainable inputs, GIS offers a path toward a future where development is inclusive, informed, and resilient, realizing both the 2030 Agenda for Sustainable Development and the African Union 2063 Agenda. Figure 2.7 illustrates the dire conditions under which the vulnerable groups survive, as seen in school-going children walking back home in the semi-arid and arid areas in North Horr, Kenya.

Using GPR, clean water resources can be identified for agriculture and the children do not need to carry water in plastic jugs as they go back home. With water mapping, more nutritious food can be grown, and children's growth and physical development can be improved. School feeding programs can then be sustained with water supply and cultivation of food to provide much-needed nutritious programs on the school compound. In arid and semi-arid areas, where rainfall is irregular or insufficient, water scarcity often leads to drought, food shortages, hunger, and starvation. In such contexts, mapping water resources with EO tools becomes critical in sustaining livelihoods and planning equitable water distribution. The seasonal rhythms and hydrological cycles that drive these water systems are foundational to regenerative agriculture and a more balanced, eco-centric approach to land use.

EO and GIS are revolutionizing agricultural production by offering spatially explicit data for improved planning, decision-making, and sustainability. These technologies help monitor land cover and land use changes, detect degradation in agricultural landscapes, and assess soil quality, vital elements for ensuring food and nutritional security. GIS tools also support precision agriculture by providing accurate insights into topography, soil health, and microclimates, helping farmers optimize input use, prevent yield losses, and plan resilient cropping systems. Integration of mobile and geospatial technologies allows for the real-time dissemination of early warning information on pests, diseases, weather forecasts, and market trends.

This empowers smallholder farmers, especially in remote and low-literacy regions, to make informed decisions and improve productivity. Below in Figure 2.8 is a young woman farmer, a member of the local women farmers' cooperative society.

Satellite-based monitoring enhances this by delivering hyper-localized data for early interventions, efficient resource use, and adaptation to climate change. Governments and institutions such as Ministries of Agriculture can leverage GIS and EO data to tailor extension messages, map disease and pest outbreaks, monitor weather extremes, and design programs that are sensitive to local conditions. These tools also support disaster risk reduction by guiding resource allocation based on spatial vulnerability assessments.

Since the 1990s, precision farming, once limited to commercial producers in high-income countries, has increasingly become accessible in developing regions

FIGURE 2.8 Young woman in Chili garden in Rwanda, Photo: Aicha Hanna Agrane.

Retrieved from: www.pawisca.org

thanks to EO and GIS. Today, these technologies not only guide on-farm decision-making but also contribute to large-scale food security strategies, environmental conservation, and participatory policy development. They enable innovative services such as weather-indexed insurance for smallholders, bridging the gap between risk and resilience.

EO and GIS technologies are game-changers in natural resource management, providing precise, data-driven insights that empower farmers, planners, and policy-makers to build resilient and sustainable systems. By enabling accurate mapping and monitoring of landscapes, these tools help anticipate environmental and production risks. In agriculture, GIS-powered farm registration has proven especially impactful in allowing for the detailed profiling of farmers and their agro-ecological zones, which in turn supports tailored services like soil degradation tracking and fertilizer blending. This precision approach enhances productivity, safeguards ecosystems, and drives smarter, equity-focused agricultural interventions. A tree planting Agro-foresting activity by the youth is illustrated in Figure 2.9.

GIS IN DECISION SUPPORT SYSTEMS

GIS plays a transformative role in agriculture and environmental sustainability by enabling data-driven decision-making, precise resource allocation, and inclusive stakeholder engagement. It supports decision support systems for risk assessment related to climate change, pests, and disease outbreaks by integrating spatial data with predictive modeling. Ministries of Agriculture can leverage localized weather

FIGURE 2.9 Tree planting activity for youth in Haiti (2011).

Retrieved from www.un.org PHOTO: Un Photo/Logan Abassi

patterns and disease forecasts to tailor extension services, while smallholder farmers use GIS insights to better prepare for changing conditions and adopt adaptive practices. Hyper-localized data also enables timely disaster risk reduction and response, including targeted interventions and climate insurance schemes for vulnerable communities.

A biocentric, ecocentric, and cosmocentric approach emphasized by initiatives like the Kunming-Montreal Global Biodiversity Framework and UN Resolution Harmony with Nature A/RES/79/210 in 2024, which aligns with SDG2 Target 2.4 and Indicator 2.4.1 promoting holistic, regenerative land management practices. GIS tools help visualize biodiversity, monitor soil health, optimize land use, and mitigate degradation, ensuring food security while protecting the planet's ecosystems. Participatory GIS platforms empower communities, especially women and youth who represent 50–80% of the agricultural labor force in developing countries, by integrating their voices into local planning. For instance, mapping water resources can reduce the time women spend fetching water by up to 80%, directly improving education outcomes and livelihoods. Programs like the Manyatta Pilot (Jackson, 2020) demonstrate how GIS-driven, community-centered approaches enhance multi-stakeholder collaboration, strengthen local ownership, and promote long-term sustainability.

Figure 2.10 illustrates the small-holder women farmers sharing their produce and knowledge.

FIGURE 2.10 Women farmers near Bulawayo,

Courtesy of Morage Whatt

REFLECTIONS AND CONCLUSION

GIS plays a transformative role in advancing agriculture and environmental sustainability by enabling data-driven decision-making, efficient resource allocation, and inclusive stakeholder participation. Through the integration of spatial data and predictive modeling, GIS supports robust decision support systems for assessing climate-related risks, pest outbreaks, and disease spread. Ministries of Agriculture can utilize localized climate and pest forecasts to refine extension services, while smallholder farmers benefit from GIS insights that improve their capacity to adapt to changing conditions. Hyper-localized geospatial data also strengthens disaster risk reduction efforts and facilitates timely responses, including climate-indexed insurance and targeted resource deployment for at-risk communities. From an anthropological perspective, life on Earth is fundamentally interconnected; people and planet belong together in a shared ecological system. This vision aligns with the aspirations of the SDGs and the upcoming Summit of the Future, echoing the spirit of the Rio+20 declaration: *The Future We Want*. It is a future that leaves no one behind, and is rooted in inclusivity, ecological harmony, and systems that sustain both humanity and nature. In agriculture, food sovereignty is challenged when key crops must be sourced externally due to environmental degradation or disrupted supply chains. For example, during the COVID-19 pandemic, global fertilizer supply was severely affected when maritime shipments from major exporters like Ukraine came to a halt. Over-reliance on synthetic fertilizers not only exposes vulnerabilities in food systems but also contributes to environmental harm such as polluting rivers and degrading soils. Equipping smallholder farmers, especially women, with timely information through mobile technology, extension services, and local media enables better choices such as switching to affordable, nature-based fertilizers that restore rather than deplete ecosystems. Moreover, shared knowledge platforms help communities develop seed banks, promote regenerative practices, and enhance food diversity and nutrition.

The agenda for the future of humanity and Mother Earth must be purposely incorporated into development with social development as an equalizer. The Mother Earth Value System ensures a balanced approach between people, planet and environment (A/RES/79/210). In this manner, we promote harmony in nature and preserve biodiversity, ensuring the sustainability and heritage of Planet for future generations. When we take care of the Earth through regenerative practices and holistic approaches such as ecosystem-based practices, Mother Earth in turn feeds us, helps her children grow and develop into healthy adults who can fully participate in their development and that of their community, and the planet at large. The communities are strengthened and respect for all forms of life on the planet is enshrined in the harmonious existence of the Earth to unite us to the source of all life. In conclusion, the use of GIS systems and tools can transform the lives of communities right where they are, providing sustainable livelihoods that are regenerative and ecosystem-centric, ensuring a people and planet holistic approach for *The Future we Want,* the 2030 Agenda and the African Union 2063 Agenda for *The Africa we Want!*

REFERENCES

African Agriculture Adaptation Atlas (AAAA): (https://adaptationatlas.cgiar.org/

Africa Soil Information Service (AfSIS): https://www.isric.org/projects/africa-soil-information-service-afsis: https://social.desa.un.org/sites/default/files/inline-files/OTIENO_PPT.pdf

AgroClimate: http://agroclimate.org/

Bottom Up Economic Transformation Agenda (BETA), (2023). Kenya Ministry of Agriculture: (https://eos.com/blog/gis-in-agriculture/

Centre for International Forestry Research (CIFOR): https://www.cifor-icraf.org/

Comprehensive Africa Agriculture Development Programme (CAADP): https://caadp.org/

Crop Monitor: https://statistics.kilimo.go.ke/en/m_s_crops/

De Groote, H., Kimenju, S. C., Munyua, B., Palmas, S., Kassie, M., & Bruce, A. (2020). Spread and impact of fall armyworm (*Spodoptera frugiperda* JE Smith) in maize production areas of Kenya. Agriculture, Ecosystems & Environment, 292, 106804.

Desert Locust Control Organization for Eastern Africa (DLCO-EA): https://dlco-ea.org/

Earth observation in support of 2030 Agenda for Sustainable Development: 2017 https://unstats.un.org/unsd/statcom

Global Monitoring for Environment and Security and Africa (GMES): https://www.esa.int/About_Us/Ministerial_Council_2012/Global_Monitoring_for_Environment_and_Security_GMES

International Centre for Tropical Agriculture (CIAT): https://www.ifpri.org/partnership/international-center-tropical-agriculture-ciat/

Jackson, E.D. (2020). The Role of Geospatial Information and Effective Partnerships in Implementation of International Agenda for Sustainable Development: https://aura.antioch.edu/etds/578 and https://storymaps.arcgis.com/stories/e500ffa099b347c18eaf7f85d64443cf

Kenya Integrated Agriculture Management Information System (KIAMIS): https://statistics.kilimo.go.ke/en/kiamis/

Kuning-Montreal Global Biodiversity Framework (GBF) Retrieved from Https://www.cbd.int/gbt

Michelle Miller, Farm Babe: Drones in agriculture: What are they used for? http://thefarmbabe.com

New Partnership for Africa (NEPAD): https://au.int/en/organs/nepad

Risk Management System for Plants Pests and Diseases: http://rcmrd.org/en/?option=com_sppagebuilder&view=page&id=54

Route to Food. Agro ecology as a transformative solution: A path to the right to healthy, affordable and accessible food" Retrieved from www.routetofood.org

Sustainable Agriculture Foundation–Africa (SAF-A): susagfoundation.org

UNDP, Human Development Report 2023/2024, pages 58,59, 125 and 126 – Examining how to enhance collective action.

UNFAO (2023). Agro-Ecological Zone (AEZ): https://www.fao.org/4/w2962e/w2962e-03.htm

UN General Assembly. (2024). Harmony with Nature. Resolution A/RES/79/210

World, Bank. ICT in Agriculture (Updated Edition): Connecting Smallholders to Knowledge, Networks, and Institutions, World Bank Publications, 2017. Module 15.

World Bank Group: Agriculture in Africa: Telling Facts from Myths: Publication: Women, Agriculture and Work in Africa.

3 SDG#3: Good Health
The Heart of UN SDGs Enabled by GIS

Shirley Orr

INTRODUCTION

Achieving UN Sustainable Development Goal #3 with Application of GIS Tools

The United Nations Sustainable Development Goals (SDGs) present a global framework aimed at fostering equity and prosperity by 2030 (UNDP, 2024). Specifically, Goal #3: Good Health and Well-Being seeks to bridge health disparities and enhance healthy living by improving access to care, health outcomes, and the social determinants of health (Figure 3.1). Central to each target is the principle of health equity, eliminating preventable health disparities caused by social, environmental, and economic factors. Despite notable advancements worldwide, healthcare inequalities in access and outcomes persist, particularly among rural and marginalized populations. These disparities are significantly impacted by variables such as education, housing, transportation, and community infrastructure, which influence how individuals access and receive healthcare. Traditional health systems often neglect these spatial and structural factors, resulting in policies that do not effectively meet the needs of the most vulnerable groups.

GIS offer a powerful solution to this challenge. GIS analyzes geographic and demographic patterns in health data (Chandran & Roy, 2024), allowing for the identification of underserved areas, tracking disease outbreaks, mapping service access, and facilitating more equitable planning and decision-making. This chapter explores the application of GIS in addressing six key SDG 3 targets: Maternal mortality, newborn and childhood mortality, communicable diseases, non-communicable diseases (NCD), sexual and reproductive health, and universal health coverage (WHO, 2022). It demonstrates how GIS supports data-driven planning, presents real-world case studies, and concludes with actionable policy recommendations for the development of sustainable and inclusive health systems.

OVERVIEW OF SDG #3 TARGETS

Target 3.1 – Reduce Maternal Mortality

The World Health Organization (2023) identifies Target 3.1 as reducing maternal mortality to less than 70 per 100,000 live births. Among the critical markers of the

 DOI: 10.1201/9781003586319-4

FIGURE 3.1 SDG#3 – ensure health lives and promote well-being for all at all ages. (The content of this publication has not been approved by the United Nations and does not reflect the views of the United Nations or its officials or member States. https://www.un.org/sustainabledevelopment).

health system and gender equality, maternal mortality is a key marker. There are improvements at the global level, but there is preventable maternal mortality, with the bulk of it still happening in middle- and lower-income countries. There are barriers to the availability of skilled birth attendants, antenatal care, emergency obstetric care, and postpartum services. Too often, factors such as rural isolation, socioeconomic disparity, and inadequate health infrastructure combine to heighten a mother's risks of harm. The high maternal mortality areas, service delivery gaps, and emergency planning response can be mapped using GIS technology and assessed alongside healthcare facility location analysis.

Target 3.2 – End Preventable Deaths of Newborns and Children Under 5

Target 3.2 focuses on reducing this mortality rate to fewer than 12 deaths each year for newborns and reducing deaths to children under five to 25 out of every thousand births. The three main dangers to children under five are pneumonia and both diarrhea and birthing complications, according to global health assessments. The root causes of these issues include lack of clean water, proper vaccination, and access to primary health services (World Health Organization, 2022). GIS will facilitate map production of areas of high mortality in children, location of health resources, and services such as vaccine distribution and skilled pediatric health personnel, and proximity of high-risk populations to critical services.

Target 3.3 – End Epidemics of Communicable Diseases

Target 3.3 aims to halt the pandemics of tuberculosis, malaria, AIDS, and neglected tropical diseases (NTDS) by 2030 and combat hepatitis, waterborne diseases, and other infectious diseases (WHO, 2023). Such diseases target the most vulnerable populations, and are exacerbated by inadequate hygiene, a lack of health education, and a lagging surveillance mechanism. The same was witnessed during the COVID-19 crisis, which showed how quickly a disease could spread from one part of the world to another. GIS-supported technologies such as contact tracing, outbreak forecasting, hotspot identification, and travel activity pattern analysis have been utilized to control and quell the spread of infectious diseases. Real-time mapping allows for targeting interventions, vaccine campaigns, and health communication within local jurisdictions.

Target 3.4 – Reduce Non-Communicable Diseases (NCDs) and Promote Mental Health

The goal of Target 3.4 is to halt the rising prevalence of NCD, including cardiovascular diseases, cancers, diabetes, and chronic respiratory tract diseases (World Health Organization, 2023b, 2023c). These diseases reflect the leading causes of death that affect worldwide populations because of lifestyle and economic circumstances, in addition to environmental conditions and healthcare access opportunities. The second target also covers mental health issues including depression, anxiety, and the rising number of suicides are global challenges which are prevalent in the world. Risk factors for tobacco and alcohol sales, levels of pollution, and the presence of recreation facilities can be identified using GIS. GIS data also allows for an assessment of the provision of healthcare services aimed at the treatment of the above diseases and areas of high disease burden with low service penetration.

Target 3.7 – Ensure Access to Sexual and Reproductive Health Services

Target 3.7 aims to ensure the availability of human reproductive health services, including family planning, education, and information. The availability of these services supports prevention of unplanned pregnancies, lowered maternal and neonatal mortality rates, and empowerment of girls and women. Some of the barriers to service access include policy barriers, cultural taboos, and inadequately equipped health facilities. GIS may be employed to identify underserved populations, map the distribution of contraceptives, analyze the geospatial accessibility of reproductive health facilities, and support educational outreach through the visualization of service deserts where action needs to be taken.

Target 3.8 – Achieve Universal Health Coverage (UHC)

Target 3.8 supports increased access to affordable and key health services for all individuals (World Health Organization, 2024b). This includes prevention, treatment, rehabilitation, palliative services, and affordable and effective therapy, along with vaccine access. Achieving UHC calls for better primary healthcare systems, financial protection, and equity in the distribution of resources. GIS enables policymakers to analyze the equity of the distribution of health services, map the gaps in coverage, estimate the population health facility densities, and create outreach strategies to target hard-to-reach populations. GIS may also be applied to health insurance planning and disaster planning to maximize the delivery of services in times of crisis.

Reducing Maternal Mortality (Target 3.1)

Understanding the Challenge

Health professionals worldwide consider maternal mortality as an essential public health concern. In 2023, the World Health Organization stated that approximately 260,000 women died in the course of childbearing, primarily women from low and lower-middle-income countries. Most of these forms of death are considered to be avoidable by health experts because they are a result of hemorrhage, infection, hypertensive conditions, and unsafe abortion. The absence of quality healthcare and care obstacles from limited resources intensify these conditions, particularly in areas with minimal resources. The main obstacles include extended travel routes, transportation constraints, financial expenses, and the lack of experienced medical personnel and materials. These factors align with the "three delays model:" delay in seeking care, reaching care, and receiving adequate treatment. To achieve SDG Target 3.1, the global maternal mortality rate should be reduced to less than 70 per 100,000 live births; health systems must therefore identify who is at risk and where the most critical gaps exist (World Health Organization, 2023a). GIS can provide that spatial clarity.

GIS for Maternal Health Planning

The use of GIS provides policymakers with the ability to view spatial disparities in maternal health outcomes and services. GIS helps identify underserved areas to prioritize interventions and to map distances to emergency obstetric and neonatal care (EmONC) facilities and distances to road networks, along with population data to identify priority areas. It can also reveal inequalities in the urban setting through spatial mapping. For example, women in urban neighborhoods may live close to hospitals but still lack access due to overcrowding or systemic neglect. GIS adds nuance to these insights, supporting targeted and equitable planning in rural and urban contexts (CDC, 2024a).

Case Study: Uganda's GIS Mapping Approach

Uganda offers a practical example of GIS in maternal health. The Ministry of Health, with international partners, mapped EmONC facility access using road, population, and health facility data. Findings showed that many women lived beyond the recommended two-hour travel time, particularly in mountainous and northern regions (Birabwa et al., 2024). In response, the government prioritized these areas for new facility construction and mobile outreach services. This approach enabled more efficient use of resources and reduced geographic disparities in care. This highlights how GIS can align strategic investment with real spatial needs.

GIS IN EMERGENCY RESPONSE AND MONITORING

GIS is also valuable in time-sensitive contexts such as obstetric emergencies. GPS-based ambulance dispatch systems reduce travel times, while real-time spatial data can identify emerging hotspots of maternal deaths. Maternal Death Surveillance and Response (MDSR) systems increasingly use GIS to visualize where interventions are needed most (CDC, 2024). Public dashboards that track maternal outcomes by region empower local leaders and communities to push for improvements. GIS-driven transparency enhances accountability and engagement, which are essential for sustained change.

Challenges and Ethical Considerations

Although GIS has great promise, implementation challenges remain. Few countries can access disaggregated and geocoded maternal data (Macharia et al., 2025). Implementing GIS in national systems takes investment, technical training, and regular maintenance. Additionally, ethical issues of privacy and data protection have to be dealt with, especially when mapping sensitive reproductive health data. Anonymization and data aggregation should be done when small-scale data are used to protect individuals from identification, particularly in small or marginalized communities.

The Way Forward

GIS empowers health systems to identify, rank, and address maternal health needs more efficiently and enables targeted interventions, infrastructure planning, and quick response coordination. For future development, GIS should be coupled with investments in midwifery training, emergency transport, education for the general public regarding health issues, and health finance, to establish a strong, equity-based maternal health system. Incorporating GIS into maternal health policy is not just a technical addition; it is a paradigm shift toward eliminating avoidable maternal mortality and implementing the global promise of SDG Target 3.1.

Ending Preventable Deaths of Newborns and Children (Target 3.2)

Understanding the Scope of the Problem

The SDG Target 3.2 establishes a dual target to achieve newborn mortality of fewer than 12 per thousand births and under-five mortality of 25 per thousand births by 2030, according to UNICEF (2024). Statistically, the worldwide population recorded substantial child survival improvements, although approximately five million young children under five years of age died in 2021 (United Nations, 2021). The leading causes of newborn death result from preterm birth complications, birth asphyxia, infections, pneumonia, diarrhea, and inadequate nutrition. Readily accessible healthcare services could prevent the majority of these health issues. The death rate of children before their fifth birthday is 15 times greater in sub-Saharan Africa compared to high-income nations, according to current data. Health planners require additional assessment methods beyond national averages because they need to be able to visualize the locations where healthcare delivery falls short. This is where GIS is indispensable.

GIS in Child Survival Planning

GIS facilitates mapping child mortality, health infrastructure, immunization rates, sanitation coverage, and social determinants (Root et al., 2020). High-risk zones are identified by superimposing data, and the location-based design of the interventions is used. For instance, GIS can identify zones of lower vaccine coverage or limited availability of skilled pediatric services. Healthcare professionals may use GIS to better target direct immunization drives and mobile medical services. GIS applications can also enhance the efforts of health systems by continuously gathering updated data that enables quick responses to emerging health threats such as disease outbreaks and areas of inadequate nutrition.

Case Study: Mapping Child Health in Nigeria

Nigeria, holding one of the world's highest under-five mortality ranks, has employed GIS to enhance planning under the direction of the National Primary Health Care Development Agency (NPHCDA). With the assistance of UNICEF, they mapped areas of population concentration, immunization rates within geographic areas, and the location of health facilities (UNICEF, 2024b). The analysis pinpointed the underserved areas in the northern states. New healthcare access points were strategically positioned, and mobile outreach services were instituted. Cold chain investment followed needs determined by GIS and helped enhance vaccine delivery while minimizing avoidable deaths among children.

Integrating WASH and Nutrition Indicators

The survival of children depends significantly on adequate water, sanitation, and hygiene (WASH), and proper nutrition. GIS allows researchers to link various health outcomes with WASH indicators such as clean drinking water availability, open defecation locations, and sanitary facilities. GIS in Ethiopia was utilized to identify concurrent challenges of inadequate sanitation, impaired growth among children, and high child mortality (Beminate Lemma Seifu et al., 2024). This knowledge guided multisectoral responses like clean water programs, supplementary feeding, and sanitation campaigns, which were all geospatially targeted from spatial data.

Monitoring and Equity Tracking

GIS helps not only with planning but also with tracking progress. Comparison of data across time allows governments to identify whether interventions are narrowing mortality gaps or leaving many behind. Such maps facilitate equitable policy to ensure that the gains benefit areas not previously advantaged. Visual maps enhance transparency and public and donor communication. Monitoring improvements in the form of improved facilities, births or decreased mortality in individual jurisdictions enhances trust and maintains momentum.

Combating Communicable Diseases (Target 3.3)

Understanding the Challenge

Target 3.3 of the SDGs works to battle AIDS along with tuberculosis, and malaria alongside NTDs through 2030. The goal is the elimination of these epidemics, hepatitis, and water-borne diseases, along with other contagious diseases (WHO, 2023). The global interventions that have reduced overall infectious disease burdens have not eliminated these diseases from causing substantial preventable deaths because low- and middle-income countries experience the most significant prevalence of communicable diseases. Globally, there were 1.6 million tuberculosis deaths, together with 619,000 malaria deaths and 650,000 deaths related to HIV (UNSD, 2021) during 2021. Increased regional prevalence of disease is noted when communities have inadequate sanitation systems, limited clean water access, inadequate healthcare services, and high population densities or active conflict. Social stigmas and insufficient public health education create barriers to disease control efforts, also underfunded surveillance systems, disrupted supply chains, and limited resources. Health resource inequality develops strategic gaps through which infectious disease outbreaks escape detection until they reach uncontrollable stages. The success of this target faces major impediments due to ineffective local disease pattern monitoring systems and rapid response capabilities. This is where GIS come to the forefront, allowing health authorities to plot the spatial patterns of disease spread and map targeted localized responses.

GIS Applications in Disease Surveillance and Control

GIS has transformed the science of infectious disease epidemiology by enabling real-time geographic surveillance, disease modeling, and outbreak mapping. Health agencies and global partners now depend on GIS to:

- Map disease prevalence and incidence by location;
- Monitor disease progression and clustering over time;
- Identify environmental conditions conducive to transmission;
- Assess health infrastructure readiness and coverage;
- Track population movement and migration patterns.

The primary utilization of GIS technologies happens through epidemic warning systems. The combination of meteorological data from rainfall and humidity alongside satellite imagery enables GIS to predict disease risk areas for cholera, malaria, and dengue since these diseases have clear environmental correlations. Using GIS, the healthcare community can conduct pre-emergent vector control activities, education campaigns, and resource allocation decisions before large outbreaks occur. During the COVID-19 crisis, governments and researchers could utilize GIS extensively to map case clusters, monitor virus spread across borders, and analyze the capacity of hospital and ICU beds (Ahasan et al., 2022). Such activities demonstrate the key role of spatial awareness in curtailing rapidly spreading health crises.

Case Study: Malaria Control in Zambia

Zambia provides a rich case of GIS applications in controlling infectious diseases (Larsen et al., 2020). Under its national malaria control program, the Zambian government partnered with organizations such as PATH and the President's Malaria Initiative to establish spatial decision-making systems. With GIS, malaria incidence, breeding sites of mosquitoes, household locations, and healthcare facility catchment areas were mapped at the district level. Such fine-scale mapping facilitated the implementation of indoor residual spraying (IRS), delivery of insecticide-treated bed nets, and seasonal malaria chemoprevention in a very targeted manner. The health workers also used GPS to register the households they covered to achieve complete geographic reach and minimize missed populations. The incorporation of GIS in malaria control facilitated the shift of Zambia from blanket to precision public health. Malaria prevalence in critical areas was notably reduced, illustrating the role of geospatial tools in maximizing the effectiveness and efficiency of programs.

Supporting Contact Tracing and Resource Allocation

In addition to vector-transmitted diseases, GIS enables contact tracing and containment measures for airborne and water-borne diseases (Blanford, 2024). GIS assist in tuberculosis outbreaks by showing drug-resistant tuberculosis locations with contact maps for patient contact screening and treatment delivery. When cholera outbreaks occur through contaminated water supply, GIS helps identify the source of contamination together with exposed communities in the affected areas. Furthermore, GIS enables resource optimization by identifying where to optimally position temporary clinics, deploy vaccination teams, or distribute medical supplies (Krzysztofowicz & Osińska-Skotak, 2021). It supports cold chain logistics by mapping optimal routes and ensuring that medications and vaccines reach remote locations without being compromised. By creating a visual overview, GIS helps different nations collaborate on disease occurrence patterns, which frequently extend across country borders in case of communicable disease outbreaks. Spatial dashboards shared by regional health bodies can inform synchronized actions among neighboring countries.

Challenges in Implementation

Despite its benefits, integrating GIS into communicable disease control efforts has obstacles (Delmelle et al., 2022). Real-time disaggregated data remains scarce throughout numerous low-resource locations, particularly in rural or disadvantaged regions. The quality of reporting and data transmission delays may cause GIS models to become less effective. Healthcare staff may find translating GIS analytical outputs into appropriate actions complex because they need further education about GIS interpretation methods. The development of essential capacities requires investments to proceed effectively. Government institutions, alongside their partners, must identify funding for the creation of spatial data infrastructure and training of epidemiologists and decision-makers on how to use GIS technology while working to establish system interoperability between different platforms and electronic health records along with national disease surveillance mechanisms.

Reducing NCD and Promoting Mental Health (Target 3.4)

Understanding the Challenge

A significant number of annual global fatalities stem from diseases that belong to the NCD group, which includes cardiovascular disease along with cancer, diabetes, and chronic respiratory illnesses. Population health research indicates that NCDs are associated with 41 million deaths yearly, with most of the fatalities involving low- and middle-income countries (Pan American Health Organization, 2022). The addition of mental health disorders along with physical NCDs in SDG Target 3.4 occurs because they both cause major impacts on work performance and quality of life. Multiple behavioral, environmental, and social factors lead to both NCDs and mental illnesses, according to Jeong et al. (2021) in their discussion of disease origin differences from infectious diseases. Excessive alcohol and tobacco use, along with unhealthy eating habits, polluted air, and insufficient health services, are among the main causes of NCDs. The risks have been perpetuated because of urbanization alongside lifestyle changes. Several of the NCDs and mental health conditions, unfortunately, remain underdiagnosed in communities in which routine surveillance of health status at the population and facility level is inadequate. GIS have a potential role in mapping high-risk environments and service gaps and pinpointing targeted interventions.

GIS and the Spatial Epidemiology of Non-Communicable Diseases (NCDs)

GIS combines health data with environmental, infrastructural, and social factors. Using the example of food deserts, GIS visualizes limited areas where people lack access to fruits and vegetables. These areas also contain multiple fast food outlets, and reliance on the less healthy foods provided through those outlets contributes to weight gain and ultimately heart disease (Muzenda et al., 2022). Using GIS enables scientists to analyze environmental conditions, including air pollution and industrial waste proximity levels, because these factors are known to lead to respiratory disease and cancer development. The GIS facilitates the combination of spatial data sets containing suicide rates with depression prevalence, service utilization rates, poverty data, joblessness statistics, and housing condition metrics. GIS informs the planning of community-based mental health services and resource targeting.

Case Study: Heart Disease and Diabetes Mapping in the United States

Through GIS mapping, the US Centers for Disease Control and Prevention (CDC) displayed heart disease and diabetes incidence patterns across the counties in the US

(CDC, 2024b). Analysis of the stroke belt area in the South applied GIS technology to reveal this region as one that bore the highest disease burden. The created maps led to focused policy implementation, including Medicaid expansion and funding for nutrition programs and mobile health unit deployments. GIS also facilitated real-time performance tracking to optimize strategies and minimize disparities.

Addressing Mental Health Gaps with GIS

A significant obstacle to mental healthcare services is unequal service distribution. Rural and urban areas of lower income frequently have inadequate access to psychiatric professionals. GIS identifies areas of need and services that must be aligned with that need (Sadler et al., 2022). GIS, for instance, was employed to identify areas of high suicide and areas with restricted psychiatric services in India (Oka et al., 2014). The gathered data enabled the proper deployment of mobile counseling centers combined with local mental health awareness initiatives for different areas disturbed by crop failure and impoverished conditions. Protective trauma services benefit from GIS for proper coordination during natural disasters as the system demonstrates both vulnerable areas and accessible resources within those locations.

Challenges and Ethical Considerations

GIS applications in NCDs and mental health face several challenges. Accurate, disaggregated data is often lacking – especially for stigmatized conditions like mental illness (Banks et al., 2024). Many health systems do not routinely geocode records, limiting spatial analysis. Interpreting GIS data also requires contextual understanding, as correlations may not imply causation. Privacy is another concern. The mapping process for health-sensitive data must combine data aggregation and anonymization techniques to safeguard personal identities, especially when working with minorities or vulnerable groups.

Ensuring Access to Sexual and Reproductive Health (SRH)- (Target 3.7)

Understanding the Challenge

Target 3.7 of SDG #3 focuses on assuring global access to SRH services that include family planning together with information and education, while incorporating reproductive health programs into national plans (World Health Organization, 2024a). Global progress in enhancing accessibility has not resolved existing inequalities between population groups. The World Health Organization reports that 257 million women worldwide cannot meet their reproductive healthcare needs with safe contraception, and millions of young people fail to obtain accurate reproductive health information (World Health Organization, 2023a).

Sexual and reproductive health services have vital importance in reducing maternal death rates as well as neonatal death rates while simultaneously protecting against sexually transmitted infections (STIs), fighting gender-based violence, and strengthening gender equality (WHO, 2024a). The combination of social stigma, together with gender norms, as well as restrictive laws along with insufficient infrastructure, provider deficits and misinformation, these perpetually restricts access to these services, particularly in underserved rural areas, impoverished locations, and disenfranchised communities. The insufficient availability of localized health service data stands as one major obstacle in achieving Target 3.7 because it hinders healthcare professionals from understanding areas that need service improvements and educational campaigns (Cowie et al., 2020). GIS becomes a strong tool that allows visual representation of sexual and reproductive health services together with population data analysis leading to spatial, strategic decisions for developing comprehensive culturally appropriate programs.

GIS and Mapping Service Gaps

GIS allows for the spatial mapping of SRH facility locations alongside population clusters, transportation networks, school distribution, and community risk profiles (Chandran & Roy, 2024). By overlaying demographic data, for example, the population density of women of reproductive age, adolescent birth rates, and STIs prevalence with clinics and outreach services locations, health planners can identify SRH service deserts, which are geographic areas where demand is high but supply is insufficient. Information plays an essential role in both urban and rural locations. Many densely populated urban locations possess SRH services, but these facilities remain underutilized due to social stigma or because patients fear being publicly recognized and because limited operational hours limit access to care. In contrast, the problem may be physical distance and lack of transportation in rural or transient populations.

GIS helps address these different barriers by:

- Determining where to site new reproductive health clinics or mobile outreach units;
- Highlighting areas that lack adolescent-friendly services;
- Planning community-based distribution of contraceptives and information materials;
- Evaluating travel times and terrain challenges for accessing SRH services.

Case Study: GIS in Family Planning in Kenya

Kenya's Ministry of Health has used GIS to strengthen its national family planning program, with support from international partners such as the United Nations Population Fund (UNFPA) and Population Services Kenya (Bouanchaud et al., 2022). By integrating health facility data with demographic and geographic information, officials identified counties with the highest unmet need for family planning. GIS maps were used to overlay contraceptive availability data, road accessibility, facility capacity, and poverty levels. Research findings demonstrated extremely low SRH service coverage in the counties with characteristics similar to Turkana and Wajir, which faced cultural and logistical barriers due to conservative practices and limited infrastructure. Due to these findings, the government began operating mobile health clinics and forming partnerships with community health volunteers to provide door-to-door services with informative materials distributed through local language resources. The tracking improvements through GIS allowed administrators to move available resources toward regions that displayed limited progress. The government achieved considerable progress in family planning, accompanied by reductions in adolescent pregnancies in targeted areas.

Addressing Adolescents and Marginalized Populations

GIS demonstrates outstanding capacity to serve equity-focused planning initiatives that focus on underserved and at-risk population groups. SRH services remain inaccessible to adolescent girls as well as sex workers and people with HIV and LGBTQ+ individuals who encounter legal, economic, and cultural constraints (Robert et al., 2020).

- GIS can support targeted strategies for these groups by:
- Mapping areas with high rates of early marriage and teenage pregnancy;
- Locating schools and overlaying SRH education program coverage;
- Identifying geographic trends in gender-based violence and STI prevalence;
- Supporting outreach in informal settlements and refugee camps.

In Bangladesh, for example, GIS was used to track the overlap between early marriage rates and regions with poor school retention among girls (Bhowmik et al., 2021). This enabled policymakers to target those areas for concurrent SRH education and community outreach programs, with specific impacts on delayed marriage and contraceptive utilization.

Integrating GIS into Health Systems and Education

Besides program planning, GIS has been used in national health information systems to enhance real-time tracking of SRH services. The system uses nationwide dashboards to display family planning item stock locations while tracking maternal and adolescent health visits as well as HIV testing destinations according to geographical regions. GIS enables comprehensive sexuality education (CSE) implementation through geographic assessments of districts and school areas that have incorporated CSE in their curricula (Ndayishimiye et al., 2020). Officials in ministries of education and health depend on this data to establish plans for improvement in adolescent health and empowerment. GIS technology with mobile digital support enables youth interaction on crowdsourcing platforms and feedback systems to make SRH service digital referrals, which helps improve reproductive health service utilization and understanding.

Overcoming Barriers and Moving Forward

Despite its promise, GIS usage in SRH programming has numerous barriers. Underreporting of sensitive topics like abortion, HIV status, and sexual orientation results from stigma or legal concerns (Bolarinwa et al., 2024). Geospatial data, hence, needs to be managed with great care, respecting privacy, ethical considerations, and data protection. Capacity constraints in GIS proficiency among SRH planners also impede implementation. Training health workers, community-based organizations, and data analysts is necessary to ensure effective collection, analysis, and interpretation of geospatial data. Additionally, funding constraints, limited internet access, and infrastructural challenges must be addressed to operationalize GIS tools consistently.

- To fully leverage GIS for Target 3.7, governments and partners should:
- Prioritize community engagement to build trust and improve data accuracy;
- Promote open data sharing between government sectors;
- Invest in ethical, gender-sensitive, and youth-responsive spatial analysis tools;
- Ensure that geospatial insights translate into real-world access, dignity, and autonomy for all individuals.

Achieving Universal Health Coverage (UHC) (Target 3.8)

Understanding the Challenge

SDG Target 3.8 focuses on UHC, which guarantees necessary access to essential health services for everyone without a financial burden to individuals and families. All healthcare services essential for health needs must be accessible to each person across the complete spectrum, including health protection, disease prevention and recovery services, clinical care, and palliative care. The World Health Assembly officially adopted UHC, yet widespread support did not result in equivalent progress because achievement depends on widely varying local conditions (World Health Organization, 2025a). Comprehensive health services remain inaccessible for over 50 percent of the global population, with 100 million people pushed into poverty through healthcare expenses each year (Kiros et al., 2020). The most susceptible groups, which are remote populations, disadvantaged urban areas, or regions of conflict, are the least likely to access health services. Structural inefficiencies, inadequate infrastructure, human resource shortages, limited data for health decision-making, and fragmented financing models also impede the achievement of UHC. To achieve UHC's ambition, the extent of services to be delivered must be expanded, and the geographic accessibility, affordability, acceptability, and quality of the services need to be ensured (World Health Organization, 2025a). GIS provide a unique lens to evaluate, visualize, and close the spatial inequalities in health service coverage.

GIS in Mapping Service Availability and Accessibility

GIS facilitates the mapping of health facility distribution and access, enabling the identification of areas of service gaps. The travel times and areas with poor access to primary, emergency, and specialist services can be estimated by superimposing health facility data, population densities, transport networks, and terrain data. For example, GIS catchment area analysis determines whether communities lie within practical travel distances (i.e., 5 km or 30 minutes of travel time) of healthcare facilities (Mostafa et al., 2024). This information becomes particularly pertinent in rural and mountain areas where the inaccessibility of services is multiplied by geographic barriers.

GIS also supports health system improvement by:

- Classifying facilities by service level (e.g., primary, secondary, tertiary);
- Mapping human resource distribution (e.g., doctors, nurses, midwives per capita);
- Tracking availability of essential medicines and equipment;
- Analyzing referral networks and service readiness across the care continuum.

Through spatial visualization, decision-makers can better prioritize investments in facility upgrades, health worker deployment, mobile clinics, and community health programs.

Case Study: Universal Health Coverage (UHC) Service Planning in Thailand

Thailand has often been cited as a world leader in UHC implementation (Tangcharoensathien et al., 2020). One factor that has made it successful is using GIS to achieve equitable geographic distribution of healthcare resources. The Ministry of Public Health, universities, and partners have worked to create GIS-based health maps to inform service coverage across provinces (Macharia et al., 2025). With GIS, the government mapped areas of inadequately covered services and added new primary care facilities near underserved communities. GIS also facilitated the equitable allocation of capitation-based funding under the universal coverage plan to support budget disbursement based on identified local health burdens and geographic challenges. This geospatial process allowed Thailand to address financial concerns related to healthcare, increase utilization levels of preventive services, and enhance health outcomes, most notably among the rural poor and the working population.

Financial Protection and Risk Pooling

GIS extends to the analysis of physical accessibility but also plays a role in financial sustainability planning by illuminating socioeconomic status, health insurance, and patterns of healthcare expenditures (Kim et al., 2022). Governments may create targeted subsidies or community-based health insurance plans for areas of high out-of-pocket costs and low health insurance penetration. The National Hospital Insurance Fund (NHIF) enrollment in Kenya received GIS-based spatial measurement by integrating poverty indicators with facility proximity assessment. Utilization of spatial understanding helped policymakers to distribute specific premium support toward informal and rural settlements, which boosted insurance accessibility. Similarly, in Nigeria, GIS helped design health insurance benefit packages by visualizing disease burden distribution and household ability to pay, ensuring the inclusion of high-priority services and populations in national coverage strategies.

Monitoring Equity and Service Quality

Achieving Universal Health Coverage (UHC) is not solely about extending coverage but also about closing equity gaps. GIS enables continuous monitoring of disparities in service utilization, patient outcomes, and health worker distribution (World Health Organization, 2025a). When linked to electronic medical records and national health information systems, GIS dashboards can provide real-time feedback on where and how well services are being delivered.

Spatial dashboards can be used to:

- Track antenatal care visits by region;
- Visualize immunization coverage among children under five;
- Monitor non-communicable disease treatment compliance;
- Analyze the quality of care indicators (e.g., wait times, supply shortages, user satisfaction).

The Ministry of Health uses GIS-integrated health information platforms in Rwanda to guide the performance-based financing model. Facilities are incentivized based on service volume and quality, with spatial monitoring tools ensuring that rural areas are not left behind in performance assessments.

Digital Health and GIS Integration

The intersection of digital health and GIS is a rapidly expanding field supporting UHC goals. Mobile health (mHealth) platforms with built-in GPS capabilities allow health workers to deliver care and record data during outreach visits (Huisman et al., 2022). Such services allow mapping that helps organizations analyze service delivery patterns to detect notable service trends and delivery gaps along with new service opportunities. The Ayushman Bharat program uses GIS technology to monitor Ayushman Bharat facilities and beneficiary locations as well as hospital numbers worldwide under the world's largest public health insurance initiative. Geographic distribution analysis powered by this system lets policymakers modify their strategies while the data is live. The combination of GIS systems, artificial intelligence (AI), and machine learning technologies allows improved predictive analysis for epidemic outbreaks and resource allocation prediction and service utilization forecasting to improve proactive UHC responses.

POLICY RECOMMENDATIONS

To accelerate progress toward SDG #3 using GIS tools, governments must prioritize the institutional integration of spatial data systems within public health planning and decision-making processes (Hariyani et al., 2025). Ministries of Health should invest in nationwide GIS infrastructure, including routine geotagging of health facilities, services, and population health indicators. This foundational data will allow policymakers to detect disparities, optimize resource allocation, and monitor real-time outcomes. Cross-sectoral collaboration is also essential. Health authorities must work with transportation, education, water, and urban planning departments to develop multi-layered GIS applications that address the social determinants of health. Policies should support inter-ministerial data-sharing agreements and ethical data governance. Non-governmental organizations (NGOs) and civil society can contribute by expanding community-based GIS mapping efforts, especially for difficult-to-reach communities. International development partners providing support in the form of training, software licensing, and funding for research should support capacity-building. Finally, equity must remain at the forefront. GIS tools should not merely be used to enhance efficiency but to balance historically asymmetric access to care. Participation of marginalized communities in GIS-based health strategies will also enhance the relevance, building of trust, and accountability. With political will and inclusive policy, GIS has the potential to reshape health systems into more equitable, evidence-based drivers of development.

CONCLUSION

SDG #3 Good Health and Well-Being requires countries to improve service delivery and develop innovative approaches to improve the overall health status of vulnerable populations. Health systems can utilize GIS to drive data-based transformations that lead to higher system inclusiveness and responsiveness. Governments and partners can use GIS to pinpoint mortality hotspots, while directing intervention programs more effectively and optimizing resource distribution and result evaluation. This chapter has demonstrated how GIS support six core targets of SDG 3, which address maternal mortality, newborn and child mortality, communicable diseases, non-communicable diseases and mental health, sexual and reproductive health, and universal access to health services. Each target benefits from GIS because it turns complex information into valuable intelligence, which reveals hidden populations and creates effective policy options. Sustained health gains require governments to include GIS technology in national health plans, strengthen data infrastructure, and establish collaborative programs between governmental departments and agencies. The involvement of communities and non-government organizations in data collection and analysis will support the sustainability and acceptance of implemented interventions.

REFERENCES

Using geographic information systems (GIS) to identify donors – fundsforNGOs. FundsforNGOs - Grants and Resources for Sustainability. https://www2.fundsforngos.org/articles-searching-grants-and-donors/using-geographic-information-systems-gis-to-identify-donors/

Ahasan, R., Alam, M. S., Chakraborty, T., & Hossain, M. M. (2022). Applications of GIS and geospatial analyses in COVID-19 research: A systematic review. *F1000Research*, *9*(2), 1379. https://doi.org/10.12688/f1000research.27544.2

Banks, D. E., Paschke, M., Rashmi, G., & Sanders, L. (2024). Benefits and challenges of geographic information systems (GIS) for data-driven outreach in black communities experiencing overdose disparities: Results of a stakeholder focus group. *BMC Public Health*, *24*(1). https://doi.org/10.1186/s12889-024-19541-3

Bhowmik, J., Biswas, R. K., & Hossain, S. (2021). Child marriage and adolescent motherhood: A nationwide vulnerability for women in Bangladesh. *International Journal of Environmental Research and Public Health*, *18*(8), 4030. https://doi.org/10.3390/ijerph18084030

Birabwa, C., Beňová, L., Olmen, J., Semaan, A., Waiswa, P., & Banke-Thomas, A. (2024). Emergency obstetric care access dynamics in Kampala City, Uganda: Analysis of Women's self-reported care-seeking pathways. *Global Health Science and Practice*. https://doi.org/10.9745/ghsp-d-24-00242

Blanford, J. I. (2024). Managing vector-borne diseases in a geoAI-enabled society. Malaria as an example. *Acta Tropica*, 107406. https://doi.org/10.1016/j.actatropica.2024.107406

Bolarinwa, O. A., Odimegwu, C., Ajayi, K. V., Oni, T. O., Sah, R. K., & Akinyemi, A. (2024). Barriers and facilitators to accessing and utilising sexual and reproductive health services during the COVID-19 pandemic in Africa: A systematic review and meta-analysis. *BMC Health Services Research*, *24*(1). https://doi.org/10.1186/s12913-024-12028-2

Bouanchaud, P., Macharia, P. M., Demise, E. G., & Nakimuli, D. (2022). Comparing modelled with self-reported travel time and the used versus the nearest facility: Modelling geographic accessibility to family planning outlets in Kenya. *BMJ Global Health*, *7*(5), e008366. https://doi.org/10.1136/bmjgh-2021-008366

CDC. (2024a). *Maternal death surveillance and response*. Global Reproductive Health. https://www.cdc.gov/global-reproductive-health/php/maternal-death-surveillance-response/index.html

CDC (2024b, August 28). *Estimates of the percentage of adults with diagnosed diabetes. Heart Disease and Stroke Maps*. https://www.cdc.gov/heart-disease-and-stroke-data/data-vis/diagnosed-diabetes.html

Chandran, A., & Roy, P. (2024). Applications of geographical information system and spatial analysis in Indian health research: A systematic review. *BMC Health Services Research*, *24*(1). https://doi.org/10.1186/s12913-024-11837-9

Cowie, J., Nicoll, A., Dimova, E. D., Campbell, P., & Duncan, E. A. (2020). The barriers and facilitators influencing the sustainability of hospital-based interventions: A systematic review. *BMC Health Services Research*, *20*(1). https://doi.org/10.1186/s12913-020-05434-9

Delmelle, E. M., Desjardins, M. R., Jung, P., Owusu, C., Lan, Y., Hohl, A., & Dony, C. (2022). Uncertainty in geospatial health: Challenges and opportunities ahead. *Annals of Epidemiology*, *65*, 15–30. https://doi.org/10.1016/j.annepidem.2021.10.002

Hariyani, D., Hariyani, P., & Mishra, S. (2025). Digital technologies for the Sustainable Development Goals. *Green Technologies and Sustainability*, 100202. https://doi.org/10.1016/j.grets.2025.100202

Huisman, L., van Duijn, S. M., Silva, N., van Doeveren, R., Michuki, J., Kuria, M., Otieno Okeyo, D., Okoth, I., Houben, N., Rinke de Wit, T. F., & Rogo, K. (2022). A digital mobile health platform increasing efficiency and transparency towards universal health coverage in low- and middle-income countries. *Digital Health*, *8*. https://doi.org/10.1177/20552076221092213

Jeong, N. J., Park, E., & del Pobil, A. P. (2021). Effects of behavioral risk factors and social-environmental factors on non-communicable diseases in South Korea: A national survey approach. *International Journal of Environmental Research and Public Health, 18*(2), 612. https://doi.org/10.3390/ijerph18020612

Kim, J., Kim, D. H., Lee, J., Cheon, Y., & Yoo, S. (2022). A scoping review of qualitative geographic information systems in studies addressing health issues. *Social Science & Medicine, 314*, 115472. https://doi.org/10.1016/j.socscimed.2022.115472

Kiros, M., Dessie, E., Jbaily, A., Tolla, M. T., Johansson, K. A., Norheim, O. F., Memirie, S. T., & Verguet, S. (2020). The burden of household out-of-pocket health expenditures in Ethiopia: Estimates from a nationally representative survey (2015–16). *Health Policy and Planning, 35*(8), 1003–1010. https://doi.org/10.1093/heapol/czaa044

Krzysztofowicz, S., & Osińska-Skotak, K. (2021). The use of GIS technology to optimize COVID-19 vaccine distribution: A case study of the City of Warsaw, Poland. *International Journal of Environmental Research and Public Health, 18*(11), 5636. https://doi.org/10.3390/ijerph18115636

Larsen, D. A., Martin, A., Pollard, D., Nielsen, C. F., Hamainza, B., Burns, M., Stevenson, J., & Winters, A. (2020). Leveraging risk maps of malaria vector abundance to guide control efforts reduces malaria incidence in Eastern Province, Zambia. *Scientific Reports, 10*(1). https://doi.org/10.1038/s41598-020-66968-w

Macharia, P. M., Beňová, L., Ray, N., Semaan, A., Musau, M. M., Kipterer, J. K., Herringer, M., Snow, R. W., & Okiro, E. A. (2025). Putting health facilities on the map: A renewed call to create geolocated, comprehensive, updated, openly licensed dataset of health facilities in sub-saharan African countries. *BMC Medicine, 23*(1). https://doi.org/10.1186/s12916-025-04023-z

Mostafa, M. A., Oluwaseun Ogunmuyiwa, J., Appleby Tenney, K., Tip, S., Zegarra Zamalloa, C. O., Blossom, J. C., & Mpo, T. (2024). Health for all: Primary care facility localization in Lesotho using qualitative research and GIS. *Global Transitions, 6*, 123–135. https://doi.org/10.1016/j.glt.2024.05.002

Muzenda, T., Dambisya, P. M., Kamkuemah, M., Gausi, B., Battersby, J., & Oni, T. (2022). Mapping food and physical activity environments in low- and middle-income countries: A systematised review. *Health & Place, 75*, 102809. https://doi.org/10.1016/j.healthplace.2022.102809

Ndayishimiye, P., Uwase, R., Kubwimana, I., Niyonzima, J., de la, C., Dzekem Dine, R., Nyandwi, J. B., & Ntokamunda Kadima, J. (2020). Availability, accessibility, and quality of adolescent sexual and reproductive health (SRH) services in urban health facilities of Rwanda: A survey among social and healthcare providers. *BMC Health Services Research, 20*(1). https://doi.org/10.1186/s12913-020-05556-0

Oka, M., Kubota, T., Tsubaki, H., & Yamauchi, K. (2014). Analysis of impact of geographic characteristics on suicide rate and visualization of result with geographic information system. *Psychiatry and Clinical Neurosciences, 69*(6), 375–382. https://doi.org/10.1111/pcn.12254

Pan American Health Organization (2022). *Noncommunicable diseases - PAHO/WHO | Pan American Health Organization.* Www.paho.org. https://www.paho.org/en/topics/noncommunicable-diseases

Robert, K., Maryline, M., Jordan, K., Lina, D., Helgar, M., Annrita, I., Wanjiru, M., & Lilian, O. (2020). Factors influencing access of HIV and sexual and reproductive health services among adolescent key populations in Kenya. *International Journal of Public Health, 65*(4), 425–432. https://doi.org/10.1007/s00038-020-01373-8

Root, E. D., Bailey, E. D., Gorham, T., Browning, C., Song, C., & Salsberry, P. (2020). Geovisualization and spatial analysis of infant mortality and preterm birth in Ohio, 2008-2015: Opportunities to enhance spatial thinking. *Public Health Reports, 135*(4), 472–482. https://doi.org/10.1177/0033354920927854

Sadler, R. C., Wojciechowski, T. W., Buchalski, Z., Smart, M., Mulheron, M., & Todem, D. (2022). Validating a geospatial healthfulness index with self-reported chronic disease and health outcomes. *Social Science & Medicine, 311*, 115291. https://doi.org/10.1016/j.socscimed.2022.115291

Seifu, B. L., Tesema, G. A., Fentie, B. M., Yehuala, T. Z., Moloro, A. H., & Mare, K. U. (2024). Geographical variation in hotspots of stunting among under-five children in Ethiopia: A geographically weighted regression and multilevel robust Poisson regression analysis. *PloS One, 19*(5), e0303071. https://doi.org/10.1371/journal.pone.0303071

Tangcharoensathien, V., Tisayaticom, K., Suphanchaimat, R., Vongmongkol, V., Viriyathorn, S., & Limwattananon, S. (2020). Financial risk protection of Thailand's universal health coverage: Results from series of national household surveys between 1996 and 2015. *International Journal for Equity in Health, 19*(1). https://doi.org/10.1186/s12939-020-01273-6

UNDP. (2024). *Sustainable development goals.* United Nations Development Programme; United Nations. https://www.undp.org/sustainable-development-goals

UNICEF (2024a). *Child survival and the SDGs.* UNICEF DATA. https://data.unicef.org/topic/child-survival/child-survival-sdgs/

UNICEF. (2024b). *NGA - UNICEF DATA.* UNICEF DATA. https://data.unicef.org/country/nga/

United Nations. (2021). *A child or youth died once every 4.4 seconds in 2021 – UN report | United Nations.* United Nations. https://www.un.org/tr/node/193074

United Nations. (2025). *Goal 3: Ensure healthy lives and promote well-being for all at all ages.* United Nations Sustainable Development Goals. https://www.un.org/sustainabledevelopment/health/

UNSD (2021). *SDG indicators.* Unstats.un.org. https://unstats.un.org/sdgs/report/2023/Goal-03/

WHO (2022). *Targets of Sustainable Development Goal 3.* World Health Organisation. https://www.who.int/europe/about-us/our-work/sustainable-development-goals/targets-of-sustainable-development-goal-3

WHO (2023). *SDG target 3.3 communicable diseases.* Www.who.int. https://www.who.int/data/gho/data/themes/topics/sdg-target-3_3-communicable-diseases

WHO (2024a). *Extending sexual and reproductive health and rights to future generations through science and evidence.* https://iris.who.int/bitstream/handle/10665/379004/9789240099975-eng.pdf

World Health Organization (2022, January 28). *Child mortality (under 5 years).* Who.int; World Health Organization: WHO. https://www.who.int/news-room/fact-sheets/detail/child-mortality-under-5-years

World Health Organization (2023a). *Contraception.* Www.who.int. https://www.who.int/health-topics/contraception

World Health Organization (2023b). *SDG target 3.1 maternal mortality.* Www.who.int. https://www.who.int/data/gho/data/themes/topics/sdg-target-3-1-maternal-mortality

World Health Organization. (2023c). *SDG target 3.4 non-communicable diseases and mental health.* Www.who.int. https://www.who.int/data/gho/data/themes/topics/sdg-target-3_4-noncommunicable-diseases-and-mental-health

World Health Organization (2024a). *SDG target 3.7 sexual and reproductive health.* Www.who.int. https://www.who.int/data/gho/data/themes/topics/sdg-target-3_7-sexual-and-reproductive-health

World Health Organization (2024b). *SDG target 3.8 | Achieve universal health coverage, including financial risk protection, access to quality essential health-care services and access to safe, effective, quality and affordable essential medicines and vaccines for all.* Www.who.int. https://www.who.int/data/gho/data/themes/topics/indicator-groups/indicator-group-details/GHO/sdg-target-3.8-achieve-universal-health-coverage-(uhc)-including-financial-risk-protection

World Health Organization (2025a, March 26). *Universal health coverage (UHC).* World Health Organization. https://www.who.int/news-room/fact-sheets/detail/universal-health-coverage-(uhc)

4 SDG #4—Mapping Learning Beyond Indicators

GIS Impact for Achieving SDG 4—Quality Education for All

Patricia Solís, Dara Carney-Nedelman, and Natalia da Silveira Arruda

INTRODUCTION

Education suffers from an *externality* paradox. Society generally recognizes the potential for change and progress that educational investment potentially brings about, but unfortunately, many societies struggle to significantly and adequately invest in universal quality education for everyone, as evidenced by the fact that SDG4 has not yet been met by all nations. This may be due to the inherent fact that investment in educational systems, particularly public universal education, benefits society, while investments are often made by taxpayers who may not always have attending learners but disproportionately bear the costs of coverage. In many countries, educator salaries are abysmally low, despite public sentiment to the contrary. Outcomes are very divergent, diffused, and accrued over the long term, and are not easily attributable to particular reforms or interventions. These make educational programming quite difficult to assess, despite that it is universally accepted that a literate and knowledgeable population drives the economic development and well-being of societies, particularly underpinning the aim of SDG4, where inclusive and equitable quality education is ensured, and lifelong learning opportunities for all are promoted. As geographers and geospatial experts, we cannot help interpreting that quality education for all can only mean quality education *everywhere*. Increasingly, it is critical to simultaneously ensure that everyone is represented on the map, all educational institutions can be found, and internet access and electricity are available ubiquitously. This intricately links SDG4 to many other development goals in mutually reciprocal and paradoxical ways.

In the context of SDGs broadly, we have long been interested in what learning about geospatial technologies and the geographic concepts they are based upon can

DOI: 10.1201/9781003586319-5

help us introduce to learners to help them understand our world for achieving sustainable development (Solís, 2005; Solís & Zeballos, 2023). Beyond indicators and targets, in this chapter, we also envision some of the ways quality education for all could mean by participating in powerful learning about the world that GIS education itself could bring about to impact progress towards the Goals as a whole. In this chapter, we thus explore SDG4 through two interconnected perspectives. First, we consider several ways GIS as a powerful tool could enhance delivery, monitoring, and targeting data for SDG4 at subnational scales, going beyond national scale reporting. Second, we examine GIS as a subject of education in itself, one with the capacity to support not only SDG4 but also the broader suite of SDGs. By framing geospatial literacy as both an enabler of informed action and a vehicle for deeper engagement with sustainability issues, we aim to highlight its foundational role in sustainable development efforts. In this context, learning about GIS can help students, educators, and policymakers alike better understand the spatial dimensions of global challenges, identify key intervention areas, and foster place-based awareness and solutions. We review a few key trends over recent decades in GIS education from both scholarly research and real-world practice that have intersected with relevance to the SDG landscape, illustrating the intersection of pedagogy, policy, and technology. These include data-driven, competency-focused, and student-centered concerns. While not a comprehensive review of all scholarships in GIS education literature, this theoretical discussion aims to ground a rationale for our making the case for greater attention and investment in going beyond indicators.

Since the particular focus of SDG4 is given to the conditions of access, whereby the aim is quality education for all, and as we interpret it, *everywhere,* we present a detailed case study involving youth-led participatory GIS mapping as evidence of this possibility, grounded in the pedagogical scholarship approach we reviewed. Reflecting on this experience, we present some lessons learned. Moving beyond this in-depth example, we consider additional transformative case studies where GIS education has had tangible impacts on community development and youth engagement, in the spirit of SDG4, concluding with strategic recommendations for leveraging geospatial capacity building more effectively in pursuit of quality education worldwide. In some ways, but not all, teaching and learning GIS and geospatial technologies in formal and informal educational spaces may help to erode some of the paradoxical impasse given the potential for empowerment and engagement. As both a tool for monitoring data about the world but also learning about and understanding the world to make it more sustainable, GIS potentially holds one key to unlocking the SDG4 investment in capacity for sustainable development, as this chapter grapples with some potential impact of doing just that.

INDICATORS FRAMEWORK FOR SDG4: ROLES FOR GIS AND GIS EDUCATION

The official source of data to monitor progress among nations for the United Nations, as adopted in the Education 2030 Framework for Action is the premier educational organization of the UN, the United Nations Educational, Scientific and Cultural

Organization, UNESCO, through their Institute for Statistics (UIS). According to the UNESCO Institute for Statistics (n.d.), this mandate aligns with each of the indicators and targets of SDG4 for every nation, and "reflects the trust of the international community in UIS data and its proven track record in methodological work and standard setting with national statistical offices, line ministries and technical partners in every region." This monitoring activity is surely complex, as there are many subcategories of statistical groups to report, including by sex, wealth, disability, and their categories, together with a wide range of indices as defined by participating institutions, and as defined by how each country in turn collects data. A number of key initiatives inform the collection and reporting of official data, which including a Technical Cooperation Group, which informs the definition and implementation of the global and thematic indicators; the Global Alliance to Monitor Learning, which sets standards for good practices in learning assessments with education stakeholders; the Inter-Agency Group on Education Inequality Indicators, which sets standards on household survey data, especially regarding indicators on education equity. These technical experts include data scientists who are knowledgeable about the spatiality of data, even if the output is not GIS-based. Output of the UIS include many products, but none of which, upon review by the authors, are available in GIS format, given the fact that the official reporting aggregates to the national country level, namely, the SDG4 Data Book and Data Digest, SDG4 Country Profiles, the UIS Global Education Database, and a downloadable eAtlas for Education 2030, which are all available from their main webpage.

It is somewhat disappointing that GIS is not integral to the official monitoring output of SDG4 at UIS, as seen in Figure 4.1, largely because the reporting process relies on nationally aggregated data, and even so, for many indicators, it is

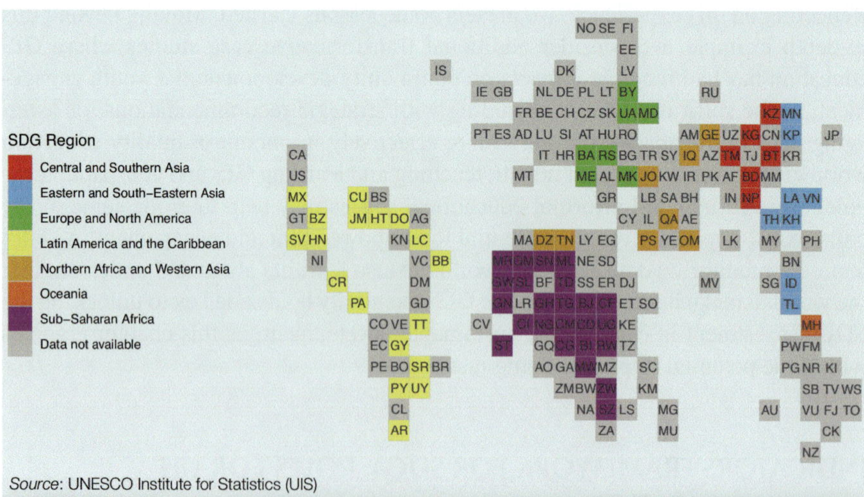

FIGURE 4.1 Reproduction of data availability map of global indicator 4.2.1 (2010–2019), proportion of children under five years of age who are developmentally on track.

Source: UIS SDG4 Data Guide, 2019, p. 20.

not even available at a national scale. Countries serve as the primary units for both national and international SDG monitoring, and national authorities – such as statistical offices and education ministries – control what level of detail is shared with global agencies. Although disaggregated data could provide more localized insights and broader utility (see Sajjad et al., 2022), most SDG4 indicators are derived from national-level administrative records or broad household surveys. These datasets are often collected through standardized questionnaires and international survey programs like Demographic and Health Survey (DHS) and the Multiple Indicator Cluster Survey (MICS), or from the World Bank, which presumably contain location data at the original source. We assume from experience working with local and national governments that while the official datasets for some areas may be well represented in spatial detail, not all regions or places in a country would always be well represented in terms of data collection, and not all countries have quality data collection or data representation. Therefore, the datasets would not be expected to be typically spatially detailed enough to benefit from GIS analysis at the regional or global level. As a result, despite the potential of GIS to reveal regional disparities and guide targeted educational interventions, its use is limited by the generalized nature of SDG4 reporting frameworks at the UIS or UN reporting levels. Nevertheless, knowing any, each, or all these indicators at a subnational level supports equity and can pinpoint areas of greater need for investment or attention, since unevenness is an unsurprising characteristic of any geography of a mapped data variable. For education, this is particularly expected with respect to remote rural areas, underserved urban or peri-urban areas, and can often coincide with locations of interest for other sociodemographic attention. Location-based analysis, hot-spot analysis, optimization analysis, and any other number of geospatial analysis of various factors or educational characteristics of interest such as absence of primary school locations, low educational attainment and/or correlation with other sociodemographic variables is easily possible with any GIS capability of virtually any open or proprietary software available and can reveal inequities where official data is available at such resolution. Yet lack of official data still hampers humanitarian and development efforts for what Marshak et al. (2023) recently called "data missingness": Missing observations, missing variables, missing respondents, missing populations, and, we would add as geographers and geospatial scientists, missing locations.

Marrying interests of official data with civil society sectors through GIS educational efforts may be one way to break this impasse, and GIS education holds examples of where such efforts have enabled advances to expanding official spatial data capacity via multi-sector partnerships for education (Parcher Wintemute et al., 2021; Arriaga et al., 2023). And fortunately, spatial data already increasingly exists to leverage in-country, subnational analyses, via the massive efforts of OpenStreetMap (OSM) (www.openstreetmap.org) as a community of communities producing open missing data, such as the Humanitarian OpenStreetMap Team (www.hotosm.org); and in the academic realm, YouthMappers, a student-led, faculty-mentored chapter-based network on more than 420 campuses in over 80 countries (www.youthmappers.org) (Solís, 2022; see also Brovelli et al., 2019). Data, particularly on educational statistics, is becoming increasingly available as well, such as via Esri's Living Atlas, where a search on "education or school or university" alone reveals no less than

1,007 distinct geospatial content resources (https://livingatlas.arcgis.com). Related to GIS data, it is important to note that Earth observation (EO) data can also be important in many cases and is increasingly available at low to no cost. Despite that, there have been nearly no dedicated EO educational resources for SDG4 specifically commissioned, per Metternicht et al. (2020), despite many available to other SDGs in the landscape of Digital Earth resources. The availability of imagery for humanitarian mapping, augmented GeoAI built from EO resources, can help identify missing populations, missing school infrastructure or assistance in disasters, determining feasibility of flooded roads *en route* to schools. And also assist with analyses such as site suitability for microgrid solar electricity installation near educational facilities. The availability of EO data basically anywhere in the world for incorporation into geospatial analyses related to facilities, infrastructure, or disaster recovery has certainly played a role in achieving the goal of universal educational access.

Certainly, there are many unofficial ways that the broader GIS community, including the GIS educational community, is already contributing to advancing SDG4, unmonitored, unreported via UNESCO, and may further potentially amplify such contributions in the future with further attention and investment. A close read of a few SDG4 targets and indicators, [https://unstats.un.org/sdgs/metadata/?Text=&Goal=4&Target=] reveals existing GIS institutions and programming for education are already dedicated and fully committed to advancing data and resources in extra-official capacity. For example, as *Indicator 4.1.1* is concerned with universal learning outcomes with respect to mathematics, Arlinghaus and Kerski (2013) creatively explore the intersection of spatial thinking and mathematical concepts through mapping and open an incredible space to advance interdisciplinary geo-literacy. *Indicator 4.3.1* opens up potential attention to non-formal geospatial educational pathways, an incredible opportunity for public participation GIS (PPGIS, Sieber, 2006) and learning modalities evolved from that tradition to advance the equitable participation rate of all women, men, youth, and adults in affordable and quality technical, vocational, and tertiary learning opportunities that include mapping skills, which can be transferrable and productive to the sustainable development goals. In fact, according to *Indicator 4.4.1,* in many countries, GIS can be considered as an information and communications technology, or ICT skill, thus advancing multiple indicators often simultaneously. Per *Indicator 4.5.1,* especially when engaging communities affected by development or crises, such as humanitarian mapping programming, some efforts can advance targets and indicators that aim to support populations with parity status, such as persons with disability status, indigenous peoples, and conflict-affected. As we will explore in the in-depth case study in this chapter, *Indicator 4.a.1,* the proportion of schools offering basic services, such as electricity and access to the internet, this is very important for GIS education that is of the highest quality, despite advances in offline data collection. Expansion of access, especially to universities in rural or remote regions, is essential for the global GIS community to continue to amplify contributions to SDG4.

With respect to the most valuable assets in the educational landscape, our human resources, the amount of direct investment in learners and educators are measured principally in terms of the volume of official development assistance flows for scholarships through *Indicator 4.b.1.* This metric is not tracked specifically in the geospatial industry, but doing so would be something that the GIS community would benefit from to assess growth and expansion for the industry, and not just the public sector response to the SDGs. For example, programs like Erasmus Mundus, Fulbright, the YouthMappers Leadership Fellowships, individual university scholarships, and other regional association programs are known for quality internationally sourced talent pipelines, but the demand greatly outstrips the supply for globally available assistance to build the GIS workforce. Similarly, *Indicator 4.c.1*, measuring the proportion of teachers with the minimum required qualifications, by education level, has grown by official programming, which is currently seeing limited support in many countries from public governmental means. These numbers could be greatly expanded for GIS through supplemental efforts from the GIS and geography communities, nationally or globally through such initiatives like the ISPRS open educational catalog (Coetzee et al., 2020), GeoAI in support of the SDGs (Brovelli, 2025), American Geographical Society Teacher Fellows in the United States, or the promotional work of the IGU Commission on Geographic Education, with support for bespoke activities. Additionally, the PanAmerican Institute for Geography and History Geography Commission indirectly raises the profile of teachers through supportive resolutions to the respective Member State Ministries of Education, as well as the actions of many other sister institutions working in this field (see also Solís & Sinton, 2023).

In short, while the possibilities for GIS as a data technology directly playing a role within the official reporting framework of national scale SDG4 target and indicator infrastructure may be constrained, there is ample potential and even realized efforts and actors working in the landscape to promote SDG4 quality education for all, everywhere, and GIS education adoption for SDGs wherever GIS is being taught and learned, with room to grow.

LITERATURE ON SDG EDUCATION AND ON GIS EDUCATION FOR THE SDGs

These actions for SDG4 aims have played out in an intellectually purposeful and rich landscape of thoughtful approaches to how to improve education generally, and GIS and geographic education specifically. While it is not our purpose to survey this field in its entirety, we would like to point out a few key trends in recent years that have enabled and shaped some of the roles for GIS to speak to SDG4, knowing that there are certainly many others that deserve additional voicing. We also recognize that such scholarship has been drawn from deserving critiques, and certainly deserves further critique and evolution as sustainability, development and GIS as well as other concepts should not be taught uncritically without careful attention to feminist, majoritarian, and other ethical understandings (Beynaghi et al., 2016; Warren et al., 2019; Kopnina, 2020; McCusker et al., 2022). In this brief overview, we attempt to speak to literature that has incorporated these perspectives, and to summarize general trends that have enabled practical action to amplify the ability of advancing educational goals broadly, in ways that can be measured, with the main purpose to introduce a detailed case study followed by some additional illustrative cases on the role of GIS for SDG4.

Our brief review thus takes as an opening standpoint, a definitive recent systematic review of education for sustainable development (ESD) for all the SDGs by Chien and Knoble (2024). Their study conducted a meta-analysis of 571 ESD papers published between 2016 and 2022, aiming to systematically categorize the key research trends and emerging issues in the field since the adoption of the SDGs. Searching in Google Scholar and Web of Science with keywords: [education OR quality education] and [sustainability OR sustainable development OR sustainable development goals] and [teaching OR learning OR skill OR training OR school OR college OR university] in titles, they looked at changes over time, global distributions in article production, key topics, most cited papers, and most active journals. Chien and Knoble identified five major issues that recent ESD studies address: (1) Education, Pedagogy, and Learning; (2) Developing Trends of Educational Institutions; (3) Industry-Academia Collaboration and Out-of-School Education; (4) Educational Policy and Governmental Institutions; and most actively (5) SDG 4 Targets, Indicators, and Achievements. They furthermore note how the literature set reveals interdependencies between education and other SDGs, advocating for policymakers to "adopt novel theoretical frameworks which consider these factors in order to implement ESD and other dimensions of sustainability for all." They particularly call out strong connections with health and well-being (SDG 3), no poverty (SDG 1), and decent work and economic growth (SDG 8), but not the potential to leverage interrelationships with neglected areas such as water (SDGs 6 and 14). From a subject matter content standpoint, GIS Education is well-situated to respond to these research opportunities in the ESD literature at large, as well as with practical approaches with ample ready K-16 curriculum (see Kerski et al., 2023a, 2023b, 2023c, 2023d, 2024a, 2024b, 2024c, 2025).

Cross-cutting research gaps per Chien and Knoble's review (2024) address issues familiar to GIS. Wu and Shen (2016) discuss the role of ICT for SDGs with respect to ESD research gaps, a space serendipitously explored recently in depth by Nuumbosho (2024) in a study of ICT tools in schools and improving internet accessibility and connectivity in Namibia. The review also noted the frequently cited ESD study by Annan-Diab and Molinari (2017) calling for greater research on interdisciplinarity, an approach well-known by geography educators, and well adapted to GIS techniques for problem-based learning (Solís et al., 2017; Silviariza et al., 2024). Finally, the ESD review identified a set of frequently cited articles that included ESD as the sole topic, where University and School Development is a critical dimension of change for progress toward SDG4. This includes the need to develop reference frameworks for and agree upon key competencies in sustainability in higher education (Brundiers et al., 2021), then connect these competencies and pedagogical approaches within institutions and industry (Lozano et al., 2017). Geospatial workforce competencies have been developed and evolved over the past decades alongside the development of SDG educational approaches, contributing in measured and unmeasured ways (DiBiase et al., 2010).

From a pedagogical approach, trends in GIS and geographic education relevant to SDG4 advances can be summarized along at least three critical pillars, relative to the examples we wish to highlight. These focus on: (1) data-driven, (2) competency-focused, and (3) participant-centered scholarship. This literature is not mutually exclusive to these categories; instead, these themes are central to the emergence of a responsive and responsible, ethical GIS educational approach that is robust for addressing inclusive access for all. The collected body of scholarship demonstrates a sustained and multifaceted contribution to advancing SDG 4 – ensuring inclusive and equitable quality education and promoting lifelong learning opportunities for all – through the lens of GIS and spatial thinking. These works coalesce around a central commitment to using GIS not just as a technical tool but as a transformative medium for equity, engagement, and global educational access.

Data-driven approaches recognize the disparity of world regions where populations are disproportionately missing in representation in cartographic, demographic, and statistical databases. The cornerstone of GIS educational scholarship recognizes the need to engage open mapping and participatory GIS approaches to foster global citizenship and education among all youth, abilities, ages, and locations. Youth at universities are uniquely positioned to implement the SDGs (El-Jardali et. al, 2018). Students active within the geospatial community are particularly well situated to do so within a data-driven perspective that can speak to impacts that can move metrics. A strong example of this is YouthMappers. The YouthMappers network is a global youth movement of university students who use open geospatial data and tools to address international development, community resilience, and disaster response. YouthMappers is a community within the open mapping and geospatial communities (Solis, 2022). Due to the global reach and locally centered impact of the network, students grow and promote the SDGs throughout their activities. Launched in 2015, the YouthMappers initiative is examined as a global learning model that builds spatial data literacy while contributing to real-world humanitarian challenges, highlighting its hybrid nature as both a pedagogical and civic movement (Anderson et al., 2023; Solís et al., 2018, 2022). The key to such approaches is creating local data on OSM for use and re-use *in situ*, driven by attention to quality, completeness, validation, and ethical production of research. This vision is expanded in analyses of students' own experiences of their learning (Coetzee et al., 2018). Over the past ten years, YouthMappers have edited more than 25 million buildings, over 1 million kilometers of new roads, and created data in the places where 67 million people on the planet live, as shown in Figure 4.2.

The content for capacity building in GIS is further explored as a mechanism for authentic, standards-based, competency-linked learning. This links to workplace ambitions and long-term impacts on learners (Solís et al., 2019). These efforts position GIS not only as a technical discipline but also as a participatory and equity-oriented platform for global education as citizen-minded learners, but also individuals preparing for work (Solís et al., 2022). Research into project-based learning frameworks demonstrates how integrated geography education can support core competencies while addressing pressing global issues such as climate change (Solís et al., 2017). Specific attention to youth engagement is also evident in studies focused on secondary education, such as a GIS-based learning design for high school girls in climate change education, which underscores the importance of inclusion in STEM pathways (Solís et al., 2018). The application of GIS in formal and informal educational settings is extended through investigations into student engagement and affective responses. For instance, the impact of contextual information on student

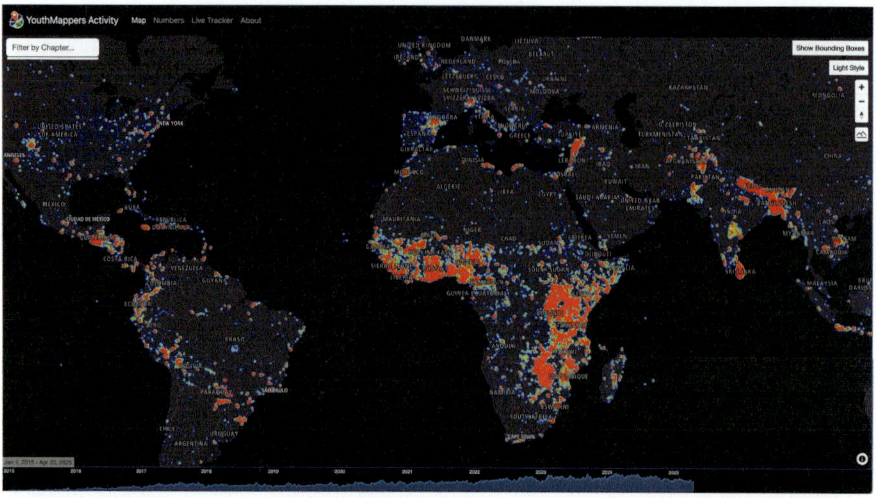

FIGURE 4.2 Data creation activity of the YouthMappers network for OpenStreetMap.

Source: https://activity.youthmappers.org/numbers.html.

motivation and performance in humanitarian mapping contexts demonstrates the potential for introducing GIS and real-world applications in learning for new audiences who are not GIS learners (Solís & DeLucia, 2019). Similarly, reflective writing analyses from YouthMappers reveal how GIS projects foster non-technical competencies like identity, empathy, and agency in students (Hite et al., 2018). International exchange and global exposure have also been pivotal themes. The development of global-minded scientists through international GIS-focused exchanges demonstrates how spatial learning can support cross-cultural understanding and global competencies (Huynh et al., 2020).

Foundational critiques on the epistemologies embedded in GIS invite deeper reflection on whose knowledge is represented and how GIS education can be structured to reflect diverse worldviews (Warren et al., 2019). These theoretical inquiries are key to ensuring that GIS contributes to SDG4 not merely through content delivery but through critical, inclusive, and reflective pedagogies. As mentioned earlier, PPGIS (Sieber, 2006) spawned some of the earliest practical conversations during the era of the millenium development goals (MDGs), influencing an earlier age of GIS for sustainable development (Solís, 2005), and has since undergone waves of critique and sophistication just as the technologies have evolved in the past two decades, to community geography, powerful geography, and beyond (Rees et al., 2021; Solem, Boehm, & Zadrozny, 2024). Enduring within this scholarship is that participants themselves are centered as powerful actors. Together, this collection of scholarships in GIS, geospatial and geography education, and ESD forms a set of literature situating how GIS can inform educational equity, empowerment, and sustainability, particularly SDG4 access to learning itself. It illustrates how spatial tools, when paired with inclusive pedagogical design and community-oriented values, can serve as powerful instruments in achieving the promise of quality education for all. This body of thought frames our in-depth example we present next.

PILOT MAPPING OF SCHOOLS BY STUDENTS AND THEIR CONNECTIVITY IN BRAZIL

Giga (2025) is an initiative by UNICEF and the International Telecommunication Union (ITU), the United Nations specialized agency for digital technologies (ICTs). ITU is made up of a membership of 193 Member States and more than 1000 companies, universities and international and regional organizations headquartered in Geneva, Switzerland, and with regional offices on every continent (https://www.itu.int/en/about/Pages/default.aspx.) Giga has been working tirelessly to achieve its ambitious goal of universal school connectivity, and is using emerging technologies such as AI, satellite imagery, blockchain, and Web 3.0. Giga's Project Connect has already mapped 2.2 million schools across 143 countries. However, they face challenges in obtaining accurate and up-to-date information on factors such as school classification and student enrollment, which are critical for mapping the magnitude of the need for connectivity. This is also compatible with the aims of SDG4, necessary to know for the goal of subnational knowledge about access to quality education for everyone, when the interpretation includes that knowledge should be *everywhere*.

To address these challenges, Giga uses Non-Fungible Tokens (NFTs) to build a decentralized and open-source database of schools worldwide, called the NFT 2.0 collection. The NFT 2.0 collection can be compared to a Wikipedia page for every school. It allows anyone in the world to contribute, validate, and update data for any school, thereby gathering precise and current information on schools globally. The collection rewards constructive actors who contribute accurate and useful data with in-game points that can be redeemed for collector-NFTs. The use of NFTs will also serve as a visual representation of the school and its connectivity status, like the Patchwork Kingdoms NFT collection, raising awareness about the digital divide and creating a sense of ownership and pride for the schools and their surrounding communities.

YouthMappers were invited in 2023 to play an important role in helping Giga achieve its ambitious goal NFT2 as a step toward universal school connectivity, while also empowering local communities to take ownership of the digital future of their schools in several ways:

1. **Conduct local mapping**: YouthMappers leveraged existing mapping experiences to collect and update data for NFT2 on school locations, classifications, and enrollment figures in a set of three pilot regions in Brazil. This was done through on-the-ground mapping, using global positioning system (GPS) devices, and by collaborating with local school administrators and communities.
2. **Verification of data**: As members of the local community, YouthMappers played a crucial role in verifying the accuracy of the information collected for NFT2, they cross-checked data with on-the-ground realities and

provided valuable feedback to ensure the authenticity of the NFT 2.0 data collection.

3. **Community engagement**: YouthMappers helped raise awareness about the NFT 2.0 collection within their communities, encouraging local stakeholders to contribute and update data about their schools.
4. **Testing of the NFT2 platform**: YouthMappers acted as beta testers for the NFT 2.0 platform, providing feedback on the user experience and functionality. Their expertise in mapping and familiarity with the target communities proved invaluable in ensuring the platform's user-friendliness and effectiveness.
5. **Refinement of the NFT2 participatory procedures**: YouthMappers leveraged significant expertise not only in mapping, but also a reputation for understanding how to execute mapping projects, resulting in providing valuable feedback to the design and execution of the program to inform Giga's commitment to the connectivity initiative. Giga presented certificates of appreciation to the participating students for their volunteer efforts. Expenses were reimbursed by the program, and the time of students, faculty, and mentors was largely volunteered or leveraged by educational and development programming.

The participatory methodology deviated in some ways from the YouthMappers' usual process to add OSM attributes scheme, although the platform utilizes OSM for location data via the Kobo Collect app. Table 4.1 presents the data schema to capture information for each school.

To frame the partnership, the Giga and YouthMappers teams together developed in advance a number of best practices for how to present the pilot work with the many different organizations involved in the volunteer effort. Therefore, a set of explicit *rules of engagement* for the UNICEF Country Office in Brazil, the Giga team, and YouthMappers chapter volunteers, which included students, faculty, and mentors, was identified. This included gaining understanding about the use of institutional names, logos, time commitment of volunteers and coordinators, whom to reach out to and when, who was collecting data on behalf of whom, with whom data was being shared and when and how, how to communicate with municipalities and schools about the project and who was supporting whom, and how to ensure the proper designated contacts and volunteers would be the right and proper person(s) to reach out to and with which kinds of updates, etc. Most importantly, it was critical to ensure that schools are not contacted multiple times for the same information.

The testing plan for the NFT2 Pilot Project outlined the successful collection and validation of data for a target of up to 100 schools using the third-party application, KoboToolBox (Kobo Collect App), to organize the effort. The main objectives of the testing plan were to ensure the accuracy, consistency, and reliability of collected and submitted school data, and to structure the communication and coordination between Giga and YouthMappers. And, to standardize procedures among the participating YouthMappers chapters, and especially, as a pilot test, to identify areas of improvement for scaling up and replicating the methodology.

The three universities, the YouthMappers chapters and locations participating in the pilot were Universidade Federal de São João del Rei, Unificar Ações e

TABLE 4.1
Data Schema for YouthMappers Giga Pilot Project in Brazil

Field Name	Type	Description
Location	object	Object containing location details of the school
Latitude	number	Latitude of the school's location (between -90 and 90 degrees)
Longitude	number	Longitude of the school's location (between -180 and 180 degrees)
Number of classrooms	integer	Number of classrooms the school has
Has Satellite Building	bool	Does school have satellite buildings
Phone	string	Phone number of the school
Email	string	Email address of the school (should be in email format)
Connectivity	object	Object containing connectivity details of the school
Number of internet links	integer	Number of (independent) Internet links which the school has
Coverage Availability	string	Does the school have internet coverage
2G Coverage	bool	Does the school have 2G coverage
3G Coverage	bool	Does the school have 3G coverage
4G Coverage	bool	Does the school have 4G coverage
Online Presence	object	Object containing links to the school's online presence
Website	string	URL of the school's website
TwitterX	string	URL of the school's Twitter page
YouTube	string	YouTube channel (or video link) of the school
Wikipedia	string	URL of the school's Wikipedia page
GoogleMap	string	link to Google map of the school
Custom	object	Object containing any additional custom data that the user wants to add to the schema

Informações Geoespaciais – known as the UAIGeo YouthMappers chapter in São João del – Rei, Brazil; Universidade Federal de Uberlândia, or UFU Mappers chapter in Uberlândia, Brazil; and the Instituto Federal de Educação, Ciência e Tecnologia do Pará – IFPA, campus, known as the YouthMappers IFPA chapter located in Castanhal, Brazil. Their areas of interest (AOI) for mapping in the project are depicted in Figures 4.3–4.5, respectively.

The methodology for data collection and validation testing involved verifying the accuracy of the data collected about schools in Brazil, including location, classification, and enrollment figures. YouthMappers were responsible for collecting this data via KoboToolbox, a data collection tool to collect, analyze, and manage data for surveys https://www.kobotoolbox.org) while Giga team members were responsible for cross-checking the data for accuracy. A YouthMappers Steering Committee Team Member provided coordination, and another team member provided oversight and minor resources for gap needs. The procedure for training the participants followed a sign-up orientation and rudimentary training on KoboToolbox, which proved largely unnecessary given the prior experience of chapter volunteers, except to set parameters. A set of school locations on Google Maps was provided for the AOIs, such as the example shown in Figure 4.6, for each of the teams.

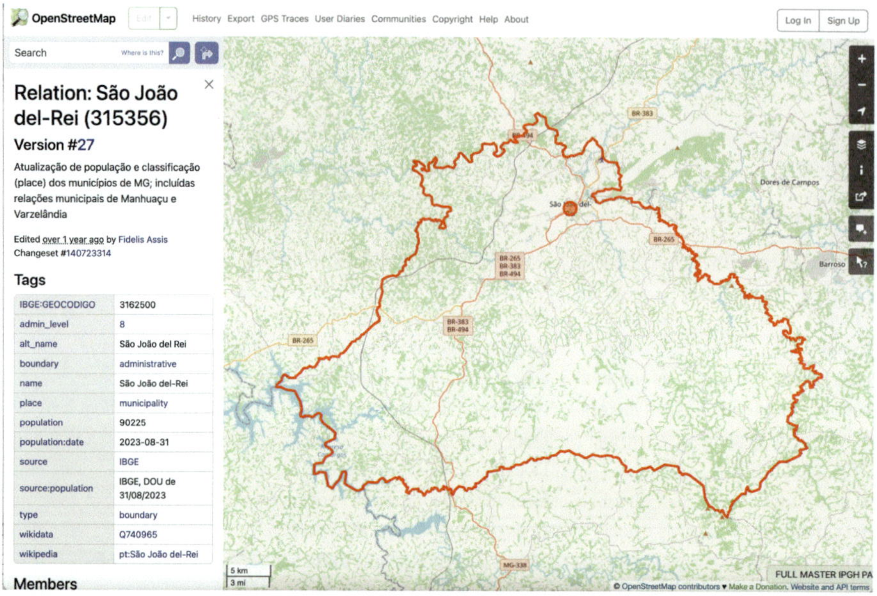

FIGURE 4.3 AOI location in São João del – Rei, Brazil for the UAIGeo YouthMappers chapter.

Source: https://www.openstreetmap.org/relation/315356#map=11/-21.2379/-44.2818

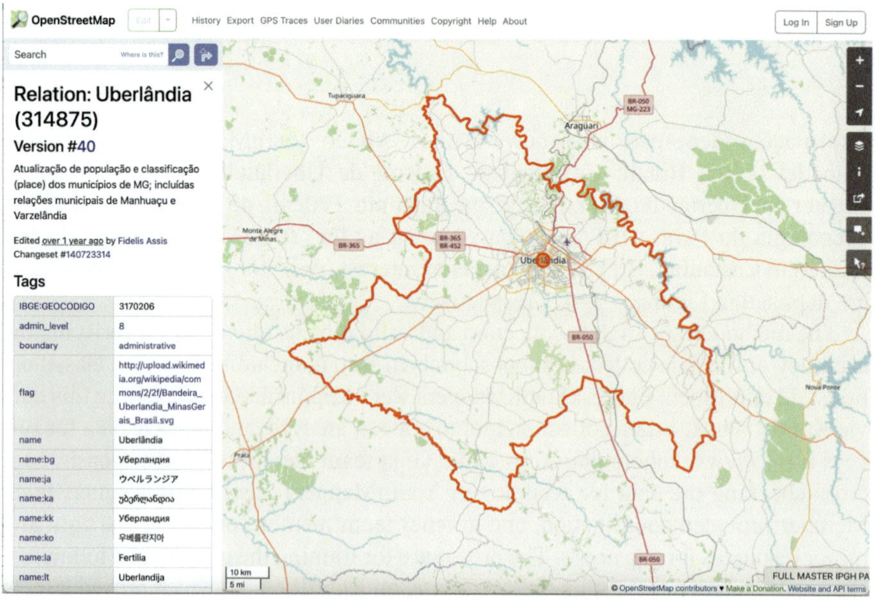

FIGURE 4.4 AOI location in Uberlândia, Brazil, for the UFU Mappers chapter.

Source: https://www.openstreetmap.org/relation/314875#map=10/-19.0303/-48.2671

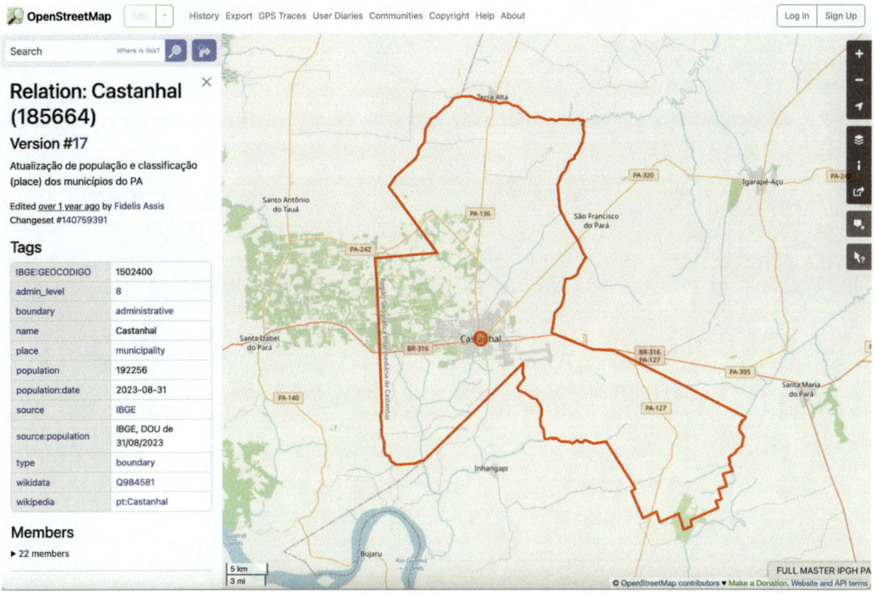

FIGURE 4.5 AOI location in Castanhal, Brazil for the YouthMappers IFPA chapter.

Source: https://www.openstreetmap.org/relation/185664#map=11/-1.2644/-47.8510

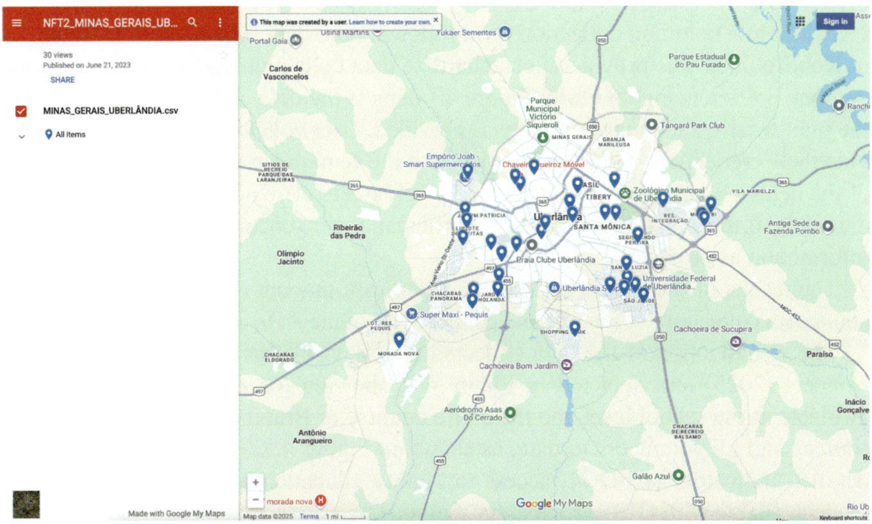

FIGURE 4.6 Sample location for YouthMappers training in pilot program using Google Maps.

Source: Authors, Google Maps.

FIGURE 4.7 Sample data collection form for YouthMappers training in pilot program using KoboToolbox.

Source: Abby Johnson for Arizona State University.

A link to a common collection form following the data scheme seen in Table 4.1 in KoboToolbox was provided to standardize data collection and allow for accurate geospatial referencing on OSM. Figure 4.7 depicts the data collection form used for the pilot, where a language option was provided. User volunteers were asked to provide their credentials, OSM usernames, and the remainder of the data fields. None of the fields was marked as mandatory, so volunteers were not required to enter any data if the information was unknown. Participants were encouraged to enter as much information as possible.

The results of these efforts yielded a significant amount of data and information about data collection on schools and connectivity in the AOI, as reflected in Table 4.2.

The visualization of the data collected is now visible on the GigaMaps Project Connect (2025) and depicted in Figure 4.8, showing the detailed data attributes available at the school level and the entire country of Brazil collected to date by all sources and all volunteers to date, as shown in Figure 4.9. These are exceptionally detailed indicators of SDG4, access to quality education at a spatial granularity, made available with GIS participatory methods.

In addition to data on the schools, data on the process were also collected to evaluate the pilot itself. Response times, efforts, communication and coordination testing, and all manner of assessing the effectiveness of execution, communication quality, frequency, and channels between Giga and YouthMappers, and among

TABLE 4.2
Results of Data Collection from YouthMappers Giga Pilot Project in Brazil.

YouthMappers Chapter	UAIGeo	UFU	IFPA
Site	*São João del – Rei*	*Uberlândia*	*Castanhal*
Average number of schools mapped by a volunteer	6 schools	6 schools	5 schools
Average time a volunteer dedicated to mapping	3 days	1 day	2 days, 3 hours/day
Sample disaggregation of time spent	Organization 2 weeks; mapping 15 minutes	Surveying calls for 15 minutes	Traveling 10 – 20 minutes; actual mapping: 30 min
Number of volunteers engaged	5 total (4 students, 1 advisor)	8 total (7 students, 1 advisor)	7 total (6 students, 1 advisor)
Total number of schools mapped	23 schools	28 schools	31 schools
Number of days/hours actively dedicated to mapping schools	14 days	7 days	10 days
Average number of schools mapped per day	1.53 schools/day	5.5 schools/day	15 schools/day
Cost per chapter	USD $50 (Gasoline and car travel)	*Unknown* (Personal use of mobile data)	USD $75 (Transportation, life insurance, internet)
Total number of students in AOI mapped	8,683 students	37,158 students	23,136 students
Total number of teachers in AOI mapped	551 teachers	1,817 teachers	1,154 teachers

YouthMappers chapters, and problem-solving during the pilot phase were critical to provide feedback for the possibility of scaling up and replicating for other potential future sites. This assessment followed three successive rounds of onboarding, training, data collection, data validation, focus group feedback, compilation of findings, and feedback reporting. This schedule allowed for an iterative process to encapsulate all learnings, achievements, and recommendations for future phases of the project. Overall, the responses revealed various challenges and recommendations, including communication obstacles, logistics, as well as technical issues for future resolution. These important lessons learned are summarized as follows:

Communication challenges: Difficulties in reaching the most responsible person, often leading to a series of referrals of one person passing the call on to another. Sometimes, participants were asked to call later. In other situations, the person willing to respond often lacked complete information. The language of the questionnaire was also considered a barrier, with students having to explain questions to school principals. Recommendations include ensuring that local languages are used in the survey instruments and platforms.

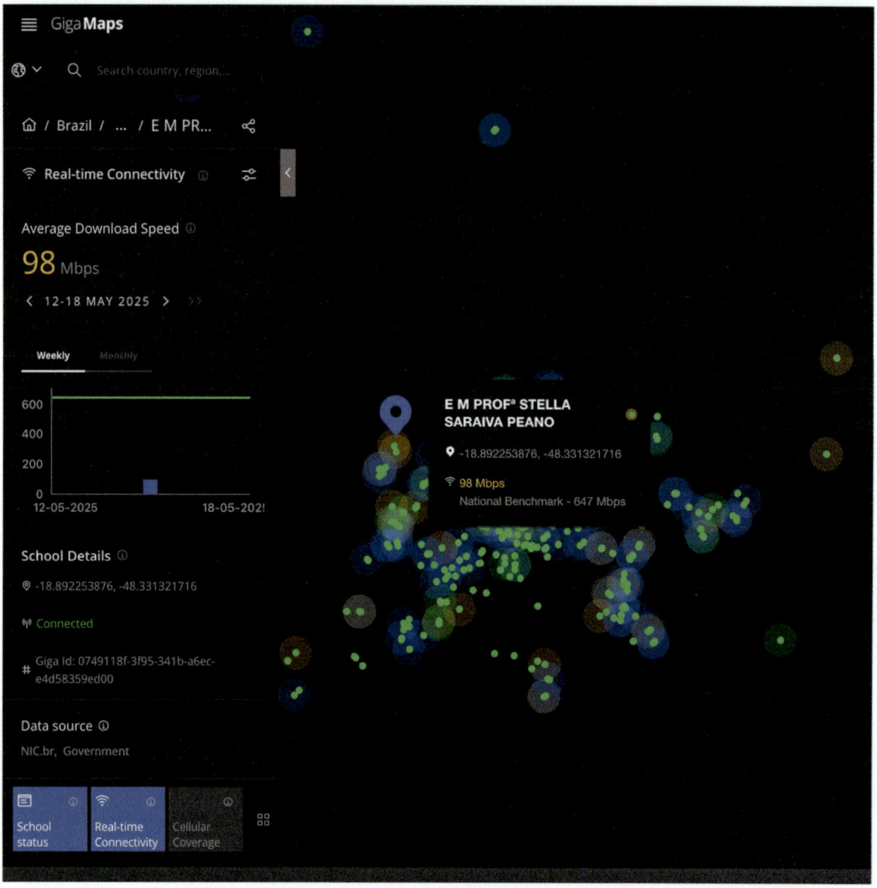

FIGURE 4.8 Example of the attributes available for data collected from one of the schools in Uberlândia, added by YouthMappers using KoboToolbox during the 2023 participatory GIS pilot project, now available on GigaMaps.

Source: https://maps.giga.global/map/schools?country=BR&school_ids=45806

***Technical and connectivity issues*:** During the survey, some participants faced problems with their cellular signal, as the signal in certain areas was not strong enough. So, even having received funds to buy data, participants had difficulty completing the survey by using the URL link without inserting it into the Kobo Collect App. Participants also re-emphasized the lack of confirmation from Kobo and the inability to access the data uploaded into the platform as barriers encountered. Recommendations include adequate and generous funding for volunteers to ensure connectivity and data backups, and redundancy for collection.

Logistical barriers: Students mentioned logistical challenges they faced with the sprawl of schools to be visited across large AOIs. Visiting the schools in person is a time-consuming activity. Some school directors were reserved

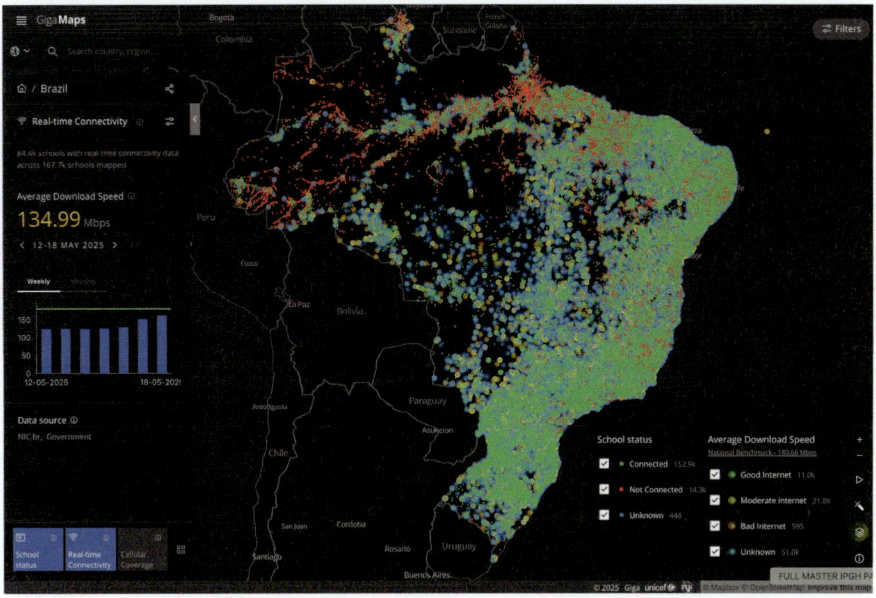

FIGURE 4.9 View of the country connectivity status for Brazil on GigaMaps, 2025.

Source: https://maps.giga.global/map/country/br

or unavailable, requiring scheduling. Recommendations include a design that recognizes the realities of student life, that leaves ample time for collection and validation; smaller and more targeted areas assigned, or larger groups of volunteers available to participate.

Timing issues: The pilot study began while students were still on vacation, and they had an extended holiday in the middle of the data collection period. However, once students learned the best strategies and established good communication with schools, survey completion time improved.

Gender and inclusion issues: One female student mentioned negative experiences with phone calls, receiving hang-ups, while a male colleague who followed up on the calls received all the answers from the same schools. Recommendations include training and support for all participants to ensure team support for members experiencing negativity when working with community members.

***Local knowledge integration*:** Students observed that some data could be more easily and readily validated using local data sites, such as the QEdu (https://qedu.org.br/), an educational data portal, for basic information, especially in larger cities with more up-to-date websites. Also, contacting municipalities first, who might have data to assist the project, could expedite achieving project goals. Integrating tools, more commonly used in daily school operations, local technology, may also speed up operations, as Kobo was perceived as somewhat distant from schools' routines.

After participating in the pilot project design, observing the chapter members during their data collection phase, and listening to their focus groups, the YouthMappers team added recommendations from a GIS ESD perspective as well. Beyond the specific lessons learned, it is important broadly for such participatory GIS partnerships to always ensure that the data collected is actually used within real world contexts, so that the time and effort of participants is honored, and, in turn, the participants involved in collecting the data are later informed about how the data is used and what is the impact. Partnering with programs like YouthMappers can bring incredibly valuable expertise and resources to projects like Giga and other SDG efforts. To facilitate such partnerships, incorporating GIS mapping tasks advances not only the particular SDG, but also advances the state of cartography on OSM or GIS, broadly creating a mutual win and aligns with the mission and enhances the project's overall impact and scope.

PARTICIPATORY MAPPING CASE STUDIES FOR SDG IMPACT BEYOND INDICATORS

Behind each SDG indicator lies a story of transformation and impact. The case studies in this section expand our thesis to further illustrate the real-world context of using geospatial tools in participatory ways to support education and sustainable development for impact. While these often meet indicators and targets, in many ways, they go beyond SDG4 and contribute to other goals, as can be seen in the following examples. These examples show how SDG targets 4.4 and 4.7 are implemented through students' involvement in geospatial projects through experiential learning. As a result of their project creation and technical skill building in GIS, participants enhance their employability and gain a better understanding of sustainable development. With the following examples of YouthMappers, the National 4-H GPS/GIS Leadership Team, and localized OpenStreetMap projects in this section, we illustrate the value and impact of these indicators for how GIS can enhance lives and shape futures.

GIS as an Experiential Learning Tool for Youth

The student-centered and peer-to-peer design of the YouthMappers' network ensures students choose how and what skills they will build (Gidudu et al., 2023). For the more geographically focused students, these can be technical skills in utilizing OSM, Quantum Geographic Information System (QGIS), and various open-source tools popular within the open geospatial profession. While other students may choose to learn how to pilot drones, manage social media, or design websites, YouthMappers form a team to carry out their activities, which allows students to build the expertise of their choosing, including soft skills. Due to the team nature of chapters, almost all participants will develop their leadership, teamwork, logistical, and event planning skills.

A powerful example of capacity building in both soft and technical skills is the rural electrification project implemented in Sierra Leone by Tommy Charles, 2020–2022 YouthMappers Regional Ambassador. During this project, Tommy coordinated power grid mapping activities among five universities across Sierra Leone

(Charles, 2023). Rooted in peer-to-peer support, YouthMappers alumni guidance, and OSM Sierra Leone community expertise, the power grid mapping project led to 69,988 buildings and 831 km of roads being added to the map. But more importantly, the volunteer mappers developed new knowledge in at least five geospatial tools and gained project management experience. The power grid mapping project is evidence of how SDG 4.4 can be met through experiential learning and leadership development within a GIS project. (See also documentary at https://youtu.be/xWOqYtv6skI).

Another youth organization capitalizing on utilizing GIS tools for youth's experiential learning is the National 4-H Geospatial Leadership Team (national4hgeospatialteam.us, 2025). 4-H is a national service-learning organization in the United States, and the Geospatial Team is a primary example of its technical initiatives. Students in grades 8–12 are equipped with training in GPS, ArcGIS, StoryMaps, Survey 123, and ArcGIS Hub either through their local 4-H club's activities or their involvement on the national team. The students directly engage in the GIS industry through conference participation and presentations, such as at the Esri User Conference. As a result of volunteering in their local 4-H club and the larger national organization, students develop tangible experience in GIS tools, geospatial project creation, and professional development within the geospatial industry (Carney-Nedelman, 2020). These opportunities prepare students for careers in and outside the geospatial industry while instilling civic engagement, which is a key outcome of SDG 4.4.

GIS in Action for Sustainable Development and Global Citizenship

YouthMappers and the National 4-H Geospatial Leadership Teams are two organizations founded on capitalizing on GIS as a tool for education and capacity building for the workforce. GIS can also be an outlet for global citizenship and application of the SDGs. An impressive method utilizing GIS to meet the SDGs is open mapping through OSM (Clark et al., 2019). OSM was created in 2004 as a global and publicly accessible map that anyone can contribute to, no matter their GIS expertise (OpenStreetMap, n.d.). A global community of open mappers has emerged, and a large component of the contributions are localized mapping projects. These projects are designed and implemented by the local community volunteer mappers, which leads to capacity building for the community members and data for sustainable development.

By using OSM as a tool for sustainable development, open mapping projects have been meeting and promoting the SDGs. Two examples of open mapping initiatives focusing on Gender Equality are EverywhereSheMaps and She Leads She Inspires. EverywhereSheMaps is a YouthMappers' program focused on increasing women's participation in the geospatial community and increasing data relevant to women's needs (Sharma et al., 2023). She Leads She Inspires is a project through the Humanitarian OpenStreetMap Team's (HOT) Asia Pacific-Hub to train women to be leaders in the OSM space (Tamura, n.d.). Both projects amplify women's voices in the geospatial industry and their local communities. They also increase their professional development and technical geospatial skills through training. The global nature of the geospatial industry and the OSM ecosystem enables participants to act as global citizens and apply their skills across borders. These initiatives apply SDG 4.7 by promoting inclusive education and supporting gender equality.

Global citizenship is a key pillar in the YouthMappers ecosystem (Carney-Nedelman & Clark, 2023). This has been achieved through the global network and the creation of Research Fellowships, Leadership Fellowships, Regional Ambassadorships, and remote mapping campaigns. YouthMappers 2018 Research Fellow and 2020-2021 Regional Ambassador Maria Fernanda Peña Valencia, wrote a blog titled *The World is a Handkerchief*, in which she expressed how attending the global fellowship in Washington, D.C. taught her how alike people are no matter their cultural and language differences; that through YouthMappers she was able to make friends and professional connections around the world (2019). Her experiences illustrate how geospatial education champions intercultural collaborations and global citizenship, key aspects of SDG 4.7.

Regional Action for Global Change

Finally, YouthMappers have shown that by working together, so much more can be accomplished than working alone. In a world growing hotter due to rising temperatures, staying cool is critical to combat the deadliest hazards on the planet. Reliable electricity in health facilities for functions like cooling, medicine storage, and heat-related care is essential. However, global data on the location of the nearest facilities and their electrification is often lacking, especially in high-need, hottest areas of the world. To address this, 75 students and faculty from 10 YouthMappers chapters in universities across Nigeria and the US collected and validated geographic and electrification data of health facilities. The effort included capturing street-level imagery and applying spatial analysis to identify priority sites for cooling centers. The study demonstrated that student-led mapping can provide valuable data for infrastructure planning, while also offering educational benefits and promoting youth leadership for good health and well-being, SDG 3 and climate action, SDG 13 (Solís et al., 2025).

Together, these initiatives demonstrate the power of geospatial education where peer-to-peer learning and participant-centered modalities are the focus. As a result of education in GIS, participants engaged in the aforementioned projects can address local and global challenges, equipped with tools for change. They become data users and contributors in their path to sustainable development for their communities. All the initiatives mentioned in this section have been volunteer-based and have had an experiential learning emphasis, which has underscored their success in building a more equipped workforce through addressing sustainable development. These stories exemplify the sentiment that when youth are empowered with geospatial tools, they don't just learn – they lead *everywhere*.

CONCLUSION AND RECOMMENDATIONS FOR THE FUTURE OF GIS EDUCATION FOR SDG4

As this chapter has shown, the intersection of GIS and education offers a compelling pathway toward achieving not only SDG4, but a broader agenda for sustainable development. By examining GIS both as an instrumental tool for planning and delivering education more equitably and as a subject of learning that fosters critical spatial awareness, we have illustrated the multifaceted role geospatial literacy can play in transforming educational landscapes. The barriers to accomplishing the former in official means – indicators gathered at national spatial levels notwithstanding – there remain ample ways the geospatial and EO communities are contributing and have further potential to contribute to university quality education for all – *everywhere*. Further investment and strategic coordination would only amplify the kinds of inroads made thus far in this journey. Our review of the global trends in this landscape, including in ESD, data-driven, competency-based, and GIS-specific, real-world grounded, youth-driven participatory mapping initiatives – highlights how amplifying geo-literacy and GIS education broadly could further bridge gaps in access, empower learners and communities, and foster data-informed action. With these considerations in mind, we reflect once more on the broader implications and recommend the following areas as focal points for future investments of attention and resources to optimize the full potential of GIS education as a systematic catalyst for inclusive and equitable quality education, everywhere and for sustainable development.

- ***Equity-focused interventions***: Because GIS highlights disparities by geography, gender, income, and powerful analytical capabilities combining socioeconomic data with educational attainment and other outcomes of other SDGs
- ***Community engagement***: Especially utilizing Participatory GIS or related types of methodologies in local school planning or filling educational gaps given its excellence as a methodology for addressing representation in place
- ***Expanding spatial data literacy*:** Particularly among education and SDG4 stakeholders themselves and the need for spatial literacy among students and lifelong learners, particularly given the need to improve the granularity of indicators and targets within nations
- ***Bridging the digital divide*:** The divide we are concerned with includes the public but also in geospatial capacity for education ministries/agencies to combat "data missingness" and improve local decision-making
- ***Advancing GeoAI*:** And predictive modeling in education domains, which is not far off in terms of what students will need to learn but also operationally, how and where schools will be located in areas of growing populations, how resources will be allocated, and generally leapfrogging the capacity to learn the latest knowledge, everywhere, now.
- ***Education for all*:** Interpreting "for all" as "everywhere" but also as tending toward "open", which means encouraging public investment in open, interoperable geospatial infrastructures in official cartographies and statistical systems
- ***Data-driven policymaking***: Because spatial evidence can inform national education plans, aligning with indicators SDG 4.7.1, especially the integration with statistical data linking census with school locations, electrification,

and internet connectivity, to meet the quality dimension of educational attainment goals

- ***Education is a responsibility of all*:** We recommend promoting multi-sectoral data integration, public-private-academic collaborations and workforce education alliances for 21st century GIS for the SDGs and beyond, especially in partnership, for SDG17.

ACKNOWLEDGMENTS

We would like to thank all of the YouthMappers around the world, but especially Bruna Kaely Souza da Silva, Gabriela Denize Silva Silva, Jameles Silva de Sousa, Maria Dolores Reis do Nascimento, Maria Luana Ferreira Ramos, Marcos Vinicius Cordeiro Conceição, Prof. Tatiana Pará, Ana Luísa Teixeira, Rafael Quetz Marques Oliveira, Julia Gabriela, Fernandes Teixeira, Hugo Frizzoni Candian, Prof. Silvia Ventorini, Deborah Bonuti Gonçalves, Ygor Henrique Bento Guimaraes, Iago Piccello, Maurício Rodrigues de Sousa, Maria Cecília Gomes Lopes, Renato Zamariolli Filho, Lorrane Alves Serafim, Prof. Rodrigo Freire, Tommy Charles, Shraddha Sharma, Courtney Clark, Sandhya Dhakal, Saugat Nepal, Mikko Tamura, Maria Fernanda Peña Valencia, Adele Birkenes, Ifeoluwa I.A Balogun, Olabanji Odunayo Aladejana, Ryan Avery, Rory Nealon, Zola Manyungwa, Chad Blevins, Abdulrahman Alsanad, Ebenezer Nana Kwaku Boateng, and Adamu Muhammad Adamu.

REFERENCES

Anderson, J., Blevins, C., Cowan, N., Carney-Nedelman, D., Clark, C., Crino, M., Engstrom, R., Hinton, R., Mann, M., McCusker, B., Nealon, R., Solís, P., & Zeballos, M. (2023). Reflecting on the YouthMappers movement. In P. Solís & M. Zeballos (Eds.), *Open mapping towards sustainable development goals* (pp. 366–370). Springer. https://doi.org/10.1007/978-3-031-05182-1_28

Annan-Diab, F., & Molinari, C. (2017). Interdisciplinarity: Practical approach to advancing education for sustainability and for the Sustainable Development Goals. *The International Journal of Management Education*, *15*(2), 73–83. https://doi.org/10.1016/j.ijme.2017.03.006

Arlinghaus, S. L., & Kerski, J. J. (2013). *Spatial mathematics: Theory and practice through mapping*. CRC Press.

Arriaga, V., Birkenes, A., Council, D., Jones, M., Lay Soler, E. K., McCarley, J. S., Wulf, E., Zhang, C., Parcher Wintemute, J., Aguirre, N., & Solís, P. (2023). Open mapping with official cartographies in the Americas. In P. Solís & M. Zeballos (Eds.), *Open mapping towards sustainable development goals* (pp. 295–304). Springer Nature. https://doi.org/10.1007/978-3-031-05182-1_26

Beynaghi, A., Trencher, G., Moztarzadeh, F., Mozafari, M., Maknoon, R., & Leal Filho, W. (2016). Future sustainability scenarios for universities: Moving beyond the United Nations Decade of Education for Sustainable Development. *Journal of Cleaner Production*, *112*, 3464–3478. https://doi.org/10.1016/j.jclepro.2015.10.117

Brovelli, M. (2025, January 22). *GEO AI in support of United Nations Sustainable Development Goals* [Lecture recording]. Department of Civil Engineering & Geomatics, Cyprus University of Technology. https://engineer-twinning.eu/2025/01/22/lecture-part-i-geo-ai-in-support-of-united-nations-sustainable-development-goals/

Brovelli, M. A., Ponte, M., Shade, S., & Solís, P. (2019). Citizen science in support of digital earth. In H. Guo, M. Goodchild & A. Annoni (Eds.), *Manual of digital earth* (pp. 593–622). Springer. https://doi.org/10.1007/978-981-32-9915-3

Brundiers, K., Barth, M., Cebrián, G., Cohen, M., Diaz, L., Doucette-Remington, S., Dripps, W., Habron, G., Harre, N., Jarchow, M., Losch, K., Michel, J., Mochizuki, Y., Rieckmann, M., Parnell, R., & Walker, P. (2021). Key competencies in sustainability in higher education—Toward an agreed-upon reference framework. *Sustainability Science, 16*(1), 13–29. https://doi.org/10.1007/s11625-020-00838-2

Carney-Nedelman, D. (2020, July 2). What can I do during a pandemic when my internship or summer camp is cancelled? Well map of course! *OpenStreetMap US News.* https://openstreetmap.us/news/2020/07/OSMUS-community-guest/

Carney-Nedelman, D., & Clark, C. (2023). The ecosystem where YouthMappers live and thrive. In P. Solís & M. Zeballos (Eds.), *Open mapping towards sustainable development goals* (pp. xx–xx). Springer. https://doi.org/10.1007/978-3-031-05182-1_29

Charles, T. G. D. (2023). Power grid mapping in West Africa. In P. Solís & M. Zeballos (Eds.), *Open mapping towards sustainable development goals* (pp. xx–xx). Springer. https://doi.org/10.1007/978-3-031-05182-1_11

Chien, S.-C., & Knoble, C. (2024). Research of education for sustainable development: Understanding new emerging trends and issues after SDG 4. *Journal of Sustainability Research, 6*(1), e240006. https://doi.org/10.20900/jsr20240006

Clark, C., Maron, M., Patel, D., Radford, T., Soden, R., & Uithol, P. (2019). *Open mapping for the SDGs: A practical guide to launching and growing open mapping initiatives at the national and local levels.* Humanitarian OpenStreetMap Team. https://documents1.worldbank.org/curated/en/967511563521548873/pdf/Open-Mapping-for-the-SDGs-A-Practical-Guide-to-Launching-and-Growing-Open-Mapping-Initiatives-at-the-National-and-Local-Levels.pdf

Coetzee, S., Minghini, M., Solís, P., Rautenbach, V., & Green, C. (2018). Towards understanding the impact of mapathons: Reflecting on YouthMappers experiences. *International Archives of Photogrammetry, Remote Sensing and Spatial Information Sciences, XLII-4/W8*, 35–42. https://doi.org/10.5194/isprs-archives-XLII-4-W8-35-2018

Coetzee, S., Rautenbach, V., Çöltekin, A., Pettit, C., Madden, M., Christophe, S., & Lkhamjav, O. (2020). Towards SDG 4: Trade-offs for geospatial open educational resources. *International Archives of the Photogrammetry, Remote Sensing and Spatial Information Sciences*, 199–205. XLIII-B5-2020. https://doi.org/10.5194/isprs-archives-XLIII-B5-2020-199-2020

DiBiase, D., Corbin, T., Fox, T., Francica, J., Green, K., Jackson, J., Jeffress, G., Jones, B., Mennis, J., Schuckman, K., Smith, C., & Van Sickle, J. (2010). The new geospatial technology competency model: Bringing workforce needs into focus. *URISA Journal, 22*(2), 55–72.

El-Jardali, F., Ataya, N., & Fadlallah, R. (2018). Changing roles of universities in the era of SDGs: Rising up to the global challenge through institutionalising partnerships with governments and communities. *Health Research Policy and Systems, 16*, 38. https://doi.org/10.1186/s12961-018-0318-9

Gidudu, A., de Newbill, M. A., Little, J., Brovelli, M. A., & Coetzee, S. (2023). Mentoring experiences in YouthMappers chapters. In P. Solís & M. Zeballos (Eds.), *Open mapping towards sustainable development goals* (pp. xx–xx). Springer. https://doi.org/10.1007/978-3-031-05182-1_28

Giga. (2025). GigaMaps. Retrieved May 25, 2025, from https://maps.giga.global/map

Hite, R., Solís, P., Wargo, L., & Larsen, T. B. (2018). Exploring affective dimensions of authentic geographic education using a qualitative document analysis of students' YouthMappers blogs. *Education Sciences, 8*(4), 173. https://doi.org/10.3390/educsci8040173

Huynh, N. T., Solís, P., Huot, P., Menkiti, N., Brown, A.-N., Carney-Nedelman, D., Coronado, A., & Soren, K. (2020). Becoming global-minded scientists: The short- and long-term impacts of international exchange. *Research in Geographic Education, 22*(1), 17–42. https://rge.grosvenor.txstate.edu/Issues/Volume-22-Number-1.html

Kerski, J. J. (2023a). Fundamentals of mapping and GIS. In Clifford, Cope, & Gillespie (Eds.), *Key methods in geography* (pp. xx–xx). Sage Publications. https://community.esri.com/t5/education-blog/16-new-lessons-as-part-of-a-fundamentals-of-gis/ba-p/1482056

Kerski, J. J. (2023b). Teaching sustainability in the 21st century with a year-long university-K–12–industry initiative. *Esri Education Blog.* https://community.esri.com/t5/education-blog/teaching-sustainability-in-the-21st-century-with/ba-p/1291745

Kerski, J. J. (2023c). Using WebGIS to teach climate resilience. *Esri Education Blog.* https://community.esri.com/t5/education-blog/using-webgis-to-teach-climate-resilience/ba-p/1266780

Kerski, J. J. (2023d). *A hands-on GeoAI workshop. Esri Education Blog.* https://community.esri.com/t5/education-blog/a-hands-on-geo-ai-workshop/ba-p/1487417

Kerski, J. J. (2024a). *Interpreting our world: 100 discoveries that revolutionized geography.* Bloomsbury Press. https://www.bloomsbury.com/us/interpreting-our-world-9798765120828/

Kerski, J. J. (2024b). Teaching and learning about geo-ethics. In Wyss & Peppoloni (Eds.), *Geoethics: Ethical challenges and case studies in earth sciences* (pp. xx–xx). Elsevier. https://storymaps.arcgis.com/stories/034813a3e60640428d7ffed5261facd7

Kerski, J. J. (2024c). Teaching and learning about conservation with GIS approaches, tools, and spatial data. *Esri Conservation Blog.* https://community.esri.com/t5/conservation-blog/teaching-and-learning-about-conservation-with-gis/ba-p/1556148

Kerski, J. J. et al (2025). Integrating web mapping and geospatial technologies. In F. Solem (Ed.), *Thriving in an academic career* (pp. xx–xx). Routledge. https://storymaps.arcgis.com/stories/caa10527808a417e9e7ca1b66aa66105

Kopnina, H. (2020). Education for the future? Critical evaluation of education for sustainable development goals. *The Journal of Environmental Education, 51*(4), 280–291. https://www.tandfonline.com/doi/full/10.1080/00958964.2019.1710444

Lozano, R., Merrill, M. Y., Sammalisto, K., Ceulemans, K., & Lozano, F. J. (2017). Connecting competences and pedagogical approaches for sustainable development in higher education: A literature review and framework proposal. *Sustainability, 9*(10), 1889. https://doi.org/10.3390/su9101889

Marshak, A., Young, H., & Naumova, E. N. (2023). Data on humanitarian crises: Who and what are we missing? *Food and Nutrition Bulletin, 44*(2_suppl), S124–S126. https://doi.org/10.1177/03795721231162429

McCusker, B., Ahmed, W., & Ramutsindela, M., & Solís, P. (Eds.). (2022). *The routledge handbook of development and environment.* Routledge.

Metternicht, G., Mueller, N., & Lucas, R. (2020). Digital earth for sustainable development goals. In H. Guo, M. Goodchild, & A. Annoni (Eds.), *Manual of digital earth* (pp. 443–472). Springer. https://doi.org/10.1007/978-981-32-9915-3_13

National 4-H Geospatial Team. (n.d.). About us. Retrieved May 20, 2025, from https://www.national4hgeospatialteam.us/pages/about-us

Nuumbosho, J. (2024). *Promoting Sustainable Development Goal 4 through ICT integration into Geography Advanced Subsidiary curriculum in selected secondary schools, Ohangwena Region* (Master's thesis, University of Namibia). University of Namibia Institutional Repository. https://repository.unam.edu.na/server/api/core/bitstreams/fa293e45-1a55-458f-bfe5-271c2e55a3a5/content

OpenStreetMap. (n.d.). What is the history of OSM? Retrieved May 21, 2025, from https://welcome.openstreetmap.org/about-osm-community/history-of-osm/

Parcher Wintemute, J., Solís, P., Aguirre, N., Arriaga, V., Birkenes, A., Zhang, C., Council, D., Wulf, E., Lay Soler, E. K., Jones, M., & McCarley, J. S. (2021). *Volunteered mapping and data governance in the Americas*. Pan-American Institute of Geography and History. https://bit.ly/VGIamericas

Peña Valencia, M. F. (2019, September 10). The world is a handkerchief. *YouthMappers Blog*. https://www.youthmappers.org/post/2019/09/10/the-world-is-a-handkerchief

Rees, A., Hawthorne, T., Scott, D., Spears, E., & Solís, P. (2021). Toward a community geography pedagogy: A focus on reciprocal relationships and reflection. *Journal of Geography*, *120*(1), 36–47. https://doi.org/10.1080/00221341.2020.1841820

Sajjad, M., Munir, H., Kanwal, S., & Naqvi, S. (2022). Spatial inequalities in education status and its determinants in Pakistan: A district-level modelling in the context of Sustainable Development Goal 4. *Applied Geography*, *140*, 102665. https://doi.org/10.1016/j.apgeog.2022.102665

Sharma, S., Clark, C., Dhakal, S., & Nepal, S. (2023). Seeing the world through maps: An inclusive and youth-oriented approach. In P. Solís & M. Zeballos (Eds.), *Open mapping towards sustainable development goals* (pp. xx–xx). Springer. https://doi.org/10.1007/978-3-031-05182-1_9

Sieber, R. (2006). Public participation geographic information systems: A literature review and framework. *Annals of the Association of American Geographers*, *96*(3), 491–507.

Silviariza, W. Y., Sumarmi, Utaya, S., Bachri, S., Marhaini, Malisi, F. F., Bobinsyah, M. F., & Hariani, V. (2024). Supporting SDG-4-education 2030: Collaboration of teaching models with technology (SPB-FC-l development). *International Journal of Technology Enhanced Learning*, *16*(3), 313–348. https://doi.org/10.1504/IJTEL.2024.139716

Solem, M., Boehm, R. G., & Zadrozny, J. (Eds.). (2024). *Powerful geography: International perspectives and applications* (2024th ed.). Springer. https://doi.org/10.1007/978-3-031-54845-1

Solis, P., Balogun, I. A., Aladejana, O. O., Avery, R., Nealon, R., Manyungwa, Z., Blevins, C., Al Alsanad, A. A., Boateng, E. N. K., & Adamu, A. M. (2025). Engaging local university students in participatory mapping for heat resilience to electrify health facilities in Africa. *International Journal of Sustainability in Higher Education*. https://doi.org/10.1108/IJSHE-09-2024-0644

Solís, P. (2005). Geographic learning for sustainable development. In W. L. Filho (Ed.), *Handbook of sustainability research* (pp. 679–704). Peter Lang GmbH.

Solís, P. (2022). YouthMappers: A hybrid movement design for the OpenStreetMap community of communities. In M. Minghini, P. Liu, H. Li, A. Y. Grinberger, & L. Juhász (Eds.), *Proceedings of the academic track at state of the map 2022, Florence, Italy, 19-21 August 2022*. Zenodo. https://doi.org/10.5281/zenodo.7004448

Solís, P., Anderson, J., & Rajagopalan, S. (2020). Open geospatial tools for humanitarian data creation, analysis, and learning through the global lens of YouthMappers. *Journal of Geographical Systems*, *23*(4), 599–625. https://doi.org/10.1007/s10109-020-00339-x

Solís, P., Huynh, N. T., Carpenter, D., Adames de Newbill, M., & Ojeda, L. (2017). Using an authentic project-based learning framework to support integrated geography education linked to standards and geospatial competencies. *Research in Geographic Education*, *19*(2), 36–65.

Solís, P., McCusker, B., Menkiti, N., Cowan, N., & Blevins, C. (2018). Engaging global youth in participatory spatial data creation for the UN sustainable development goals: The case of open mapping for malaria prevention. *Applied Geography*, *98*, 143–155. https://doi.org/10.1016/j.apgeog.2018.07.013

Solís, P. & DeLucia, P. (2019). Exploring the impact of contextual information on student performance and interest in open humanitarian mapping. *Professional Geographer*, *71*(3), 523–535. https://doi.org/10.1080/00330124.2018.1559655

Solís, P., Rajagopalan, S., Villa, L., Mohiuddin, M. B., Boateng, E., Nakacwa, S. W., & Peña Valencia, M. F. (2022). Digital humanitarians for the sustainable development goals: YouthMappers as a hybrid movement. *Journal of Geography in Higher Education*, *46*(1), 80–100. https://www.tandfonline.com/doi/full/10.1080/03098265.2020.1849067.

Solís, P., Balogun, I. A., Aladejana, O. O., Avery, R. H., Nealon, R., Manyungwa, Z., Blevins, C., Alsanad, A., Boateng, E. N. K., & Adamu, A. M. (2025). Engaging local university students in participatory mapping for heat resilience to electrify health facilities in Africa. *International Journal of Sustainability in Higher Education*, *26*(6), 1410–1433. https://doi.org/10.1108/IJSHE-09-2024-0644

Solís, P., & Sinton, D. (2023). Geospatial organizations and programs, internationally-based or with a non-US focus. In J. P. Wilson (Ed.), *The Geographic Information Science & Technology Body of Knowledge* (2nd Quarter 2023 Edition). https://doi.org/10.22224/gistbok/2023.2.2

Solís, P., & Zeballos, M. (Eds). (2023). *Open mapping towards sustainable development goals*. Springer. https://doi.org/10.1007/978-3-031-05182-1.

Tamura, M. (n.d.). She leads and she inspires: Aspiring for a representative movement in the asia-pacific region. Humanitarian OpenStreetMap Team Our Work. Retrieved May 21, 2025, from https://www.hotosm.org/projects/she-leads-and-she-inspires/

UNESCO Institute for Statistics (n.d). *Sustainable development goal 4*. UNESCO. Retrieved May 26, 2025, from https://uis.unesco.org/en/topic/sustainable-development-goal-4

Warren, S., Harris, T. M., Goodchild, M. F., & Solís, P. (2019). Commentaries on "Evaluating the Geographic in GIS." (Who is putting whose Geography into whose GIS, and how? Solís). *Geographical Review*, *109*(3), 308–320. https://www.tandfonline.com/doi/full/10.1111/gere.12359

Wu, Y. C. J., & Shen, J. P. (2016). Higher education for sustainable development: A systematic review. *International Journal of Sustainability in Higher Education*, *17*(5), 633–651. https://doi.org/10.1108/IJSHE-01-2015-0004

Part 2

Innovations, Safe Communities, and Life of Dignity

5 SDG #5—Leveraging Spatial Dimensions in Advancing Gender Equality

Nashon J. Adero

INTRODUCTION

Sustainable Development Goal 5 (SDG 5) aims to achieve gender equality and empower all women and girls. It addresses a range of interlinked issues such as ending discrimination, eliminating violence, ensuring equal participation in leadership, and granting access to sexual and reproductive health and rights. While the goal is ambitious and multifaceted, spatial dimensions offer a powerful but underutilized lens to analyze, monitor, and address gender disparities across geographies. Closing the global gender gap has been a towering challenge, a target the 2025 World Economic Forum's Global Gender Gap Report estimated to be more than 120 years away at the current rates. By 2025, the global gender gap was reported to have narrowed to 68.8% (World Economic Forum, 2025). The report on the global gender gap revealed the unsettling fact that, assuming the business-as-usual scenario, 123 more years would be needed to attain gender parity with the gender gap in STEM being even more skewed against women, even in Europe and North America. Though European tech companies post that women are approaching parity in employment representation overall, they still hold just 22 % of technical roles, such as developers and data engineers (World Economic Forum, 2020; Ellingrud et al., 2025).

Historically, women have been disadvantaged by skewed representation in natural resource ownership, economic opportunities, scientific leadership, and top management positions. Over decades, the dominant epistemological responses have taken the form of feminist geography and feminist GIS, domains which have evolved over time as critical and justice-oriented lenses for addressing gender-differentiated spatial inequalities and injustices (Kwan, 2002; McLafferty, 2005; Leszczynski & Elwood, 2015; Falahatkar, 2024).

WHY PLACE MATTERS FOR SDG 5

Approximately one-fifth of all SDG indicators can be measured wholly or partly using geospatial data (UN-GGIM, 2017, 2021). Aggregate statistics conceal spatial patterns of gender discrimination, which location-based intelligence can reveal while

DOI: 10.1201/9781003586319-7

supporting an all-inclusive dialogue with shared and geospatially mapped evidence. Though geospatial data makes up a substantial share of the century's big data revolution, research into how geospatial data can help leverage the attainment of gender equality remains limited. By leveraging spatial dimensions and extracting actionable geospatial intelligence, stakeholders can better understand the specific challenges and barriers faced by women and girls in different geographical locations, allowing for more targeted and effective interventions.

Key examples of geographically clustered burdens and discriminations that disproportionately affect women can be cited. For example, the geography of high rates of gender-based violence, unequal access to education or healthcare, disinheritance of land and property, or unequal economic opportunities for women can be mapped to scale for democratized visualization and remedial action using GIS technology. Mapping and visualizing the spatial trends and patterns of gender-related disputes, deprivation, destitution, dispossession, discrimination, and disenfranchisement empowers policymakers, organized and institutionalized advocacy groups, non-governmental organizations (NGOs), the private sector, and other stakeholders to better prioritize and allocate resources to address these skewed outcomes. In addition, gendered geospatial metrics enable location-specific monitoring of progress toward gender equality by tracking changes in key indicators over time and across different regions. Holding governments and other stakeholders accountable for their commitments, responsibilities, and mandates in promoting gender equality and empowering women and girls becomes more objective, inclusive, and traceable when spatial and temporal parameters are applied. Thus, the integration of geospatial information and analysis holds a high potential for closing the gender data gap and advancing SDG 5. Understanding the spatial and temporal dimensions of gender inequality can lead to effective strategies and interventions that finally empower women and girls and help create a more inclusive and equitable society for all.

CONTEXT AND MOTIVATION

Gender inequality in education and at workplace has been pervasive, especially in low-income countries. This study has drawn motivation from two case studies, both conducted in Kenya. First, a STEM-education audit of eight schools across three Kenyan counties showed that delivering quality science education to a girl transitioning from primary school to secondary school would require a budget of USD 750 per annum, versus USD 410 per annum for a boy, because girls have more sanitary and extra-maintenance needs (Adero, 2021a). The rural-urban divide was also evident, with wider disparities in rural areas. The findings confirmed the financial barriers girls face in accessing quality education and transitioning into STEM careers, particularly in rural areas and informal urban settlements.

Second, a Kenya-wide youth-skills survey conducted in 2021 by Adero (2021b) as part of the African Centre for Career Enhancement and Skills Support (ACCESS) Idea Competition on addressing youth unemployment returned spatially skewed results, with lower scores concentrated in rural wards and informal urban settlements. A key observation in these cases was that spatially clustered deprivation metrics reveal the extent and depth of inequality in access to educational resources and opportunities. Urbanized counties such as Nairobi, Kiambu, and Nakuru showed narrow gender gaps in post-secondary education attainment, but rural counties showed steep male leads in post-secondary education and STEM careers.

This chapter aims to:

1. Map SDG 5 targets to spatially explicit indicators
2. Demonstrate geospatial technologies and methods that aid in exposing, quantifying, and monitoring gender gaps
3. Recommend a scalable model of strategic and high-leverage entry points for enhancing spatial dimensions in advancing gender equality

CONCEPTUAL AND ANALYTICAL FRAMEWORK

Geospatial information and analysis can play a critical role in identifying spatial disparities and informing decision-making processes. By leveraging spatial dimensions in understanding the challenges faced by girls and women in accessing opportunities, decision makers can work towards creating a more inclusive and equitable society where all individuals have the opportunity to thrive and succeed. The conceptual basis of this chapter is grounded in identifying SDG 5 targets and exploring their inherent linkages with spatially explicit indicators. The theory of change is that, if gendered spatial parameters can be precisely mapped and tracked over time, then progress on achieving SDG 5 can be accelerated, complete with geographically verifiable outcomes. Geography reveals the spatial aspects of gender justice that are otherwise concealed in aggregate statistics. The spatial statistics derived from the spatially enabled approach lay the basis for the analytical framework with solid parameters and models used to quantify progress indicators for rational monitoring, verification, reporting, evaluation, and learning. It is possible to adopt a feminist-GIS lens, align with the United Nations Committee of Experts on Global Geospatial Information Management (UN-GGIM) Geospatial SDGs Roadmap, and integrate indicators from the UN-Women's 2022–2025 Integrated Results and Resources Framework (IRRF) (UN-GGIM, 2021; UN Women & UN Stats Division, 2022; UN Women, 2022).

SDG 5 TARGETS AND SPATIAL INDICATORS

Spatial indicators such as the gender gap in access to services, proximity to legal aid centers, women's mobility patterns, and exposure to environmental risks provide critical insights into the lived realities of women across different contexts. Several SDG 5 targets are inherently spatial. For instance, **Target 5.2** seeks to eliminate violence against women and girls in public and private spaces. The prevalence of violence is often unevenly distributed and can be mapped geographically using crime data, health facility reports, or survey data (Friedberg et al., 2020). **Target 5.4**, which recognizes unpaid care and domestic work, can be analyzed spatially by comparing infrastructure distribution, such as the presence of childcare services and public transportation within easy reach and the associated rural-urban divide (Watch, 2022).

Target 5.a, aimed at equal access to ownership and control over land, is directly spatial. The requisite property ownership data and information, land registry maps, and cadastral surveys showcase significant gender disparities in land tenure. The targets 5.2 (violence), 5.a (economic resources), and 5.b (ICT access) have explicit spatial dimensions, such as mapped violence hotspots by physical and human geographical characteristics, and travel-time inequality between women and men in search of economic opportunities. In addition, there are unequal exposure risks for women and men due to male-dominated work cultures and the design of occupational tools. In communication technologies, there is an interesting case of mobile-signal gender parity, a simple but powerful ratio showing how close women are to enjoying the same mobile-connectivity benefits as men. It can be used as a guide for gender-responsive mobile network rollout. Integrating spatial indicators into monitoring frameworks helps locate where gender inequalities are most entrenched and where interventions may have the most impact, hence the high-leverage intervention points. These indicators support location-relevant policymaking, allowing governments and organizations to tailor responses to regional disparities and foster spatial justice alongside gender justice. By embedding spatial considerations into SDG 5 tracking, stakeholders can more precisely address gaps, enhance accountability, and design more effective, place-based gender interventions (Shah, 2023).

FEMINIST-GIS AND INTERSECTIONALITY

It is worth noting that space is socially constructed and experienced differently based on gender, race, class, and other axes of identity. Feminist-GIS is a methodological and epistemological framework that critiques traditional spatial analysis for often being gender-neutral or even gender-blind. Feminist-GIS principles ensure that power relations embedded in spatial data are exposed and mitigated. Intersectional layers such as age, disability, or livelihood are treated as analytical axes rather than mere attributes. By incorporating feminist theories into spatial data practices, feminist-GIS challenges how space, power, and identity intersect and are represented. A central concept in feminist-GIS is intersectionality, coined by legal scholar Kimberlé Crenshaw. Intersectionality acknowledges that women are not a homogenous group and that their experiences of inequality are shaped by multiple overlapping social identities. For example, an indigenous woman in a remote area may face compounded marginalization due to her ethnicity, location, and gender. Mapping intersectional data—such as overlapping indicators of poverty, health access, environmental vulnerability, and education—provides nuanced insights into where and how gender inequalities manifest (Watch, 2022).

Feminist-GIS methods include participatory mapping, qualitative spatial narratives, and counter-mapping. Participatory mapping engages communities, especially women and marginalized groups, in creating spatial data that reflects their realities—such as places where they feel unsafe or excluded (Friedberg et al., 2020). These practices democratize spatial knowledge production and help challenge dominant narratives and power structures embedded in traditional cartography.

Integrating intersectionality into spatial analysis not only improves data richness but also ensures that interventions are equitable. For example, in urban planning, understanding how mobility differs for elderly women, migrant women, or women with disabilities can lead to the design of safer public spaces and inclusive transportation systems (Collins et al., 2023). Ultimately, feminist-GIS and intersectionality shift the focus from "where gender inequality happens" to understanding why and how it happens—offering a more grounded, empathetic, and actionable spatial framework for achieving SDG 5.

GEOSPATIAL DATA CUBE INTEGRATION

The United Nations Global Geospatial Information Management (UN-GGIM) Geospatial SDGs Roadmap process includes: (i) Data audit, (ii) cube assembly, (iii) indicator derivation, and (iv) decision-support services. Geospatial Data Cubes (GDCs) represent a transformative approach in handling large, multidimensional spatial datasets. They integrate time-series data across various geographies and themes—such as population, land use, infrastructure, health, and education—into a cohesive analytical environment. Africa, under the Digital Earth Africa (DEA) initiative for analysis-ready data (ARD), democratizes satellite data processing and analysis. The Africa Regional Data Cube (ARDC) promotes the Open Data Cube (ODC) initiative for open-source technology and structured geospatial data with interoperability, mainly from time-series satellite imagery. Thus, access to open and analysis-ready satellite data on a continental-scale platform has been feasible. By leveraging GDCs, spatial analysts and policymakers can investigate patterns, monitor progress, and identify inequalities across multiple layers and over time, offering unprecedented potential for advancing SDG 5. A key practical application of GDCs in SDG 5 efforts can be realized in monitoring women's economic participation and land tenure rights. For example, combining geospatial agricultural data with land registry records empowers analysts to visualize gender disparities in land ownership, cropland access, and financial resources. A spatial overlay of such information with policy zones or investment initiatives can reveal whether women are being actively included in economic development efforts or inadvertently left behind and disenfranchised (Watch, 2022).

GDCs support data disaggregation, which is essential for gender-responsive analysis and integrates satellite imagery with census data disaggregated by gender and enables analysis of access to essential services—such as electricity, clean water, or the Internet—across rural and urban areas, broken down by the sex of the household head (Shah, 2023). Furthermore, GDCs can link environmental data with health and demographic datasets to uncover correlations between climate-related extreme events, e.g., floods and droughts, and increased gender-based vulnerabilities such as school dropout rates or domestic violence incidents. GDCs also enable spatiotemporal trend analysis of the long-term impact of infrastructure investments, such as road development or new public transit routes. Geospatial data analysis aids in investigating women's labor force participation, commuting times, and safety perceptions (Ababneh et al., 2025). Thus, GDCs make it easier to evaluate policy impacts over time, identifying how gender equality interventions are producing sustained improvements in targeted regions. GDCs are interoperable with AI and machine learning tools. These predictive models can forecast areas at risk of increased gender disparities due to external shocks such as pandemics or migration, enabling proactive or predictive planning and resource allocation (Ababneh et al., 2025).

This chapter demonstrates how spaceborne, airborne, and ground-based sensors aid in capturing large volumes of geospatial data, based on which artificial intelligence (AI) and GIS technology can be utilized to enhance gender-focused decision support. AI is singularly potent as a data processing and analysis tool for revealing gender-differentiated access, risk, and opportunity landscapes. The second section presents the conceptual and analytical framework. The spatial data ecosystem and methods needed to set up a replicable model for leveraging spatial dimensions and advancing gender equality are highlighted in the third section. The fourth section presents and discusses the key findings and lessons. The chapter then closes with policy pathways and spatial data governance checklist for scaling geospatial gender analytics.

SPATIAL DATA ECOSYSTEM AND GENDER-RESPONSIVE MODEL

Meeting gender equality goals as envisaged in SDG 5 calls for adequate spatial data ecosystems that can support robust gender-responsive methodologies. The following spatial datasets and their accompanying characteristics should inform a suitable data acquisition and quality assurance strategy. This section also presents the recommended methodological approach and model for timely achievement of geodata-driven gender parity outcomes.

SPATIAL DATA ECOSYSTEM ELEMENTS

In Table 5.1 are displayed the key elements of the spatial datasets that should inform the optimal utilization of EO/geospatial data and GIS technologies to achieve the relevant SDG 5 targets.

METHODS FOR A GENDER-RESPONSIVE MODEL

The following GIS-based methods are recommended for monitoring progress toward achieving the SDG 5 targets.

1. **Service-Area Modeling** (Network Analyst) to compute walking-time buffers (15, 30, and 45 min) around service centers with gender-responsive disaggregation.
2. **Kernel-Density Estimation** (KDE) on harassment reports, followed by Getis-Ord Gi* to identify statistically significant clusters.
3. **Random-Forest Classification** of Sentinel-2 composites to delineate informal settlements where women's safety risks are elevated, e.g., urban slums.
4. **Geographically Weighted Regression (GWR)** linking GREI (Gender-Responsive Expenditure Index) to school dropout rates. GREI and dashboard prototypes can also illustrate how location-based intelligence can guide gender-responsive budgeting and help monitor Target 5.a on equal rights to economic resources and 5.b on ICT empowerment.

TABLE 5.1
Spatial Data Ecosystem Matrix

Spatial dataset	Source	Nominal spatial/ temporal resolution	Illustrative gender-lens use-case	Primary access/ custodian	Relevant SDG 5 target(s)†
EO – Optical, multispectral and panchromatic					
1	**Sentinel-2 MSI**	10 m/5-day revisit	Land-use proxies for basic service reach (e.g., clinics, water points) by sex-disaggregated demand	ESA Copernicus (SciHub)	5.1, 5.4, 5.a
2	**Landsat 8/9 OLI–TIRS**	30 m/16-day	Historical urban-expansion trends affecting women's commuting time	USGS (EarthExplorer)	5.4
3	**PlanetScope**	3 m/daily	Fine-grain mapping of street-lighting gaps linked to women's night-time safety	Planet Labs	5.2, 5.b
4	**Maxar WorldView-3/4**	0.31 m/tasked	Locating informal markets dominated by women vendors for tenure-security advocacy	Maxar SecureWatch	5.a, 5.5
5	**Pléiades-Neo**	0.30 m/tasked	Monitoring post-disaster shelter conditions for female-headed households	Airbus OneAtlas	5.2
6	**MODIS (Terra/Aqua)**	250 m–1 km/daily	Fire-count and drought stress are affecting women's time collecting fuel and water	NASA LAADS	5.4
EO – Radar, thermal and night-lights					
7	**Sentinel-1 SAR**	10 m/6–12 day	Flood-extent mapping for gender-sensitive emergency response	ESA Copernicus	5.1, 5.c

(Continued)

TABLE 5.1 *(Continued)*
Spatial Data Ecosystem Matrix

Spatial dataset	Source	Nominal spatial/ temporal resolution	Illustrative gender-lens use-case	Primary access/ custodian	Relevant SDG 5 target(s)†
8	**VIIRS Day/Night Band**	500 m/nightly	Night-time-lights proxy for electrification and women's evening study hours	NOAA NCEI	5.b, 5.4
EO – Population and infrastructure layers					
9	**WorldPop gridded population**	100 m–1 km (annual)	Sex-ratio maps informing location of women's resource centers	University of Southampton	5.1
10	**Microsoft/Meta building footprints and HRSL**	1–2 m polygons (updates quarterly)	Enumerating structurally safe shelters for women in disaster zones	Microsoft & Meta (open)	5.2
EO – Very-high-resolution low-altitude					
11	**UAV orthomosaics**	2–10 cm (on demand)	Audit of school infrastructure (separate latrines, safety fences)	Operator flight logs	5.1, 5.2
12	**Airborne LiDAR**	0.5–2 m (campaign)	Barrier-free access mapping for expectant mothers and persons with disability	National mapping agencies	5.2, 5.4
Survey, census and crowd-sourced data					
13	**Demographic and Health Survey (DHS)**	Cluster centroids jittered ≤ 2 km (rural) /≤ 5 km (urban); 5-year cycle	Maternal-health, GBV and ICT-ownership indicators with geotags	ICF/USAID	5.2, 5.6, 5.b
14	**Multiple Indicator Cluster Survey (MICS)**	Similar to DHS, 3- to 5-year	Girls' school-attendance & water-collection burden	UNICEF	5.4

15	**Living Standards Measurement Study – ISA**	HH GPS (jittered); 2- to 3-year	Women's agricultural land rights and asset control	World Bank	5.a
16	**Population and Housing Census**	Sub-location polygons; decennial	Sex-ratio, literacy, labor-force participation by place	National Bureau of Statistics	5.1, 5.5
17	**Labor-Force and Time-Use Surveys**	County or cluster-level; 3- to 5-year	Unpaid-care workload mapping	National Bureau of Statistics	5.4
18	**Call-Detail Records (CDR, anonymized)**	Cell-tower polygons; sub-daily	Women's mobility patterns vs public-transport routes	Mobile network operators	5.2, 5.b
19	**OpenStreetMap (OSM**)	Varies (crowd-sourced)	Mapping gender-sensitive amenities (maternal clinics, GBV centers)	OSM Community	*5.2, 5.6*

† Key SDG 5 targets abbreviated: **5.1** (end discrimination), **5.2** (eliminate violence), **5.4** (recognise unpaid care and domestic work), **5.5** (leadership and economic participation), **5.a** (land and economic resources), **5.b** (ICT access), **5.c** (enabling policies), **5.6** (SRHR).

5. **Participatory mapping sessions** with women's groups to ground-truth model outputs and add qualitative layers. Ethical protocols to follow the UN-Women gender-data guidelines unwomen.org.

KEY FINDINGS AND LESSONS FROM CASE STUDIES

Several case studies on gender-differentiated opportunities and outcomes have demonstrated the power of geospatial data and technologies in enhancing the understanding of how location influences the revealed disparities. The following examples serve to showcase the dynamic atlas of place-based gender nuances.

Gendered Transition to STEM Careers in Kenya

The findings of the 2021 study on gender disparity in STEM education for eight schools across three counties in Kenya showed that rural girls require annual learning and sanitary budgets 2–3 times higher than boys, if they are to reach parity in transitioning to secondary and tertiary education and pursuing STEM careers (Adero, 2021a, see Table 5.2 and Figure 5.1).

The map in Figure 5.1 shows the spatial context of the selected Gender-Responsive Expenditure Index (GREI) values. The following key spatial messages emerge:

1. **Nairobi** (private, urban schools) sits at the **top of every resource metric**; its parity in enrollment shows that once cost barriers are removed, girls can compete equally.
2. **Kisumu** (public, urban schools) occupies a middle tier—better than rural public schools, yet still below private schools, and it retains the **3:2 boy advantage** characteristic of public provision.
3. **Homa Bay** (rural schools) exhibits the **deepest dual disadvantage**: the lowest STEM and sanitary budgets and the same gender imbalance seen in other public schools.

Evident Gendered Resource Gaps

1. **STEM Learning Inputs**: Girls in public rural schools study with **less than half** the material budget of their private-school sisters, reinforcing the STEM gender gap at its root.
2. **Sanitary Provision**: Though girls' needs are recognized everywhere (budgets > boys by a factor ≈ 6), the **absolute level** in rural public schools (≈ USD 80 yr^{-1}) is still **below the USD 200 Standard** identified in project benchmark.
3. **Teacher Incentives**: Where mean monthly top-ups exceed USD 10 (all private schools), teachers are more willing to run girl-centered after-school motivation clubs, cited locally as key to retention.

What the Results Mean for SDG 5 Targets at County Scale

1. **Target 5.1 (End discrimination)**: Rural Homa Bay schools need direct bursaries for girls if the 3:2 imbalance is to be closed.

TABLE 5.2

Results of 2019 Budget (USD) Requirements for Girls and Boys in Critical Transition Stages from Primary to Secondary School in Kenya. "Factor" Represents the Ratio of the Mean of Private to Public School Budget Requirements

Budget estimates in USD as of July 2019

Primary school type	Class 7 annual budget for learning materials	Class 8 annual budget for learning materials	Class 7 Boy sanitary material annual budget	Class 7 Girl sanitary material annual budget	Class 8 Boy sanitary material annual budget	Class 8 Girl sanitary material annual budget	Class 7 teacher motivation payments monthly	Class 8 teacher motivation payments monthly	Boy's extra maintenance annual budget	Girl's extra maintenance annual budget
Public, rural	120.00	150.00	0.00	84.00	0.00	84.00	2.00	3.00	24.00	72.00
Private, rural	250.00	350.00	50.00	240.00	50.00	240.00	15.00	22.00	60.00	240.00
Private, urban	180.00	180.00	20.00	100.00	20.00	100.00	5.00	7.00	36.00	120.00
Public, rural	80.00	130.00	0.00	80.00	0.00	80.00	1.50	1.50	24.00	60.00
Public, urban	120.00	150.00	40.00	240.00	40.00	240.00	6.00	6.00	120.00	600.00
Mean, public	**106.67**	**143.33**	**13.33**	**134.67**	**13.33**	**134.67**	**3.17**	**3.50**	**56.00**	**244.00**
Mean, private	**215.00**	**265.00**	**35.00**	**170.00**	**35.00**	**170.00**	**10.00**	**14.50**	**48.00**	**180.00**
Factor	**2.02**	**1.85**	**2.63**	**1.26**	**2.63**	**1.26**	**3.16**	**4.14**	**0.86**	**0.74**

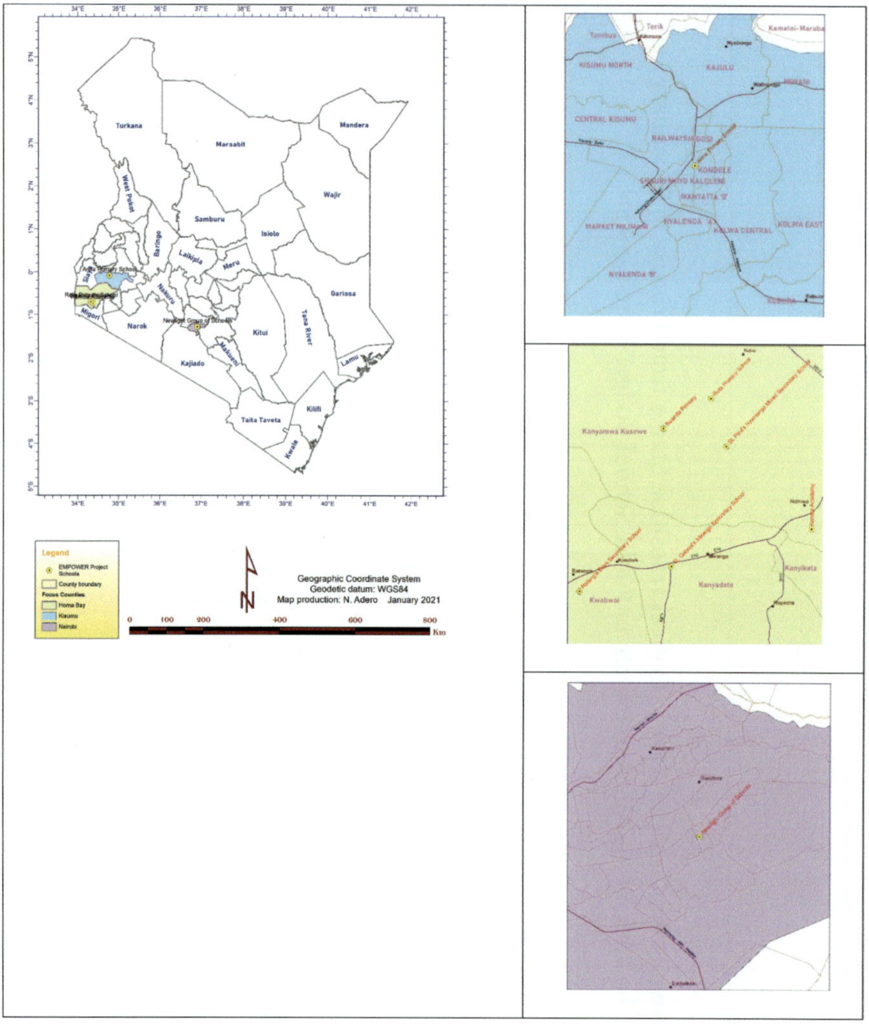

FIGURE 5.1 Spatial distribution of eight schools across three counties in Kenya surveyed in 2019 for gendered access to quality STEM education.

2. **Target 5.2 (Violence & safety)**: Better-lit compounds in resource-rich private sites underscore how street-lighting investments can help raise perceived safety for girls (PlanetScope imagery at 3 m resolution can provide visual evidence).
3. **Target 5.4 (Unpaid care and domestic work)**: Underfunded sanitary budgets in rural publics push girls into absenteeism and domestic chores; bridging the gap can help tackle this hidden workload.

4. **Target 5.b (ICT & STEM access)**: The USD 300 STEM-learning resource benchmark achieved only in urban private schools should become the **irreducible minimum** for public allocations if counties aspire to achieving parity in digital fluency.

The key lesson here is that geography amplifies gender. A girl in a rural public school endures a "triple deficit" of lower STEM inputs, meager sanitary support, and residual enrollment bias. Conversely, urban private environments show that with adequate funding, gender parity is immediately attainable—a compelling argument for spatially redistributive education budgets.

These spatially distributed results show that gender disparities in Kenya remain highly spatialized: Counties only a few kilometers apart can exhibit diametrically opposite patterns. Thus, tailoring interventions to micro-geographies should be more cost-effective than nationwide one-size-fits-all schemes. A GIS dashboard can serve as a baseline for county governments, TVET institutions, and private-sector partners to track progress toward more equitable, digitally enabled, and STEM-oriented labor markets.

The country-wide youth-skills survey conducted in 2021 by Adero (2021b) as part of the ACCESS Idea Competition on addressing youth unemployment found that only **24%** of young Kenyans graduating from colleges felt market-ready, with their digital-literacy scores averaging **16%**. The scores were spatially skewed, with lower scores concentrated in rural wards and informal urban settlements. Of notable significance was that spatially clustered deprivation metrics do reveal the extent and depth of inequality in access to educational resources and opportunities. Urbanized counties such as Nairobi, Kiambu, and Nakuru showed narrow gender gaps in post-secondary education attainment, while rural counties showed steep male leads in post-secondary education and STEM careers. STEM-qualified respondents were 1.4 × more likely to report being employed, and this premium was slightly larger for women; therefore, women in STEM can secure more career progression advantages. Women are also more exposed and vulnerable, and they bear the greater burden of occupational health and safety concerns, a fact well demonstrated by recent findings in the artisanal gold mining industry, in which women are more exposed than men to mercury poisoning in the process of extracting gold from the mercury-gold amalgam (Adero, 2025). However, women earn 2-3 times lower wages than their male counterparts. Overlaying UAV-mapped mine pits with mercury-vapour risk rasters can inform high-hazard zones and derisking mechanisms.

The use of GDCs for SDG 5 requires addressing data gaps, ensuring ethical data governance, and enhancing local capacity. Many regions still lack gender-disaggregated spatial data, and without community input, the insights may not reflect ground realities. Thus, integration must be paired with inclusive data practices and gender-responsive frameworks to ensure that spatial technologies genuinely empower women and marginalized communities. The findings underscore the importance of addressing both gender disparities and spatial inequalities in education and skills development. By understanding the unique challenges faced by girls and youth in specific geographic locations, policymakers and stakeholders can design targeted interventions to bridge these gaps and ensure equal access to quality education and skills training for all.

POLICY INTEGRATION

Several policy recommendations can arise from the review of case studies.

1. **Gender-Responsive Budgeting**: Earmark "equity top-ups" of **USD 200/learner/year** for wards scoring GREI < 60.
2. **Open-Street-Lighting Program**: Prioritize road segments frequented by female users.
3. **Digital-skills Accelerators** for rural women, leveraging ACCESS survey evidence of low digital literacy (mean 16%)
4. **Equity-weighted Capitation Grants:** The Ministry of Education to introduce an equity-weighted capitation formula that allocates at least USD 300 per learner per year for STEM learning materials in all public primary schools, with an automatic 50% top-up for rural schools in counties whose enrollment ratio of girls to boys is below 0.9.
5. **Ring-fenced Menstrual-Health Budget:** County governments in Kenya, through their education budgets, need to ring-fence a minimum of USD 200 per girl per year for menstrual-health and sanitary materials in every public primary school, disbursed directly to school boards and audited under the Public Finance Management Act.
6. **Performance-linked Teacher Motivation Payments:** A teacher-motivation scheme needs to be established that rewards schools with monthly incentive payments of USD 10–20 per teacher when they (i) achieve gender parity in the transition grades and (ii) maintain ≥ 95% attendance of girls, with higher weightings for rural and low-income urban zones.

CHALLENGES AND FUTURE DIRECTIONS

1. Data-privacy safeguards for sensitive gender-based incident reports.
2. Need for higher-resolution Earth Observation (EO) products (< 1 m) to resolve informal-settlement micro-grids.
3. Digital twins for gender-inclusive smart-city planning remain under-tested.

CONCLUSION

Spatial intelligence makes gender injustice visible and measurable. By embedding gender-disaggregated indicators in national and sub-national geospatial platforms, governments can locate, prioritize, and track interventions that close equality gaps. The Kenyan case studies show that bridging a mere 15-km service-area deficit or a USD 200 budget gap can halve dropout rates and improve safety for thousands of women and girls.

Resource endowments rise sharply with urbanity and private ownership, and every line-item shortfall—STEM materials, sanitary support, teacher incentives—translates directly into lower female retention in rural public settings. Yet the near-perfect gender parity observed in well-resourced urban private schools demonstrates that the gap is not a cultural destiny but a funding deficit. By pairing equity-weighted capitation, ring-fenced menstrual-health budgets and performance-linked teacher incentives, county and national authorities can transform the "triple deficit" faced by rural girls into a springboard for achieving SDG 5 targets and unlocking full human-capital dividend.

ACKNOWLEDGMENTS

The author acknowledges the significant roles of Impact Borderless Digital (IBD), the Institute for Conscious Global Change, GraFA Empower Program of TU Bergakademie Freiberg, and the ACCESS Idea Project in supporting the gender-focused studies that have yielded the case study results presented in this chapter. Beda Ogola and Billford Otieno contributed to data collection and preliminary analyses, which were crucial inputs for the overall form and contents of the chapter.

REFERENCES

Ababneh, A., Lok, K. L., Abdeyazdan, H., Opoku, A., & Chen, C. (2025). Gender equality in smart sustainable cities: Literature review. *Frontiers in Sustainable Cities*, *7*, 1535561.

Adero, N. J. (2025). *The toxic tools – A poisoned ecosystem and the gender toll: Policy paper on artisanal and small-scale gold mining (ASGM) in Migori County, Kenya*. Fr iedrich Naumann Foundation for Freedom. https://www.freiheit.org/publikation/toxic-tools-poisoned-ecosystem-and-gender-toll

Adero, N. J. (2021a). *Enhancing the quality and equity of access to STEM education in rural Kenya by optimising transition rates from primary to secondary school. GraFA EMPOWER Project Report.*

Adero, N. J. (2021b). *Addressing youth unemployment by matching lifelong skills development needs with talents and labour-market demographics* (ACCESS Idea Report).

Collins, T., Di Clemente, R., Gutiérrez-Roig, M., & Botta, F. (2023). Spatiotemporal gender differences in urban vibrancy. *Environment and Planning B: Urban Analytics and City Science*, *51*(7), 1430–1446.

Ellingrud, K., Yee, L., & Martínez, M. M. (2025, June 22). *Closing the gender gap in engineering for a digital-first future*. McKinsey & Company. https://www.mckinsey.com/featured-insights/themes/closing-the-gender-gap-in-engineering-for-a-digital-first-future

Falahatkar, H. (2024). A feminist GIS approach for identifying, mapping and evaluating gender inclusive features in urban public spaces (Master's thesis, University of Calgary, Calgary, Canada). Retrieved from https://prism.ucalgary.ca.

Friedberg, R., Sarnquist, C., Nyairo, G., Amuyunzu-Nyamongo, M., & Baiocchi, M. (2020). Understanding the spatial burden of gender-based violence: Modelling patterns of violence in Nairobi, Kenya through geospatial information. *arXiv preprint arXiv:2002.06710.*

Kwan, M. P. (2002). Feminist visualization: Re-envisioning GIS as a method in feminist geographic research. *Annals of the Association of American Geographers*, *92*(4), 645–661.

Leszczynski, A., & Elwood, S. (2015). Feminist geographies of new spatial media. *The Canadian Geographer/Le Géographe Canadien*, *59*(1), 12–28.

McLafferty, S. (2005). Women and GIS: Geospatial technologies and feminist geographies. *Cartographica: The International Journal for Geographic Information and Geovisualization*, *40*(4), 37–45.

Shah, C. (2023) *Leveraging Geospatial technology to achieve SDG-5: Promoting gender equality and empowering women…, Medium.* Available at: https://medium.com/%40shahchancy/leveraging-geospatial-technology-to-achieve-sdg-5-promoting-gender-equality-and-empowering-women-c482e5f1f613

UN Women. (2022). *Strategic Plan 2022-2025: Integrated Results and Resources Framework.* https://www.unwomen.org(PDF).unwomen.org

UN Women & UN Stats Division. (2022). *Progress on the Sustainable Development Goals: The Gender Snapshot 2022.* https://unstats.un.org.unstats.un.org

UN-GGIM. (2021). *The Geospatial SDGs Roadmap.* https://ggim.un.org(PDF).ggim.un.org

United Nations Committee of Experts on Global Geospatial Information Management (UN-GGIM) (2017). *The role of geospatial information in the Sustainable Development Goals: 2030 Agenda for Sustainable Development.* United Nations.

Watch, O. D. (2022). *Advancing gender data and statistics in Africa.*

World Economic Forum. (2020). *The future of jobs report 2020.* WEF.

World Economic Forum (2025). *Global gender gap report 2025.* WEF.

6 SDG #6 – Earth Observation for SDG 15 & SDG 6 Reporting in North African Countries

Fatma Trabelsi, Salsebil Bel Hadj Ali, Lamia Khemiri, El Hadi Gashut, Mustapha Mimouni, and Luis Evence Zoungrana

INTRODUCTION

Water and natural resources (WNRs) constitute foundational components of socio-economic systems across all scales, from local to global, and are critical determinants of human well-being, environmental integrity, and economic prosperity (IRP, 2019). However, unsustainable patterns of natural resource utilization have precipitated widespread environmental degradation, elevating anthropogenic activities to the primary driver of global Earth system changes and exerting significant pressure on multiple planetary boundaries (Rockström et al., 2009). Given the indispensable role of WNRs in economic activities (Bakshi et al., 2011), a global consensus underscores the imperative for their sustainable management. This approach aims to ensure the planet's capacity to meet the needs of both present and future generations (UN, 2015), while concurrently mitigating excessive disruption to local and global environmental systems (IRP, 2019).

In 2015, the international community launched the 2030 Agenda for Sustainable Development. This comprehensive action plan addresses critical challenges related to environmental degradation and resource depletion, which exacerbate global human challenges. The 2030 Agenda comprises seventeen integrated and indivisible sustainable development goals (SDGs), 169 associated targets, and 234 indicators. These are designed to catalyze action across areas of paramount importance for humanity and the planet during the 2015–2030 timeframe (UN, 2017). For the development and monitoring of these indicators, remotely sensed Earth Observations (EO), geospatial data, and derived information play an increasingly crucial role. They provide a consistent framework for measuring spatial and temporal attributes pertinent to the SDGs, offering valuable insights for target monitoring, strategic planning, progress tracking, and informing national and stakeholder decision-making, planning, and adaptive adjustments necessary for SDG attainment. When integrated with demographic and statistical data, these sources empower nations to conduct comprehensive analyses and

DOI: 10.1201/9781003586319-8

modeling, generate maps and other visualizations, assess cross-sectoral and regional impacts, monitor changes over time in a standardized manner, and enhance accountability. Key advantages associated with the application of EOs for SDG monitoring include the continuous availability of observational data, the existence of multi-annual time series facilitating change detection, their capacity to provide stable observations, and the cost-effectiveness of EOs for monitoring remote or otherwise inaccessible areas where ground-based surveys are impractical. Despite these benefits, significant challenges persist regarding the effective utilization of EO-derived information for SDG monitoring. These include the necessity for improved integration with national statistics and the requisite technical capacity for EO data analysis. Such capacity, particularly for developing nations, could be bolstered through targeted capacity-building initiatives and enhanced access to cloud computing facilities for online data processing.

A series of pilot projects is actively applying and testing the utility of EOs in supporting the assessment and tracking of SDGs. These initiatives demonstrate a clear alignment between specific SDG Goals and distinct types of EO and geospatial information products. An exemplary project is *Earth Observation for Sustainable Land and Water Management in North Africa* (GMES&Africa), implemented by the Sahara and Sahel Observatory (OSS) in collaboration with regional and national partners specializing in remote sensing. These partners include ASAL (Algeria), DRC (Egypt), LCRSSS (Libya), CRTS (Morocco), Nouakchott University AL-Aasriya (Mauritania), and CNCT (Tunisia) at the national level, and CRTEAN and CRASTE-LF at the regional level. The overarching objective of the GMES&Africa project is to facilitate decision-making in the sustainable management of natural resources and water through the provision of products and services derived from EO data and techniques. This chapter presents the findings of a regional study conducted by the Regional Center for Remote Sensing of North African (NA) States (CRTEAN) within the framework of the GMES&Africa project, spearheaded by the OSS. This study aims to assess EO use for SDGs reporting and decision-making in NA countries. It reviews the state of art in the use of EO for operational management of WNRs and for decision-making. It discusses the key technical and operational considerations to integrate the EO dimension in decision support and policy making. It highlights the significant progress that has been made in recent years by NA countries for operational EO integration at high level. Then, the study explores some opportunities of using EO in WNR management, assesses the suitability of EO in each legal and institutional framework, and recommendations on the way forward to implement the EO in a new era through a roadmap for WNR managers, policy makers, and stakeholders.

METHODOLOGY

Study Area – Within the African continent, North Africa distinguishes itself as the most economically developed region, also exhibiting a high degree of religious and linguistic homogeneity. This region demonstrates superior performance in the context of sustainable development, with four of its six constituent countries ranking within the top six performers continent-wide, culminating in a regional score of 61.87. Overall, North Africa appears well-positioned to maintain its status as a leading performer in SDG achievement, evidenced by moderate improvements or 'on

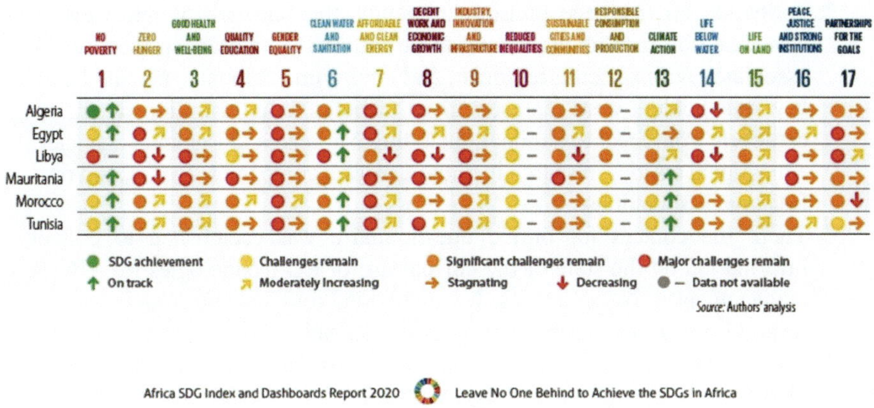

FIGURE 6.1 SDGs monitoring in NA countries – Africa SDG Index and Dashboards Report 2020. (The content of this publication has not been approved by the United Nations and does not reflect the views of the United Nations or its officials or Member States. https://www.un.org/sustainabledevelopment).

track' ratings for 52% of its SDG indicators. However, it is important to note that progress on numerous goals remains insufficient to meet the ambitious SDG targets. In this context, EO data and derived products represent a valuable asset for African nations, with NA countries being particularly well-suited to leverage these resources. EO provides a continental-scale data infrastructure that democratizes the capacity for processing and analyzing satellite data. This infrastructure delivers analysis-ready information, facilitating more informed, strategic, and inclusive decision-making through detailed tracking of changes in water bodies, land use, infrastructure development, and vegetation cover across national boundaries. The insights generated from EO can be effectively applied to address a broad spectrum of critical issues, including but not limited to water scarcity, land degradation, and food security challenges.

METHODOLOGY AND DATA

This study is conducted in four phases, adopting the following methodology as seen in Figure 6.2:

1. The design of the main elements of the study was based on a retrospective analysis of national studies conducted in Algeria, Libya, Mauritania, and Tunisia and done by the OSS consortium in the context of GMES & Africa project to plan the main phases of fieldwork in terms of questions that needed to be asked and potential stakeholders who should be approached. It was recommended that a more open or grounded approach should be adopted as far as possible to allow for the emergence of insights regarding the use of EO data in Algeria, Libya, Mauritania, and Tunisia.

2. Following on from the preliminary study, the selection of interviewees was based on institutions known to be the most important in their sector associated with agriculture, forest, and environmental management. These key stakeholders were representatives of public administrations at different levels from local to national, including National and Regional agencies, educational entities, research institutes, private companies, NGOs, and professional associations.
3. Then, the content for online questionnaire was established to collect information on the state of the current use of EO technologies for WNRs management in NA countries. It was divided into 7 sections and took the respondents around 15 to 20 minutes to complete it. All answers to mostly 'closed' questions were processed anonymously and information was handled according to the rules of the Data Protection Directive (95/46/EC) and national legislation. The data collected were summarized for each country and comparatively analyzed, resulting in lists of key requirements. In total, 76 completed questionnaires out of 506 questionnaires distributed were returned, resulting in an overall response rate for the entire region of 15%. The average response for the entire NA region was 15 for each country, ranging from 2 to 34 responses. For a web survey, this response rate was satisfactory with significant differences among the countries ranging from 3% (Mauritania) on the lower side and 42% (Tunisia) on the higher side, as seen in Figure 6.1.
4. The elaboration of critical study showcase the state of art of EO for SDG's reporting in North Africa as well as the current use of EO for WNR management in North Africa; the potential of EO-based operational services in support to decision-making for WNRs management such as forest monitoring, agriculture monitoring and yields forecasts, land degradation assessment, assessment of groundwater abstractions for

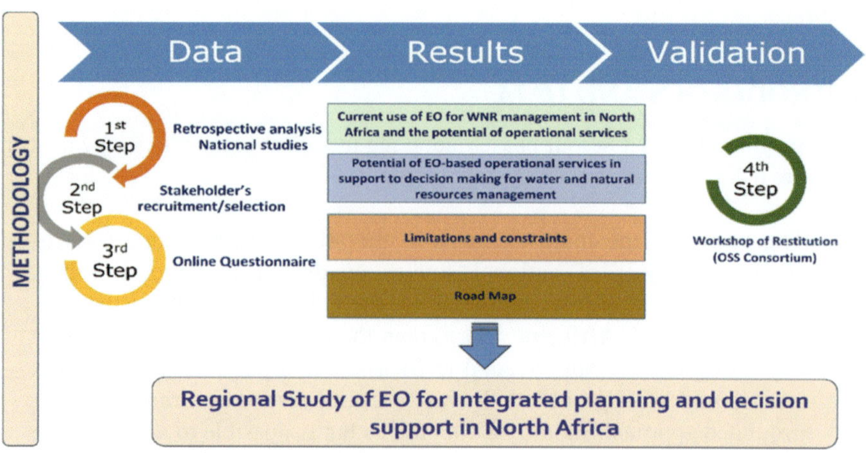

FIGURE 6.2 Chart of methodology.

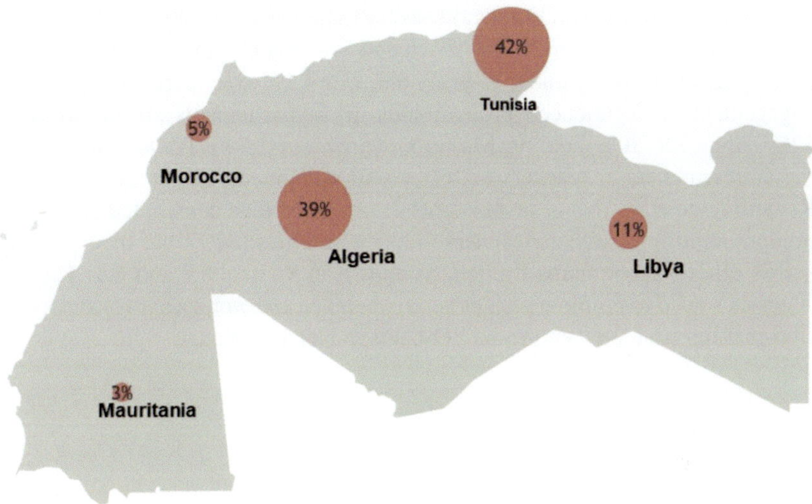

FIGURE 6.3 Distribution of survey responses among the NA regions.

agriculture use present the main limitations and constraints. And identified related gaps between the potential use of EO data and techniques and actual use by national administrations in charge of environment, water and agriculture monitoring. This resulted in a proposal for a roadmap for EO integration in WNRs management in North Africa based on the study outputs.

5. The study outputs were discussed during the workshop of restitution, when a discussion took place about the proposed solutions to bridge the gap between the potential use of EO data and techniques and their actual use by national administrations and to validate the proposed roadmap as seen in Figures 6.2 and 6.3.

RESULTS AND ANALYSIS

General Information

This section presents insights from survey questions about respondents' organizations, roles, and demographics. The participants predominantly work in the public sector (95%), with 17% in academia, and significant representations at the national (32%) and local (20%) administrative levels, while only 5% worked with regional authorities, private sector, or NGOs, as shown in Figure 6.4. Most respondents (~34%) are involved in the agriculture sector, mainly in Algeria and Tunisia, followed by higher education (16%), forestry (11%), environment and natural protection (8%), and EO-related downstream (4%) and midstream (4%) activities; other sectors such as construction, ICT, mapping, and security were less cited. Organizations primarily focus on improving EO use for agriculture, research, and cross-country activities, while upstream EO sectors such as satellites, launchers, and data reception remain underdeveloped in North Africa and as seen in Figure 6.5. Demographically, 36% of respondents are under 35 years old, with 64% over 35, reflecting the targeted engagement of experienced professionals. Most participants are highly qualified: 32% engineers, 30% holding Master's degrees, and 34% PhDs, with many serving as researchers or teachers (42%) or senior engineers (17%) seen in Figure 6.6. Regarding gender, 67% of respondents are men; women constituted 33%, and were predominantly researchers/teachers (42%) with PhDs (34%), but their participation across other roles remains limited, as shown in Figures 6.7 and 6.8. The analysis suggests a need to promote greater gender balance and women's involvement in decision-making positions within the EO fields in North Africa.

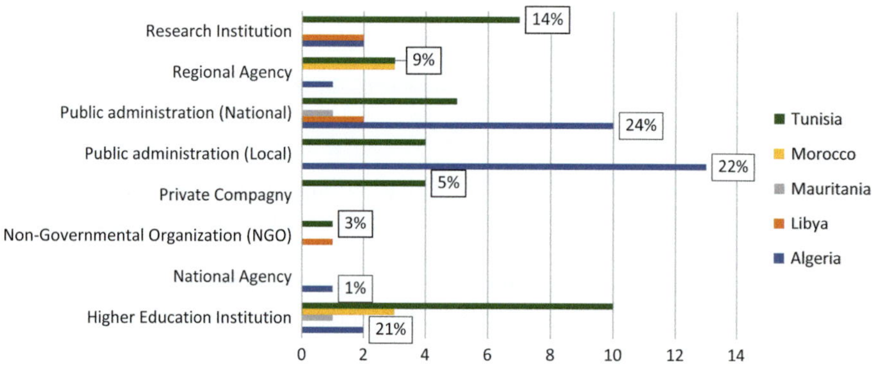

FIGURE 6.4 Type of responding organizations among the NA regions.

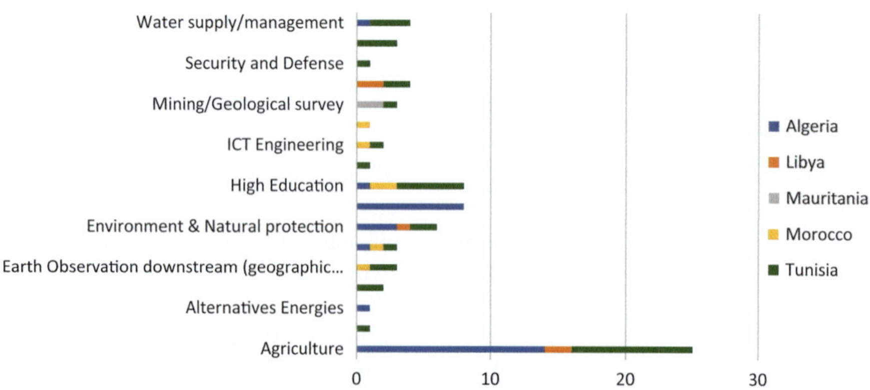

FIGURE 6.5 Sector of activity by country.

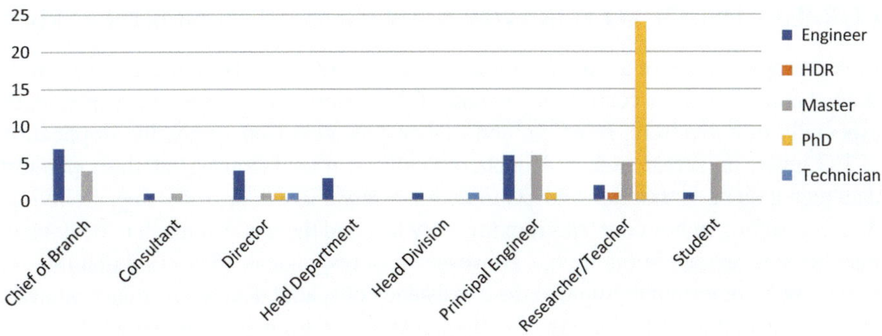

FIGURE 6.6 Position of respondents per educational background.

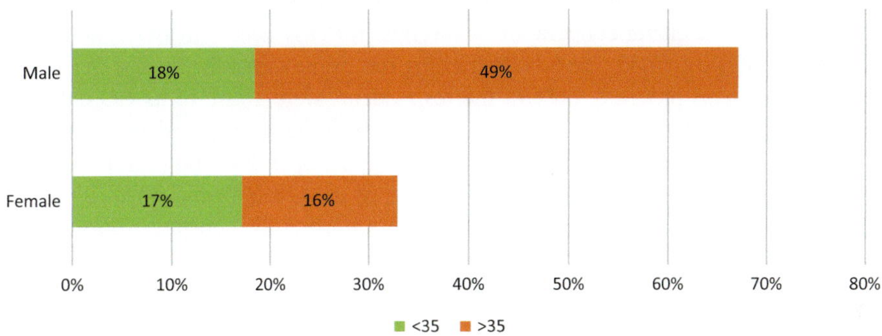

FIGURE 6.7 Gender & age repartition.

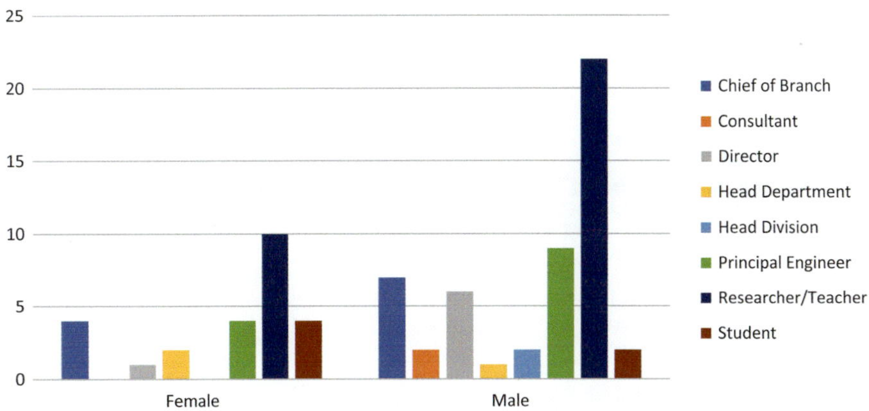

FIGURE 6.8 Position of respondents per gender.

CURRENT USE OF EO FOR WNR MANAGEMENT IN NORTH AFRICA

Concerning the level of awareness of and stance on EOs in NA countries, the results show that the EO play a central role in most of the sectors that involved the respondents, especially in agriculture, research, and higher education Conversely, the implication of EO upstream sector such as launchers/satellites, manufacturing/satellite operator/data reception or distribution and EO midstream sector like data processing, archiving, delivery, selling in NA countries remains very low and there is still much to be done in these sectors as seen in Figure 6.9. However, most respondents (53%) find that the EO is extremely important to improve the consistency of spatial data across their countries and at a second level to increase the frequency and currency of spatial as shown in Figure 6.10. The main aspects that they preferred to have available from EO use were to create maps for inclusion in the reports (72%), and to model and predict events and phenomena (66%); to provide general background information (57%) and to calculate the data on which the reports are based like inventory reporting (55%). Moreover, the majority consider emerging EO data flows as future strategic assets for WNR management, and that global cooperation is essential for easy access to EO systems and open data (84%) seen in Figure 6.11. Regarding the awareness of the European Union's (EU's) Copernicus program, 66% of the respondents were aware of its EO data and thematic services and most of them (72%) had already integrated its data, while 28% of all respondents expressed their willingness to use Copernicus program data in the future shown in Figure 6.12. Regarding the use of the Geoinformation & EO in the NA countries, it shows that 70% of respondents are using the geospatial information system (GIS) at different scales: Regional (20%), national (35%), local (31%), and organization (14%). Where GIS is used mainly are:

- Agricultural Information System (29%),
- Water Information System (24%),
- Soil Information System (25%),
- Forestry & Pastoral Information System (20%).

When asked about the level of access and upgrade of these GIS, the respondents felt it was still moderate (45%) to low (28%), and 28% of respondents considered it as

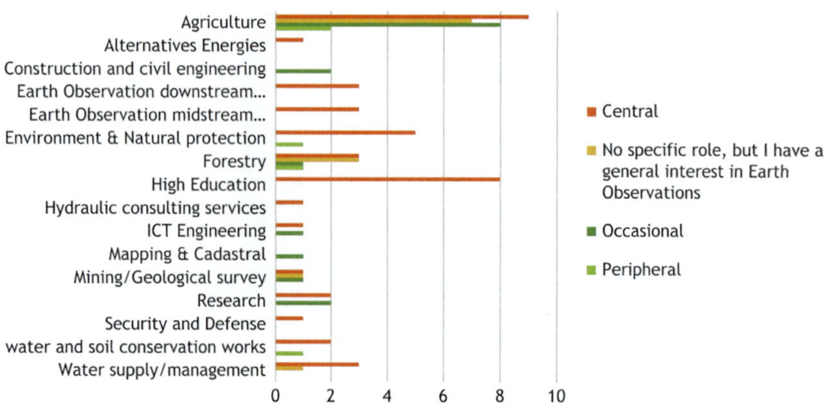

FIGURE 6.9 Role of EO per sector of activity.

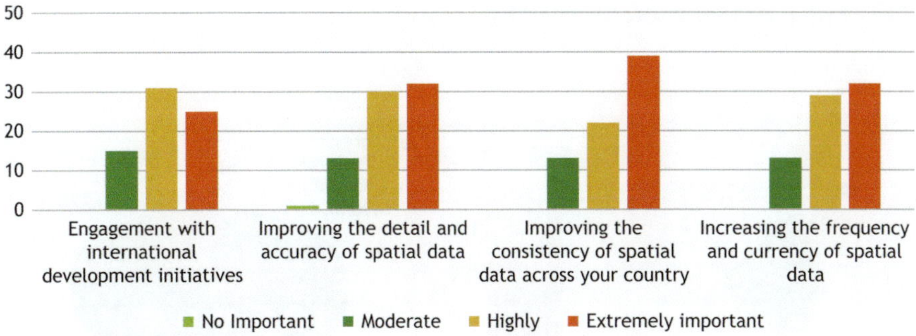

FIGURE 6.10 The need for EO according to different issues.

high. Furthermore, the majority stated they used open-source software (42%), while 22%, 16%, 18% of them respectively were using software with license, geospatial platform, and image processing system. They indicated that having different sources of spatial data, such as Sentinel data (1 & 2) are the most used (39%) followed by Landsat data (29%) while MODIS and SPOT data are equally used by only 9% of the interviewed persons. The use of other satellite data types, NOAA, AVHRR, AIRBUS, and QuickBird, remains very limited (1 to 5%) and can be seen in Figure 6.13.

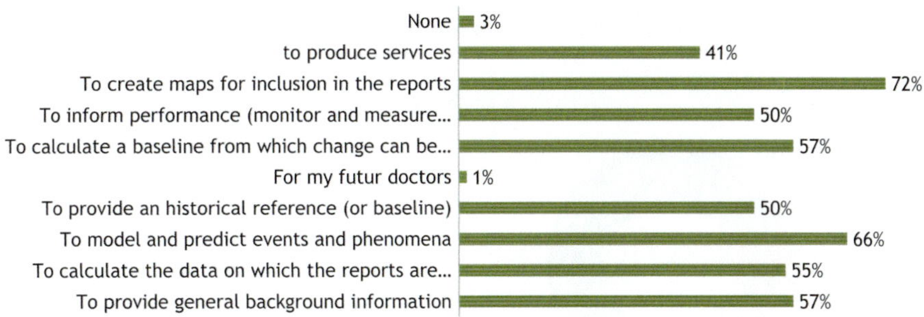

FIGURE 6.11 Preferences from survey respondents regarding the use of EO technologies.

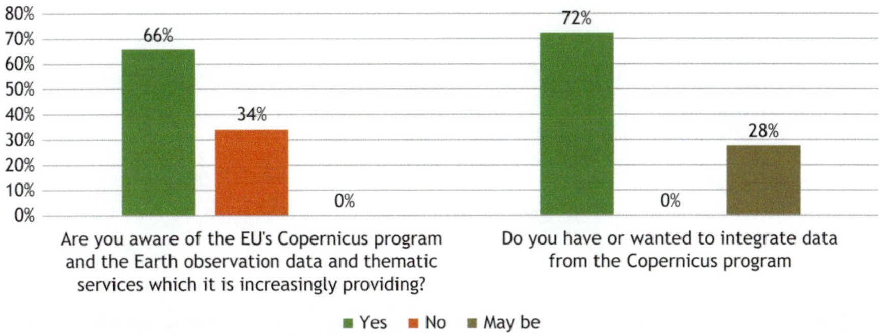

FIGURE 6.12 Awareness level about the EU's Copernicus program and providing EO data and thematic services.

Geospatial Information System Use

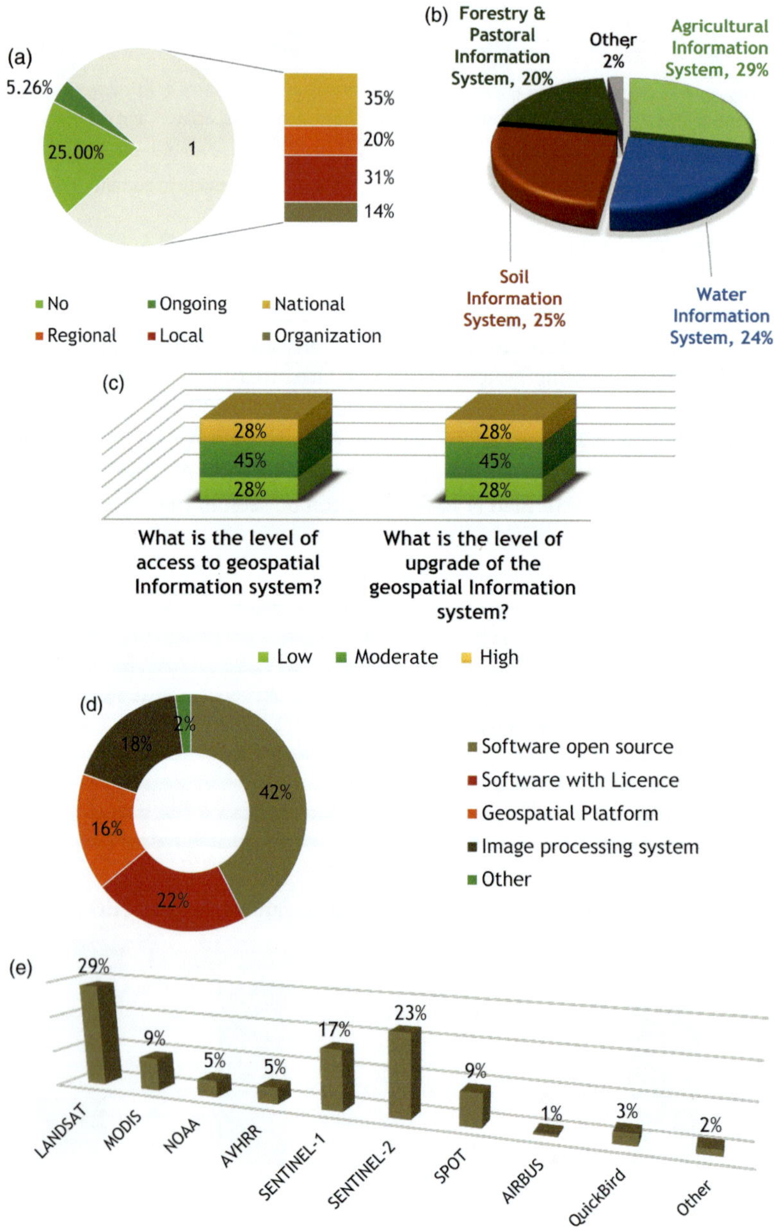

FIGURE 6.13 (a) Level use of GIS. (b) The fields of application of GIS. (c) The level of access and upgrade of geospatial information system types of softwares and satellite data used. (d) Types of software used by respondents' organizations. (e) Types of spatial data used by respondents' organizations.

FROM POTENTIAL TO PRACTICE: CONCRETE CASE STUDIES FOR SDG 6 MONITORING

While the survey provides a broad overview of EO usage and perceptions, a deeper analysis of specific applications reveals how this potential is being translated into practice. The following case studies provide concrete evidence of how NA countries are leveraging EO to directly calculate, monitor, and report on specific SDG 6 targets and indicators, moving from sparse in-situ data to comprehensive, spatially explicit assessments. This evolution highlights a clear methodological synergy. The same satellite platforms, and often similar data processing workflows, are used to assess both the quality of water within a pixel and the classification of that pixel as water for extent mapping. This suggests that capacity-building efforts should focus on creating a core competency in aquatic remote sensing, which can efficiently serve multiple monitoring needs simultaneously. The case studies reveal a *data maturity ladder,* which monitors water extent (Indicator 6.6.1) is a more mature and directly quantitative application of EO, while monitoring water quality (Indicator 6.3.2) often relies on proxies that require more intensive local calibration and validation. This progression offers a logical pathway for countries looking to adopt and expand their EO capabilities for water management.

CASE STUDY: LARGE-SCALE WATER QUALITY MONITORING IN THE NILE DELTA AND LAKE NASSER, EGYPT

Assessing Ambient Water Quality (Indicator 6.3.2)

The Nile River and its delta are the lifeblood of Egypt, but the sheer scale of these water bodies makes comprehensive in-situ monitoring exceptionally challenging (Fitoka et al., 2020; EOMAP). To address this, Egyptian institutions use a multi-sensor EO approach. Data from the Sentinel-3 OLCI provides frequent, medium-resolution coverage for large-scale phenomena across Lake Nasser, while higher-resolution Sentinel-2 and Landsat-8/9 data are used for detailed analysis of inland delta lakes. Standard algorithms derive key water quality proxies like chlorophyll-a (Chl-a) and Total Suspended Matter (TSM) (El-Sadek et al., 2022; Spyrakos et al., 2025). Platforms like the UNESCO IIWQ World Water Quality Portal (UNESCO, 2018) visualize these parameters, providing synoptic maps that are impossible to generate from point-based sampling alone and directly informing the national calculation of Indicator 6.3.2. This work is supported by international collaborations, including an initiative led by the United Nations Environment Programme (UNEP) and funded by the German Agency for International Cooperation (GIZ). The technological framework is also supported by the UNESCO International Initiative on Water Quality (IIWQ) and its World Water Quality Portal.

CASE STUDY: INTEGRATED RIVER MONITORING IN THE SEBOU BASIN, MOROCCO

The Sebou Basin is subject to severe pollution from industrial and agricultural sources. Researchers and water agencies in Morocco are pioneering a hybrid approach that integrates traditional in-situ sampling (for parameters like BOD5, COD, and Total Phosphorus) with advanced EO analysis using Sentinel-2 and Landsat imagery. Machine learning models, such as Multivariate Adaptive Regression Splines (MARS), are used to establish robust relationships between spectral reflectance and measured water quality parameters (Kanga Idé et al., 2021; El Bilali et al., 2023). This allows for the generation of predictive maps of a composite Water Quality Index (WQI) across the entire river basin, transforming point data into comprehensive maps that directly support the reporting requirements of Indicator 6.3.2.

CASE STUDY: MONITORING COASTAL WETLAND DYNAMICS IN TUNISIA AND MOROCCO

Tracking the Spatio-Temporal Extent of Water-Related Ecosystems (Indicator 6.6.1)

North Africa is home to internationally important wetlands, such as Ichkeul Lake in Tunisia and the Merja Zerga Lagoon in Morocco, both of which are threatened by upstream water management and pollution (Hamed et al., 2018; El Mahrad et al., 2023). To monitor these ecosystems, national agencies use the long-term archives of Landsat and the higher-resolution Sentinel-2 data. Time-series analysis of spectral indices like the Normalized Difference Water Index (NDWI) accurately delineates open water areas, while classification algorithms map key vegetation such as the *Zostera noltei* seagrass beds in Merja Zerga. This application provides a direct, quantitative measurement (in hectares) of *the change in extent* for Indicator 6.6.1, which is crucial for national reporting and for developing targeted conservation strategies. At Ichkeul Lake, monitoring efforts have been supported through financing from the Global Environment Facility (GEF) and the Kreditanstalt für Wiederaufbau (KfW), Germany. In the case of Merja Zerga, a tidal lagoon on the Atlantic coast of Morocco, this work is primarily governed by Morocco's national commitments under the Ramsar Convention.

CASE STUDY: ASSESSING GROUNDWATER-DEPENDENT ECOSYSTEMS IN THE ALGERIAN SAHARA

The oases of the Algerian Sahara are classic groundwater-dependent ecosystems (GDEs) reliant on the vast but non-renewable Northwestern Sahara Aquifer System (NWSAS). Direct monitoring of this aquifer is impractical, but EO provides a powerful proxy by monitoring the health and extent of the surface vegetation that depends on this groundwater. Using the historical archive of Landsat imagery and classification algorithms, researchers can track the expansion of date palm plantations and assess vegetation health (via NDVI) as a direct indicator of water availability and use (Bouarfa et al., 2024). This approach, operationalized through platforms like MISBAR, which was developed as part of the GMES & Africa program, is a joint initiative financed by the EU and the African Union Commission (AUC). It provides critical data on the pressures on these finite resources, addressing the GDE component of Indicator 6.6.1 and highlighting alarming water stress levels.

CASE STUDY: THE FAO WaPOR PROJECT IN Tunisia

Quantifying Water-Use Efficiency and Water Stress (Indicators 6.4.1 and 6.4.2)

With a water stress level of 98%, improving agricultural water efficiency is a national survival strategy for Tunisia. The FAO's WaPOR portal provides open-access data on actual evapotranspiration (ETa) and Net Primary Production (NPP) derived from MODIS, VIIRS, and Landsat sensors. The ratio NPP/ETa yields a direct physical measure of water productivity (kg/m3). This data is actively integrated into management tools like the "IREY" mobile application, which provides farmers with real-time irrigation schedules, closing the loop from high-level monitoring for Indicator 6.4.1 to on-farm decision support (Bouselmi et al., 2023). The WaPOR (Water Productivity through Open-access of Remotely sensed derived data) is a project funded by the Ministry of Foreign Affairs of the Netherlands and is implemented by the Food and Agriculture Organization (FAO) with key partners including the IHE-Delft Institute for Water Education and the International Water Management Institute (IWMI).

CASE STUDY: MOROCCO'S NATIONAL DROUGHT EARLY WARNING SYSTEM

Drought is a recurrent hazard in Morocco, and the Royal Centre for Remote Sensing (CRTS) has developed a sophisticated Composite Drought Indicator (CDI) to provide early warning. The CDI integrates four satellite-derived parameters: The Standardized Precipitation Index (SPI) from CHIRPS, NDVI anomalies from MODIS, Land Surface Temperature (LST) anomalies from MODIS/VIIRS, and ET anomalies (Bijaber et al., 2018). The resulting monthly maps provide a dynamic, national-scale assessment of drought conditions, which is a direct result of water stress and provides a critical layer of information for reporting on Indicator 6.4.2. The development of the CDI was part of the Land Data Assimilation Systems-Maroc (LDAS-Morocco) project and was financed by the GEF and managed by the World Bank, with additional support from the United States Agency for International Development (USAID) and technical assistance from NASA.

CASE STUDY SMART IWRM MEDJERDA – *SMART INTEGRATED WATER RESOURCES MANAGEMENT IN TUNISIA*

The SMART IWRM Medjerda project, implemented by the Higher School of Engineers of Medjez El Bab (ESIM) in collaboration with the USGS and funded by USAID (PEER7), addresses the critical issue of groundwater overexploitation and quality degradation in the lower Medjerda Basin, Tunisia. The project integrates EO technologies with in-situ sensing and advanced modelling to support the implementation of Integrated Water Resources Management (IWRM), in line with SDG 6. A real-time Decision Support System (DSS) was developed, combining a network of IoT-based sensors monitoring aquifer dynamics with a geospatial platform (Geoportail) that centralizes and visualizes multi-source data, including EO-derived land use, vegetation cover, climatic variables, and groundwater vulnerability maps. These datasets are used to assess spatial patterns of nitrate and pesticide pollution, salinization processes, and to model groundwater recharge zones under climate change scenarios (Trabelsi et al, 2023). Furthermore, the project applied remote sensing and machine learning algorithms to predict piezometric trends and to inform sustainable abstraction strategies (Trabelsi et al, 2022). Participatory governance was strengthened through multi-stakeholder workshops and in collaboration with CIRAD, facilitating the co-construction of local groundwater management approaches (Hassenforder et al, 2023). The initiative also emphasized capacity building and gender inclusion through targeted training events and scientific outreach. SMART IWRM Medjerda contributes directly to several SDG 6 indicators, notably 6.3.2: Ambient water quality; 6.4.1: Water-use efficiency; 6.5.1: IWRM implementation level; 6.6.1: Ecosystem extent and condition, and 6.b.1: Community engagement in water governance, which demonstrates the potential of EO to support science-based, participatory water management in data-scarce regions.

CASE STUDY: THE GMES & AFRICA MISBAR PLATFORM FOR IRRIGATION MONITORING

The MISBAR platform, led by the OSS and financed by the European Union (EU) and the AUC, is a diagnostic tool designed to quantify agricultural water withdrawal, the primary driver of water stress in the region. Using Sentinel-1 and Sentinel-2 data, the platform maps irrigated areas and combines this with crop type maps and ET models to estimate the total volume of water withdrawn for irrigation (Hadria et al., 2021). This provides a direct, EO-based methodology for calculating the agricultural water withdrawal component of Indicator 6.4.2, a critical data point that is very difficult to measure using traditional methods.

CASE STUDY: THE GMES & AFRICA MISLAND PLATFORM FOR LAND DEGRADATION NEUTRALITY

Monitoring Land Degradation (Indicator 15.3.1)

To address the challenge of land degradation and desertification in North Africa, a global hotspot that threatens food security and livelihoods (Akhtar-Schuster et al., 2011), the Monitoring Integrated Services for Land Degradation (MISLAND) platform was developed by the OSS within the GMES & Africa program. Designed as a DSS, MISLAND supports Algeria, Egypt, Libya, Morocco, Mauritania, and Tunisia with their national reporting for the United Nations Convention to Combat Desertification (UNCCD) and SDG Indicator 15.3.1, a framework that requires robust, large-scale monitoring capabilities to be effective (Sims et al., 2019). The platform utilizes time-series EO data to assess degradation via the three official sub-indicators: Land Productivity Dynamics, Land Cover Change, and Soil Organic Carbon. By integrating this data to classify land as "degraded" or "not degraded" and providing supplementary indicators on vulnerability, MISLAND delivers a standardized, evidence-based tool through a user-friendly web dashboard and QGIS plugin, enabling these nations to effectively monitor degradation on a national scale, report on Land Degradation Neutrality (LDN) targets, and prioritize interventions.

POTENTIAL OF EO-BASED OPERATIONAL SERVICES IN SUPPORT OF DECISION MAKING FOR WATER AND NATURAL RESOURCES MANAGEMENT

In the last years, EO data has increasingly been used in NA countries as advances in the technology make it more easily accessible to scientists, researchers, and other users. Its applications range from surveying, weather assessment and forecasting, water resources management, vegetation monitoring, early warning activities such as drought, flood and crop yield forecasting, detection of pollution, and all aspects of land resource planning.

Analysis undertaken in this chapter identified potential of EO-based operational services in support of decision-making for WNRs management. In fact, the survey identifies different uses of EO in WNR services, where the most used are Agriculture management service (28%), Assessment of Land suitability and Land use/cover service (23%), and Monitoring & Surveillance service (19%). Then, at moderate level, they used EO in Assessment of Natural Hazards service (8%), Early Warning Systems (7%), Land Administration and Conflict Resolution (6%), in Set up of National Information System (7%), in Insurance and Financial Market Sectors (1%) and in other services (1%) (Figure 6.14).

Relating to the source of data used in these services, most respondents (76%) indicated that they used hybrid production, while 14% produce their own data and 14% used external data and reported their need to update these products monthly (37%). Concerning the agriculture management services, most of the respondents were strongly interested in applying, in the future, applications for technical advice on water irrigation, and those on information about fertilizer applied in the past.

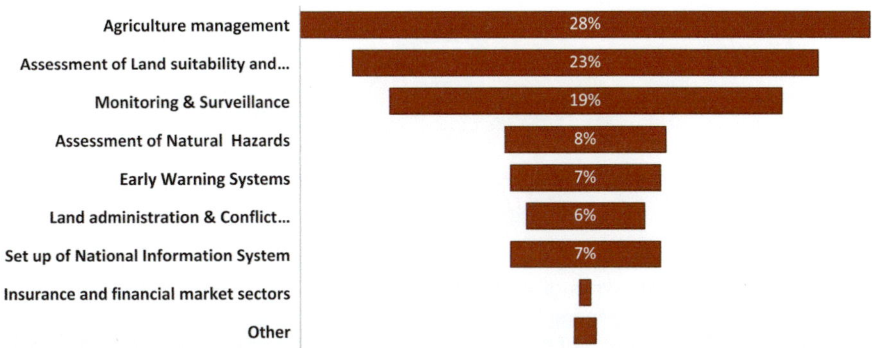

FIGURE 6.14 Scope of services using EO data in the responding organizations.

Conversely, they reported they used applications in the past that were essentially for weather forecast (10%), information about fertilizer (9%), pest control and diseases (8%), technical advice on sowing/planting and harvesting (12%) applied in the past, technical advice on irrigation water (7%), precision agriculture tools such as remote sensors, yield monitors, variable rate technology, GPS, etc. (16%). Additionally, these applications were mostly applied at national levels (22%), then at regional level (19%), at the scale of watershed (17%), Agro-climatic zone (17%) (Figure 6.15b). When we asked survey participants about the source of data for their applications, most of them (68%) mentioned hybrid production; 13% of organizations produced their own data and 19% of them use external data (Figure 6.15c). Finally, (38%) of the respondents indicated the need to update these products monthly (Figure 6.15d).

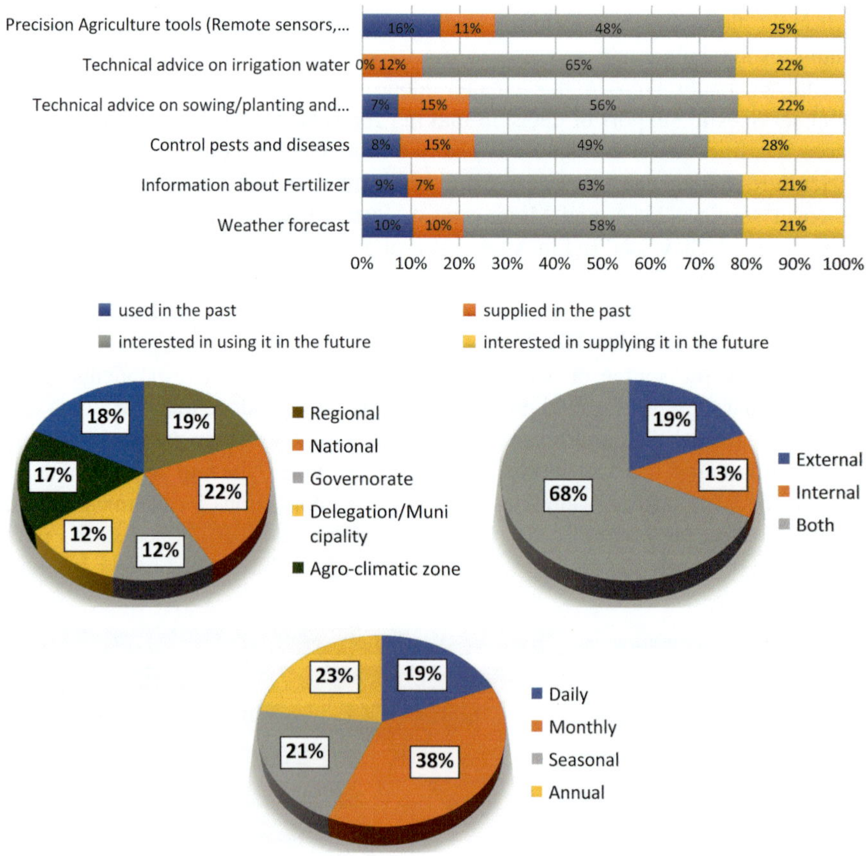

FIGURE 6.15 (a) What types of use/fields of application are in the sector of agriculture management? (b) At what geographic scale would you/your organization plan to use the EO applications? (c) What sources of data? (d) How often would you need to update or distribute EO information products? (a–d) Geoinformation & EO applications used in the agriculture management.

Regarding the assessment of land suitability and land use and land cover services, the respondents were largely interested in applying applications for crop selection (64%), assessment of soil salinity (63%), assessment of soil nutrient content (60%), soil moisture assessment, and modeling (58%). Furthermore, the used or applied applications in the past by respondents were essentially for land evaluation and site assessment (21%), assessing natural capital and ecosystem services (7%), and assessment of water resources (7%) (Figure 6.16).

For monitoring and surveillance services, most of the respondents are mostly interested in using applications for monitoring water quality (60%), soil moisture

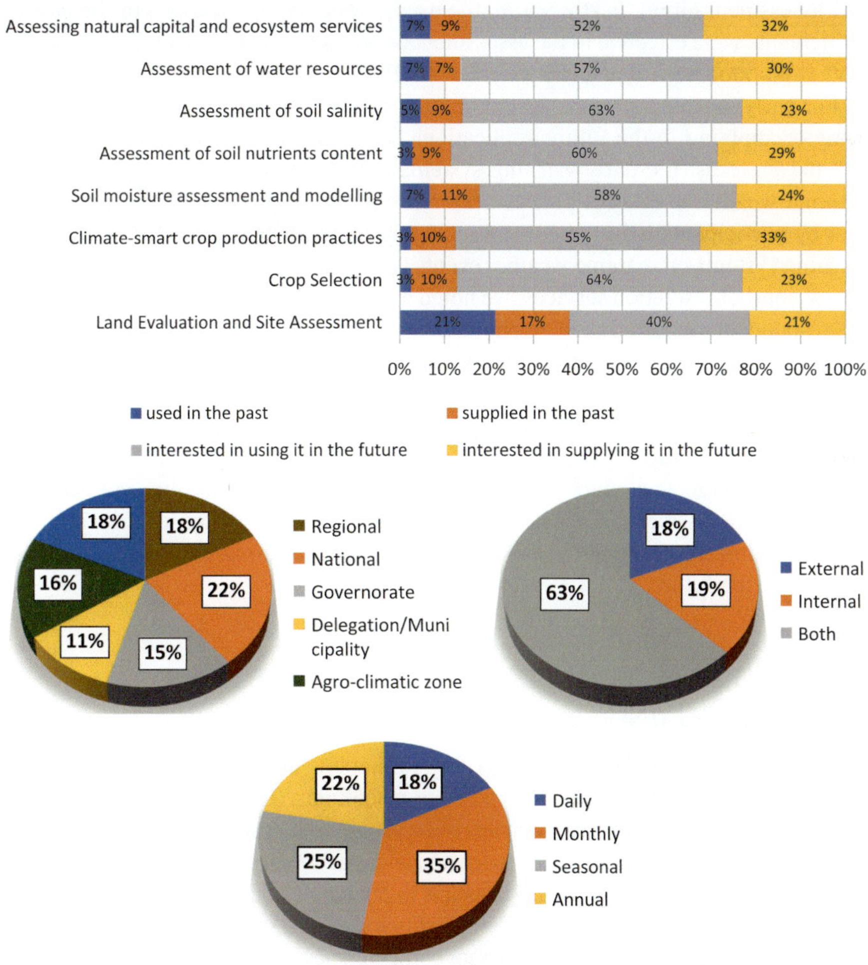

FIGURE 6.16 (a) What types of use/fields of application in the sector of assessment of land suitability and land use/cover? (b) At what geographic scale would you/your organization plan to use the EO applications? (c) What sources of data? (d) How often would you need to update or distribute EO information products? (a–d) Geoinformation & EO applications used in the assessment

assessment and modeling (54%), soil degradation and erosion monitoring (51%), identification and monitoring of water body extent (50%), agricultural control and compliance monitoring (50%), and flood monitoring system (smart-IoT) (49%). In addition, the used or applied applications in the past by respondents were essentially in land use; land-use change and forestry monitoring (18%), soil degradation and erosion monitoring (18%), forest fire monitoring (13%), and drought monitoring (13%). For assessment of natural hazards services, most of the respondents were strongly interested in the use or to apply applications in the future for disaster risk assessment and monitoring (49%), assessment of drought hazard vulnerability and risk (49%), modeling adaptation monitoring of climate change (47%), flood hazard and risk assessment (46%). Furthermore, the used or applied applications in the past by participants were effective in the assessment of drought hazard vulnerability and risk (16%) and flood hazard and risk assessment (15%).

For early warning systems services, most of the respondents were interested in using or applying in the future applications for early warning systems to the flood risk (67%), forest fire early warning systems (65%), and early warning systems on extreme weather events (61%). Furthermore, the use of applied applications in the past by participants is in the early warning systems for flood risk (18%) and early warning systems on extreme weather events (15%). Concerning the land administration & conflict resolution services, we noted a lesser use than in the past (3 to 6%) but the respondents reported their strong interest in using these specific applications in the future. Regarding the setup of National Information System services, we identified the use of EO data in different fields such as the services of National Agricultural Information Systems, National Water Information Systems, and National Natural Resources Information Systems. Finally, the application of EO data in the services of insurance and financial market sectors was as follows: Agricultural Index Insurance, used in the assessment of access and use of agricultural market information, was lightly used in the past and most of the respondents are strongly interested to use or to apply in the future.

LIMITATIONS AND CONSTRAINTS OF EO USE FOR WNR MONITORING

This regional study deals with identifying limitations and constraints of the use of EO technologies, their causes, and possible resolutions. The survey analysis identifies the existence of three main barriers such as the high cost of the EO products and services, and also the limited knowledge and awareness of the innovative EO products and services. People are generally resistant to change. Concerning the competencies and needs as regards the use of EO data, technologies and services, the survey participants were asked to evaluate their current competencies in different geomatics fields:

- Image acquisition,
- Image processing and calibration,
- Map and model production,
- Map validation against field data,

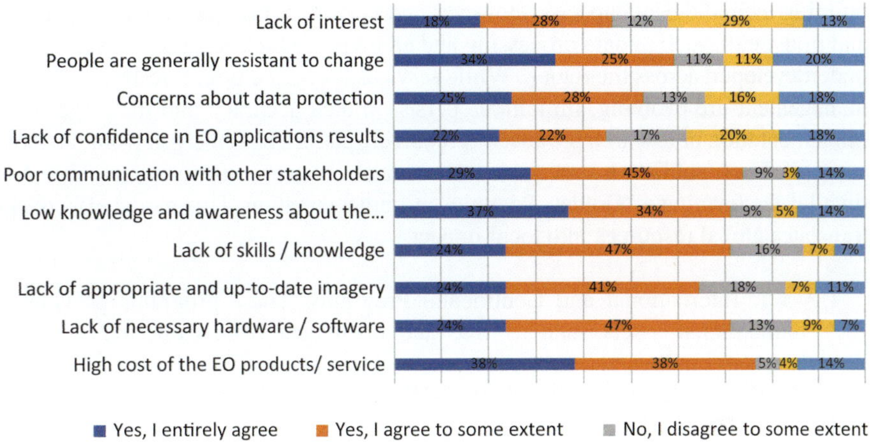

FIGURE 6.17 Barriers to exploit EO data and technologies in the NA countries

- Big data management,
- Spatial Data Infrastructure & Platforms,
- Data manipulation, analysis, design, and modeling.

Most of the listed competences are largely necessary for the participants according to their job and specialty. In general, there appears to be a greater need for competencies related to Big Data management (30%) and Spatial Data Infrastructure & Platforms (20%). It should be noted that most of the respondents are familiar with the competencies of map validation against field data, maps, and model production that are mainly used for routine reporting purposes. In general, for all listed competences, the participants stated their need to improve their skills through continued training since non-expert users lack training and tutorials needed to become familiar with EO data. This issue of expertise is especially pertinent for local authorities and users. It can be concluded that, despite the great demand for remote sensing in NA countries, its use has been hampered by problems ranging from financial constraints, gaps in technology, lack of enough trained personnel. These constraints need to be resolved if advances in space technology applications for operational management of WNRs and for decision-making are to be achieved.

DISCUSSION

Current Use and Institutional Landscape of EO in North Africa

NA countries have increasingly integrated EO products into national and regional projects, reflecting a growing recognition of space technology's potential. This is supported by national agencies such as Algeria's ASAL, Libya's LCRSSS, Morocco's CRTS, and Tunisia's CNCT, alongside regional centers like CRTEAN and universities specializing in geospatial technologies. Several nations have launched their own satellites, including Algeria, Morocco, and Tunisia, highlighting regional progress in building technological capacity. However, the upstream, launchers, manufacturing, and midstream such as data processing and archiving, which are EO sectors, remain underdeveloped across the region. While services for agriculture, forestry, and water management are evolving, limitations persist in data accuracy and the large-scale implementation of research products. A recurring pattern has emerged from the most successful case studies: A synergistic collaboration between institutions at three distinct levels, forming a "Golden Triangle" of implementation. This model effectively translates global resources into local impact.

Level 1: International and Continental Programs: These programs provide foundational assets such as free and open data from Copernicus and Landsat, funding, and strategic frameworks from initiatives like the EU-AU GMES&Africa program and the FAO's WaPOR project.

Level 2: Regional Hubs of Excellence: Organizations like the OSS and CRTEAN act as crucial intermediaries, transforming raw data into regionally adapted tools (e.g., MISBAR, MISLAND) and fostering cross-border cooperation.

Level 3: National Implementing Agencies: These are the *last-mile* actors like CRTS, CNCT, and ASAL who domesticate, validate, and deliver EO products to end-users, building crucial trust with government ministries.

POTENTIAL OF EARTH OBSERVATION-BASED OPERATIONAL SERVICES FOR WNR MANAGEMENT (SDG 6 & SDG 15)

The GMES/Copernicus program aims to establish a sustainable infrastructure for environmental monitoring, supporting informed decision-making. Within this framework, the GMES North Africa project promotes EO-based decision support for sustainable resource management, with services such as Water Abstraction Surveillance, Land Degradation Monitoring, and Agriculture Seasonal Monitoring and Early Warning. For water management (SDG 6), EO-enabled applications provide significant value in crop monitoring, soil moisture assessment, optimized irrigation scheduling, and drought monitoring. Precision agriculture stands out as a domain where EO can have a profoundly beneficial influence, optimizing the use of water and fertilizers to enhance productivity and environmental protection. For land degradation (SDG 15), North Africa is a global hotspot. EO offers a robust and reliable source for monitoring land cover change and biomass activity over extended periods, which is critical for meeting the requirements of SDG indicator 15.3.1: Proportion of land that is degraded over total land area. Initiatives like the Copernicus Global Land Service and the increasing availability of free high-resolution data such as Landsat, Sentinel, now allow for accurate and regular monitoring of land degradation at national scales.

OVERCOMING BARRIERS AND EMBRACING INNOVATION

The successful case studies demonstrate tangible strategies for overcoming the constraints identified in the survey. The cost barrier has been largely dismantled by free and open data policies such as Copernicus, Landsat and open-source software.

The expertise gap is bridged through *platformization*, where complex data processing is handled by platforms like WaPOR and MISBAR, delivering user-friendly products to non-specialist end-users.

Resistance to change is mitigated by demonstrating clear value and co-designing solutions with users, as seen in the WaPOR/IREY project in Tunisia and the IRRISAT project in Morocco. A technical hallmark of these mature applications is the sophisticated integration of data from multiple satellite sensors. This *fit-for-purpose* approach combines high spatial resolution data like Sentinel-2, Landsat) and high temporal resolution data such as MODIS, VIIRS, Sentinel-3, and all-weather radar data, Sentinel-1, to create superior, more reliable information products. Morocco's CDI is a prime example, fusing precipitation, thermal, and vegetation data into a single robust indicator.

ROADMAP FOR EARTH OBSERVATION INTEGRATION IN WNR MANAGEMENT

This study presents a roadmap to increase the use of, and confidence in, EO data for sustainable development and monitoring the SDGs in NA countries. The proposed actions are based on the results of the GMES project studies, literature review, the regional survey, and outcomes from a regional restitution workshop.

Pillar 1: Strengthen the "Golden Triangle" Institutional Framework. Formalize and deepen partnerships between international, regional, and national bodies. This includes establishing clear protocols for data sharing and co-development of tools to prevent duplication of effort and ensure interoperability of platforms (e.g., MISBAR and WaPOR).

Pillar 2: Accelerate the trend toward open data and citizen science. Empower citizens through EO-related citizen science and engagement in the collection of WNR data. This can supplement official monitoring and increase local ownership and awareness.

Pillar 3: Implement stronger coordination and Public-Private Partnerships (PPPs). Establish a NA forum for ongoing dialogue with European and local businesses. Set up PPPs for Earth observation to facilitate stakeholder engagement and the uptake of services by the private sector, based on a full, free, and open data policy.

Pillar 4: Boost downstream application development and awareness. Increase awareness about Copernicus EO services in local companies, especially SMEs and start-ups. Boost the demonstration and promotion of new business opportunities based on mobile applications tailored to local needs for non-expert users.

CONCLUSION AND RECOMMENDATIONS

EO stands as a pivotal asset for North Africa, offering extensive satellite data crucial for monitoring and managing the continent's natural resources and the impacts of human activity and climate change. This technology provides a continental-scale data infrastructure that democratizes the ability to process and analyze satellite imagery, facilitating more informed, strategic, and inclusive decision-making relevant to numerous SDGs, particularly SDG 6 and SDG 15. Concrete case studies from across the region provide compelling evidence of this maturation. In Morocco, a sophisticated national drought indicator provides critical early warning of water stress: Indicator 6.4.2, while innovative projects deliver plot-level irrigation advice to improve efficiency: Indicator 6.4.1. In Tunisia, the FAO's WaPOR platform is being used to conduct robust water productivity assessments and empower farmers with data-driven tools. In Egypt, multi-sensor satellite systems provide synoptic views of water quality in the vital Nile system: Indicator 6.3.2. Regionally, platforms like MISBAR are being deployed to monitor shared aquifers, and the long-term archives of Landsat and Sentinel are used to track the changing extent of critical wetlands from the coasts of Morocco to the lakes of Tunisia: Indicator 6.6.1. These successes are not isolated; they are the product of a functioning ecosystem that connects international programs, regional technical hubs, and national implementing agencies. The key to unlocking the full potential of EO for achieving water security in North Africa lies not in a quest for new technology alone, but in the deliberate strengthening of this collaborative institutional architecture. By scaling up these proven applications, investing in targeted, user-centric capacity building, and continuing to embrace innovation, the nations of North Africa can further harness the power of EO to build a more resilient and sustainable future. To further maximize the benefits of EO in North Africa, several aspects warrant sustained attention:

Enhance User Satisfaction and Information Platforms: Improve the robustness of user satisfaction surveys and ensure the widespread adoption of information platforms across core EO services.

Invest in User Uptake and Impact Studies: Continue investment in user adoption within both institutional and private sector client groups and strengthen the development of impact case studies that clearly demonstrate the benefits of Earth Observation.

Strengthen Governance and Coordination: Increase interaction with Member States and regional authorities to improve coordination, create greater synergy, and enhance international cooperation.

Support Innovation in Core Services: Provide ongoing support for innovation within core EO services and their platforms, focusing on greater utilization of data linking and big data methodologies.

Boost Downstream Application Development: Substantially increase support for developing downstream applications, including providing incentives for smaller businesses and ensuring public institutions have the capacity to provide access to services and data for Research and Development (R&D) and demonstration projects.

REFERENCES

Akhtar-Schuster, M., Thomas, R. J., Stringer, L. C., Chasek, P., & Seely, M. (2011). Improving the enabling environment to combat land degradation: Institutional, financial, legal and science-policy challenges and solutions. *Land Degradation & Development*, *22*(2), 299–312.

Allen, C., Metternicht, G., & Wiedmann, T. (2019). Prioritising SDG targets: Assessing baselines, gaps and interlinkages. *Sustainable Science, 14*(2), 421–438.

Anderson, K., Ryan, B., Sonntag, W., Kavvada, A., & Friedl, L. (2017). Earth observation in service of the 2030 agenda for sustainable development. *Geo-Spatial Information Science, 20*(2), 77–96.

ASAL. (n.d.). *Algerian Space Agency.* Retrieved from official agency documentation.

Bakshi, B. R., et al. (2011).

Bijaber, N., El Hadani, D., Saidi, M., Svoboda, M. D., Wardlow, B. D., Hain, C. R., Poulsen, C. C., Yessef, M., & Rochdi, A. (2018). Developing a remotely sensed drought monitoring indicator for Morocco. *Geosciences, 8*(2), 55.

Bouarfa, S., Benaradj, A., & Nacer, A. (2024). Agricultural oasis expansion and urban growth in Tolga, southeastern Algeria: A geospatial study. *Journal of Arid Land, 16*(5), 635–653.

Bouselmi, A. et al. (2023). *Utility of WaPOR platform outputs for evapotranspiration mapping in Tunisia.* Hybrid Symposium of NENA Regional Network for Evapotranspiration.

Bullock, E. L., Woodcock, C. E., & Olofsson, P. (2020). Monitoring tropical forest degradation using spectral unmixing and Landsat time series analysis. *Remote Sensing of Environment, 238*, 111798.

Chasek, P. S., Akhtar-Schuster, M., Orr, B. J., Luise, A., Ratsimba, H. R., & Safriel, U. (2019). Land degradation neutrality: The science-policy interface from the UNCCD to national implementation. *Environmental Science & Policy, 92*, 182–190.

CNCT. (n.d.). *National Center for Cartography and Remote Sensing (Tunisia).* Retrieved from official agency documentation.

Cochran, F., Daniel, J., Jackson, L., & Neale, A. (2020). Earth observation-based ecosystem services indicators for national and subnational reporting of the sustainable development goals. *Remote Sensing of Environment, 244*, 111796.

Committee on Earth Observations Satellites. (2018). *Satellite Earth Observations in Support of the Sustainable Development Goals.* Retrieved from http://eohandbook.com/sdg/.

Copernicus Market Report. (2019). Issue 2. Prepared by PwC. ISBN 978-92-79-98973. doi:10.2873/011961.

Cowie, A. L., Orr, B. J., Sanchez, V. M. C., Chasek, P., Crossman, N. D., Erlewein, A., Louwagie, G., Maron, M., Metternicht, G. I., Minelli, S., Tengberg, A. E., Walter, S., & Welton, S. (2018). The Scientific Conceptual Framework for Land Degradation Neutrality. A Report of the Science-Policy Interface. *United Nations Convention to Combat Desertification (UNCCD)*, Bonn, Germany.

CRTS (n.d). *IRRISAT-Morocco project.* Space for Climate Observatory (SCO). Retrieved from SCO project documentation.

CRTS. (n.d.). *Earth observation for monitoring drought in Morocco.* United Nations Office for Outer Space Affairs (UNOOSA). Retrieved from UNOOSA Space for SDGs documentation.

DESA, United Nations. (2016). *Transforming our World: The 2030 Agenda for Sustainable Development.*

DESA, United Nations. (2019). *Synthesis of Voluntary National Reviews.* Retrieved from https://sustainabledevelopment.un.org/content/documents/2523/2019_VNR_Synthesis_Report_DESA.pdf.

DESA, United Nations. (2020). *Sustainable Development Knowledge Platform.* Retrieved from https://sustainabledevelopment.un.org/vnrs/.

Dhu, T., Giuliani, G., Juárez, J., Kavvada, A., Killough, B., Merodio, P., Minchin, S., & Ramage, S. (2019). National open data cubes and their contribution to country-level development policies and practices. *Data, 4*(4), 144.

Earth Observations for Sustainable Development Goals (EO4SDG). (2020–2024). *Strategic Implementation Plan.* Retrieved from http://eo4sdg.org/wp-content/uploads/2019/09/EO4SDG-Strategic-Impl.-Plan-2020-2024.pdf.

El Bilali, A. et al (2023). A novel approach for water quality index monitoring using a combination of sentinel-2 data and artificial intelligence models. *Remote Sensing*, *15*(9), 2359.

El Mahrad, B. et al (2023). Mapping and monitoring of Zostera noltei in the Merja Zerga Lagoon (Morocco) using remote sensing. *Journal of Marine Science and Engineering*, *11*(4), 882.

El-Sadek, A. et al (2022). Cross-sensor analysis of Sentinel-3 OLCI, Sentinel-2A MSI, and Landsat-8 OLI for water quality retrieval in the Nile Delta coast. *Egypt. Remote Sensing*, *14*(4), 593.

EOMAP. (n.d.). *IIWQ World Water Quality Portal Demonstrations*. Retrieved from EOMAP documentation.

FAO (2024, February 12). *WaPOR team visits high-resolution area in Jendouba, Tunisia*. Food and Agriculture Organization of the United Nations.

FAO. (n.d.). *WaPOR: Water productivity through open-access of remotely sensed derived data*. Food and Agriculture Organization of the United Nations. Retrieved from https://www.fao.org/in-action/remote-sensing-for-water-productivity/en/.

Fitoka, E., Tompoulidou, M., Hatziiordanou, L., Apostolakis, A., Höfer, R., Weise, K., & Ververis, C. (2020). Water-related ecosystems' mapping and assessment based on remote sensing techniques and geospatial analysis: The SWOS national service case of the Greek Ramsar sites and their catchments. *Remote Sensing of Environment*, *245*, 111795.

Forkuor, G., Zoungrana, J. B. B., Dimobe, K., Ouattara, B., Vadrevu, K. P., & Tondoh, J. E. (2020). Above-ground biomass mapping in West African dryland forest using Sentinel-1 and Sentinel-2 datasets—a case study. *Remote Sensing of Environment*, *236*, 111496.

Ghebrezgabher, M. G., Yang, T., Yang, X., & Wang, X. (2019). Assessment of desertification in Eritrea: Land degradation based on Landsat images. *Journal of Arid Land*, *11*(3), 319–331.

Gilbey, S. E., Davies, J., Metternicht, G., & Magero, C. (2019). Operationalising land degradation neutrality: The role of a national land degradation monitoring system. *Environmental Science & Policy*, *94*, 170–179.

GIZ (2021). *Using water from the Nile more efficiently*. Deutsche Gesellschaft für Internationale Zusammenarbeit. Retrieved from GIZ project database.

GMES & Africa (n.d). *Programme information*. African Union & European Union. Retrieved from various GMES & Africa sources.

Group on Earth Observations. (2017).

Hadria, R. et al (2021). MISBAR geoservice under OSS/North Africa Consortium of GMES&Africa program in support of agriculture monitoring. *Journal of Science and Space Technologies, Special Edition*, 46–47.

Hamed, Y. et al (2018). Spatial-temporal variation of surface water quality in the Ichkeul Lake and rivers ecosystem, Northern Tunisia. *Arabian Journal of Geosciences*, *11*(18).

Hassenforder, Emeline & FERJANI, Arij & Trabelsi, Fatma. (2023). Participatory modeling of past, current and future groundwater governance An experiment in Aousja Ghar El Melh, Tunisia. Futures. 155. 103281. 10.1016/j.futures.2023.103281.

Honeck, E., Castello, R., Chatenoux, B., Richard, J. P., Lehmann, A., & Giuliani, G. (2018). From a vegetation index to a Sustainable Development Goal indicator: Forest trend monitoring using three decades of earth observations across Switzerland. *ISPRS International Journal of Geo-Information*, *7*(12), 455.

International Resource Panel (IRP). (2019). Land restoration for achieving the Sustainable Development Goals. A think piece of the International Resource Panel. United Nations Environment Programme, Nairobi, Kenya.

Kanga Idé, S., Naimi, M., & Chikhaoui, M. (2021). Assessment of surface water quality using indices and geographic information system in the Sebou River basin, Morocco. *European Scientific Journal*, *17*(10), 249.

Kupss, J. A. et al (2023). Monitoring inland water via Sentinel satellite constellation: A review and perspective. *Geo-Spatial Information Science*, *26*(4), 525–546.

Metternicht, G., Akhtar-Schuster, M., & Castillo, M. L. (2019). (As cited in Trabelsi et al., 2024).

Olthof, I., Butson, C., & Fraser, R. H. (2005). Signature extension through space for northern landcover classification: A comparison of radiometric correction methods. *Remote Sensing of Environment*, *95*(4), 446–461.

OSS (n.d). *MISBAR platform*. Sahara and Sahel Observatory. Retrieved from http://misbar.oss-online.org/.

OSS (n.d). *MISLAND platform*. Sahara and Sahel Observatory. Retrieved from http://misland.oss-online.org/.

Pekel, J.-F., Cottam, A., Gorelick, N., & Belward, A. S. (2016). High-resolution mapping of global surface water and its long-term changes. *Nature*, *540*(7633), 418–422.

Rockström, J., et al. (2009).

SDG 6 Data Portal. (2023). *Country profiles for Algeria*. UN-Water. Retrieved from https://sdg6data.org/.

Sims, N. C., et al. (2019).

Spyrakos, E. et al. (2025). Unlocking the global benefits of earth observation to address the SDG 6 in situ water quality monitoring gap. *Frontiers in Remote Sensing*.

The Sustainable Development Goals Center for Africa and Sustainable Development Solutions Network. (2020). *Africa SDG index and dashboards report 2020*. Kigali and New York: SDG Center for Africa and Sustainable Development Solutions Network.

Trabelsi, F., Bel Hadj Ali, S. (2022). Exploring Machine Learning Models in Predicting Irrigation Groundwater Quality Indices for Effective Decision Making in Medjerda River Basin, Tunisia. Sustainability 2022, 14, 2341. https://doi.org/10.3390/su14042341

Trabelsi, F., Bel Hadj Ali, S., Lee, S. (2023). Comparison of Novel Hybrid and Benchmark Machine Learning Algorithms to Predict Groundwater Potentiality: Case of a Drought-Prone Region of Medjerda Basin, Northern Tunisia. Remote Sens. 2023, 15, 152. https://doi.org/10.3390/rs15010152

UN. (2015). *Transforming our world: The 2030 Agenda for Sustainable Development*. A/RES/70/1.

UN. (2017).

UNEP (n.d). *SDG indicator 6.3.2: Proportion of bodies of water with good ambient water quality*. United Nations Environment Programme. Retrieved from https://www.unep.org/goal-6-clean-water-and-sanitation/indicator-632.

UNEP (n.d). *SDG indicator 6.6.1: Change in the extent of water-related ecosystems over time*. United Nations Environment Programme. Retrieved from https://www.unep.org/goal-6-clean-water-and-sanitation/indicator-661.

UNEP & GIZ. (2024, February 21). *UNEP, Nile Basin Initiative partners strive to better monitor pollution in rivers*. United Nations Environment Programme.

UNESCO. (2018). *IIWQ world water quality portal*. UNESCO International Initiative on Water Quality. Retrieved from https://wwqa.info/.

UN-Water. (2017). Integrated Monitoring Guide for SDG 6.

World Bank. (2018). *Beyond scarcity: Water security in the Middle East and North Africa*. MENA Development Report. Washington, DC.

7 SDG #9—Blueprint Strategies for Resilient Infrastructure, Inclusive Industrialization, and Innovation

Lincoln T. Satkunarajah

INTRODUCTION

CONTEXT AND SIGNIFICANCE OF SDG#9

Sustainable Development Goal 9 (SDG#9) represents one of the most transformative elements of the United Nations 2030 Agenda, focusing on three interconnected pillars that form the backbone of modern economies: infrastructure, industrialization, and innovation. This triad constitutes not merely standalone objectives but fundamental enablers that catalyze progress across the entire sustainable development framework. Infrastructure provides the physical and digital foundations upon which societies function; industrialization drives economic growth and employment creation; and innovation fuels productivity improvements and solutions to pressing global challenges. Together, these elements create a powerful nexus that can accelerate human development while promoting environmental sustainability. The significance of SDG#9 lies in its cross-cutting nature and multiplier effects. Reliable transportation networks connect remote communities to markets and services, addressing poverty (SDG#1), and hunger (SDG#2). Modern energy infrastructure supports clean energy transitions (SDG#7) and climate action (SDG#13). Digital connectivity enables quality education (SDG#4) and gender equality (SDG#5) by democratizing access to information and opportunities. Industrial development creates decent work (SDG#8) and can reduce inequalities (SDG#10) when properly managed. Meanwhile, innovation generates new solutions for health challenges (SDG#3), water management (SDG#6), and sustainable consumption (SDG#12). This interconnectedness makes SDG#9 a cornerstone of sustainable development, with progress—or lack thereof—reverberating throughout the entire 2030 Agenda.

The formal articulation of SDG#9 calls for *building resilient infrastructure, promoting inclusive and sustainable industrialization, and fostering innovation* (United Nations, 2015). Each component deserves careful consideration. Resilient infrastructure implies systems that can withstand shocks, whether from natural disasters,

DOI: 10.1201/9781003586319-9

economic crises, or pandemics—while continuing to provide essential services. *Inclusive and sustainable industrialization* emphasizes that industrial growth must benefit all segments of society while minimizing environmental harm. *Fostering innovation* recognizes that technological advancement and knowledge creation are essential for addressing complex development challenges. These three dimensions are operationalized through eight specific targets and twelve indicators that provide a framework for measuring progress. As we approach the midpoint between the 2015 adoption of the SDGs and the 2030 target date, global progress on SDG#9 presents a mixed picture characterized by significant disparities. High-income regions have maintained or even increased their industrial output, with manufacturing value added per capita in Europe and North America reaching record levels. Meanwhile, Least Developed Countries (LDCs) have seen stagnation or even regression in manufacturing output, with per capita figures remaining extremely low. Similar divides exist in infrastructure access, with rural communities in developing regions still lacking basic connectivity, and in innovation capacity, with research and development investments heavily concentrated in a handful of countries. These disparities highlight the challenge of ensuring that no one is left behind in the pursuit of SDG#9.

Post-Covid-19 Landscape

The COVID-19 pandemic represented an unprecedented shock to global systems, with profound implications for SDG#9 implementation. The crisis exposed vulnerabilities in infrastructure, disrupted industrial production, and simultaneously accelerated certain aspects of innovation. Understanding this transformed landscape is essential for recalibrating strategies to achieve SDG#9 targets by 2030. The pandemic's immediate impact on infrastructure was multifaceted. Transportation networks experienced dramatic usage reductions as mobility restrictions took effect—global air passenger travel plummeted by over 60% in 2020, public transit ridership collapsed in cities worldwide, and road traffic declined significantly during lockdowns (ICAO, 2021). While this temporarily reduced carbon emissions, it also strained the financial viability of transportation systems and delayed maintenance and expansion projects. Simultaneously, the crisis underscored critical infrastructure gaps, particularly in healthcare facilities, water and sanitation systems, and digital connectivity. Many communities lacking reliable internet access found themselves unable to participate in remote education, telemedicine, or digital commerce, exacerbating existing inequalities.

Industrial sectors faced severe disruptions from the pandemic. Global manufacturing output contracted by 8.4% in 2020, with particularly sharp declines in automotive, aerospace, and apparel industries (UNIDO, 2021). Supply chain vulnerabilities became glaringly apparent as production shutdowns in one region cascaded through global value chains, causing shortages of essential goods from medical supplies to semiconductor chips. Small and medium enterprises (SMEs) were disproportionately affected, with many lacking the financial reserves to weather extended closures. The International Labour Organization estimated that 114 million jobs were lost globally in 2020, with manufacturing employment declining in most regions (ILO, 2021). These setbacks have made the target of doubling industry's share in

LDCs' economies by 2030 increasingly challenging to achieve. Paradoxically, the pandemic also catalyzed unprecedented innovation and digital transformation. Necessity drove rapid adoption of digital technologies across sectors—from remote work platforms and e-commerce to telemedicine and online education. Companies that might have taken years to digitalize operations did so in weeks or months. The crisis spurred remarkable scientific collaboration, most visibly in the development of COVID-19 vaccines at record speed. Digital payment systems expanded rapidly in regions previously reliant on cash transactions. According to the International Telecommunication Union, internet usage increased by 10% globally in 2020, the largest annual increase in a decade (International Telecommunication Union, 2021). This accelerated digital transformation presents opportunities to leapfrog traditional development pathways, particularly for developing economies.

The post-COVID recovery phase has brought new dynamics relevant to SDG#9. Many governments have launched infrastructure-centered stimulus packages to revitalize economies and create jobs. The European Union's €750 billion recovery fund allocated substantial resources to green and digital infrastructure. The United States passed a $1.2 trillion Infrastructure Investment and Jobs Act in 2021, representing the largest infrastructure investment in generations. These initiatives could accelerate progress on SDG#9 targets if effectively implemented. Simultaneously, there has been renewed emphasis on industrial policy and supply chain resilience, with countries seeking to rebuild domestic manufacturing capabilities for critical goods and reduce vulnerability to future disruptions. However, the recovery remains uneven, with many developing countries facing fiscal constraints that limit their ability to invest in infrastructure and industrial development. The pandemic has widened existing gaps between advanced and developing economies in digital infrastructure, manufacturing capacity, and innovation ecosystems. Rising inflation, supply chain bottlenecks, and geopolitical tensions further complicate the post-COVID landscape for SDG#9 implementation. These challenges underscore the need for enhanced international cooperation, innovative financing mechanisms, and targeted capacity-building efforts to ensure that all countries can advance toward SDG#9 targets in the pandemic's aftermath.

Theoretical Framework: Measuring SDG#9 Progress

SDG#9 Targets and Indicators Framework

The United Nations' SDG 9 framework consists of eight targets—five outcome targets (9.1–9.5) and three implementation targets (9.a–9.c)—each with specific indicators to guide resilient infrastructure, inclusive industrialization, and innovation measurement. This structure underpins progress tracking, gap identification, and evidence-based policymaking.

Target 9.1 addresses the development of quality, reliable, sustainable, and resilient infrastructure and Indicator 9.1.1 measures the share of rural inhabitants within 2 km of an all-season road using geospatial analyses like the Rural Access Index (World Bank, 2016). Indicator 9.1.2 captures passenger- and freight-kilometers by transport mode, relying on standardized methodologies from the International Transport Forum and UNSD (United Nations Statistics Division, 2020).

Target 9.2 promotes inclusive and sustainable industrialization by tracking manufacturing value added (9.2.1) and employment share (9.2.2). Data derived from national accounts, enterprise surveys, and labor force surveys are standardized by UNIDO (UNIDO, 2020) and ILO (ILO, 2020). Informal-sector activity can be estimated via satellite imagery, night-time lights, and mobile phone data. Target 9.3 focuses on small-scale enterprises' roles by measuring their contribution to industry value added (9.3.1) and access to credit (9.3.2). Definitions typically follow UNSD employment thresholds (United Nations Statistics Division, 2018), with data from economic censuses, enterprise surveys, and financial inclusion studies complemented by digital finance records. Target 9.4 seeks to upgrade infrastructure and industries for sustainability, using CO_2 emissions per unit of value added (9.4.1) as the sole indicator. Emissions are estimated with IPCC guidelines (Intergovernmental Panel on Climate Change, 2019) and combined with national accounts. Remote sensing increasingly validates reported data. Target 9.5 fosters research and technological capability by tracking Research and Development (R&D) expenditure (9.5.1) and researcher density (9.5.2). Surveys following the Organization for Economic Co-operation and Development (OECD Frascati Manual; Organization for Economic Co-operation and Development, 2015) provide core data, supplemented by bibliometrics and patent analyses. Means of implementation with measurement tools and systems include:

- Target 9.a's measurement of official international infrastructure support (9.a.1) via the OECD's Creditor Reporting System (Organization for Economic Co-operation and Development, 2021);
- Target 9.b's share of medium- and high-technology manufacturing
- (9.b.1) classified by UNIDO (United Nations Industrial Development Organization, 2019); and
- Target 9.c's mobile network coverage indicator (9.c.1) from ITU datasets (International Telecommunication Union, 2020).

Despite its comprehensiveness, the SDG 9 framework faces data gaps in LDCs, informal economies, and subnational contexts. Variations in definitions, collection frequency, and mapping resources hinder comparability. Strengthening national statistical offices and leveraging innovative sources—satellite imagery, machine learning, mobile phone records, and digital platforms—are critical to narrowing these measurement gaps.

Geospatial Science as a Measurement Tool

Geospatial methods integrate location-based data with advanced analytics to enrich SDG 9 monitoring. Remote sensing, including satellite imagery, aerial photography, and LiDAR, maps transportation, energy, water, and industrial infrastructure across vast areas. Sentinel satellites exemplify frequent, high-resolution imagery for monitoring road construction and industrial expansion (ESA, 2021).

For indicator 9.1.1, machine learning algorithms extract road networks from imagery and combine them with population grids, as in the World Bank's Spatial Development Infrastructure Framework (World Bank, 2019). In Tanzania, this method revised rural road access estimates from over 50% to 33% (World Bank, 2019).

Beyond roads, nighttime lights data serve as proxies for electrification and economic activity (Henderson et al., 2018), while radar interferometry monitors infrastructure health. Geospatial analytics by UNIDO's Industrial Analytics platform reveal industrial clusters and supply-chain linkages (UNIDO, 2021).

Subnational disparities emerge through spatial disaggregation: the World Bank's mapping of mobile network coverage (indicator 9.c.1) across African districts highlights connectivity gaps masked by national averages (World Bank, 2020). Temporal sequences of Landsat imagery track infrastructure growth over decades and inform trend analyses (Southworth et al., 2011). Visualization tools like UNOOSA's Space4SDGs democratize data through interactive maps and dashboards (UNOOSA, 2020). Challenges include uneven data quality, limited geospatial expertise, and privacy concerns when revealing sensitive infrastructure details. International collaborations via GEO (GEO, 2020), open-source GIS tools, and training programs can address these barriers. Integrating satellite data with ground-based sensors, surveys, and administrative records offers comprehensive measurement systems. For example, combining thermal infrared imagery with factory energy consumption data enriches carbon intensity assessments for indicator 9.4.1.

AI-Enhanced Measurement Systems

Artificial intelligence accelerates SDG 9 measurement by automating data extraction and analysis. Convolutional neural networks detect roads, buildings, power lines, and industrial facilities in imagery with over 80% accuracy (Cadamuro et al., 2019). Facebook's AI lab mapped 300,000 miles of African roads in days (Facebook AI Research, 2019), underpinning more frequent Rural Access Index updates (World Bank, 2019). Computer vision models assess infrastructure condition—identifying cracks or corrosion—and predict maintenance needs using sensor and street-level imagery (American Society of Civil Engineers, 2020). In industrial monitoring, AI estimates production levels, emissions profiles, and energy efficiency from satellite observations and operational data.

Night-time lights combined with daytime imagery enable GDP predictions and manufacturing output proxies surpassing traditional methods (Jean et al., 2016). The World Bank's Night Light Index now embeds AI for near-real-time economic activity tracking (World Bank, 2021).

Natural language processing of patents, technical reports, and scientific publications uncovers innovation trends and green technology diffusion (World Intellectual Property Organization, 2020). Predictive AI models forecast infrastructure demand based on demographic growth and climate scenarios, as applied by the Asian Development Bank (Asian Development Bank, 2020).

Big data sources like mobile phone metadata for transport patterns, e-commerce transactions for small enterprise market access, and social media sentiment on infrastructure services expand measurement granularity between formal reporting cycles. UN Global Pulse projects illustrate these capabilities (United Nations Global Pulse, 2020). Data scarcity, model biases, technical complexity, interpretability, and privacy risks constrain AI adoption. Capacity building through the UN Technology Innovation Labs (United Nations Technology

Innovation Labs, 2021), open-source AI frameworks, explainable AI, and ethical guidelines are essential for responsible deployment. The most robust measurement ecosystems combine AI insights with expert judgment and institutional processes, ensuring that technological methods serve human-centered sustainable development objectives.

Integrated Measurement Frameworks

Effective SDG 9 monitoring demands frameworks that fuse traditional statistics, geospatial analyses, AI methods, and other innovative techniques. A conceptual model should map data inputs such as surveys, administrative records, satellite imagery, sensor networks through analytical processes to SDG 9 indicators. The UN Global Working Group on Big Data outlines guidelines for integrating diverse data sources, emphasizing quality assessment, transparency, and validation (UN GWG, 2020). Cross-sectoral systems like the UN SEEA link economic and environmental accounts for holistic industrial and infrastructure measurement (UN SEEA, 2021).

Spatial integration via GIS, as demonstrated by the EO4SDG initiative, combines satellite, ground, and socioeconomic data for indicators such as rural road access (GEO, 2020). Temporal integration such as backcasting, gap-filling, and nowcasting supports consistent time series for targets, as implemented by the World Bank's Development Data Hub (World Bank, 2021).

Methodological integration orchestrates statistics, geospatial analytics, AI, economic modeling, and qualitative insights into cohesive measurement approaches, following examples from UNDP's SDG Integration initiative (UNDP, 2020). Institutional integration requires clear mandates and collaboration among statistical offices, infrastructure and industry agencies, research institutions, and civil society. The Praia Group offers a governance model for such coordination (UN Praia Group, 2020). Real-time monitoring by leveraging mobile data, social media streams, and rapid satellite imaging provides early warnings and responsive insights during crises or rapid development shifts (UN Global Pulse, 2020). Challenges in integration include data interoperability, methodological complexity, institutional coordination, and technical infrastructure demands. Solutions encompass data standards, metadata systems, cloud platforms, capacity building, and governance frameworks to sustain adaptive, learning-oriented measurement systems that drive progress toward resilient infrastructure, inclusive industrialization, and innovation.

Infrastructure Implementation Through Geospatial Intelligence

Infrastructure Gap Analysis

Understanding infrastructure gaps, which are the differences between existing provision and needs, is essential for SDG 9 implementation. Traditional aggregate statistics often mask spatial variations in presence, quality, capacity, accessibility, and resilience. Geospatial intelligence offers precise, location-specific insights, revealing where infrastructure is absent or underperforming and guiding targeted investments. Global assessments estimate a $94 trillion investment requirement from 2016 to 2040, leaving a $15–20 trillion gap that disproportionately affects developing regions (Global Infrastructure Hub, 2020). Geospatial mapping by the World Bank's Infrastructure Spatial Database highlights that sub-Saharan Africa has only 43% electricity access, 70% improved water access, and 30% sanitation coverage, while South Asia exhibits pronounced rural service deficits (World Bank, 2021). High-resolution satellite imagery such as Sentinel-1 radar and Sentinel-2 multispectral data forms the baseline for mapping roads, energy systems, water facilities, and telecommunications, even under cloud cover (ESA, 2021). Machine learning models have identified over 300,000 km of previously unmapped roads in Nigeria (Development Seed, 2019). Synthetic aperture radar and multitemporal analyses detect infrastructure deterioration, while overlaying WorldPop population grids identifies communities lacking proximity to services, such as 67% of rural Tanzanians beyond 2 km of all-season roads (WorldPop, 2021; World Bank, 2019). Network-based travel-time models further refine access assessments, revealing multi-hour journeys to hospitals for more than 90% of some populations (Malaria Atlas Project, 2020).

Temporal insights from decades of Landsat imagery expose whether gaps are narrowing or widening, informing progress evaluation (Google, 2021). Integrating hazard maps of floods, landslides, cyclones with infrastructure layers reveals resilience vulnerabilities, such as 36% of Nepal's strategic roads at high landslide risk (GFDRR, 2021). Coupling infrastructure maps with socioeconomic indicators clarifies how deficits correlate with poverty and growth patterns (World Bank, 2020). Georeferenced financing records show misaligned investment flows that favor developed areas over those with greatest needs (Aid Data, 2021). Participatory mapping in informal settlements uncovers locally relevant gaps, like inadequate drainage and last-mile connectivity (Missing Maps, 2021). Interactive dashboards from humanitarian platforms democratize gap insights for stakeholders (OCHA, 2021), while initiatives by the Global Partnership for Sustainable Development Data support capacity building for robust gap analyses (GPSDD, 2021).

Geospatial Planning for Resilient Infrastructure

Resilient infrastructure planning integrates spatial data on hazards, vulnerabilities, interdependencies, and lifecycle dynamics. The 5D modeling framework combining 3D location, time, and contextual factors enables simulation of infrastructure performance from construction through operation and adaptation. Singapore's Virtual Singapore digital twin exemplifies this, merging detailed 3D models with sensor feeds and climate projections for scenario testing (National Research Foundation Singapore, 2021). Similarly, the UK's National Digital Twin Programme links infrastructure network twins to assess cascading risks (Centre for Digital Built Britain, 2021), and Rotterdam's port twin-maintained operations under COVID-19 disruptions (Port of Rotterdam, 2020).

Geospatial planning supports climate-resilient design by overlaying climate projections with infrastructure inventories via platforms like the World Bank's Resilient Infrastructure Geospatial Platform (World Bank, 2020). Analyses of past hazard impacts such as Stanford's examination of hurricane damage patterns yield design recommendations for stronger materials and siting (Stanford University, 2020). Open-source tools from the Global Earthquake Model guide seismic retrofitting (GEM, 2021), while USACE's spatial optimization of structural and nature-based flood controls maximizes protection and ecological co-benefits (US Army Corps of Engineers, 2020). Network analysis identifies critical nodes for redundancy enhancements, as demonstrated in Switzerland's infrastructure protection strategy (ETH Zurich, 2020). Equity mapping, such as Georgetown Climate Center's toolkit, ensures resilience investments target vulnerable communities (Georgetown Climate Center, 2020), and participatory GIS by the Red Cross fosters community-driven disaster risk reduction (Red Cross Red Crescent Climate Centre, 2020). Nature-based solutions identified via geospatial methods like wetland restorations, urban green spaces, and coastal buffers deliver multiple benefits beyond resilience (The Nature Conservancy, 2021). Scenario planning with geospatial simulations under varied climate, demographic, and policy futures has guided major city resilience plans, notably New York's $20 billion climate strategy (NYC Mayor's Office of Resiliency, 2020). Continuous monitoring systems such as the Philippines' NOAH program track infrastructure performance during extreme events, offering feedback loops for adaptive management (DOST-NOAH, 2021). Challenges include data sparsity in hazard mapping, limited technical capacity, institutional fragmentation, and financing models that undervalue resilience. Collaborations like the Coalition for Disaster Resilient Infrastructure are improving access to data and tools (CDRI, 2021), and cloud-based geospatial platforms lower technical barriers, enabling more integrated, participatory, and finance-aligned resilience planning.

USA CASE STUDY: SMART INFRASTRUCTURE INITIATIVES

The Infrastructure Investment and Jobs Act allocates $1.2 trillion over five years, embedding geospatial decision-making via ARPA-I and mandating a National Infrastructure Atlas for assets, conditions, and vulnerabilities (U.S. Department of Transportation, 2022). The Department of Transportation's GIS-based prioritization framework integrates economic, equity, resilience, and connectivity metrics to restore community linkages disrupted by past projects (U.S. Department of Transportation, 2022).

The Digital Twin Consortium fosters standards for interoperable twins across transportation, energy, and water, underpinning implementations like the Port of Los Angeles twin for optimized logistics (Digital Twin Consortium, 2021; Port of Los Angeles, 2022). USACE's Engineering With Nature program applies LiDAR-driven floodplain restorations on the Mississippi, delivering 15–30% cost savings and environmental co-benefits (USACE, 2021; 2022). Colorado DOT's risk-based asset management uses hazard, climate, and socioeconomic layers to prioritize resilience upgrades, demonstrating improved outcomes following the 2013 floods (Colorado DOT, 2021). Public-private efforts such as the RISE Consortium's CIVE platform analyze interdependencies between energy and water for strategic resilience investments (RISE Consortium, 2022). Urban sensor networks from Chicago's Array of Things enable real-time air quality and traffic optimizations (City of Chicago, 2021), while NOAA's Digital Coast offers coastal planners high-resolution elevation and sea-level projections (NOAA, 2022). The USGS 3DEP LiDAR initiative generates foundational elevation data with an estimated $13 billion annual economic benefit (USGS, 2022).

Geospatial prioritization under the IIJA has increased investments in underserved communities by 35% (U.S. Department of Transportation, 2022). Digital twins report 10–15% maintenance cost reductions and 20–25% reliability gains (Digital Twin

Consortium, 2022). Nature-based flood solutions under Engineering With Nature save 15–30% compared to conventional methods (USACE, 2022). Obstacles include cross-sector data integration, privacy concerns, uneven technical capacities, and sustainable financing for ongoing geospatial systems. Key lessons emphasize building shared geospatial datasets, leveraging public-private partnerships, and integrating technical and institutional strategies. Frontier initiatives by NSF's Civic Innovation Challenge and DOE's Grid Modernization are advancing participatory and energy resilience applications (NSF, 2022; DOE, 2022).

SRI LANKA CASE STUDY: POST-CONFLICT INFRASTRUCTURE DEVELOPMENT

Post-2009 recovery in Sri Lanka's Northern Province faced 60% road damage, 50% electricity deficits, and 20% water access (Asian Development Bank, 2020). The Sustainable Fisheries Development Project used satellite damage assessments and community participatory mapping to reconstruct harbors, landing sites, and markets for equitable resource access (Ministry of Fisheries, Sri Lanka, 2019; Ministry of Fisheries, Sri Lanka, 2021).

The nationwide iRoad program prioritizes roads via a GIS-based multi-criteria framework including conflict sensitivity, enhancing ethnic cohesion and service access (Road Development Authority, Sri Lanka, 2020). Mobile geotagging apps tracked construction and maintenance in real time, improving on-schedule completion from 60% to 85% (Road Development Authority, Sri Lanka, 2021). The Northern Rural Road Connectivity Project designed resilient roads with LiDAR-informed terrain and flood analyses, achieving 90% passability during monsoons versus 40% for older roads (Ministry of Highways, Sri Lanka, 2020; 2021).

The Dry Zone's Climate Resilient Integrated Water Management rehabilitated cascade tank systems through remote sensing and hydrological models, boosting agricultural water availability by 40% and cutting flood damage by 60% (Ministry of Irrigation, Sri Lanka, 2020; 2022). Rebuilding foundational geospatial datasets via combined satellite, aerial, and ground surveys was critical after map loss (Survey Department, Sri Lanka, 2020). Capacity programs trained 500+ practitioners in GIS and remote sensing (ICTA, Sri Lanka, 2021). The Northern and Eastern Coordination Mechanism used a shared geospatial database to reduce duplication and save 15–20% in costs (Ministry of National Policies, Sri Lanka, 2020). Participatory GIS approaches improved design legitimacy in diverse communities (Ministry of National Integration, Sri Lanka, 2019). Future directions include IoT sensor deployments under Smart Sri Lanka and interoperable national spatial data infrastructure (Survey Department, Sri Lanka, 2022; ICTA, Sri Lanka, 2022).

GLOBAL BEST PRACTICES IN INFRASTRUCTURE IMPLEMENTATION

Rwanda's National Feeder Road Program applied a GIS prioritization combining agricultural potential, population, poverty, and market data to expand all-season roads from 41% to 69% rural coverage, boosting productivity and reducing transport costs (Ministry of Infrastructure, Rwanda, 2019; 2021). Community smartphone monitoring improved maintenance response by 60% (Rwanda Transport Development Agency, 2020). Colombia's Territorial Renewal Agency integrated participatory mapping and spatial analysis to align post-conflict investments with peacebuilding goals, achieving 78% on-time project completion via a transparent Obras PDET dashboard (Agencia de Renovación del Territorio, Colombia, 2020; 2022).

The Netherlands' Room for the River used hydrological models and elevation data to create floodplain space rather than higher dikes, reducing flood risks and delivering ecological benefits during the 2018 and 2021 high-water events (Rijkswaterstaat, Netherlands, 2020; Rijkswaterstaat, Netherlands, 2021). Indonesia's One Map Policy harmonized spatial data across 19 agencies, slashing planning times by 40% and reducing land conflict by 30% (Geospatial Information Agency, Indonesia, 2020; 2021; Ministry of National Development Planning, Indonesia, 2022). Ethiopia's Low Volume Road Manual optimized rural road designs for terrain and materials, constructing 36,000 km of all-weather roads and raising rural coverage from 22% to 41% (Ethiopian Roads Authority, 2020; Ethiopian Roads Authority, 2022). Singapore's National Digital Twin provides a city-scale, sensor-integrated virtual environment for predictive maintenance and scenario planning, supported by a collaborative geospatial platform (Singapore Land Authority, 2021; Singapore Land Authority, 2022; Land Transport Authority, Singapore, 2022). Sharing of methodologies and tools through South-South cooperation networks, such as UNOSSC, facilitates adaptation of geospatial best practices across diverse contexts (UNOSSC, 2021).

Industrial Innovation in the Digital Era

Industry 4.0 and SDG#9

Industry 4.0, encompassing cyber-physical systems, IoT, advanced robotics, AI, additive manufacturing, cloud computing, and big data analytics, enables *smart factories* that are highly flexible, efficient, and responsive to market dynamics (World Economic Forum, 2020). By bridging digital and physical production worlds, these technologies generate vast operational data and support adaptive manufacturing processes. However, realizing inclusive and sustainable industrialization under SDG 9 depends on governance frameworks that ensure equitable access and address potential technology-driven disparities. The impact of Industry 4.0 on Target 9.2, inclusive industrialization, is demonstrated by productivity gains of 15–20% and 20–50% reductions in product development time reported by adopters (McKinsey Global Institute, 2021). Nonetheless, automation risks displacing routine roles even as it creates demand for digital-savvy jobs in data analysis and human-machine collaboration (International Telecommunication Union, 2021). Policies must therefore support reskilling and ensure that digital industrial transformation benefits workers across skill and demographic groups.

Industry 4.0 also directly advances Target 9.4 by improving resource efficiency such as real-time monitoring and optimization of energy, water, and materials leads to energy savings of 10–30% and material reductions of 20–25% in digitally enabled factories (UNIDO, 2020). Circular economy practices like digital product passports, predictive maintenance, and additive manufacturing minimize waste and support multiple product lifecycles, potentially cutting industrial waste by up to 80% in key sectors (Ellen MacArthur Foundation, 2021). Smart factory exemplars vary by context: high-income nations retrofit existing plants to achieve near-perfect quality and major productivity boosts, as seen in Siemens' Amberg facility (World Economic Forum, 2021). Emerging economies target new industrial parks with integrated digital infrastructure, exemplified by China's Made in China 2025 strategy (Ministry of Industry and Information Technology, China, 2020). LDCs benefit from selective, low-barrier technologies like mobile quality control, simple IoT tools, and cloud design services guided by frameworks like the UN Technology Bank's Appropriate Digital Manufacturing model (UN Technology Bank, 2021).

Effective governance of Industry 4.0 requires regulatory and standardization efforts to protect data sovereignty, ensure cybersecurity, and foster interoperability. Participation in standards bodies such as ISO's smart manufacturing committee is crucial for reflecting diverse development priorities (ISO, 2021). Concurrently, education systems must evolve; technical and vocational training should be aligned with emerging Industry 4.0 competencies, which are essential, given projections that most future jobs will require skills not yet prevalent today (World Manufacturing Foundation, 2021). The COVID-19 pandemic accelerated digital adoption but also risks widening the gap between digital leaders and laggards, underscoring the need for inclusive, future-oriented strategies (World Economic Forum, 2021).

AI-Driven Industrial Transformation

AI accelerates industrial transformation by enabling predictive maintenance, supply chain intelligence, and automated quality control, which are key to resilient and sustainable manufacturing systems. Predictive maintenance leverages sensor data and machine learning to forecast equipment failures, reducing downtime by 30–50% and extending machine life by 20–40% (World Economic Forum, 2021). Siemens' platform processes over 30 million data points daily to achieve failure prediction accuracies above 95% (Siemens, 2021), while UNIDO's starter kits support SMEs in deploying cost-effective AI maintenance solutions (UNIDO, 2021). Supply chain intelligence powered by AI analyzes real-time data like shipping logs, weather forecasts, news feeds, and social media to predict disruptions and optimize logistics. During the COVID-19 crisis, companies like Unilever reconfigured production and distribution networks swiftly using AI, maintaining product availability and reducing waste (Unilever, 2021). Comprehensive AI adoption across global supply chains could reduce CO_2 emissions by 1.5 gigatons annually by 2030 through route and inventory optimization, aligning with resource-efficiency goals (World Economic Forum, 2022). Automated quality control uses computer vision and intelligent analytics to inspect products at scale with high accuracy. BMW's AI system inspects over 5,000 components daily with 99% precision, detecting defects as small as 0.2 mm (BMW Group, 2021). UNIDO's AI Quality Control Competence Centers in developing countries help local industries adopt these technologies, achieving material waste reductions of 15–25% (Manufacturing Technology Centre, 2021; United Nations Industrial Development Organization, 2022).

AI drives new industrial business models with predictive analytics, which underpin servitization by linking service outcomes to equipment performance, digital twins enable virtual prototyping, and generative design algorithms to optimize products for functionality and sustainability. However, governance frameworks covering transparency, data ownership, and ethical deployment are vital to ensure AI benefits align with public interest and inclusivity (IEEE, 2021). Realizing AI's potential across diverse contexts requires addressing skill gaps and data barriers. Human-centered AI strategies emphasize workforce upskilling, while innovations like *small data* techniques, transfer learning, and federated learning reduce dependency on large datasets and protect privacy. ILO guidance on human-centered AI highlights the need for education and training systems that prepare workers for new roles, ensuring AI-driven transformation supports SDG 9's inclusive vision (ILO, 2022; United Nations Conference on Trade and Development, 2021).

USA CASE STUDY: ADVANCED MANUFACTURING INITIATIVES

The Manufacturing USA network, with 16 institutes formed under the Revitalize American Manufacturing and Innovation Act, drives public-private collaboration on advanced manufacturing technologies, from robotics to sustainable biofabrication (Manufacturing USA, 2022). With over $1 billion in federal funding matched by industry and academia, these institutes accelerate R&D, technology transfer, and workforce development.

Chicago's Digital Manufacturing and Design Innovation Institute (DMDII) has completed over 60 applied projects with 300 partners, producing tools and methodologies that boost productivity by 15–25%, improve energy efficiency by 10–20%, and reduce defects by 30–40% (DMDII, 2021). Its Digital Manufacturing Commons offers open-source tools for SMEs, while deployment programs provide hands-on support, and democratizing access to digital manufacturing innovations (DMDII, 2022). The Department of Energy's Advanced Manufacturing Office (AMO) supports energy-efficient and low-carbon technologies through $300 million annual R&D and the Better Plants program. Over 250 companies have pledged 25% energy intensity reductions, collectively saved 1.9 quadrillion BTUs and avoided 116 million tCO_2 emissions (U.S. Department of Energy, 2022). Traditional industries are also transforming: General Motors' $2.2 billion Factory ZERO in Detroit combines robotics, IoT, digital twins, and 100% renewable power to produce electric vehicles with optimized quality and efficiency (General Motors, 2021; 2022). General Motors' retraining programs and diverse apprenticeships exemplify inclusive workforce approaches. Workforce development is supported by the Manufacturing Extension Partnership (MEP), which offers training, apprenticeships, and education partnerships (NIST MEP, 2022). Certifications like MSSC's Production Technician 4.0 align skills with Industry 4.0, and NSF-funded community college programs train technicians in IoT and smart manufacturing (MSSC, 2022; Lorain County Community College, 2022). Yet adoption gaps persist: advanced capabilities remain concentrated in select regions, and SMEs and underrepresented groups lag, underscoring the need for targeted inclusion initiatives (Brookings Institution, 2021).

SRI LANKA CASE STUDY: INDUSTRIAL MODERNIZATION EFFORTS

Sri Lanka's industrial base with 75% SMEs contributing 52% of GDP, aims to raise manufacturing's GDP share from 16% to 25% by 2030 (Ministry of Industry and Commerce, Sri Lanka, 2021; Board of Investment, Sri Lanka, 2022). The National

Export Strategy (NES) targets six priority sectors: IT/BPM, spices, boat building, processed food, electronics, and wellness tourism, with tailored technology, quality, and market development pathways (Export Development Board, Sri Lanka, 2022). Sectoral technology programs include the Electrical and Electronic Components Upgrading initiative, assisting 45 firms in adopting automated testing and design software. Participants report 15–25% productivity gains and export-enabling quality certifications (Export Development Board, Sri Lanka, 2021). In textiles, leaders like MAS Holdings implemented digital planning, automation, and energy management to achieve 30% productivity and 25% energy improvements, along with carbon neutrality via renewables and offsets (MAS Holdings, 2021).

SME digitalization, through the SME Technology Adoption Program, provides matching grants, mentorship, and technical assistance for entry-level automation and digital tools. Over 300 SMEs achieved 12–18% productivity and 8–15% cost reductions (Ministry of Industry and Commerce, Sri Lanka, 2022). Complementary export-innovation programs, like Nature's Secrets' digital formulation and automated production for natural cosmetics, demonstrate value-addition success, exporting to 35 countries at premium price points (Nature's Secrets, 2022).

The Sri Lanka Institute of Nanotechnology (SLINTEC)—a public-private R&D partnership—drives applied innovation in smart textiles, nano fertilizers, and advanced materials, with technologies transferred to industry partners for commercial use (SLINTEC, 2022). Institutional barriers such as finance, skills, and weak research-industry linkages limit SMEs' adoption, but collaborative testing facilities and capacity programs like Asian Development Bank (2021), ICTA, Sri Lanka, 2021 are strengthening the ecosystem. Policy frameworks such as the National Industrial Development Policy and fiscal incentives for automation and R&D align modernization with SDG 9 targets (Ministry of Industry and Commerce, Sri Lanka, 2020). Sustainability initiatives, including the National Cleaner Production Centre and the Sustainable Manufacturing Hub, deliver 15–20% energy savings and integrated resource-recycling infrastructure, advancing environmentally sound industrialization (National Cleaner Production Centre, Sri Lanka, 2022; Board of Investment, Sri Lanka, 2022).

GLOBAL BEST PRACTICES IN SUSTAINABLE INDUSTRIALIZATION

Circular economy practices are central to sustainable industrialization. Denmark's strategy combining extended producer responsibility, landfill bans, and economic incentives enabled a 9% reduction in material consumption, 15% increase in secondary material use, and creation of 7,000 circular economy jobs (Danish Ministry of Environment, 2021; Danish Circular Economy Hub, 2022). China's Circular Economy Promotion Law mandates resource targets and producer take-back in key sectors, leading to 20% cost savings and 16,000 remanufacturing jobs in automotive (National Development and Reform Commission, China, 2021; China Association of Automobile Manufacturers, 2022). Rwanda's e-waste strategy underpinned by regulatory fees and public-private facilities, processed over 3,000 tons of waste, creating 400+ green jobs in materials recovery (Rwanda Environment Management Authority, 2021; Enviroserve Rwanda, 2022).

Green manufacturing technologies reduce environmental footprints through advanced process innovations. Germany's Green Factory initiative developed continuous flow reactors that cut energy use by 40–60% and hazardous waste by 50%, supported by demonstration hubs for SME adoption (Fraunhofer Society, 2021; 2022). India's Green Manufacturing Initiative tailored semi-automated dyeing and energy measures in textiles, achieving 30–40% water savings and 20–30% chemical reductions with payback periods of one to three years (Ministry of Heavy Industries, India, 2022; Textile Ministry, India, 2021). Vietnam's Eco-Industrial Manufacturing Program advanced biogas, heat recovery, and water reuse in food processing, delivering 15–25% resource cost savings for moderate investments (Vietnam Cleaner Production Centre, 2022). Industrial symbiosis turns waste streams into inputs across co-located operations. Denmark's Kalundborg network with over 30 exchange relationships saves 24 million m^3 of water and avoids 275,000 tCO_2 annually, generating €24 million in economic benefits (Kalundborg Symbiosis, 2022). China's Eco-Industrial Parks legislation fostered 217 designated zones with centralized waste, energy, and water infrastructures; Suzhou Industrial Park achieves >75% water recirculation and >90% solid waste utilization (Ministry of Ecology and Environment, China, 2022; Suzhou Industrial Park Administrative Committee, 2022). Morocco's cluster approach in olive oil processing recovered compost and extracted valuable compounds, reducing pollution by 80% and raising producer incomes by 15–20% (Morocco Cleaner Production Centre, 2022).

Enabling policy frameworks such as the EU's Sustainable Products Initiative, combined with ecodesign regulations, digital product passports, extended producer responsibility, and green procurement to drive lifecycle sustainability (European Commission, 2022). Looking ahead, advanced materials, distributed manufacturing, digital commons, and renewable energy integration offer further pathways to align industrial growth with environmental and social imperatives, fulfilling SDG 9's vision of resilient, inclusive, and sustainable industrialization.

Capacity Building for SDG#9 Implementation

Education and Training Approaches

Realizing SDG 9's targets hinges on a workforce equipped to design, build, and manage modern infrastructure and industry. Higher education curricula integrating sustainability, systems thinking, and digital competencies are foundational. MIT's Civil and Environmental Engineering department restructured its program around a New Engineering Education Transformation, embedding sustainability and digital skills across courses (MIT, 2022). Its Sustainable Infrastructure Systems course uses project-based learning where students collaborate on climate-resilient transport, water systems in informal settlements, and renewable energy integration—to cultivate both technical mastery and contextual implementation abilities (MIT Department of Civil and Environmental Engineering, 2022). Globally, universities adapt these models to local contexts. India's IITs offer Sustainable Manufacturing Systems programs emphasizing frugal innovation that are affordable, resource-efficient solutions for constrained environments, yielding advances in decentralized renewables and water-efficient processes (Indian Institute of Technology Delhi, 2021). Kenya's PAUSTI integrates engineering with environmental assessment and community engagement, training graduates to plan infrastructure that is technically sound, socially inclusive, and environmentally sustainable (Pan-African University, 2022).

Vocational training delivers essential hands-on skills for infrastructure and industry roles. Germany's dual system blends classroom theory with workplace learning, now incorporating "Green Skills" modules for energy-efficient construction, renewable installations, and circular manufacturing practices (Federal Institute for Vocational Education and Training, Germany, 2022). India's Industrial Training Institutes similarly offer Green Industrial Skills, focusing on energy savings in small manufacturers and maintenance of renewables (Ministry of Skill Development and Entrepreneurship, India, 2021). In East Africa, EACREEE's technician training in renewable system installation and maintenance has certified over 2,000 practitioners since 2018, bolstering both sustainable energy rollout and job creation (East African Centre for Renewable Energy and Energy Efficiency, 2022).

Digital competencies underpin modern infrastructure and manufacturing. The ITU's Digital Skills for Sustainable Development framework outlines basic, intermediate, and advanced levels of digital literacy (International Telecommunication Union, 2021). Rwanda's Digital Ambassadors have trained 5 million citizens in basic digital skills, improving inclusion in e-government and digital business services, especially in rural areas (Ministry of ICT and Innovation, Rwanda, 2022). For advanced skills, coding bootcamps like Andela in Africa equip software engineers to lead digital infrastructure projects (Andela, 2022). Continuing professional development ensures existing workforces adapt to evolving demands. The Institution of Civil Engineers' Engineering for the *SDGs* program offers modules on sustainable planning, resilience, and digital transformation via blended online and workshop formats (Institution of Civil Engineers, 2021). Sector-specific initiatives like Malaysia's green construction training and Brazil's Sustainable Manufacturing Training Network train practitioners across roles in sustainable building techniques, waste management, and efficiency measures, achieving 15–25% resource savings (Construction Industry Development Board, Malaysia, 2021; Brazilian Cleaner Production Network, 2022).

Alternative learning pathways broaden access. Open educational resources (MIT OpenCourseWare, SDG Academy) and MOOCs (IDB's Sustainable Infrastructure course on edX) offer modular, self-paced study to tens of thousands worldwide (SDG Academy, 2022; Inter-American Development Bank, 2022). Communities of practice such as the World Bank's Global Infrastructure Community enable peer exchange of tacit knowledge through forums, webinars, and collaborative projects (World Bank, 2021). Mentorship programs like Women in Sustainable Infrastructure link early-career professionals with sector leaders to grow technical and leadership skills (Global Infrastructure Basel Foundation, 2022). Hackathons, exemplified by the African Development Bank's Africa Infrastructure Hackathon, foster multidisciplinary teams to co-develop contextually viable infrastructure innovations under time-bound, intensive learning conditions (African Development Bank, 2022). Measuring training effectiveness requires moving beyond enrollments to assess how skills translate into practice. The OECD's Skills for a Sustainable Future framework evaluates not only knowledge acquisition but also application capacity and contributions to development outcomes (Organization for Economic Co-operation and Development, 2021). Common success factors include multidimensional curricula blending technical, environmental, and social content; experiential learning that builds real-world implementation skills; and contextual relevance ensuring applicability across diverse resource levels.

PUBLIC–PRIVATE PARTNERSHIPS FOR TECHNOLOGY TRANSFER

Public–private partnerships (PPPs) are vital for transferring technologies and building local capacity for sustainable infrastructure and industry. Effective PPPs combine public agencies' developmental mandate with private firms' technical expertise and academia's research strengths, ensuring technologies are not only deployed but also adapted, maintained, and improved locally.

The World Bank's Sustainable Infrastructure Partnership Program illustrates a successful PPP model. In Vietnam, international engineering firms, local contractors, government agencies, and universities co-deliver climate-resilient bridges, water systems, and renewable installations as demonstration projects. Structured knowledge transfer such as training, mentorship, and progressive local leadership in subsequent project transitions local partners from learners to independent implementers, embedding resilient design standards within government planning processes (World Bank, 2022; Ministry of Construction, Vietnam, 2022).

UNIDO's Global Cleantech Innovation Programme fosters industrial PPPs for clean manufacturing. In Morocco, the program assessed manufacturers' processes, then collaborated with technology providers and local firms to install energy-efficient equipment and recycling systems. Concurrently, it has developed a support ecosystem training local suppliers, equipping technical institutes, and engaging financiers to offer clean-tech loans. Forty-five companies saw 25–30% energy and 30–40% water savings; over 200 additional firms adopted similar solutions, and fifteen local service providers now specialize in clean manufacturing (United Nations Industrial Development Organization, 2022).

Research and Development (R&D) partnerships focus on context-appropriate technology development. The UN Technology Bank's Affordable Technologies Network teamed up with local partners in Bangladesh to design solar-powered irrigation pumps for small farmers. These pumps are inexpensive, easy to keep in working order, and built with parts that can be manufactured or sourced locally. Over 2,000 systems serve 15,000 farmers, reducing diesel use; five local firms produce key parts and as a result, university research capacity for rural energy has grown significantly (Bangladesh Agricultural Development Corporation, 2022).

Digital PPPs leverage open-source models like The Digital Public Goods Alliance's Sierra Leone program that built a geospatial asset management platform, mobile condition-monitoring apps, and analytics dashboards. Released under open licenses and paired with developer and administrator training, these tools now track 5,000 assets nationally, and local software firms have established dedicated digital infrastructure practices (Digital Public Goods Alliance, 2021; Ministry of Information and Communications, Sierra Leone, 2022). Governance and financing determine PPP success; effective models feature balanced steering committees, transparent monitoring of both technology deployment and capability development, and blended finance such as concessional, private, and grant funds aligned to project milestones. Intellectual property arrangements, modeled on pharmaceutical patent pools or CTCN licensing, can balance innovation incentives with local adaptation needs (Climate Technology Centre and Network, 2022).

Success factors include demand-driven project selection, adaptive implementation allowing iterative learning, and ecosystem development addressing skills, supply chains, finance, and regulations. Emerging trends such as South-South cooperation, digital collaboration platforms, open innovation, and triangular partnerships promise more diverse and context-sensitive technology transfer models for SDG 9 delivery.

KNOWLEDGE NETWORKS AND COMMUNITIES OF PRACTICE

Knowledge networks and communities of practice accelerate SDG 9 implementation by enabling peer learning and sharing of explicit and tacit insights across institutions and borders. Unlike formal hierarchies, these voluntary networks leverage relationships to disseminate best practices and catalyze innovation. The Global Infrastructure Forum's Climate Resilient Infrastructure Working Group exemplifies a structured network that connects multilateral banks, governments, private investors, and researchers. Through annual meetings, thematic working groups, and online repositories, it fosters exchange of resilience methodologies and case studies. Its Contextual Resilience Framework guides adaptation of resilience strategies to specific risk profiles and resource contexts, resulting in 85% of members applying shared approaches and 70% reporting policy or practice changes (Global Infrastructure Forum, 2022).

Communities of practice like the Sustainable Manufacturing Practitioners Network focus on practitioner-level exchange. Through implementation clinics and dialogues, members collaboratively solve challenges such as reducing water use or defect rates. Over 90% of active practitioners have improved environmental metrics using shared insights, and the network has published assessment protocols adopted by 500+ companies (Sustainable Manufacturing Practitioners Network, 2022).

Regional networks address context specificity. ASEAN's Sustainable Infrastructure Network harmonizes standards while allowing national flexibility, aligning infrastructure policies across Member States and guiding over 30 cities in sustainable urban programs benefiting 15 million residents (ASEAN Secretariat, 2022). Digital platforms expand reach. The Sustainable Development Solutions Network (SDSN) Knowledge Platform's SDG 9 community of 50,000 users from 180 countries combines resource libraries with discussion forums and case databases. Eighty percent of active users implement methods learned, and 65% share resources further, demonstrating digital networks' scale (Sustainable Development Solutions Network, 2022).

South-South and triangular networks, such as the South-South Global Thinkers, promote peer learning among developing countries. Its Sustainable Industrialization in Africa initiative informed policy changes in seven nations and spurred technology adoption, such as mobile finance models adapted across borders (United Nations Office for South-South Cooperation, 2022).

Academic networks like the Global Research Alliance for Sustainable Infrastructure integrate research with practice. Its Climate Resilient Infrastructure Observatory co-creates risk assessment and resilience investment tools with practitioners, influencing planning in 18 countries and standards at five development banks (Global Research Alliance for Sustainable Infrastructure, 2022). Key challenges include ensuring inclusive participation, overcoming language barriers, balancing power dynamics, and capturing tacit knowledge. Success factors are focused purpose, stakeholder diversity, skilled facilitation, digital tool integration, and sustained resourcing. Emerging trends like AI-driven knowledge discovery, immersive virtual exchanges, and transparent blockchain-based governance will further enhance global learning communities, driving progress toward SDG 9's vision of resilient, inclusive, and sustainable infrastructure and industrial development.

Policy Recommendations and Implementation Pathways

Integrated Policy Frameworks for SDG#9

Achieving SDG 9 targets requires policies that link infrastructure, industrial, and innovation strategies across sectors, governance levels, and timeframes. The European Green Deal's Integrated Sustainable Industrial Policy exemplifies regional integration: a cross-Directorates-General Task Force unites climate targets, circular economy standards, digital priorities, and industrial competitiveness. Instruments like the EU Taxonomy for Sustainable Activities, circular design standards, and Skills for the Green Transition program ensure coherent incentives for resource efficiency and innovation (European Commission, 2021; 2022).

Colombia's National Policy for Sustainable Infrastructure and Production, developed through inclusive stakeholder dialogues, aligns transport, energy, water, and digital connectivity with industrial and environmental goals. A Sustainable Development Coordination Commission including ministries, regions, business, and civil society oversees coherence and supervises 23 Territorial Sustainable Development Plans. Since adoption, 70% of public infrastructure spending follows integrated plans, and industrial resource efficiency has improved 12%, with stronger gains in advanced regions (National Planning Department, Colombia, 2021; 2022). In resource-constrained contexts, Rwanda's Integrated Development Program links infrastructure investments, economic transformation, and environmental management within Vision 2050. Sector Working Groups and an Inter-Ministerial Committee coordinate thematic strategies, while district performance contracts ensure accountability for integrated outcomes. This approach reduced inter-agency coordination costs by 25% and cut project completion times by 30%, delivering a 45% rise in households with both electricity and clean water (Ministry of Finance and Economic Planning, Rwanda, 2021; 2022). Key enablers are shared visions, cross-sector bodies with clear mandates, flexible instruments aligned across regulations, standards, procurement, skills, and finance, plus multi-level planning that tailor national objectives to local contexts. Digital platforms such as Estonia's X-Road data exchange can further bind siloed agencies by enabling secure information sharing for planning, service delivery, and monitoring (e-Estonia Briefing Centre, 2022).

FINANCING MECHANISMS FOR SUSTAINABLE INFRASTRUCTURE AND INDUSTRY

Closing SDG 9's funding gap that is estimated at $15 trillion globally by 2040 demands both more capital and better-aligned finance. Public budgets, when guided by wellbeing frameworks like New Zealand's Living Standards approach, channel spending to investments that boost natural, human, social, and physical capital, increasing sustainable transport, renewables, and circular projects with transparent cost-benefit analysis (New Zealand Treasury, 2021; 2022).

For constrained fiscal environments, Ghana's Public Investment Management System enhances project appraisal and monitoring across economic, social, environmental, and priority-alignment criteria. A dedicated Public Investment Division oversees standardized appraisals and tracks outcomes, raising project completion rates from 65% to 85% and improving sustainability in public works (Ministry of Finance, Ghana, 2021; 2022).

Development finance institutions mobilize finance and expertise such as the AIIB's Sustainable Infrastructure Framework that integrates climate and social standards and through its Project Preparation Special Fund has readied 35 green projects totaling $8 billion while Brazil's BNDES directs credit lines and technical support to renewables, energy efficiency, and clean industry, financing 15 GW of renewables and 2,500 km of urban rail (Asian Infrastructure Investment Bank, 2021; 2022; BNDES, 2022).

Private capital flows via green bonds now a $1.5 trillion market fund, Chile's $7.5 billion sovereign green bond program for renewables, low-carbon transport, and water infrastructure with rigorous transparency (Ministry of Finance, Chile, 2022) are examples. Sustainability-linked bonds, like Suzano's $1.25 billion issuance tied to emissions targets, align industry-wide decarbonization incentives (Suzano, 2021). PPPs such as the Philippines' Green Infrastructure PPP Program blend procurement, finance, and sustainability criteria, delivering $4.5 billion in resilient water, transport, and energy assets with contractual incentives for environmental and social outcomes (Public-Private Partnership Center, Philippines, 2022). Blended finance structures such as Climate Investor One's three-tiered fund, mobilizing $850 million for renewables and digital finance models like M-KOPA's pay-as-you-go solar systems illustrate catalytic approaches for distributed infrastructure and small-scale industry (Climate Fund Managers, 2022; M-KOPA, 2022). Domestic resource mobilization, through green bond markets and local capital development as in Rwanda's Sustainable Finance Initiative strengthens self-reliant funding pathways that are aligned with SDG 9 (Capital Market Authority, Rwanda, 2022).

MONITORING AND EVALUATION FRAMEWORKS

Robust Monitoring and Evaluation (M&E) systems track SDG 9's multidimensional progress, integrating global, regional, national, subnational, and project-level metrics. Global UN indicators: transport coverage, manufacturing share, R&D spend, and network access provide comparability but require local adaptation for data gaps and context nuances. Regionally, the African Union's Agenda 2063 monitoring adds connectivity, value addition, and indigenous innovation measures, complementing SDG metrics to reflect African priorities (African Union Commission, 2021). National systems like Colombia's Sustainable Development Indicator System merge SDG, development plan, and sectoral indicators on a unified platform, enhancing data quality and cross-sector analysis through statistical capacity-building and a central data hub (National Administrative Department of Statistics, Colombia, 2021; 2022).

City-level indices, such as India's Sustainable Cities Index, assess infrastructure, economy, environment, and governance across 100 cities, linking online dashboards, peer learning, and performance-based grants to improve services for 100 million urban residents (Ministry of Housing and Urban Affairs, India, 2021; 2022). Project evaluations use multidimensional frameworks like IDB's Sustainable Infrastructure Assessment that rates economic, environmental, social, institutional, and impact dimensions in over 200 projects (Inter-American Development Bank, 2021; Inter-American

Development Bank, 2022), while UNIDO's Green Industry Progress framework tracks resource productivity, environmental performance, and competitiveness across 500 facilities, identifying $450 million in efficiency gains and 2.3 million tCO_2 avoided annually (UNIDO, 2021; United Nations Industrial Development Organization, 2022).

Innovation system assessments from the Global Innovation Index to Africa's Inclusive Innovation Monitor combine traditional R&D and patent metrics with grassroots and social inclusion measures, capturing technologies serving marginalized populations (WIPO, 2022; African Centre for Technology Studies, 2021).

Advanced M&E methods include theory of change mapping, contribution analysis for complex attribution, participatory monitoring engaging beneficiaries, and geospatial techniques for remote infrastructure and environmental impacts. Digital tools such as mobile surveys, IoT sensors, and big data analytics enhance timeliness and granularity. Tanzania's Rural Water Monitoring System, integrating mobile reports, sensors, and GIS for 130,000 pumps, raised functionality from 54% to 78% in five years (Ministry of Water, Tanzania, 2022). Key enablers for effective SDG 9 M&E are capacity building for data systems, open data platforms, methodological guidance, and knowledge networks that share evaluation innovations. Emerging trends such as AI-driven analytics, citizen science, integrated SDG frameworks, and real-time digital monitoring promise more responsive, inclusive evidence bases to guide sustainable infrastructure, industrial, and innovation pathways.

CONCLUSION

SDG 9's integrated vision linking infrastructure, industrialization, and innovation demands tools that reveal their interdependence. Geospatial methods enhance measurement by mapping access gaps, monitoring industrial distribution, and tracking innovation diffusion. Advanced planning techniques such as gap analysis and digital twins improve resilience and equity in diverse contexts, while Industry 4.0 technologies boost resource efficiency but require governance to ensure inclusivity. Capacity building through formal education, vocational training, PPPs, and knowledge networks underpins sustainable implementation. Integrated policy frameworks align objectives across sectors and levels; blended and market-based finance mobilize public, private, and international capital; and multidimensional M&E systems from global SDG indicators to geospatial monitoring provide evidence for adaptive management.

Key research priorities should include exploring SDG 9 synergies and trade-offs; disaggregating distributional impacts; advancing resilience assessment methods; systematizing technology transfer; valuing non-market benefits; studying governance innovations; empirically assessing digitalization's net effects; designing finance for underserved markets; improving knowledge management; and developing integrated metrics that balance global comparability with local relevance. Midway to 2030, SDG 9 implementation faces fiscal, climate, and geopolitical headwinds yet benefits from proven approaches worldwide. Accelerating progress demands deeper integration, stronger inclusion, built-in resilience, and deliberate digital governance. Sustained international cooperation, guided by geospatial insight and human-centered values, can transform infrastructure and industry into pillars of equitable, low-carbon, and innovative development.

REFERENCES

A Digital Twin for the Port of Rotterdam by Esri. DITTO. Retrieved from https://ditto-oceandecade.org/use-cases/a-digital-twin-for-the-port-of-rotterdam-by-esri/

Aid Data. (2021). *Tracking Chinese Development Finance: An Application of AidData's TUFF 2.0 Methodology* [Dataset and report]. Williamsburg, VA: AidData at William & Mary. Retrieved from https://www.aiddata.org/datasets

American Society of Civil Engineers. (2020). *2020 Report Card for America's Infrastructure*. Reston, VA: ASCE. https://infrastructurereportcard.org/

ASEAN Secretariat. (2022). ASEAN Annual Report 2022–2023. Jakarta: ASEAN Secretariat. Retrieved from https://asean.org/wp-content/uploads/2023/07/NEW-FINAL-ASEAN-Annual-Report-2022-2023.pdf

Asian Development Bank. (2020). *Asian Development Outlook 2020: What Drives Innovation in Asia?* Manila: ADB. https://www.adb.org/publications/asian-development-outlook-2020-innovation-asia

Asian Development Bank. *Sri Lanka: Integrated Road Investment Program. South Asia Innovation*. Retrieved from https://www.adb.org/sites/default/files/publication/879306/saipb-003-sri-lanka-integrated-road-investment-program.pdf

Better Evaluation. (n.d.). *Contribution Analysis*. Retrieved from https://www.betterevaluation.org/methods-approaches/approaches/contribution-analysis

BioMADE: Revitalizing U.S. Biomanufacturing and Fostering Collaboration. BioMADE. Retrieved from https://www.biomade.org/news/biomade-revitalizing-us-biomanufacturing-and-fostering-collaborationnbspnbsp

BMW Group. (2021). *BMW Group Report 2021*. Munich: BMW AG. https://www.bmwgroup.com/en/investor-relations/reports.html

Cadamuro, G., et al. (2019). *Mapping Infrastructure Using Deep Learning*. Proceedings of the IEEE International Conference on Computer Vision Workshops (ICCVW). https://openaccess.thecvf.com/content_ICCVW_2019/html/AI4CE/Cadamuro_Mapping_Infrastructure_Using_Deep_Learning_ICCVW_2019_paper.html

Capital Market Authority Rwanda. (2022). *Annual Report 2021/2022*. Kigali: CMA Rwanda. Retrieved from https://www.cma.rw/publications/annual-reports/

Coalition for Disaster Resilient Infrastructure (CDRI). (2021). *Infrastructure for Resilient Island States (IRIS) Initiative*. New Delhi: CDRI. Retrieved from https://cdri.world/IRIS

Construction Industry Development Board Malaysia. (2021). *Construction 4.0 Strategic Plan (2021–2025)*. Kuala Lumpur: CIDB. Retrieved from https://www.cidb.gov.my/wp-content/uploads/2022/12/CR4_web_compressed.pdf

Department of Environmental Conservation, NY. *New York State Adaptation and Resilience Plan*. Retrieved from https://dec.ny.gov/environmental-protection/climate-change/climate-change-adaptation-resilience-planning/new-york-state-adaptation-and-resilience-plan

Digital Twin Consortium. (2022). *Digital Twin System Interoperability Framework* [White paper]. Boston, MA: Digital Twin Consortium. Retrieved from https://www.digitaltwinconsortium.org/resources/whitepapers.htm

ECOWAS Centre for Renewable Energy and Energy Efficiency. (2022). *Annual Report 2022*. Praia: ECREEE. Retrieved from https://www.ecreee.org/wp-content/uploads/2024/03/ECREEE-ANNUAL-REPORT-2022-EN.pdf

European Commission. (2022). *Decision (EU) 2022/2481 establishing the Digital Decade Policy Programme 2030. Official Journal of the European Union*, L 323, 19 Dec 2022. Retrieved from https://eur-lex.europa.eu/legal-content/EN/TXT/?uri=CELEX:32022D2481

European Space Agency. (2022). *Earth Observation for Sustainable Development*. Paris: ESA. https://eo4sd.esa.int/

Facebook AI Research. (2019). *Mapping Africa's Roads with AI*. Facebook Engineering Blog. https://ai.facebook.com/blog/mapping-africas-roads

Group on Earth Observations. (2016). *GEO Strategic Plan 2016–2025: Implementing GEOSS*. Geneva: GEO Secretariat. Retrieved from https://old.earthobservations.org/documents/GEO_Strategic_Plan_2016_2025_Implementing_GEOSS.pdf

Group on Earth Observations (GEO). (2020). *EO4SDG Annual Report 2020*. Geneva: GEO Secretariat. Retrieved from https://eo4sdg.org/reports/

Global Environment Facility. (2021). *GEF Monitoring Report 2021*. Washington, DC: GEF. Retrieved from https://www.thegef.org/council-meeting-documents/gef-monitoring-report-2021

Global Facility for Disaster Reduction and Recovery (GFDRR). (2021). *Risk Data Library Standard (RDLS)* [Technical framework]. Washington, DC: World Bank. Retrieved from https://gfdrr.org/risk-data-library

Global Infrastructure Hub. (2020). *InfraCompass 2020: Set your infrastructure policies in the right direction*. Sydney: Global Infrastructure Hub. Retrieved from https://infra-compass.gihub.org/

Global Partnership for Sustainable Development Data (GPSDD). (2021). *Data Values Project—What We Heard in 2021*. Washington, DC: GPSDD. Retrieved from https://www.data4sdgs.org/resources/data-values-project-what-we-heard

Google LLC. (2021). *Google COVID-19 Community Mobility Reports* [Dataset]. Retrieved from https://www.google.com/covid19/mobility/

GW Prime. *Virtual Singapore - Building a 3D-Empowered Smart Nation*. Retrieved from https://geospatialworld.net/prime/case-study/national-mapping/virtual-singapore-building-a-3d-empowered-smart-nation/

Henderson, J. V., Storeygard, A., & Weil, D. N. (2018). The development of the African manufacturing sector: Evidence from night lights. *Journal of Economic Growth*, 23(3), 217 -244. https://doi.org/10.1007/s10887-018-9151

IEEE Standards Association. (2021). *Ethically Aligned Design: A Vision for Prioritizing Human Well-being with Autonomous and Intelligent Systems*. IEEE. https://standards.ieee.org/industry-connections/ec/autonomous-systems.html

International Labour Organization (ILO). (2021). *World Employment and Social Outlook 2021: The Role of Digital Labour Platforms in Transforming the World of Work*. Geneva: ILO. Retrieved from https://www.ilo.org/global/research/global-reports/weso/2021/WCMS_771749/lang--en/index.htm

Intergovernmental Panel on Climate Change. (2019). *2019 Refinement to the 2006 IPCC Guidelines for National Greenhouse Gas Inventories*. Geneva: IPCC. https://www.ipcc-nggip.iges.or.jp/public/2019rf/index.html

International Telecommunication Union. (2020). *Measuring Digital Development: Facts and Figures 2020*. Geneva: ITU. https://www.itu.int/en/ITU-D/Statistics/Documents/facts/FactsFigures2020.pdf

Institution of Civil Engineers. (2021). *Introduction to Sustainable Development*. London: ICE. Retrieved from https://www.icetraining.org.uk/courses/environment-and-sustainability/introduction-to-sustainable-development

International Finance Corporation. (2022). *Andela: Launching Africa's Tech Talent*. Washington, DC: IFC. Retrieved from https://www.ifc.org/en/insights-reports/2022/andela

International Telecommunication Union. (2021). *Digital Skills Toolkit*. Geneva: ITU. Retrieved from https://academy.itu.int/sites/default/files/media2/file/2401045_1f_Digital%20Skills%20Toolkit_compressed_0.pdf

Inter-American Development Bank. (2022). *Sustainable Infrastructure: An Integrated Vision for Development*. Washington, DC: IDB. Retrieved from https://cursos.iadb.org/en/indes/sustainable-infrastructure-integrated-vision-development

Jean, N., Burke, M., Xie, M., Davis, W. M., Lobell, D. B., & Ermon, S. (2016). Combining satellite imagery and machine learning to predict poverty. *Science*, 353(6301), 790 -794. https://doi.org/10.1126/science.aaf7894

Malaria Atlas Project (MAP). (2020). *Global Malaria Risk and Intervention Data* [Dataset]. University of Oxford. Retrieved from https://malariaatlas.org/

Manufacturing Technology Centre. (2021). *Impact of Artificial Intelligence in Manufacturing: Study Report.* Coventry, UK: MTC. https://www.the-mtc.org/resources

McKinsey Global Institute (MGI). (2021). *The Future of Work After COVID-19.* New York: McKinsey & Company. Retrieved from https://www.mckinsey.com/mgi/our-research/the-future-of-work-after-covid-19

Ministry of ICT and Innovation. (2022). *Digital Ambassadors Programme (DAP).* Kigali: Government of Rwanda. Retrieved from https://www.minict.gov.rw/projects/digital-ambassadors-programme

Missing Maps. (2021). *The Missing Maps Project.* Retrieved from https://www.missingmaps.org/

M-KOPA. (2022). *Annual Impact Report 2022.* Nairobi: M-KOPA. Retrieved from https://www.m-kopa.com/impact/

National Research Foundation Singapore (NRF). (2021). *Virtual Singapore: A Dynamic 3D City Model.* Singapore: NRF. Retrieved from https://www.nrf.gov.sg/programmes/virtual-singapore

Organization for Economic Co-operation and Development. (2015). *Frascati Manual 2015: Guidelines for Collecting and Reporting Data on Research and Experimental Development.* Paris: OECD Publishing. https://doi.org/10.1787/9789264239012-en

Organization for Economic Co-operation and Development. (2021). *Creditor Reporting System (CRS) Aid Activities.* Paris: OECD. https://stats.oecd.org/Index.aspx?DataSetCode=CRS1

Pereira, F. A., & Sant'Anna, A. M. O. (2016). *The State of Research on Cleaner Production in Brazil. Revista de Administração de Empresas*, 56(1), 88–103. Retrieved from https://www.scielo.br/j/rae/a/KByHLYFV5xjDMKjmKgfMrkr/

Port of Rotterdam Authority. (2020). *Digital Twin Port and Smart Infrastructure Programme.* Rotterdam: Port of Rotterdam Authority. Retrieved from https://www.portofrotterdam.com/en

Praia Group on Governance Statistics (United Nations). (2020). *Handbook on Governance Statistics.* New York: United Nations. Retrieved from https://unstats.un.org/unsd/methodology/data-stat/praia/

Rwanda Dispatch. *Feeder Roads Transform Lives of Farmers in Gatsibo District. Rwanda Dispatch.* Retrieved from https://rwandadispatch.com/feeder-roads-transform-lives-of-farmers-in-gatsibo-district/

Southworth, J., Munroe, D. K., & Nagendra, H. (2011). Land cover change and landscape fragmentation --Comparing the utility of continuous and discrete analyses for a western Honduras region. *Applied Geography*, 31(1), 94 -102. https://doi.org/10.1016/j.apgeog.2010.03.003

Sustainable Development Solutions Network. (2022). *Sustainable Development Report 2022.* Cambridge: Cambridge University Press. Retrieved from https://sdsn.eu/2022-sustainable-development-report/

SDG Academy. (2022). *Open Education Resources (OER).* New York: Sustainable Development Solutions Network. Retrieved from https://sdgacademy.org/open-education-resources/

United Nations Committee of Experts on Environmental-Economic Accounting (UN SEEA). (2021). *System of Environmental-Economic Accounting --Ecosystem Accounting 2021.* New York: United Nations. Retrieved from https://seea.un.org/ecosystem-accounting

United Nations Conference on Trade and Development. (2021). *Technology and Innovation Report 2021: Catching Technological Waves.* Geneva: UNCTAD. https://unctad.org/webflyer/technology-and-innovation-report-2021

UNDP. (2020). *Climate Resilient Integrated Water Management Project.* GCF, Sri Lanka. Retrieved from https://www.adaptation-undp.org/sites/default/files/resources/gcf_sri_lanka_summary_document_v2.pdf

United Nations Global Working Group on Big Data for Official Statistics (UN GWG). (2020). *Big Data for Official Statistics: Annual Report 2020.* New York: United Nations Statistics Division. Retrieved from https://unstats.un.org/bigdata/

United Nations Industrial Development Organization. (2019). *Industrial Development Report 2020: Industrializing in the Digital Age.* Vienna: UNIDO. https://www.unido.org/

resources-publications-flagship-publications-industrial-development-report-series/industrial-development-report-2020

United Nations Industrial Development Organization (UNIDO). (2021). *Industrial Development Report 2022: The Future of Industrialization in a Post-Pandemic World*. Vienna: UNIDO. Retrieved from https://www.unido.org/resources-publications-flagship-industrial-development-report-2022

United Nations Industrial Development Organization. (2022). *AI for Industrial Development: Progress Report*. Vienna: UNIDO. https://www.unido.org/resources-publications

United Nations Global Pulse. (2020). *Big Data for Development: A Primer*. New York: UN Global Pulse. https://www.unglobalpulse.org/document/big-data-for-development-a-primer/

United Nations Statistics Division. (2018). *Handbook on Statistical Organization: The Operation and Organization of a Statistical Agency*. New York: UN. https://unstats.un.org/unsd/dnss/hb/

United Nations Statistics Division. (2020). *Handbook on Governance Statistics*. New York: UN Statistics Division. Retrieved from https://unstats.un.org/unsd/publication/seriesf/SeriesF_117E.pdf

United Nations Technology Innovation Labs. (2021). *Annual Report 2021*. UN. https://until.un.org/sites/default/files/2021-12/UNTIL_Annual_Report_2021.pdf

United Nations Office for the Coordination of Humanitarian Affairs (OCHA). (2021). *Humanitarian Data Exchange (HDX)*. New York: OCHA. Retrieved from https://data.humdata.org/

United Nations Office for Outer Space Affairs (UNOOSA). (2020). *Space for the Sustainable Development Goals (Space4SDGs)*. Vienna: UNOOSA. Retrieved from https://www.unoosa.org/oosa/en/ourwork/space4sdgs/index.html

US Army Corps of Engineers. (2020). *Coastal Infrastructure Program*. Retrieved from https://www.sam.usace.army.mil/Missions/Program-and-Project-Management/Civil-Projects/MsCIP/#:~:text=Mississippi%20Barrier%20Islands%20Restoration,from%20the%20Gulf%20of%20Mexico

U.S. Army Corps of Engineers (USACE). (2020). *Engineering With Nature: An Atlas, Volume 2*. Vicksburg, MS: U.S. Army Engineer Research and Development Center (ERDC). Retrieved from https://ewn.erdc.dren.mil/publications.html

U.S. Army Corps of Engineers (USACE). (2022). *Engineering With Nature: An Atlas, Volume 3* (ERDC SR-24-2). Vicksburg, MS: U.S. Army Engineer Research and Development Center (ERDC). https://doi.org/10.21079/11681/48453

U.S. Department of Transportation (USDOT). (2022). *National Roadway Safety Strategy*. Washington, DC: U.S. Department of Transportation. Retrieved from https://www.transportation.gov/NRSS

World Intellectual Property Organization. (2020). *World Intellectual Property Indicators 2020*. Geneva: WIPO. https://www.wipo.int/edocs/pubdocs/en/wipo_pub_941_2020.pdf

World Bank. (2017). *Public-Private Partnerships: Reference Guide Version 3.0*. Washington, DC: World Bank. Retrieved from https://openknowledge.worldbank.org/entities/publication/e64a94b9-b0cb-51a3-b8dd-36bd7944284f

World Bank. (2019). *Methodology for the Rural Access Index*. Washington, DC: World Bank. https://documents.worldbank.org/en/publication/documents-reports/documentdetail/093241468181510003/methodology-for-the-rural-access-index

World Bank. (2020). *World Development Report 2020: Trading for Development in the Age of Global Value Chains*. Washington, DC: World Bank. Retrieved from https://www.worldbank.org/en/publication/wdr2020

World Bank. (2021). *World Development Report 2021: Data for Better Lives*. Washington, DC: World Bank. Retrieved from https://www.worldbank.org/en/publication/wdr2021

WorldPop. (2021). *WorldPop Open Spatial Demographic Data* [Dataset]. University of Southampton. Retrieved from https://www.worldpop.org/

8 SDG #11—Geospatial Information for Sustainable Cities

Rosario Casanova

INTRODUCTION

The Sustainable Development Goals (SDGs) require the collection and analysis of reliable, detailed and disaggregated data, often involving the use of information from various sources. Geospatial data provides the spatial context necessary to interpret these data more comprehensively. This data, which includes information on the location and distribution of geographic and social phenomena, enables a better understanding of global challenges and provides informed solutions to complex problems related to sustainability, social justice, and economic well-being. Geospatial information is fundamental for measuring and monitoring the SDGs.

Indeed, through advanced technologies like Geographic Information Systems (GIS), remote sensing, and Earth observation, geospatial information provides a detailed view of the real local conditions affecting human development and the environment. This information allows for the spatial visualization of problems such as poverty, food security, access to drinking water, and climate change, facilitating more precise decision-making adapted to local realities and needs.

Sustainable Development Goal 11 (SDG 11), part of the 2030 Agenda for Sustainable Development, focuses on making cities and human settlements inclusive, safe, resilient, and sustainable. This goal has become one of the most urgent priorities of the global community, as more than half of the world's population lives in urban areas, a figure that continues to rise. While urban growth presents economic and development opportunities, it also poses significant challenges in terms of sustainability, access to basic services, and the quality of life in cities. The rapid expansion of cities and population densification in urban areas have created pressures on natural resources, infrastructure, transportation, housing, and the environment. As cities grow, more efficient planning and management are needed to ensure they are environmentally sustainable and resilient. This is where geospatial information plays a key role.

Integrating geospatial data with other types of information is fundamental to addressing territorial challenges. GIS and other geospatial data technologies allow for mapping, analyzing, and understanding urban phenomena with a level of precision and detail that is difficult to achieve with other methods. These data not only allow for identifying and localizing problematic areas, but also for generating and visualizing potential solutions, providing decision-makers with the necessary tools to act in an informed and efficient manner.

 DOI: 10.1201/9781003586319-10

This chapter examines how geospatial data is being used to monitor and measure the progress of SDG 11, particularly through geospatial-based indicators that allow for evaluating access to adequate housing, air quality, access to public transport, and the availability of public spaces, among other key aspects. Through effective integration of geospatial and statistical data, a more comprehensive view of urban dynamics can be obtained, facilitating the creation of more effective public policies. It also helps in simulating future scenarios, as well as different potential incidents based on the application of public policies onto the territory. This chapter aims to explore the importance of geospatial data in monitoring SDG 11 indicators and how these related technologies not only help to improve urban management but also contribute to building smarter, more resilient, and sustainable cities. Furthermore, it will discuss how geospatial data can help overcome the challenges that many cities face on their path toward sustainability, and how they can be used to adapt public policies to the changing needs of urban populations.

IMPORTANCE OF SDG 11: SUSTAINABLE CITIES AND COMMUNITIES

SDG 11 stands out for its comprehensive approach to the well-being of people living in cities and urban communities. As urbanization continues to accelerate, with estimates suggesting that approximately 60% of the global population will live in urban areas by 2030, this goal becomes critically relevant to ensure that cities not only grow in number but do so sustainably, inclusively, and resiliently. SDG 11 is designed to address a range of challenges faced by urban areas, such as providing adequate housing, access to public transportation, managing air quality, accessibility to basic services like drinking water and electricity, and safety in public spaces. These are key issues for many cities, especially in developing countries, where urban areas are expanding rapidly without adequate planning, resulting in informal settlements and precarious living conditions.

SDG 11 seeks to transform these cities into places where natural resources and infrastructure are managed efficiently, people have access to essential services, and social inclusion is promoted. The indicators of SDG 11 cover multiple areas, from measuring housing quality and public space safety to managing urban pollution and access to transportation services. All these aspects are crucial to ensure that cities are not only livable but also contribute to the health and well-being of their inhabitants.

In this context, the integration of geospatial information is essential. Geospatial data allows for a deeper understanding of how different factors affect cities and their inhabitants, providing a spatial perspective that helps identify patterns and make informed decisions. For example, maps of public transportation accessibility, the distribution of green areas, or air quality monitoring are crucial for assessing the sustainability of cities and the quality of life of their residents.

Furthermore, geospatial data plays a key role in sustainable urban planning. By enabling the visualization of the spatial distribution of resources, problems, and opportunities, decision-makers can implement more effective strategies to improve access to services, mitigate risks, and ensure equity in the distribution of resources. The ability to integrate geospatial information with data on social inequality, poverty distribution, and vulnerabilities allows for creating public policies that directly address access gaps and ensure that no community is left behind. Additionally, SDG 11 aligns with the urgent need to adapt to climate change and strengthen urban resilience. Smart cities, which leverage technology to improve the quality of life for citizens and the efficiency of urban services, are essential for achieving sustainability goals. The use of geospatial technologies allows for monitoring and managing urban growth, waste management, energy consumption, and resilient infrastructure in real-time, providing a platform for intelligent city management that is sustainable in the long term. Therefore, the importance of SDG 11 lies not only in creating cities that are sustainable and resilient but also in ensuring that these places are inclusive and balanced, providing all inhabitants with the opportunity to access basic services fairly and equitably. Geospatial information is the fundamental tool for making these problems visible and transforming cities into environments that are both sustainable and livable for all.

THE IMPORTANCE OF GEOSPATIAL DATA FOR THE SDGS

In the context of SDG 11, the application of geospatial technologies and the participatory engagement of communities are essential for advancing toward the creation of sustainable and resilient cities. The use of GIS and GeoDesign enables citizens to actively participate in urban planning and resource management. This collaborative approach is crucial to ensure that development decisions are inclusive, transparent, and felt as their own by the population.

The value of community participation in urban planning is emphasized in the approach of the Toolkit for Activating SDG 17. Using GeoDesign, citizens can directly participate in the decision-making process by providing geospatial data about their needs and priorities. Furthermore, the integration of geospatial technologies allows for visualizing potential solutions and evaluating scenarios in real-time, making it easier to adopt measures to mitigate negative territorial impacts.

GeoDesign, a key tool in this context, extends the analytical capabilities of GIS to include factors such as risks, vulnerability, capacity, proximity, accessibility, and other critical elements that inform design decisions. By integrating geospatial analysis with design technologies and community participation, a more holistic approach adapted to local realities can be achieved, ensuring that development is both sustainable and resilient.

Moreover, effective partnerships between governments, international organizations, the private sector, academia, and communities are essential to successfully implementing the SDGs, including SDG 11. Strategic partnerships and international cooperation enable the sharing of technologies, knowledge, and resources, overcoming technological barriers and ensuring the participation of all relevant stakeholders in the sustainable development process.

CONTRIBUTION OF GEOSPATIAL INFORMATION TO MONITORING THE SDGS

It is increasingly recognized that geospatial information could contribute significantly to the implementation and monitoring of the SDGs. For example, in SDG 6 on clean water and sanitation, geospatial data helps identify areas with water scarcity or without access to adequate sanitation facilities. By integrating this data with other social and economic indicators, a more accurate view of inequalities in access to these resources can be obtained, facilitating the design of more effective public policies tailored to local needs.

Similarly, in SDG 11 on sustainable cities, geospatial data allows for monitoring urban growth, public transportation accessibility, pollution distribution, and the safety of public spaces, among other aspects. Without the integration of geospatial information, it would not be possible to have an accurate view of areas that require intervention, making efforts toward sustainable urban development less effective or just. Urban infrastructure management, service planning, and risk mitigation depend on geospatial data to ensure that all communities have access to basic services, and that urban growth is managed efficiently.

GEOSPATIAL INFORMATION IN SUPPORT OF DECISION-MAKING TOOLS

One of the most important aspects of geospatial information is its ability to support decision-making. Instead of relying solely on aggregated or statistical data, geospatial data offers an additional dimension, allowing for mapping phenomena, analyzing patterns, and evaluating solutions in real-time. This type of analysis is essential for addressing challenges such as climate change, social inequality, and natural resource management, which require a spatial approach to identify geographic patterns, vulnerable areas, and zones requiring urgent intervention. For example, air quality data collected through remote sensing enables public institutions to monitor pollution in real-time, facilitating the implementation of public health policies in urban areas. Likewise, in the context of SDG 13 on climate action, Earth observation provides crucial data on sea level, global temperature, and ecological habitats, helping to model future scenarios and develop adaptation strategies for regions vulnerable to climate change. Moreover, geospatial data is also essential for monitoring SDG indicators. SDG 3 (health and well-being), SDG 4 (quality education), and SDG 5, gender equality, for example, benefit from the integration of geospatial data to measure their indicators and map the distribution of healthcare services, schools, and analyze the safety of public spaces, which in turn enhances the planning and monitoring of progress in these sectors.

FACILITATING PLANNING AND STRATEGY IMPLEMENTATION

Urban planning, natural resource management, and public service distribution are just some of the areas where geospatial data plays a crucial role. In SDG 11, geospatial analysis of housing distribution and access to safe public spaces allows governments to identify areas with urgent needs, plan infrastructure expansion, and design inclusive urban development strategies. Without this detailed information, efforts to create sustainable and resilient cities would be much less effective. Geospatial data enables decision-makers to create more efficient strategies and policies adapted to local realities. For example, through the integration of advanced technologies like artificial intelligence (AI) and predictive analysis, future problems such as uncontrolled urban expansion or infrastructure deterioration can be anticipated, and proactive public policies can be designed rather than reactive ones.

Some key benefits of using geospatial data for producing information associated with SDG indicators include: The ability to derive SDG indicators that would otherwise be complex to calculate (technically or financially), reducing the frequency of surveys and associated costs by providing information at a higher level of disaggregation, as well as providing breakdowns and granularity of the indicators, with geospatial enablement and guidance. Therefore, geospatial data is not only crucial for monitoring progress toward the SDGs but also serves as a fundamental tool for the sustainable planning and management of resources, enabling governments and international organizations to implement more effective and balanced strategies to achieve sustainable development at the global level.

INTEGRATION OF GEOSPATIAL INFORMATION WITH STATISTICAL INFORMATION IN SUPPORT OF THE SDGS

The integration of geospatial and statistical data is a powerful tool that allows for a more accurate understanding of the spatial distribution of problems and resources related to the SDGs. This integration provides a solid foundation for creating more effective public policies, as it allows decision-making based on a comprehensive view of territories and the needs of their inhabitants.

THE VALUE OF INTEGRATING GEOSPATIAL AND STATISTICAL DATA

Statistical data provides information about the size, composition, and trends of populations, while geospatial data allows us to visualize how these phenomena are distributed in space. The integration of these two types of data provides a multidimensional view that is essential for measuring and monitoring the SDGs. For example, while statistical data may indicate that an area has a high poverty rate, geospatial data allows for mapping areas of higher vulnerability, identifying marginalized urban areas, and planning the distribution of resources and services more efficiently.

The integration of statistical and geospatial data is crucial not only for the decision-making process and data sharing but also for generating new knowledge and added value based on relationships, analysis, and geostatistical data modeling. It also enables continuous monitoring at different scales of the SDG indicators. For SDG 11, which aims to create sustainable cities and communities, data integration is essential to measure indicators such as access to adequate housing, public transport, air quality, and safe public spaces. Geospatial data allows us to identify high-risk areas (such as areas with high pollution or lack of access to services), while statistical data provides details about the social and economic conditions of people living in those areas. This integration facilitates a localized and disaggregated approach, which is crucial to ensure that interventions do not leave anyone behind.

PRACTICAL APPLICATIONS FOR DATA INTEGRATION

A concrete example of how the integration of geospatial and statistical data can be applied to achieve the SDGs is in monitoring access to public transportation (indicator 11.2.1). By combining data on the location of public transportation stations with information about population density and the transportation needs of different social groups (e.g., people with disabilities, elderly, etc.), it is possible to identify underserved urban areas and design policies to improve accessibility to these services. This approach also allows for analyzing geographical and socioeconomic disparities in access to public transportation, ensuring that solutions are inclusive and equitable.

In Figure 8.1, the different realiteies of the Colombian cities of Medellin, and Cartagena (based on the information of National Administrative Department of Statistics, DANE), as represented by the geographic indicator 11.2.1, provide a basis for a comparative analysis of these urban contexts. This, in turn, would help inform the decision-making process for defining differentiated public policies tailored to the specific needs of each city.

Another example is Indicator 11.6.2 on air quality in urban areas. Earth observation provides data on the concentration of pollutants in different areas of a city, while statistical data can link this information with public health trends (such as the increase in respiratory diseases in certain areas). Integrating these data helps identify risk factors and design intervention policies to improve air quality, thereby protecting public health. In the context of SDG 6 on clean water and sanitation, the integration of geospatial and statistical data can facilitate the identification of rural or urban areas without access to drinking water. Using geospatial technologies, water sources, existing infrastructure, and the population's needs can be mapped, and with statistical data, the most vulnerable groups can be identified. This not only improves water resource management but also ensures that interventions reach communities with the greatest water needs.

CHALLENGES IN DATA INTEGRATION

Despite the opportunities presented by the integration of geospatial and statistical data, there are several challenges that must be overcome to fully realize their potential. One of the main challenges is the availability and quality of geospatial data in many regions, especially in developing countries. The high costs of obtaining and processing geospatial data, as well as the lack of technological infrastructure in certain areas, can make it difficult to collect accurate and up-to-date data.

Another challenge is the lack of standardization in the methods used to collect and analyze geospatial and statistical data. Harmonizing data is crucial to ensure that indicators are comparable and reliable. Without a common methodology, the

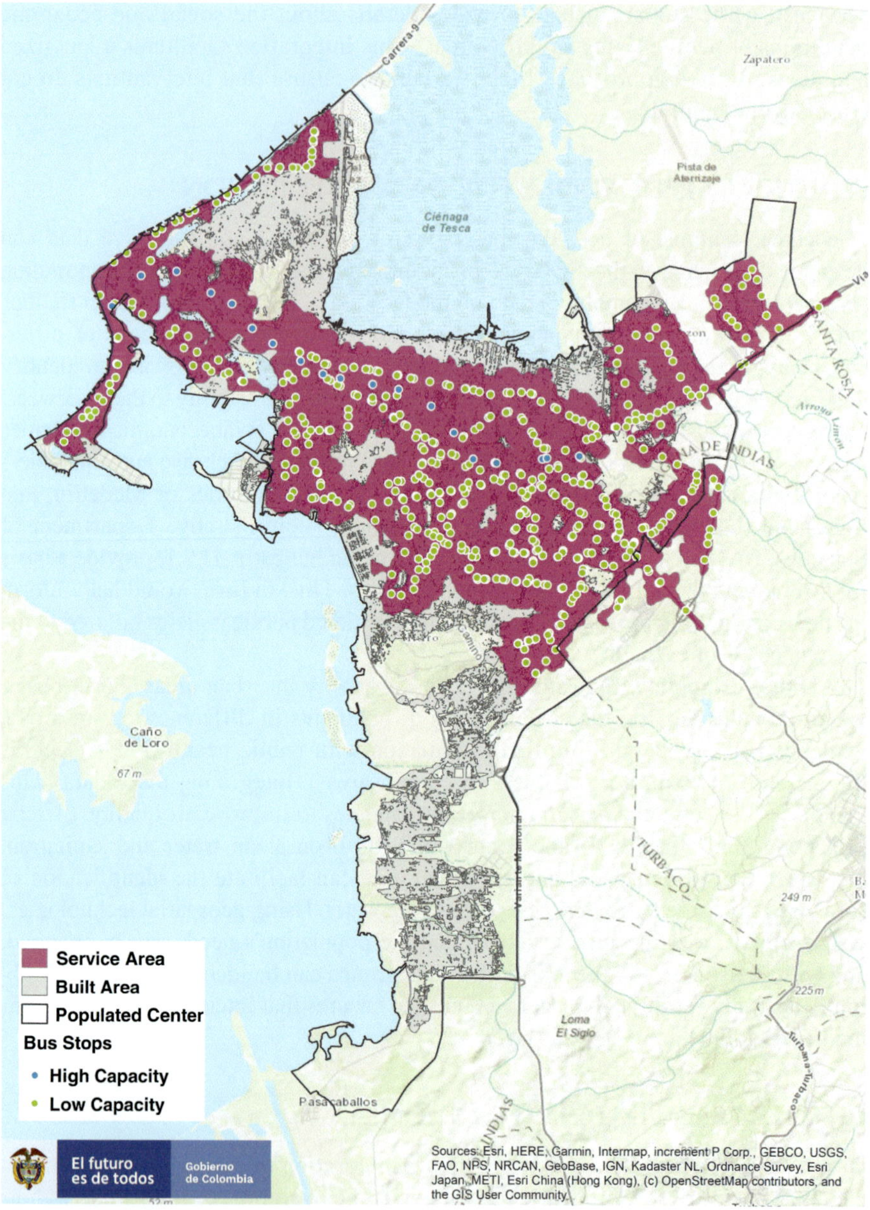

FIGURE 8.1 Maps of indicator 11.2.1 of different cities in Colombia. (*Continued*)

Source: DANE, 2024.

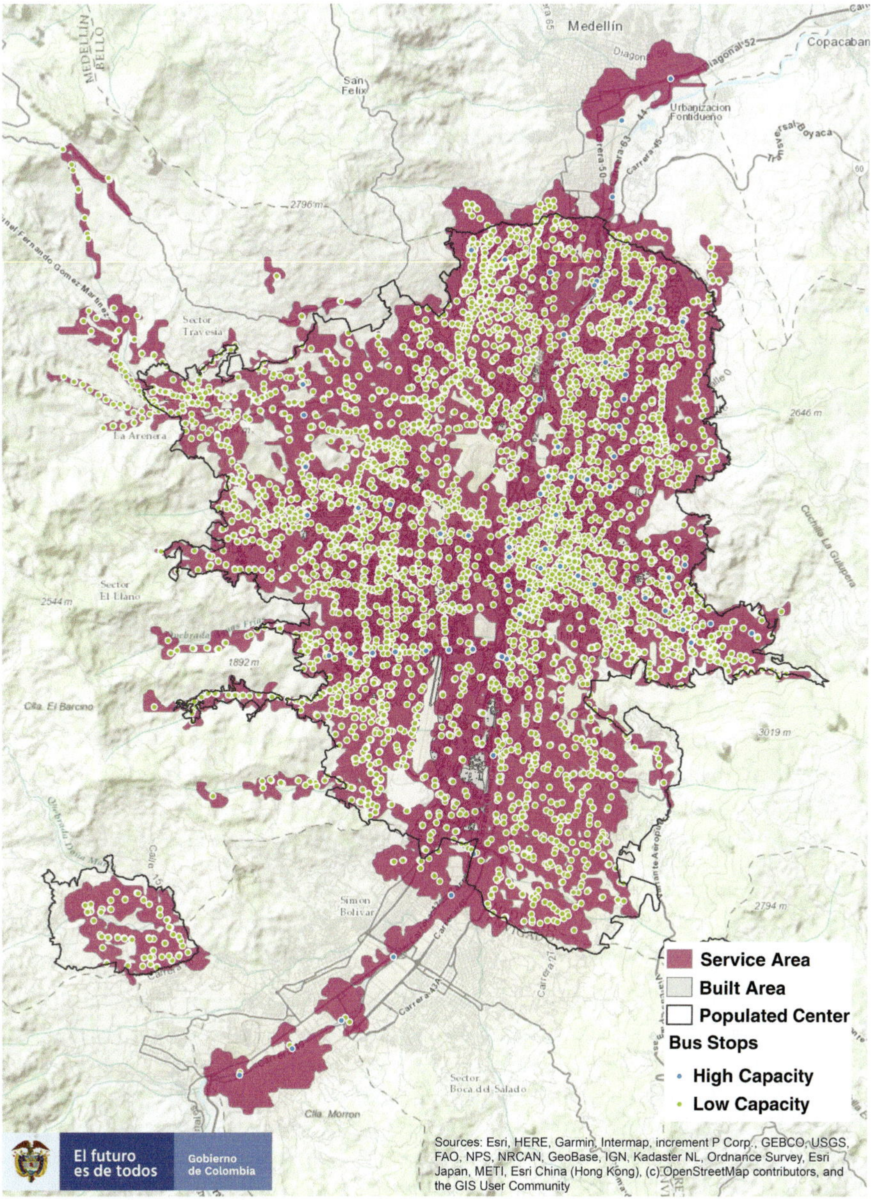

FIGURE 8.1 (Continued)

integration of data from different sources can result in inconsistent results, which makes it harder to make informed decisions.

OPPORTUNITIES TO OVERCOME THE CHALLENGES

Despite these challenges, there are multiple opportunities to improve the integration of geospatial and statistical data. Advanced technology is facilitating the creation of collaborative platforms that enable more efficient data collection, storage, and analysis. Remote sensing, smart sensors, and geospatial big data are tools that can help overcome many of the current limitations in terms of access to and quality of data.

Moreover, international cooperation between countries, non-governmental organizations, universities, and the private sector can facilitate joint data collection and training in the use of geospatial technologies, especially in low-income countries. Initiatives such as the Global Geospatial Information Framework (UN-GGIM) and the Global Statistical and Geospatial Framework (GSGF) are working on standardizing data collection and analysis methods, which makes integration easier and improves the quality of geospatial indicators.

CHALLENGES AND OPPORTUNITIES IN THE PRODUCTION OF GEOSPATIAL SDG INDICATORS

The creation and monitoring of geospatial indicators for the SDGs offer great opportunities to improve the effectiveness of public policies, but they also present a range of technical, institutional, and financial challenges that must be overcome to maximize their impact. These challenges affect both the quality of geospatial data and its accessibility, and they are especially relevant for developing countries, where technological capabilities and data infrastructure may be limited.

Challenges in the Production of Geospatial Indicators

One of the biggest challenges in producing geospatial indicators is data quality. Geospatial data must be accurate, up-to-date, and consistent to ensure that indicators accurately reflect ground realities. However, data quality varies considerably depending on the region and the source. In many areas, especially in developing countries, the lack of geospatial data collection infrastructure and inadequate geographic coverage can lead to data gaps or inaccurate information. Another challenge is the lack of standardization in data collection, processing, and analysis methods. Geospatial data comes from various sources, such as satellites, remote sensors, geospatial censuses, and crowdsourcing platforms, which can lead to discrepancies in data quality and format. The absence of global standards for how to collect and process these data makes integration and comparability difficult. Without a common methodology, the production of geospatial indicators becomes fragmented, hindering their effective use in global and national public policy formulation.

Inequality in access to advanced geospatial technologies and technical capabilities among countries is a significant obstacle to the production of quality geospatial indicators. While developed nations have access to Earth observation satellites, high-precision sensors, and advanced GIS platforms, many developing countries lack these resources and, in some cases, even basic geospatial data infrastructure. This creates a significant gap in training and data collection capacity, which in turn limits the accuracy and consistency of indicators at the global level.

The development, maintenance, and analysis of geospatial data is costly. Earth observation technologies, the use of satellites, remote sensors, and the storage of large amounts of geospatial data require significant investments in technological infrastructure and skilled human resources. In many cases, developing countries do not have the financial resources to invest in the right technology or in training their teams to create and maintain geospatial indicators. Data governance is another critical challenge. In many countries, geospatial data is not managed in a centralized way, leading to fragmentation in data collection and usage efforts. Additionally, there are challenges related to data ownership and accessibility, limiting their use for policy decision-making. Open data access and transparency in geospatial data management are essential for making geospatial indicators useful and effective in planning and formulating public policies.

Opportunities To Overcome The Challenges

Despite these challenges, there are multiple opportunities to improve the production and use of geospatial indicators for the SDGs. Emerging technologies, such as AI, predictive analysis, and geospatial big data, are helping improve the quality and precision of geospatial indicators. AI and machine learning can process large volumes of data efficiently, allowing the detection of patterns and trends that would not be visible with traditional methods. Furthermore, the use of more accessible remote sensors and satellites reduces costs and increases the accuracy of available data, opening new possibilities for countries with limited resources.

International cooperation is key to overcoming technological inequalities between countries. International agreements can promote the transfer of technology and technical training in developing countries, improving their capacity to collect and use geospatial data. Initiatives like the Global Geospatial Information Management (UN-GGIM) and the GSGF are working on standardizing data collection and analysis methods, which facilitates integration and improves the quality of geospatial indicators. Investment in infrastructure for geospatial data collection and analysis is a key opportunity to improve the accuracy and effectiveness of indicators. Creating national geospatial data infrastructures (NSDI) will allow countries to better manage spatial data, promoting interoperability and coherence between different data sources. These infrastructures can facilitate coordination between government agencies, the private sector, and international organizations, improving data collection and its integration into SDG monitoring systems. Establishing international standards for geospatial data collection and analysis is crucial to ensure the comparability and compatibility of indicators at the global level. Initiatives like the GSGF are working to standardize data collection methods and create frameworks that allow for the effective integration of geospatial data with other data sources, such as national statistics and population censuses. Opening up geospatial data is one of the most promising opportunities. The use of open data platforms and crowdsourcing allows local communities and international organizations to contribute to real-time SDG monitoring. Initiatives such as OpenStreetMap (OSM) and UN mappers enable citizens to collect geospatial data to improve the accuracy of local indicators and support the implementation of public policies at the community level.

Examples of Geospatial Information Use in Calculating SDG 11 Indicators

SDG 11 aims to make cities and human settlements inclusive, safe, resilient, and sustainable. To measure progress toward this goal, geospatial indicators are used, which rely on precise and detailed geospatial data. Below are some examples of how geospatial information is specifically used to calculate SDG 11 indicators and how this data contributes to sustainable urban planning.

Indicator 11.1.1: Proportion of Urban Population Living in Informal Settlements, Slums, or Marginalized Neighborhoods

One of the main urban challenges is the expansion of informal settlements. In many cities, unplanned areas with precarious living conditions, known as slums or marginalized neighborhoods, grow rapidly due to rural-to-urban migration and lack of access to adequate housing. Geospatial information is essential to measure this indicator, as it allows for mapping the location of informal settlements in relation to formal urban areas. By integrating geospatial data from urban censuses and remote sensors, it is possible to identify high-risk areas where informal settlements are growing, as well as risk factors such as lack of access to basic services such as drinking water, sanitation, electricity or vulnerability to natural disasters. Furthermore, by using GIS platforms and satellite data, governments can monitor the expansion of informal settlements over time, which helps forecast and manage urban growth, planning public policies that encourage affordable housing development and inclusive urban planning. Figure 8.2 illustrates the phases of the process applied in Colombia to measure indicator 11.1.1 (2018), including the integration of census data

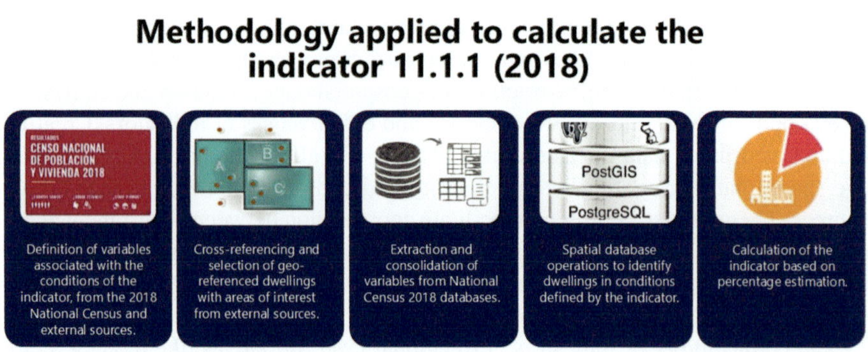

FIGURE 8.2 Methodological process applied in Colombia for the calculation of SDG indicator 11.1.1 (2018)

Source: DANE, 2024.

with geospatial analysis and external sources for estimating the proportion of dwellings in adequate conditions.

From the spatial distribution of the values obtained in each of the cities of Colombia, it can be observed, in Figure 8.3, that the highest percentages of the indicator are concentrated in cities located in the Caribbean and Pacific regions. In contrast, cities located in the interior of the country show the lowest percentages.

Based on the calculations made for the selected cities, the national value for indicator 11.1.1 shows that 11.4% of the urban population lives in inadequate housing, informal settlements, or slums. This spatial variation highlights the territorial inequality in access to adequate housing conditions and underscores the need for differentiated public policies that address the specific challenges faced by cities in coastal regions. The indicator provides crucial evidence for prioritizing investment in housing programs, infrastructure, and services in areas where housing vulnerability is most acute.

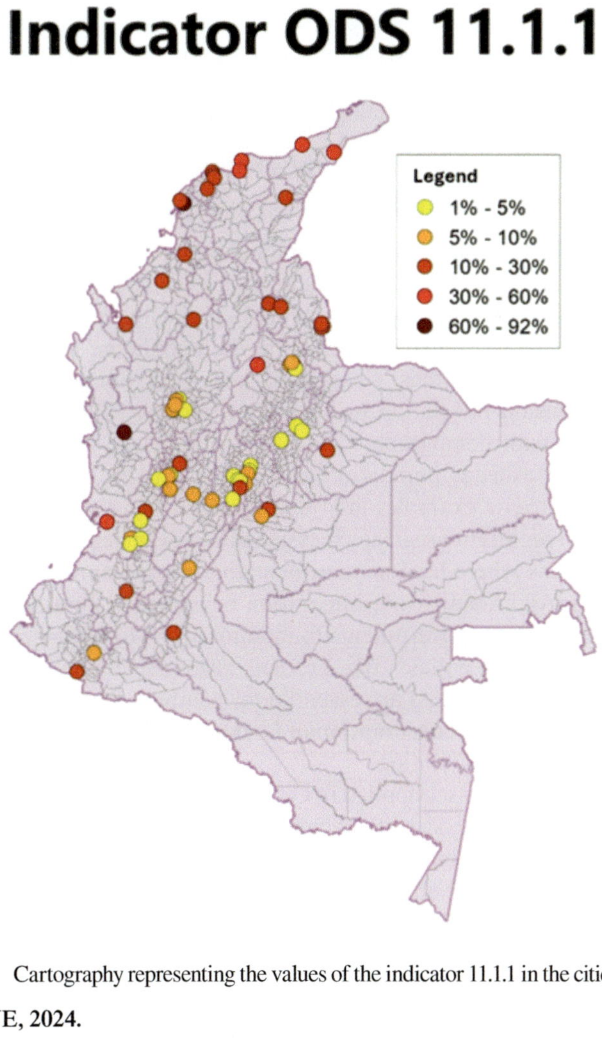

FIGURE 8.3 Cartography representing the values of the indicator 11.1.1 in the cities of Colombia.

Source: DANE, 2024.

Indicator 11.2.1: Proportion of Urban Population with Access to Public Transportation

Access to public transportation is a key factor for the sustainability of cities, as it helps reduce congestion, pollution, and socioeconomic disparity. This indicator measures the proportion of the population with access to efficient public transportation systems, especially in densely populated urban areas. Geospatial data allows for mapping the distribution of transportation routes and stations in relation to residential areas. By combining this information with data on population density and demographics, it is possible to identify areas of the city with limited access to public transportation and design strategies for expanding transportation infrastructure. For example, accessibility maps generated using GIS technologies can identify critical points where public transportation is insufficient, allowing for the planning of new routes or the improvement of existing infrastructure.

Furthermore, the use of real-time monitoring technologies, such as GPS sensors on transportation vehicles, enables governments to collect continuous data on the efficiency of routes and user satisfaction, helping to optimize routes and improve public transportation services.

Indicator 11.3.1: Ratio of Land Consumption Rate to Population Growth Rate

The ratio of land consumption rate to population growth rate is a crucial indicator for assessing urban sustainability, as it helps to evaluate the balance between urban expansion and population growth. This indicator measures the extent to which land is being consumed in relation to the population increase, especially in rapidly growing urban areas. Geospatial data allows for mapping urban land use and its expansion in relation to population growth patterns. By combining this information with demographic data, it is possible to identify areas where land consumption exceeds population growth, potentially indicating unsustainable urban sprawl. For example, land consumption maps generated using GIS technologies can highlight areas of excessive urban expansion, enabling the identification of regions where land management and urban planning strategies need to be improved. Additionally, the integration of real-time data on land use and population trends allows governments to monitor urban development more effectively, helping to design policies that promote sustainable land management and urban planning. These insights are essential for planning more efficient and resilient cities, preventing overconsumption of land while ensuring that urban growth meets the needs of the population.

The methodological phases applied in Colombia to calculate indicator 11.3.1 integrate satellite imagery analysis with demographic projections to determine the relationship between land consumption and population growth.

Below, in Figure 8.4, there are two images showing the indicator values for the cities of Medellín and Baranoa in Colombia, in the period between the years 2018 and 2023.

As shown in the previous figure, the comparison between Medellín and Baranoa in Colombia reveals significant differences in the land consumption to population growth

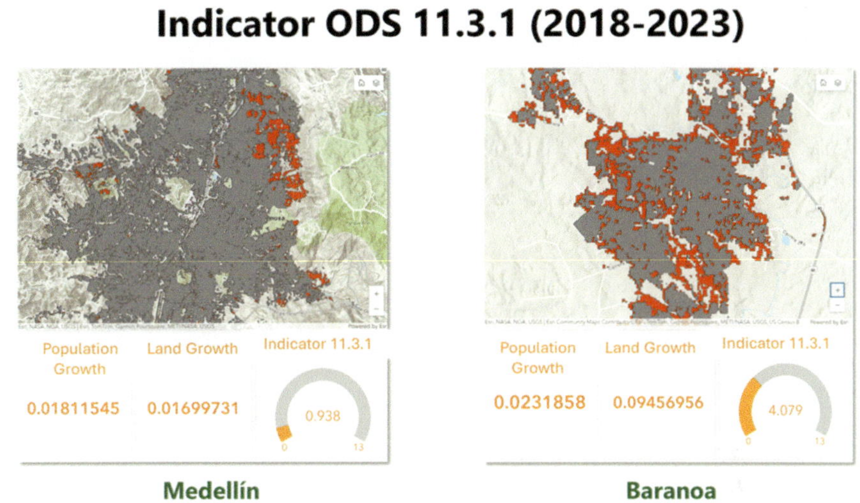

FIGURE 8.4 Land consumption and population growth in Medellín and Baranoa for indicator 11.3.1 (2018 -2023)

Source: DANE, 2024

ratio (indicator 11.3.1). Medellín presents a land consumption rate (0.0169) slightly below its population growth rate (0.0181), resulting in an indicator value of 0.938. This suggests that the city is expanding at a rate that is relatively aligned with its population growth, indicating a more controlled and sustainable urban development pattern.

In contrast, Baranoa shows a land consumption rate (0.0945) that far exceeds its population growth rate (0.0231), yielding a much higher indicator value of 4.079. This disparity suggests that urban expansion in Baranoa is disproportionate to its population increase, potentially reflecting inefficient land use or urban sprawl. Such comparative analysis using geospatial data allows decision-makers to identify where land management policies may need to be strengthened to ensure more balanced and sustainable urban growth. Following the example presented in the previous figure, which displays the calculation of the indicator at the city level, the following Figure 8.5 illustrates its application in Uruguay at the national scale.

At the national level, Uruguay presents a land consumption to population growth ratio of 1.8, indicating that land is being consumed at nearly twice the rate of population growth. This reflects a territorial expansion of urban centers that exceeds demographic growth and therefore points to increasing inefficiency in land use. When interpreting these results, it is important to consider that, although Uruguay has experienced almost no population growth in recent years, there is an ongoing shift in household composition. This trend directly impacts the demand for land and contributes to the observed increase in land consumption. At the departmental level, the data reveal heterogeneous patterns across the country. Florida, Montevideo, and Rocha stand out as the departments with the greatest disparity between land consumption and population growth, highlighting the need for differentiated planning and land management strategies.

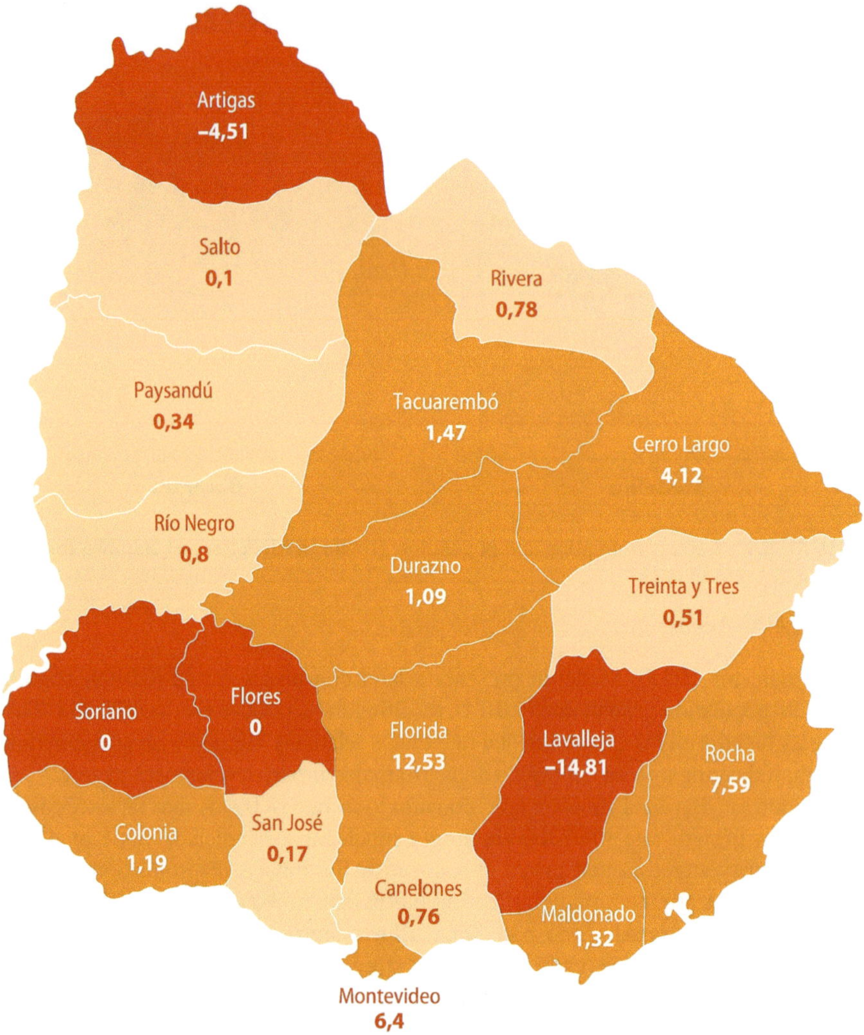

	<1	Urban densification: land consumption rate is lower than the population growth rate
	>1	Urban expansion: land consumption rate is higher than the population growth rate.
	≤ 0	Urban explansion with population decline.

FIGURE 8.5 Ratio between land consumption rate and population growth rate by state in Uruguay, between the years 2011 and 2015.

Source: National Direction of Urban Planning, 2018.

Indicator 11.6.2: Proportion of Urban Areas that Monitor and Manage Air Quality

Air quality in urban areas is one of the main environmental issues affecting the health of inhabitants. This indicator measures the proportion of urban areas that have air quality monitoring systems and management policies to reduce air pollution. Geospatial data provided by air quality sensors and satellites allow for real-time monitoring of pollutant concentrations. These geospatial data are integrated with statistical information on public health to identify critical areas where pollution is higher, enabling the implementation of local policies to reduce pollution and protect the health of residents. Air pollution maps generated by remote sensors and Earth observation technologies are valuable tools for governments to carry out effective interventions, such as limiting vehicle emissions, improving public transportation, and promoting the use of clean energy.

Indicator 11.7.1: Proportion of the Population with Access to Safe and High-Quality Public Spaces

Access to safe public spaces is fundamental for urban quality of life, especially for vulnerable groups such as women, children, and the elderly. This indicator measures the proportion of the population with access to high-quality public spaces, such as parks, squares, and recreational areas. The use of geospatial information to map the distribution of public spaces in cities is essential for identifying areas lacking access to recreational spaces and green zones. By integrating this information with data on population density and demographics, urban planners can propose solutions to improve access to these spaces, promoting social inclusion and urban well-being. Moreover, geospatial coverage analysis using sensors and drones allows for monitoring the quality of public spaces and security in high-traffic areas, helping implement public safety measures and improve city infrastructure to make them more accessible and livable. Figure 8.6 illustrates the methodological

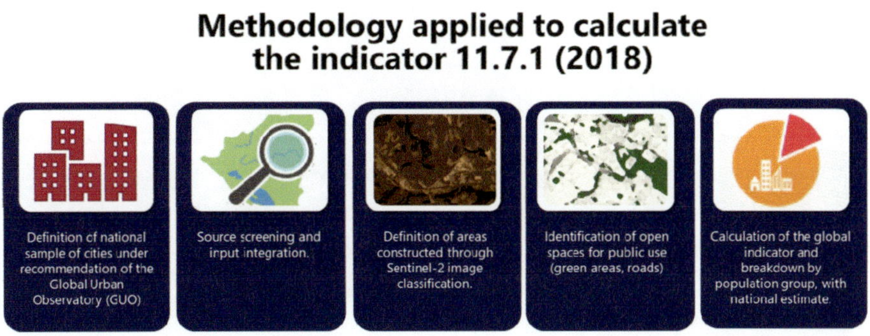

FIGURE 8.6 Methodological process applied in Colombia for the calculation of SDG indicator 11.7.1 (2018).

Source: DANE, 2028.

process applied in Colombia to calculate indicator 11.7.1 in 2018. The procedure includes the definition of a national sample of cities, integration of data sources, classification of satellite imagery, and identification of open spaces for public use, concluding with the calculation of the indicator and its disaggregation by population group.

The comparison of Medellín and Cartagena in Colombia regarding indicator 11.7.1 reveals significant differences in access to public open spaces. Medellín registers a higher value, with 42.1% of its built-up area dedicated to public open spaces, while Cartagena reports only 23.7%. When disaggregated by population groups, Medellín also demonstrates greater inclusiveness, with over 88% of women, children under 14, and persons with disabilities having access to public spaces. In contrast, Cartagena shows substantially lower accessibility: 60.5% for women, 57.4% for children under 14, and 64.1% for persons with disabilities. These differences highlight the importance of spatial and social equity in urban planning. The data suggest that while Medellín has made significant progress in ensuring inclusive access to public spaces, Cartagena faces important challenges that require targeted policies and investment to improve spatial justice and urban livability.

As shown in Figure 8.7, Uruguay demonstrates a generally strong performance in indicator 11.7.1, with a national average of 36% of the built-up area dedicated to public open spaces. Most departments report values at or above this average,

State	Proportion of open spaces to built-up area
Artigas	34%
Canelones	38%
Cerro Largo	38%
Colonia	34%
Durazno	38%
Flores	38%
Florida	35%
Lavalleja	27%
Maldonado	35%
Montevideo	34%
Paysandú	35%
Río Negro	38%
Rivera	32%
Rocha	37%
Salto	32%
San José	40%
Soriano	37%
Tacuarembó	38%
Treinta y Tres	39%
Total del país	**36%**

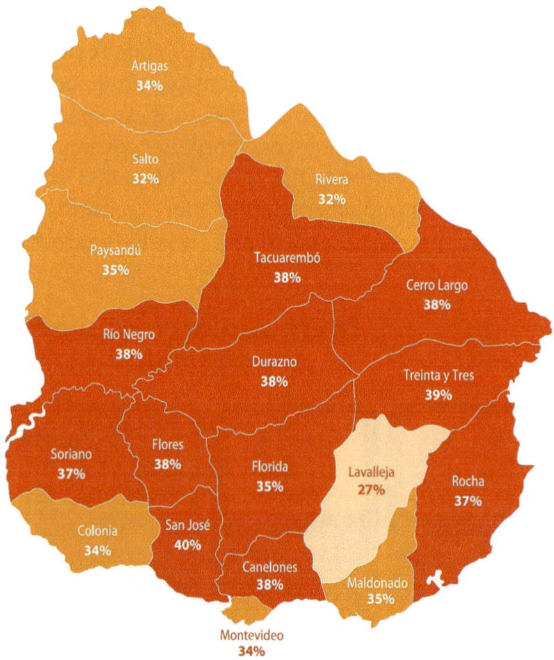

FIGURE 8.7 Values of indicator 11.7.1 per each state in Uruguay, year 2015.

Source: National Direction of Urban Planning, 2018.

highlighting a favorable national landscape in terms of urban planning for accessibility and livability.

Departments such as San José (40%), Treinta y Tres (39%), Cerro Largo, Durazno, Flores, and others with 38%, stand out for their high proportions of public space, suggesting robust infrastructure supporting social interaction, mobility, and recreation.

On the other hand, Lavalleja (27%) presents the lowest proportion, which may indicate limited availability or accessibility of public open spaces and suggests the need for targeted interventions to improve urban quality and equity. Departments such as Rivera, Salto, and Montevideo also fall below the national average, pointing to opportunities for urban redevelopment or policy adjustments to enhance public space provision.

Overall, this distribution reinforces Uruguay's commitment to ensuring inclusive and accessible cities. However, continued monitoring, along with qualitative assessments of the functionality, distribution, and accessibility of these spaces, especially for vulnerable groups, will be essential to advance toward the broader goals of SDG 11.7.1.

CASE STUDIES ON THE USE OF GEOSPATIAL DATA IN SDG 11

Case Study One: New York City—Transformation Toward a Smart and Sustainable City

New York City has been a pioneer in the use of geospatial technologies to manage its urban growth and promote sustainability. Through a series of smart city initiatives, the city has integrated geospatial data into urban planning, public transportation, waste management, and air quality. One of the most notable examples is the use of remote sensors and GIS technologies to measure and manage air quality in high-traffic areas. Thanks to geospatial data, high-pollution areas have been identified, and sustainable transportation policies, such as promoting the use of electric vehicles and expanding public transportation, have been implemented. Additionally, accessibility maps generated through GIS technologies have helped identify areas in the city where access to public spaces and green zones is limited. This has led to the creation of new urban parks and the rehabilitation of public spaces in vulnerable neighborhoods. The combination of geospatial data with public health statistics has enabled New York City to make more informed decisions to improve the quality of life for its residents, significantly advancing toward a more inclusive and sustainable SDG 11

Case Study Two: Medellín, Colombia—Sustainable Urban Growth Management

Medellín, one of Colombia's major cities, has undergone a remarkable transformation in the last two decades, and part of its success lies in the innovative use of geospatial technologies to manage urban growth and improve the quality of life for its residents.

The city has used geospatial data to improve access to adequate housing and public transportation. In Indicator 11.1.1 (proportion of the population living in informal settlements), Medellín has significantly reduced the number of informal settlements thanks to geospatial data-based urban planning. The city implemented a public housing system that has helped regulate urban expansion and provide affordable housing in high-demand areas. On the other hand, the use of GIS technologies for public transportation has been key in creating efficient transportation systems and improving connectivity throughout the city. The Metrocable, a cable car system connecting the most impoverished areas with the city center, was planned and optimized using geospatial information to ensure it reached the neediest areas, reducing congestion and CO_2 emissions. Additionally, air quality sensors and the use of remote sensing to monitor air pollution have enabled Medellín to implement policies to reduce pollution levels and promote the use of electric vehicles and sustainable public transportation. This comprehensive approach has helped the city make significant progress in creating a more resilient and sustainable urban environment, in line with SDG 11.

Case Study Three: Rural Areas in Sub-Saharan Africa - Monitoring and Management of Drinking Water

In many rural areas of Sub-Saharan Africa, access to drinking water and adequate sanitation is an ongoing challenge. However, the use of geospatial technologies is helping to overcome some of these challenges by monitoring water resources and managing water sustainably. Geospatial data and remote sensing are being used to map the distribution of water sources and assess water infrastructure in rural areas. In countries such as Ethiopia, Uganda, and Kenya, governments and NGOs are using GIS platforms to identify areas with water scarcity and prioritize the construction of wells and rainwater harvesting systems. Geospatial information also helps identify areas vulnerable to droughts or floods, allowing for more efficient management of water resources and long-term planning to improve access to drinking water. Additionally, geospatial data is integrated with statistical data on population and demographics, facilitating the distribution of resources based on community needs. This approach to geospatial resource management not only improves access to drinking water in rural areas but also contributes to building more resilient communities against natural disasters and climate change, helping to achieve the goals of SDG 6 and SDG 11.

REFLECTIONS ON CASE STUDIES

The case studies presented highlight how the use of geospatial information has transformed urban management and sustainable planning in different parts of the world. From New York and Medellín to rural areas of Sub-Saharan Africa, geospatial data is essential for addressing some of the major urban and rural challenges faced by cities and communities on their path toward sustainability.

These examples demonstrate how the integration of geospatial technologies not only facilitates the monitoring of SDG 11 indicators but also offers practical solutions to improve urban quality of life, promote access to basic services, and reduce the impacts of climate change. The ability to visualize and analyze spatial data in real-time allows governments and organizations to implement more effective and balanced policies aligned with SDGs.

CONCLUSION

The use of geospatial information is essential for achieving the SDGs, particularly SDG 11, which aims to create sustainable cities and communities. Throughout this chapter, we have shown how the integration of geospatial and statistical data enables informed decision-making, improving urban planning, natural resource management, and the provision of public services in cities. However, despite progress in some regions, significant challenges persist that must be addressed to ensure these geospatial data are used appropriately and effectively.

One of the crucial points emerging from the discussion about geospatial information is the urgent need for governments to become aware of the importance of reliable georeferenced data. This data is not just a technical resource; they are essential source of information for sound decision-making and the formulation of appropriate public policies. When governments lack accurate and up-to-date data on resource distribution, service accessibility, and environmental challenges, effective planning and implementation of sustainable development strategies become difficult.

However, in many developing countries, there is little awareness of the relevance of these geospatial data and their value for improving urban and rural governance. This is reflected in the lack of resource allocation for the capture, processing, and analysis of these data. Investment in geospatial infrastructure and capacity development such as training of specialized professionals, is still insufficient in many regions, resulting in significant gaps in the availability of reliable geospatial data.

In other cases, although geospatial data already exists, it is inaccessible due to interoperability issues, restrictive regulations, or lack of accessible platforms. Privacy, data ownership, and technological barriers can hinder access to information, preventing governments and organizations from using these data to improve citizens' quality of life. This reinforces the need to create national strategies to improve geospatial data governance, fostering open access and interoperability between databases and systems.

The underuse or mismanagement of geospatial and statistical data can, in part, be attributed to the lack of trained professionals in these areas. Training in geospatial technologies and spatial statistics remains limited in many countries, creating a knowledge gap that prevents governments and institutions from using these tools effectively. It is essential that global education systems prioritize training in these areas, not only to close the skills gap but also to empower future generations of decision-makers at national, regional, and local levels. Furthermore, in some cases, the lack of awareness of the value of geospatial data is also due to fragmented efforts within governments. Without a coordinated approach and a comprehensive strategy for collecting and using geospatial data, different government departments may work in isolation, reducing the efficiency and effectiveness of public interventions.

In summary, the proper management of geospatial data is a key factor in achieving the SDGs, particularly SDG 11. To move toward sustainable and resilient cities, it is essential that governments recognize the value of this data and allocate the necessary resources for its capture, processing, and analysis. International collaboration, standardization of data collection methods, and fostering the training of skilled professionals are fundamental steps to bridge the data gap and ensure that all countries, regardless of their level of development, have access to reliable information that enables efficient and sustainable management of their cities and communities.

BIBLIOGRAPHY

Global Taskforce of Local and Regional Governments, UNDP, & UN-Habitat. (2016). *Roadmap for localizing the SDGs: Implementation and monitoring at subnational level.* https://www.uclg.org/sites/default/files/roadmap_localizing_sdgs_0.pdf

Global Partnership for Sustainable Development Data (GPSDD). (2022). *Data Interoperability Guidebook: Connecting Data to Improve Lives.* https://www.data4sdgs.org

Jackson, E. D. (2020). *The role of geospatial information and effective partnerships in the implementation of the International Agenda for Sustainable Development*: https://aura.antioch.edu/etds/578/ (Doctoral dissertation, Antioch University).

Kraak, M.-J., Roth, R. E., Ricker, B., Kagawa, A., & Le Sourd, G. (2020). *Mapping for a sustainable world.* United Nations & International Cartographic Association. https://digitallibrary.un.org/record/3898826

National Administrative Department of Statistics (DANE). (2024, November). *Geospatial information to support the monitoring of the Sustainable Development Goals* [PowerPoint presentation]. 5th United Nations World Data Forum, Medellín, Colombia.

OECD. (2023). *Smart Cities and Inclusive Growth: Building on the Outcomes of the OECD Programme on Smart Cities and Inclusive Growth.* https://www.oecd.org/

Office of Planning and Budget (OPP), & National Directorate of Land Planning (DINOT). (2018). *Voluntary National Review: Uruguay 2018 – Sustainable Development Goal 11.* Government of Uruguay. https://hlpf.un.org/sites/default/files/vnrs/2018/Uruguay_VNR2018.pdf

Scott, G., & Rajabifard, A. (2019). SDGs roadmap. In *Sustainable Development Goals Connectivity Dilemma* (pp. 3–11). CRC Press.

UN-Habitat. (2023). *World Cities Report 2022: Envisioning the Future of Cities.* United Nations Human Settlements Programme. https://unhabitat.org/wcr/

United Nations Committee of Experts on Global Geospatial Information Management. (2021). *The SDGs Geospatial Roadmap.* United Nations. https://ggim.un.org/

United Nations Committee of Experts on Global Geospatial Information Management. (2024). *The Future Geospatial Information Ecosystem* (E/C.20/2024/8/Add.1). United Nations Economic and Social Council. https://ggim.un.org/meetings/GGIM-committee/14th-Session/documents/E_C20_2024_8_Add_1_e.pdf

United Nations Geospatial Network. (2021). *Geospatial in action: Data and insights for the Sustainable Development Goals.* United Nations. https://ggim.un.org/

United Nations Statistical Commission. (2024). *Report of the Inter-Agency and Expert Group on Sustainable Development Goal Indicators* (E/CN.3/2025/6). United Nations Economic and Social Council. https://unstats.un.org/sdgs/iaeg-sdgs/

World Bank. (2023). *Data for Better Lives: World Development Report 2021 Update.* https://www.worldbank.org/en/publication/wdr2021

Part 3

Climate, Food Consumption, and Environmental Sustainability

9 SDG #12—Responsible Consumption and Production

Spatially Enabled Assessment of Resource Consumption, Waste Generation, and Production Patterns

Cecilia A. Wandiga, Shem O. Wandiga, and Nashon J. Adero

INTRODUCTION

The United Nations Sustainable Development Goal (SDG#12): Responsible Consumption and Production aims to create a world where resources are utilized efficiently, waste is minimized, and industrial production aligns with environmental sustainability. Achieving this goal requires data-driven decision-making and the ability to monitor resource flows, assess environmental impacts, and design effective policies. Geographic information systems (GIS) offer powerful tools for tracking resource consumption, waste generation, and production patterns, providing policymakers and industries with valuable insights to promote sustainability.

The above Figure 9.1 is a generalized framework for linking workers, indicator variables, pollution data, and tasks pertaining to ensuring sustainable consumption and production of Artisanal small-scale mining (ASM) products in cities and communities. One sector where these issues are particularly evident is mining, a critical economic activity that has driven industrial development and economic growth in many countries, including Kenya. Mining provides essential raw materials for various industries, from construction to technology, and is a significant source of

DOI: 10.1201/9781003586319-12

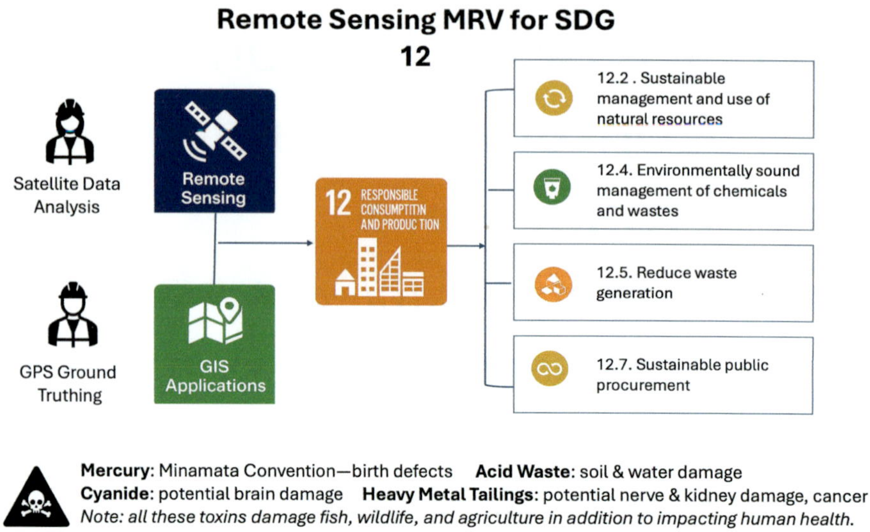

FIGURE 9.1 SDG 12—a generalized framework

employment. However, if not managed sustainably, mining can lead to severe environmental degradation, resource depletion, and health hazards for local communities. A key concern in mining operations is waste generation, which includes mine tailings, chemical runoff, and heavy metal contamination. Among these pollutants, mercury contamination from Artisanal and Small-scale Gold Mining (ASGM) is a pressing issue, as mercury is commonly used to extract gold but often leaches into nearby water sources and soils, endangering both ecosystems and human health. Sustainable mining practices, therefore, play a crucial role in balancing economic benefits with environmental protection. GIS technology helps address these challenges by enabling precise mapping of mining sites, monitoring changes in land use, and assessing the spread of pollutants over time. By integrating satellite imagery/remote sensing, on-the-ground sensor data, and predictive models, GIS can provide valuable insights into how mining waste is generated, where it accumulates, and how it impacts surrounding communities. GIS and remote sensing integration play a crucial role in Measurement, Reporting, and Verification (MRV), as seen in Figure 9.2. This information supports decision-makers in implementing policies that promote responsible resource extraction, encourage the adoption of cleaner technologies, and mitigate the harmful effects of unsustainable mining practices.

This chapter will explore how geospatial data can be applied to resource consumption, waste generation, and production inefficiencies in Kenya, using mercury contamination from artisanal mining as a case study. By leveraging GIS, stakeholders, including policymakers, researchers, and industry leaders, can develop strategies that support responsible mining, safeguard natural ecosystems, and align industrial production with the principles of sustainable development.

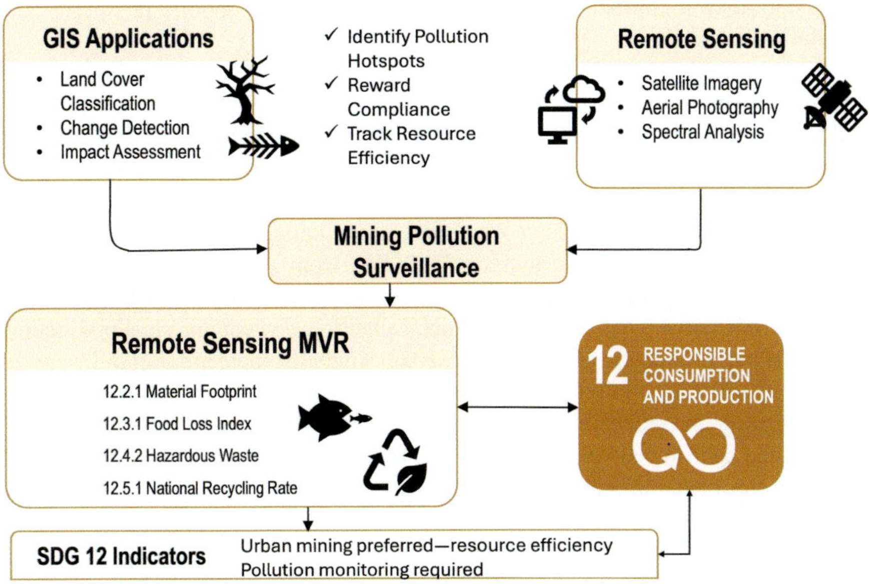

FIGURE 9.2 SDG 12—mining surveillance geospatial data linkages

UNDERSTANDING THE CONTEXT: ARTISANAL MINING, MERCURY CONTAMINATION, AND FRESHWATER SYSTEMS

ASM in Kenya is a significant economic activity, providing livelihoods for thousands but posing severe environmental and health risks. In 2022, ASM accounted for over half of Kenya's mining output with an ASM economic value of $224 million (Wario, 2024). Similar trends exist across Africa with a key characteristic being the linkages between ASM and farming activities (Hilson et al., 2025). Mercury, commonly used in gold extraction, contaminates freshwater systems, soil, and crops irrigated with polluted water (Kohio et al., 2024; Mulenga et al., 2024). This directly threatens SDG Target 12.4, which aims for the environmentally sound management of chemicals and waste throughout their life cycle. By enabling precise mapping of mining sites, monitoring changes in land use, and assessing the spread of pollutants over time, GIS technology offers a transformative approach to managing ASM's environmental and health risks. Unlike traditional regulatory efforts that often rely on limited field inspections and reactive measures, GIS enables proactive interventions by identifying high-risk zones, tracking deforestation linked to mining activities, and modeling the movement of contaminants in soil and water systems. In Kenya, this means local governments, environmental agencies, and mining cooperatives could use real-time geospatial data to enforce sustainable mining policies, promote safer extraction techniques, and designate protective buffer zones to safeguard agriculture and freshwater sources. The application of GIS in mining governance has already shown success in other regions, where spatial analysis has guided land reclamation efforts and optimized resource use, proving that a data-driven approach to mining can lead to both economic prosperity and environmental resilience.

Mercury, used as an affordable chemical to extract gold, poses a serious risk to human and ecological health. Mercury causes birth defects, miscarriages, cancer, and a chronic syndrome known as Minamata disease because it was identified in Japan. These health risks occur from direct exposure to mercury or from eating contaminated fish. Mercury contamination therefore intersects critically with SDG 2 (Zero Hunger) and SDG 6 (Clean Water and Sanitation). It is important to note that while this chapter focuses on mercury pollution from ASM, there are numerous sources of Mercury pollution including landfill waste (Mercury filament light bulbs and electronic parts), coal combustion, metal smelting, cement and steel production, chlorine alkali production (Teng & Altaf, 2022). Globally, there are over 500 years of accumulated mercury pollution from human activities (Streets et al., 2019). Polluted freshwater systems reduce agricultural productivity, harming food security and livelihoods, and contribute to water scarcity and quality degradation. In addition to SDG 6, Freshwater contamination also directly impacts and relates to SDG 2 indicators as follows, including:

SDG 2 Indicators:

1. **Indicator 2.4.1:** Proportion of agricultural area under productive and sustainable agriculture. Mercury-contaminated irrigation water hinders sustainable practices.
2. **Indicator 2.1.1:** Prevalence of undernourishment. Contaminated water and crops exacerbate food insecurity.
3. **Indicator 2.3.2:** Average income of small-scale food producers, by sex and indigenous status. Farmers near contaminated sites face declining yields and incomes.

SDG 6 Indicators:

1. **Indicator 6.1.1:** Proportion of population using safely managed drinking water services. Mercury contamination in water systems undermines safe drinking water availability.
2. **Indicator 6.3.1:** Proportion of domestic and industrial wastewater safely treated.
3. ASM Mercury discharge into water systems highlights gaps in wastewater treatment and pollution control.
4. **Indicator 6.3.2:** Proportion of bodies of water with good ambient water quality. Mercury contamination directly reduces water quality, affecting ecological and human health.
5. **Indicator 6.6.1:** Change in the extent of water-related ecosystems over time. Mercury pollution contributes to the degradation of freshwater ecosystems.

These indicators are also closely linked with SDG 12 targets and indicators:

1. **Indicator 12.a.1:** Installed renewable energy-generating capacity in developing countries (in watts per capita). GIS-enabled research on Mercury pollution aligns with this indicator by advancing sustainable practices for

renewable energy use in mining and its impact on the availability of clean energy to communities surrounding mining sites.

2. **Indicator 12.b.1:** Implementation of standard accounting tools to monitor the economic and environmental aspects of tourism sustainability. While typically applied to tourism, this indicator highlights the role of geospatial tools in tracking environmental impacts of economic activities such as ASM and craft tourism sales of metallic ornaments.
3. **Indicator 12.6.1:** Number of mining companies publishing sustainability reports. Geospatial data can provide a foundation for mining and agricultural companies to report on their environmental impact. Most importantly, independent studies are needed to benchmark the veracity of company-reported data.
4. **Indicator 12.7.1:** Number of countries implementing sustainable public procurement policies and action plans. Policies integrating GIS-based data ensure that mining procurement processes address environmental sustainability.

GIS plays a pivotal role in tracking mercury pollution by enabling the integration, visualization, and analysis of spatial data from diverse sources. Reports on ASM mining activities by location can be integrated with satellite data, field measurements, and environmental models to compare company report data by pinpointing areas with high mercury concentrations. GIS visualizations can communicate complex data in an accessible way, fostering public awareness and stakeholder collaboration. Policymakers and researchers can use GIS-based maps to prioritize interventions, allocate resources efficiently, and design targeted strategies to mitigate pollution.

The critical importance of toxic mineral and mining pollution surveillance under Indicators 6.6.1: Change in the extent of water-related ecosystems over time and 12.6.1: Number of mining companies publishing sustainability reports) is poignantly present when disasters occur due to lack of early warning systems. On February 18, 2025, the tailings dam in Kitwe Zambia, collapsed releasing 50 million liters of copper mining acid waste, heavy metals, and dissolved solids in the Kafue River thereby affecting approximately 12 million people (Kille & Zimba, 2025). The spillage floated at least 100 km downstream with locals complaining of dead fish (Kille & Zimba, 2025). Trace leakages are likely to have been evident in the Kafue River before the tailings dam collapsed. Satellite time series surveillance data has been effectively used to monitor artisanal and small-scale mining sites in South-Western Ghana (Forkuor et al., 2020). Assuming surveillance was active, an early warning system could have been triggered to repair the tailings dam on time, as seen in Figure. 9.3.

Figure 9.3 above also shows change in the extent of water-related ecosystems over time and such as pollution from a tailings dam collapse, which should be reported under Indicator 12.6.1: Number of mining companies publishing sustainability reports. Human workers are shown entering field observations into tablets linked to an online analytical portal that can capture data in real-time or when the tablets are returned to the office.

FIGURE 9.3 Geospatial real-time data reporting under indicators 6.6.1. Adapted from Figure 9.1 in Forkuor, Ullmann & and Griesbeck (2020) under a Creative Commons License. The photos are generated by NightCafé Studio Flux AI Algorithm (shared as Creative Commons Photos 1 and 2).

REGULAR GIS VS. HYPERSPECTRAL IMAGING

While traditional GIS provides robust tools for spatial analysis and data management, hyperspectral imaging adds a layer of specificity by capturing detailed spectral information from the environment. The key differences include:

Data Collection:

- Regular GIS relies on multispectral satellite imagery, drone data, and ground-based measurements, which are often limited to visible or infrared bands.
- Hyperspectral imaging captures hundreds of narrow spectral bands across the electromagnetic spectrum, enabling the identification of specific substances like Mercury.

Pollution Detection:

- Regular GIS maps land-use changes, water flows, and general environmental indicators.
- Hyperspectral imaging, preferred for granular aspects of materials science, mineral exploration, and environmental chemistry, helps to detect unique spectral signatures of mercury and other contaminants, making it invaluable for pinpointing contamination sources.

Cost and Complexity:

- Regular GIS tools are more widely accessible and cost-effective, for routine wide-area environmental monitoring, while hyperspectral imaging requires specialized sensors and analytical expertise

- For detailed environmental and natural resource studies targeting focused and in-depth scientific investigations.

Combining the strengths of regular GIS and hyperspectral imaging allows for comprehensive tracking and management of mercury pollution, enhancing both the precision and breadth of environmental monitoring.

EXAMPLES OF KENYAN GEOSPATIAL MAPPING PROJECTS TRACKING ASM MERCURY

Kenya has implemented several notable geospatial mapping projects to monitor Mercury contamination from ASM activities. These projects employ advanced techniques and ground-truthing methodologies to enhance data accuracy and reliability:

1. **Historical Baseline Data on Mercury contamination in the Lake Victoria Basin** (Campbell et al., 2003a; Campbell et al., 2003b)**:**

   This case study shows the multi-faceted aspects of collecting data for SDG 2.4.1: Agricultural production area and Indicator 6.3.2: Proportion of water bodies with good water quality. Mercury toxicity accumulates over time; hence, one-time random samples of fish, agricultural crops, or water are not enough to understand the accumulation of mercury in the human body. The case study exemplifies the current state of data collection as limited to static geospatial information that exists in separate databases but is not yet integrated across studies and locations. It is important to note that the data was collected by academic researchers, meaning ASM groups and companies have not begun to submit sustainability reports on Indicator 12.6.1. High total mercury (THg) content in hair samples is an important test for Mercury contamination because it provides a long-term, cumulative measure of exposure, unlike water testing, which only captures a snapshot of contamination at a single point in time. This case study showcases the use of hair sample data in mercury pollution monitoring but lacks adequate geospatial tagging (Campbell et al. 2003). Table 9.1 has latitude/longitude coordinates but no mention of the GIS source. Campbell et al. (2003) do not have GIS data but do have evidence of high THg content hair samples. Collecting hair samples is simple, non-invasive, and cost-effective compared to repeated water testing, making it ideal for large-scale health studies. Mercury accumulates in the human body over weeks to months, and hair, which grows at a relatively constant rate (~1 cm per month), provides a timeline of exposure. Mercury, particularly methylmercury, bio-accumulates in organisms and biomagnifies up the food chain, meaning people may ingest high doses even if water concentrations seem low. Most importantly, hair analysis can differentiate between methylmercury (organic Mercury) and inorganic mercury. This distinction is critical in regions with artisanal mining, where workers may be exposed to inorganic Mercury vapor during gold extraction, while the general population may be consuming methylmercury-contaminated fish.

Techniques Used:

- Remote sensing tools to link latitude and longitude coordinates to samples collected.

Ground Truthing:

- Soil and water samples were collected from suspected contamination hotspots to confirm the presence of Mercury.
- Water chemistry analysis using EPA Method 1631 (U.S.E.P.A. 1999).

Even though hair collection sampling is non-invasive, ethical approval and consent are required for human subject testing. Care must be taken to ensure no physical or ethical damage is inflicted on participants.

2. **Mercury Levels in Groundwater near Artisanal Small-Scale Gold Mines in Migori County, Kenya (Adero, 2025; Omondi et al. 2023):**

Two recent studies illustrate the relevance of repeat independent testing for confirmatory analysis of reporting under SDG 6.3.2: Bodies of water with good ambient water quality, SDG: 6.6.1: Ecosystem changes over time and SDG 12.6.1: Companies publishing sustainability reports. Migori county has three water bodies: Lake Victoria, Kuja River, and Migori River. All three are affected by Mercury from artisanal mining activities. These independent studies provide a good baseline against which to assess sustainability reports from mining companies in the area. Over the period of October to December 2024, a comprehensive study across the expansive Migori County, an active gold mining district in Kenya, interviewed 424 artisanal and small-scale gold miners (ASGM), collected and tested 28 samples of water, soil, tailings, and fish for Mercury contamination, and further engaged 57 key informants and diverse stakeholders in a consultative forum. The main aim was to establish the spatial patterns and levels of Mercury-related public health and environmental risks in the gold mining business with an emphasis on the exposure risks to the miners, especially women (Adero, 2025). The study applied GIS technology to deliver on the geospatial precision needed for parametric policy decision support. One salient outcome of the analysis was that the findings helped to recommend a buffer zone of 100 m and 400 m from the gold mining hotspots, over which the government should ban the extraction of water for domestic use, and limit agricultural activities, respectively. Indicator 2.4.1: Agricultural area under sustainable production. This is because the mines are point sources of Mercury contamination, which was confirmed from the test results showing mercury levels exceeded the safe WHO limits by up to 100 times. Figure 9.4 shows how GIS was applied to sample the point sources of mercury contamination and measure their proximity to water resources and other critical ecosystem components.

Spatial data in Figure 9.4 shows GIS application in geographical sampling of Mercury contamination and proximity analysis for gold mines and exposed ecosystems in Migori, Kenya (Adero, 2025). Regular site sampling is more effective when combined with longitudinal data collection.

FIGURE 9.4 Indicator 6.3.2—Proportion of bodies of water with good ambient water quality

GIS equipment can be set up for real-time reporting over time. The study Adero (2025) was preceded by a 2023 study focused on determining if mercury contamination concentrations are higher during drought seasons, that is, concentration in water and soil is less diluted because there is less rain. Pertaining to Indicator 6.3.2: Bodies of water with good ambient water quality; groundwater samples were analyzed from 15 boreholes within 6 km of five Migori mines: Masara, Osiri Matanda, Macalder, Kitere, and Kehancha (Omondi et al., 2023). Additionally, at each mine site groundwater was collected during wet vs dry seasons from 20 pan-pond water and soil tailing samples. Although wet season samples were within Kenya Bureau of Standards (KEBS) safety limits, dry season samples showed borehole groundwater Mercury levels higher than the recommended KEBS limit of 0.001 mg/L. Similar trends were found in pan-pond water and soil tailing samples. Dry season mercury levels were above the National Environment Management Authority (NEMA) effluents discharge standards of 0.05 mg/L (pan-pond water) and 0.05 mg/Kg (soil tailings). The findings confirm drought conditions correlate with increased concentration of mercury because there is less water dilution. Dry season readings are important for tracking Indicator 6.6.1: Ecosystem changes over time because fish are still consumed when there is drought and higher concentrations of Mercury. Human hair samples during wet and dry seasons can be used to determine the accumulation of Mercury within the human body. ASM workers are also likely to be exposed to Mercury emissions in the air, which also build up in human hair. Although the World Health Organization (WHO) standards set a safety level of 45 mcg/day of daily exposure to mercury, the 2013 Minamata Convention established there is no safe level of mercury exposure (UNEP, 2017; Mayo Clinic Laboratories, 2024). Hyperspectral imaging can be used to capture Mercury levels in air, soil and water across different seasons—Indicator 6.6.1: Ecosystem changes over time.

Techniques Used:

- GIS mapping identified artisanal mining sites based on topographical changes and human activity footprints.
- Spectral analysis from hyperspectral remote sensing distinguished mercury-contaminated areas.

Ground-Truthing:

- Field surveys involved collaboration with local communities to validate mining site locations and contamination data.
- Biomonitoring techniques, including analysis of mercury levels in fish and crops, helped assess downstream impacts.

3. **The Taita Taveta Mining Environmental Monitoring Initiative**

   Indicator 12.7.1 tracks the number of countries implementing sustainable public procurement policies and action plans. Sub-national reporting is aggregated into national reporting tallies. Frameworks are essential tools for understanding what is tracked. The Taita Taveta Sustainable Mining Action Plan was finalized in January 2021 as a collaborative effort aimed at addressing the environmental impacts of mining activities in Taita Taveta County, Kenya (Muthuri et al., 2021). This initiative brings together local communities, government agencies, and academic institutions to monitor and mitigate the adverse effects of both large-scale and artisanal mining operations. Multi-stakeholder input was gathered between August 2019

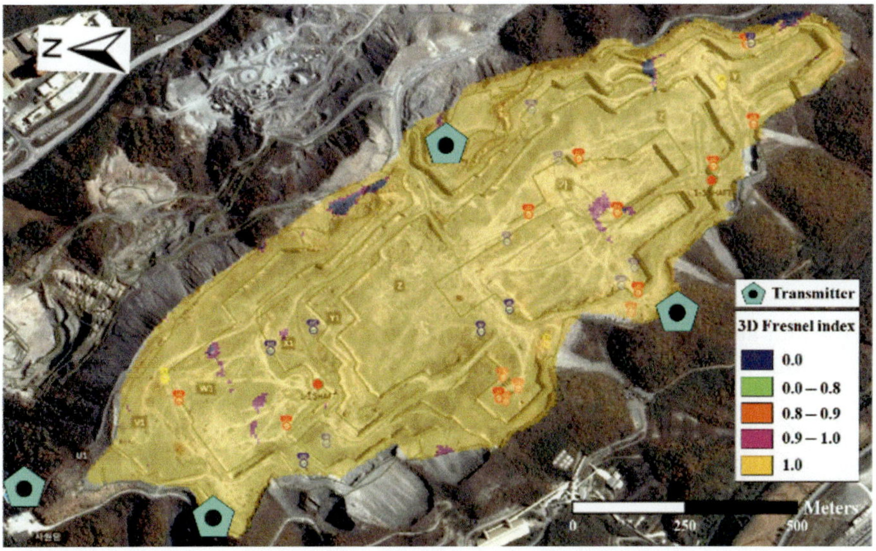

FIGURE 9.5 Indicator 6.6.1—Open-pit mine transmitter locations (Baek & Choi, 2018, Figure 9.14)

and December 2020. The plan delineates targets for each of the 17 SDGs making GIS a potential solution for creating a progress dashboard seen in Figure 9.1. Input layers for selecting candidate transmitter locations in the open pit mine: (a) pit area; (b) road; (c) slope gradient; and (d) vegetation seen in Figure 9.5 below.

Aligned with best practice are SDG 12 targets in the Taita Taveta Sustainable Mining Action Plan highlight the need to track hazardous chemical processing waste, pollutants and land degradation resulting from ASM mineral processing. An example of a geospatial mining dashboard platform is the ESRI ArcGIS Storymap on GIS-Based Mining Operations (Berberian, 2022). Open-pit mines are typical in ASM operations.

4. **Open Pit Mining and Tailings Surveillance Opportunities**

Indicator 6.6.1 tracks data on the changes in the extent of water-related ecosystems over time. The tailings dam collapse that massively impacted the Kafue River basin in Zambia highlights the fact that ecosystem changes can happen instantaneously. For reporting Indicator 6.6.1: Ecosystem changes over time, adding transmitters around an open pit mine makes it possible to track emissions, climate, slope gradients and vegetation (Baek & Choi, 2018). GIS surveillance systems can be used as early warning systems to prevent such disasters. Communication viewshed analysis and topographic data were used in to calculate the 3D Fresnel Index for open pit mines in Samcheok-si, Gangwon-do, South Korea, shown in Figure. 9.5. A 3D Fresnel Index is a GIS calculation measuring the line of sight between communication transmitters in complex terrain (Baek & Choi, 2018). The same transmitters can be deployed around a tailings dam to detect leaks and risks of collapse if one or more transmitters fall below the line of sight, that is an indication that a portion of the pit wall is starting to collapse. If filled with acid waste, changes in vegetation can also be detected. Figure 9.6 depicts hyperspectral satellite images for surveillance of water pipelines near open pit mines on the Salar de Atacama salt flat in Chile (Maus et al., 2020 Figure 9.2, Creative Commons License), global tracking of open pit and ASM locations (Tang & Werner, 2023 Figure 9.1, Creative Commons License), landside direction of phyllite materials from a collapsed mine wall at the Yanqianshan iron mine in China (Kolapo et al., 2022, Figure 9.12, Creative Commons License), and Spatiotemporal variation of gold ASM areas near the Ofin River in Ghana (Tang & Werner, 2023, Figure 9.5, Image A. Image © 2023 CNES/Airbus and Maxar Technologies, visualized in Google Earth Pro. Creative Commons License).

Locations with a high probability of acid waste pit mining water leakages flowing downstream can be tracked using data like these input layers for selecting candidate transmitter locations in the open-pit mine in Samcheok-si, Gangwon-do, South Korea: (a) pit area; (b) road; (c) slope gradient; and (d) vegetation. (Baek & Choi, 2018, Figure 9.6. Creative Commons License). This South Korean case study exemplifies the use of sensors linked to GIS surveillance systems and the real-time detection of ground movement that can trigger emergency crews to install reinforcements before a catastrophe occurs.

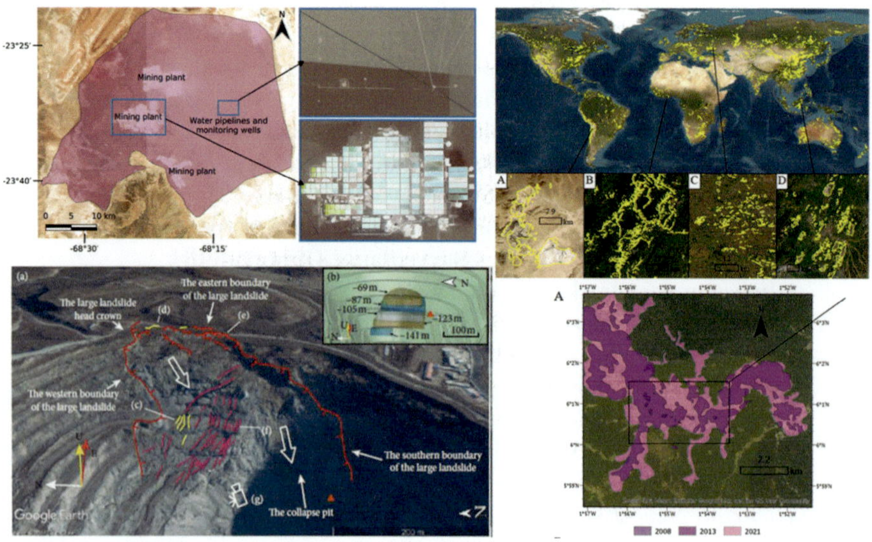

FIGURE 9.6 Indicator 6.1.1—Multiple perspectives on the use of Hyperspectral Satellite-2 data (clockwise left to right: Maus et al., 2020, Figure 9.2; Tang & Werner, 2023, Figure 9.1; Kolapo et al., 2022, Figure 9.12; Tang & Werner, 2023, Figure 9.5)

Techniques Used:

- Synthetic Aperture Radar (SAR) was applied to detect ground disturbances and waterbody changes linked to mining activities.
- 3D Fresnel Index to calculate changes between ground-level communication transmitters.
- Time-series analysis tracked environmental degradation over months.

Ground Truthing:

- Water sampling along key rivers and reservoirs identified Mercury levels.
- Interviews with artisanal miners provided insights into mercury usage patterns and disposal practices.

At a larger scale, satellite spectral data linked to a support vector machine classifier can be calibrated to detect the ecological impacts of ASM sites on the surrounding communities, Indicator 6.6.1: Ecosystem changes over time. In southeastern Senegal, Ngom et al. (2020) combined Google Earth images with Sentinel 2 data to track the following:

- Normalized difference vegetation index (NDVI)
- Soil moisture
- Normalized difference water index (NDWI)
- Surface material brightness index (BI)
- Surface material color index (CI)
- Surface material refractive index (RI)
- Occurrence of bare ground using the normalized difference built-up index (NDBI) and bare soil index (BSI)
- Normalized built-up area index (NBAI)
- Band ratio for built-up areas (BRBA)

SDG reporting by mining companies requires an evaluation of sustainability impacts. These indices such as NDVI, NDWI, BI, and NDBI can be used to assess changes in agricultural productive areas surrounding ASM sites per Indicator 2.4.1, which changes in water-related ecosystems over time and Indicator: 6.6.1, and the effectiveness of reported sustainable mining action plans in Indicator 12.7.1 over time. The charts in Figure 9.7 showcases two combined indices from the study, the spectral evolution over time for BI, NDVI, NDWI, and NDBI in the Bantakokouta village adjacent to the ASM in southeastern Senegal. The circled part on each chart corresponds to the period from April to June 2017. Such analysis is useful to establish correlations between the start and stop of ASM activities and changes in surrounding land use.

The photos in Figure 9.7 combine—Sentinel Level 2-A imagery from ASM site and surrounding village conditions at Bantakokouta village between February 2017 and 2018. Black arrows point to the location of the pictures in the Sentinel level-2A image. (Ngum et al., 2020, Figures 9.5 & 9.11. Creative Commons License). Photographs are needed for ground truthing of the satellite data and are an integral part of good reporting practices.

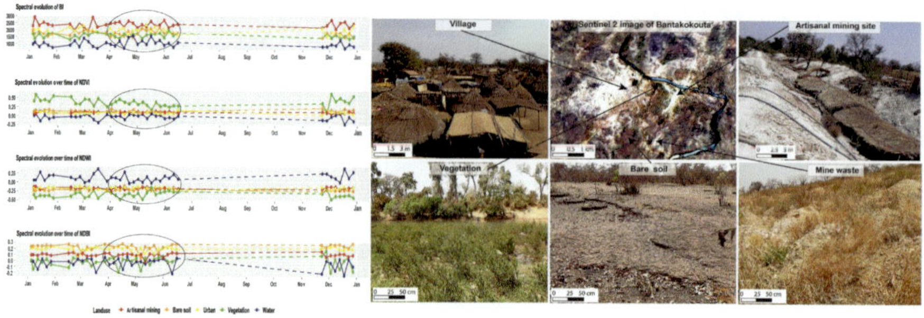

FIGURE 9.7 Showcasing hyperspectral indicator 12.6.1 for NDBI changes over time

GAPS IN KENYA'S DATA COLLECTION COMPARED TO GLOBAL STANDARDS

The case studies in this chapter highlight the extent to which water bodies carry and spread chemical contamination from mining sites. Indicator 12.b.1 calls for tools to monitor the impact of sustainable tourism. This is highly relevant in the case of Migori county, which has several tourism sites including natural landscapes, wildlife sanctuaries, and historical sites. Key attractions include Ibencho Hills Sanctuary, Gogo Falls, and the Thimlich Ohinga archaeological site. Additionally, there are various lodges and resorts like Treat House Resort, Migori Hillview Hotel, and Migori Country Lodge. All these sites are exposed to air, water and soil contamination from mining. The same is true for Taita Taveta county, which houses the Taita Hills Wildlife Sanctuary, Tsavo East National Park, and the Voi Commonwealth War Graves. Additionally, the county boasts attractions like Mzima Springs in Tsavo West National Park, Ngangao Forest, and Lake Chala, a transboundary lake share with Tanzania. Kenya's existing data collection systems for monitoring Mercury pollution face notable gaps when compared to global platforms like the Environmental Defense Fund (2025) (EDF) and Chemical Exposure Action Map. Although the map does not track mercury, it does track methylene chloride, which is used as a solvent in some industrial-scale gold extraction processes, particularly in lab-based gold recovery methods and certain chemical separation techniques, as seen in Figure 9.7. Methylene chloride is also sometimes used in analytical chemistry when testing gold ore samples. If an ASM site is located near industrial waste sites, chemical storage facilities, or illegal gold refining operations, methylene chloride could be present in water sources or air emissions. When inhaled in confined spaces, methylene chloride can cause dizziness, nausea, liver and kidney damage, and is classified as a potential carcinogen.

The search results for facilities releasing methylene chloride are shown in Figure 9.8. The search zoomed into a facility near a water body and similar analysis can be done to identify the risks of industrial facilities near ASM sites and water. These images do not represent the position, statements or endorsement of the EDF.

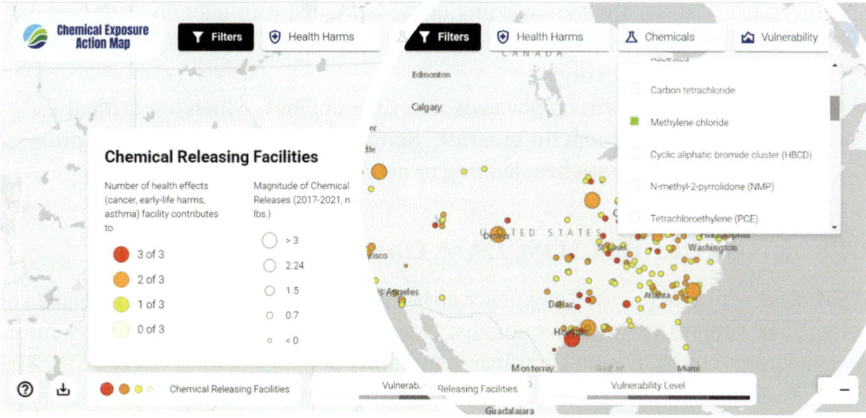

FIGURE 9.8 Indicator 12.b.1 potential chemical risks can affect tourism—Environmental Defense Fund (Accessed February 17, 2025)

A more common pollutant found in ASM is cyanide, which is used as an alternative to Mercury. When cyanide reacts with chlorine in water treatment processes or under high-temperature conditions, it releases cyanogen $(CN)_2$ and cyanogen chloride (CNCl) metabolites. These metabolites can form secondary pollutants in air and water, where the effect is amplified through the formation of other cyanide byproducts. Cyanide is toxic, and both cyanogen $(CN)_2$ and cyanogen chloride (CNCl) metabolites cause severe respiratory and neurological effects if inhaled.

Compared to the EDF's Chemical Action Map, ASM GIS mapping gaps in Kenya include:

1. **Granularity and Accessibility of Data:**
   - The EDF's Chemical Action Map offers high-resolution, detailed data on chemical hotspots globally, whereas Kenya's systems lack similar granularity. Many datasets are not updated frequently or are inaccessible to stakeholders.
2. **Integration Across Sectors:**
   - Kenya's geospatial data systems are fragmented, with limited collaboration among mining, agriculture, health, and environmental sectors. This fragmentation hampers holistic analysis and response.
3. **Pollutant Coverage:**
   - While Kenya's efforts focus on Mercury, broader chemical tracking—including pesticides, industrial effluents, and other hazardous substances—remains insufficient. The EDF platform's multi-chemical tracking approach could serve as a model.
4. **Visualization and Public Engagement:**
   - EDF's platform excels in providing intuitive visualizations that are accessible to non-technical users, fostering public awareness and

participation. Kenyan systems often lack such user-friendly interfaces, limiting engagement.

5. **Real-Time Monitoring:**
   - The EDF platform emphasizes real-time updates, which are critical for timely interventions. In contrast, Kenyan data collection systems often rely on periodic studies, leading to delays in addressing pollution.

OPPORTUNITIES TO CLOSE DATA GAPS

ASM plays a critical role in Kenya's economy, providing livelihoods for thousands of miners and supporting local economies. However, the sector remains largely informal and unregulated, posing significant environmental, health, and safety risks. The widespread use of Mercury and cyanide in gold extraction, coupled with inadequate waste management practices, threatens water sources, agricultural land, and the well-being of mining communities. Addressing these challenges requires accurate, real-time data to monitor pollution, land use, and mining activities, which is where Geospatial Information Systems (GIS) mapping becomes an essential tool. GIS technology can significantly improve the safety and quality of ASM operations by enabling spatial tracking of pollutant dispersion, land degradation, and health risks associated with mercury and other hazardous chemicals.

To enhance its geospatial capabilities, Kenya could:

1. **Adopt Open Data Standards:**
   - Make geospatial data more accessible and shareable across sectors.
2. **Leverage International Partnerships:**
   - Collaborate with organizations like the EDF to adopt the best practices in chemical monitoring and data visualization.
3. **Invest in Training and Infrastructure:**
   - Build technical capacity and invest in state-of-the-art geospatial tools, including hyperspectral imaging.
4. **Establish a Centralized Platform:**
   - Develop a national platform like the EDF's Chemical Action Map to integrate, analyze, and disseminate data on pollutants.
5. **Encourage Multi-Stakeholder Engagement:**
   - Engage communities, industry players, and policymakers through accessible tools and clear visualizations to drive collective action.

By addressing these gaps, Kenya and countries like Zambia can enhance its ability to track and manage mercury and other pollutants, advancing progress toward SDG 12 and related goals.

COSTS AND FINANCING GAPS IN ADDRESSING MERCURY CONTAMINATION

Through satellite imagery, remote sensing, and on-the-ground data collection, GIS provides actionable insights that policymakers, industry leaders, and local communities can use to promote safer and more sustainable mining practices. For instance, mapping high-risk zones for Mercury contamination can help prioritize intervention efforts, while real-time monitoring of deforestation and soil erosion near mining sites can inform environmental restoration strategies. To maximize these benefits and close existing data gaps, Kenya must strengthen its geospatial capabilities by adopting open data standards, leveraging international partnerships, and investing in technical training and infrastructure. A centralized GIS-based platform could integrate data from multiple sources, enabling authorities to track pollutant exposure levels, enforce regulations, and promote cleaner technologies in ASM. The average costs associated with geospatial monitoring and mitigation of Mercury and related chemical processing waste from ASM gold extraction contamination in Kenya include:

1. **Geospatial Data Acquisition and Analysis:**
   - Satellite imagery subscriptions (e.g., Landsat, Sentinel) cost approximately $5,000 to $10,000 annually for high-resolution data.
   - Unmanned Aerial System (UAS).
   - surveys (e.g., using sensors aboard drones) for localized assessments range from $2,000 to $5,000 per survey area.
   - Software and GIS tools, including training, can cost $15,000 to $20,000 annually.
2. **Ground Truthing and Field Studies:**
   - Soil and water sampling for mercury analysis costs $50 to $100 per sample, with a typical study requiring hundreds of samples.
   - Portable XRF devices cost approximately $25,000 each, with maintenance adding annual costs of $5,000.
3. **Community Engagement and Capacity Building:**
   - Workshops and training sessions for local communities and officials cost $10,000 to $20,000 per program.

FINANCING GAPS:

Effective environmental monitoring and pollution mitigation in ASM require sustained financial investment in geospatial tools, chemical analysis, and enforcement mechanisms. However, Kenya faces significant financing gaps that hinder the adoption of advanced monitoring technologies and sustainable mining practices. Public sector funding remains insufficient, leaving key agencies without the necessary resources to deploy real-time data collection and GIS-based tracking systems. At the same time, private sector participation is limited, as mining and agricultural companies often lack financial incentives to invest in pollution control measures. While international grants from organizations like UNEP and the Global Environment Facility (GEF) exist, challenges in accessing, managing, and scaling these funds persist, preventing Kenya from fully leveraging global financing opportunities. Additionally, the absence of blended finance models, which combine public, private, and philanthropic capital further restricts the development of integrated solutions

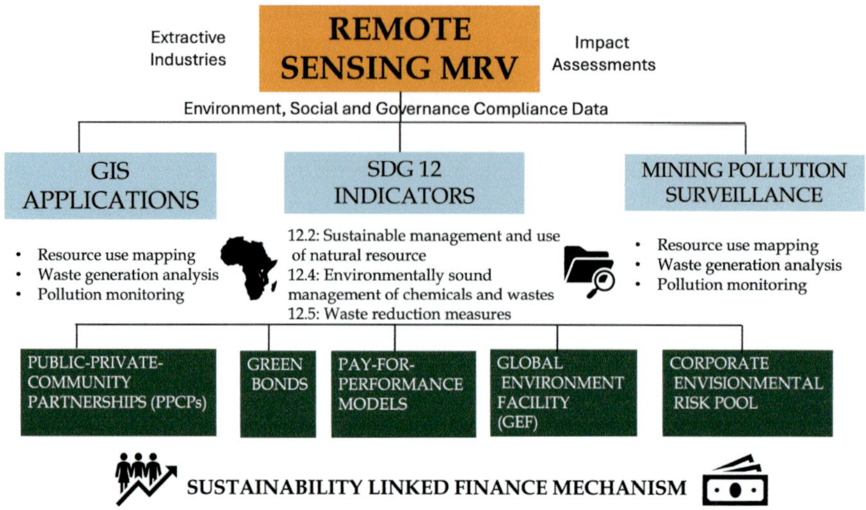

FIGURE 9.9 SDG 12—GIS—Mining surveillance and finance linkages for sustainable reporting by ASM and national planning resource allocation

for sustainable ASM. Remote sensing monitoring reporting and verification (MRV) platforms are a cost-effective solution to only use physical surveillance. Figure 9.9 summarizes these issues into an SDG 12 framework. Addressing these financing challenges is crucial to ensuring that environmental monitoring systems are adequately funded, operationalized, and scaled, ultimately supporting Kenya's commitment to SDG 12 on responsible consumption and production.

RECOMMENDED FINANCIAL MODELS

Figure 9.9 presents a framework for integrating mining pollution surveillance data, SDG 12 indicator tracking statistics, and impact assessment reports. Integrated data facilitates the identification of both the size and magnitude of mining-related chemical pollutants and their corresponding environmental degradation impacts. Financing of mining operations should not be restricted to equipment and operations. Instead, financing should reward improvements in regulatory compliance, governance, as well as public health impacts (social) and biodiversity conservation (environmental), otherwise known as ESG incentives. To address ESG performance, innovative financial models can be adopted:

1. **Public-Private-Community Partnerships (PPCPs):**
   - Establish partnerships between the government, mining companies, and agricultural stakeholders to co-finance geospatial monitoring projects.
   - Example: A PPP to fund a centralized geospatial platform for tracking mercury hotspots and mitigating contamination to affect fisheries and farms.

2. **Green Bonds:**
   - Issue green bonds to finance pollution monitoring and sustainable agricultural practices.
   - Example: A $50 million green bond could fund satellite-based monitoring, field studies, and community capacity building.
3. **Pay-for-Performance Models:**
   - Investors provide upfront capital, with returns linked to the achievement of measurable outcomes, such as reduced Mercury levels in water systems
   - Executive pay should be linked to SDG performance, not just profits.
   - Example: Collaborations with impact investors to fund mercury-free mining technologies and accelerated salary reductions in relation to environmental disasters that occur.
4. **International Grant Programs:**
   - Strengthening access to grants from UNEP, the GEF, and other multilateral organizations.
   - Example: A proposal targeting funding for geospatial tools under the Minamata Convention on Mercury.
5. **Blended Finance:**
   - Combining concessional funding from development banks with private capital to de-risk investments in sustainable technologies.
   - Example: The African Development Bank co-financing Mercury mitigation initiatives with private mining companies.
6. **Corporate Sustainability Funds:**
   - Encourage mining and agricultural companies to allocate portions of their Corporate Social Responsibility (CSR) budgets for geospatial monitoring and sustainable practices. The funds would be allocated into an interest-bearing shared corporate environmental risk pool for the mining sector in a company or region. Large payouts can be linked to catastrophe bonds that direct compensation to government and affected populations.

Establishing financial mechanisms will not be effective if the financial mechanisms are not directly linked to independently verified MRV platforms. Each of the SDG 12 indicators can be tracked using Key Performance Metrics (KPIs) that are common to the green and impact finance sector, as seen in Figure 9.10.

OPPORTUNITIES FOR IMPROVEMENT

System integration does not occur simply because GIS surveillance data is linked to impact assessments and financial incentives. There is a need to ensure data collected is real-time, accurate, and received by the authorities able to enforce penalties or safeguards. Regulatory action to curtail the slow accumulation of Mercury pollution in the Lake Victoria Basin is very different from the high-speed and large cost of regulatory action needed to increase safety after an acid waste spillage from a tailings dam affecting the Kafue River in Zambia.

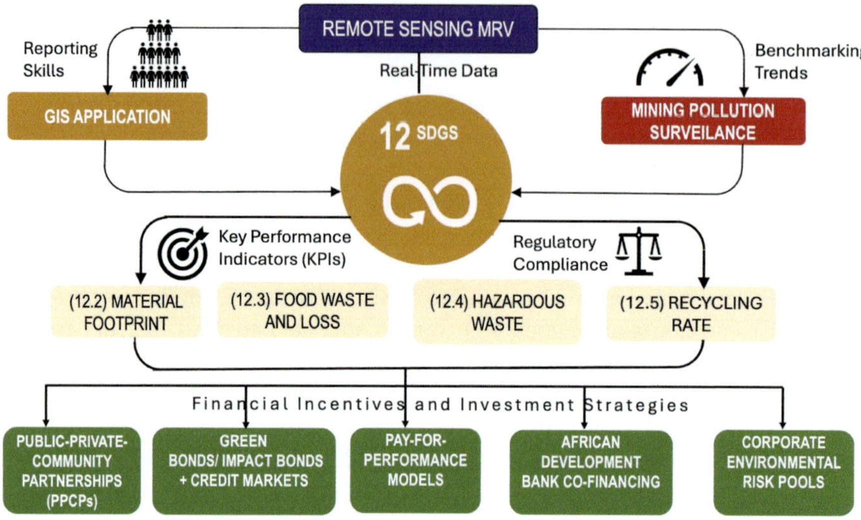

FIGURE 9.10 SDG 12—GIS—Mining surveillance and finance key performance indicators (KPIs) for transboundary coordination and planning of sustainable ASM production and consumption.

Lake Victoria pollution monitoring is a transboundary regulatory intervention. Zambia's Kafue River is not transboundary, but the resources needed might exceed national capacity to deal with the crisis. Data on similar accidents by the same company and the same type of tailing dam collapse helps to accelerate damage assessments as well as fines that should be assessed. By establishing a unified system of national frameworks in Target 12.1, transboundary assessments and regulatory coordination become easier and this is particularly true with units of measurement. If one country measures pollution exposure using affected hectares while another country simply reports number of sites with a risk, there is no mechanism for financial evaluation of relevant priorities or rewards for compliance.

To bridge these framework integration gaps, Kenya can adopt the following strategies:

1. **Satellite-Based Monitoring:**
    - High-resolution satellite imagery can reduce the costs and logistical challenges of data collection, enabling continuous monitoring of Mercury contamination hotspots.
    - Open-access platforms like Copernicus and Landsat provide valuable datasets for environmental monitoring.
    - And data standardization.
2. **Centralized Data Platforms:**
    - Develop a centralized, publicly accessible geospatial platform similar to the EDF's Chemical Action Map that can facilitate

multi-stakeholder collaboration and real-time monitoring of chemical and waste hotspots.

3. **Capacity Building:**
   - Develop training programs for local government officials and researchers can enhance the technical capacity to analyze and interpret geospatial data.
   - Collaboration with universities and international experts can foster knowledge exchange and innovation.
   - And on measurements and risks.
4. **Policy Integration:**
   - Embed geospatial data into 10-year policy frameworks that ensure evidence-based decision-making, addressing Target 12.1 by creating actionable policies and plans to ensure ASM compliance with sustainable consumption and production.
   - The indicator states: Implement the 10-Year Framework of Program on Sustainable Consumption and Production Patterns, all countries taking action, with developed countries taking the lead, considering the development and capabilities of developing countries
   - Integrate frameworks to facilitate benchmarking between countries regarding mining resource efficiency versus reductions in risks to human and environmental health.

Both MRV data collection and green financing mechanisms require enabling regulatory and policy frameworks that are aligned with local GIS surveillance skills capacity. In other words, the missing gaps or challenges in data collection and analysis also drive the determination of skills training needed for effective implementation.

EMPOWERING POLICYMAKERS WITH GEOSPATIAL DATA

In summary, to accelerate progress toward SDG 12 and its intersections with SDG 2, government policymakers must embrace geospatial data as a cornerstone of sustainable development strategies. Key actions include:

1. **Adopting Open Data Policies:** Ensuring geospatial data is accessible to stakeholders, fosters collaboration and innovation.
2. **Investing in Geospatial Infrastructure:** Expanding data collection networks, including satellite ground stations and remote sensing technologies, strengthens national capacity.
3. **Fostering Cross-Sectoral Collaboration:** Aligning efforts across agriculture, mining, water management, and environmental protection ensures holistic solutions to complex challenges.
4. **Establishing Monitoring and Evaluation Systems:** Integrating geospatial data into SDG indicator tracking frameworks ensures accurate, real-time progress assessments.
5. **Implementing Target 12.1:** Developing comprehensive and actionable 10-year national frameworks for sustainable mining using geospatial data to

assess how changes in the consumption and production impacts of mining production simultaneously impact water, fisheries, and agriculture in areas surrounding ASMs. This goes beyond simple resource efficiency benchmarks of the amount of minerals extracted. Geospatial data guides policies, action plans, and international collaboration to address transboundary environmental and resource challenges.

CONCLUSION

The integration of geospatial data into Kenya's efforts to achieve SDG 12 represents a paradigm shift toward evidence-based, sustainable decision-making through ASM sustainable action plans and reports. By addressing resource consumption, waste generation, and production inefficiencies through spatially enabled assessments, policymakers can protect ecosystems, promote public health, and foster economic resilience. The example of mercury contamination from artisanal mining underscores the urgent need for geospatial tools to identify and mitigate environmental risks. With targeted investments in geospatial capacity and innovative financial models, Kenya can lead the way in leveraging technology for responsible consumption and production, contributing to a sustainable future for all, while safeguarding food systems and freshwater resources as part of a broader commitment to SDG 2 (zero hunger by eliminating ASM-related pollution of agricultural production areas and freshwater ecosystems).

REFERENCES

Adero, N. J. (2025). *The toxic toleols, a poisoned ecosystem and the gender toll.* Policy paper Potsdam-Babelsberg: Friedrich Naumann Foundation for Freedom.

Baek, J., & Choi, Y. (2018). A new GIS-based algorithm to support initial transmitter layout design in open-pit mines. *Energies, 11*(11), 3063. https://doi.org/10.3390/en11113063

Berberian, L. (2022, December 2). *GIS-based mining operations: The usage of GIS systems in the sustainable exploration and operation stages of mining.* ESRI Storymaps. Retrieved Feb 16, 2025 from https://storymaps.arcgis.com/stories/7b9ad44bdf2a40ba8e34c7208763f857

Campbell, L. M., Dixon, D. G., & Hecky, R. E. (2003a). A review of mercury in Lake Victoria, East Africa: Implications for human and ecosystem health. *Journal of Toxicology and Environmental Health, Part B, 6*(4), 325–356.

Campbell, L. M., Hecky, R. E., Muggide, R., Dixon, D. G., & Ramlal, P. S. (2003b). Variation and distribution of total mercury in water, sediment and soil from northern Lake Victoria, *East Africa. Biogeochemistry*, 65(2), 195–211.

Environmental Defense Fund. (2025). *Chemical Exposure Action Map.* Accessed Feb 17, 2025. https://chemicalactionmap.edf.org/

Forkuor, G., Ullmann, T., & Griesbeck, M. (2020). Mapping and monitoring small-scale mining activities in Ghana using Sentinel-1 time series (2015–2019). *Remote Sensing, 12*(6), 911. https://doi.org/10.3390/rs12060911

Hilson, G., Laing, T., Hilson, A., Arnall, A., & Mondlane, S. (2025). How does small-scale mining stabilize rural livelihoods in sub-Saharan Africa? The case of Mozambique. *World Development, 185*, 106761. https://doi.org/10.1016/J.WORLDDEV.2024.106761

Kille, R., & Zimba, J. (2025, March 15). A river 'died' overnight in Zambia after an acidic waste spill at a Chinese-owned mine. *AP News*. https://apnews.com/article/mining-pollution-china-zambia-environment-93ee91d1156471aaf9a7ebd6f51333c1

Kohio, E. N. B., Karoui, H., Sossou, S. K., & Yacouba, H. (2024). Review of pollution trends and impacts in artisanal and small-scale gold mining in sub-Saharan Africa: Advancing towards sustainable practices through equitable redistribution of gold spin-offs. *Journal of Cleaner Production*, *476*, 143754. https://doi.org/10.1016/J.JCLEPRO.2024.143754

Kolapo, P., Oniyide, G. O., Said, K. O., Lawal, A. I., Onifade, M., & Munemo, P. (2022). An overview of slope failure in mining operations. *Mining 2022*(2), 350–384. https://doi.org/10.3390/MINING2020019

Maus, V., Giljum, S., Gutschlhofer, J., da Silva, D. M., Probst, M., Gass, S. L. B., Luckeneder, S., Lieber, M., & McCallum, I. (2020). A global-scale data set of mining areas. *Scientific Data*, *7*(1), 1–13. https://doi.org/10.1038/S41597-020-00624-w

Mayo Clinic Laboratories. (2024). *Mercury, Blood | Neurology Catalog*. https://neurology.testcatalog.org/show/HG

Mulenga, M., Ouma, K. O., Monde, C., & Syampungani, S. (2024). Aquatic mercury pollution from artisanal and small-scale gold mining in sub-Saharan Africa: Status, impacts, and interventions. *Water*, *16*(5), 756. https://doi.org/10.3390/W16050756

Muthuri, J. N., Gul, M., & Mwagandi, S. M. (2021). *Taita Taveta Sustainable Mining Action Plan*. https://www.researchgate.net/publication/371905174_Taita_Taveta_Sustainable_Mining_Action_Plan

Ngum, N. M., Mbaye, M., Baratoux, D., Baratoux, L., Catry, T., Dessay, N., Faye, G., Sow, H., & Delaitre, E.. (2020). Mapping artisanal and small-scale gold mining in Senegal using Sentinel 2 data. *Advancing Earth and Space Sciences – GeoHealth*, *4*(12). https://doi.org/10.1029/2020GH000310

Omondi, G. Z. O., Nyamari, J., & Mugo, J. (2023). Mercury levels in groundwater near artisanal small-scale gold mines in Migori County, Kenya. *East African Journal of Environment and Natural Resources*, *6*(1), 421–431. https://doi.org/10.37284/EAJENR.6.1.1543

Streets, D. G., Horowitz, H. M., Lu, Z., Levin, L., Thackray, C. P., & Sunderland, E. M. (2019). Five hundred years of anthropogenic mercury: Spatial and temporal release profiles. *Environmental Research Letters*, *14*(8). https://doi.org/10.1088/1748-9326/AB281F

Tang, L., & Werner, T. T. (2023). Global mining footprint mapped from high-resolution satellite imagery. *Communications Earth and Environment*, *4*(1), 1–12. https://doi.org/10.1038/S43247-023-00805-6

Teng, H., & Altaf, A. R. (2022). Elemental mercury (Hg0) emission, hazards, and control: A brief review. *Journal of Hazardous Materials Advances*, *5*, 100049. https://doi.org/10.1016/J.HAZADV.2022.100049

UNEP. (2017). *Minamata Convention on Mercury at a glance | Minamata Convention on Mercury*. https://minamataconvention.org/en/resources/minamata-convention-mercury-glance

Veiga, M., & Baker, R. (2004). *Protocols for environmental and health assessment of mercury released by artisanal and small-scale gold miners | Global Mercury Project*. https://www.unep.org/globalmercurypartnership/resources/guidance/protocols-environmental-and-health-assessment-mercury-released-artisanal-and

Wario, H. (2024, February 20). Gold and governance provide hope for Kenya's artisanal miners | ISS Africa. *ISS Today*. https://issafrica.org/iss-today/gold-and-governance-provide-hope-for-kenyas-artisanal-miners

10 SDG#13: Global South Climate Change Protection Through GIS Study and Coordinated Sustainable Development Measures

Williams S. Ebhota and Pavel Y. Tabakov

INTRODUCTION

The Intergovernmental Panel on Climate Change (IPCC), in its Fifth Assessment Report (AR5), provides unequivocal evidence that "anthropogenic greenhouse gases (GHG) emissions have increased since the pre-industrial era, driven predominantly by economic development and population growth, and are now at their highest recorded levels" (IPCC, 2014a). This increase has resulted in atmospheric concentrations of carbon dioxide (CO_2), methane (CH_4), and nitrous oxide (N_2O) that are unprecedented in at least the past 800,000 years (EPA, 2023). Today, the term "climate change" is commonly associated with anthropogenic (human-caused) global warming, driven by the increased concentration of GHG, such as carbon dioxide (CO_2), methane (CH_4), and nitrous oxide (N_2O) in the atmosphere due to activities like burning fossil fuels, deforestation, and industrial processes (Yoro & Daramola, 2020). Therefore, climate change refers to long-term alterations in the Earth's climate patterns, primarily due to natural processes and human activities, as depicted by Figure 10.1. It involves significant changes in temperature, precipitation, wind patterns, and other aspects of the climate system over extended periods, decades to millions of years. This occurrence is attributed to GHG resulting from several human socioeconomic activities, such as industrialization, deforestation, urbanization, and agricultural practices. Burning fossil fuels for energy is the main contributor and since the pre-industrial era, human-induced GHG emissions have caused global temperatures to rise by approximately 1.1°C, with profound consequences for the planet.

According to the IPCC, the effects of GHG alongside other anthropogenic influences have been detected across the climate system and are "extremely likely" to be the dominant cause of the observed warming since the mid-20th century

DOI: 10.1201/9781003586319-13

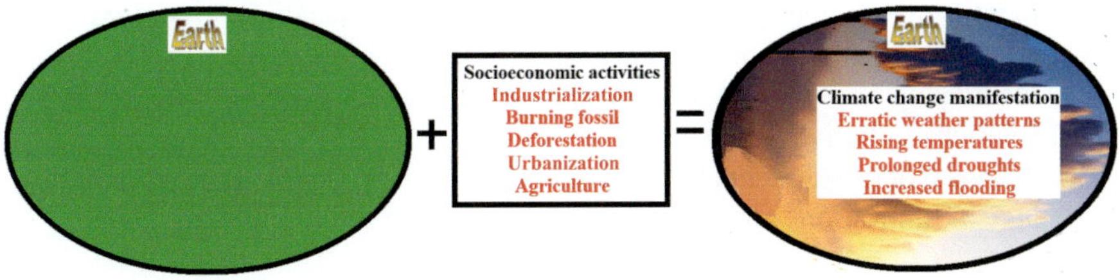

FIGURE 10.1 Schematic of causes and manifestations of climate change.

(IPCC, 2014b). Although a 1°C increase in global mean surface temperature may seem marginal, it is crucial to recognize its profound implications (Robinson et al., 2021). Such warming is sufficient to intensify hydrological cycles, accelerate glacier and polar ice melt, raise sea levels, and increase the frequency and severity of extreme weather events. Climate change poses an existential threat to the Global South (GS), with its severe impacts manifesting through rising temperatures, erratic weather patterns, prolonged droughts, and increased flooding (Mohan, 2025). Furthermore, the spatial distribution of warming is not uniform. Regional temperature anomalies can be significantly higher than the global average, compounding the local impacts on ecosystems and human systems. Data from the Berkeley Earth 2019 global temperature report illustrates these disparities, revealing notable regional variations in temperature increases compared to the 1951–1980 reference period (Allen et al., 2018). These variations underscore the complexity of climate change, where seemingly modest changes in global averages translate into severe localized consequences. These challenges are exacerbated by socio-economic vulnerabilities, limited adaptive capacity, and a lack of resources to mitigate the risks.

Climate change has emerged as the defining crisis of our time, threatening ecosystems, economies, and the well-being of billions. With impacts ranging from extreme weather events to food insecurity, it disproportionately affects vulnerable populations, particularly in the GS. As the IPCC highlights, urgent climate action is necessary to limit global warming to 1.5°C and prevent catastrophic consequences (IPCC, 2021). These challenges require efforts taken at individual, community, national, regional, and global levels to address and reduce the causes of climate change and mitigate its impacts. Therefore, this chapter aims to x-ray the GS climate change protection through Geographic Information Systems (GIS) study and coordinated sustainable development instruments. To achieve the study's goal, therefore, the objectives will include:

1. Assessment of the impact of climate change on socio-economic and environmental systems, particularly in the GS.
2. Evaluation of climate action strategies in addressing climate change impacts, including mitigation and adaptation measures.
3. Examination of the role of GIS in climate change mitigation and adaptation efforts.

4. Identifying of policy frameworks and financial mechanisms required for effective climate resilience.
5. Identifying research and implementation gaps in GIS-driven climate change solutions initiatives in the GS.
6. Proposal of recommendations for enhancing climate action through technological innovations and international cooperation.

The study discusses climate action and identifies implementation tools and GIS as tools that facilitate climate action, which will be at the centre point. Further, it addresses climate action and implementation gaps in the context of the GS.

The remaining sections of this research are structured as follows:

- Reviews the literature on the foundation and application of the GHG Protocol framework.
- Outlines of the methodology for this systematic literature review (SLR).
- Key research topics are examined, data from the selected studies are synthesized, and the findings are presented, finally,
- Conclusion of the study by discussing its strengths and limitations, as well as outlining future research directions.

Methodology: This study employs a SLR approach to analyze existing research on climate action, mitigation, adaptation, and the application of GIS in climate resilience. The methodology involves:

1. **Literature Review:** Examining peer-reviewed articles, IPCC reports, and policy documents on climate change and GIS applications.
2. **Data Analysis:** Synthesizing findings from case studies and empirical research to understand climate change trends and mitigation strategies.
3. **GIS-Based Analysis:** Evaluating spatial data utilization in climate adaptation planning, renewable energy siting, and disaster risk management.
4. **Comparative Assessment:** Analyzing different mitigation and adaptation strategies across regions to determine best practices for climate action.

CLIMATE ACTION

Climate Action, which is United Nations Sustainable Development Goal (UN-SDG) 13, refers to efforts taken at individual, community, national, and global levels to reduce the causes of climate change and mitigate its impacts (Sorooshian, 2024). The goal of climate action is to protect the planet, promote sustainable development, and ensure a healthy environment for future generations. Climate Action highlights the importance of urgent steps to combat climate change and its impacts. It recognizes that climate change is a global challenge that affects every aspect of life, from food security and water resources to public health and economic stability. Climate Action is built on two pillars: *Mitigation*, aimed at reducing greenhouse gas emissions, and *adaptation*, focused on enhancing resilience to inevitable climate impacts.

Mitigation: Preventing or Reducing GHG Emissions

Mitigation is the cornerstone of limiting global warming. Mitigation focuses on addressing the root causes of climate change by reducing GHG emissions or enhancing carbon sinks, improving efficiency, and adopting sustainable practices across sectors. It involves the moving away from fossil fuels to renewable energy transition (RET), energy efficiency, sustainable agriculture and forestry, as a central strategy. The global energy transition is being driven by the rapid deployment of solar and wind technologies. The IPCC identifies mitigation as essential to keeping global temperature rise below 1.5°C above pre-industrial levels.

Mitigation Motivation

Mitigation plays a crucial role in addressing climate change, as failing to implement significant efforts could lead to a global temperature rise of more than 2°C by 2050. Such an increase would intensify extreme weather events, accelerate sea level rise, and contribute to widespread biodiversity loss. The consequences would be particularly severe for regions in the GS, which, despite contributing the least to GHG emissions, remain among the most vulnerable to climate change impacts. Beyond preventing environmental and social devastation, effective mitigation strategies offer substantial economic and health benefits. Transitioning to cleaner energy sources and enhancing energy efficiency can improve air quality and reduce respiratory diseases caused by pollution. Moreover, investments in renewable energy and sustainable industries can drive job creation and economic growth, ensuring a more resilient and secure future for communities worldwide. By prioritizing mitigation, nations can simultaneously combat climate change while fostering sustainable development and improving overall well-being.

The energy sector is the largest contributor to global GHG emissions, primarily due to the burning of fossil fuels such as coal, oil, and natural gas. To effectively mitigate climate change, a fundamental shift towards renewable energy sources is essential. By replacing fossil fuels with clean and sustainable alternatives, it is possible to significantly reduce emissions while ensuring a stable energy supply. Among the most impactful solutions are solar and wind energy, which have seen rapid advancements in efficiency and affordability. Expanding the use of solar photovoltaic (PV) systems and wind farms helps decrease dependence on carbon-intensive energy sources. Additionally, hydropower and geothermal energy provide low-emission and reliable alternatives that contribute to a balanced and diversified energy mix.

KEY STRATEGIES FOR GHG MITIGATION

Comprehensive GHG mitigation requires a multi-faceted approach involving renewable energy adoption, energy efficiency improvements, sustainable transportation, agricultural reform, and carbon capture technologies. Successful case studies from countries like Kenya, Norway, and the European Union demonstrate that large-scale emission reductions are feasible with strong policies, investment, and innovation. Effective GHG mitigation requires a combination of strategies across multiple sectors, ranging from energy and transportation to agriculture and industrial processes (Andrei, 2023). Countries that adopt these strategies can significantly reduce emissions while fostering economic growth and environmental sustainability.

TRANSITIONING TO RENEWABLE ENERGY

Shifting from fossil fuels to renewable energy is one of the most impactful ways to lower GHG emissions. Investments in solar, wind, hydro, and geothermal power provide cleaner alternatives to coal, oil, and natural gas, reducing reliance on carbon-intensive energy sources. A prime example of successful energy transition is Kenya, which generated over 90% of its electricity from renewable sources in 2022, demonstrating the potential of sustainable energy systems in reducing emissions (Rotich et al., 2024).

ENERGY EFFICIENCY AND SUSTAINABLE INDUSTRY

Enhancing energy efficiency is one of the most cost-effective and immediate strategies for reducing GHG emissions while promoting sustainability. By optimizing energy use across various sectors, significant environmental and economic benefits can be achieved. One of the most impactful measures is building efficiency, which involves the construction of green buildings with better insulation, LED lighting, and smart energy management systems. These innovations can reduce overall energy consumption by 30–50%, making buildings more sustainable and cost-effective in the long run (PlanRadar, 2024). The effectiveness of these initiatives is demonstrated by the European Union's Energy Efficiency Directive, which led to a 12% reduction in final energy consumption (FEC) and a 19% decrease in primary energy consumption (PEC) between 2005 and 2023 (EEA, 2025a). This achievement marks a significant improvement over historical energy efficiency trend. Moreover, the pace of reduction in PEC and FEC over the past three years indicates that, if sustained, the EU remains on track to meet its 2030 energy efficiency targets (EEA, 2025b).

SUSTAINABLE TRANSPORTATION

The transportation sector accounts for approximately 74% of global CO_2 emissions in 2022, making it a crucial area for climate mitigation efforts (Carlin & Arshad, 2024). A shift toward low-carbon and sustainable mobility solutions is essential to reduce dependence on fossil fuels and decrease emissions from road, air, and maritime transport. One of the most effective strategies is electrification, which involves the widespread adoption of electric vehicles (EVs) and the development of an extensive charging infrastructure. By phasing out gasoline-powered cars and replacing them with battery-electric and hydrogen-powered alternatives, significant reductions in transport-related emissions can be achieved. In addition to vehicle electrification, public transport and active mobility play a crucial role in cutting emissions. Investing in buses, trains, cycling lanes, and pedestrian-friendly urban planning reduces car dependency and promotes sustainable commuting options. Well-designed public transportation networks can provide affordable, efficient, and low-emission mobility solutions for urban populations.

Furthermore, addressing emissions from aviation and shipping is critical for global sustainability. The integration of biofuels, hydrogen-based fuels, and energy efficiency improvements in logistics and freight transportation helps curb emissions from these traditionally high-polluting sectors. Sustainable innovations in aircraft and ship design, coupled with improved fuel efficiency, contribute to long-term carbon reduction. A notable success in transportation decarbonization is Norway, where over 80% of new cars sold in 2023 were electric. The country has set an ambitious target for all new vehicles to operate at zero emissions by 2025, demonstrating the potential of strong policies and investment in accelerating the transition to clean transportation (Adomaitis, 2025). By implementing electrification, public transport enhancements, and sustainable aviation and shipping practices, the transport sector can significantly contribute to global GHG reduction goals, fostering a cleaner and more resilient future.

SUSTAINABLE AGRICULTURE AND LAND USE

Agriculture is a significant contributor to GHG emissions, particularly methane CH_4 from livestock and N_2O from fertilizer use. Implementing sustainable agricultural practices and responsible land management is crucial to reducing emissions while maintaining food security. One effective strategy is regenerative agriculture, which enhances soil health and carbon sequestration. Techniques such as crop rotation, no-till farming, and the use of organic fertilizers improve soil quality while reducing reliance on synthetic inputs. These practices restore natural carbon sinks, which help to offset agricultural emissions. Advancements in precision farming further optimize resource use. By leveraging GIS and Artificial Intelligence (AI), farmers can minimize water and fertilizer waste, leading to lower emissions and improved crop yields. These data-driven approaches enhance efficiency and sustainability across the agricultural sector. Reducing emissions from livestock farming is another critical aspect of sustainable agriculture. Innovations such as methane-reducing feed additives and improved livestock management techniques help lower CH_4 emissions from ruminant animals. Additionally, better waste management systems can further mitigate environmental impacts. Beyond agriculture, forest conservation and restoration play a vital role in absorbing carbon dioxide (CO_2) and restoring degraded ecosystems. Large-scale projects like Africa's Great Green Wall, which aims to restore 100 million hectares of degraded land, serve as natural carbon sinks while combating desertification.

CARBON CAPTURE, UTILIZATION, AND STORAGE

While reducing emissions at the source is a priority, carbon capture technologies provide an additional mechanism to address hard-to-decarbonize sectors such as steel and cement production. One promising approach is Direct Air Capture (DAC), which removes CO_2 directly from the atmosphere for storage or repurposing (Hanson et al., 2025). Carbon sequestration, another key method, involves storing captured CO_2 in underground geological formations or using it for enhanced oil recovery. Additionally, carbon utilization technologies convert captured emissions into synthetic fuels, construction materials, and other industrial products, preventing them from re-entering the atmosphere. A successful example of carbon storage is Norway's Sleipner project, which has safely stored over 16 million tons of CO_2 in undersea rock formations since 1996 (Dickson, 2024). This project demonstrates the viability of long-term CO_2 sequestration as a tool for climate mitigation

Adaptation: Strengthening resilience against climate change

While mitigation efforts aim to reduce GHG emissions, adaptation strategies ensure that societies, economies, and ecosystems withstand and recover from climate-related shocks with significant socio-economic impacts (Abbass et al., 2022). The shocks, which threaten food and water security, and shift climatic patterns include rising sea levels, extreme weather events, such as droughts, floods, and heatwaves. Adaptation is critical for regions already experiencing the adverse effects of climate change, and this requires adjusting to current or anticipated climatic conditions and their impacts (Weiskopf et al., 2020). In human systems, the aim is to reduce vulnerabilities and mitigate potential harm while also leveraging any beneficial opportunities that may arise. Adaptation requires significant investment in climate-resilient infrastructure, agriculture, and disaster preparedness and strengthening adaptation requires the following:

BUILDING CLIMATE-RESILIENT INFRASTRUCTURE

Developing robust infrastructure is a key component of climate adaptation, enabling communities to withstand environmental stresses. Coastal protection systems and flood defences are essential in mitigating the effects of rising sea levels and storm surges. Urban areas can adopt climate-smart planning approaches, such as green roofs, permeable pavements, and improved drainage systems, to reduce urban flooding (Dharmarathne et al., 2024). Additionally, resilient transportation and energy systems designed to operate under extreme weather conditions can help maintain essential services and prevent disruptions.

ENHANCING AGRICULTURAL RESILIENCE

Climate change threatens global food security, necessitating the adoption of resilient agricultural practices. The development of drought-resistant and climate-resilient crop varieties can help maintain agricultural yields despite unpredictable weather patterns. Sustainable water management techniques, including rainwater harvesting and efficient irrigation systems, can optimize water usage and reduce dependence on erratic rainfall (Roohi et al., 2025). Furthermore, agroforestry and soil conservation practices can protect land from erosion and degradation, ensuring long-term agricultural productivity.

PROTECTING AND RESTORING NATURAL ECOSYSTEMS

Natural ecosystems serve as critical buffers against climate change, providing protection against extreme weather events while supporting biodiversity. Restoring mangroves and wetlands can shield coastal communities from storm surges and erosion (K et al., 2024). Forest conservation and reforestation help regulate local climates and prevent land degradation, while biodiversity protection ensures ecosystem stability, contributing to food and water security (Sunkur et al., 2023).

STRENGTHENING COMMUNITY AND SOCIAL RESILIENCE

Communities, particularly in climate-vulnerable regions, must be equipped to respond effectively to climate-related disruptions. Implementing early warning systems for extreme weather events allows for timely evacuations and preparedness measures. Disaster risk reduction (DRR) programs can educate communities on climate adaptation strategies, equipping them with the necessary skills to respond to environmental challenges ("Key Risks across Sectors and Regions," 2023). Additionally, climate-responsive healthcare systems must be developed to address emerging health threats, including heat stress, vector-borne diseases, and waterborne illnesses.

CLIMATE ADAPTATION FINANCING AND POLICY SUPPORT

Significant investments in climate-resilient infrastructure, agriculture, and disaster preparedness are essential for effective adaptation but are often hindered by financial and policy barriers (Ishiwatari & Surjan, 2019). Addressing these challenges requires increased climate financing to support resilience-building initiatives in developing nations. Integrating adaptation strategies into national policies ensures that climate risks are considered in long-term development planning. Furthermore, fostering international collaboration and knowledge-sharing can accelerate the adoption of innovative and scalable adaptation solutions.

THE URGENCY OF ADAPTATION

As climate change continues to impact ecosystems, economies, and human well-being, adaptation has become an urgent necessity rather than an option. Strengthening resilience at all levels—from infrastructure to communities—will ensure that societies can withstand climate shocks, minimize losses, and secure a sustainable future for generations to come.

GEOGRAPHICAL INFORMATION SYSTEM (GIS) IN CLIMATE CHANGE MITIGATION AND ADAPTATION

Study Concept

Global South Climate Change Protection through Geographical Information System (GIS) Study and Coordinated Sustainable Development Measures refers to a comprehensive initiative aimed at addressing the impacts of climate change in the GS using GIS studies and implementing coordinated sustainable development measures. The key components of this concept are:

THE GS

The term "Global South" typically refers to countries located in the Southern Hemisphere, particularly those in Africa, Latin America, Asia, and Oceania. These regions often face uniquc challenges related to climate change, including extreme weather events, rising sea levels, and threats to biodiversity.

Climate Change Protection and GIS Study

This aspect of the initiative focuses on safeguarding the GS from the adverse effects of climate change. This may include mitigation efforts to reduce GHG emissions and adaptation strategies to cope with the impacts that are already occurring.

GIS is a technology that involves capturing, analyzing, and interpreting spatial and geographic data. The GIS study component implies using this technology to comprehensively analyze and understand the geographical aspects of climate change in the GS. This may involve mapping vulnerable areas, assessing climate-related risks, and identifying potential solutions. GIS technologies have become essential tools for climate change research and action, offering data-driven solutions for risk assessment, resource management, resource potential assessment and policy planning. This technology supports efficient water management and crop planning, which are critical in regions affected by water scarcity. Some of the GIS tools and case studies that have successfully addressed some of the key research gaps in climate adaptation and sustainable development in the GS are presented in Table 10.1.

TABLE 10.1
Some of the GIS Tools and Case Studies

GIS Tool	Use and Impact	Case study
Google Earth Engine (GEE)	Provides free access to satellite imagery and environmental data for climate monitoring. Used by researchers and NGOs in Africa and Asia to monitor deforestation, droughts, and urban heat islands.	Deforestation Monitoring in the Congo Basin: GEE and Global Forest Watch were used to track illegal logging and forest loss in Central Africa, helping policymakers enforce conservation laws (Masolele, 2023).
Digital Earth Africa	Provides open-access satellite data tailored for African climate challenges. Used in drought monitoring, water resource management, and sustainable agriculture.	Drought Prediction in Kenya: Digital Earth Africa helped identify drought-prone areas, allowing early interventions for farmers and herders (Kropff et al., 2023).
QGIS (Open-source GIS Software)	Provides a free alternative to commercial GIS tools for local governments and communities. Used in various Global South countries for land-use planning, disaster risk mapping, and environmental monitoring.	Community-Led Climate Resilience in India: NGOs trained local farmers in using QGIS for flood risk assessment, improving disaster preparedness in flood-prone regions (Shrestha et al., 2025).

(Continued)

TABLE 10.1 *(Continued)*
Some of the GIS Tools and Case Studies

GIS Tool	Use and Impact	Case study
SERVIR (A NASA & USAID Program)	Provides technical training and tools for using GIS in climate resilience. Active in Africa, Asia, and Latin America, supporting governments and NGOs	Flood Prediction System in Bangladesh: SERVIR helped develop GIS-based flood models to improve early warning systems, reducing flood-related losses (Shah & Ai, 2024).
Participatory GIS (PGIS)	Combines indigenous climate knowledge with geospatial mapping to improve local adaptation strategies. Used in Latin America and Africa for land-use planning and DRR.	Indigenous Climate Adaptation in the Peruvian Andes: Farmers used PGIS to map historical climate patterns and adjust agricultural practices accordingly (Mantas & Caro, 2023).
African Flood and Drought Monitor (AFDM)	Provides near-real-time flood and drought forecasting using satellite and ground data and is used in East and West Africa to improve disaster response	Flood Mitigation in Ghana: AFDM data was integrated with local hydrological models to improve early flood warnings (Ampofo et al., 2024).
Climate Hazards Group InfraRed Precipitation with Station Data (CHIRPS)	Provides 40+ years of rainfall data to monitor drought trends. Used in water management and food security planning across the Global South	Drought Response in Ethiopia: CHIRPS data was used to predict severe droughts, leading to better food aid distribution (Bojer et al., 2024).
Global Solar Atlas (by World Bank)	Identifies high-potential solar energy sites using geospatial analysis. Used by governments and investors to expand solar energy infrastructure.	Solar Energy Expansion in Nigeria: The Nigerian government used Global Solar Atlas to map rural electrification projects (Okedu et al., 2024).
FAO's Land Resources Information System (LRIS)	Supports sustainable land management by mapping soil quality and land degradation risks. Used to prevent desertification and improve agricultural practices.	Combatting Desertification in the Sahel: GIS-based land assessments guided reforestation efforts in the Great Green Wall project (Negede, 2024).
UN Biodiversity Lab	Maps biodiversity and climate risks to support environmental policymaking. Helps nations meet climate finance and conservation goals.	Carbon Offset Planning in Brazil: GIS data was used to prioritize reforestation areas for carbon credit investments (Blanton et al., 2024).
Climate Policy Mapping by Climate Watch	Tracks how countries integrate GIS data into climate policies. Informs decision-makers on best practices for GIS-driven climate governance.	GIS in Climate Action Plans in South Africa: GIS-supported climate policy helped improve city-level emissions reduction targets (Anggara et al., 2025).

These GIS tools and case studies highlight solutions that bridge data gaps, strengthen local capacity, integrate indigenous knowledge, improve real-time climate monitoring, and support sustainable energy and land use.

In rapidly urbanizing areas of the GS, GIS is used to analyse heat island effects, assess urban infrastructure vulnerabilities, and promote green infrastructure development. Sustainable urban planning can mitigate temperature increases and reduce the impact of extreme weather events. GIS is widely deployed across various sectors for spatial analysis, mapping, and decision-making. The deployment of GIS continues to expand into new sectors as technology advances and data availability increases. Some of the key areas where GIS is commonly used are classified in this study into climate-related and other areas, as presented in Table 10.2.

TABLE 10.2
GIS Application in Climate-Related and Other Areas

Climate-related areas	Description
Climate change monitoring—involves analyzing temperature changes, sea level rise, and GHG emissions.	The aim of studies under this category is to highlight the critical role of GIS in analysing climate change phenomena, specifically temperature variations, sea level rise, and GHG emissions. The objectives include using GIS to—assess methodologies; identify vulnerable regions; evaluate data sources; understand climate-health links; and support policy and mitigation strategies.
Disaster management and mitigation – encompasses mapping flood-prone areas, wildfire risks, and extreme weather events	The aim of these is to highlight the essential role of GIS in disaster management and mitigation by showcasing its applications across various disaster scenarios. This analysis underscores the versatility of GIS in strengthening disaster preparedness, response, and mitigation efforts. Specifically, the objectives of these include to—analyse GIS integration; enhance disaster preparedness; support coastal and flood risk management; leverage community knowledge, inform policy and planning.
Environmental management – concerns tracking deforestation, soil erosion, and land degradation	GIS plays crucial role in environmental management studies as it enables researchers to analyse spatial relationships and patterns in environmental data; integrate and manage large datasets from various sources, including satellite imagery, aerial photography, and field surveys; and assess the potential impacts of proposed projects on the environment. Several studies used GIS to assess GIS techniques, monitor land use and deforestation; improve waste management strategies, enhance urban and rural planning, and support conservation and climate action.
Renewable energy planning—Identifying optimal locations for solar, wind, and hydro energy projects	GIS is pivotal in the assessment and planning of renewable energy projects, offering spatial analysis tools to identify optimal locations and evaluate resource potentials. Several studies underscore the versatility and effectiveness of GIS in renewable energy assessment and planning, facilitating informed decision-making for sustainable energy development. The objectives of these studies on the application of GIS in renewable energy assessment and planning include— evaluating renewable energy resources, local energy planning, case studies and market analysis, sustainable deployment, multi-criteria decision analysis, dynamic energy potential mapping, and data quality assessment.

(Continued)

TABLE 10.2 *(Continued)*
GIS Application in Climate-Related and Other Areas

Climate-related areas	Description
Agriculture and precision farming—Monitoring soil health, crop conditions, and irrigation efficiency	GIS have become integral to modern agriculture, particularly precision farming, by enabling data-driven decisions that enhance productivity and sustainability. Several notable studies and applications illustrate the impact of GIS in this field and underscore its pivotal role in advancing precision agriculture, leading to more efficient and sustainable farming practices. The overarching objective of these studies is to examine the role of GIS in precision agriculture and its impact on improving agricultural efficiency, sustainability, and decision-making. The specific objectives include access to GIS applications in agriculture, develop sustainable farming strategies, enhance decision-making in farming, improve agricultural productivity, and leverage AI and remote sensing for smart farming.
Water resource management—Mapping watersheds, groundwater reserves, and drought-prone regions	GIS have become essential tools in water resource management, enabling the analysis and visualization of hydrological data to enhance decision-making processes. Notable studies and applications demonstrating the use of GIS in this field. These studies and applications underscore the critical role of GIS in advancing water resource management practices, offering robust frameworks for analysing hydrological data and implementing effective management strategies. The primary objective of these studies is to explore the role of GIS in managing and analysing water resources for sustainable and efficient use. The specific objectives include assessing hydrological processes, enhance watershed management, improve water quality monitoring, optimize water supply and distribution, develop flood and drought risk assessment models, support sustainable groundwater management, integrate remote sensing and GIS for water resource mapping, and enhance decision-making for policy and planning.
Biodiversity conservation—tracking wildlife habitats, migration patterns, and ecosystem health	GIS have become indispensable tools in biodiversity conservation, enabling researchers and conservationists to analyse spatial data, monitor species distributions, and identify critical habitats. Notable studies and applications in this field. These studies and applications underscore the critical role of GIS in advancing biodiversity conservation efforts, providing essential tools for effective analysis, planning, and management. The primary objective of these studies is to leverage GIS to enhance biodiversity conservation efforts by providing spatial analysis, monitoring tools, and decision-making support. The specific objectives include assessing species distribution and habitat mapping, identify and monitor critical biodiversity hotspot, analyse land use and habitat fragmentation, support the design and management of protected areas, monitor climate change effects on biodiversity, enhance wildlife corridor planning, integrate remote sensing for ecosystem monitoring, develop decision support tools for conservation policy, and support biodiversity informatics and open data initiatives.

(Continued)

TABLE 10.2 *(Continued)*
GIS Application in Climate-Related and Other Areas

Climate-related areas	Description
Urban green planning - designing climate-resilient cities with sustainable land use strategies	GIS plays a pivotal role in urban green planning by facilitating the analysis and visualization of spatial data to enhance the development and management of green infrastructure. Notable studies and applications in this field. These studies and applications highlight the integral role of GIS in urban green planning, facilitating data-driven decisions that promote sustainable urban development and environmental resilience. The primary objective of these studies is to utilize GIS to enhance urban green planning by supporting sustainable development, improving environmental quality, and optimizing land use. The specific objectives include assessing and map urban green spaces, support sustainable urban development, optimize green infrastructure design, enhance climate adaptation and mitigation, improve urban flood management, analyse land use and urban expansion, monitor and manage tree canopy coverage, develop walkable and liveable cities, support decision, and enhance community engagement in urban greening.
Other GIS applications	
Urban & regional planning; Transportation & logistics; Health & epidemiology; Defence & security; Real estate & land management; Telecommunications Tourism & recreation Mining & geology	GIS have been extensively applied across various sectors to analyse spatial data, enhance decision-making, and improve operational efficiency. Below is a curated list of studies and applications demonstrating the use of GIS in these fields. GIS have been extensively applied across various sectors to analyse spatial data, enhance decision-making, and improve operational efficiency. Below is a curated list of studies and applications demonstrating the use of GIS in these fields. These studies underscore the versatility and critical role of GIS in various sectors, facilitating data-driven decisions and promoting efficiency across diverse applications.

THE ROLE OF GIS TECHNOLOGY IN PV SYSTEM ADVANCEMENT

GIS technology plays a crucial role in assessing photovoltaic (PV) potential by integrating spatial data, environmental factors, and solar energy models. It enables efficient site selection, resource estimation, and decision-making for solar energy projects. Through its application at different stages—from planning and design to real-time monitoring and management—GIS enhances accuracy, improves decision-making, and optimizes solar energy deployment.

***Site selection and resource assessment*:** GIS enables the precise selection of suitable locations for solar PV installations by analysing solar radiation, land use, and environmental constraints. It helps in identifying high solar potential areas using solar radiation mapping; avoiding environmentally sensitive areas and optimizing land use; and evaluating proximity to grid infrastructure to minimize transmission losses.

***Energy yield estimation and system simulation*:** GIS integrates real-time and historical climate data to simulate PV system performance under different conditions. It supports accurate energy output forecasting using solar radiation models; optimization of PV panel orientation and tilt angles to maximize efficiency; 3D modeling of shading effects to minimize energy losses.

***Grid integration and infrastructure planning*:** GIS supports grid connection analysis and load balancing by mapping the shortest and most cost-effective routes for transmission lines; assisting in microgrid planning for rural electrification; optimizing energy storage placement, and distribution networks (Ye et al., 2023).

***Real-time monitoring and maintenance*:** GIS-powered Internet of Things (IoT) sensors and AI-driven analytics enable efficient system monitoring by detecting system faults and performance inefficiencies; using drones and thermal imaging to identify malfunctioning panels, and predicting maintenance needs through AI-based analytics (Jamroen et al., 2023).

KEY GIS ALGORITHMS USED IN SOLAR PV SYSTEMS

GIS technology relies on various algorithms to analyze spatial data, optimize solar PV panel placement, and enhance solar PV system management. These algorithms are presented in Table 10.3.

TABLE 10.3
GIS Algorithms Used in Solar PV Systems

Models	Description of GIS algorithm	PV software application deployed
Solar radiation estimation models	Pérez Model, Bird's Clear Sky Model, and Solar Position Algorithm (SPA) for solar position and irradiance estimation	PVGIS, SolarGIS, and Helioscope for solar potential assessment and site selection
Terrain and shading analysis models	Digital Elevation Models (DEM), Hill shade Algorithm, Sky View Factor (SVF), and Ray Tracing for obstruction and shadow impact assessment.	ArcGIS Solar Analyst, GRASS GIS, and QGIS use these algorithms for PV system layout optimization.
Energy yield estimation models	PVWatts, Sandia PV Performance Model, and AI-based forecasting techniques like Artificial Neural Networks (ANNs).	PVGIS, RETScreen, and SAM (System Advisor Model) for solar PV feasibility studies and energy output simulation
Grid optimization algorithms	Dijkstra's Algorithm for transmission planning and Genetic Algorithms for grid balancing.	HOMER Pro, OpenDSS, and ETAP use these algorithms for solar grid integration studies
Predictive maintenance models	Machine learning models like Random Forest and Recurrent Neural Networks (RNNs) for performance monitoring	SCADA systems, AI-based maintenance platforms, and UAV (drone) inspections use these algorithms for solar PV farm monitoring
Multi-Criteria Decision Analysis (MCDA) models	Analytical Hierarchy Process (AHP) and Weighted Overlay Analysis for combining multiple spatial datasets to create suitability maps for solar power projects.	ArcGIS ModelBuilder, ILWIS, and Google Earth Engine use these algorithms for solar site selection studies.

COORDINATED SUSTAINABLE DEVELOPMENT MEASURES FOR THE GS

While GIS provides the technological foundation for addressing climate change, it must be integrated with broader sustainable development strategies to ensure long-term success. The following coordinated measures are essential for climate protection in the GS:

1. *Policy integration and governance*: Climate action must be embedded in national development plans and governance frameworks. Cross-sectoral coordination between environmental, agricultural, energy, and urban development agencies is critical for coherent climate strategies.
2. *Community-based approaches*: Local communities must be at the center of adaptation measures. Engaging indigenous knowledge and local practices enhances the relevance and effectiveness of interventions, ensuring that they are context-specific and culturally appropriate.

***Capacity building and knowledge transfer*:** Many countries in the GS face a lack of technical expertise in GIS and climate adaptation. Capacity-building initiatives, such as training programs and knowledge-sharing networks, can bridge this gap and empower local institutions.

***Access to climate finance*:** Adequate funding is essential for scaling up GIS-based climate protection projects. Mechanisms like the Green Climate Fund and international development programs must prioritize investment in GIS capacity and sustainable development projects in vulnerable regions.

***Technology transfer and innovation*:** Partnerships with global technology providers can facilitate access to advanced GIS tools and remote sensing technologies. Affordable solutions tailored to the specific needs of the GS can drive innovation and local adaptation efforts.

PROJECTS OF GIS-DRIVEN CLIMATE SOLUTIONS

Several successful GIS-based climate initiatives demonstrate how technology can transform climate adaptation and sustainable development in the GS as presented in Table 10.4.

RESEARCH GAPS IN GIS-DRIVEN CLIMATE SOLUTIONS IN THE GS

Despite the transformative potential of GIS in climate adaptation and sustainable development, several research and implementation gaps hinder their full impact in the GS (Gorelick et al., 2017). These include data limitations, technical capacity deficits, weak integration of indigenous knowledge, inadequate real-time climate monitoring, underutilization in renewable energy planning, and policy shortcomings. Addressing these gaps requires concerted efforts from governments, academia, and international organizations to harness GIS's potential in combating climate challenges effectively.

TABLE 10.4
GIS-based Climate Initiatives in the GS

GIS-based climate initiatives	Description
The SERVIR Program	SERVIR stands for "Regional Visualization and Monitoring System" in Spanish. It's a joint initiative of National Aeronautics and Space Administration (NASA) and United States Agency for International Development (USAID) to support countries in the GS by providing satellite data, maps, and tools to address challenges like climate change, disaster response, agriculture, and water resources management (Deval et al., 2024).
Digital Agriculture Platforms in East Africa	GIS-driven platforms help smallholder farmers access climate forecasts, optimize crop yields, and manage risks associated with climate variability (Abate et al., 2023).
Mangrove Restoration Projects in Southeast Asia	GIS tools are used to map and monitor mangrove ecosystems, which act as natural buffers against coastal erosion and rising sea levels. These projects enhance biodiversity while protecting vulnerable coastal communities (Syahid et al., 2023).
Africa Risk Capacity (ARC)– Africa	Uses GIS and remote sensing for drought and disaster risk financing and helps African governments access insurance payouts for climate-related disasters (Chamberlain & Bernards, 2024)
Digital Earth Africa– Africa	Uses open-source satellite data to monitor deforestation, flooding, and water resources and aims to improve environmental decision-making across African nations (Kerner et al., 2024).
The Global Mangrove Watch (GMW) – GS	Tracks mangrove ecosystems using GIS to support conservation efforts and helps mitigate coastal erosion and carbon sequestration loss (Rosenqvist, 2023).
Climate Risk and Early Warning Systems (CREWS) – Small Island Developing States & Africa	Uses GIS to enhance early warning for extreme weather and climate hazards, which strengthens preparedness in vulnerable regions (Avalon-Cullen et al., 2023).
Global Forest Watch (GFW) – Latin America, Africa, Asia	Uses GIS and remote sensing to monitor deforestation and illegal logging and provides real-time forest loss alerts to policymakers and conservationists (Richter et al., 2024).

Limited high-resolution climate data: Many regions lack localized climate, land-use, and hydrological data, relying on global datasets that fail to capture microclimates. Restricted access to satellite imagery and remote sensing data further hampers climate planning.

Capacity gaps and technical expertise: Government agencies and local organizations often lack the GIS skills needed for climate adaptation. Limited training opportunities and poor integration into decision-making reduce GIS effectiveness in agriculture and disaster management.

Weak integration of traditional knowledge: Indigenous climate adaptation methods remain underutilized in GIS models, creating a disconnect between technology and local adaptation strategies. Participatory GIS approaches can help bridge this gap.

Challenges in real-time climate monitoring: Sparse meteorological stations and weak data-sharing frameworks hinder early warning systems for floods, droughts, and extreme weather, increasing climate vulnerability.

INADEQUATE GIS USE IN RENEWABLE ENERGY PLANNING

Despite abundant solar and wind resources, GIS-based site assessments are lacking, leading to inefficient energy investments. Improved geospatial analysis can support sustainable energy transitions.

Policy and governance barriers: Weak GIS integration into climate policy and governance limits adoption, with insufficient financial incentives for geospatial technology use in key sectors like water management and urban planning.

WAY FORWARD: STRENGTHENING GIS-DRIVEN CLIMATE SOLUTIONS IN THE GS

To address the identified gaps in GIS-driven climate action, the following strategic steps are recommended:

1. Invest in localized, high-resolution climate data collection. This can be achieved by governments and research institutions collaborating to improving access to high-resolution spatial datasets tailored for regional climate challenges.
2. Develop capacity-building programs for GIS adoption at community and policy levels by developing training programs for policymakers, urban planners, and local communities can enhance the practical use of GIS in climate adaptation.
3. Integrate indigenous knowledge with geospatial technologies by combining traditional environmental knowledge with GIS tools can improve local adaptation strategies.
4. Strengthen real-time climate monitoring infrastructure by investing in satellite-based remote sensing, IoT sensors, and AI-driven climate models can enhance early warning systems.
5. Embed GIS in climate finance and policy frameworks. Governments and climate finance institutions must integrate GIS-based risk assessments into funding mechanisms.

CONCLUSION

Climate change presents an urgent global challenge, with disproportionate effects on vulnerable regions. Effective climate action requires integrating mitigation efforts to reduce GHG emissions and adaptation strategies to enhance resilience. GIS emerges as a crucial tool for climate planning, offering insights into risk assessment, resource management, and sustainable development. Coordinated policy frameworks, increased financial investment, and technological advancements are essential for fostering climate resilience. Strengthening international collaboration and adopting innovative solutions can accelerate the transition toward a sustainable and climate-resilient future.

ACKNOWLEDGMENTS

The authors hereby acknowledge the Research and Postgraduate Support Directorate, Institute for Systems Science, and the Management of Durban University of Technology, South Africa, for their continuous support

REFERENCES

Abate, G. T., Abay, K. A., Chamberlin, J., Kassim, Y., Spielman, D. J., & Tabe-Ojong, M. P. J. (2023). Digital tools and agricultural market transformation in Africa: Why are they not at scale yet, and what will it take to get there? *Food Policy*, *116*, 102439.

Abbass, K., Qasim, M. Z., Song, H., Murshed, M., Mahmood, H., & Younis, I. (2022). A review of the global climate change impacts, adaptation, and sustainable mitigation measures. *Environmental Science and Pollution Research*, *29*(28), 42539–42559. Available: https://doi.org/10.1007/s11356-022-19718-6

Adomaitis, N. (2025). *In Norway, nearly all new cars sold in 2024 were fully electric*. Retrieved from Reuters, London, UK: https://www.reuters.com/business/autos-transportation/norway-nearly-all-new-cars-sold-2024-were-fully-electric-2025-01-02/

Allen, M. R., Dube, O. P., Solecki, W., Aragón-Durand, F., Cramer, W., Humphreys, S., & Zickfeld, K.. (2018). *Framing and context. In: Global warming of 1.5 C*. Retrieved from Special Report on the impacts of global warming of 1.5 C of The Intergovernmental Panel on Climate Change (IPCC), Cambridge University Press, Cambridge, UK and New York, NY, USA: https://www.ipcc.ch/sr15/chapter/chapter-1/

Ampofo, J. A., Sekyere, E., & Adongo, R. (2024). Effectiveness of flood mitigation strategies used by residence in the Tamale metropolis of the Northern region, Ghana. *Educ Adm Theory Pract, 30*, 3062–3069.

Andrei, D.-M. (2023). The energy efficiency issue in the European Union: Perspectives, objectives and challenges. *Romanian Journal of European Affairs*, *23*(1).

Anggara, E. D., Aditya, T., & Sutanta, H. (2025). PGIS for tiered participatory development planning. *International Review for Spatial Planning and Sustainable Development*, *13*(1), 136–154.

Avalon-Cullen, C., Caudill, C., Newlands, N. K., & Enenkel, M. (2023). Big data, small island: Earth observations for improving flood and landslide risk assessment in Jamaica. *Geosciences*, *13*(3), 64.

Blanton, A., Mohan, M., Galgamuwa, G. P., Watt, M. S., Montenegro, J. F., Mills, F., & Pons, J. (2024). The status of forest carbon markets in Latin America. *Journal of Environmental Management*, *352*, 119921.

Bojer, A. K., Biru, B. H., Al-Quraishi, A. M. F., Debelee, T. G., Negera, W. G., Woldesillasie, F. F., & Esubalew, S. Z. (2024). Machine learning and remote sensing based time series analysis for drought risk prediction in borena zone, Southwest Ethiopia. *Journal of Arid Environments*, *222*, 105160.

Key Risks across Sectors and Regions. (2023). In C. Intergovernmental Panel on Climate (Ed.), *Climate change 2022 – Impacts, adaptation and vulnerability: Working group II contribution to the sixth assessment report of the Intergovernmental Panel on Climate Change* (pp. 2411–2538). Cambridge: Cambridge University Press Retrieved from https://www.cambridge.org/core/product/0E18FB2DE3A051281267D5C9C047986F

Carlin, D., & Arshad, M. (2024). *Climate risks in the transportation sector.* Retrieved from United Nations Environment Programme, Geneva: https://www.unepfi.org/wordpress/wp-content/uploads/2024/05/Climate-Risks-in-the-Transportation-Sector.pdf

Chamberlain, J., & Bernards, N. (2024). Insurance and the contradictions of the climate-development-finance nexus: The case of the African risk capacity. *Competition & Change*, 10245294241226985.

Deval, C., Anderson, E., Kruskopf, M., Limaye, A., Bhandari, B., Markert, A., & Shrestha, M. (2024). Connecting science to sustained impact: SERVIR's journey in earth observations for water resources management and future pathways. *Authorea Preprints.*

Dharmarathne, G., Waduge, A. O., Bogahawaththa, M., Rathnayake, U., & Meddage, D. P. P. (2024). Adapting cities to the surge: A comprehensive review of climate-induced urban flooding. *Results in Engineering*, *22*, 102123. Available: https://www.sciencedirect.com/science/article/pii/S2590123024003773

Dickson, D. (2024). *Norway's Sleipner: Where CO2 has been buried in the rock since 1996.* Retrieved from Vattenfall, Stockholm, Sweden: https://group.vattenfall.com/press-and-media/newsroom/2024/norways-sleipner-where-co2-has-been-buried-in-the-rock-since-1996

EEA. (2025a). *Primary and final energy consumption in the European Union.* Retrieved from European Environment Agency (EEA), Denmark: https://www.eea.europa.eu/en/analysis/indicators/primary-and-final-energy-consumption

EEA. (2025b). *Share of energy consumption from renewable sources in Europe.* Retrieved from European Environment Agency (EEA), Denmark: https://www.eea.europa.eu/en/analysis/indicators/share-of-energy-consumption-from

EPA. (2023). *Climate change indicators: Atmospheric concentrations of greenhouse gases.* Retrieved from United States Environmental Protection Agency (EPA), Washington, D.C.: https://www.epa.gov/climate-indicators/climate-change-indicators-atmospheric-concentrations-greenhouse-gases

Gorelick, N., Hancher, M., Dixon, M., Ilyushchenko, S., Thau, D., & Moore, R. (2017). Google Earth Engine: Planetary-scale geospatial analysis for everyone. *Remote Sensing of Environment*, *202*, 18–27. Available: https://www.sciencedirect.com/science/article/pii/S0034425717302900

Hanson, E., Nwakile, C., & Hammed, V. O. (2025). Carbon capture, utilization, and storage (CCUS) technologies: Evaluating the effectiveness of advanced CCUS solutions for reducing CO2 emissions. *Results in Surfaces and Interfaces*, *18*, 100381. Available: https://www.sciencedirect.com/science/article/pii/S2666845924002010

IPCC. (2014a). *Climate change 2014: Impacts, adaptation, and vulnerability.* Retrieved from Cambridge University Press, Cambridge, United Kingdom and New York, USA: https://www.ipcc.ch/site/assets/uploads/2018/03/ar5_wgII_spm_en-1.pdf

IPCC. (2014b). *Climate change 2014: Impacts, adaptation, and vulnerability. Part B: Regional aspects.* Retrieved from United Kingdom and New York, NY, USA: https://www.ipcc.ch/site/assets/uploads/2018/02/ar5_wgII_spm_en.pdf

IPCC. (2021). *Sixth assessment report.* Retrieved from Intergovernmental Panel on Climate Change (IPCC), Switzerland. https://www.ipcc.ch/assessment-report/ar6/

Ishiwatari, M., & Surjan, A. (2019). Good enough today is not enough tomorrow: Challenges of increasing investments in disaster risk reduction and climate change adaptation. *Progress in Disaster Science*, *1*, 100007. Available: https://www.sciencedirect.com/science/article/pii/S2590061719300079

Jamroen, C., Yonsiri, N., Odthon, T., Wisitthiwong, N., & Janreung, S. (2023). A stand-alone photovoltaic/battery energy-powered water quality monitoring system based on narrowband internet of things for aquaculture: Design and implementation. *Smart Agricultural Technology, 3*, 100072.

Kerner, H., Nakalembe, C., Yang, A., Zvonkov, I., McWeeny, R., Tseng, G., & Becker-Reshef, I. (2024). How accurate are existing land cover maps for agriculture in sub-Saharan Africa? *Scientific Data, 11*(1), 486.

Kropff, W., Ortiz Crespo, B., Steinke, J., Tesfaye, L., Ramirez-Villegas, J., Ghosh, A., & Tablia, A. (2023). *Mainstreaming digital approaches for adaptation in agriculture in Kenya*. The Global Center on Adaptation, The Alliance of Bioversity International and The Center for Tropical Agriculture (CIAT), and The African Development Bank, 2023. [Online]. Available: https://hdl.handle.net/10568/138918

Mantas, V., & Caro, C. (2023). User-relevant land cover products for informed decision-making in the complex terrain of the Peruvian Andes. *Remote Sensing, 15*(13), 3303.

Masolele, R. N. (2023). *Assessing land use following tropical deforestation: Combining remote sensing and deep learning*. Wageningen University and Research.

Mohan, V. (2025). In search of consensus: Examining global South perspectives on climate security in UNSC debates. *Earth System Governance, 23*, 100231. Available: https://www.sciencedirect.com/science/article/pii/S2589811624000314

Muhammed, J., & Augustine, A. (2024). Mangroves in environmental engineering: Harnessing the multifunctional potential of nature's coastal architects for sustainable ecosystem management. *Results in Engineering, 21*, 101765. Available: https://www.sciencedirect.com/science/article/pii/S2590123024000185

Negede, O. (2024). *Faculty of Engineering*. University of Nairobi.

Okedu, K. E., Oyinna, B., Colak, I., & Kalam, A. (2024). Geographical information system based assessment of various renewable energy potentials in Nigeria. *Energy Reports, 11*, 1147–1160.

PlanRadar. (2024). *The economic benefits of green building: How sustainable architecture can save money in the long run*. Retrieved from PlanRadar Office, Vienna, Austria: https://www.planradar.com/au/economic-benefits-of-green-building/

Richter, J., Goldman, E., Harris, N., Gibbs, D., Rose, M., Peyer, S., & Velappan, H. (2024). *Spatial database of planted trees (SDPT version 2.0)*. World Resources Institute, Washington, DC, 2024, vol. 23. [Online]. Available: https://doi.org/10.46830/writn.23.00073

Robinson, A., Lehmann, J., Barriopedro, D., Rahmstorf, S., & Coumou, D. (2021). Increasing heat and rainfall extremes now far outside the historical climate. *NPJ Climate and Atmospheric Science, 4*(1), 45. Available: https://doi.org/10.1038/s41612-021-00202-w

Roohi, Kamboj, P., Soni, P. G., Rani, S., Sukirtee, Kumar, R., & Mishra, A. K. (2025). Advancing climate-resilient agriculture: Adaptive strategies and soil-centric approaches for mitigation. In A. K. Mishra, S. Sharma, A. Mishra, & A. Roy (Eds.), *Transition to regenerative agriculture: Principles and indicators of soil health management* (pp. 55–75). Singapore: Springer Nature Singapore. Retrieved from https://doi.org/10.1007/978-981-96-1421-9_3

Rosenqvist, A. (2023). *Global Mangrove Watch–GMW Mangrove Loss Alerts*. Food and Agriculture Organization (FAO), Rome, Italy, 2023. [Online]. Available: https://openknowledge.fao.org/handle/20.500.14283/cc6247en

Rotich, I. K., Chepkirui, H., & Musyimi, P. K. (2024). Renewable energy status and uptake in Kenya. *Energy Strategy Reviews, 54*, 101453. Available: https://www.sciencedirect.com/science/article/pii/S2211467X24001603

Shah, S. A., & Ai, S. (2024). Flood susceptibility mapping contributes to disaster risk reduction: A case study in Sindh, Pakistan. *International Journal of Disaster Risk Reduction, 108*, 104503.

Shrestha, A., McCrone, A., Láng-Ritter, J., Gautam, S., Taka, M., & Varis, O. (2025). Bridging gaps, saving lives: Integrating communities' voices and impact mapping into flood early warning systems in rural Nepal. *International Journal of Disaster Risk Reduction*, *118*, 105238.

Sorooshian, S. (2024). The sustainable development goals of the United nations: A comparative midterm research review. *Journal of Cleaner Production*, *453*, 142272. Available: https://www.sciencedirect.com/science/article/pii/S0959652624017207

Sunkur, R., Kantamaneni, K., Bokhoree, C., & Ravan, S. (2023). Mangroves' role in supporting ecosystem-based techniques to reduce disaster risk and adapt to climate change: A review. *Journal of Sea Research*, *196*, 102449. Available: https://www.sciencedirect.com/science/article/pii/S1385110123001181

Syahid, L. N., Sakti, A. D., Ward, R., Rosleine, D., Windupranata, W., & Wikantika, K. (2023). Optimizing the spatial distribution of Southeast Asia mangrove restoration based on zonation, species and carbon projection schemes. *Estuarine, Coastal and Shelf Science*, *293*, 108477.

Weiskopf, S. R., Rubenstein, M. A., Crozier, L. G., Gaichas, S., Griffis, R., Halofsky, J. E., & Whyte, K. P. (2020). Climate change effects on biodiversity, ecosystems, ecosystem services, and natural resource management in the United States. *Science of The Total Environment*, *733*, 137782. Available: https://www.sciencedirect.com/science/article/pii/S0048969720312948

Ye, Y., Zhu, R., Yan, J., Lu, L., Wong, M. S., Luo, W., & Wang, Y. (2023). Planning the installation of building-integrated photovoltaic shading devices: A GIS-based spatiotemporal analysis and optimization approach. *Renewable Energy*, *216*, 119084.

Yoro, K. O., & Daramola, M. O. (2020). Chapter 1 - CO2 emission sources, greenhouse gases, and the global warming effect. In M. R. Rahimpour, M. Farsi & M. A. Makarem (Eds.), *Advances in carbon capture* (pp. 3–28). Woodhead Publishing. Retrieved from https://www.sciencedirect.com/science/article/pii/B9780128196571000013

11 SDG #15—Disrupted Progress

The Impact of COVID-19 on Sustainable Development Goals in the Indian Subcontinent

Hishmi Jamil Husain, Saied Pirasteh, Tammineni Rajitha, Nasim Khonsari, and Mahdieh Shirmohammadi

INTRODUCTION

In 2015, to promote "Global Partnership" for sustainable development of the world, Agenda 2030 was launched. The Agenda aims to transform systems toward an equitable society and healthy planet and has been documented in the form of a global indicator framework consisting of 17 Sustainable Development Goals (SDGs) with 169 targets and 234 indicators for ease of monitoring and measuring (United Nations, 2022). The years between 2020 and 2030 are marked as the ***Decade of Action*** for fulfilling the agenda and achieving the 2030 goals (Zhenmin, 2020). The beginning period saw a setback due to the unfortunate spread of the novel coronavirus, thus diverting the global resources toward pandemic management (Pirasteh et al., 2020). Apart from human health, the pandemic had an impact on the environmental, economic, and social structures and a tangible impact on the progress of all the 17 SDGs and in this regard, the years 2020 and 2021 were even declared recession periods by the IMF. The extent of environmental damage and limits of planetary boundaries have also come into the global conscience and the urgency for global climate action and biodiversity conservation has brought about the release of the Sixth Assessment Report of the United Nations Intergovernmental Panel on Climate Change (IPCC) (https://www.ipcc.ch/report/ar6/wg1/), under the (UNFCCC) United Nation Framework Convention on Climate Change's Conference of the Parties (COP) 26 and Convention on Biodiversity (CBD) COP-15, which has built a base for integrating biodiversity conservation into climate action. Although the world is recovering now, aggressive measures need to be taken in the aftermath of COVID-19 to make the decade of action a success, with environmental sustainability forming the foundation for economic and social recovery.

DOI: 10.1201/9781003586319-14

In addition, the SDGs built on the earlier Millennium Development Goals (MDGs) and encompass a broader set of interlinked objectives, from eradicating extreme poverty and hunger to achieving gender equality and combating climate change. Measuring progress toward these ambitious goals is critical. To that end, various indices and reports track how countries are faring. A key metric is the SDG Index, which aggregates performance across all 17 goals on a scale of 0 to 100 and interpreted as a percentage of the overall SDG achievement (dashboards.sdgindex.org). Using such indicators, this paper examines the trajectory of SDG progress in India in comparison to global averages from 2015 through 2020. Besides, the pandemic has also presented an opportunity for the world to develop more inclusive and sustainable pathways and processes in the recovery phase, thus increasing the relevance of SDGs in the post-COVID-19 world. Hence, it is important to understand the impact of COVID-19 on the SDGs to begin rebuilding toward achieving them. The work toward achieving SDGs begins with taking the SDGs from global to local. The first step is setting the national and the subnational context of the 2030 Agenda. Once the context is set, the goals and targets must be adapted from national to local levels, indicators have to be identified, local means and structures of implementation have to be designed, and monitoring frameworks have to be created from national to local levels (Pirasteh et al., 2020; UN, 2020). With the constraint of time and resources, the moment demands the development of frameworks for global collaboration, consensus, and evidence-based decision-making in dealing with developmental challenges, including environmental issues. These challenges can be mapped and analysed in the geographical context to encourage a more holistic approach at the policy level. Along with understanding the challenges, it is important to understand what is being implemented at the national level and how to measure and monitor it. Data-driven geospatial information can provide science and time-based monitoring solutions for these challenges (Alaguraja et al., 2010; Mihai, 2013; Cho et al., 2017; Shekhar & Aryal, 2019; Moomen et al., 2019; Pirasteh & Varshosaz, 2019). The global indicator framework is highly dependent on diverse and reliable data to provide the evidence base for decision-making and reporting. Thus, the availability of the data and its analysis within a geographical context can help build an understanding of resource requirements for implementation at the national level.

At the starting point of the SDGs in 2015, India faced a substantial development gap relative to high-income nations, but it has since embarked on aggressive national initiatives to improve outcomes. Over the next five years, India's scores on global SDG benchmarks showed notable improvement. For instance, the Sustainable Development Report 2020, which tracked progress up to 2019, ranked India 117th out of 166 countries, with an overall SDG Index score of 61.9, compared to 84.7 for the top performer, Sweden (testbook.comtestbook.com). This indicated that by 2019, India had achieved roughly 62% of the SDG targets on average, up from less than 50% in 2015. Globally, however, progress remained modest and uneven with many countries not on track to meet the Goals by 2030 (unstats.un.org). Key human development indicators did improve in the aggregate, for example, the proportion of youth out of school fell and communicable disease rates declined, yet other trends were worrying with food insecurity rising, and inequalities remaining pervasive (unstats.un.org).

Against this backdrop, the COVID-19 pandemic that struck in 2020, dramatically disrupted SDG efforts worldwide. This chapter focuses on the comparative outcomes of these developments and is organized into sections covering the methodology of analysis, the empirical findings on SDG progress in India, globally, pre- and post-2020, a discussion interpreting these findings, recommendations for policy action, and a conclusion. Through this structured approach, the analysis provides insights into how far India and the world have come in the first five years of the 2030 Agenda and how the crisis in 2020 altered the trajectory.

METHODOLOGY

This study employs a Comparative Trend Analysis using secondary data from authoritative SDG monitoring sources. The primary quantitative measure used is the SDG Index score published in the annual Sustainable Development Reports by the Sustainable Development Solutions Network (SDSN). As noted, the SDG Index is a composite indicator (0–100) summarizing a country's overall progress toward the 17 goals (dashboards.sdgindex.org). India's annual index scores from 2015 through 2020 are compared against the global average SDG Index over the same period to gauge relative performance. In addition, disaggregated data on individual goals are considered, especially to contrast the pre-pandemic period (2015–2019) with the pandemic's first year (2020). The analysis integrates information from United Nations SDG Progress Reports, including the Sustainable Development Goals Report 2020, which provided a snapshot of global trends before and during the initial phase of COVID-19 (unstats.un.orgunstats.un.org). For India-specific insights, the study references national SDG index reports (e.g., NITI Aayog's SDG India Index) and other research highlighting country-level progress and challenges. No primary data were collected; instead, the paper synthesizes existing statistics and findings from the above sources. Visual charts have been produced to illustrate key data: one line graph shows India vs. world SDG Index scores from 2015–2020, and a bar chart depicts comparative global performance on each SDG in 2018–2019 vs. 2020. These figures were derived from the compiled data and helped visualize the pace of change. All citations are provided as footnotes in APA style. Different reports sometimes use varying methodologies, for example, changes in indicators or index computation, so the analysis emphasizes broad trends and percentage changes rather than exact fine-grained values. By triangulating multiple reputable sources, the methodology ensures a comprehensive and credible assessment of SDG progress over the period in question.

IMPACT OF THE COVID-19 PANDEMIC ON SDGs

For the scope of this chapter, SDG 6, 7, 11, 12, 13, 14, and 15 are in focus as they directly (SDG 13, 14, and 15) or indirectly (SDG 6, 7, 11, and 12) pertain to the environmental sustainability seen in Table 11.1.

There is a marked decrease in progress in SDG 6 (Clean Water and Sanitation) and SDG 13 (Climate Action). SDGs 11, 12, and 7 (Affordable and Clean Energy)

TABLE 11.1
Impact of COVID-19

Sustainable Development Goal	Pre-Covid Era		Covid-Era	Change
	2018	2019	2020	
1 - No Poverty	54	50	60	Increase
2 - Zero Hunger	48	35	47	Increase
3 - Good Health and Well Being	52	61	74	Increase
4 - Quality Education	58	58	57	Decrease
5 - Gender Equality	36	42	48	Increase
6 - Clean Water and Sanitation	63	88	83	Decrease
7 - Affordable and Clean Energy	51	70	92	Increase
8 - Decent Work and Economic Growth	65	64	61	Decrease
9 - Industry, Innovation and Infrastructure	44	65	55	Decrease
10 - Reduced Inequalities	71	64	67	Increase
11 - Sustainable Cities and Communities	39	53	79	Increase
12 - Responsible Consumption and Production	0	55	74	Increase
13 - Climate Action	0	60	54	Decrease
14 - Life Below Water	0	0	0	NA
15 - Life on Land	90	66	66	No Change
16 - Peace, Justice, And Strong Institutions	71	72	74	Increase

have shown a significant increase with the world pushing toward renewable energy, such as green hydrogen, phasing down coal, and increasing investments in technological advancements. Unfortunately, SDG 14 (Life Below Water) and SDG 15 (Life on Land) show no change due to a lack of enough data.

SDG Index score is the aggregate score of all the goals, which helps determine the progress of a nation with respect to achieving the 2030 Agenda, as shown in Figure 11.1, with India scoring lower than the global average. The dip in COVID-19 is prominently showing a regression in SDG status at the national level due to the pandemic.

Further, as all the SDGs are interlinked as shown in Table 11.2, so is their impact. This cascading impact on each SDG is important for policymakers to understand so that they can take an interdisciplinary approach to develop solutions. Following scoring, the technique was developed to map the cascading impact.

Based on the above scoring technique, the impact on the SDGs is documented and scored accordingly in Table 11.3. An impact map has been created with a focus on SDGs directly or indirectly pertaining to environmental sustainability.

In the following diagram, Figure 11.2, the impact of the COVID-19 pandemic on sustainable goals has been mapped along with the interaction between different sustainable goals, since each goal is affected by the progress of other goals. The 17 SDGs do not exist in isolation, and there is a synergy among all goals. This synergy has been depicted through various linkages on the map. There are no data from SDGs 14 and 17. Data for SDG 15 has also not been tracked in the pandemic era, due

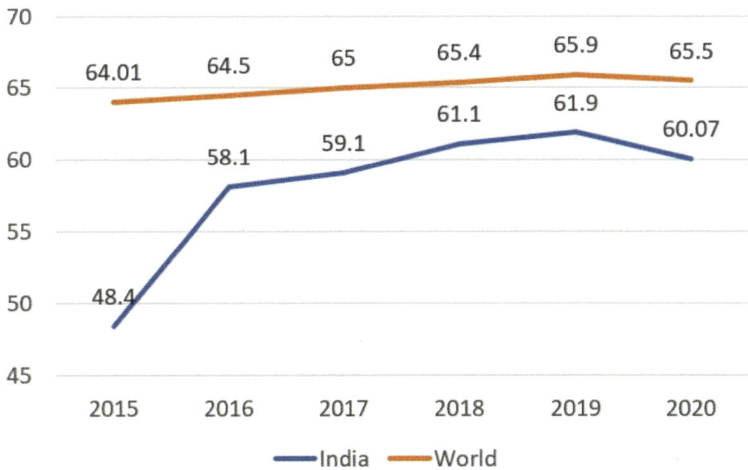

FIGURE 11.1 SDG index score of India and global average.

to which accurate status and impact for these goals have not been mapped. It gives an incomplete and obscure picture of environmental sustainability.

FINDINGS

The comparative trends in SDG Index scores from 2015 to 2020 reveal several key patterns. Globally, the SDG Index showed a slow but steady increase during the first five years of SDG implementation. The world's average index score was around 64.0 in 2015 and inched up to approximately 65.9 by 2019, indicating modest overall

TABLE 11.2
SDGs Interlinked

Score	Impact	Explanation
+3	Benefiting	A positive impact on one target caused a direct positive impact on another target
+2	Accelerating	Positive impact on one target leads to an increased pace of progress on another target
+1	Fostering	Positive impact on one target created conditions suitable for progress in another target
0	No Impact	No Impact
-1	Inhibiting	Negative impact on one target stalled the progress of another target.
-2	Decelerating	Negative impact on one target slowed the progress of another target.
-3	Detrimental	Negative impact on one target reversed the progress of another target.

Source: Designed by the author with the Relationship scorecard mentioned in the guide to interactions as the reference (http://pure.iiasa.ac.at/id/eprint/14591/1/SDGs-Guide-to-Interactions.pdf).

TABLE 11.3
Impact Scores

Goals	Targets	Key Impact in India	Impact Score
1<-->2	1.2, 1.3, 1.4--> 2.1, 2.2, 2.3	Disruption in social protection schemes.	-3
1<-->3	1.3-->3.1, 3.2	Diversion of resources.	-2
1-->4	1.3-->4.1, 4.3, 4.5	Inequal access to resources.	-3
1-->5	1.4-->5.a	Inequal resource distribution.	-3
1-->10	1.2, 1.4, 1.3-->10.1, 10.2, 10.4	Inequal access to resources.	-1
2-->3	2.1,2.2-->3.1,3.2,3.7	Disruptions in social protection schemes.	-3
2-->5	2.3-->5.1	Reduced opportunities	-3
2-->8	2.3-->8.5	Reduced opportunities	-1
3-->4	3-->4	Closure of educational institutions	-1
3-->5	3.7-->5.2	Loss of access to resources	-3
3-->8	3-->8	Economic instability	-3
3-->10	3-->10	Inequal access due to affordability	-1
4-->2	4.1-->2.1, 2.2	Disruption of social protection schemes	-3
4<-->5	4.5<-->5	Inequal access to technology	-1
4-->8	4-->8	Reduction in opportunities	-1
4-->10	4.1, 4.3, 4.5--> 10.1, 10.2	Increased inequalities	-2
5<-->8	5<-->8	Economic instability	-1
6-->3	6.1, 6.2 --> 3.3	Increased pressure on resources	-3
7-->13	7-->13	Increased demand for renewable energy	1
8<-->9	8.1, 8.3, 8.5 <--> 9.2	Economic instability	-3
8-->10	8.5-->10.1	Increased financial inequalities	-1
11-->13	11.6-->13	Improved environment	1
11-->14	11.6-->14.1	Improved water environment	1
12<-->11	12.5<-->11.6	New technologies and processes	1
12-->14	12.4-->14.1	No impact	0
13<-->14	13<-->14	Degrading air and water quality	-3
13-->15	13-->15	Decreasing terrestrial life diversity on Earth	-3
17-->16	17-->16	Increased partnerships and global cooperation	1

improvement. This reflects incremental progress on many goals worldwide but also underscores that the rate of change was not rapid enough to meet the 2030 targets, even before the pandemic as no country was on a fully sustainable trajectory. By contrast, India's SDG Index performance started from a considerably lower base in 2015 but improved faster over the same period. India's score climbed from about 48.4 in 2015 to 61.9 in 2019, a remarkable rise of over 13 points in four years. This gain reflects substantial progress in multiple domains, for example, increases in

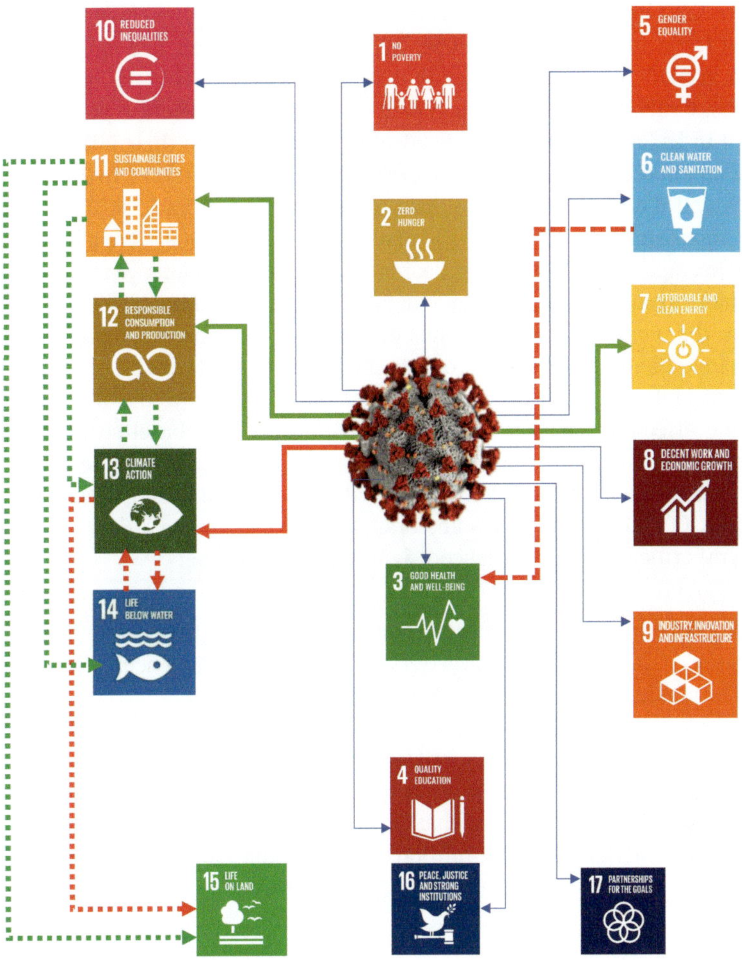

Map Key

Line	Relationship
	Negative Impact
	Negative Impact
	Positive Impact
	Positive Impact

FIGURE 11.2 Impact of COVID-19 pandemic on SDGs 2030. (The content of this publication has not been approved by the United Nations and does not reflect the views of the United Nations or its officials or member States.

Source: Created by qualitative analysis of collated information on the impact of COVID-19.

access to sanitation, education, and electricity in India, which helped narrow the gap between India and the global average. By 2019, India had reached roughly the global median in SDG achievement, though it still trailed far behind the best-performing countries. Crucially, however, the year 2020 marked an inflection point. Both globally and in India, SDG Index scores declined slightly from their 2019 peak, representing an unprecedented setback. India's index fell to about 60.1 in 2020 from 61.9 in 2019, and the global average dipped to roughly 65.5 from 65.9. An independent analysis confirmed that 2020 was the first time since the SDGs were adopted that the world's average SDG Index score decreased from the previous year, a drop largely attributable to surges in poverty and unemployment caused by the COVID-19 crisis. In short, the quantitative index trends demonstrate clear pre-pandemic gains and a sharp pandemic-era reversal in progress shown in Figure 11.3.

Figure 11.4 provides a goal-by-goal perspective on how the first year of the pandemic affected SDG outcomes, using 2018–2019 as a baseline for recent pre-pandemic progress. Across nearly all the 17 goals, the data show either stalled progress or outright declines in 2020 (represented by the red bars) compared to the immediate pre-COVID values (yellow and orange bars for 2018 and 2019).

Several critical areas experienced significant backsliding:

- No Poverty (SDG 1): After decades of steady decline in extreme poverty rates, 2020 saw global poverty rise for the first time since 1998 (un.org). The pandemic's economic fallout pushed an estimated 71 million additional people into extreme poverty worldwide (un.org). This is reflected in Figure 11.2 by a drop in the SDG 1 indicator, a clear regression from 2019. In India, which had been reducing poverty steadily, the poverty rate also spiked. One World Bank study suggests that of the tens of millions newly impoverished globally, a staggering 56 million were in India alone (fortuneindia.com), underscoring the pandemic's concentrated impact on Indian livelihoods.

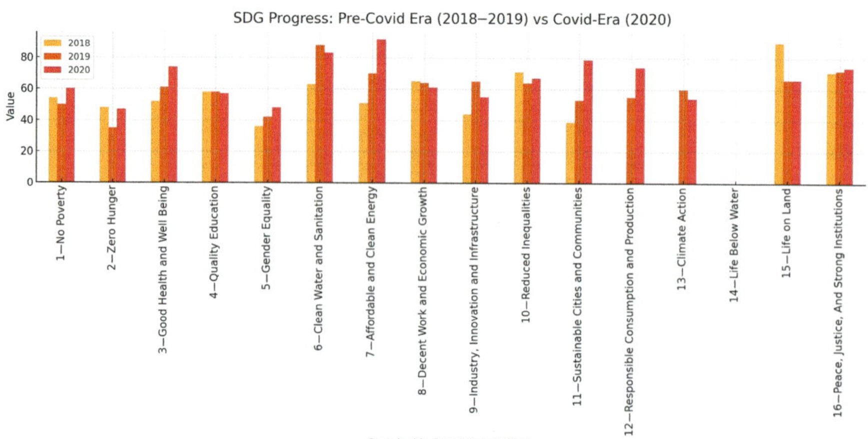

FIGURE 11.3 SDG progress (2018–2020)—global trends.

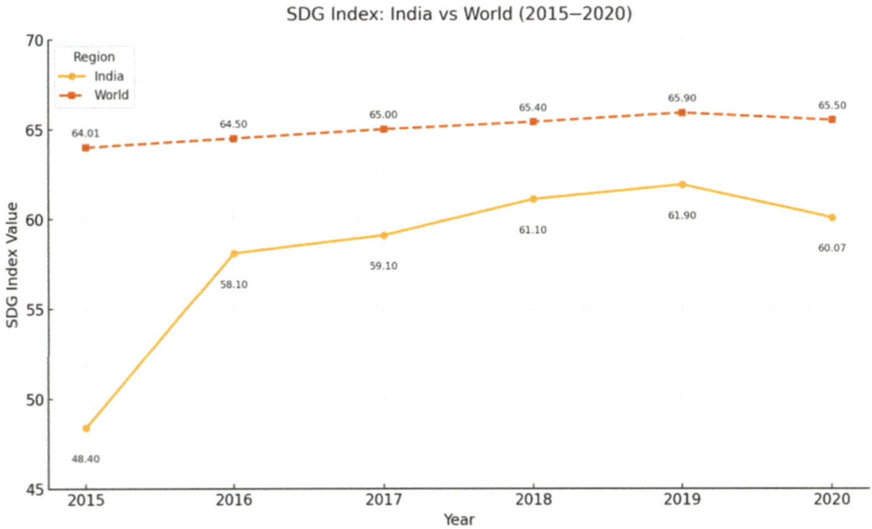

FIGURE 11.4 Comparison of global SDG performance by goal—Pre-COVID (average 2018–2019) vs. COVID-impacted (2020). Bars represent composite scores/values for each goal.

- Zero Hunger (SDG 2): Hunger and food security were already problematic as the number of undernourished people had been increasing in the late 2010s (unstats.un.org), and 2020 exacerbated this issue. Food supply disruptions, income losses, and other shocks led to tens of millions more people facing undernourishment. The Goal 2 bar in 2020 consequently fell below its 2019 level, indicating a rise in global hunger. India likewise struggled with surges in hunger and malnutrition, wiping out previous gains in reducing these rates.
- Good Health and Well-Being (SDG 3): Health outcomes deteriorated in 2020 due to both direct and indirect effects of COVID-19. Millions of COVID cases and related deaths, plus overwhelmed health systems, contributed to higher mortality and reduced access to other healthcare services. Essential health programs such as vaccinations and maternal care were interrupted in many countries. The global health goal metrics in 2020 reflected these setbacks. In India, the strain on the healthcare system was severe, and combined with pandemic-related service disruptions, it impeded progress on health indicators.
- Quality Education (SDG 4): Perhaps one of the most visibly impacted sectors, education saw an almost unprecedented disruption. By April 2020, over 1.6 billion students were out of school due to worldwide school closures (unstats.un.org). Remote learning alternatives were uneven and often inaccessible to poorer communities, leading to significant learning losses. The 2020 bar for SDG 4 shows a decline accordingly, and India experienced prolonged school closures in 2020; despite efforts at online education, large

digital divides meant many children effectively received no formal schooling for months.

- Decent Work and Economic Growth (SDG 8): The global economy contracted sharply in 2020, the worst recession since World War II, adversely affecting employment and incomes. Global working-hour losses translated to the equivalent of hundreds of millions of full-time jobs lost. The crisis severely affected the livelihoods of roughly half the global workforce, particularly those in informal or vulnerable employment (unstats.un.org). Consequently, SDG 8 indicators, which include unemployment and GDP growth rates, plummeted in 2020 and India's situation mirrored this trend after years of robust economic growth, which lifted millions out of poverty with India's GDP shrinking by about 7% in 2020, and unemployment and poverty rose sharply.
- Industry, Innovation and Infrastructure (SDG 9) and Sustainable Cities (SDG 11): These goals also saw setbacks as infrastructure projects were delayed and urban economies hit hard. In India, for example, numerous infrastructure and industrial activities stalled during lockdowns, affecting progress on SDG 9 and 11 targets like improved urban services and connectivity.
- Reduced Inequalities (SDG 10): The pandemic tended to hit hardest the most vulnerable populations within and between countries, thus widening many inequalities (unstats.un.org). While the SDG 10 indicators are complex, 2020 likely halted or reversed any progress by reducing income and opportunity disparities. India witnessed widening gaps as well; for instance, urban vs. rural divides in access to healthcare and education were accentuated during the crisis.

On a slightly positive note, a few environmental indicators saw short-term improvements in 2020. For example, global carbon emissions dropped temporarily, and air quality improved in many regions during lockdowns. This might reflect marginal upticks in SDG 13 (Climate Action) or SDG 11 (urban air quality) for 2020. However, these gains were largely transient and came at the cost of economic and social turmoil. By and large, the 2020 data show a broad-based regression in SDG progress. The pandemic erased several years of gains on multiple goals in a span of months (unstats.un.org). Both the global community and India were left in a more precarious position relative to the SDG targets at the end of 2020 than in 2019.

In summary, the findings can be divided into two phases: (1) 2015–2019, a period of gradual progress, with India making above-average strides in improving its SDG index standing, though still facing serious gaps in many goals, and (2) 2020, a sharp downturn in progress for both India and the world due to the COVID-19 shock. By 2020, India remained off-track on the majority of the goals and it continued to face "major challenges" on ten-plus SDGs, notably including zero hunger, good health, clean water and sanitation, among others, according to SDG preparedness assessments (downtoearth.org.in). The global setbacks of 2020 similarly left virtually every country with widened shortfalls. These empirical observations set the stage for a deeper discussion of the reasons behind the trends and their implications.

DISCUSSION

The UN2019 progress report on SDGs stated that while it is convenient to track the progress of Agenda 2030 through the aggregated indicators, these indicators did not completely disclose the impact of varying situations at the national, regional, and local geographical levels. There is a lack of sufficient and good-quality data generated from the local grassroots levels in many parts of the world, which hinders the appropriate tracking of SDG progress (Bird et al., 2010; Li et al., 2019). The COVID-19 pandemic has taught us how the developmental challenges are all integrated from the regional to the global level in the globalized world. The scale and range of challenges needed to be addressed for a sustainable world with a healthy environment and equitable society demand an integrated and cooperative approach by the nations. Data collection, big data processing, information sharing, and collaborative research played a significant role in helping the world cope with the pandemic. Also, previous experience in tracking MDGs has shown the world that data forms the core of any effective development initiative. Similarly, data plays a significant role in the achievement of SDGs as well, for it will form the basis of evidence-based decision-making and policy formulation. In addition, geographic information systems (GIS) are central in data hosting, sharing, imaging processing, simulation, and decision-making (Li et al., 2019; International Development Research Centre (IDRC), Canada, 2017; Yuan, 2021). The study addresses the challenge of determining the comparative process for measuring SDG progress at the global level, with realities at the local level.

The consistent tracking of SDG goals, targets, and indicators is necessary to get an accurate picture of the SDGs' progress over the years across geographies. It will help bring comparability across nations and help globally harmonize the national data collected at local levels. Thus, the location aspect is going to play a major role in data management and usage for SDG tracking. In addition, the findings highlight a complex narrative of partial success and significant shortfall in SDG implementation up to 2020. It is instructive to analyze the factors that drove progress in the early years and those that contributed to the setbacks, and compare India's trajectory with global patterns in context. Pre-Pandemic Progress (2015–2019): Globally, the moderate gains during this period can be attributed to sustained economic growth in many developing countries, technological improvements, and concerted efforts in areas like public health. However, progress was uneven across goals and regions. For example, poverty reduction was faster in East Asia than in sub-Saharan Africa; health and education improved, but hunger increased due to conflicts and climate-related shocks affecting food security. Environmental degradation such as biodiversity loss and greenhouse gas emissions, continued largely unabated despite some localized improvements. This unevenness meant that by 2019, the world was not on track to achieve the SDGs, a reality acknowledged by the UN at the SDG Summit in 2019 (unstats.un.org). India's strong improvements from 2015 to 2019 stand out, and they were driven by several national initiatives aligned with the SDGs. One notable success was in sanitation (Goal 6): the Government of India's Swachh Bharat Mission Clean India Campaign built nearly 90 million toilets between 2014 and 2019 and effectively eliminated open defecation in rural areas (en.wikipedia.org). As a result, India achieved the SDG target for sanitation (SDG 6.2) well ahead of the 2030 schedule, an accomplishment recognized as reaching the goal *11 years early*. This had positive ripple effects on health and well-being. Progress was also significant in clean energy (Goal 7); India undertook one of the world's largest renewable energy expansion programs, dramatically increasing solar and wind power capacity, which contributed to better performance on climate-related metrics and energy access. Additionally, poverty rates in India fell substantially during the late 2010s, thanks in part to economic growth and targeted poverty alleviation schemes. Over the decade of 2006–2016, India lifted an estimated 271 million people out of multidimensional poverty, highlighting the scale of improvement possible with sustained effort.

Despite these advances, India in 2019 still lagged on numerous SDGs, mirroring many global challenges but often at a larger scale due to India's population and development stage. Areas of concern included nutrition, *SDG 2*, where India continued to have one of the highest rates of child malnutrition and anemia in the world, indicating slow progress on hunger and food security. Gender equality, *SDG 5*, was another persistent gap, as evidenced by low female labor-force participation and continued gender disparities in health, education, and political representation. Quality education, *SDG 4*, remained a challenge in terms of learning outcomes, even though school enrollment was high. Economic inequality, *part of SDG 10*, in India was rising in the late 2010s, with wealth concentration increasing. Moreover, several environmental goals such as *SDGs 13, 14, 15* posed difficulties, for instance, as India struggled with severe air pollution and biodiversity losses. In short, by the eve of the pandemic, India exemplified the broader global scenario, notable progress in some areas, offset by stagnation or insufficient movement in others, leaving a significant distance to cover to attain the SDGs.

Impact of COVID-19 in 2020: As captured in the findings, the pandemic's disruptive effects on SDG progress were profound. The discussion here highlights why the impacts were so severe. Public health systems worldwide were unprepared for a shock of this magnitude; consequently, *SDG 3*, Health, suffered as countries diverted resources to fight COVID-19 at the expense of other health services and because the virus directly caused millions of deaths. Economically, the lockdowns and downturn led to a spike in unemployment and poverty—a reversal of years of poverty alleviation. The estimated addition of 71 million people to extreme poverty in 2020 means *SDG 1*, No Poverty, temporarily moved backward. This single year increase effectively erased the global poverty reduction achievements of the previous several years. For India, the economic contraction, which was about 7% of GDP in 2020, and the massive job losses in the informal sector translated into millions of newly poor households, as noted earlier. Social inequalities were laid bare: migrant workers, for example, faced enormous hardship during India's lockdown, illustrating the intersecting failures in urban planning *SDG11*, decent work *SDG 8*, and reduced inequalities *SDG 10*.

Education disruptions *SDG 4* in 2020 have long-term implications: the loss of an academic year for many students, especially in developing countries like India, could lead to lower lifetime productivity and earnings, potentially reinforcing cycles of poverty and inequality. Likewise, *SDG 2* Hunger setbacks are alarming because undernutrition early in life causes irreversible damage to human capital. The pandemic has thus created cascading effects that threaten to slow future SDG progress even after the immediate crisis. Another aspect is how the crisis exposed weaknesses in infrastructure and institutions, from digital divides impacting remote learning and work, to a lack of social safety nets to protect the most vulnerable. All these factors explain why the SDG index scores fell and why the global progress reversal in 2020 was so widespread.

India vs. Global Comparison: In broad terms, India's SDG progress has been a microcosm of global trends, but with some differences in scale and speed. India's improvements pre-2020 were relatively faster, in percentage terms, than many higher-income countries, reflecting the potential for catch-up growth and the implementation of some high-impact programs, as discussed: sanitation and renewable energy. However, by 2020, India still ranked low in absolute terms, 117th globally, indicating that other countries also progressed, and many were far ahead in overall SDG achievement. One reason is that developed countries had already largely met basic targets on poverty, hunger, etc., whereas India was still working to fully address these. Thus, India's rapid gains were narrowing a historical gap. The global context also includes countries that stagnated or regressed due to conflicts or other crises in the late 2010s, which India avoided during that period. The pandemic's impact on India was somewhat harsher than on many other countries, especially economically given India's GDP drop in 2020 was among the steepest of major economies. Consequently, the magnitude of India's 2020 SDG setback was particularly large in areas like poverty increase. On the other hand, some wealthy nations also saw significant SDG challenges in 2020, for instance, certain European countries who had high COVID mortality impacting *SDG 3*, and the United States who saw rising inequalities and hunger. So, 2020 was a global shock felt across the board; the difference is that countries like India, with more developmental deficits to begin with, had more ground to make up post-pandemic.

In summary, the discussion underscores that while there were important gains in 2015–2019, neither India nor the world was on a trajectory to achieve all SDGs by 2030, even before COVID-19. The pandemic then dealt a serious blow, amplifying pre-existing weaknesses. The comparative perspective reveals that accelerating progress, especially for countries lagging behind, like India, building resilience against such shocks is critical moving forward. The next section outlines recommendations to address these issues.

RECOMMENDATIONS

Considering the above analysis, it is clear that extraordinary efforts will be required through 2030 to recover lost ground and hasten progress on the SDGs. The following recommendations are proposed for India and the global community, focusing on strategic areas to bolster SDG achievement in the post-2020 context:

1. **Strengthen Health Systems and Social Protection:** Governments should invest in robust healthcare infrastructure and universal health coverage to ensure resilience against future pandemics and health crises. This includes enhancing disease surveillance, emergency response, and vaccine

manufacturing/distribution capacities. Alongside health systems, expanding social safety nets such as unemployment insurance, cash transfers and food security programs are vital to protect the most vulnerable from economic shocks. Such measures would directly support *SDG 3, Health* and *SDG 1, No Poverty*, creating a buffer that prevents people from falling into poverty during crises.

2. **Inclusive and Green Economic Recovery:** As countries implement recovery plans, they must prioritize inclusive, sustainable growth that creates jobs while accelerating the transition to a low-carbon economy. Stimulus packages and infrastructure investments should be directed toward renewable energy, energy efficiency, clean transport, and climate-resilient agriculture, aligning with SDG 7 (Affordable & Clean Energy) and SDG 13, Climate Action. At the same time, emphasis on decent job creation and support for small businesses can help reduce inequalities, *SDG 8 and SDG 10.* An inclusive recovery also means focusing on education catch-up programs and digital inclusion like broadband access and digital skills so that the gains of growth are broadly shared and human capital losses from 2020 are mitigated.
3. **Mobilize Financing and Partnerships**: Achieving the SDGs by 2030 will require significant financial resources, especially in developing countries. The international community should enhance support through grants, concessional loans, and debt relief initiatives to free up fiscal space in low and middle-income countries. Strengthen global partnerships, SDG 17, is crucial: developed countries, multilateral banks, and the private sector all have roles in scaling up investment in sustainable development. For India, increased collaboration with global funds for climate, health, etc. and public-private partnerships that bring in technology and funding. Additionally, within-country resource mobilization must improve, for instance, better tax collection and curbing illicit financial flows can generate domestic revenues for SDG programs. In sum, a coordinated multilateral response is needed so that countries like India are not constrained by a lack of finance in pursuing SDG targets.
4. **Integrate SDGs into Policy Planning and Data Systems:** To ensure sustained focus, nations should embed SDG targets into their national development plans, budgets, and monitoring frameworks. As noted in one analysis, India's national and state budgets did not explicitly align with SDG priorities as of 2022; this needs to change. Mainstreaming the SDGs in policy means each ministry and state aligns projects with relevant goals and regularly reports on progress. Enhancing statistical capacities is also important: investing in data collection like timely surveys and SDG indicator tracking that will help identify gaps and guide interventions. Data-driven policy can better target underserved areas and groups, thereby improving effectiveness. In addition, transparency and citizen participation in SDG monitoring can bolster accountability. By institutionalizing the SDGs in governance, governments essentially make *SDG achievement a core governance mandate,* countries can then more systematically drive progress.

By pursuing these recommendations, India and other countries can aim not only to get back on track but to "build back better," creating more resilient and equitable systems. Indeed, as the UN Secretary-General urged, the world's recovery from COVID-19 must be leveraged to reshape economies and societies to be more inclusive and sustainable than before. The above measures resonate with that vision, addressing immediate weaknesses revealed by the pandemic and laying a stronger foundation for achieving the 2030 Agenda.

CONCLUSION

The degree of impact might vary, but every single goal of the 17 SDGs has been positively or negatively affected by the COVID-19 pandemic. Environmental sustainability-related goals have been at the center of global discussions in the post-pandemic world. Even as the impact varies across geographies, the COVID-19 experiences from various countries indicate that the more information we can access and analyze together, the more effective nations we have. This collaboration of partners requires several key components, such as existing and accessible geospatial information and building the spatial data infrastructure (SDI). Geospatial information requires the integration of statistics and is used in various geographies on a local, national, and global scale. We need technologies to demonstrate our geographic dimension on a global scale, and consider information that can be conveyed clearly and in a more useful way than statistical data. We suggest empowering the implementation of geospatial technologies and methods for effective monitoring of SDG progress from a global to a local scale. Moreover, from 2015 to 2019, the world, including India, made measurable but insufficient progress toward the SDGs. India's case demonstrates both the possibilities and the challenges inherent in the SDG journey: targeted initiatives can yield rapid improvements as seen in sanitation, energy, and poverty reduction, yet deeply entrenched issues like malnutrition and gender inequality require much longer-term, systemic efforts. The global comparison highlights that many countries shared similar patterns of partial progress, and that no region was fully on pace to meet all goals by 2030. In 2020, the COVID-19 pandemic emerged as a colossal setback, abruptly halting or reversing gains on multiple fronts. The first five-year "window" of SDG implementation thus closed with a sobering reality, instead of accelerating, progress in 2020 regressed for the first time, leaving the world farther from the targets than it had been a year earlier.

As of 2020, India and the global community have entered the United Nations-declared ***Decade of Action*** with a clear imperative: dramatically scale up and speed up SDG efforts. The findings of this analysis make plain that "business as usual" will not suffice. Recovering from the pandemic's impacts and surpassing the pace of previous progress is essential if the SDGs are to be salvaged by 2030. This will entail implementing the recommendations discussed, such as fortifying healthcare, pursuing green growth, financing the goals, and improving governance for SDGs thus and fostering a sense of shared urgency and commitment. It is heartening that the ethos of ***leave no one behind***, which underpins the SDGs, has gained renewed emphasis in the wake of the pandemic's unequal toll. Going forward, policies must consciously aim to reduce disparities, be they economic, social, or digital, so that development gains reach all segments of society. In conclusion, the period 2015–2020 offers valuable lessons. It showed that progress is possible, even in challenging contexts, when sound policies and adequate resources are applied, as evidenced by several success stories in India and elsewhere. It also laid bare the vulnerabilities and gaps that must be addressed, especially after the catastrophic shock of COVID-19. The road to 2030 is now shorter and steeper, yet the SDG vision can still be approached with resolute action, innovation, and international solidarity. In the words of the UN Secretary-General, everything done in the recovery from this crisis should be oriented toward *building more equal, inclusive, and sustainable economies and societies* that are resilient to future crises (unstats.un.org). India and the world at large have no time to lose in translating this vision into reality—the sustainable future envisioned in the 2030 Agenda depends on the choices and commitments made in this decisive decade.

ACKNOWLDGMENT

This chapter of the book is dedicated to students who may work on the new transformative 2030 agenda for SDGs 2030 and beyond. The authors appreciate the reviewers of this chapter for their remarkable comments.

REFERENCES

Alaguraja, P., Yuvaraj, D., & Sekar, M. (2010). Remote sensing and GIS approach for water pollution and management in Tiruchirappalli taluk, Tamil Nadu, India. *International Journal of Environmental Science*, *1*, 66–70.

Bird, K., Higgins, K., & Harris, D. (2010). Spatial poverty traps: An overview. Overseas Development Institute 111 Westminster Bridge Road London SE1 7JD. ISBN 978 1 907288 11 1.

Cho, N.-W., Maeng, J. H., & Lee, M.-J. (2017). Use of environmental geospatial information to support post-management of environmental impact assessment. *Korean Journal of Remote Sensing*, *33*(5-3), 799–807 (in Korean with English abstract). https://doi.org/10.7780/kjrs.2017.33.5.3.4.

International Development Research Centre (IDRC), Canada. (2017). Big data and SDGs: The state of play in Sri Lanka and India. International Development Research Centre (IDRC), Canada. p. 23.

Li, G., Cai, Z., Liu, J., Liu, X., Su, S., Huang, X., & Li, B. (2019). Multidimensional poverty in rural China: Indicators, spatiotemporal patterns and applications. *Social Indicators Research*, *144*(3), 1099–1134.

Mihai, V. (2013). The role of geospatial technologies, geographic information and ICT in promoting rural communities sustainable development in Transylvania. *Academica Science Journal, Geographica Series*, *3*(2013), 90.

Moomen, A. W., Bertolotto, M., Lacroix, P., & Jensen, D. (2019). Inadequate adaptation of geospatial information for sustainable mining towards Agenda 2030 sustainable development goals. *Journal of Cleaner Production*, *238*, 117954., ISSN 0959-6526, https://doi.org/10.1016/j.jclepro.2019.117954.

Pirasteh, S., Husain Hishmi, J., & Tammineni, R.. (2020). COVID-19 pandemic challenges and impacts on the SDGs 2030: Indian perspective. CRC Press, Taylor & Francis Group ISBN 9780367775315, 544 Pages 148. P.455-468.

Pirasteh, S., & Varshosaz, M. (2019). *Sustainable development goals connectivity dilemma land and geospatial information for urban and rural resilience*. CRC Press, Taylor & Francis Group, ISBN:13:978-0-367-25935-8, P.93-108.

Shekhar, S., & Aryal, J. (2019). Role of geospatial technology in understanding urban green space of Kalaburagi City for sustainable planning. *Urban Forestry & Urban Greening*, *46*, 126450, ISSN 1618-8667. https://doi.org/10.1016/j.ufug.2019.126450.

United Nations (2022). SDG Indicators: Global indicator framework for the Sustainable Development Goals and targets of the 2030 Agenda for Sustainable Development. Retrieved on 27 July.

United Nations High-Level Political Forum for Sustainable Development (2020). Decade of action: taking SDGs from Global to Local, India Voluntary National Review 2020. Retrieved on 27 July 2020.

Yuan, M. (2021). Geographical information science for the United Nations' 2030 agenda for Sustainable Development. *International Journal of Geographical Information Science*, *35*(1), 1–8.

Zhenmin, L. (2020). Finding transformative pathways in turbulent times. The Sustainable Development Goals Report 2020. Retrieved on 27 July. Sixth Assessment Report (ipcc.ch).

Part 4

Strong Institutions and Sustainable Development

12 SDG#16—The Role of Spatially Enabled Governance in the Implementation of the SDGs

Jeanne-Marie Col

INTRODUCTION

Indicators for SDG 16—Peace, Justice, Strong Institutions, and Inclusiveness—can be measured with geospatial data. And geospatial data can be aligned to government jurisdictions, providing a basis for greater accountability. SDG 16 involves the characteristics of government jurisdictions that can facilitate reduction in violence, ensure fairness and equal access to justice, and establish government institutions and processes that can make decisions, provide services to all citizens, and ensure inclusiveness. But concepts such as fairness, equal access, inclusiveness, and effectiveness are vague, controversial, difficult to measure, and often politically sensitive and governance concepts are defined differently in different societies, and some governments are reluctant to share their data with their citizens and while governments are happy to take credit for success, they are reluctant to be blamed for lack of progress. It is not surprising, therefore, that measurement of SDG 16 governance indicators has lagged other Sustainable Development Goal (SDG) indicators. Many governments also do not want to publicize information showing the incidence of violence or unfairness. However, using geospatially enabled indicators for governance concepts provides a framework for identifying conditions and progress in specific government jurisdictions. A high level of public transparency can lead to accountability, as well as providing a platform for better identification of problems and their solutions.

Geospatially engaged indicators bring a geographic lens to the measurement of progress towards peace, justice, strong institutions, and inclusiveness that contribute to accomplishment of the SDGs in communities. Geography is important to societies, which divide their territory into geographic jurisdictions such as provinces, counties, municipalities, and wards. Geospatial data facilitates data aggregation and disaggregation to all levels of government, thereby adding accuracy and precision in tracking progress towards the SDGs.

One of the best ways to learn how to strengthen SDG performance is to compare performance in different jurisdictions. Analysis of geospatial data facilitates the

DOI: 10.1201/9781003586319-16

discovery of exemplary practices in other jurisdictions, encouraging mutual learning among jurisdictions. With comprehensive geospatial data, jurisdictions can compare their causal factors with their impact measures. Geospatial data is also essential to "leaving no one behind" as it enables public servants and citizens to locate geographical pockets of marginalization.

Indicators for the various targets of the 17 SDGs were proposed by a wide range of civil society organizations, national statistical offices, UN agencies and other stakeholders. This discussion of SDG 16 includes indicators chosen through this process, as well as other widely recognized governance indicators. At the inception of the SDGs in 2015, because there is a wide variety of governance practices throughout the world, there was disagreement on the definition of elements such as peace, justice, participation, and inclusion among the many stakeholders. Among the variety of government models around the world, there are many different governing frameworks, such as institutional arrangements and especially the role of citizens in governing processes. While the UN has established many conventions about proper civic behavior, the UN has been careful to avoid promoting institutional arrangements and excels in appreciating the variety of experiences among societies. The following ideographic Figure 12.1 illustrates some SDG 16 targets.

Measures of the effectiveness of legislatures, bureaucracies, and judiciaries, and especially inclusiveness, are still hotly debated. Many of these measures lack established measurement methodologies and are not regularly produced by National Statistical Offices. These limitations served as a call to governments and civil society organizations to explore ways to measure these governance-related concepts. Besides the agreed-upon minimal indicators, there is a wider range of governance indicators developed by organizations such as the Worldwide Governance Indicators (World Bank Institute, 2024), and Gothenburg's Quality of Governance Standard Data Set (Teorell et al., 2025). These more complex measures of governance allow

FIGURE 12.1 SDG16 targets (The content of this publication has not been approved by The United Nations and does not reflect The views of The United Nations or its officials or Member States https://www.un.org/sustainabledevelopment)

researchers, officials, and citizens to compare their jurisdiction's progress and processes with other jurisdictions, within the same county, country or globally. With the addition of geospatial data—where the data is geo-tagged—people can easily make comparisons to national and global standards and between jurisdictions. Where the data is also time-stamped, jurisdictions can also track progress from past conditions to current data. These data characteristics provide effective monitoring of progress by government jurisdictions, a framework for future planning and priority-setting, as well as a citizen-window to oversee and evaluate government performance.

When the data is made accessible to everyone, including ordinary citizens, the data promotes transparency, accountability, and participation. Open data systems allow government officials and citizens alike to view, analyze, and use data. The process of opening datasets to all stakeholders and facilitating the participation of citizens in providing, validating, visualizing, and using data also leads to governance effects, such as greater inclusiveness and citizen empowerment. This chapter reviews the theory and practicality of using geospatial data to measure governance indicators, as well as promoting the widespread sharing and use of data for sustainable development. Several case studies will illustrate these processes and results.

THEORY OF GOVERNANCE

Governance encompasses both the structure of a society and the behavior of the people within that society. "Governance is the framework by which [governments] are directed and controlled. It identifies who can make decisions, who has authority to act on behalf of the [government] and who is accountable for how government and its people behave and perform" (Chartered Governance Institute cgi.org.uk). In most societies, there is a government structure with people—civil servants—carrying out functions within that structure. This structure is established in the country's constitution and the processes which are set out in its laws. The preeminence of the constitution and laws is expressed in the concept "rule of law". One of the most important aspects of a constitution is the description of the relationship between the government and the people. In *The Narrow Corridor*, Acemoglu and Robinson (2019), state that "Both state and society must be strong. A strong state is needed to control violence, enforce laws and provide public services that are critical for a life in which people are empowered to make and pursue their choices. A strong, mobilized society is needed to control and shackle the strong state. It is in this corridor that the state and society balance each other out" (Acemoglu and Robinson, 2019). So, for accomplishing the SDGs, countries and societies need a state apparatus that can deliver services and a society with enough voice to provide advice, monitor progress and evaluate outcomes. The constitution also provides a framework for distributing responsibilities—especially among legislative, executive, and judicial functions. Some countries have parliamentary systems in which the national leader, the prime minister, is chosen from the elected legislators. Other countries have a separation of powers, whereby the leader, president, is elected independently. Some countries fuse the two systems but, in all cases, the judiciary is appointed but expected to decide cases and principles without undue influence from the executive and legislative branches. Among actual nations, there is a wide variety in the strength of the different branches and the independence of one branch from the others. The three functions exist at the national and local levels. For instance, in a devolved system, there is a county leader with their executive, who is separately elected from the elected legislators in a county assembly. One measure of governance is the degree of independence between the executive and the legislative branches, as well as the degree of independence of the judiciary at all levels. This variable is difficult to measure and is affected by politics. Further, the constitution provides a framework for intergovernmental relations. In a unitary state, the national government controls all decisions and actions, down to the lowest local level. In a federal state, lower-level governments—often states, provinces or counties—have specified powers and resources, and directly interact with citizens. There is a wide variety of examples of centralization and decentralization among nations. Recently, some societies have changed their governmental systems to devolve more responsibilities and resources to local governments. For example, Kenya devolved authority and responsibility for many development functions to its 47 counties (Kenya Law Review Commission, *Constitution*, Chapter 11: Devolved Government). Successful implementation of the SDGs involves capable local governments that collaborate closely with their citizens. There are many models of government-citizen relations, ranging from mere consultation to perfunctory participation to robust empowerment. How can a society be more powerful in relation to the state? Society can be characterized by how inclusive it is. That is, which citizens are informed, involved and powerful? And which citizens are excluded from governmental processes? These questions are politically sensitive. Only the most inclusive of governments reach out to citizens who are most remote, less educated, and with the lowest income. If the theme is "leave no one behind", then the government needs to expand the space for public participation and provide access to accurate information—to even the most marginalized people.

Most importantly, citizens need access to data, preferably in real time and in understandable forms. Equally important is that the citizens cannot be isolated as individuals but must be gathered regularly to review community needs and progress of the SDGs. Local governments can facilitate involving citizens in data gathering, analysis and use and with greater involvement of citizens in decisions and evaluation, the government officials will have more accurate information on the needs and the impact of government activities. Furthermore, by examining needs and progress, citizens develop a more holistic and comprehensive view of development and participate in deciding priorities and trade-offs. While participating in these sessions, citizens experience efficacy, trust, and mutual respect. While there are many sectoral forums for discussing strategies for development, it is only in general-purpose governments like counties or municipalities where whole-of-society priorities can be set and trade-offs between priorities can be debated and decided for whole-of-government strategies. Through these processes, the citizens build more inclusive government and community processes.

The United Nations Committee of Experts on Public Administration (2018) decided on eleven principles of effective governance that could contribute to accomplishment of the SDGs. The eleven principles are divided into three categories: Effectiveness, accountability, and inclusiveness. Effectiveness is measured by competence, such as expertise, resources, and tools and sound policy making, which includes strengthening national statistical systems and open data sharing, together with collaboration among all levels and sectors of government, including non-governmental organizations and citizens. Accountability is measured by integrity, honesty, fairness, moral values, and transparency including public scrutiny of government operations and sharing of open data. And independent oversight with a degree of independence of oversight of institutions and the inclusion of citizens in oversight functions. Inclusiveness is measured by leaving no one behind and considering the needs and aspirations of all people, non-discrimination in access to public services, and protection of human rights. This includes participation where people are actively involved in matters that directly affect them, and they have a chance to influence policy. Subsidiarity, which focuses on decentralization allows central authorities to perform only those tasks which cannot be performed more effectively at lower local levels. Intergenerational equity, as another way to ensure fairness, balances short-term needs of today's generation with the long-term needs of future generations. Measuring these variables is controversial and often politicized. More important is the need to measure these variables with geospatially influenced measures so that local jurisdictions can be held accountable and can make improvements to their performance.

The United Nations Economist Network (2025) noted that "the desirable future of governance could be shaped by technological advancement, participatory governance and decentralization". To facilitate participatory governance and decentralization, technological advancement is likely to be characterized by open geospatial data. With geospatial data and data visualization, ordinary citizens can perceive data in picture form to create understanding about the natural and built environment and perceive strategies for improvement. In a decentralized government system—following the principle of subsidiarity—functions are performed at the lowest level that is practical and with the participation of local level stakeholders. Geospatial data facilitates subsidiarity.

PRACTICALITY OF SDG INDICATORS OF GOVERNANCE

It is possible to align indicators of effective governance with the UN Guidelines for SDG 16 indicators (Inter-Agency and Expert Group on SDG Indicators (AIEG-SDGs), as well as supplemental governance indicators. The official UN SDG Indicators were formulated by the AIEG-SDGs and agreed upon by the United Nations Statistical Commission (UNSC) in 2017. These indicators are updated annually by the UNSC. Table 12.1 compares UNDESA's Effective Governance Indicators with those established for the SDG 16 by the AIEG-SDGs and with supplemental suggestions.

Given the wide diversity of definitions of governance and its indicators, we recall that SDG 16 focuses on peace, justice, inclusiveness, and institutional strengthening. According to the SDG indicators, peace is measured by the reduction in violence, and justice is measured by access to and fairness in judicial processes. Targets 16.1 and 16.2 specifically relate to the outcomes of reducing or eliminating violence, especially against children, and deaths in society. Target 16.3 refers to the outcome of promoting the rule of law, which includes equal treatment and access to justice.

TABLE 12.1
Comparing Governance Concepts and Indicators (Comments in Italics)

Effective Governance Indicators (UNDESA, 2023)	Established SDG 16 Indicators (UN AIEG-SDGs and UNSC, 2021)	Supplemental Indicators and *comments in italics*
Effectiveness		
Competence (expertise, resources and tools)	**16.2.1** The proportion of children aged 1–17 years who experienced any physical punishment and/or psychological aggression by caregivers in the past month; **16.2.2** Number of victims of human trafficking per 100,000 population, by sex, age and form of exploitation; and **16.2.3** Proportion of young women and men aged 18–29 years who experienced sexual violence by age 18. **16.6.1** Primary government expenditures as a proportion of original approved budget, by sector (or by budget codes or similar); and **16.6.2** Proportion of population satisfied with their last experience of public services.	*Data on the incidence of social violence can be captured though law enforcement statistics and checked by community awareness of social violence incidents.* For the local level, one could ask "to what extent are community assemblies capable of questioning, investigating and exercising oversight over the operations of the county execution?" and "to what extent do members of the public perceive that county institutions to be responsive to their concerns in a transparent manner?" (TAP Network)
Sound Policy Making (strengthening national statistical systems, open data sharing	**16.10:** Ensure public access to information	Adding geospatial dimension to national statistical systems, along with open data sharing (Teorell et al., 2025)
Collaborations (all agencies, all sectors, all levels, NGOs, citizens)	**16.1.1** Number of victims of intentional homicide per 100,000 population, by sex and age; 16.1.2 Conflict-related deaths per 100,000 population, by sex, age, and cause; **16.1.3** Proportion of population subjected to (a) physical violence, (b) psychological violence and (c) sexual violence in the previous 12 months; and 16.1.4 The proportion of the population that feel safe walking alone around the area they live. **16.4.1** Total value of inward and outward illicit financial flows (in current United States dollars); **16.4.2** Proportion of seized, found or surrendered arms whose illicit origin or context has been traced or established by a competent authority in line with international instruments.	*Data on the incidence of social violence can be captured though law enforcement statistics and checked by community awareness of social violence incidents.*

(Continued)

TABLE 12.1 ***(Continued)***
Comparing Governance Concepts and Indicators (Comments in Italics)

Effective Governance Indicators (UNDESA, 2023)	Established SDG 16 Indicators (UN AIEG-SDGs and UNSC, 2021)	Supplemental Indicators and *comments in italics*
Accountability		
Integrity (honesty, etc. morality	**16.5.1** Proportion of persons who had at least one contact with a public official and who paid a bribe to a public official, or were asked for a bribe by those public officials, during the previous 12 months; **16.5.2** Proportion of businesses that had at least one contact with a public official and that paid a bribe to a public official, or were asked for a bribe by those public officials during the previous 12 months	*Targets 16.4 and 16.5 refer to outcomes that reflect a reduction in corruption, whereby all money allocated to development is applied to development rather than being mis-directed to fraud, waste and abuse.*
Transparency (share information, public scrutiny of government operations	**16.10.2** Number of countries that adopt and implement constitutional, statutory and/or policy guarantees for public access to information	*Need data on ease of access and understandability.* *Geospatial dimension offers greater specificity and accuracy*
Independent Oversight (build trust through independent institutions	Targets **16.4 and 16.5** refer to outcomes that reflect a reduction in corruption, whereby all money allocated to development is applied to development rather than being misdirected to fraud, waste and abuse.	*Disaggregating this data by jurisdiction facilitates the identification of responsible government authorities. At the lowest level, ordinary people are more likely to be aware that development projects are not reaching fruition, and that money has disappeared, and where anti-corruption measures could be most effective. Robust data gathering and visibility can create ways to trace the use and abuse of money.*
Inclusiveness		
Leaving No One Behind (consider the needs and aspirations of all people)	**16.7.1** Proportions of positions in national and local institutions, including (a) the legislatures; (b) the public service; and (c) the judiciary, compared to national distributions, by sex, age, persons with disabilities and population groups; and 16.7.2 Proportion of population who believe decision-making is inclusive and responsive, by sex, age, disability, and population group.	*While representation and sentiment are generally important, it is more effective to disaggregate these and other measures to the local level, where public participation can be facilitated.* For example, "are county and ward residents regularly consulted by government officials and within official government processes?"; or "the proportion of the county and ward populations that have access to data and documents involved in decision-making, as well as monitoring and evaluating county programs" (TAP Network)

(Continued)

TABLE 12.1 *(Continued)*

Comparing Governance Concepts and Indicators (Comments in Italics)

Effective Governance Indicators (UNDESA, 2023)	Established SDG 16 Indicators (UN AIEG-SDGs and UNSC, 2021)	Supplemental Indicators and *comments in italics*
Non-discrimination (access to public services and human rights protections)	**16.3.1** Proportion of victims of violence in the previous 12 months who reported their victimization to competent authorities or other officially recognized conflict resolution mechanisms; **16.3.2** Unsentenced detainees as a proportion of overall prison population; **16.3.3** Proportion of the population who have experienced a dispute in the past two years and who accessed a formal or informal dispute resolution mechanism, by type of mechanism	Supplemental measures could include the capacity of criminal justice systems to investigate and adjudicate and the accessibility, impartiality and effectiveness of criminal justice systems. Surveys could measure sentiment about fairness in the justice system and trust in police and judicial institutions. Use measures such as the speed with which cases and trials are dealt with, and citizen satisfaction with the justice process, including the feeling that the law is being applied fairly to all cases. Disaggregate justice data into geographic jurisdictions in order to identify the areas where these variables are better to serve as examples and where these variables are worse to provide guidance for improvements.
Participation (all people involved in matters that affect them and have a chance to influence policy, including community-driven development)	**16.7.1** Proportions of positions in national and local institutions, including (a) the legislatures; (b) the public service; and (c) the judiciary, compared to national distributions, by sex, age, persons with disabilities, and population groups; **16.7.2** Proportion of population who believe decision-making is inclusive and responsive, by sex, age, disability and population group.	*Participation should be measured by citizen participation in local development activities.*
Subsidiarity (all tasks should be performed effectively at the lowest possible level of government)	At the lowest level, such as county or ward, ordinary people are more likely to be aware that development projects are not reaching fruition, and that money has disappeared. Robust data gathering and visibility can create ways to trace the use and abuse of money.	*By 2025, most forums were focusing on local government arenas as the most promising place to focus on SDG implementation.*

The indicators for violence in society include 16.1.1: Number of victims of intentional homicide per 100,000 population, by sex and age; 16.1.2: Conflict-related deaths per 100,000 population, by sex, age and cause; 16.1.3: Proportion of population subjected to (a) physical violence, (b) psychological violence, and (c) sexual violence in the previous 12 months; and 16.1.4: The proportion of the population that feel safe walking alone around the area in which they live. Data on the incidences of social violence can be captured through law enforcement statistics and checked by community awareness of social violence incidents. Various other measures relating to sub-sets of crimes and sub-sets of victims could be identified and geo-coded for location. While it is interesting to have these data sets at the national level, it is even more useful to be able to disaggregate the data to the local level, thereby engaging the local government to work to reduce the problems.

Indicators for reducing violence and trafficking children include: 16.2.1: The proportion of children of ages 1–17 years who experienced any physical punishment and/or psychological aggression by caregivers in the past month; 16.2.2: Number of victims of human trafficking per 100,000 population, by sex, age, and form of exploitation; and 16.2.3: Proportion of young women and men aged 18–29 years who experienced sexual violence by age 18. Violence against children is often underreported because much of it takes place in domestic settings. Supplemental measures for crimes against children could include the number and location of child soldiers and other categories of crimes against children.

Indicators for the justice system include 16.3.1: Proportion of victims of violence in the previous 12 months who reported their victimization to competent authorities or other officially recognized conflict resolution mechanisms; 16.3.2: Unsentenced detainees as a proportion of overall prison population; and 16.3.3: Proportion of the population who have experienced a dispute in the past two years and who accessed a formal or informal dispute resolution mechanism, by type of mechanism. Supplemental measures could include the capacity of criminal justice systems to investigate and adjudicate, and the accessibility, impartiality and effectiveness of criminal justice systems. In addition, surveys could measure sentiment about fairness in the justice system and trust in police and judicial institutions. The functioning of the justice system can be measured by the speed with which cases and trials are dealt with, and citizen satisfaction with the justice process, including the feeling that the law is being applied fairly to all cases. It is important to disaggregate justice data into geographic jurisdictions to identify the areas where these variables are better to serve as examples and where these variables are worse to provide guidance for improvements.

Another dimension of justice is the reduction in corruption, which takes money away from development and into the hands of people who are involved in activities such as bribery, fraud, or self-dealing. Targets 16.4 and 16.5 refer to outcomes that reflect a reduction in corruption, whereby all money allocated to development is applied to development rather than being misdirected to fraud, waste and abuse. Disaggregating this data by jurisdiction facilitates the identification of responsible government authorities. Unfortunately, these data are usually measured at the national level and seldom disaggregated to the lowest levels, where anti-corruption initiatives could be most effective. At the lowest level, such as the administrative unit "county" or "ward", ordinary people are more likely to be aware that development projects are not reaching fruition, and that money has disappeared. Robust data gathering and visibility can create ways to trace the use and abuse of money. However, data on corruption measures are usually underreported. Also, where corruption is more prevalent, there is a tendency to ignore instances of fraud, waste and abuse, as just part of normal life. With open geospatial data on corruption variables such as bribery, institutions can create stronger anti-corruption initiatives in government jurisdictions.

Furthermore, data on corruption is not only at the national level but also at the international level, due to the huge influence of international crime. Indicators for international crimes include: 16.4.1: Total value of inward and outward illicit financial flows (in current United States dollars); and 16.4.2: Proportion of seized, found, or surrendered arms whose illicit origin or context has been traced or established by a competent authority in line with international instruments. These indicators are important to understanding factors that can increase the overall loss of funding for development programs, but their solution is found in strengthening international regulations of banking and arms dealing.

At the more local level, instances of acts such as bribery and self-dealing are apparent to local stakeholders. Indicators for bribery include: 16.5.1: Proportion of persons who had at least one contact with a public official and who paid a bribe to a public official, or were asked for a bribe by those public officials, during the previous 12 months; and 16.5.2: Proportion of businesses that had at least one contact with a public official and paid a bribe to a public official, or were asked for a bribe by those public officials during the previous 12 months. Due to lack of concrete data on corruption, it is common to measure, instead, perception of corruption. For example, supplemental measures for bribery might include the percentage of businesses that believe organized crime imposes costs on businesses in their country.

The rest of the SDG 16 Targets 16.6 to 16.10 refer to strengthening institutions, especially increasing inclusive citizen participation in planning, tracking, and evaluating government activities. For example, Target 16.6: Develop effective, accountable, and transparent institutions at all levels is measured by the indicator 16.6.1: Primary government expenditures as a proportion of original approved budget, by sector (or by budget codes or similar); and indicator 16.6.2: Proportion of population satisfied with their last experience of public services. But these two indicators are very general and need to be addressed at the local level, as most service delivery operations are at the local level. To disaggregate to the local level, these measures should be defined geospatially. For the local level, one could ask "to what extent are community assemblies capable of questioning, investigating and exercising oversight over the operations of the county execution?" and "to what extent do members of the public perceive that county institutions should be responsive to their concerns in a transparent manner?" (TAP Network).

For performance data to be useful in governance at the local level, these variables need to be mapped geospatially.

CASE STUDY: MAPPING MODI GAUPALIKA IN NEPAL

An example of increasing the effectiveness of local institutions, "Mapping Modi Gaupalika" recruited volunteers to this local jurisdiction in Nepal. They used a GPS Survey, along with self-developed mobile applications to collect the data. Through this survey, the local government—Modi Gaupalika—was able to facilitate the construction and maintenance of roads and monitor service delivery in several key sectors. "Because these maps existed in an open spatial platform rather than relying on the tabular information in the municipal profile documents, there was accountability and transparency to the users and the community" (Sharma et al., 2023, pp. 107–109).

Target 16.7: Ensure responsive, inclusive, participatory, and representative decision-making at all levels. According to the agreed upon indicators, Target 16.7 is measured by the indicator 16.7.1: Proportions of positions in national and local institutions, including (a) the legislatures; (b) the public service; and (c) the judiciary, compared to national distributions, by sex, age, persons with disabilities and population groups; and the indicator 16.7.2: Proportion of population who believe decision-making is inclusive and responsive, by sex, age, disability, and population group. While representation and sentiment are generally important, it is more effective to disaggregate these and other measures to the local level, where public participation can be facilitated. For example, "are county and ward residents regularly consulted by government officials and within official government processes?"; or "the proportion of the county and ward populations that have access to data and documents involved in decision-making, as well as monitoring and evaluating county programs" (TAP Network).

CASE STUDY: USING OPENSTREETMAP IN UGANDA

An example of measuring and including geospatial data related to participatory decision-making is a flood mapping project in Kampala, Uganda, using open geospatial information. In two informal settlements—Bwaise and Kalerwe—volunteers from Makerere University documented conditions of drainage systems and rubbish dumps. They used the app KoboCollect for OpenStreetMap (OSM) to record data from 562 data locations, focusing on how rubbish dumps block drainage channels. This information was used by the local councils in these two slums to conduct mitigation plans. The key to success was that this project involved direct engagement with residents who were suffering from regular flooding in their low-lying slums (Kintu and Bulley, 2023, pp. 209–219). Figure 12.2 shows flood vulnerability data collected for Bwaise and Kalerwe slums, as visualized on an OSM building footprint basemap. Figure 12.3 shows that rubbish dumps are situated very close to drainage canals, and some are evidently blocking, leading to stagnant water.

FIGURE 12.2 Flood vulnerability shows data collected for Bwaise and Kalerwe slums.

Solis, Patricia and Marcela Zeballos, (eds). Open mapping towards sustainable development goals: Voices of YouthMappers on community engaged scholarship. Switzerland: Springer, 2023.

FIGURE 12.3 Rubbish dumps are situated very close to drainage canals.

Solis, Patricia and Marcela Zeballos, (eds). Open mapping towards sustainable development goals: Voices of YouthMappers on community engaged scholarship. Switzerland: Springer, 2023.

CASE STUDY: LOCAL PEOPLE AS MAPPERS IN BANGLADESH

Another example of local community involvement in geospatial data collection occurred in Bangladesh on Khasbaroshimul, Sirajgang, a small flood-prone island. Volunteers from Dhaka College trained local people to work as mappers. Together, they collected GPS and field papers, as well as working door-to-door to collect fine-grained data. Because the community members participated in the collection of this data, the data accuracy level was 100% (Andal et al., 2023, pp. 85–91).

Target 16.10: Ensure public access to information, inter alia, is measured by the indicator 16.10.2: Number of countries that adopt and implement constitutional, statutory and/or policy guarantees public access to information. While access to information needs to be guaranteed, it is useful to detect the extent to which access to data and information exists at local levels, especially in more rural jurisdictions. Supplemental measures could be *the percentage of budget, procurement, and natural resource concessions are publicly available and easily accessible in open data format; and the availability of up-to-date open data in user friendly and/or machine-readable format on strategic issues for all public institutions* (TAP Network). If the data are geo-tagged and time-stamped, in a geospatial ecosystem, citizens in local communities can engage with government officials and each other in online and in-person joint monitoring of progress in achieving development goals.

CASE STUDY: OPEN GEOSPATIAL DATA IN COLOMBIA AND MEXICO

Freedom of information can be increased effectively with geospatial data. Of the seven countries involved [in the study], only Mexico and Colombia have legislation to allow free and open access to official geographic spatial information. They allow the public to directly download data in GIS format from their open web portals (Arriaga et al., 2023, p. 299).

Many governments have some degree of open information laws, but usually not about open data and open geospatial data, however, the implementation of even these minimal open information laws are fraught with problems, such as fees, delays, and retractions. Some governments are reluctant to share data with citizens to avoid citizen feedback. When citizens have accurate data, they might be able to compare their jurisdiction to neighboring jurisdictions. Some regional organizations—for example, the African Union—have made efforts to embed review governance indicators into regional processes.

CASE STUDY: AFRICA PEER REVIEW MECHANISM FOR GOVERNANCE INDICATORS

For example, the African Union created the African Peer Review Mechanism (APRM) in 2003, as a voluntary arrangement among African states to systematically review and assess governance at Heads of State level to self-monitor their governance performance. There are five general themes: Democracy and political governance, economic governance and management, corporate governance, socio-economic development, and state resilience to shocks and disasters. Currently, 38 African countries participate in the process, which involves a base review, periodic reviews, early warning reviews and targeted reviews. At the national level, there is a national focal point, a national commission or council, and a national secretariat. After the base review or other review, the national authorities create a national program of action, short-term, medium-term, and long-term, to begin to implement recommendations from the review. As recently as 2024, the APRM produced a sample survey of 10 countries based on indicators such as the rule of law, transparency and accountability, participation and socio-economic governance and included several cross-cutting issues, such as corruption. The APRM process more finely defines and measures governance indicators than those in the UN's SDG indicator guide. It is important to note that while belonging to the APRM is voluntary, there is considerable pressure from civil society organizations, government officials and ordinary citizens on governments to join in the APRM process. The APRM, in association with the African Development Bank, has produced the *African Union Governance Atlas,* which uses geospatial and other data as background for information on such things as infrastructure, resource distribution, and population density, which can be mapped and overlaid with governance data. Currently, this information is used mostly at the national level, but it would provide more actionable intelligence if the data were systematically disaggregated to local levels within countries and made available to local officials and local citizens. Also, in this geospatial arena is the Regional Center for Mapping of Resources for Development (RCMRD), with its GMES and Africa GeoPortal, in Nairobi, Kenya. RCMRD has created and analyzed geospatial data for specific African 23 governments in Eastern and Southern Africa. Usually, this data is sector-specific and used for specialized development projects. So far, most of governments are not using geospatial data for general planning and monitoring purposes in their local government jurisdictions.

There are several non-governmental organizations (NGOs) that analyze governance factors in various countries. However, some governments do not trust the results of NGO analyses. The *Afrobarometer* (2022), based in Ghana, measures public opinion on governance issues, including 12 countries in Africa in 1999 and increasing to 38 countries by 2022. In each country, they conduct face-to-face interviews with a randomly selected nationally representative sample of 1200 to 2400 adult citizens. They measure opinions on accountability, bribery, corruption, courts of law and justice, decentralization, government performance, local government, parliaments, state capacity, transparency and trust. The *Latinobarometro*, based in Chile, conducts similar analyses for countries in the Latin America region.

Based in Germany, the Bertelsmann Stiftung's *Bertelsmann Transformation Index BTI 2024: Governance in International Comparison* covers governance issues, such as performance, capacity and accountability of political decision-makers, in countries throughout the world. Among other things, it measures the performance of government leaders in steering capacity, resource efficiency and consensus building. To create a more realistic environment for their measurements, the BTI adds in a measure of difficulty based on the degree of complexity of the national environment, such as the incidence of natural disasters and the complexity of their civic order. The measurement of governance factors embedded in SDG 16 would be improved by a greater variety of indicators, beyond those of the UN's SDG indicator guidance, and a greater inclusion of geospatially enabled indicators.

IMPORTANCE OF OPEN GEOSPATIAL DATA FOR CITIZEN ENGAGEMENT WITH THE SDGs

An important aspect of access to data regarding Target16.10 is that the information should be available to everyone to access, reuse and redistribute without restrictions (UN DPIDG/DESA, 2013, p. 6). Bringing this requirement up to date suggests that timely, relevant and comprehensive information and data should be available on the Internet, and that this information should be available in a useable and understandable manner. This can mean that the data should be machine-readable for analysis and that the data should be visualized so that ordinary people can understand the meaning. With these conditions, the local people can exercise their right to engage in policy making and policy implementation (UN DPIDG/DESA, 2013, p. 6).

"Citizen engagement is seen as the interaction between government and citizens in order to share information and data in policy making processes, including but not limited to public service provision and delivery, and more specifically in defining the issues that affect them, identifying possible solutions, and developing priorities for action, often jointly with the government and other governance actors" (UN DPIDG/DESA, 2013, p. 17). In determining a society's readiness for open geospatial data for citizen engagement, the assessment often includes:

- Political commitment and appropriate policies
- Capacity of civil society, the media and other users
- Legislative and regulatory framework
- Institutional framework and organizational conditions
- Cultural and human resources conditions
- Financial conditions
- Technical infrastructure
- Data and information systems
- Legal openness of open government data
- Technical openness of open government data (UN DPIDG/DESA, 2013, *p. 51*)

Not mentioned above is the capacity of the national and local energy system to support the electrical needs of these information systems, as well as the telecommunications needs of mobile phone networks. Energy capacity can limit the reliability and widespread use of the open geospatial data system. Energy reliability is especially problematic in remote, rural locations. When the SDGs were first established in 2015, SDG implementation was limited by that current level of technological conditions. Now, in 2025, with greater advancement of the internet, geospatial information systems and artificial intelligence, open data systems are more dependable, more scalable, and more interoperable. These more advanced systems are not only effective for citizen engagement but also for assisting government officials to better coordinate and integrate development programs. Open data systems contribute to increasing government transparency and accountability, increasing participation and engagement, increasing government efficiency and effectiveness, and improved coordination between government agencies and various levels of government and between government and stakeholders, such as civil society organizations, private sectors and citizens (UN DPIDG/DESA, 2013, p. 57).

INSIGHTS INTO THE FUTURE OF GEOSPATIALLY ENABLED GOVERNANCE INDICATORS

The role of geospatially enabled governance indicators in implementation of the SDGs depends upon several current trends in societies: Identification of responsible jurisdictions, localization of the SDGs, and increasing citizen engagement as a key element of governance. As the world moves closer to the SDG deadline of 2030, and the SDGs are not yet nearing accomplishment, there is more interest in identifying exactly which jurisdictions are making progress and which are falling behind in the effort. At the same time, it has been noticed that most citizens notice the extent to which development has occurred in their local areas, and that more emphasis needs to be placed on SDG implementation at the local level. Both trends point out the importance of geography and in parallel, the importance of geospatially enabled indicators. Furthermore, development can best emerge into relevant successes when citizens have unrestricted access to up-to-date information as a platform for citizens to be involved in prioritizing activities across many sectors and monitoring progress. Geospatial data fulfills these conditions for more robust development, leading to fulfillment of the SDGs. Currently, the official SDG 16 indicators of governance make less use of the geo-spatial lens. Some regional organizations, such as the African Union, are encouraging the greater use of geospatial data and at the same time have also increased its focus on governance factors, as represented in the African Peer Review Mechanism. The APRM provides a framework for African governments, as well as civil society organizations and citizens, to compare their performance on governance measures and to strategize plans to improve their national and local development results. To the extent that governance is best measured locally, there is a call for governments to use more geospatially enabled data that can be disaggregated to lower levels of government. There is greater emphasis on the local level when dealing with the SDGs. A recent UNDESA webinar emphasized the importance of *Localization of the SDGs* (DPIDG-DESA, 2024). Further focus on ground-level development takes the form of training local citizens to co-produce open data through various iterations of OSM and other relevant applications. Governments are more likely to pursue useful development activities where local citizens participate in setting local priorities. And these activities are more likely to lead to progress where citizens have access to useable data on development progress, often leading to cooperation between local government officials and local citizens. With close citizen scrutiny of government operations, there are more opportunities to find instances of corruption in terms of bribery, fraud, or self-dealing.

Another way to measure performance in government institutions is to focus on the inputs (funds, etc.) and outputs (programs and projects) for development.

With a robust geospatial database, budgets and outcomes can be aligned with geographic jurisdictions, leading to measurement of effectiveness in these government operations, as well as finding anomalies that indicate missing money, missing assets and corruption. These measurements can lead to greater effectiveness in a wide variety of sectors, including the accomplishment of other SDGs—such as education, health or decent work, not just those belonging to SDG 16. For example, it is common to track levels of educational attainment, clean water, and decent employment, but these measurements of service delivery depend upon a functioning bureaucracy to undertake projects and an engaged citizenry to oversee government operations. A high level of government effectiveness contributes to citizen satisfaction with community services, leading to greater citizen trust in government, in this regard, adding geospatial information to data provides a way to increase the effectiveness of planning and implementing the SDGs. While it might be unrealistic to add finely tuned geo-tags and timestamps to current data, it is quite possible to establish a geospatial database and to geo-tag and time-stamp new data, going forward. Making these databases open to all officials and all members of the public adds transparency and accountability and overtime this data will provide a framework for increasing institutional effectiveness, reducing corruption, and strengthening citizen engagement.

REFERENCES

Acemoglu, D. and Robinson, J.A. *The Narrow Corridor: States, Societies and the Fate of Liberty*. New York: Penguin Press, 2019

African Peer Review Mechanism. *An Innovative Self-Assessment Tool for Governance in Africa*. Midrand, South Africa, 2024.

African Union and African Development Bank. *The African Governance Atlas*. Midrand, South Africa, 2024.

Andal, F., Islam, M., Shaheen, A.R., and Anderson, J. "Understanding YouthMappers' Contributions to Building Resilient Coummunities in Asia" in Patricia Solis and Marcela Zeballos (eds.) *Open Mapping towards Sustainable Development Goals: Voices of YouthMappers on Community Engaged Scholarship*. Switzerland: Springer, 2023, pp. 85–91.

Arriaga, V., Birkenes, A., Council, D., Jones, M., Lay Soler, E.K., McCarley, J.S., Wulf, E., Zhang, C., Wintemute, J.P., Aguirre, N., and Solis, P. "Open Mapping with Official Cartographies in the Americas" in Patricia Solis and Marcela Zeballos (eds.) *Open Mapping towards Sustainable Development Goals: Voices of YouthMappers on Community Engaged Scholarship*. Switzerland: Springer, 2023, p. 299.

Bertelsmann Stiftung. (2024) *Bertelsmann Transformation Index 2024: Governance in International Comparison*. Berlin: Bertelsmann Stiftung, 2024

Committee of Experts on Public Administration (CEPA). *Principles of Effective Governance for Sustainable Development*. Endorsed by the United Nations Council for Economic and Social Development (ECOSOC), 2018.

Division of Public Institutions and Development Management of the UN Department of Economic and Social Affairs (DPIDG/DESA), *Guidelines on Open Government Data for Citizen Engagement*. New York: DPIDG/DESA, 2013. ST/ESA/PAD/SER.E/177.

Division of Public Institutions and Development Management of the UN Department of Economic and Social Affairs (DPIDG/DESA), 2024. *Webinar on Localization the SDGs*. New York: DPIDG/DESA, 2024.

Goal 16—The Indicators We Want. *Sourcebook on Measuring Peace, Justice and Effective Institutions.* New York: *United Nations Development Program UNDP).* 2016.

Habitat. *Urban Observatory Setup Guide.* Available at https://data.UNHabitat.org

Kaufmann, D., Kraay, A., and Mastruzzi, M. *Governance Matters VIII: Aggregate and Individual Governance Indicators, 1996-2008.* Washington DC: World Bank Institute, 2009.

Kenya Law Review Commission, *Constitution*, Chapter 11: Devolved Government. Nairobi, Kenya: Government of Kenya, 2005.

Kintu, I.M., Bulley, H.N.N. (2023). "Youth Engagement and Participation in Mitigating Perennial Flooding in Kampala, Uganda Using Open Geospatial Data" in Solís, P., Zeballos, M. (eds) *Open Mapping towards Sustainable Development Goals. Sustainable Development Goals Series.* Springer, Cham. https://doi.org/10.1007/978-3-031-05182-1_18

Kintu, I.M. and Bulley, H.N.N. (2023) "Youth Engagement and Participation in Mitigating Perennial Flooding in Kampala, Uganda; Using Open Geospatial Data" in Patricia Solis and Marcela Zeballos (eds.) *Open Mapping towards Sustainable Development Goals: Voices of YouthMappers on Community Engaged Scholarship.* Springer, Switzerland, pp. 209–219.

Latinobarometro (Chile). *Public Opinion Surveys. Chile: Latinobarometro*, 2022.

Migdal, J. *State-Society Relations and State Capacities in the Third World.* Princeton: Princeton University Press, 1988.

Regional Center for Mapping Resources for Development (RCMRD). *GMES and Africa GeoPortal.* Nairobi: RCMRD, 2025.

International Institute for Democracy and Electoral Assistance (IDEA). *SDG 16 DATA Initiative Report on Complementary Global Indicators.* Stockholm: IDEA, 2022.

Sharma, S., Clark, C., Dhakal, S., and Nepal, S. "Seeing the World Through maps: An Inclusive and Youth Oriented Approach" in Patricia Solis and Marcela Zeballos (eds.) *Open Mapping towards Sustainable Development Goals: Voices of YouthMappers on Community Engaged Scholarship.* Switzerland: Springer, 2023, pp. 107–109.

Solis, P. and Zeballos, M. (eds.). *Open Mapping towards Sustainable Development Goals: Voices of YouthMappers on Community Engaged Scholarship.* Switzerland: Springer, 2023.

TAP Network (Transparency, Accountability and Participation for the 20230 Agenda). *SDG 16+ Civil Society Tool Kit. https://www.sdg16toolkit.org/guides/sdg16-indicators-guide/*

Teorell, J., Sundström, A., Holmberg, S., Rothstein, B., Alvarado Pachon, N., Dalli, C.M., Valverde, R.L., Phiri, V.S., and Gerber, L. *The Quality of Government Standard Dataset, Version Jan25.* University of Gothenburg: The Quality of Government Institute, 2025. https://www.gu.se/en/quality-government. doi:10.18157/qogstdjan25

Transparency International Global Corruption Barometer. *Corruption Perception Index.* Transparency International, Berlin, Germany, 2023.

United Nations Department of Economic and Social Affairs (UN DESA). Available at un.org/sustainabledevelopment/peace-justice/

United Nations Economist Network. *Governance Futures: Key Insights and Policy Implications.* New York, UN Department of Economic and Social Affairs, 2025.

United Nations Geospatial and Geographic Information Management (GGIM) of the UN Department of Economic and Social Affairs (UN DESA). New York: United Nations Statistical Division (UNSD). ggim.un.org

United Nations Secretary-General. UN Sustainable Development Solutions Network (among other things, producing new geospatial indicators

World Bank Institute. *Worldwide Governance Indicators.* Washington, DC: World Bank, 2024.

13 Activation of SDG #17: Implementation through Partnerships and Technology

Etta D. Jackson

INTRODUCTION

A revolutionary aspect of the 2030 Agenda for Sustainable Development is the integrated comprehensive and holistic way all 17 goals, their related 169 targets and 234 indicators were constructed. This was made clear in the Outcome Document *The Future We Want* produced by the Rio+ 20 conference in Rio de Janeiro in the summer of 2012 (UNDESA, 2012, A/RES/66/288). At this meeting of Heads of States and high-level representatives, at which there was full participation of civil society, business, academia, and philanthropic organizations, they renewed and committed to sustainable development and to promoting social, economic and environmentally sustainable development for everyone, everywhere by the year 2030. The overarching goal would be the eradication of extreme poverty for everyone everywhere. The goals produced in the Outcome Document of 2014 and adopted in 2015: *Transforming Our World: The 2030 Agenda for Sustainable Development* achieved the desired outcome and are integrated and indivisible, global in nature and universal in application (UNDESA, 2015). This was intentional. The idea of sustainability is at its core and meant to allow for the eradication of extreme poverty and the understanding that this can only be achieved if the three pillars on which they were constructed: Social, economic, and environmental sustainability, are the outcome of their implementation. The wisdom of SDG#17: Means of Implementation becomes apparent in this regard. It was intended that in implementing the SDGs each country's situation and level of development would be considered while at the same time taking into consideration the priorities each country has for its own development.

SDG#17: MEANS OF IMPLEMENTATION TARGETS

The mandates laid out in the targets under SDG #17 for the successful implementation of the whole agenda make clear that because all 17 goals are interconnected, and that only an integrated approach would suffice to achieve the intended objective of eradicating extreme poverty. This would be achieved while strengthening global partnerships. The goal states that the following elements, illustrated in Figure 13.1, must be given great consideration.

DOI: 10.1201/9781003586319-17

FIGURE 13.1 SDG#17: Means of implementation through partnerships. (The content of this publication has not been approved by the United Nations and does not reflect the views of the United Nations or its officials or Member States. https://www.un.org/sustainabledevelopment).

They are:

1. *Finance:* The mandate of the 2030 Agenda is that domestic resources should be mobilized with support from the international community to developing countries, which include assistance in improving domestic capacity for collecting taxes and other revenues. In addition, developed countries are to contribute .07% of their revenues to, and consider giving debt relief to developing countries. (UNDESA, 2014, p. 30)
2. *Governments*: The implementation of the whole agenda should be country-led. This is a significant departure from a project-oriented approach of the past. (UNDESA, 2014, 31).
3. *Citizen Participation.* The sustainability of any implementation requires a feeling of ownership by the citizens of any community who must be invested in its success. (UNDESA, 2014, 31).
4. *Technology Transfer*: Giving access and making available technology needed to achieve the SDGs through knowledge sharing in a North-South and South-South triangular and regional cooperation. Give access to science, technology and innovation needed to build capacity in the countries, which include providing environmentally sound technologies. A recommendation offered was that by 2017, a technology bank, a science and technology bank, science and technology and innovation capacity building mechanism, especially in information technology, would be operational to facilitate the achievement of the SDGs. (UNDESA, 2014, 31). Imaz and Sheinbaum (2017) further emphasized that if technology transfer is not recognized, the SDGs will continue to have only marginal success.

5. *Capacity-building:* To achieve this mandate, developing countries need the support from the international community to help them achieve their national development plans. It is recognized that human resource development and training, knowledge sharing, and transfer are essential (UNDESA, 2014, 31). The Bali Strategic Plan for Technology Support and Capacity Building adopted by the United Nations Environmental Programme offers a good reference (UNEP/GC,.23/6/Add.1 and Corr.1, annex)
6. *Trade*: A rules-based open and non-discriminatory multi-lateral trading system would be promoted under the World Trade Organization (WHO) and the negotiations agreed upon under Doha Development Agenda. (UNDESA, 2014, 31)
7. *Systems Issues*: Each country's policy space and leadership should be respected in the enhancement of macroeconomic stability and policy coherence to achieve sustainable development in developing countries. (UNDESA, 2014, 31–32)
8. *Multi-Stakeholder Partnership*: A strong emphasis in the implementation of the 2030 Agenda, multi-stakeholder partnerships are seen as key to the successful outcome of the whole agenda. This partnership should include strong civil-society, public-private, and public partnerships in which knowledge, expertise, technology, and financial resources are shared to achieve the SDGs in all countries but especially the developing countries. (UNDESA, 2014, p. 32)
9. *Data, Monitoring, and Accountability:* It is widely understood by the UN community that the successful achievement of the SDGs is not possible without the use of accurate data.

In this regard, it was hoped that by 2020, increased high-quality and reliable disaggregated data by gender, age, ethnicity, migratory status, income, disability, and geographic location would be available in least developed and small island developing countries, and as is relevant, in the national context of each country. The hope was that by 2030, those measurements could be built upon. According to the 2020 SDG Report by the Secretary-General, the COVID-19 crisis has rendered a setback to many of these efforts. Out of this commitment of a strong partnership framework as the glue to achieving this ambitious Agenda has emerged the 5Ps of the SDGs: Peace and Prosperity for People and Planet through Partnerships.

THE UNITED NATIONS TECHNOLOGY FACILITATION MECHANISM

Since the inception of the SDGs, technology has been seen as an enabler for their effective implementation. When the 2030 Agenda was adopted at the United Nations Sustainable Development Summit in September 2015, Science, Technology, and Innovation (STI) was positioned as a key means of implementation of the SDGs, and it was in this regard that the United Nations Technology Facilitation Mechanism (TFM) was launched. The primary fora to discuss topics about TFM that are of common interest among Member States and STI stakeholders in the context of the 2030 Agenda, have been the Annual Multi-Stakeholder Forum for STI. This is supported by the Inter-Agency Task Team on STI for the SDGs (IATF). At the Third Annual STI Forum in 2018, action-oriented outcomes included the establishment of a work-stream dedicated to supporting discussions and initiatives on STI Roadmaps. The objective of the work-stream was to outline an inter-sessional work program to inform and enrich STI Forum discussions on STI Roadmaps, focusing on tangible impacts to be achieved over the next 12 months. The primary focus was to shape consensus on common guidance, principles and possible frameworks/methodologies for country-level STI Roadmaps. The second priority established at the 2018 forum was initiating discussions on possible elements of a global initiative or international assistance mechanisms to facilitate the development of such roadmaps. To accomplish these outcomes, the workstream would support a collective effort by a group of volunteering countries and would also facilitate a multi-stakeholder collaborative approach among the Inter-Agency Task Force (IATF), other UN agencies, non-UN partners, and stakeholders. One of the notable deliverables of the roadmap workstream that IATF made to the STI Forum in 2018 was an online information repository platform. It was created by compiling and making available a selection of STI or STI-related roadmaps from organizations and agencies both within the UN system and beyond. The purpose of this platform was to facilitate efficient and effective access to, and exchange information among stakeholders on the most current work being done by each organization in support of the development of STI roadmaps for SDGs. In this context, the platform will become an online resource portal on the state of STI policy-setting frameworks and will contribute to enhancing the capacity of Member States in integrating STI in development processes. Another outcome of the meeting was the proposition of the United Nations Inter-Agency Task Team to establish a Global Pilot Programme on Science, Technology and Innovation Roadmaps for the SDGs. A call to action was issued to countries to volunteer to have such a pilot conducted based on the guidelines outlined in the TFM roadmap.

UNITED NATIONS GLOBAL GEOSPATIAL INFORMATION MANAGEMENT (UNGGIM)

In 2011, the Economic and Social Council (ECOSOC) established the United Nations Committee of Experts on Global Geospatial Information Management (UN-GGIM) as the first inter-governmental body to set the global agenda on the use and management of geospatial information. It coordinates among countries and international organizations and other stakeholders, and develops strategies to build geospatial capacities of nations, especially in developing countries (UN E/2016/L.28). In 2017, the United Nations and the World Bank agreed to collaborate on a joint vision to promote growth and prosperity through creating and strengthening geospatial information capacity and development (United Nations Committee of Experts on Global Geospatial Information Management, 2018). The objective of the collaboration was to develop an Integrated Geospatial Information Framework that countries can use to develop and enhance their own geospatial information management. To this end, an Integrated Geospatial Information Framework (n.d.) was developed, consisting of three parts: An Overarching Strategic Framework, an Implementation Guide, and the Country-level Action Plans. The Overarching Strategic Framework is intended to give high-level policy and decision makers, institutions, and organizations within and across governments a framework within which to work to develop a national plan that would integrate existing development plans with the SDGs in alignment with the mission and vision of each country and according to their needs, priorities and circumstances. The framework consists of seven underpinning principles, eight goals, and nine strategic pathways to be used to create a national approach. The principles are as follows:

- Transparency and accountability
- Information accessibility and ease of use
- Strategic enablement
- Collaboration and cooperation
- An integrative solution
- Sustainable and valued, and,
- Consistent with the direction given for "how" the agenda is to be implemented, one of collaboration in partnership, sustainable and integrated in delivery.

The Implementation Guide's eight goals are as follows:

- Effective geospatial information management
- Leveraging international cooperation and partnerships
- Increased capacity, capability and knowledge transfer
- Sustained education and training programs
- Integrated geospatial information systems and services
- Economic return on investment
- Enhanced stakeholder engagement and communication
- Enriched societal value and benefits.

The goal of this guide is to provide guidance for governments to establish holistic nationally integrated geospatial information frameworks in countries in such a way that transformational change is "enabled, visible and sustainable" (United Nations Committee of Experts on Global Geospatial Information Management, 2018, p. 4). The Implementation Guide's nine strategic pathways provide the "how, when and who?" (the subtitle of the document, Country-level Action Plans) and are intended to assist countries prepare and implement their own country-level action plans while taking into consideration national circumstances and priorities. The country-level action plans are to include elements such as the economic impact and value of geospatial information systems, identification of investment needs, priorities, analysis of socio-economic benefits and potential funding sources. According to the United Nations Committee of Experts on Global Geospatial Information Management (2018), the plans are to have the following components:

- *Governance and Institutions:* Establishes the leadership, governance model, institutional arrangements and a clear value proposition; it includes commitment to achieving an Integrated Geospatial Information Framework (p. 7).

- *Policy and Legal*: Establishes framework to institute appropriate national geospatial information legislation and policy (p. 19).
- *Financial:* Aims to help countries see the benefits of investment, business models, financial partnerships, and opportunities realized (p. 20).
- *Data:* Aims at data curation and delivery and data supply chain interlinkages in which custodianship, acquisition, management, and fundamental geospatial data themes are emphasized (p. 21).
- *Innovation*: Helps to bridge the digital divide and promote innovation and creativity through innovation, improvement of process, technology, and technological advancement as possible.
- *Standards:* Aims to create technical, semantic, data and legal interoperability and establish best practices standards (p. 23).
- *Partnerships:* Aims to strengthen international collaboration and community participation; also, industry partnerships, joint ventures, cross-sectoral and interdisciplinary corporations (p. 24).
- *Capacity and Education:* Includes professional development, workplace training, entrepreneurship, formal education, and the raising of awareness. Establishes long-lasting capacity-building programs and education systems (p. 25).
- *Communication and Engagement*: Establishes monitoring and evaluation, planning and execution, integrated engagement strategies and stakeholder identification (p. 26).

THE CASE FOR GIS AND GEODESIGN FOR ACTIVATION OF SDG#17

As stated before, the SDGs were carefully designed so that all 17 goals and their targets are interlinked and meant to be implemented in a comprehensive and integrated way. Given this mandate, it becomes important to use a platform and technology that would provide for such integration and Geographic Information Systems (GIS) technology provides the best option for this. A GIS is generally defined as a system that captures, stores, manipulates, analyzes, manages, and presents geographic or spatial data. GIS applications, on the other hand, are tools that allow users to create interactive questions, analyze spatial information, edit data in maps, and present the results of these queries. The volume of data needed to be stored and manipulated gives GIS a great advantage and brings into focus the characteristics of Big Data that apply well to implementing the SDGs. The access and storage of such large amounts of data became defined, initially, as the three Vs: Volume, Velocity, and Variety became a challenge. These evolved to 5Vs: Volume, Velocity, Variety, Value, and Veracity and now even to 7Vs: Volume, Velocity, Variety, Value, Veracity, Variability, and Visualization and are the main characteristics that define Big Data (Evans, Owda, Crockett and Vilas, 2019). It is easy to see why a system like GIS is most appropriate. Geospatial information science provides increased transparency and accountability for citizens. With the combination of the cloud for remote storage of data, sharing of data, maps, and pertinent information, GIS is allowing increased engagement with citizens. GIS can succinctly show many kinds of trends in the community visually, enabling constructive and informed conversation for government ministries, citizens and other stakeholders.

PARTICIPATORY GIS

To fully engage government and stakeholders in the process of imagining and planning for *the future they want* for their communities, participatory GIS (PGIS) becomes a very useful tool in achieving that end. It is an innovative approach increasingly being used by governments to support data collection and management operations and is being introduced for consideration, evaluation, and adoption. PGIS explores aspects of the control and ownership of geographical information; additionally, it offers the representation of local and indigenous knowledge and the ability to scale and scale up web-based approaches. It also offers a fragile transitory potential to explore a world where conventional GIS has a stronger foothold and through information communication technology geographic information can become more mainstream (Dunn, 2007).

PGIS, as shown in Figure 13.2, allows citizens and all other parties to work together for the mutual benefit of the sustainable development of the country and can help to reduce the workload of governments while empowering citizens and communities. Given that some governments lack the resources to collect the data

FIGURE 13.2 Participatory geospatial mapping in Kisumu, Kenya

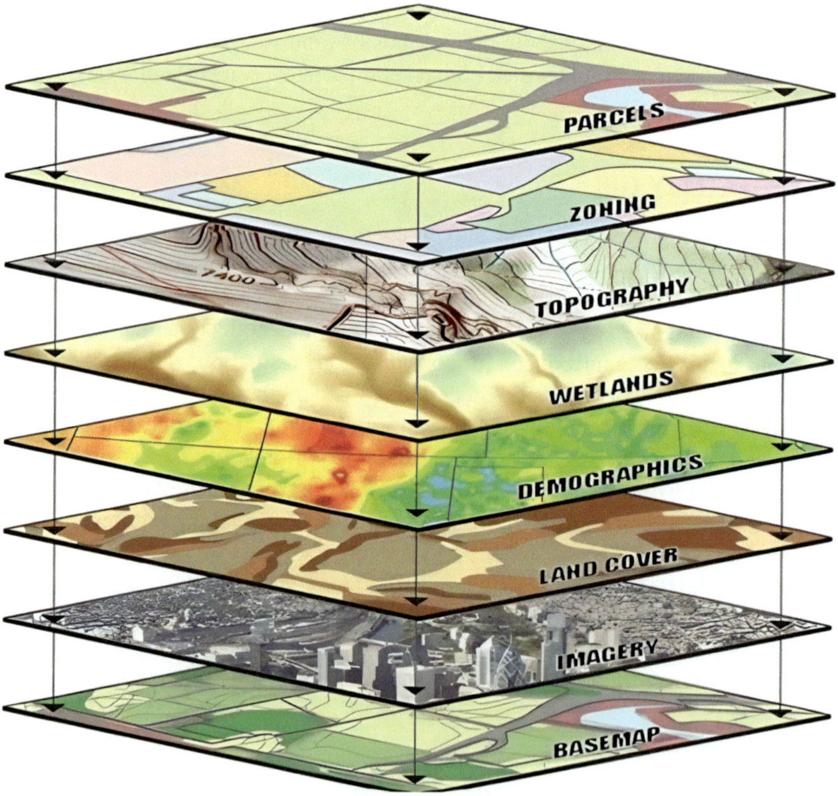

FIGURE 13.3 Illustration of typical GIS data layers.

From "GIS Data Layers Visualization," by U.S. Geological Survey

required to support operations, some data sets reside in the private sector. A strong government-citizen partnership for data collection and use can be of great benefit to any community.

As seen in Figure 13.3, Geographic information systems or geographic information science (GIS) deal with boundaries—what is "in" and what is "out;" to be able to do this, the discussion itself must have some boundaries (Gold, 2006). Geospatial information science provides increased transparency and accountability for citizens. With the combination of the cloud for remote storage of data, sharing of data, maps, and pertinent information, GIS is allowing increased engagement with citizens. GIS can succinctly show many kinds of trends in the community visually, enabling constructive and informed conversation for government ministries, citizens, and other stakeholders. Social, economic, and environmental sustainability are the three pillars on which the 2030 Agenda (United Nations General Assembly, 2015b) is based, and geographic information science provides the capacity for the most appropriate response for its implementation. Given that groups are fundamental units underlying intra-organizational, organization-wide, and inter-organizational activity in society, social behavioral studies on group use of GIS can help in understanding the social implications of GIS (Nyerges et al., 2002). PGIS is, therefore, used when group communication technology is integrated with basic GIS capabilities, leading to an enhanced application of GIS. In some emergency conditions, focus groups may use GIS fed by field data to create community mapping, which, even though its quality may not be high, has the advantage to be rapidly performed making it useful for an early response (Laaribi & Peters, 2019). The knowledge that citizens have of the area in which they live and any maps, however simple they may have been created, can become very helpful in emergency situations to re-create the area and inform decision-making.

GeoDesign

GeoDesign is a method that brings geographic analysis into a collaborative process allowing governments, citizens, designers, planners, geographers, and civil engineers to find and design resilient solutions to 21st-century challenges like population growth, decreasing resources, disaster mitigation, climate change, and many more. Flaxman (2010) explained the workflow of GeoDesign as a design and planning method which tightly couples the creation of a design proposal with impact simulations informed by a geographic context. In an ideal case, he stated, a planner or designer receives real-time guidance on performance at every phase of design, from early site visit or conceptual sketch to final detail. The use of contextual geographic information means that design performance can be evaluated relative to local conditions, and that evaluation can and should consider off-site impacts. The focus is on supporting a 'human in a loop' kind of design, providing continuous feedback on multiple aspects of performances and producing "designs-in-progress rather than on a post-hoc evaluation" (p. 29).

GeoDesign is a technology that has emerged as an essential tool providing the perfect formula for a data-driven integrated and holistic approach to implementation of the 2030 Agenda as stated. The GeoDesign method shown in Figure 13.4 consists of four elements:

- Geographic Information Science (GIS)
- Information technology
- Design technology, and most importantly,
- The People of the Place (stakeholder engagement).

The most effective method for collecting this data is the use of the Global Navigation Satellite System (GNSS), chief among them being the Global Positioning System (GPS), used with hand-held devices. With this technology, ordinary citizens can participate in data collection and research about their situated villages or towns and conditions. More specifically, this technology assists in massive point-based data collection, at the individual level, ensuring that no one is left behind. The GIS system is uniquely designed to allow for the storage of extremely large and complex amounts of data as is needed for the 17 SDGs, 169 targets, and 234 indicators involved in the

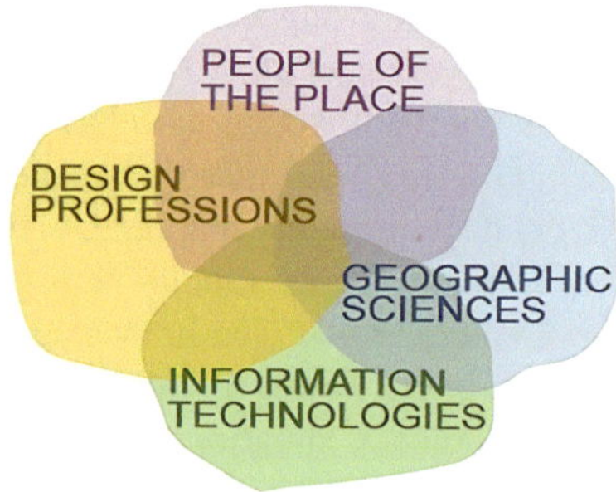

FIGURE 13.4 Components of the GeoDesign method.

From A Framework for GeoDesign by Carl Steinitz (2012)

2030 Agenda. GIS and GeoDesign are meant to be utilized by communities in the following ways:

1. Integrated into school curricula
2. Develop government agencies' capacity
3. Visualize the current landscape and future development plans
4. Improve citizens' attitude toward future development
5. Engage citizens in sustainable planning
6. Enable timely reporting on local SDG indicators

Flaxman (2010) argued that participatory GeoDesign allows for the involvement of all parties with a stake in the sustainable development and growth of an area. This ensures that everyone's voice is heard and that all have a hand in the design and construction of their future. The combination of expertise and voices involved in GeoDesign also ensures a holistic process. The four essential categories of GeoDesign are: information technologies—GIS, 3D, and 2D design programs; geographic information sciences (geology, hydrology, etc.); design professionals like architects, urban planners and designers; and the people of the place including local citizens, in collaboration with governments, the UN, business, academia, philanthropic organizations, etc. Wilson (2011) emphasized that data is central to geographical technologies and provides the pathways in which geographic investigations are brought forward. Wilson discussed the importance of mattering of data; He explains mattering as data products that result from specific practices, adding that there is an affective aspect of data which is imaginative, generative, and evocative. Batty (2013) explained what GeoDesign is not: namely, a return to the old ways of overlay mapping, but instead is a way of combining, using, and adapting the tools of geospatial science to very different contexts. At the heart of GeoDesign lies participation in the process and an understanding of the science of geography, which is both a prelude as well as an afterward to design. At the same time, design is regarded as a constant and continuing process of reaching out for solutions and useful responses to urgent problems. Below is a schematic in Figure 13.5 of model development and the roles of stakeholders in GeoDesign.

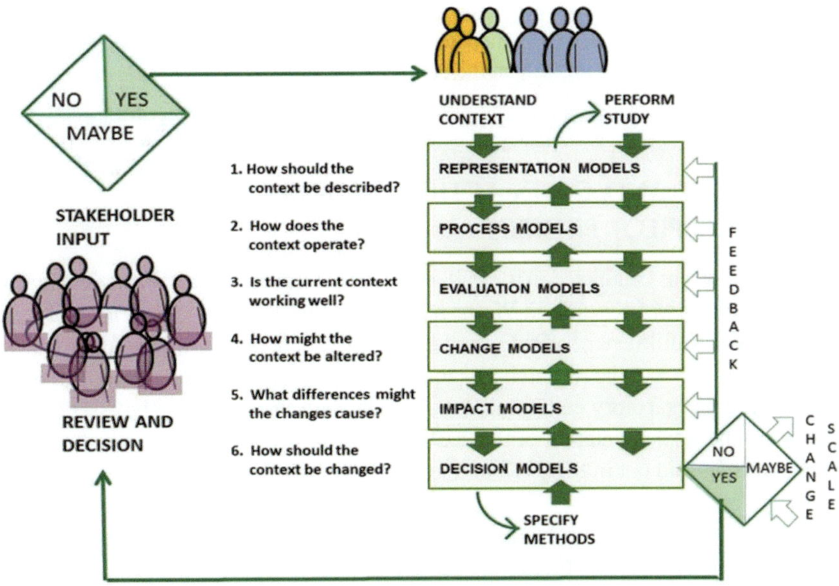

FIGURE 13.5 GeoDesign framework of development of models and roles.

From a framework for GeoDesign by Carl Steinitz (2012), p. 25.

According to Steinitz (2012), the GIS and GeoDesign application in the change process addresses the following questions by first looking at the world as it is:

- How can we describe geography as it relates to data inventory?
- How does geography work? It is addressed by developing process models
- How can we alter geography?—answered by developing capacity sustainability models.
- What are the alternative scenarios?—which relate to the world as it could be and are addressed by the creation of designs and sketches
- What are the consequences of change?—evaluation and analysis
- How should geography be changed?—Looking at decisions and values of the community.

Wilson (2015) described GeoDesign as both matter and meaning, both material and discursive. He explained that to practice GeoDesign is to draw upon both forces, which would be the concretization of both ideas and effects. Thus, GeoDesign is part of a broader techno-scientific endeavor that draws together investment with development and fashions a problem space while in the process of providing specific solutions. An important point made in the Outcome Document of the Brussels Conference on STI Forum held in 2018:

> Innovative, market-ready integrated technological solutions that can substantially contribute to the achievement of the SDGs already exist. These technologies are

> not sufficiently deployed on a global scale because policy and decision makers are unaware of their existence or their potential to contribute to achieving the SDGs or lack the political will to implement them.
>
> (G-STIC, as cited in Dodds, 2018, para. 9).

PARTNERSHIP AND GIS IN ACTION: MANYATTA GEODESIGN PILOT PROJECT

A research pilot was conducted in the Fall of 2019 in the informal settlements of Manyatta A and B in the city of Kisumu in Western Kenya on the shores of Lake Victoria. It was done in partial fulfillment of a Ph.D. by the author in a scholar-practitioner doctoral program. The objective of the research was to show how geospatial information technology together with participatory engagement tools, to include academia, can be essential contributors to the implementation of the 2030 Agenda and could most effectively respond to the mandates laid out in SDG#17, and included the following:

1. Engagement with the leadership of the County and City of Kisumu for direction in helping them and their community fulfill their development aspirations. The Ministry of Planning and the Office of City Planning provided direct input and involvement throughout the process.
2. Brought together the multi-stakeholder groups to create a shared vision for the future of the settlement that included:
   - Farmers
   - Women and Girls
   - The Business Community
   - Youth
   - Religious leaders
   - Persons with Disabilities
   - The Chiefs (National representatives)
   - Ward Administrators (County representatives)
   - Community Organizers
   - Community-Based Organizations
   - Academia
   - Politicians
   - Opinion Leaders
   - United Nations-African Union—The Regional Center for Mapping Resources for Development (RCMRD).
3. *Capacity Building*: It was important that the Assistant City Planner, the Chiefs, Ward Administrators of both settlements together with representatives of the major groups mentioned, participated. It should be noted that all the research assistants who participated were young adults, all university graduates, from Kenya and gained first-hand knowledge of the 2030 Agenda and "how" it can be implemented.
4. *Knowledge Transfer*: The intense representative participation of members from very diverse groups ensured that no one was being left behind in the technical and non-technical knowledge shared.

5. *Technology Transfer*: The Household survey conducted revealed that a significant percentage of the population had heard of geospatial information technology and for the first time many had the opportunity to work with it and see how it applied directly to their daily lives and the potential to provide essential information to improve their experience in their community.

PARTNER-GENERATED DATA SOURCES

Mapping and geo-visualization techniques were applied to ensure inclusive stakeholder representation during data collection. This is an essential tenet in, and a prerequisite for, supporting scalability in acquiring actionable location-based intelligence. Though mapping is an old practice with significant implications for development studies across spatial scales, its utility has yet to be fully exploited through active citizen engagement. GeoDesign is an emergent and integrated methodological framework for visually facilitating multi-stakeholder inclusion at scale. The leading concept guiding this study is the collaborative design and realization for the optimal solution for spatial challenges in the built and natural environments. Manyatta is a peri-urban estate on the eastern outskirts of Kisumu, Kenya's third most important urban center on the shores of Lake Victoria. It is home to one of the major slums in the city. The slums consist largely of informal semi-permanent housing for the poor, many of whom migrated from rural areas in pursuit of economic opportunities in the city. The slums have grown for decades skirting around the center and suburbs of Kisumu since independence. The other well-known slums in Kisumu include: Kondele, Nyawita, Obunga, and Nyalenda. The socioeconomic, environmental, and spatial properties of Manyatta qualify it as a suitable candidate for research on participatory development planning. Manyatta slum is made up of two main administrative units: Manyatta A and Manyatta B, each acting as a Ward or Location at the same time. This implies each administrative unit has a Chief representing the National Government and a Ward Administrator representing the County Government. Manyatta A is about twice as populated as Manyatta B with each further divided into smaller units. Manyatta A has six units and 10 sub-divisions, namely, Flamingo Lower, Flamingo Upper, Gonda Lower, Gonda Upper, Magadi Upper, Magadi Lower, Kondele Lower, Kondele Upper, Metameta, and Kona Mbuta. Manyatta B is made up of three sub-divisions: Kuoyo, Upper Kanyakwar, and Lower Kanyakwar.

The sources of data used to GeoDesign the settlement included:

1. ***A 500 Geo-referenced Household Survey*** as seen in Figure 13.6 using GNSS with hand-held GPS devices allowed for the maximum representative inclusion of all citizens of all units in the settlements. As intended, 500 households were interviewed, hence reaching the intended target of 100 percent. Households were mapped for scalability and visualization of the spatial spread. In addition to this, other points of interest such as water points were also mapped to make up a total of 513 mapped points. The full household data captured showed that less than half (44.4) of the Manyatta slum community displayed a total lack of information about the SDGs,

while a quarter (25.6) had only a faint understanding of the same. Only a tenth of the population (10.6) had clear information on the objectives of the SDGs, the targets against which to assess their performance, and the strategies in place for their implementation. The HH questionnaire asked about problems the households faced, followed by what they would want changed for a better Manyatta. When asked what problems they faced, the majority of the HH (87%) surveyed indicated poor sanitation and/or poor waste management as their greatest concern and is consistent with the results of a Problem Tree Analysis discussion conducted. This location-based data helps to ensure no one is left behind and given the overarching goal of the 2030 Agenda is to achieve the eradication of extreme poverty for everyone.

2. ***Key Informant Interviews (KII).*** There were 11 key informants—seven males (63.6%) and four females (36.4%), thereby meeting the one-third gender rule representation as required on any engagement set by the Kenya

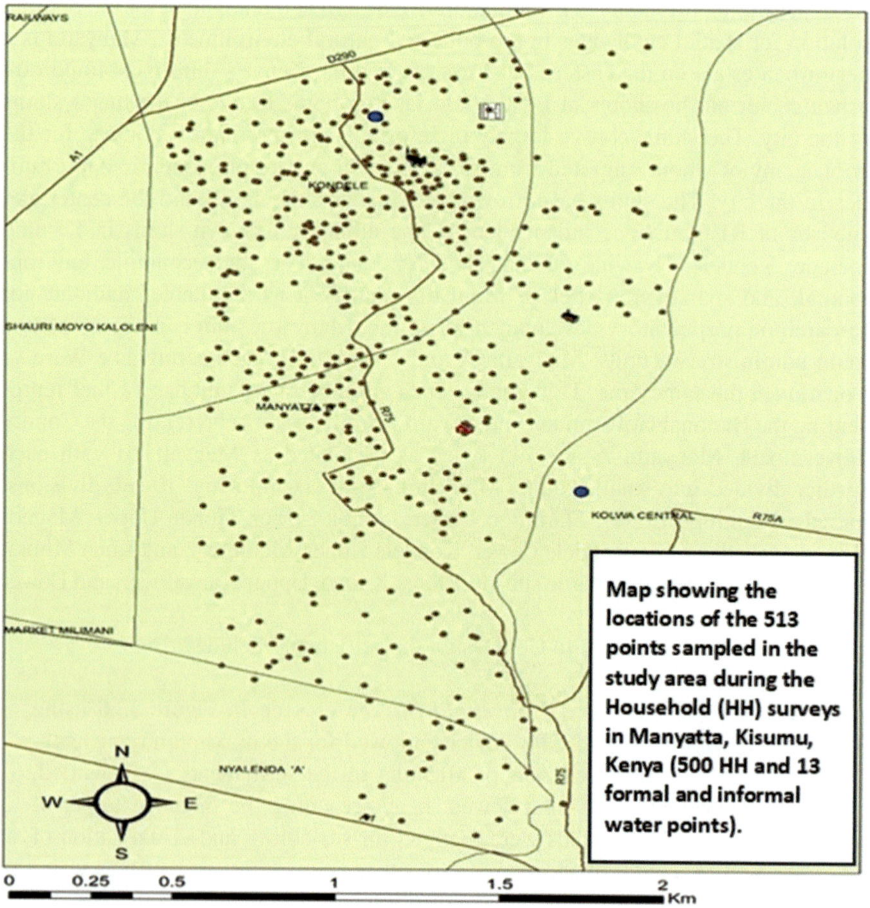

FIGURE 13.6 Geo-referenced household surveys in Manyatta A and B, Kisumu, Kenya

2010 constitution. The majority, 63.6% and 18.2% respectively, had worked with Manyatta informal settlements for between 11 and 20 years and over 20 years. Only one participant (9.1%) had spent less than five years working in the community. This by extension could be interpreted as most key informants having extensive knowledge of the slum. All participants had acquired at least secondary school education, with five (45.0%) being at university level, and three at non-university tertiary education. Unlike the general population of the Manyatta slum, the leadership and stakeholders are well-informed groups. They have a very good understanding of the United Nations Sustainable Development Goals and are aware of the implications of their proper implementation. A cross-tabulation of awareness of SDG and education levels revealed levels of awareness positively related to education levels with only one person—the one at the lowest level of education, secondary, giving a negative response. With respect to gender, there was no evidence showing any of the gender categories as having greater awareness of the SDGs.

3. ***Focus Group Discussion.*** Thirteen participants shared their opinions on how to improve the Manyatta community in a discussion setting. The discussion, which included a GIS demonstration, revealed a general agreement by the participants about the power of GIS and GeoDesign, mainly due to and in relation to concerns associated with livelihood opportunities as well as community development. The discussion yielded valuable insights into the underlying hopes of the people in Manyatta for improving development.

   Key focus group discussion points

   The stakeholders spoke about the following matters:

   - The feeling of how things operate in Manyatta and what could be done to address them effectively
   - The quality of life in the Manyatta settlement in terms of the planning and design of critical amenities
   - The settlement package foreseen for residents in Manyatta to build family houses, toilets, schools, healthcare facilities, and water supply on the allocated area; and
   - Identifying the likely community-based support approaches to effective planning in Manyatta.

   It should be noted that most residents of Manyatta have lived in the place for more than 20 years and have served the residents in different capacities such as church leaders, chiefs, youth leaders, businesspeople, among others.

4. A ***Problem Tree Analysis*** as shown in Figure 13.7 was conducted to assist participants in identifying what they considered to be the core problem in the community and to engage the group of enthusiastic participants. This problem and solution tree approach, as suggested by Snowdon et al. (2008), is used to work through layers of determinants and then develop potential interventions for a specific issue using available data and expertise. Participatory research is more of an orientation than a set of methods, which emphasizes the importance of the knowledge and views of the community (Leung et al., 2004). This approach focuses on visualizations and

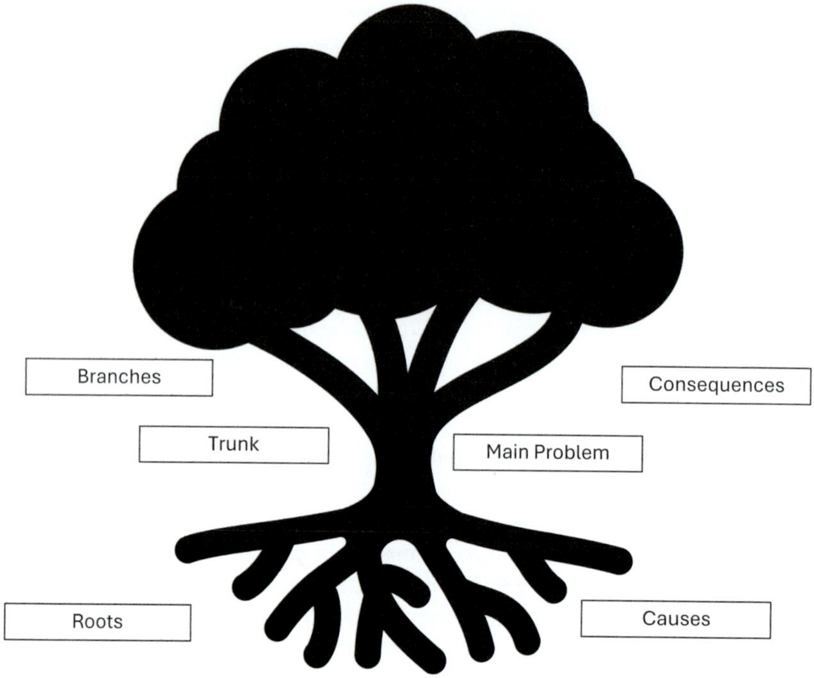

FIGURE 13.7 Problem tree analysis

is commonly used to assist the process and usually includes mapping and diagrams.

Figure 13.7 here – created in Microsoft Word

The participants were guided through the problem tree approach and asked to identify the core problem in Manyatta as well as the causes and effects of that problem. In the process of identifying what is referred to as the core problem, the participants made inventories of their core problem, which they identified as *poor sanitation* and *poor waste management.*

5. *The Stakeholder Analysis* matrix in Figure 13.8 shows the next data gathering tool aimed at determining those in the community having the *interest, power,* and *influence* needed to implement the identified changes. The participants in this discussion were the same as those who participated in the problem tree analysis. They received full instructions on the stakeholder analysis to help them comprehend various opinions of the stakeholders in their community. Freeman (1984) described stakeholders as individuals who affect or may be affected by decisions and actions that fit with the organization's objectives. He also proposed a framework of three levels for stakeholder analysis: rational, process, and transactional and suggested that the level from which each stakeholder is operating should be identified. The results of the stakeholder analysis can be seen in Figure 13.9.
6. ***Situational Analysis***: The concept of *situation* is key, argued Clarke (2005) who was inspired by several scholars to include Thomas's (1923/1978)

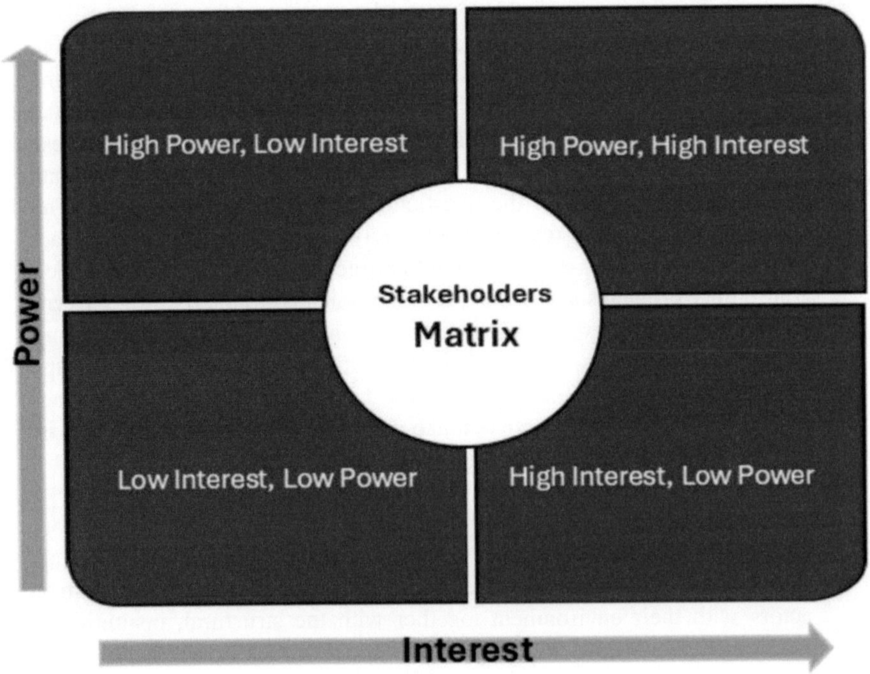

FIGURE 13.8 Stakeholder analysis matrix

theorem from the 1920s, which states that: "If situations are perceived as real, they are also real in their consequences." This theorem is at the heart of social constructionism and symbolic interactionism, which is foundational for Situational Analysis (SA), and citing others, Clarke concluded that the key point is that in SA, the situation itself becomes the fundamental unit of

		INTERESTS	
		Low	**High**
POWER	**Low**	Religious Leaders	Business Communities
	High	Residents Local Administration County Administration Politicians Development Partners Academia Professional Bodies	Opinion Leaders Civil Society members Government Agencies

FIGURE 13.9 Manyatta responses to stakeholder analysis

analysis. The SA is inextricable from the data harvested from the Household Survey, Key Informants Interviews, the Focus Group Discussions, and Participatory Action Research/GeoDesign discussions. The analysis of the data gathered from the household surveys illustrated the current economic, cultural, environmental, and infrastructural realities in Manyatta. These prevailing phenomena informed the engagement with the key informants and were encapsulated in the questionnaires administered to them. The key informants, being relevant community authorities, regularly interact and solicit feedback from the public and act as channels through which community problems and needs get communicated to the relevant governmental bodies for development assistance. On the other hand, focus group discussions enabled specific problems identification from diverse groups in Manyatta. Waltner-Toews (2017) felt that stories and dialogue impact how political and public leaders make decisions, which were confirmed in speaking with the leadership who participated in the discussions and who expressed a commitment to help solve the issues together in partnership.

Geospatial Data Describing Situation: This study sought to support one of the premises of SA made by Charmaz (2005) and Clarke et al. (2018); it explores the relational ecologies between human and non-human actors with their environment together with the structural, institutional, social, socio-political, cultural, and historical worlds in which they live. The extensive household surveys conducted with the aid of GPS devices, coupled with recorded observation and description of the physical landscape, allowed for the capturing of location data. The data provided import and information of the human and non-human situation in the location both from the perspective of the interviewer and the data provided by the interviewee. Key informant interviews provided data from the perspective of respected community leaders regarding the situation in Manyatta. The focus group discussion gave voice to the situation as they experienced it but additionally what is needed to change the situation. The data gathered through problem tree analysis provided rich information about poor sanitation and waste management, which contributed to many health issues to produce a troubling social, environmental, and economic situation for all. The participatory mapping activities included members of all previous groups and provided the opportunity to apply the nascent aspects of GeoDesign; to help make visible the physical situation in an interactionist approach (Genat, 2009) as part of the participatory action research approach meant to inform social policy decisions that would remedy the reality in the neighborhood. The situational data gathered was meant to change the geographic features on the landscape that have not served the community well and to create a future that more accurately reflects a vision of the community that is viable.

Situational Analysis Considerations for GeoDesign. Various design elements were incorporated to ensure a better, livable, compatible, harmonious, and sustainable environment in Manyatta. They include but not limited to space and its definition, such as elements of form, urban functions like living, working, leisure, mobility and administration, urban fabric, and environmental factors like precipitation, temperatures, humidity, wind, and lighting. This was done to ensure that there was functional efficiency, improved environmental conditions, ease of movement, community integration and character (a place with its own identity). The spatial design of the Manyatta slum belt was also primarily informed by SA of the social, economic, environmental, and infrastructural landscape. Some of the structuring elements that greatly contributed to the design included, but were not limited to, existing developments, typology and topography of soil, road networks, rivers, lagas, soils and terrain; lagas, also called ephemeral streams, are shallow wet gullies that drain the surface run-off within a target area (one exists in Manyatta B).

7. A **Participatory-GeoDesign-Situational Analysis Discussion**

Data collected involved: Earth Observation, GNSS to include GPS handheld devices, household (HH) surveys, key informant interviews (KII), focus group discussions (FGD), participatory mapping, problem tree analysis, stakeholder analysis, data points for health and education facilities, formal and informal water points, land use, etc. All these data informed the GeoDesign of Manyatta important to the GeoDesign process to achieve the social, economic, and environmental objectives of the community; and its understanding and approach to land use policy in allocating and managing land to improve the lives of citizens along the following principles:

1. Continual learning and adaptive management that is dynamic
2. Common concern as an entry point
3. Multiple scales: Many systems of influence, feedback and constraints that affect management
4. Multifunctionality: Multiple uses of landscapes have value in different ways to different stakeholders
5. Multiple stakeholders: To ensure an equitable outcome in decision-making about land use, the process should be ethical and inclusive
6. Negotiated and Transparent- There should be trust among stakeholders to avoid conflict
7. Clarification of rights and responsibilities: The rights and responsibilities of the different actors should be defined so they can be better accepted by all stakeholders
8. Participatory and user-friendly monitoring- A system that integrates different kinds of information should be in place
9. Resilience: A system-level resilience to allow for threats and vulnerabilities
10. Strengthen stakeholder capacity: The participation assumes certain skills and abilities such as social, cultural, financial (Sayer et al., 2013).

The two maps below show the before and after scenarios of the Manyatta informal settlement and make clear the importance of data: geospatial, statistical, situational, participatory that was contributed by citizens, government, multi-stakeholders, that included:

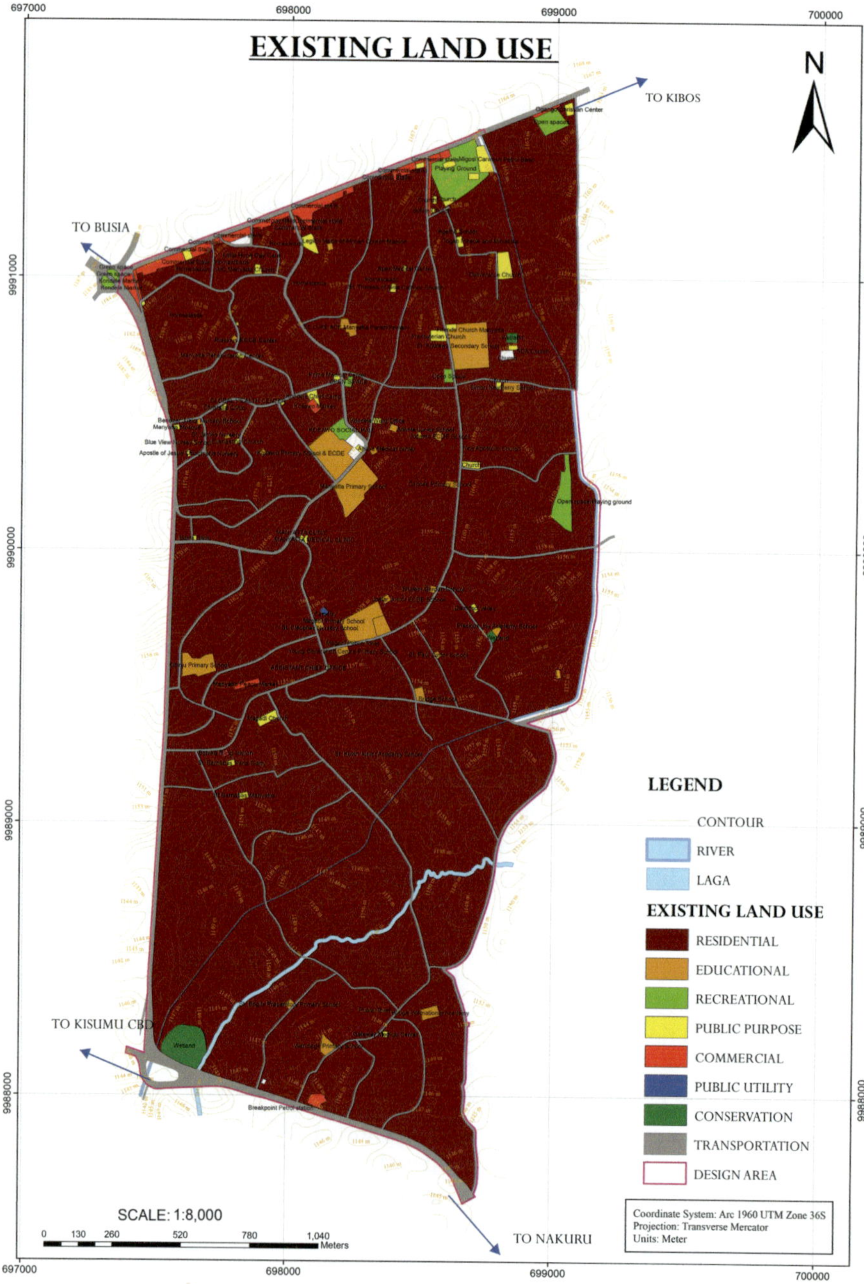

FIGURE 13.10 Current land use in Manyatta A and B

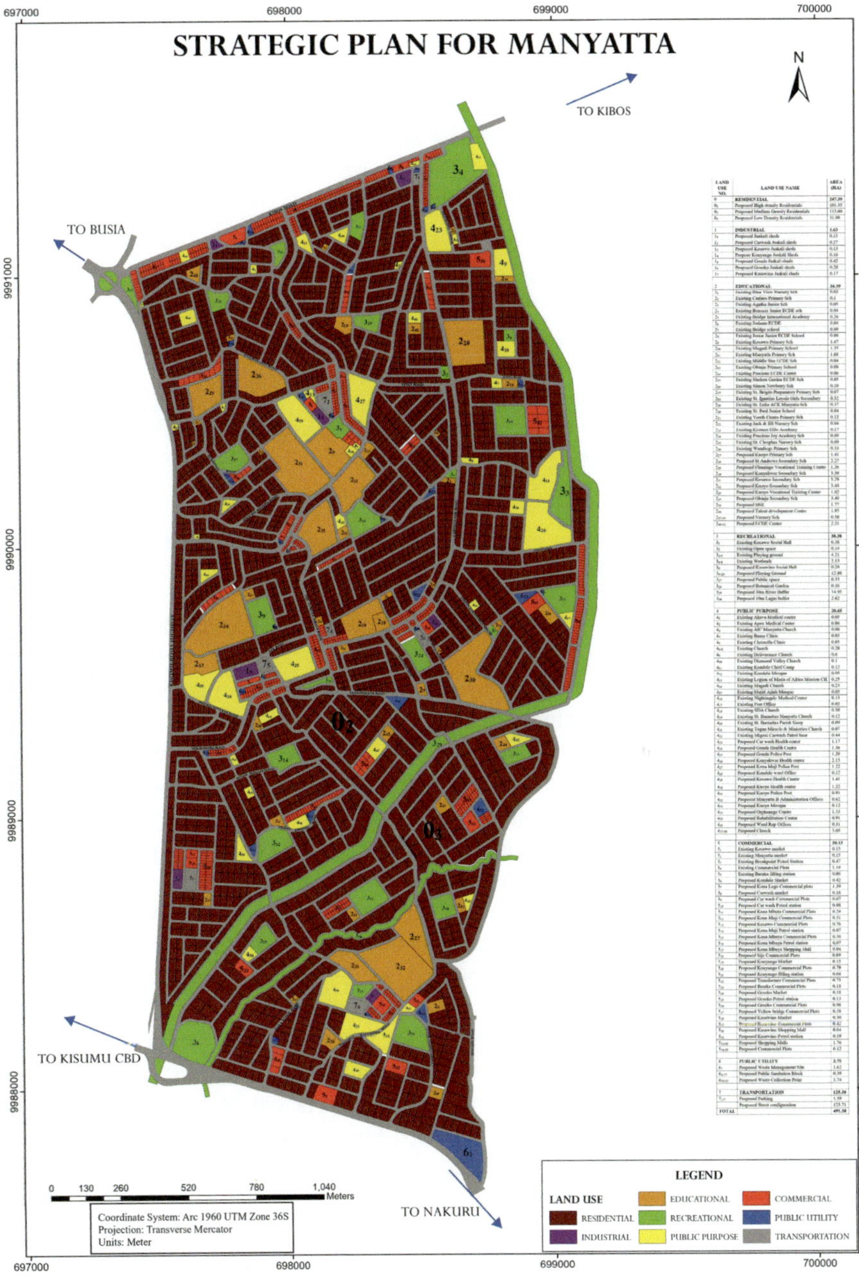

FIGURE 13.11 Detailed strategic plan for Manyatta A and B

- Persons with Disability
- Business
- Academia
- Government (City Planner
- Chiefs of both Wards
- Ward Administrators)
- Religious community
- Farmers
- Youth
- Women
- Community-Based Organizations (CBOs).

The first, Figure 13.10 reflects the current land use in both Manyatta A and B, while Figure 13.11 shows the comprehensive integrated and holistic develop plan of the two communities reflecting the mandate of SDG#17, one that is county-led with community engagement to reflect ***the future they want.***

CONCLUSION

The research pilot study of Manyatta using geospatial information and with the assembly of a strong effective partnership to address the initial overarching question of the research, which was "The Role of Geospatial Information and Effective Partnerships in the Implementation of the International Agenda for Sustainable Development" (Jackson, 2020) overwhelmingly demonstrated how essential this combination is to the successful implementation of the 2030 Agenda for Sustainable Development. The study further makes clear how crucial the intermediate step of integrated, comprehensive, and holistic planning is to any successful implementation that would eradicate poverty and leave no one behind, sustainably. A data-driven approach to a process, which brings all stakeholders to the table, was facilitated by the collaborative platforms that participatory geographic information systems and GeoDesign afforded. The findings and results of this research pilot demonstrate **"the how"** for the implementation of global United Nations 2030 Agenda. The outcome of this was the GeoDesign of the community to reflect the hopes and aspirations of the communities with the engagement of the Ministry of Planning and reflects "the future they want" and can be seen below in a very detailed presentation in the ArcGIS Online Storymap: https://storymaps.arcgis.com/stories/e500ffa099b347c18eaf7f85d64443cf

REFERENCES

Batty, M. (2013). Editorial: Defining GeoDesign. *Environment and Planning B: Planning and Design*, *40*, 1–2. https://doi.org/10.1068/b4001ed

Charmaz, K. (2005). Grounded theory in the 21st century: Applications for advancing social justice studies. In N. Denzin & Y. Lincoln (Eds.), *The SAGE handbook of qualitative research* (3rd ed., pp. 507–537). SAGE.

Clarke, A. (2005). *Situational analysis: Grounded theory after the postmodern turn.* SAGE.

Cornwall, A., & Jewkes, R. (1995). What is participatory research? *Social Science and Medicine, 41*(1), 1667–1676.

Dodds, F. (2018, December 20). Connecting technological innovation to decision making for sustainability: G-STIC 2018 co-host statement. Retrieved from http://blog.felixdodds.net/2018/12/connecting-technological-innovation-to.html

Dunn, C. E. (2007). Participatory GIS—A people's GIS? *Progress in Human Geography, 3*(5), 616–637. https://doi.org/10.1177/0309132507081493.

Evans, L., Owda, M., Crockett, K., & Vilas, A. F. (2019). Big data fusion model for heterogeneous financial market data (FinDf). In K. Arai, S. Kapoor, & R. Bhatia (Eds.), *Intelligent systems and applications. IntelliSys 2018. Advances in intelligent systems and computing* (pp 1085–1101). Springer.

Flaxman, M. (2010, May 22). Fundamentals of GeoDesign. In E. Buhmann, M. Pietsch, & E. Kretzler (Eds.), *Digital landscape architecture 2010* (pp. 28–41). Retrieved from http://www.kolleg.loel.hs-anhalt.de/landschaftsinformatik/fileadmin/user_upload/_temp_/2010/Proceedings/Buhmann_28-41.pdf

Freeman, R. E. (1984). *Strategic management: A stakeholder approach.* Pitman.

Genat, B. (2009). Building emergent situated knowledge in participatory action research. Action Research, 7(1), 101–115. https://doi.org/10.1177/1476750308099600

Gold, C. M. (2006). What is GIS and what is not? *Transactions in GIS, 10*(4), 505–519. https://doi.org/10.1111/j.1467-9671.2006.01009.x

Imaz, M., & Sheinbaum, C. (2017). Science and technology in the framework of the Sustainable Development Goals. *World Journal of Science, Technology & Sustainable Development, 14*(1), 2–17. https://doi.org/10.1108/WJSTSD-04-2016-0030.

Jackson, E. D. (2020). The Role of Geospatial Information and Effective Partnerships in the Implementation of the International Agenda for Sustainable Development (https://aura.antioch.edu/etds/578/)

Laaribi, A., & Peters, L. (2019). *GIS and the 2020 Census–Modernizing official statistics.* ESRI.

Leung, M. W., Yen, I. H., & Minkler, M. (2004). Community based participatory research: A promising approach for increasing epidemiology's relevance in the 21st century. *International Journal of Epidemiology, 33*(1), 499–506. https://doi.org/10.1093/ije/dyh010

Nyerges, T., Jankowski, P., & Drew, C. (2002). Data-gathering strategies for social-behavioural research about participatory geographical information system use. *International Journal of Geographical Information Science, 16*(1), 1–22.

Sayer, J., Sunderland, T., Ghazoul, J., Pfund, J. L., Sheil, D., Meijaard, E., & Van Oosten, C. (2013). Ten principles for a landscape approach to reconciling agriculture, conservation, and other competing land uses. Proceedings of the National Academy of Sciences, 110(21), 8349–8356. https://doi.org/10.1073/pnas.1210595110

Snowdon, W., Schultz, J., & Swinburn, B. (2008). Problem and solution trees: A practical approach for identifying potential interventions to improve population nutrition. *Health Promotion International, 23*(4), 345–353. https://doi.org/10.1093/heapro/dan027.

Steinitz, C. (2012). *A framework for GeoDesign: Changing geography by GeoDesign. ESRI.*

United Nations Committee of Experts on Global Geospatial Information Management. (2018). Integrated geospatial information framework: A strategic guide to develop and strengthen national geospatial information management. Retrieved from http://ggim.un.org/meetings/GGIM-committee/8th-Session/documents/Part%201-IGIF-Overarching-Strategic-Framework-24July2018.pdf

United Nations General Assembly. (2012, July 27). The future we want. A/RES/66/288. New York, NY. Retrieved from https://www.un.org/en/development/desa/population/migration/generalassembly/docs/globalcompact/A_RES_66_288.pdf

United Nations General Assembly (2015b). Transforming our world: The 2030 Agenda for Sustainable Development. A/RES/70/1. Retrieved from https://www.un.org/en/development/desa/population/migration/generalassembly/docs/globalcompact/A_RES_70_1_E.pdf

Waltner-Toews, D. (2017). Zoonoses, one health, and complexity: Wicked problems and constructive conflict. *Philosophical Transactions of the Royal Society B*, 372, 1–9. doi:10.1098/rstb.2016.0171

Wilson, M. W. (2011). Data matter(s): Legitimacy, coding, and qualifications-of-life. *Environment and Planning D: Society and Space*, *29*(1), 857–872. https://doi.org/10.1068/d7910

Index

Note: Pages in *italics* and **bold** refer to figures and tables, respectively.

H

I

K

L

M

N

O

P

R

S

For Product Safety Concerns and Information please contact our EU representative GPSR@taylorandfrancis.com Taylor & Francis Verlag GmbH, Kaufingerstraße 24, 80331 München, Germany

1 Biofuel Significance in a Today's Transforming World

*Aditya Arvind Yadav, Mithilesh Kumar Jha, Sachin Kumar, and Meenu Hans**

1.1 INTRODUCTION

The discovery of fossil fuels such as coal, oil, and natural gas has been a stepping stone in the history of mankind. The complete need for energy in the world is currently fulfilled by these fossil-based fuels from the past centuries. Fossil-based fuels have been the backbone of the Industrial Revolution. Fossil fuels, after combustion, emit excessive carbon dioxide, nitrogen oxides, and sulfur oxides in variable amounts. The major sectors where fossil fuels are extensively used worldwide are power generation, industries, and the transportation sector. Carbon dioxide, which is one of the major pollutants from the combustion of fossil fuels, is responsible for climate change and global warming. Fossil fuels such as crude oil, coal, and natural gas are highly supportive energy items of trade for any country in today's world. Another concern of these fossil fuels is the challenging and variable pricing range in international markets and the uneven distribution of fossil fuels throughout the world, which raises the issue of importing fossil fuels from one nation to another nation to meet their respective energy needs. India is primarily dependent upon Gulf countries such as Iran, Iraq, Qatar, Indonesia, and Middle East nations for conventional fuel imports to meet its energy demands (Liu et al., 2021). Nearly 95% of transportation fuel is imported into India. This raises concerns about energy security due to fluctuating prices of fossil fuels and ever-increasing energy demands. In India, the energy demand has exponentially increased over the past few years and is expected to increase by a rate of 4.8% and become the world's third-largest user of fossil fuels by the end of the year 2030 (Liu et al., 2021). As far as coal is concerned, India ranks fifth in the world in terms of coal reserves but still faces the scarcity of coal every year. The primary reasons for coal scarcity are improper storage during monsoon season, extensive growing energy demands, and a shift towards urbanization and a modern lifestyle.

Due to the growing concern about climate change and global warming, the unstable pricing of fossil fuels and the depletion of conventional fuels have encouraged society to search for renewable energy sources that can cater to the needs of mankind in the coming centuries. Biofuels are primarily produced fuels from renewable sources such as biomass that could be a sustainable option for liquid fuels such as bioethanol and biodiesel that can substitute and minimize the consumption of

* Corresponding author

DOI: 10.1201/9781003485636-1

fossil-based liquid fuels such as gasoline and diesel. These liquid biofuels have a wide range of benefits, such as being environmentally friendly clean energy alternatives for liquid fossil fuels. Biofuels are considered environmentally friendly due to the presence of high oxygen content, which promotes complete combustion with the insignificant exhaust of gases. The oxygen content in the biofuels ranges from 15% to 45%, which makes them a premier choice over petroleum-based fossil fuels (Valli, 2019).

Bioethanol, when blended with gasoline, creates a lean mixture that enhances the air–fuel ratio and improves engine performance by complete combustion of fuel (Hans et al., 2019). These enhance the performance of the engine and reduce the emission of unburnt hydrocarbons, carbon monoxide, carbon dioxide, nitrogen oxides, and sulfur oxides, subsequently reducing greenhouse gas (GHG) and the concern of climate change. The fuel methanol can also be used as an alternative fuel for internal combustion engines due to its high octane rating, high energy density, and clean combustion characteristics. Biodiesel can also be blended with conventional diesel and used for demand substitution. Gaseous fuels such as methane, also known as compressed natural gas (CNG), can be used as fuel for gasoline-based engines with certain modifications in the engines (Abdulkareem-Alsultan et al., 2020).

1.2 WHAT ARE BIOFUELS?

Biofuels are liquid fuels that are extracted from plant biomass. The biofuels can be categorized into three primary forms, namely, solid biofuels in the form of biomass briquettes and dried animal waste; liquid biofuels in the form of ethanol, biodiesel, and bio-oil; and gaseous fuels such as syngas, biogas, and bio-hydrogen. They are considered environmentally friendly compared to fossil fuels, as they can reduce significant GHG emissions. The liquid biofuels can be classified into four generations based on the feedstock used: first generation using edible feedstock, second generation using non-edible feedstock, third generation using algal feedstock, and fourth generation using genetically modified organisms.

Biomass is an excellent renewable resource that can be used as a feedstock for biofuel production. The biomass can broadly be divided into waste products such as agricultural residue, generated waste, forest residues, and energy crops for specific applications (Kumar et al., 2015). It is estimated that agriculture and forest residue can provide nearly 100 to 150 exajoules of energy per year (Subramaniam et al., 2020). The biomass, being a primary feedstock for the production of biofuels, basically consists of three key biomolecules: 35–45% cellulose, 25–30% hemicellulose, and 15–20% lignin (Hans et al., 2019, 2021). Depending upon biomass composition, the appropriate conversion technique needs to be implemented, as it may affect biofuel productivity. Figure 1.1 elaborates on the generation of biofuels based on the type of feedstock.

1.2.1 First-Generation (1G) Biofuels

The 1G biofuels include biodiesel and bioethanol. The primary feedstock for 1G biofuels includes edible sugar- and starch-rich food crop resources and their by-products,

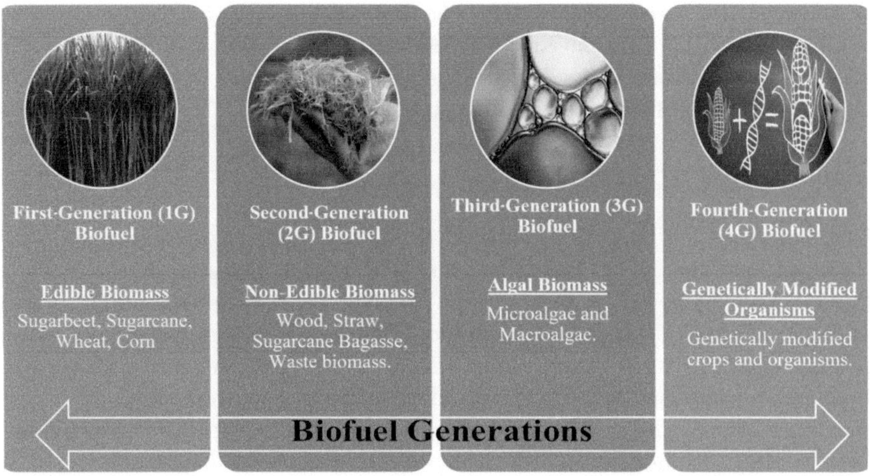

FIGURE 1.1 Generations of biofuels.

such as molasses from the sugar industry (Hans et al., 2021). Bioethanol was the first 1G biofuel produced from corn and sugarcane juice using fermentation technology. Bioethanol is one of the renowned biofuels produced by maturing sugar from sugarcane juice in Brazil, corn in the U.S.A., and molasses in India (Ambaye et al., 2021). The other feedstock for 1G bioethanol includes sugar beet, wheat, barley, etc. Bioethanol can be utilized in internal combustion engines and can be blended with conventional gasoline in variable proportions to be used as a conveyance fuel. Brazil is one of the ethanol-powered countries; Brazil and the U.S. together contribute 89% of worldwide ethanol production (Hans et al., 2019). It can be blended with gasoline at low rates, ranging from 10% ethanol and 90% gasoline (E10) to 85% ethanol and 15% gasoline (E85). The government of India has mandated the blending of ethanol in gasoline up to 20% presently and targeted 30% blending by 2030 under the Ethanol Blending Program (EBP) started in 2006 (National Biofuel Policy, Govt. of India). Ethanol has been used as a conveyance fuel since its inception. The significant advantage of using bioethanol as a fuel is that it can be used in variable proportions with gasoline as an oxygenated additive due to its 35% oxygen content and also can be used as neat ethanol, thereby reducing the demand for conventional nonrenewable gasoline (Hans et al., 2019).

The biodiesel can be produced from vegetable oils, wherein the fundamental fatty acids are linoleic and oleic acids originating from plant-based biomass. Another component is glycerides, which are unsaturated fats that can be further converted into biodiesel. Biodiesel production depends upon oil yields from various biomass, and nearly 75% of production cost depends on raw material cost (Abdulkareem-Alsultan et al., 2020). The commonly used food crop sources for 1G biodiesel production are oil crops such as sunflower, soybean, castor, corn, rice, and olive. The conversion of raw feedstock, i.e., vegetable oil, into biodiesel is called transesterification. Biodiesel can be effectively blended with conventional diesel and utilized in the transportation sector for internal combustion engines. The major problem with biodiesel is improper mixing or blending with conventional diesel fuel due to differences in densities. To overcome these issues, various emulsification agents can be used to settle the high thickness of biodiesel with conventional diesel (Saleh et al., 2022).

1.2.2 Second-Generation (2G) Biofuels

The major problem associated with 1G biofuels is the utilization of food-based feedstock, leading to food insecurity. Thus, to overcome this issue, the 2G biofuels existed. The 2G biofuels are also known as advanced biofuels. The primary feedstock for 2G biofuel includes lignocellulosic biomass such as agricultural residue, forest residue, and various energy crops (Lloyd & Kossmann, 2021). The 2G biofuel, biodiesel, has been developed from non-edible feedstock such as Jatropha, neem, and Karanja oil seeds. The physio-chemical and thermochemical processes are involved in converting oils into biodiesel. The physio-chemical process involves transesterification, whereas the thermochemical process involves gasification followed by fuel blending.

On the other hand, 2G ethanol has been developed using non-edible feedstock such as sugarcane bagasse, corn stover, rice straw, and wheat straw (Mat Aron et al., 2020). The 2G biofuel feedstock is a non-consumable waste obtained after processing the edible feedstock, such as agro-residues. The significant advantage of 2G feedstock is that non-consumable materials are cost-effective and economical and do not impact the food chain cycle. The only drawback of 2G feedstock is the complex process consisting of three steps: pretreatment, saccharification, and fermentation, which makes the conversion process cost-intensive. The lignocellulosic biomass is a predominant feedstock for 2G ethanol production and consists of cellulose, hemicellulose, lignin, and organic components in variable proportions. Cellulose and hemicellulose are significant carbohydrate fractions that can be utilized as sugar sources in fermentation for ethanol production. The initial step for conversion of 2G ethanol requires biomass pretreatment and enzymatic hydrolysis of cellulose and hemicellulose fractions, which are broken down into simple sugars, which further get converted into ethanol using various microorganisms.

1.2.3 Third-Generation (3G) Biofuel

Pretreatment and enzymatic hydrolysis in 2G biofuel production are cost-intensive steps that stop the flourishing of 2G technology at the ground level. The 3G technology utilizes photosynthetic microorganisms like cyanobacteria and microalgae as a source of biofuel production. The lipid content of algal species lipids could be exploited for biodiesel, whereas cellulosic/hemicellulosic fractions could be utilized for bioethanol, etc. The cultivation of these algal species requires light to perform photosynthesis. The significant advantages of using microalgae as a feedstock for biofuel production are high photosynthetic transformation, fast development, less land requirement, and the ability to grow in saline, even in wastewater. Nearly 15,000 algal species have been investigated to date worldwide (Abdulkareem-Alsultan et al., 2020).

The various steps involved in bioethanol production from microalgae include cultivation, harvesting, cell disruption, conversion, and downstreaming for ethanol production. The cultivation of microalgae includes media preparation under approximate temperature, pH, and light density conditions. The harvesting step involves filtration, centrifugation, and flocculation (Cavelius et al., 2023). The cell interruption requires a physical or mechanical breakdown of microalgal cells using various chemical and biological techniques. The major advantage of cell interruption is that no waste is generated through the process. The conversion process involves steps like saccharification and fermentation. The microalgae have a long chain of unsaturated fats, an essential component for biodiesel production. These long chains of unsaturated fatty acids are called lipids. The extracted oils from the microalgae are used as a feedstock for transesterification to obtain 3G biodiesel. Various other conversion processes are also investigated to obtain various products from microalgae, such as the biochemical conversion of microalgae to methane via anaerobic digestion. The other conversion route involves thermochemical conversion, such as pyrolysis for biochar and biohydrogen production, gasification for biomethane and biohydrogen production, and combustion for charcoal and bioelectricity production (Patnaik et al., 2020).

1.2.4 Fourth-Generation (4G) Biofuels

The 4G process is one of the latest technologies in biofuel production. It uses genetic engineering to enhance the characteristics of particular microorganisms for specific biofuel production. It involves the utilization of multiple sugars like pentose and hexose for fermentation by a single set of microorganisms, as well as higher lipid synthesis, lower lignin, and higher cellulose/hemicellulose formation in plant biomass and higher photosynthetic microorganisms. Two pathways have been employed, namely, developing pathways in native producers through optimization and use of various carbon sources, while another pathway is reconstruction/introduction of the pathways in natural producers such as bacteria, yeast, and algae (Cavelius et al., 2023). In 4G technology, the metabolic pathway of these microorganisms is so versatile that they can produce a wide variety of biofuels. For example, in the butanol pathway, genes from *Clostridia sp.* were introduced to *E. coli*, *Pseudomonas putida,* and *Bacillus sub*tilis strain*s* for enhanced biofuel production (Ambaye et al., 2021).

The microalgae can also be used as a feedstock for bio-butanol production. The various microalgal species that can be utilized for biobutanol production are *T. subcordifornis*, *Chlorella sp.*, *Chlorella vulgaris,* Chlorella *Reinhardt, and S. obliquus*, which have very high carbohydrate content. Bio-butanol can be produced from the acetone–butanol–ethanol (ABE) fermentation process. The substrate used for fermentation was *the C. acetobutylicum* bacterium for anaerobic digestion. The bacterium produces butyric acid and acetic acid in a process called acidogenesis, followed by solventogenesis, where ethanol and butanol are produced (Yeong et al., 2018). The significant substrates and microbes used in bio-butanol production are traditional substrates such as cane, corn, millet, wheat, rice, and potatoes; the agricultural residues such as corn straw, corn fibers, wheat straw, and Miscanthus can also be effectively used for bio-butanol production. The primary microbial strains utilized for bio-butanol production are *Clostridia*. The butanol pathway genes isolated from *Clostridia* are introduced in *P. putida, E. coli, and S. cerevisiae.* The significant challenges associated with bio-butanol production are the unavailability of cheaper feedstock, improved fermentation strategies, sustainable and economical solvent recovery, and water recycling (Rathour et al., 2018).

Similarly, genetically modified algae can yield a wide range of products such as biomethane, biohydrogen, biodiesel, and bioethanol (Adebami et al., 2022). The photosynthetic efficiency of microalgae can be enhanced by advanced antenna systems that can absorb a broader range of the light spectrum. The 4G biofuel provides a wide range of more sustainable products by combining carbon capture with the help of genetic engineering and nanotechnology. The genetics and metabolic engineering of yeast strains for ethanol production play a vital role in 4G biofuel (Cao et al., 2014). The yeast strain improvement techniques can be broadly categorized into randomized mutagenesis, whole genome sequencing, evolutionary engineering, protein engineering, quantitative trait locus, genome editing, promoter engineering, and metabolic engineering. The primary yeast strains reviewed for ethanol production were *Yarrowia lipolytica* and *S. cerevisiae.* The major challenge associated with 2G ethanol production is the formation of toxins during process conversion, as chemical pretreatment is majorly involved in the preliminary stages. The stress response of yeast plays a vital role in efficient conversion. The recombinant DNA technology for *S. cerevisiae* of the mutant strain has faster cellulose utilization for bioethanol production than wild strains (Kim et al., 2013). Genetic engineering can play a vital role in economic bioethanol production. The stress-tolerant genes were introduced to *S. cerevisiae* (Sharma et al., 2016). The xylose utilization by *S. cerevisiae* by xylose reductase (XYL 1) and xylitol dehydrogenase (XLY2) genes for NADPH-linked XR and NAD-linked XDH from *P. stiptis* on *S. cerevisiae* (Han et al., 2021). The effect of xylitol by-product for recombinant *S. cerevisiae is* due to high production of NADPH rather than NADH (Sharma & Arora, 2020).

1.3 ENVIRONMENTAL BENEFITS

Biofuels are advantageous and disadvantageous at the same time, considering environmental, social, and economic parameters related to biofuel production. The major drivers in biofuel production are GHG emission reduction and rural development. Continuously growing demand for various agricultural yields over a period of time has impacted 1G biofuel production on a large scale. The 2G biofuel production has also been impacted due to economic constraints, due to higher pretreatment and conversion processes. The 3G biofuel production could certainly overcome food production issues, as the microalgae can be grown on any land and even on wastewater. However, on the other hand, the production of 3G biofuels is energy-intensive, and high cost is generally involved during the conversion process (Malode et al., 2022).

Qin et al. (2016) reported spatially and modeled carbon emission's influence on ethanol production through corn and cellulosic feedstock. The study was conducted for the USA, the effects of biofuel production on soil organic carbon resulting in life cycle GHG. From the results obtained, switch grass and miscanthus show positive results in terms of land use change GHG emission, whereas corn and corn stover have increased land use change GHG emission. The range of values for land use change GHG emission is as follows: 2.1 to 9.3 for corn, –0.7 for corn stover, –3.4 to 12.9 for switch grass, and –20.1 to –62 for miscanthus grams of CO2 equivalent per MJ for ethanol production. The soil depth of nearly 30 cm to 100 cm was affected due to soil organic carbon change, land use change, and GHG emission. The total life cycle GHG emissions were 59 to 66 for corn ethanol, 14 for stover ethanol, 18–26 for switch grass ethanol, and –7 to 0.6 for miscanthus ethanol. Koponen et al. (2013) reported the GHG emissions for various liquid transportation fuels in Finland with EU sustainability criteria. The EU has set a renewable energy directive (RET) for increased use of renewable energy with conventional energy. The GHG emission reduction targets for transportation biofuels and other bio-liquids were estimated per EU sustainability criteria. An uncertainty and sensitivity analysis was performed on liquid biofuels. From the results, it is unpredictable whether the biofuel chain cycle passes the emission-saving limit suggested by EU standards. The major emission sources include nitrogen oxide from soil and nitrogen fertilization, as well as the emission of heat production and soil carbon stock in biomass processing. Haus et al. (2020) reported the GHG emission reduction obligation system for Swedish sawdust-to-ethanol production. Sawdust ethanol production is one of the low life cycle GHG emissions, with 93% emission savings as compared to fossil fuel. The sawdust-based ethanol production is one of the low life cycle GHG emission feedstocks for Sweden, which will increase the country's economic competitiveness in the road transportation sector. Additional economic incentives are required to make it possible at the commercial level.

The significant advantages of biofuels are food security, biomass availability, climate change, and social benefits, which enhance sustainable development. As per the United Nations Sustainable Development Goals (SDGs), goal 7 reflects affordable and clean energy for all (Cavelius et al., 2023). The other 17 goals are also linked with biofuel production in one way or another. As per the net zero carbon emission targets of 2050, biofuel technology is the major contributor to achieving the aim. The production of biofuels will enhance the carbon capture and storage (CCS) mechanism, which will enhance the net zero emission targets. The United Nations Framework Convention on Climate Change (UNFCCC) has set a conference of parties (COP) to implement net zero emissions by 2050 and reduce GHG emissions (Gheewala, 2023). The Paris Agreement, COP 26, held in Glasgow, has thrown light on GHG emission reduction and has set up the goal of reducing global average temperature by 1.5°C. The latest meeting of COP27 was held in Sham El-Sheikh, where the company adopted climate finance and adaptation to reduce GHG emissions on a large scale. The significant factors for the adoption of biofuels in the energy transition are directly linked to political and regional support. The biofuels alone cannot reduce the total GHG but indeed can play a crucial role in sustainable energy transition towards net zero targets by 2050.

The transportation sector alone is a very energy-intensive sector and requires barely 560 exajoules of energy every year. The transportation sector accounts for nearly 30% of the world's fossil fuels. Now replacing these sets of non-renewable-based energies with renewable energy is the major challenge, as most of the renewable energy, such as solar and wind, is intermittent in nature. Biofuels are one of the

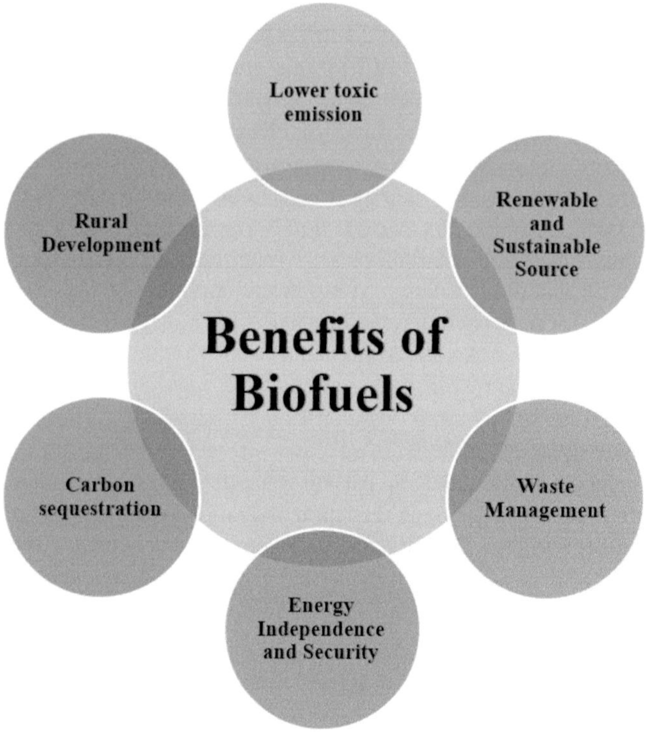

FIGURE 1.2 Benefits of biofuels.

promising and sustainable options that can replace carbon-intensive fossil fuels. The combustion of one liter of conventional diesel accounts for nearly 2.67 kg of carbon dioxide emission, whereas the combustion from biofuels does not affect the long-term carbon cycle, as it originated from renewable sources of energy (Gheewala, 2023). Figure 1.2 represents various benefits of biofuels in terms of usage.

1.4 CHALLENGES AND POTENTIAL BARRIERS IN BIOFUEL PRODUCTION

1.4.1 Economic Implications

The major factor that affects the conversion of biomass to biofuels is the cost involved in performing various steps of the process. The economic challenges associated with biofuel production are as follows:

1.4.2 Equipment and Technology Cost

The significant expenditure associated with the conversion of biomass to biofuels is the cost of reactors, equipment, accessories, and technology employed in the process.

For example, 4G biofuel production uses genetically modified organisms that can be highly invested in technology. The high-end process equipment can also hamper the economy on a large scale. The significant costs involved in biofuel production are reactor costs and separation and purification systems such as distillation units for bioethanol and biodiesel production. The other major cost components are storage and handling, such as tanks, piping, and pumping systems. The other auxiliary costs associated with equipment are energy supply, control system, instrumentation, and monitoring of equipment (Malode et al., 2022).

1.4.3 Material Cost

Material cost is one of the important parameters taken into consideration during the conversion economics process. The significant cost involved is the cost of enzymes, feedstock, chemicals used as pretreatment agents, and other operational costs. The other material costs are water supply and energy costs; the water supply costs depend on the availability of water near the production unit and the treatment of raw water for processing. The energy costs include the production of energy for process heating and cooling applications at local rates (Sahay, 2022).

1.4.4 Waste Disposal

The waste generated from the process needs to be treated before disposal, which further increases the cost. The closed technology cycle can enhance the entire process of economics (Jeswani et al., 2020). The closed-loop waste disposal cycle is a system that produces no waste; the waste in the closed-loop cycle has a circular life cycle, i.e., the waste is recycled into a new product that is environmentally friendly and returned to the environment as biodegradable waste.

1.4.5 Infrastructure Cost

The cost of infrastructure, such as electricity, water, and other utilities, is the major factor. The infrastructure costs of biofuel production involve facility construction, such as production units, storage tanks, and equipment installation costs. The other costs associated with infrastructure are utilities and transportation costs. The utility costs include water supply and treatment and energy supply infrastructure. The transportation costs associated with infrastructure include road infrastructure and loading and unloading facilities at production units for feedstock and biofuel (Malode et al., 2022).

1.4.6 Cultivation and Harvesting Costs

It is one of the most neglected factors for consideration of expenditure on biofuel technology. The economic viability depends upon the cultivation and harvesting of feedstock in the case of 2G, 3G, and 4G biofuel processes. The cultivation costs include costs associated with seed and planting biomass, soil preparations, fertilizer and pesticide costs, irrigation and water costs, and labor costs. The harvesting costs include equipment and machinery costs, labor costs, and transportation and storage costs (Mat Aron et al., 2020). These economic challenges can be overcome by proper research and development in each sector of biofuels. The proper economics can make the biofuel industry feasible on a commercial scale.

1.5 BIOMASS RECALCITRANCE

The primary challenge faced by the 2G biofuel production is the presence of lignin in the feedstock. The complex structure of the lignin hinders the conversion process. Various advanced pretreatment technologies are effectively used for efficient conversion. Another major challenge faced by biofuel production is the cost associated with the production, which is much higher than that of conventional fuels. Various cost-effective enzymes have been developed to overcome this issue (Arora et al., 2015; Hans et al., 2019).

The lignocellulosic biomass is recalcitrant to various microbial strains and enzymes used in process conversion, which is the major obstacle at the commercial level. The factors that affect the enzymatic hydrolysis are the presence of cellulose, hemicellulose, acetyl groups, and lignin in the biomass. The cellulose is most abundantly available in biomass, constituting 60–70% by weight of total biomass. It consists of β-glucopyranose units linked together by β-(1,4) glycosidic bonds with cellobiose as a fundamental unit. The degree of polymerization ranges from 800 to 1500 D-glucose units to form cellulose microfibrils. The major challenge of cellulose is its high crystallinity and porosity. The cellulose microfibrils are very resistant to enzymatic hydrolysis. The long cellulose chains contain more hydrogen bonds and are challenging to hydrolyze, and short cellulose chains have weak hydrogen bonds and tend to have easy enzymatic hydrolysis (Sahay, 2022).

The hemicellulose biopolymers present nearly 25–40% of the total weight of biomass. It contains xylans, xyloglucan, mannans, and glucomannans in monosaccharide subunits. The hemicellulose is amorphous with very little physical strength. It acts as a barrier for the enzymatic hydrolysis. The lignin is the most abundantly available polymer in LCB, consisting of 25–30% of the total weight of biomass. It comprises phenylpropanoid building units consisting of p-coumaryl, coniferyl, and sinapyl alcohol subunits. Hydrophobicity and structural rigidity are hindrances to enzymes. The lignin limits the accessibility of enzymes to various polysaccharides in LCB. It can also absorb a certain amount of enzymes due to its hydrophobic nature. The significant factors in LCB that impact enzyme activity are crystallinity, particle size, accessible surface area, accessible volume, pore size, and internal surface area (Zoghlami & Paës, 2019).

1.6 TECHNICAL CHALLENGES

The major technical challenge associated with 2G technology includes saccharification, which is one of the important steps in biofuel production that degrades various polysaccharides into simple sugars. Various advanced enzymes can be utilized that can convert carbohydrates to simple sugars and make saccharification more efficient; utilization of multiple feedstocks, the agro-residue, can be utilized

TABLE 1.1
Potential Barriers and Strategies in Biofuel Production

Sl. No.	Potential Barriers	Strategies to Overcome the Barriers
1	Presence of lignin in the feedstock	The various advanced pre-treatment technologies can be effectively explored to efficiently separate lignin from LCB. A combination of two or more pre-treatment techniques can effectively remove lignin. The pre-treatment techniques should have considered, taking into account the overall process economics.
2	Cost of biofuel production	The significant cost involved in biofuel production is the cost of enzymes used in process conversion. Efficient and indigenous enzymes should be developed for the effective conversion of biomass substrate into biofuels.
3	Ethical and regulatory issues in 4G biofuel production	The comprehensive strategies should be developed through various certification systems, taking into account the protection of human and environmental rights, life cycle assessment of GHG emissions, fair trade policies and benefit-sharing schemes.
4	Efficient Saccharification process	Various advanced Indigenous enzymes can be developed that can convert carbohydrates to simple sugars and make saccharification more efficient; using multiple feedstocks, the agro-residue can be utilized for biofuel production.
5	Equipment's for efficient process conversion	The high-end reactors and fermenters that can handle the various LCB feedstock are still unexplored areas in the bio-conversion system. The energy-efficient reactors and fermenters need to be developed for efficient utilization of biomass.
6	Supply chain management of LCB	The effective transportation logistics and storage of biomass should have developed, taking into account the cost and shelf life of biomass.

for biofuel production. The agro-residue has variable amounts of C5 and C6 sugars, and depending upon availability, it can be utilized with multiple feedstocks; for enzymatic hydrolysis, the various enzyme cocktails can be utilized for the efficient degradation of LCB into biofuels. The primary concern of optimizing the enzymes is still under research; the high-end reactors and fermenters that can handle the various LCB feedstocks are still an unexplored area in the bioconversion system. Table 1.1 summarizes the key barriers and strategies in the biofuels sector. The energy-efficient reactors and fermenters need to be developed for efficient utilization of biomass. The other primary concern associated with biomass is its supply chain management and the economics associated with it. The transportation logistics and storage should be cost-effective (Rame et al., 2023).

1.7 POLICY AND REGULATORY FRAMEWORK

The National Biofuel Policy (NBP) 2018 was implemented in India on May 16, 2018, to develop a sustainable biofuel economy in India (NBP, Govt. of India). Under the NBP 2018, the Indian government has set up the target of 10% ethanol blending in gasoline by the year 2022. India mainly depends on imported gasoline, which is around 80% of the crude oil import requirements. The biofuel industry in India started in 2001, with ethanol blending of 5% as a pilot-scale program. The National Biofuel Mission was adopted in the year 2003 with a biodiesel blending program with targets of 20% biodiesel blending by the year 2011–12 (MoPNG, 2018). The same national policy on biofuel was followed in 2009, and some revised mandates were implemented in 2018, with optional blending targets of 20% ethanol and biodiesel by 2017. The NBP 2009 focused more on 2G biofuel production with a major goal of the development of indigenous feedstock and its utilization for biofuel production. Later, the NBP was revised in 2018 with an indicative target of 20% ethanol blending in gasoline and 5% biodiesel blending in diesel by 2030 (NBP Amendment, Govt. of India, 2022).

The major objective of NBP 2018 was to reduce crude oil imports, increase farmers' income, utilize dry and arid lands, and generate local employment with a focus on sustainability. The United States is one of the top countries in the production of biofuel, with a production capacity of nearly 38,000 tons of oil equivalent with 40% of the share globally, followed by Brazil with 21,375 ktoe with a share of 23% globally. The Indian biofuel status is nearly 1023, with a global share of just 1.1%. Various countries have come up with their own biofuel policies, such as the USA, Canada, the EU, Australia, Brazil, Japan, and India (Sahay, 2022).

1.8 INTERNATIONAL BIOFUEL POLICIES

1.8.1 USA

The USA has launched a program named Renewable Fuel Standard (RFS) under the Environmental Protection Agency. The mandate under RFS is the utilization of a minimum quantity of fuel sourced from renewable sources such as biofuels. The utilization of RFS was in the year 2006 with blending targets of ethanol in gasoline for nearly 4 billion gallons in the year 2006; later, it was increased to 7.5 billion gallons by the year 2012. The RFS 1 was mandated under the Energy Policy Act of the US in the year 2005. The updated RFS 2 targeted nearly 9 billion gallons of biofuels for nearly 2022. The RSF has categorized biofuels into four categories based on the feedstock; the US has focused its RSF towards cellulosic 2G ethanol due to the corn market in the US.

1.8.2 Canada

Canada had mandated Renewable Fuels Legislation in the year 2010 with a 5% mandate for renewable fuel in gasoline by volume and 2% renewable fuel in conventional diesel. The gasoline and diesel blending was mandated after legislation in the year 2010–11. The other policies, such as the ecoENERGY for biofuels initiative, were active from 2008 to 2017 for biofuel production. The primary objective of ecoENERGY was to produce incentives for biofuel manufacturing in Canada.

1.8.3 Argentina

Argentina enacted the biofuels law in the year 2006 with the 5% biodiesel blending and 5% ethanol blending in diesel and gasoline, respectively. The major objective of biofuel law was to provide tax incentives for the production of biofuels.

1.8.4 Brazil

The first biofuel program in Brazil was launched under the name of Proalcool with a major focus on the oil crisis and oil price hike in 1973. The national biofuel policy of Brazil is named the RenovaBio Programme and was started in the year 2020, with targets of reducing carbon emissions by 10% by the year 2028. The main aim of the policy was to create a carbon credit mechanism market; the fuel distributors require certificates called CBIOs. The more the number of CBIOs, the more the amount of profit. The latest targets of RenovaBio are 27% ethanol blending and 11% biodiesel blending in gasoline and diesel, respectively.

1.8.5 European Union (EU)

The European Union has set a mandate under the Renewable Energy Directive (RED) to set 20% usage of renewable energy in the total energy mix of the EU. Out of 20% of the total share of energy, 10% should be dedicated to the transportation sector for biofuels. Under RED, the GHG emission reduction should be at least 35% in the transportation sector through biofuels. The blending targets set under RED were 2% biofuel blend by 2005 and 5.75% biofuel blend by 2010. The Common Agricultural Policy (CAP), launched in the year 1962, gave a boost to 2G biofuel production in the EU; it also provides subsidies to farmers.

1.8.6 Germany

The German biofuel mandates are the same as that of the EU Common Agricultural Policy (CAP). Germany has set the target of a 6% GHG reduction in the transportation sector by the year 2025. Germany is one of the leaders in the production of bioethanol and biodiesel in the EU. Germany has adopted the EU Directive Policies on biofuel.

1.8.7 China

China has categorized the entire biofuels sector into four sections, namely, denatured fuel ethanol, ethanol gasoline for motor vehicles, biodiesel blend stock, and biodiesel fuel blend. The ethanol blend was started in the year 2001, whereas biodiesel blending was started in the year 2010. China has more focus on bioethanol rather than biodiesel, as most of the vegetable oil in the country is from imports (Das, 2020; Sahay, 2022).

1.9 OVERVIEW AND CRITICAL ANALYSIS OF NBP 2018

The Indian transportation sector depends on nearly 80% of crude oil imports with a view to reducing imports and having indigenous biofuel production. The NBP was enacted on May 16, 2018, to encourage the usage of biofuel in the transportation sector in India. The earlier mandate of biofuel blending was 2% ethanol in gasoline and 0.5% biodiesel in diesel. The latest mandates under NBP 2018 are 20% ethanol and 5% biodiesel in gasoline and diesel, respectively. Under the NBP 2018, the biofuel sector is divided into three generations: 1G, basic biofuels from edible feedstock; 2G, advanced biofuels from non-edible feedstock; and 3G, algal biofuels.

The policy allows various feedstocks, such as sugarcane juice, molasses, corn, sugar beet, cassava, damaged food grains, broken rice, and potatoes, for bioethanol production through various routes. The surplus and broken or leftover food grains can be utilized for ethanol production under the supervision of the National Biofuel Coordination Committee (NBCC). The NBP 2018 is more focused on 2G ethanol production and has led the investment of about 50,000 million rupees in 2G biorefineries for six years. The primary focus on foreign investment, i.e., 100% foreign direct investment, is permitted under NBP 2018 with direct approval. The NBP has set a ban on the import and export of various biofuels in the country. With a significant focus on 2G biorefineries, the NBP has provided significant tax incentives, lower depreciation costs, and higher funding for setting up 2G biorefineries. The oil manufacturer companies will be responsible for the storage, distribution, blending, and marketing of various biofuels in India (Das, 2020).

1.9.1 Critical Analysis from NBP 2018

The critical analysis for various clauses from NBP 2018 is as follows:

Clause 6.1: The NBP 2018 mentions that no biofuel should be imported from foreign nations. The import of bioethanol and biodiesel should be allowed to avoid the shortage of biofuels in the near future.

Clause 5.9.3: NBP mentions the usage of food grains for ethanol production as per the latest guidelines from foreign nations. The food grains should be limited from usage as a feedstock for ethanol production, as the problem may arise for a food vs. fuel crisis in the near future due to various environmental issues. This may also hamper food prices over the long run.

Clause 5.10.1: The policy mentions that the ethanol production from 1G feedstock, i.e., sugarcane and molasses, has certain limitations, as the ethanol from 1G feedstock is extensively used in potable alcohol and various chemical industries. This fact will hamper the Ethanol Blending Program on a large scale, and hence, the government has shifted its focus from 1G ethanol production to 2G ethanol production, but the major part of the ethanol is still procured from 1G feedstock, which makes the EBP program successful.

Clause 5.10.4: The Oil Manufacturer Companies (OMCs) have agreed to sign an agreement on ethanol purchase from 2G feedstock for 15 years to provide a stable market for 2G ethanol in India. The major part of the ethanol is still procured from 1G feedstock, while 2G ethanol is still under the research and development phase.

Clause 5.17: The policy has given more limelight to tax incentives, advanced depreciation on plant expenditure, and financial assistance in the form of loans for setting up 2G biorefineries in India at a commercial scale. This will highly impact the 1G ethanol industry. The majority of ethanol is still procured by 1G feedstock, while the price variation for 1G and 2G ethanol will be the significant deciding factor in the future. The fact that India is the second largest producer of sugarcane and molasses means the 1G biorefinery could provide ethanol on a large scale in the near future for EBP if proper incentives are provided. No such incentives and tax benefits are proposed in NBP 2018.

1.10 FUTURE OUTLOOKS

Biomass to biofuels is one of the emerging technologies that can be a boon to the biofuel industry. The various dynamics in the process conversion can provide a well-defined track for the biofuel industry. The future outlooks for the biofuel industry are as follows: Various nanocatalysts, such as alkali homogenous catalysts, for example, NaOH, KOH, and $NaOCH_3$, heterogeneous catalysts, and solid acid catalysts, such as Y-type zeolites, could be utilized for efficient conversion of triglycerides to biodiesel for enhanced productivity (Rame et al., 2023). The enzymes play an important role in the conversion of biomass to biofuel. Various nanoparticles and nanocomposites can be used to stabilize enzyme activity, thereby improving the efficiency of enzymes. The nanotechnology can also enhance lipid extraction from algal biomass with the use of nanosolvents. The other applications of nanotechnology include biosensors for monitoring parameters such as pH, temperature, and concentration. The various biofuel blends can be tested for combustion characteristics based on the environmental outlook. The ethanol blends such as E20, E30, E50, and E80 and biodiesel blends such as B5 and B20 enhance engine efficiency, fuel quality, and environmental benefits (Hyrchenko et al., 2021). The long-term effects of various blends on the engines can also be studied in terms of the materials of the engines and other properties.

The 2G ethanol production has a large-scale viability considering the overall availability of biomass in India. Efficient conversion techniques should be developed to foster the production of 2G ethanol from various feedstocks. The environmentally friendly pretreatment technologies can help to reduce energy consumption as well as the environmental footprint of the process. It is essential to integrate 2G ethanol production with other products such as biochemicals, bioelectricity, and anaerobic digestion to improve overall process economics. Optimization of enzymes for efficient conversion of LCB to biofuel. Cost-efficient enzyme extraction and its effective utilization in biofuel conversion. With the help of genetic engineering, the modification of enzymes at the molecular level is possible to make biofuel production cost-effective. The major challenge faced in commercial biofuel production is the stability of enzymes for harsh environments such as temperature, and pH fluctuation. Enzyme stabilization using genetic engineering will be a way forward to overcome this issue. The catalysts used for transesterification should be reusable and recyclable to reduce the overall cost of production.

1.11 CONCLUSION

Biofuel technology is considered one of the sustainable energy solutions that can overcome the issues of conventional fossil fuel depletion. The biochemical conversion pathways such as fermentation, transesterification, and anaerobic digestion can effectively produce biofuels such as bioethanol, biodiesel, and biomethane, respectively. The major hurdle posed by the 2G biofuel production is the degradation of LCB into various constituent compounds efficiently in terms of economics. The 3G biofuel from microalgae is one such untapped area that can harness the needs of the biofuel industry. Biofuels will play a pivotal role in energy transition and can face particular challenges in a changing world. They can offer sustainable solutions to conventional fossil fuels, contributing to the environment and economic development of the nation. Renewable resources such as biomass can be effectively utilized for biofuel production that can reduce GHG emissions, prevent climate change, and promote rural development through enhanced agricultural productivity. However, the widespread adoption of bioenergy requires technological advancements, continuous policy frameworks, and public awareness.

ACKNOWLEDGEMENTS

Authors gratefully acknowledge Sardar Swaran Singh National Institute of Bio-Energy (SSS-NIBE), Kapurthala, India, for providing research facilities. Meenu Hans is thankful to the Guru Nanak Dev University, Amritsar, India, for providing the facility to carry out the work. Meenu Hans is also grateful to DST-SERB for providing funds to do the necessary researchvia grant no. EEQ/2023/000427. Aditya Arvind Yadav is thankful to the Ministry of Education for providing a research fellowship for a full-time PhD in the Centre for Energy and Environment at Dr B R Ambedkar National Institute of Technology, Jalandhar, India. [Roll No. 22501203]

REFERENCES

Abdulkareem-Alsultan, G., Asikin-Mijan, N., Lee, H. V., & Taufiq-Yap, Y. H. (2020). Biofuels: Past, present, future. In *Green Energy and Technology* (pp. 489–504). Springer Singapore. https://doi.org/10.1007/978-981-13-9012-8_21

Adebami, G. E., Kuila, A., Ajunwa, O. M., Fasiku, S. A., & Asemoloye, M. D. (2022). Genetics and metabolic engineering of yeast strains for efficient ethanol production. *Journal of Food Process Engineering*, *45*(7), e14051. https://doi.org/10.1111/jfpe.14051

Ambaye, T. G., Vaccari, M., Bonilla-Petriciolet, A., Prasad, S., van Hullebusch, E. D., & Rtimi, S. (2021). Emerging technologies for biofuel production: A critical review on recent progress, challenges, and perspectives. *Journal of Environmental Management*, *290*, 112554. https://doi.org/10.1016/j.jenvman.2021.112554

Arora, S., Behera, S., & Kumar, R. (2015). Bioprospecting thermophilic/thermotolerant microbes for production of lignocellulosic ethanol: A future perspective. *Renewable and Sustainable Energy Reviews*, *51*, 699–717. https://doi.org/10.1016/j.rser.2015.07.086

Cao, L., Tang, X., Zhang, X., Zhang, J., Tian, X., & Wang, J., et al. (2014). Two-stage transcriptional reprogramming in *Saccharomyces cerevisiae* for optimizing ethanol production from xylose. *Metabolic Engineering, 24*, 150–159. https://doi.org/10.1016/j.ymben.2014.06.007

Cavelius, P., Engelhart-Straub, S., Mehlmer, N., Lercher, J., Awad, D., & Brück, T. (2023). The potential of biofuels from first to fourth generation. *PLoS Biology, 21*(3), e3000489. https://doi.org/10.1371/journal.pbio.3000489

Das, S. (2020). The national policy of biofuels of India – A perspective. *Energy Policy, 143*, 111595. https://doi.org/10.1016/j.enpol.2020.111595

Gheewala, S. H. (2023). Life cycle assessment for sustainability assessment of biofuels and bioproducts. *Biofuel Research Journal, 10*(1), 1810–1815. https://doi.org/10.18331/BRJ2023.10.1.5

Govt. of India Ministry of Petroleum and Natural Gas. National Policy on Biofuels 2018. Gazette of India. 2018;(D):1–23. Available from: https://mopng.gov.in/files/uploads/NATIONAL_POLICY_ON_BIOFUELS-2018.pdf

Govt. of India Ministry of Petroleum and Natural Gas. National Policy on Biofuels-2018 Amendment, 2022; (D):4–7. Available from: https://mopng.gov.in/files/article/articlefiles/Notification-15-06-2022-Amendments-in-NPB-2018.pdf

Han, J., Gong, G., Wu, X., & Zha, J. (2021). Systematic metabolic engineering of Saccharomyces cerevisiae for efficient utilization of xylose. In: Liu, ZH., Ragauskas, A. (eds) *Emerging Technologies for Biorefineries, Biofuels, and Value-Added Commodities*. Springer, Cham. https://doi.org/10.1007/978-3-030-65584-6_7

Hans, M., Garg, S., Pellegrini, V. O. A., Filgueiras, J. G., de Azevedo, E. R., & Guimaraes, F. E. C., et al. (2021). Liquid ammonia pretreatment optimization for improved release of fermentable sugars from sugarcane bagasse. *Journal of Cleaner Production, 281*, 125366. https://doi.org/10.1016/j.jclepro.2020.125366

Hans, M., Kumar, S., Kumar Chandel, A., & Polikarpov, I. (2019). A review on bioprocessing of paddy straw to ethanol using simultaneous saccharification and fermentation. *Process Biochemistry, 86*, 76–84. https://doi.org/10.1016/j.procbio.2019.04.006

Haus, S., Björnsson, L., & Börjesson, P. (2020). Lignocellulosic ethanol in a greenhouse gas emission reduction obligation system—a case study of Swedish sawdust-based ethanol production. *Energies, 13*(5), 1089. https://doi.org/10.3390/en13051089

Hyrchenko, Y., Skibina, T., Us, Y., & Veckalne, R. (2021). World market of liquid biofuels: Trends, policy and challenges. *E3S Web of Conferences, 280*, 01012. https://doi.org/10.1051/e3sconf/202128005005

Jeswani, H. K., Chilvers, A., & Azapagic, A. (2020). Environmental sustainability of biofuels: A review. *Proceedings of the Royal Society A: Mathematical, Physical and Engineering Sciences, 476*(2243), 20200428. https://doi.org/10.1098/rspa.2020.0351

Kim, S.R., Park, Y-C., Jin, Y-S., Seo, J-H. (2013). Strain engineering of Saccharomyces cerevisiae for enhanced xylose metabolism. Biotechnology Advances, 31(6), 851–861. https://doi.org/10.1016/j.biotechadv.2013.03.004

Koponen, K., Soimakallio, S., Tsupari, E., Thun, R., & Antikainen, R. (2013). GHG emission performance of various liquid transportation biofuels in Finland in accordance with the EU sustainability criteria. *Applied Energy, 102*, 440–448. https://doi.org/10.1016/j.apenergy.2012.07.019

Kumar, S., Dheeran, P., Singh, S.P., Mishra, I.M., & Adhikari, D.K. (2015). Kinetic studies of two-stage sulphuric acid hydrolysis of sugarcane bagasse. *Renewable Energy, 83*, 850–858. https://doi.org/10.1016/j.renene.2015.05.033

Liu, Y., Cruz-Morales, P., Zargar, A., Belcher, M. S., Pang, B., & Englund, E., et al. (2021). Biofuels for a sustainable future. *Cell, 184*(6), 1636–1647. https://doi.org/10.1016/j.cell.2021.01.039

Lloyd, J. R., & Kossmann, J. (2021). Improving crops for a changing world. *Frontiers in Plant Science*, *12*, 690. https://doi.org/10.3389/fpls.2021.728328

Malode, S. J., Gaddi, S. A. M., Kamble, P. J., Nalwad, A. A., Muddapur, U. M., & Shetti, N. P. (2022). Recent evolutionary trends in the production of biofuels. *Materials Science for Energy Technologies*, *5*, 262–277. https://doi.org/10.1016/j.mset.2022.02.002

Mat Aron, N. S., Khoo, K. S., Chew, K. W., Show, P. L., Chen, W. H., & Nguyen, T. H. P. (2020). Sustainability of the four generations of biofuels – A review. *International Journal of Energy Research*, *44*(12), 9266–9282. https://doi.org/10.1002/er.5196

Patnaik, R., Bagchi, S. K., & Mallick, N. (2020). The refinery concept: Addressing the challenges of microalgal biodiesel production. In *Recent Developments in Bioenergy Research* (pp. 195–223). Springer. https://doi.org/10.1007/978-3-030-34175-6_9

Qin, Z., Dunn, J. B., Kwon, H., Mueller, S., & Wander, M. M. (2016). Influence of spatially dependent, modeled soil carbon emission factors on life-cycle greenhouse gas emissions of corn and cellulosic ethanol. *GCB Bioenergy*, *8*(6), 1136–1149. https://doi.org/10.1111/gcbb.12338

Rame, R., Purwanto, P., & Sudarno, S. (2023). Biotechnological approaches in utilizing agro-waste for biofuel production: An extensive review on techniques and challenges. *Bioresource Technology Reports*, *24*, 101662. https://doi.org/10.1016/j.biteb.2023.101662

Rathour, R. K., Ahuja, V., Bhatia, R. K., & Bhatt, A. K. (2018). Biobutanol: New era of biofuels. *International Journal of Energy Research*, *42*(15), 4532–4345. https://doi.org/10.1002/er.4107

Sahay, S. (2022). *Handbook of Biofuels*. Elsevier. ISBN: 978-0128242362.

Saleh, N. M., Saleh, A. M., & Mahdi, H. H. (2022). Production of biofuels from biomass as an approach towards sustainable development: A short review. *NTU Journal of Renewable Energy*, *3*(1), 9–21. https://doi.org/10.56286/ntujre.v3i1.346

Sharma, S., & Arora, A. (2020). Tracking strategic developments for conferring xylose utilization/fermentation by *Saccharomyces cerevisiae*. *Annals of Microbiology*, *70*(1), 13. https://doi.org/10.1186/s13213-020-01590-9

Sharma, V., Sharma, S., & Kuila, A. (2016). A review on current technological advancement of lignocellulosic bioethanol production. *Journal of Applied Biotechnology and Bioengineering*, *1*(2), 00011. https://doi.org/10.15406/jabb.2016.01.00011

Subramaniam, Y., Masron, T. A., & Azman, N. H. N. (2020). Biofuels, environmental sustainability, and food security: A review of 51 countries. *Energy Research and Social Science*, *68*(April), 101549. https://doi.org/10.1016/j.erss.2020.101549

Valli, C. (2019). Mitigating enteric methane emission from livestock through farmer-friendly practices. In *Global Climate Change and Environmental Policy: Agriculture Perspectives* (pp. 257–273). Springer. https://doi.org/10.1007/978-981-13-9570-3_8

Yeong, T. K., Jiao, K., Zeng, X., Lin, L., Pan, S., & Danquah, M. K. (2018). Microalgae for biobutanol production – Technology evaluation and value proposition. *Algal Research*, *31*(February), 367–376. https://doi.org/10.1016/j.algal.2018.02.029

Zoghlami, A., & Paës, G. (2019). Lignocellulosic biomass: Understanding recalcitrance and predicting hydrolysis. *Frontiers in Chemistry*, *7*(December), 599. https://doi.org/10.3389/fchem.2019.00599

2 Analytical Characterization of Biomass for Biofuel Production

*Aditya Sharma, Tanvi Bisht, Rohit Kumar Singh, Anu Chaudhary, Pushpak Dahiya, Ekta Arya, Arvind Kumar Bhatt, and Ravi Kant Bhatia**

2.1 INTRODUCTION

The global energy landscape is undergoing a profound transformation, driven by concerns over energy security, climate change, and environmental sustainability. In this context, renewable energy sources, particularly biomass-derived biofuels, have gained significant attention as viable alternatives to fossil fuels (Bhaskar et al., 2018). Biomass, sourced from agricultural residues, forestry waste, dedicated energy crops, and organic wastes, is a renewable resource that can be converted into biofuels through biochemical or thermochemical processes (Demirbas, 2008). Utilizing biomass for biofuel production offers several benefits, including reduced greenhouse gas emissions, decreased reliance on fossil fuels, and enhanced rural economic opportunities (Ragauskas et al., 2006). However, the success of biomass-based biofuel production depends on effective feedstock characterization. Biomass varies widely in composition and structure based on feedstock type, geographic origin, and cultivation practices (Das et al., 2021). Analytical characterization is essential to understand these variations, enabling process optimization and ensuring product quality (Sluiter et al., 2013). Biomass composition, typically including cellulose, hemicellulose, lignin, proteins, lipids, ash, and moisture, is a major factor influencing its suitability for biofuel production (Himmel et al., 2007). Techniques such as proximate and ultimate analysis, as well as elemental analysis, are commonly employed to assess chemical composition (Demirbas, 2009). Proximate analysis provides data on moisture, volatile matter, fixed carbon, and ash, while ultimate analysis determines elemental content (C, H, N, S, O) (Rowell, 2005), supporting energy potential evaluation. Feedstock quality, such as moisture content, ash content, particle size, bulk density, and heating value, impacts storage, handling, and processing performance (Kumar & Sharma, 2017). Structural properties like particle size, surface area, porosity, and crystallinity also play critical roles in conversion efficiency (Chundawat et al., 2011; Rodrigues et al., 2022). Analytical tools such as scanning electron microscopy (SEM), transmission electron microscopy (TEM), X-ray

* Corresponding author

DOI: 10.1201/9781003485636-2

diffraction (XRD), and Fourier-transform infrared spectroscopy (FTIR) offer insights into biomass morphology and functional groups (Vanholme et al., 2010). SEM and TEM provide high-resolution imaging of surface and internal structure, while XRD identifies crystalline phases, and FTIR characterizes chemical bonds (Pereira et al., 2016). Thermal properties of biomass, including decomposition, pyrolysis, and combustion behavior, are crucial for thermochemical processes. Thermogravimetric analysis (TGA), differential scanning calorimetry (DSC), and calorimetry help evaluate thermal stability, decomposition kinetics, and heat release (Bridgwater, 2012; Cui et al., 2013). These insights guide process design and conversion system optimization. Analytical methods addressing these parameters enable informed decisions on feedstock selection and preprocessing (Pereira et al., 2016). Comprehensive analytical characterization and analysis of all these parameters is vital for efficient and sustainable biofuel production from biomass. Continued advancements in analytical techniques will further enhance process efficiency and accelerate the biofuel production for a cleaner and greener future.

2.2 COMPOSITION OF LIGNOCELLULOSIC BIOMASS

The composition analysis of biomass involves a precise and complex procedure when individual constituents are measured separately. Accurate quantification of these components is critical for determining fuel conversion yields and performing mass balance calculations. Lignocellulosic biomass (LCB) consists of polysaccharides, phenolic polymers that form the structural framework of plant woody cell walls, and ash, which is mostly constituted of inorganic substances such as silica (Sluiter et al., 2013). Its complex structure includes cellulose, which is encased in a matrix of hemicellulose and lignin, as shown in Figure 2.1. Biochemical processes use LCB

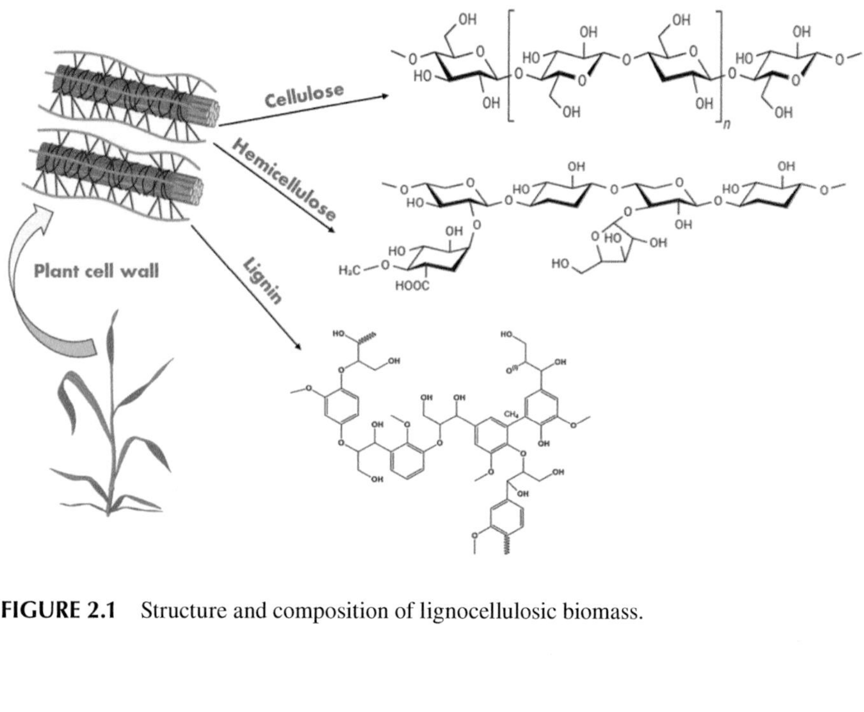

FIGURE 2.1 Structure and composition of lignocellulosic biomass.

with a carbon-to-nitrogen ratio below 30 and moisture content above 30%. These processes involve chemical reactions facilitated by enzymes, fungi, and microorganisms. When LCB has a C/N ratio above 30 and moisture content below 30%, thermochemical methods become a viable alternative. Recent advancements have led to the development of innovative biofuels from LCB, including biohydrogen (bio-H_2) and butanol (Yousuf, 2012). Despite challenges like natural degradation resistance, LCB remains a promising raw material due to its abundant availability, cost-effectiveness, and environmental benefits. Various components of LCB are described as given below.

2.2.1 Cellulose

Cellulose is an organic polysaccharide made up of a long chain of D-glucose molecules connected by β-(1-4) glycosidic bonds, with the chemical formula (C6H10O5). It serves as the primary structural component of the cell walls in green plants, algae, and oomycetes (Rathour et al., 2025). A notable feature of cellulose is its crystallinity, which can be altered. For industrial purposes, higher concentrations of cellulose are found in wood pulp and cotton. It is commonly used to produce paperboard and paper products (Chaudhary et al., 2022). For example, when cellulose is processed under specific conditions at a pressure of 25 MPa and a water temperature of 320°C, it can transform into an amorphous solid. Additionally, cellulose can be converted into several eco-friendly biofuels from sources like agricultural residues and energy crops (Vaca-Medina et al., 2013).

2.2.2 Hemicellulose

Hemicellulose is a branched heteropolymer with 500–3,000 sugar units, primarily composed of pentoses like xylose and arabinose, hexoses like mannose, glucose, rhamnose, and galactose, and acetylated sugars (Gibson, 2012). Its structural heterogeneity and difficult enzymatic hydrolysis make it complex to utilize. Nevertheless, hemicellulose offers greater potential for regioselective chemical and enzymatic modifications, owing to its diverse sugar constituents, glycosidic linkages, variable structure of glycosyl side chains, and the presence of two reactive hydroxyl groups on the xylose repeating unit (Chaudhary et al., 2022).

2.2.3 Lignin

Lignin is a crucial component of lignocellulosic biomass (LCB), providing structural integrity to vascular plants and algae (Martone et al., 2009). Its complex polymeric structure, composed of cross-linked phenolic chemicals, makes plant tissues durable and resistant to degradation, protecting cellulose and hemicellulose. Lignin's amorphous (noncrystalline) nature contributes to the structure's complexity and variability, making it more resistant to degradation (Sluiter et al., 2008). As a result, lignin protects the more easily degradable cellulose and hemicellulose from microbial attack and enzymatic degradation. However, characteristics of lignin pose both obstacles and opportunities in industrial applications. Lignin's resistance to breakdown challenges biofuel production and value-added products but can be used for bioplastics, adhesives, and other high-value materials in industrial applications (Rathour et al., 2025).

2.3 ANALYTICAL ANALYSIS OF LIGNOCELLULOSIC BIOMASS

The composition analysis of biomass involves a precise and complex procedure when individual constituents are measured separately. Accurate quantification of these components is critical for determining fuel conversion yields and performing mass balance calculations. Various techniques are used in the analytical analysis of the lignocellulosic biomass, as shown in Figure 2.2 and also discussed below.

2.3.1 Moisture Content

Accurate compositional analysis of biomass begins with proper sample preparation, particularly moisture reduction and particle size adjustment. Moisture content should be reduced to below 10% by weight, and the sample should be milled to a uniform particle size suitable for analysis. When environmental conditions permit, air-drying is the simplest and most cost-effective method (Hames, 2009). It effectively lowers moisture levels with minimal impact on sample integrity. However, this method is highly dependent on ambient conditions such as temperature and humidity, making it unsuitable in high-humidity environments where samples may degrade or develop mold. An alternative approach involves drying samples in a convection

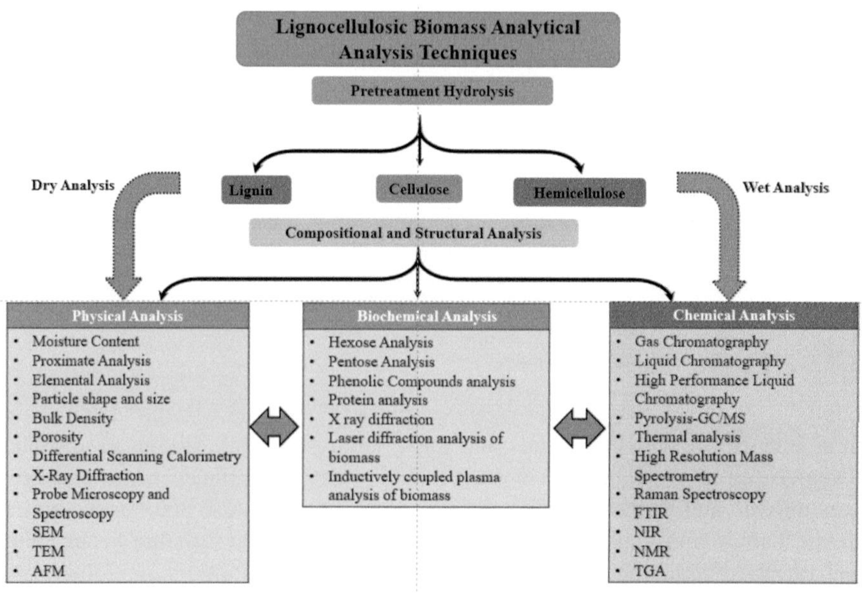

FIGURE 2.2 Various techniques used in the analytical analysis of the lignocellulosic biomass.

oven at 45°C. While more controlled, this method must be carefully managed to prevent thermal degradation, as temperatures above 45°C may damage the biomass (Williams et al., 2017). For precise moisture determination, the sample is weighed and placed in an oven set at a standard temperature, typically 105°C or 103°C ± 2°C. Drying continues until the sample reaches a constant weight, indicating that negligible mass is lost between successive weighing. Moisture content is then calculated by comparing the original and final dry weights. Analytical values are often reported on a dry weight basis, which mathematically excludes water content from the results (Vassilev et al., 2010). For example, if a biomass sample contains 50% moisture and 10% ash, the ash content expressed on a dry-weight basis would be 20%. This reporting method ensures consistency and comparability across different samples. Since moisture correction affects all other compositional values, it is one of the most critical steps in biomass analysis. Any error in moisture determination will propagate through subsequent calculations, increasing overall uncertainty (Orlando & Borja, 2020). Furthermore, due to the highly dynamic nature of biomass moisture content, which can fluctuate within minutes depending on laboratory storage and ambient conditions, it is essential to determine moisture simultaneously with each sample preparation for analysis.

2.3.2 Proximate Analysis

Proximate analysis is a thorough technique for measuring several major components of a material, including moisture content (M), fixed carbon content (FC), volatile matter (VM), and ash. This analysis is carried out by a proximate analyzer, which measures mass loss at certain temperatures. To start with, moisture is removed at around 105–110°C, which is just above the boiling point of water and solely eliminates physically bonded water (Nimmanterdwong et al., 2021). Volatile matter is then evaporated in an inert environment at 950°C at a slow rate. After the volatile materials have been removed, the residual material is burned at temperatures exceeding 700°C in an oxygen environment to determine the ash content. The fixed carbon content can be calculated using this formula:

$$\text{Fixed carbon}(\text{FC}) = 1\text{-M-Ash-VM}$$

During combustion, the minerals in the material are converted into ash, which can be examined to identify specific components such as CaO, Na_2O, K_2O, MgO, SiO_2, P_2O_5, Fe_2O_3, SO_3, and Cl. To conduct further analysis on these separated minerals or ash, they are frequently dissolved in various acids. While there are other analytical procedures available, this one is highly complex and not commonly utilized (Bridgeman et al., 2010).

2.3.3 Elemental Analysis

Elemental analysis of the organic fraction of biomass measures the mass percentages of key elements such as carbon (C), hydrogen (H), nitrogen (N), sulfur (S), and oxygen (O), with chlorine (Cl) also included in some cases. These values can subsequently be expressed on an atomic basis for further analysis. In many setups, instruments are designed to specifically measure the mass percentages of C, H, and N, while sulfur is often analyzed using a separate instrument (Huang & Lo, 2020). The analysis involves combusting the sample and examining the resulting combustion products. To determine the molecular or atomic composition of biomass from ultimate analysis data, the first step is to obtain the mass percentages of the relevant elements, typically C, H, N, S, and O, with Cl when applicable. These percentages are then converted into mass fractions by dividing each by 100, forming the basis for subsequent stoichiometric or elemental ratio calculations.

Using the atomic masses of key elements-carbon (C = 12.01 g/mol), hydrogen (H = 1.008 g/mol), nitrogen (N = 14.01 g/mol), sulfur (S = 32.07 g/mol), oxygen (O = 16.00 g/mol), and chlorine (Cl = 35.45 g/mol), the mass fractions obtained from ultimate analysis can be converted into moles by dividing each element's mass fraction by its respective atomic mass (Singh et al., 2017). The resulting mole values are then normalized to derive the simplest whole-number ratio. This is achieved by dividing each mole value by the smallest among them, yielding the empirical formula of the biomass. It is important to consider the influence of water content on the elemental composition, particularly hydrogen and oxygen. If residual water is present, it can artificially inflate the hydrogen content. To correct for this, the hydrogen and oxygen contributions from water must be subtracted from the overall percentages prior to mole conversion (Fajobi et al., 2022). For example, consider a biomass sample with the following ultimate analysis on a dry basis: 50.0% carbon, 6.0% hydrogen, 1.0% nitrogen, 0.5% sulfur, and 42.5% oxygen (calculated by difference). These percentages are first converted to mass fractions and then to moles by dividing each by the corresponding atomic mass. The mole values are then normalized using the smallest value, often that of sulfur in such cases. The result is an approximate empirical formula, such as $C_{266}H_{382}N_5S_{11}O_{171}$, which reflects the elemental composition of the biomass at the molecular level (Bridgeman et al., 2010). This empirical formula provides valuable insights into biomass composition and is a useful representation for understanding combustion characteristics, stoichiometry, and other chemical behavior based on ultimate analysis data.

2.3.4 Particle Size and Shape Analysis

Particle size and particle size distribution (PSD) are key factors influencing the properties of particulate and densified materials, affecting characteristics such as flowability, bulk density, compressibility, and durability. To maintain consistent quality, controlling the PSD of biomass particles is essential due to their varying sizes and shapes. Mechanical screening, often performed through sieve analysis, is a common method for assessing PSD in biomass (Sluiter et al., 2013). Sieve analysis is a traditional technique for determining particle size distribution, which can be expressed by either mass or volume. This laboratory method involves passing particles through a mesh sieve, with movement occurring either vertically or horizontally. The selection of sieving method manual, mechanical, dry, or wet depends on the specific characteristics of the particles and the analysis requirements (Singh et al., 2017). Manual sieving, typically conducted in environments without electricity, is mainly used for differentiating between large and small particles on-site. The mechanical sieving method, the most common in laboratories to ensure quality, is widely used today. Dry sieving is employed when the particles are in a dry state, while wet sieving is used for particles that are wet or suspended in a liquid (Sluiter et al., 2013). The analysis generally includes the percentage of particles retained on sieves with different opening sizes, the cumulative undersize distribution, and the determination of geometric and arithmetic mean values, along with their standard deviations. These elements provide a comprehensive understanding of the material's particle distribution.

2.3.5 Density Measurement of Biomass

Bulk density is a crucial physical property assessing biomass quantity and quality, calculated using three methods, including weight and volume (Irwansyah et al., 2024).

2.3.5.1 Standard Bulk Density Using ISO 17828 Method

Standard bulk density (BD_ISO) for biomass is determined using the ISO 17828 method, which provides a standardized approach for measuring how much biomass fits into a given volume. It involves collecting samples, measuring their wet weight, and drying them at 65°C to account for moisture content, providing a consistent reference for comparing bulk densities across different samples (Sultana & Kumar, 2011).

2.3.5.2 Actual Bulk Density

Actual bulk density is a method for measuring biomass density in real-world conditions. It involves weighing biomass in a collection vehicle and measuring its load volume using a platform truck scale and a laser survey instrument that captures detailed 3D profiles of the dump body and the loaded biomass. By calculating the volume of the chips and subtracting any voids, this method provides an accurate representation of the biomass's true density in practical transportation scenarios and logistics optimization (Fajobi et al., 2022).

2.3.5.3 Effective Bulk Density

This method is useful when the exact volume of the material is not measured but estimated from the container size. This method involves measuring the volume of the empty dump body and using this fixed volume to calculate the density when the vehicle is filled with biomass. While effective bulk density does not account for potential voids or overfills, it provides a straightforward and practical estimate of density that is easy to obtain (Singh et al., 2017). This measure is useful for logistics and operational purposes where simplicity and quick assessment are prioritized over precision.

2.3.6 Surface Area and Porosity

The Brunauer–Emmett–Teller (BET) method accurately determines the surface area of porous materials, including biomass (Eisenbies et al., 2019). It uses an adsorption isotherm to measure gas adsorption onto a sample's surface, providing crucial information, such as the creation of a monolayer of gas molecules on the material's surface, which is required to calculate the specific surface area. The best method for processing biomass samples involves accurately weighing and placing the sample in a clean chamber, securely storing the sample cell in a heating mantle, setting the outgassing temperature to 250°C, and degassing for three hours with the heater turned on to ensure stable temperature conditions (Meng et al., 2013). After the outgassing period is finished, turn off the heater and allow the sample cell to cool naturally within the heating mantle to avoid temperature shocks. After cooling, reweigh the biomass sample to calculate weight loss due to outgassing. This rigorous technique guarantees that the biomass sample is effectively purified of adsorbed gases and volatile chemicals, making it ideal for subsequent analytical or experimental uses (Nsude & Orie, 2022).

The equation for BET is given by:

$$\frac{1}{v\left(\frac{P_0}{P}\right)-1}=\frac{C-1}{V_m C}\left(\frac{P}{P_0}\right)+\frac{1}{V_m\, C}$$

where:

v = Volume of gas adsorbed at pressure P
v_m = Monolayer adsorbed gas quantity
P = Equilibrium pressure of adsorbing gas
P_0 = Saturation vapor pressure of the adsorbing gas
C = BET constant, related to the energy of adsorption

Surface Area Calculation: The specific surface area (SSA) is calculated using the monolayer capacity v_m obtained from the BET plot and knowing the cross-sectional area of the adsorbed gas molecule (Lowell et al., 2012).

2.3.7 Advanced Methods of Lignocellulosic Biomass Analysis

Advanced methods for lignocellulosic biomass analysis include thermogravimetric analysis (TGA), spectroscopy (FTIR, NMR), and chromatography (GC-MS, HPLC). These techniques enable detailed characterization of cellulose, hemicellulose, and lignin, as shown in Figure 2.3. Emerging tools like pyrolysis-GC/MS and high-resolution microscopy enhance understanding of structural and chemical changes during biomass conversion and biofuel production (Oladunni et al., 2023). The understanding of these various properties of biomass during conversion processes provides insights for optimizing processing conditions and improving performance and stability of biomass for various applications (Gaffney et al., 2002). Studying feedstock behavior can lead to innovative processing methods for biofuels, addressing global renewable energy demand and evaluating energy potential.

2.3.7.1 Thermogravimetric Analysis

Thermogravimetric analysis (TGA) is an essential technique for evaluating the thermal properties and decomposition behavior of biomass, offering insights into

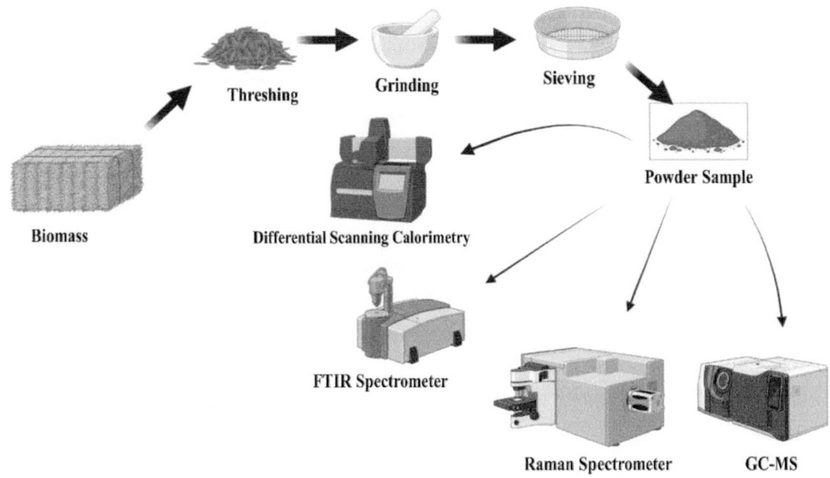

FIGURE 2.3 Various instrumentations used in analytical characterization of biomass.

its suitability for energy production. TGA measures the weight loss of a sample as temperature increases at a controlled rate, enabling the assessment of reaction rates and thermal stability under various conditions (Almusafir & Smith, 2024). Typically, a small biomass sample (2–30 mg) is heated from 25°C to 1,200°C at rates such as 5, 20, 40, or 80°C min^{-1} in inert atmospheres like nitrogen (Figure 2.4). This setup

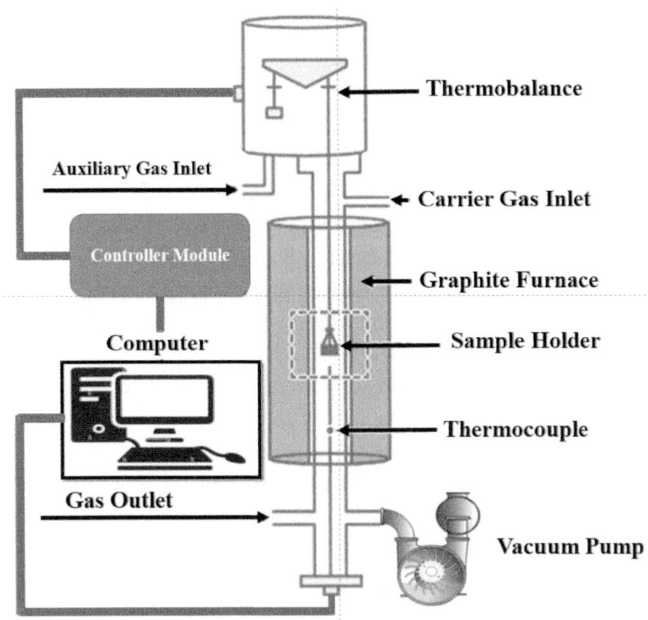

FIGURE 2.4 Simplified representation of the thermogravimetric analyzer.

allows for simultaneous monitoring of temperature, time, and mass in a dynamic environment. TGA is used to examine biomass feedstocks such as hardwood (HW), softwood (SW) pellets, and refuse-derived fuel (RDF), revealing thermal degradation patterns of key biomass components. Hemicellulose degrades over a broad range (170–690°C), cellulose between 291°C and 395°C, and lignin from 170°C to 835°C due to its complex structure (Apaydın-Varol & Mutlu, 2023). These distinct decomposition ranges help interpret biomass reactivity during thermochemical conversion. Mass fluctuations during TGA are categorized as positive, such as adsorption, oxidation, and reduction, or negative, including desorption, thermal decomposition, combustion, vaporization, and sublimation (De Blasio, 2019). Understanding these behaviors enhances the design and optimization of bioenergy systems, making TGA a valuable tool in biomass research.

TGA can be coupled with differential thermogravimetric (DTG) analysis, which provides a more detailed understanding of the decomposition process by displaying the rate of weight loss at specific temperatures. It is a novel method for determining lignocellulosic composition in biomass, offering enhanced accuracy for hemicellulose content but not suitable for lignin quantification (Carrier et al., 2011).

2.3.7.2 Differential Scanning Calorimetry (DSC)

DSC is increasingly valued in biofuel research for analyzing the thermal behavior of biomass and biofuels. It offers critical data on properties such as melting temperatures, heat capacities, and energy transitions key to optimizing biofuel production and performance. DSC is widely used to characterize lignocellulosic materials, providing insight into kinetic parameters and energetic profiles necessary for evaluating bioenergy potential (Silva et al., 2024). The technique is especially useful in assessing the thermal behavior of biofuels like biodiesel and bioethanol under varying conditions of temperature and pressure. These assessments reveal changes in heat flux and thermal conductivity, contributing to a deeper understanding of fuel stability and efficiency. DSC detects thermal events, including melting, crystallization, glass transitions, and decomposition. A notable example is the DSC thermogram of Chinese tallow tree seeds, which displayed multiple transitions, indicating the presence of lipids, fatty acids, and possibly hemicellulose (Picou & Boldor, 2012). Furthermore, DSC can monitor protein thermal behavior, such as unfolding and aggregation, which are relevant to enzyme-based biofuel processes. The presence or absence of substrates can significantly influence these thermal transitions, offering insights into enzyme activity and structural stability (Di Matteo et al., 2023). Overall, DSC is a powerful tool for examining thermal transitions in biofuels, enhancing the development of efficient and sustainable bioenergy systems.

2.3.7.3 Pyrolysis-Gas Chromatography/Mass Spectroscopy (Py-GC/MS)

The thermal decomposition of lignocellulosic substrates produces a range of volatile compounds formed from the degradation of hemicellulose, cellulose, and lignin. The Py-GC/MS technique is a powerful analytical tool for identifying the chemical components released during the pyrolysis of various biomass resources. Understanding the roles of different polymers and additives is crucial for optimizing processes, as these factors significantly influence the yield and quality of biofuels (González-Vila et al., 2001). Py-GC/MS enables in situ studies without the need for sample pretreatment. Biomass samples were introduced into a metallic receptacle, and an autosampler deposited them into a pyrolysis furnace set at 500°C for 12 seconds. The resulting vapors were directed into a gas chromatograph (GC). The GC operated with an initial 4-minute hold at 50°C, followed by a 5°C per minute ramp to 280°C, and another 5-minute hold at that temperature. The mass spectrometer was set at 70 eV and 280°C. Two hundred program peaks and their areas were extracted using TurboMass GC/MS software with a signal/noise threshold of $\geq$ 2,000 and identified against the National Institute of Standards and Technology (NIST) library (Edmunds et al., 2018).

2.3.8 Spectroscopic Analysis

Spectroscopic analysis uses electromagnetic radiation to provide qualitative and quantitative information about substances. It focuses on the ultraviolet region (wavelengths between 200 nm and 350 nm) to the far-infrared region (wavelengths from 700 nm to 50,000 nm), including visible (400–780 nm), near-infrared (780–2,500 nm), and mid-infrared spectrums (2,500–25,000 nm) (Pereira et al., 2016). Shorter wavelengths include gamma rays (approximately 0.001 nm) and X-rays (about 0.01 nm), while longer wavelengths include microwaves (around 107 nm) and radio waves (approximately 1,010 nm) (Delwiche, 2015). Spectroscopic analysis is essential in various fields, including biomass and biofuels. The techniques applied in spectroscopy offer significant advantages in analyzing biological materials, playing a crucial role in drug development, quality assurance, and numerous investigative studies. The following sections will clarify the fundamental characteristics of spectroscopic analysis. Prominent techniques include nuclear magnetic resonance (NMR), Fourier transform infrared (FTIR) spectroscopy, Raman spectroscopy, and ultraviolet-visible (UV-Vis) spectroscopy, discussed below and also summarized in Table 2.1.

2.3.8.1 FTIR Spectroscopy

FTIR spectroscopy is a cost-effective, non-destructive technique widely used to analyze the chemical structure of biomass, playing a vital role in biofuel production. It preserves sample integrity while providing accurate qualitative and quantitative data, enabling further testing and analysis (Oladunni et al., 2023). FTIR identifies functional groups, detects chemical bonds, and maps the chemical composition of biomass at the micron scale, offering insights into key components such as lipids, proteins, and carbohydrates crucial for efficient biofuel conversion. By comparing FTIR spectra with reference spectra from pure compounds like ergosterol, glucose, and chitin, researchers can assign specific bands to molecular structures (Gaffney et al., 2002). Organic molecules contain covalent bonds that act like springs, exhibiting vibrational modes such as stretching and bending. Symmetric and asymmetric stretching common in CH_2 and CH_3 groups differ in energy demand, with asymmetric stretching requiring more energy. Bending vibrations, or deformation vibrations, involve changes in bond angles and can be in-plane or out-of-plane (Khan et al., 2018).

TABLE 2.1
Summary of Different Analytical Techniques Used in Biomass Characterization

Analytical Technique	Use/Principle	Quality Assessment	Analysis Type	References
Scanning Electron Microscopy (SEM)	Uses focused beams of electrons	Surface Morphology	Non-Destructive	Vaseghi & Nematollahzadeh, 2020
Transmission Electron Microscopy (TEM)	Uses transmitted electrons	Internal Structure	Destructive	Vaseghi & Nematollahzadeh, 2020
Fourier Transform Infrared Spectroscopy (FTIR)	Measures infrared intensity vs. wavelength	Chemical Structure, Functional Group	Non-Destructive	Khan et al., 2018
Thermogravimetric Analysis (TGA)	Measures change in weight as sample heated or cooled	Characterize Materials	Destructive	Almusafir & Smith, 2024
Differential Scanning Calorimetry (DSC)	Measures heat flows associated with transitions	Material Characterization	Destructive	Silva et al., 2024
PY-GC/MS	Combines pyrolysis with GC/MS	Chemical Composition	Destructive	Edmunds et al., 2018
Raman Spectroscopy	Uses inelastic scattering of monochromatic light	Chemical Structure	Non-Destructive	Dzurendova et al., 2023
Nuclear Magnetic Resonance Spectroscopy (NMR)	Uses magnetic fields and radio waves	Chemical Composition, Molecular Structure	Non-Destructive	Elliot et al., 2019
Atomic Force Microscopy (AFM)	Uses a sharp probe to scan surfaces	Surface Structure and Behaviors	Non-Destructive	Xing et al., 2019
Gas Chromatography (GC)	Separates and analyzes compounds that can be vaporized	Component of volatile and semi-volatile Compounds	Destructive	Soria et al., 2021
Liquid Chromatography (LC)	Separates and analyzes compounds in liquid form	Identification and quantification of targeted compounds in a sample.	Non-Destructive	Jacques & Caramao, 2009

(Continued)

TABLE 2.1 (*Continued*)
Summary of Different Analytical Techniques Used in Biomass Characterization

Analytical Technique	Use/Principle	Quality Assessment	Analysis Type	References
High-Performance Liquid Chromatography (HPLC)	A type of LC with high resolution and speed.	Qualitative and quantitative Composition of the sample	Non-Destructive	Monteiro et al., 2008
Mass Spectrometry (MS)	Analyzes ionized particles based on mass-charge ratio m/z	Chemical structure, Molecular Composition	Destructive	Bahng et al., 2009
X-Ray Diffraction (XRD)	Uses X-rays	Crystallographic Structure, Chemical Composition	Non-Destructive	Waseda et al., 2011
Laser Diffraction Analysis	Measures particle size distribution using laser light scattering	particle's size and shape	Non-Destructive	Dapkunas et al., 2001
Inductively Coupled Plasma (ICP) Analysis	Uses plasma to ionize samples	Elemental composition	Destructive	Tafur-Marinos et al., 2016

2.3.8.2 Raman Spectroscopy

Raman spectroscopy is a non-invasive analytical technique widely used across agriculture, medicine, and materials science to provide detailed molecular information. It generates unique spectral fingerprints of materials, allowing real-time, sample-free analysis. Particularly in biomass and biofuel research, Raman spectroscopy is valuable for identifying functional groups, analyzing the structural arrangement of polymers, and determining the orientation and conformation of macromolecular chains. In agriculture, Raman spectroscopy enables the molecular-level assessment of plant tissues, supporting crop quality evaluation and early detection of stress—key factors for sustainable agricultural practices (Lemes, 2024). The technique is based on vibrational transitions within the ground electronic state of molecules. A monochromatic laser light interacts with the sample, polarizing the electron cloud around the atomic nuclei. The spectrometer then detects the scattered light, where the frequency of re-emitted photons shifts either up or down relative to the incident light. This frequency shift is known as the Raman effect (Lupoi et al., 2015). The process involves the excitation of molecules into oscillating dipoles, which emit light at different frequencies as they return to their original vibrational state. The electron cloud distortion during this interaction results in photon scattering, providing critical molecular insights. Online Fourier-transform Raman (FT-Raman) spectroscopy is particularly effective in monitoring biomass production and metabolite concentrations during microbial fermentation. It delivers real-time data, enhancing the optimization and control of biofuel production processes (Dzurendova et al., 2023).

$$\omega\text{Raman} = \omega\text{excite} - \omega\text{scatter} = (\text{h}/2\pi)$$

The energy difference ωRaman between the initially absorbed photon (ωexcitation) and the emitted photon (ωscattering) in Raman scattering indicates the variation in vibrational energy levels. In the equation, h represents the Planck constant, and μ denotes the reduced mass. For a process to be considered Raman active, there must be a net change in bond polarizability during the electronic transition that occurs after the absorption of an electromagnetic wave (Kim et al., 2013). While Raman spectroscopy offers significant advantages in biomass analysis, challenges remain in addressing overlapping spectral features, particularly in complex mixtures like plant fibers, which can complicate data interpretation (Agarwal & Ralph, 2023).

2.3.8.3 NMR Spectroscopy

NMR spectroscopy is a powerful tool in biofuel research, offering precise analysis of bioactive compounds in biomass. It plays a critical role in characterizing pretreatment processes and determining bio-oil compositions, ultimately improving biofuel yield and process efficiency (Cui et al., 2024). One of its key strengths is the detailed analysis of carbohydrate mixtures, providing essential structural and compositional information that supports the development of high-yield biofuels. Time-domain NMR (TD-NMR) has proven effective in evaluating biomass pretreatments by revealing how different methods impact enzymatic saccharification, the conversion of complex carbohydrates into fermentable sugars vital for second-generation ethanol production (Bianquini et al., 2020). TD-NMR helps optimize these pretreatments, boosting ethanol output. Quantitative NMR techniques also simplify analysis by accurately measuring compound concentrations without requiring purified standards. This is especially useful in monitoring low-concentration intermediates during biomass conversion (Elliot et al., 2019). Solid-state NMR offers further insights by examining the three-dimensional structures of biopolymers like cellulose, xylan, and lignin, allowing researchers to study molecular rearrangements that affect extraction efficiency during processing (Munson et al., 2022; Sarpal et al., 2015).

2.3.9 Microscopic Analysis

Imaging methods reveal morphological information and surface characteristics of lignocellulosic biomass, aiding in characterization and enhancing enzymatic and chemical hydrolysis, thereby enabling the creation of biomaterials, biofuels, and cellulose fibers (Singh et al., 2020). Microscopy uses light and electron beam emission sources for various applications. The contact of an electron-scanning probe with the surface of the material to be examined is known as probe microscopy scanning (Di Gianfrancesco, 2017). For the purpose of characterizing biomass through imaging, optical microscopy techniques like atomic force microscopy (AFM), scanning electron microscopy (SEM), and TEM are frequently used.

2.3.9.1 SEM

SEM is an essential technique for analyzing the surface morphology and porosity of biomass-derived materials like biochar. It provides high-resolution images (down to 15 nm) of particle texture, shape, and size, revealing the presence and development of micropore, mesopore, and macropore features that directly influence surface area and adsorption capacity (Vaseghi & Nematollahzadeh, 2020). SEM is particularly useful in assessing how processing conditions, such as pyrolysis temperature, affect pore formation and surface structure. When paired with energy-dispersive X-ray spectroscopy (EDX), SEM also enables the analysis of elemental composition, offering insights into the chemical and structural transformations of biomass during treatment (Zoghlami et al., 2019). Biomass surface modifications and irregularities resulting from various pretreatments can be effectively visualized using SEM imaging. Specimen preparation is critical to SEM accuracy. Fixatives like glutaraldehyde (GA) and osmium tetroxide (OsO_4) are used to stabilize ultrastructures and enhance conductivity. Dehydration is typically done using ethanol or acetone, followed by critical point drying with CO_2. However, solvents like ethanol and acetone can cause micro-ripping due to their high surface tension. To avoid such artifacts, liquid CO_2 or hexamethyldisilazane (HMDS) is preferred (Fischer et al., 2024). For improved conductivity, samples are mounted on sticky carbon discs, often with silver adhesive, and coated with a conductive material such as gold. Despite its advantages, SEM sample preparation has limitations. It requires conductive coatings, which can increase costs, and exposure to the electron beam may damage delicate biomass structures or alter their morphology (Karimi & Taherzadeh, 2016).

2.3.9.2 TEM

TEM is a powerful technique for nanoscale imaging of lignocellulosic biomass, offering detailed insights into internal structures through electron beam scattering, unlike SEM, which is limited to surface analysis (Vaseghi & Nematollahzadeh, 2020). TEM is especially valuable for studying lignification, hemicellulose deposition, and plant cell wall architecture. When combined with potassium permanganate ($KMnO_4$) staining, it can highlight lignin distribution, while immunogold labeling allows for precise identification of lignin substructures and monomeric composition. Similarly, hemicellulose deposition related to plant tissue growth and differentiation can be visualized using TEM with specific immunolabeling (Zhou et al., 2015). Sample preparation for TEM involves fixation, embedding, sectioning, and staining. Fixatives like formaldehyde or glutaraldehyde crosslink proteins and cellular components to preserve structure. Samples are typically cut into ~1 mm pieces and undergo secondary fixation with osmium tetroxide to retain lipids and enhance membrane contrast (Grams, 2022). Uranyl acetate is used for post-sectioning to bind lipids, proteins, and nucleic acids, enhancing contrast further. Dehydration is performed using ethanol or acetone, followed by embedding in a liquid resin, which is then cured into a solid block using heat or UV light. Ultra-thin sections (<100 nm) are sliced from the resin block and mounted on grids for analysis. These thin sections allow the electron beam to pass through the sample, enabling internal visualization.

However, TEM has limitations, including the need for ultra-thin sectioning, extensive sample preparation, and staining to improve contrast, which may complicate the study of organic matter (Zhou et al., 2015).

2.3.9.3 AFM

AFM is a high-resolution technique widely employed to analyze the surface morphology and mechanical properties of biomass materials. While it offers detailed surface imaging, its effectiveness in consistently monitoring nanoscale properties remains limited due to challenges in image stability and interpretation (Xing et al., 2019). Successful AFM analysis requires careful consideration of sample preparation, probe selection, and imaging conditions to ensure optimal accuracy and image quality. To obtain reliable microscopic morphology, the solid biomass surface must be smoothed, often achieved using a slide-away sectioning method and an ultramicrotome diamond knife (Lou et al., 2023). The prepared sample is affixed to a holder with clear double-sided tape. Mica is commonly used as the base substrate because of its flatness and strong interaction with biomass particles, which aids in producing clear topographical images (Qian et al., 2001). The biomass material is dispersed on a mica sheet and attached to a glass slide. Through van der Waals forces, the sample adheres to the mica, and the glass slide is then mounted onto the AFM stage for scanning. This method also improves image brightness at equal magnifications, enhancing clarity during observation (Yan et al., 2008). Probe selection is critical to ensure accurate measurements. During scanning, the AFM probe tip contacts the sample surface, and the cantilever deflects in response to weak surface forces, as shown in Figure 2.5. This deflection is measured by a position-sensitive detector, which converts the movement into an electrical signal. The system adjusts the piezoelectric scanner accordingly, constructing a detailed 3D image of the sample's surface (Lou et al., 2023).

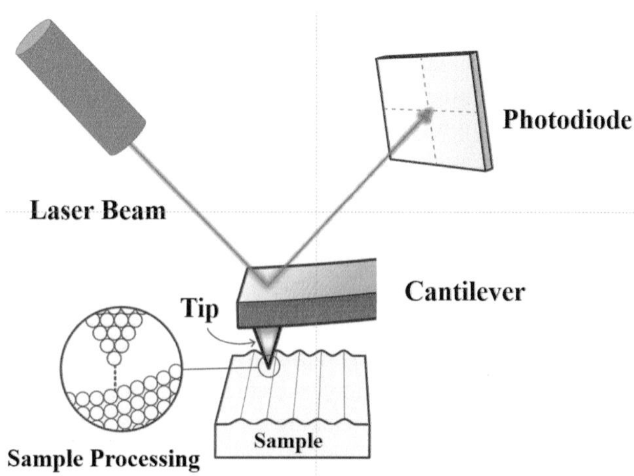

FIGURE 2.5 Schematic diagram of contact between the atomic force microscope tip and the sample surface.

AFM is effective in studying biomass materials, but test conditions can affect image quality. Incorrect settings can lead to inaccuracies in nanomechanical maps, so it's advisable to use the same probe and imaging settings whenever possible to ensure consistent findings across multiple samples.

2.3.10 Mass Spectrometric Analysis of Biomass

Mass spectrometry (MS), particularly when coupled with thermal degradation techniques like pyrolysis, is a widely utilized method for identifying byproducts from biomass. One special type of this method, called matrix-assisted laser desorption/ionization (MALDI), works really well for figuring out the molecular weight, which helps to tell apart compounds that have the same nominal mass (Bahng et al., 2009). In this technique, the sample is introduced into the mass spectrometer and ionized, typically using electrons. Depending on the nature of the sample and the ionization method, the resulting ion beam is directed through a series of electric or magnetic fields. These fields sort the ions by deflecting them according to their mass-to-charge (m/z) ratios; ions with identical m/z values follow the same path toward the detector. As ions reach the detector, their impact generates a very weak electrical signal either through induced charge or current, which is significantly amplified to produce a readable output (Mase et al., 2023). Detectors such as ion-to-photon converters, Faraday cups, or electron multipliers are commonly used. The amplified signals are plotted to create a mass spectrum, a graphical representation of ion intensity versus m/z values. Each peak in this spectrum corresponds to a specific ion. By comparing the observed spectrum with known mass profiles or identifying characteristic fragmentation patterns, researchers can accurately determine the molecular composition of the sample.

2.3.11 X-ray Diffraction

Diffraction happens when the periodic array of molecules in a crystal interacts with a collimated beam of electromagnetic waves that has a wavelength similar to interatomic distances (Waseda et al., 2011). When analyzing the structure of cellulose, XRD can be used without causing significant irradiation damage or OH/OD exchange, whether the cellulose is isolated or in its natural condition. The positions of the atoms in the crystal's unit cell, the smallest structural unit that can be generated entirely by simple translations, largely determine the positions of the diffracted beams with respect to the incident x-ray beam. The unit cell's dimension is commonly expressed using the unit lengths a, b, and c, along with the interaxial angles α, β, and γ. If unit cells have perfect three-dimensional periodicity, diffraction meets constructive interference requirements, as per Bragg's law.

$$n\lambda = 2d \times \sin\theta$$

where n is a positive integer, λ is the wavelength of x-ray, d is the spacing between atomic planes with equal electron density, and θ is the incidence and reflection angle. Consequently, a d-spacing of the planes reflecting x-rays corresponds to the diffraction peak observed at a given angle (θ). Miller indices are then used to connect the d-spacing values to the unit cell dimensions (a, b, c and α, β, γ) (Kim et al., 2013). The kinds and relative positions of the component atoms determine the diffracted beam intensities. The size and level of crystallite perfection can be correlated with the reciprocal of the breadth of the diffracted peak (Nishiyama et al., 2003). Large single crystals, often over 100 μm, are crucial for producing a two-dimensional array of diffraction spots for high-precision structural models, with cellulose crystals ideal for XRD examination to determine crystallinity index. The radiation used in most approaches is in the 2ϑ range, which is between 5° and 50°, and the duration is fixed with a step interval of 0.02° (Fitria et al., 2023).

2.3.12 Chromatographic Analysis

For the analytical community, chromatographic examination of biomass constituents poses a number of difficulties. Pretreatments were performed on complex samples in order to increase the chromatographic resolution power and concentrate or enrich portions of species of interest through the use of extraction, distillation, and column chromatography. This allows for the accurate quantification of individual components as well as the detection and characterization of numerous compound classes (Chen et al., 2020). Various chromatographic analytical techniques used in biomass biofuel analysis are summarized in Table 2.2.

2.3.12.1 Gas Chromatography (GC)

GC is a widely used analytical method for separating and analyzing volatile compounds, particularly carbohydrate monomers in biomass research. Due to the thermal instability of carbohydrate oligomers, errors may occur during analysis, making chemical derivatization essential (Soria et al., 2021). Common derivatization

TABLE 2.2
Methods for Analysis of Different Components of Biomass

Analysis type	Method	Principle	Component Analyzed	Advantages	References
Carbohydrate Analysis	NREL Acid Hydrolysis + HPLC	Hydrolysis of cellulose and hemicellulose into sugars, quantified via HPLC	Glucose, xylose, arabinose (monosaccharides)	Standard method for biomass sugars	Dashtban et al., 2010
Lignin Analysis	Py-GC-MS (Pyrolysis-Gas Chromatography-Mass Spectrometry)	Thermal decomposition of lignin followed by GC-MS analysis	Lignin monomers	Provides detailed structural info	Edmunds et al., 2018

techniques include silylation and conversion to alditol acetates, with acetyl, trifluoroacetyl (TFA), and trimethylsilyl (TMS) being widely used derivatives. These modifications stabilize the sugars, prevent incomplete derivatization, and disrupt the ring structure at the anomeric carbon to improve analysis accuracy (Sun et al., 2020). Sample preparation involves several steps: collection using gas syringes, volatilization, pre-concentration, and cryogenic condensation. In manual injection, samples are introduced into carrier gases such as hydrogen or air; inert gases like helium, argon, and nitrogen are also used. High-purity carrier gases enhance analytical performance. In automated GC systems, samples are injected by synchronizing carrier gas flow with a sample loop (Martínez et al., 2019). Separation occurs inside temperature-controlled columns packed with a molecular sieve or stationary phase material. As components elute, they pass through a detector, generating characteristic peaks on a chromatogram. The choice of detector depends on the compound type: flame ionization detectors (FIDs) for hydrocarbons, photoionization detectors (PID) for volatile organics, and thermal conductivity detectors (TCDs) for general-purpose use. For lignocellulosic biomass analysis, techniques like GC-FID and GC-MS are commonly employed due to their specificity and sensitivity in detecting and quantifying complex carbohydrate derivatives (Stilo et al., 2021).

2.3.12.2 Liquid Chromatography (LC)

LC is a physical separation technique used to identify and quantify components in complex mixtures. It works by separating compounds based on their differential interactions between a mobile phase (liquid solvent) and a stationary phase (solid or liquid supported on a solid). Accurate results depend on proper sample preparation, including solid-phase extraction (SPE), liquid-liquid extraction (LLE), and filtration. SPE uses sorbents such as silica, reversed-phase, or ion-exchange materials to selectively retain analytes and remove unwanted matrix components (Reymond et al., 2020). LLE relies on partitioning analytes between two immiscible liquid phases, typically an aqueous and an organic solvent, based on solubility. Filtration removes particulate matter, preventing clogging and protecting the chromatography system. During analysis, the composition of the mobile phase can be adjusted to modify interactions with analytes, affecting retention time. Each analyte elutes from the column based on its affinity to the stationary phase, forming peaks on a chromatogram (Jacques & Caramao, 2009). The area under each peak corresponds to the concentration of that component. Comprehensive two-dimensional liquid chromatography (LC×LC) significantly enhances resolution, especially for complex samples like bio-oils and aqueous biomass extracts. In LC×LC, compounds separated in the first dimension undergo a second separation with a different mechanism, using a high-pressure switching valve. This approach improves peak capacity, reduces co-elution, and identifies more compounds. A major advantage of LC is the ability to directly inject aqueous samples without prior extraction, streamlining analysis (Dugo et al., 2008).

2.3.12.3 High-Performance Liquid Chromatography (HPLC)

HPLC is an extensively utilized analytical method in biofuel research, particularly effective for identifying and quantifying biodiesel blend components such as fatty acids, triglycerides, diglycerides, monoglycerides, and free fatty acids. Compared to GC, HPLC typically operates at lower analytical temperatures, reducing the risk of compound degradation and often eliminating the need for derivatization, making it a faster and more direct method for many applications (Reymond et al., 2020). In HPLC, a sample is introduced into a non-polar stationary phase column and eluted using a polar mobile phase. As the sample moves through the column, compounds are separated based on their polarity: more polar substances elute faster, while non-polar components are retained longer. Compounds are then detected using various detectors, and their identities are determined by comparing their retention times to known standards (Monteiro et al., 2008). Common detectors used in HPLC for bio-fuel analysis include

HPLC-UV (Ultraviolet Detection): Employs DADs for the detection of polar compounds like aldehydes in aqueous pyrolysis samples. These compounds are often water-soluble and can be extracted with minimal recovery steps using methods like liquid-liquid extraction or solid-phase extraction (Abou-Yousef, 2014).

HPLC-RID (Refractive Index Detection): Ideal for detecting sugars in aqueous biomass samples, particularly those derived from sugar-rich sources like sugarcane straw and bagasse. RID measures the refractive index difference between the mobile phase and analytes, enabling detection of compounds that don't absorb UV or visible light.

HPLC-MS (Mass Spectrometry): This technique is used to analyze both bio-oil and aqueous fractions from biomass pyrolysis, especially targeting polar and thermolabile compounds that cannot be easily analyzed by GC. HPLC-MS can handle a wide range of polarities and both volatile and non-volatile compounds, providing high sensitivity and specificity (Chiaradia et al., 2008). Overall, HPLC is a versatile and robust tool for analyzing complex biomass-derived mixtures, offering a comprehensive approach for monitoring the chemical composition of biofuels and their intermediates.

2.3.13 Biochemical Analysis

Optimizing biomass conversion processes requires biochemical analysis for biomass characterization, with an emphasis on enzyme activity. Numerous enzymes, including xylanases and cellulases, are important in the conversion of lignocellulosic materials into fermentable sugars. In order to improve the production of biofuel, the investigation of the stability, activity, and efficiency of these enzymes under various circumstances is very crucial.

2.3.13.1 Glucose Analysis

The dinitrosalicylic acid (DNSA) method is a widely used biochemical protocol for detecting glucose and other reducing sugars released from lignocellulosic biomass hydrolysis. In this procedure, the biomass is first subjected to enzymatic hydrolysis using cellulolytic enzymes such as cellulase and β-glucosidase, typically incubated at 50°C and pH 4.8 for 24–72 hours to convert cellulose into glucose. After hydrolysis, the reaction mixture is centrifuged or filtered to obtain a clear supernatant. An aliquot of the supernatant is then mixed with an equal volume of DNSA reagent, which contains 3,5-dinitrosalicylic acid, sodium potassium tartrate, and sodium hydroxide. The mixture is heated in a boiling water bath for 5–10 minutes, resulting in the reduction of DNSA to 3-amino-5-nitrosalicylic acid, producing an orange-red color proportional to the concentration of reducing sugars. After cooling, the absorbance is measured at 540 nm using a spectrophotometer. The glucose concentration is determined by comparing the absorbance to a standard curve prepared using known glucose concentrations. This method is simple, cost-effective, and commonly used for monitoring enzymatic saccharification efficiency in biofuel research (Rinki et al., 2024).

2.3.13.2 Xylose Analysis

The detection of xylose released from lignocellulosic biomass hydrolysis can be effectively performed using colorimetric or chromatographic biochemical methods. One commonly used protocol involves hydrolyzing hemicellulosic components of the biomass using dilute acid (e.g., 1–2% sulfuric acid) or specific hemicellulolytic enzymes such as xylanase under controlled conditions (e.g., 50°C, pH 5.0) for 24–48 hours. Following hydrolysis, the liquid fraction containing the released monosaccharides is separated by centrifugation or filtration. The xylose content can then be quantified using colorimetric methods such as the orcinol assay or phloroglucinol assay, which involve reacting xylose with reagents that form a colored complex measurable by spectrophotometry (typically at 540–560 nm) (Behl et al., 2023). Alternatively, more specific quantification is achieved using high-performance liquid chromatography (HPLC) equipped with a refractive index detector or pulsed amperometric detector, using an appropriate carbohydrate column and mobile phase (often water or dilute acid). The concentration of xylose is calculated by comparing the peak area or absorbance to a standard curve generated from known xylose concentrations (Sluiter et al., 2008). This biochemical approach is essential for evaluating hemicellulose degradation and optimizing biofuel production from lignocellulosic feedstocks.

2.3.13.3 Phenolic Compound Analysis

The detection of phenolic compounds released during the hydrolysis of lignocellulosic biomass can be effectively carried out using the Folin–Ciocalteu colorimetric assay. In this protocol, the hydrolysate is first filtered to remove residual solids, and an aliquot is mixed with Folin–Ciocalteu reagent followed by sodium carbonate solution. The reaction mixture is incubated at room temperature for 30 to 60 minutes, allowing the phenolic compounds to reduce the reagent, resulting in the formation of a blue complex. The absorbance of the solution is then measured at 765 nm using a spectrophotometer, and the concentration of phenolics is quantified using a calibration curve typically prepared with gallic acid as the standard. This method provides a rapid and sensitive means of determining the total phenolic content and is widely used in biomass conversion studies due to its simplicity and reproducibility (Singleton & Rossi, 1965).

2.4 CONCLUSION

The analytical characterization of biomass for biofuel production plays a pivotal role in assessing the properties of biomass materials, refining biofuel production processes and ensuring the sustainability of bioenergy systems. This chapter underscores the significance of comprehending the chemical composition and structural attributes of biomass through analytical techniques like spectroscopy, chromatography, and microscopy. These methodologies enable researchers to quantify essential parameters such as moisture content, ash content, elemental composition, and calorific value, pivotal for assessing the energy potential and suitability of biomass feedstocks for biofuel production. By employing advanced analytical techniques and methodologies, researchers can glean valuable insights into biomass feedstock properties, thereby optimizing biofuel production processes for heightened efficiency and environmental sustainability. Persistent research and innovation in biomass characterization are imperative to propel the development of cost-effective and eco-friendly biofuel production technologies for ushering in a more sustainable energy future.

ACKNOWLEDGMENT

The authors would like to acknowledge the University Grant Commission (UGC) and Council of Scientific and Industrial Research (CSIR) for providing financial assistance to Mr. Aditya Sharma [NTA reference no. 231620036265] and Ms. Pushpak Dahiya [CSIR File no 09/0237(15947)/2022-EMR 1] respectively. Author also acknowledge National Mission on Himalayan Studies Project Management Unit (NMHS-PMU), MoEF&CC, Govt. of India, New Delhi, [Grant number: NMHS/2022-23/SG 81/01/280] for the financial support to Dr. Ravi Kant Bhatia and the Department of Biotechnology, HPU Shimla, for all necessary facilities

REFERENCES

Abou-Yousef H (2014). Efficient utilization of aqueous phase bio-oil to furan derivatives through extraction and sugars conversion in acid-catalyzed biphasic system. *Fuel*, *137*, 115–121.

Agarwal UP, and Ralph SA (2023). Raman spectra of delignified plant fibers: Exploring the impact of Xylan's presence on the spectral features of cellulose. *Fibers*, *12*(1), 5.

Almusafir R, and Smith JD (2024). Thermal decomposition and kinetic parameters of three biomass feedstocks for the performance of the gasification process using a thermogravimetric analyzer. *Energies*, *17*(2), 396.

Apaydın-Varol E, and Mutlu Ü (2023). TGA-FTIR analysis of biomass samples based on the thermal decomposition behavior of hemicellulose, cellulose, and lignin. *Energies*, *16*(9), 3674.

Bahng MK, Mukarakate C, Robichaud DJ, and Nimlos MR (2009). Current technologies for analysis of biomass thermochemical processing: A review. *Analyticachimicaacta*, *651*(2), 117–138.

Behl M, Dahiya P, Kumari D, Thakur K, Devi M, Rathour RK, and Bhatia RK (2023). Development of tailored bioprocess for pretreatment and saccharification of corn stalk into bioethanol using hydrolytic enzymes cocktail and fermentative yeasts. *Biocatalysis and agricultural biotechnology*, *53*, 102896.

Bhaskar T, Bhavya B, and Sharma A (2018). Overview of biofuels and its properties. *Materials today: proceedings*, *5*(3), 7413–7420. https://doi.org/10.1016/j.matpr.2017.11.620

Bianquini TS, Florencio C, dos Santos Garcia RH, Colnago LA, Ceccato-Antonini SR, Carmo M, and Ribeiro TS (2020). Time domain NMR spectroscopy as a fast method for probing the efficiency of biomass pretreatments for second generation ethanol production. *Biomass and bioenergy, 142*, 105734.

Bridgeman TG, Jones JM, Williams A, and Waldron DJ (2010). An investigation of the grindability of two torrefied energy crops. *Fuel, 89*(12), 3911–3918.

Bridgwater AV (2012). Review of fast pyrolysis of biomass and product upgrading. *Biomass and bioenergy, 38*, 68–94.

Carrier M, Loppinet-Serani A, Denux D, Lasnier JM, Ham-Pichavant F, Cansell F, and Aymonier C (2011). Thermogravimetric analysis as a new method to determine the lignocellulosic composition of biomass. *Biomass and bioenergy, 35*(1), 298–307.

Chaudhary S, Jain VP, and Jaiswar G (2022). The composition of polysaccharides: Monosaccharides and binding, group decorating, polysaccharides chains. In *Innovation in nano-polysaccharides for eco-sustainability* (pp. 83–118). Elsevier.

Chen WH, Eng CF, Lin YY, and Bach QV (2020). Independent parallel pyrolysis kinetics of cellulose, hemicelluloses and lignin at various heating rates analyzed by evolutionary computation. *Energy conversion and management, 221*, 113165.

Chiaradia MC, Collins CH, and Jardim IC (2008). The state of the art of chromatography associated with the tandem mass spectrometry for toxic compound analyses in food. *Química nova, 31*, 623–636.

Chundawat SP, Beckham GT, Himmel ME, and Dale BE (2011). Deconstruction of lignocellulosic biomass to fuels and chemicals. *Annual review of chemical and biomolecular engineering, 2*(1), 121–145.

Cui Y, Yan Q, Sun J, and Zhang Q (2013). Biomass thermochemical characterization and its application in energy yield assessment. *Energy conversion and management, 74*, 149–157. https://doi.org/10.1016/j.enconman.2013.05.029

Cui Y, Zeng Y, Huang C, Yang Q, Zhan Z, Feng X, and Chen Z (2024). Highly efficient analysis on biomass carbohydrate mixtures by DREAMTIME NMR spectroscopy. *Analytical chemistry, 97*(1), 912–920.

Dapkunas SJ, Jillavenkatesa A, and Lum LH (2001). Particle size characterization. In *NIST recommended practice guide* (pp. 960–961). https://nvlpubs.nist.gov/nistpubs/Legacy/SP/nistspecialpublication960-1.pdf (Accessed on 18/12/2025).

Das N, Jena PK, Padhi D, Kumar Mohanty M, and Sahoo G (2021). A comprehensive review of characterization, pretreatment and its applications on different lignocellulosic biomass for bioethanol production. *Biomass conversion and biorefinery*, 1–25.

Dashtban M, Maki M, Leung KT, Mao C, and Qin W (2010). Cellulase activities in biomass conversion: Measurement methods and comparison. *Critical reviews in biotechnology, 30*(4), 302–309.

De Blasio C (2019). Thermogravimetric analysis (TGA). *Fundamentals of biofuels engineering and technology*, 91–102.

Delwiche SR (2015). Basics of spectroscopic analysis. In *Hyperspectral imaging technology in food and agriculture* (pp. 57–79). New York, NY: Springer New York.

Demirbas A (2008). Biofuels sources, biofuel policy, biofuel economy and global biofuel projections. *Energy conversion and management, 49*(8), 2106–2116.

Demirbas A (2009). Biofuels securing the planet's future energy needs. *Energy conversion and management, 50*(9), 2239–2249.

Di Gianfrancesco A (2017). Technologies for chemical analyses, microstructural and inspection investigations. In *Materials for ultra-supercritical and advanced ultra-supercritical power plants* (pp. 197–245). Woodhead Publishing.

Di Matteo P, Luziatelli F, Bortolami M, Mele ML, Ruzzi M, and Russo P (2023). Differential scanning calorimetry (DSC) as a tool for studying thermal properties of a crude cellulase cocktail. *Chemical papers, 77*(5), 2689–2696.

Dugo P, Cacciola F, Kumm T, Dugo G, and Mondello L (2008). Comprehensive multidimensional liquid chromatography: Theory and applications. *Journal of chromatography A*, *1184*(1–2), 353–368.

Dzurendova S, Olsen PM, Byrtusová D, Tafintseva V, Shapaval V, Horn SJ, and Zimmermann B (2023). Raman spectroscopy online monitoring of biomass production, intracellular metabolites and carbon substrates during submerged fermentation of oleaginous and carotenogenic microorganisms. *Microbial cell factories*, *22*(1), 261.

Edmunds CW, Reyes, Molina EA, André N, Hamilton C, Park S, Fasina O, and Labbé N (2018). Blended feedstocks for thermochemical conversion: Biomass characterization and bio-oil production from switchgrass-pine residues blends. *Frontiers in energy research*, *6*, 79.

Eisenbies MH, Volk TA, Therasme O, and Hallen K (2019). Three bulk density measurement methods provide different results for commercial-scale harvests of willow biomass chips. *Biomass and bioenergy*, *124*, 64–73.

Elliot SG, Tosi I, Riisager A, Taarning E, and Meier S (2019). Response factors enable rapid quantitative 2D NMR analysis in catalytic biomass conversion to renewable chemicals. *Topics in catalysis*, *62*(7–11), 590–598.

Fajobi MO, Lasode OA, Adeleke AA, Ikubanni PP, and Balogun AO (2022). Investigation of physicochemical characteristics of selected lignocellulose biomass. *Scientific reports*, *12*(1), 2918.

Fischer ER, Hansen BT, Nair V, Hoyt FH, Schwartz CL, and Dorward DW (2024). Scanning electron microscopy. *CurrentProtocols*, *4*, e1034. https://doi.org/10.1002/cpz1.1034

Fitria, Liu J, and Yang B (2023). Roles of mineral matter in biomass processing to biofuels. *Biofuels, bioproducts and biorefining*, *17*(3), 696–717.

Gaffney JS, Marley NA, and Jones DE (2002). Fourier transform infrared (FTIR) spectroscopy. In *Characterization of materials* (pp. 1–33). John Wiley & Sons.

Gibson LJ (2012). The hierarchical structure and mechanics of plant materials. *Journal of the royal society interface*, *9*(76), 2749–2766.

González-Vila FJ, Tinoco P, Almendros G, and Martin F (2001). Pyrolysis– GC– MS analysis of the formation and degradation stages of charred residues from lignocellulosic biomass. *Journal of agricultural and food chemistry*, *49*(3), 1128–1131.

Grams J (2022). Surface analysis of solid products of thermal treatment of lignocellulosic biomass. *Journal of analytical and applied pyrolysis*, *161*, 105429.

Hames BR (2009). Biomass compositional analysis for energy applications. *Biofuels: methods and protocols*, 145–167.

Himmel ME, Ding SY, Johnson DK, Adney WS, Nimlos MR, Brady JW, and Foust TD (2007). Biomass recalcitrance: Engineering plants and enzymes for biofuels production. *Science*, *315*(5813), 804–807.

Huang YF, and Lo SL (2020). Predicting heating value of lignocellulosic biomass based on elemental analysis. *Energy*, *191*, 116501.

Irwansyah FS, Amal AI, Diyanthi EW, Hadisantoso EP, Noviyanti AR, Eddy DR, and Risdiana R (2024). How to read and determine the specific surface area of inorganic materials using the Brunauer-Emmett-Teller (BET) method. *ASEAN journal of science and engineering*, *4*(1), 61–70.

Jacques RA, and Caramao EB (2009). Two dimensional liquid chromatography. *Chromatographic Science*, *1*, 73–82.

Karimi K, and Taherzadeh MJ (2016). A critical review of analytical methods in pretreatment of lignocelluloses: Composition, imaging, and crystallinity. *Bioresource technology*, *200*, 1008–1018.

Khan SA, Khan SB, Khan LU, Farooq A, Akhtar K, and Asiri AM (2018). Fourier transform infrared spectroscopy: Fundamentals and application in functional groups and nanomaterials characterization. In *Handbook of materials characterization* (pp. 317–344). Cham: Springer International Publishing.

Kim SH, Lee CM, and Kafle K (2013). Characterization of crystalline cellulose in biomass: Basic principles, applications, and limitations of XRD, NMR, IR, Raman, and SFG. *Korean journal of chemical engineering*, *30*, 2127–2141.

Kumar AK, and Sharma S (2017). Recent updates on different methods of pretreatment of lignocellulosic feedstocks: A review. *Bioresources and bioprocessing*, *4*(1), 7.

Lemes EM (2024). Raman spectroscopy – A visit to the literature on plant, food, and agricultural studies. *Journal of the science of food and agriculture*.

Lou Z, Zhang Y, Li Y, and Xu L (2023). Study on microscopic physical and chemical properties of biomass materials by AFM. *Journal of materials research and technology*, *24*, 10005–10026.

Lowell S, Shields JE, Thomas MA, and Thommes M (2012). *Characterization of porous solids and powders: Surface area, pore size, and density* (Vol. 16). Springer Science & Business Media.

Lupoi JS, Gjersing E, and Davis MF (2015). Evaluating lignocellulosic biomass, its derivatives, and downstream products with Raman spectroscopy. *Frontiers in bioengineering and biotechnology*, *3*, 50.

Martínez MG, Ohra-Aho T, da, Silva, Perez D, Tamminen T, and Dupont C (2019). Influence of step duration in fractionated Py-GC/MS of lignocellulosic biomass. *Journal of analytical and applied pyrolysis*, *137*, 195–202.

Martone PT, Estevez JM, Lu F, Ruel K, Denny MW, Somerville C, and Ralph J (2009). The discovery of lignin in seaweed reveals the convergent evolution of cell-wall architecture. *Current Biology*, *19*(2), 169–175.

Mase C, Maillard JF, Marcuz S, Hubert-Roux M, Afonso C, and Giusti P (2023). Contribution of LDI and MALDI for the characterization of a lignocellulosic-based pyrolysis bio-oil. *Journal of the American society for mass spectrometry*, *34*(8), 1789–1797.

Meng X, Foston M, Leisen J, DeMartini J, Wyman CE, and Ragauskas AJ (2013). Determination of porosity of lignocellulosic biomass before and after pretreatment by using Simons' stain and NMR techniques. *Bioresource technology*, *144*, 467–476.

Monteiro MR, Ambrozin ARP, Liao LM, and Ferreira AG (2008). Critical review on analytical methods for biodiesel characterization. *Talanta*, *77*(2), 593–605.

Munson CR, Gao Y, Mortimer JC, and Murray DT (2022). Solid-state nuclear magnetic resonance as a tool to probe the impact of mechanical preprocessing on the structure and arrangement of plant cell wall polymers. *Frontiers in plant science*, *12*, 766506.

Nimmanterdwong P, Chalermsinsuwan B, and Piumsomboon P (2021). Prediction of lignocellulosic biomass structural components from ultimate/proximate analysis. *Energy*, *222*, 119945.

Nishiyama Y, Sugiyama J, Chanzy H, and Langan P (2003). Crystal structure and hydrogen bonding system in cellulose Iα from synchrotron X-ray and neutron fiber diffraction. *Journal of the American chemical society*, *125*(47), 14300–14306.

Nsude OP, and Orie KJ (2022). Microcrystalline cellulose of oil bean pod: Extraction, physicochemical, Brunauer–Emmett–Teller (BET), and flow-ability analysis. *Asian journal of applied chemistry research*, *12*(4), 1–12.

Oladunni AA, Adeleke AA, Bello S, Fajobi MO, Ikubanni PP, and Olosho AI (2023, November). Overview of Fourier Transform Infrared (FTIR) Spectroscopy for Biomass and BioFuel Characterization. In *2023 2nd International Conference on Multidisciplinary Engineering and Applied Science (ICMEAS)* (Vol. 1, pp. 1–5). IEEE.

Orlando MQ, and Borja VM (2020). Pretreatment of animal manure biomass to improve biogas production: A review. *Energies*, *13*(14), 3573.

Pereira SC, Maehara L, Machado CM, and Farinas CS (2016). Physical–chemical–morphological characterization of the whole sugarcane lignocellulosic biomass used for 2G ethanol production by spectroscopy and microscopy techniques. *Renewable energy*, *87*, 607–617.

Picou L, and Boldor D (2012). Thermophysical characterization of the seeds of invasive Chinese tallow tree: Importance for biofuel production. *Environmental science & technology, 46*(20), 11435–11442.

Qian L, Luo J, Wen S, and Xiao X (2001). The experimental rules of mica as a reference sample of AFM/FFM measurement. *Chinese science bulletin, 46*, 349–352.

Ragauskas AJ, Williams CK, Davison BH, Britovsek G, Cairney J, Eckert CA, and Tschaplinski T (2006). The path forward for biofuels and biomaterials. *Science, 311*(5760), 484–489.

Rathour RK, Behl M, Sakhuja D, Kumar N, Sharma N, Walia A, and Bhatia RK (2025). Recent updates in biomass pretreatment techniques for a sustainable biofuel industry: A comprehensive review. *Waste and biomass valorization*, 1–26.

Reymond C, Dubuis A, Le Masle A, Colas C, Chahen L, Destandau E, and Charon N (2020). Characterization of liquid–liquid extraction fractions from lignocellulosic biomass by high performance liquid chromatography hyphenated to tandem high-resolution mass spectrometry. *Journal of chromatography A, 1610*, 460569.

Rinki, Yadav P, Sharma A, Dahiya P, Kashyap A, Walia A, and Bhatia RK (2024). Upcycling of tetra pack waste cellulose into reducing sugars for bioethanol production using Saccharomyces cerevisiae. *Biotechnology for sustainable materials, 1*(1), 4.

Rodrigues RC, Rodrigues BG, Canettieri EV, Martinez EA, Palladino F, Wisniewski A Jr, and Rodrigues D Jr (2022). Comprehensive approach of methods for microstructural analysis and analytical tools in lignocellulosic biomass assessment–a review. *Bioresource technology, 348*, 126627.

Rowell RM (2005). *Handbook of wood chemistry and wood composites.* CRC press.

Sarpal AS, Teixeira CMLL, Silva PRM, Da Costa Monteiro TV, Da Silva JI, Da Cunha VS, and Daroda RJ (2015). NMR techniques for determination of lipid content in microalgal biomass and their use in monitoring the cultivation with biodiesel potential. *Applied microbiology and biotechnology, 100*(5), 2471–2485.

Silva MVB, Otaguro H, and Assunção RMN (2024). Thermal properties of biomass. In *Handbook of biomass* (pp. 349–375). Singapore: Springer Nature Singapore.

Singh RK, Pandey D, Patil T, and Sawarkar AN (2020). Pyrolysis of banana leaves biomass: Physico-chemical characterization, thermal decomposition behavior, kinetic and thermodynamic analyses. *Bioresource technology, 310*, 123464.

Singh YD, Mahanta P, and Bora U (2017). Comprehensive characterization of lignocellulosic biomass through proximate, ultimate and compositional analysis for bioenergy production. *Renewable energy, 103*, 490–500.

Singleton VL, and Rossi JA (1965). Colorimetry of total phenolics with phosphomolybdic–phosphotungstic acid reagents. *American journal of enology and viticulture, 16*(3), 144–158.

Sluiter A, Hames B, Ruiz R, Scarlata C, Sluiter J, Templeton D, and Crocker DLAP (2008). Determination of structural carbohydrates and lignin in biomass. *Laboratory analytical procedure, 1617*(1), 1–16.

Sluiter A, Sluiter J, and Wolfrum EJ (2013). Methods for biomass compositional analysis. In *Catalysis for the conversion of biomass and its derivatives* (Vol. 2). Open Access, Max Planck Institute for the History of Science.

Soria AC, Mena A, Ruiz-Matute AI, and Sanz ML (2021). Gas chromatographic analysis of carbohydrates. In *Gas chromatography* (pp. 703–726). Elsevier.

Stilo F, Bicchi C, Jimenez-Carvelo AM, Cuadros-Rodriguez L, Reichenbach SE, and Cordero C (2021). Chromatographic fingerprinting by comprehensive two-dimensional chromatography: Fundamentals and tools. *TrAC trends in analytical chemistry, 134*, 116133.

Sultana A, and Kumar A (2011). Optimal configuration and combination of multiple lignocellulosic biomass feedstocks delivery to a biorefinery. *Bioresource technology, 102*(21), 9947–9956.

Sun K, Zhang L, Zhang Z, Shao Y, Sun Y, Zhang S, and Hu X (2020). Investigation into properties of carbohydrate polymers formed from acid-catalyzed conversion of sugar Monomers/Oligomers over brønsted acid catalysts. *Energy technology*, *8*(6), 1901476.

Tafur-Marinos JA, Ginepro M, Pastero L, and Zelano V (2016). Digestion procedures for the elemental analysis of wood by inductively coupled plasma optical emission spectrometry. *Analytical letters*, *49*(11), 1722–1733.

Vaca-Medina G, Jallabert B, Viet D, Peydecastaing J, and Rouilly A (2013). Effect of temperature on high pressure cellulose compression. *Cellulose*, *20*, 2311–2319.

Vaseghi Z, and Nematollahzadeh A (2020). Nanomaterials: Types, synthesis, and characterization. In *Green synthesis of nanomaterials for bioenergy applications* (pp. 23–82).

Vassilev SV, Baxter D, Andersen LK, and Vassileva CG (2010). An overview of the chemical composition of biomass. *Fuel*, *89*(5), 913–933.

Waseda Y, Matsubara E, and Shinoda K (2011). *X-ray diffraction crystallography: Introduction, examples and solved problems*. Springer Science & Business Media.

Williams CL, Emerson RM, and Tumuluru JS (2017). Biomass compositional analysis for conversion to renewable fuels and chemicals. In *Biomass volume estimation and valorization for energy* (pp. 251–270).

Xing C, Liu L, and Wang M (2019). A new preparation method and imaging parameters of asphalt binder samples for atomic force microscopy. *Construction and building materials*, *205*, 622–632.

Vanholme R, Demedts B, Morreel K, Ralph J, and Boerjan W. (2010). Lignin biosynthesis and structure. *Plant physiology*, *153*(3), 895–905.

Yan YU, Ze-hui JIANG, Ge WANG, Dao-chun QIN, and Qiang CHENG (2008). Visualization of cellulose microfibrils of Moso bamboo fibers with atomic force microscopy. *Journal of Beijing forestry university*, *30*(1), 124–127.

Yousuf A (2012). Biodiesel from lignocellulosic biomass–prospects and challenges. *Waste management*, *32*(11), 2061–2067.

Zhou X, Ding D, Ma J, Ji Z, Zhang X, and Xu F (2015). Ultrastructure and topochemistry of plant cell wall by transmission electron microscopy. In *The transmission electron microscope–theory and applications* (pp. 285–306). InTech.

Zoghlami A, Refahi Y, Terryn C, and Paes G (2019). Multimodal characterization of acid-pretreated poplar reveals spectral and structural parameters strongly correlate with saccharification. *Bioresource technology*, *293*, 122015.

3 Biomass Pretreatment
An Analytical Approach

*Suman Salhotra#, Deepika Umrao#, Pragati Bhardwaj, and Meenu Hans**

3.1 INTRODUCTION

Global warming and limited stock of nonrenewable resources shift the social interest toward the use of renewable resources. The renewable resource, or biomass, is used for extraction of valuable products of biofuel, mainly cellulosic ethanol, as an alternative to gasoline. Demand for transportation fuel is expected to increase in coming years, which makes it difficult to produce on a large scale due to its challenges related to the complexity of the process. The United States of America and Brazil are the major contributors of bioethanol by the fermentation of sugars derived from starch (corn) and sugar (sugarcane), respectively, and termed as first-generation (1G) ethanol production (feedstock from food resources or their by-products). On a large scale, it may create the food vs fuel debate; therefore, to avoid this debate, lignocellulosic biomass (LCB), termed as second-generation ethanol (2G), is a suitable approach for bio-ethanol production on a large scale. Lignocellulosic biomass includes the feedstocks other than the food stream, like forestry waste, agricultural waste, and municipal waste. LCB can also be advantageous in creating job opportunities and new income sources in rural areas by selling their agricultural wastes to industries. The prime minister of India launched an ambitious E-100 pilot project in Pune for the production and distribution of ethanol across the country. The Government of India also sets the ethanol blending program (EBP), which was increased by 1.53% in 2013–2014 to 10% in 2021–2022. Ethanol blending increased from 38 crore liters in 2013–2014 to 433.6 crore liters in 2021–2022. The Indian government also introduces the scheme "Pradhanmantri JI-VAN Yojana," which provides financial support to integrate the bio-ethanol project for setting up 2G bio-ethanol production. The establishment of 2G bioethanol technology could also reduce the foreign exchange burden on the country. The cumulative foreign exchange impact due to the EBP program is estimated to be over USD 64,297 lakh (https://mopng.gov.in).

In spite of immense potential for ethanol production, the successful adoption of 2G ethanol technology is difficult due to the complexity of feedstock composition, i.e., LCB, which involves three steps. Biomass in its native form is challenging to use for the production of energy, so pretreatment is required to treat its refractory nature and make the biomass more efficient for conversion through various conversion routes for maximum recovery. Pretreatment of feedstock is the first step in the process, which

# Equal Contribution of the Authors
* Corresponding author

 DOI: 10.1201/9781003485636-3

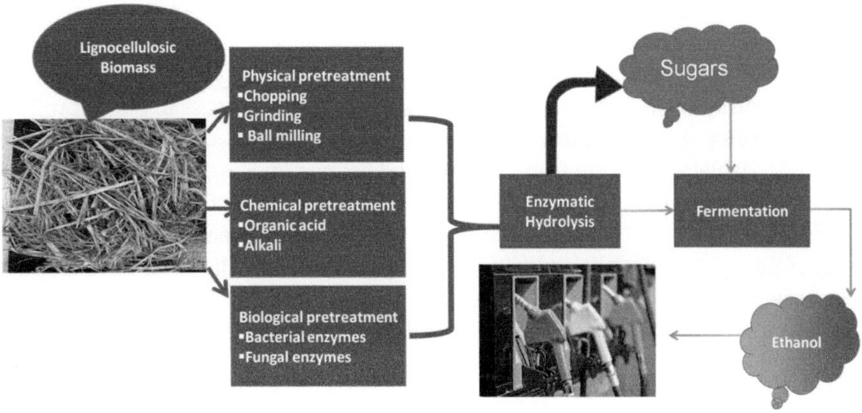

FIGURE 3.1 Schematic representation of bio-ethanol production using various pretreatment methods.

alters the structure and composition of biomass, which could be easily available for hydrolysis by enzymatic reactions and then fermentation of the sugars [C6 and C5] with the help of different microorganisms, resulting in ethanol production as the end product. Depending upon the type of LCB, different types of pretreatments are given to the biomass for the efficient recovery of the product. A schematic representation of pretreatment methods used for bioethanol production is illustrated in Figure 3.1. Further, efficiency of the treatment could be determined by the use of high-quality analytical techniques, which could provide the morphological as well as biochemical changes in biomass for the maximum recovery of the product at the end. This chapter discusses the different approaches for pretreatment and analytical techniques used for morphological and biochemical characterization of biomass after the treatment.

3.2 NECESSITY OF PRETREATMENT

Pretreatment of biomass is a critically important and expensive process of 2G biofuel technology. LCB consists of cellulose (38–50%), hemicellulose (23–32%), and lignin (15–25%) as its three major components (Mankar et al., 2021). Cellulose is a polysaccharide made up of long straight chains of β-D glucose bonded with β-1-4 glycosidic linkages to form the basic repeating units of cellobiose in the chain (Hans et al., 2019). Hemicellulose is also a polysaccharide, which is made up of predominantly pentose sugars and a number of hexose sugars, located in the primary and secondary cell walls of biomass. It has a low degree of polymerization and lower molecular weight as compared to cellulose. The common saccharides present in hemicellulose are galactomannans, glucomannans, β-D-glucans, β-D-glucan-callose, xyloglucans, and xylans (arabinoxylans and 4-O-methyl-glucuronoxylan, forming a major fraction of the polymer) (Shukla et al., 2023). Lignin is an amorphous and phenolic macromolecule derived from three monolignols: specifically, p-coumaryl alcohol, coniferyl alcohol, and sinapyl alcohol (Shukla et al., 2023). These long phenyl units are attached together by β-O-4, β-β, β-5, α-O-4, and β-1 with β-O-4 ether linkages and make the lignin degradation more difficult (Rezania et al., 2020). Lignin keeps the cellulose and hemicellulose safe from various infectious agents and natural damage by covering them under its matrix. Lignin provides toughness to the cell wall of biomass, which makes it harder to perform the pretreatment on it. Pretreatment is required to isolate the lignin from cellulose and hemicellulose (delignification of biomass) so that enzymes can easily penetrate the cell wall and access the cellulose and hemicellulose to hydrolyze them into sugars to be used in the next step. Pretreatment reduces the crystalline index of cellulose and particle size of the biomass to enhance the surface area so that enzymes could hydrolyze it efficiently. Pretreatment given to the biomass should be cost- and operationally effective and eco-friendly to be applicable to the commercial level.

3.3 TYPES OF PRETREATMENT

Different types of pretreatment are given to the biomass to achieve maximum possible delignification and high recovery of cellulose and hemicellulose and therefore release a higher amount of sugars. Various methods available are physical, physiochemical, biological, radiation, and combined, as given in Table 3.1 and discussed ahead.

3.3.1 Physical Pretreatment

This pretreatment is given to the feedstock for reducing its particle size (up to 10–30 mm), which increases the surface area to let enzymes penetrate to the deepest layers of sugar polymers. Physical pretreatment involves the usage of various tools, such as chopping, cutting, grinding, and milling, to shorten the particle size and ultimately reduce the crystallinity index (CI) of biomass (Hans et al., 2019). Grinding and Milling reduces the particle size of biomass up to 0.2–2.0 mm as it helps in reducing the CI and particle size, which expands the surface area and enhances the enzymatic action. In a study, oil palm empty fruit bunch was pretreated by ball milling for 120 min and resulted in the reduction of CI from 56.1% to 9.3% and higher sugar yields (glucose/xylose ratio as 67.5/80.1) after enzymatic hydrolysis as compared to the untreated sample (15.9/5.4) (Mankar et al., 2021). In a study, Douglas fir residuals (*Pseudotsuga menziesii*) were pretreated by hammer milling, air classifier milling, ball milling, chipping, and mashing for 7–30 mins. Ball milling decreased the cellulose crystallinity content from 40.73% to 11.70% (Jędrzejczyk et al., 2019). Ball milling treatment did not affect the chemical composition of the biomass but reduced crystallinity and increased the bioethanol yield in the study. A disadvantage of this process is that it uses high energy, which makes up about 40% of the total production cost. Three physical methods, including air classifier mill, continuous ball mill, and high-speed mill, were applied for the pretreatment of corn stover. All these methods reduced the particle size of biomass in the range of 100–230 mm. The size reduction of biomass increased the glucose in the range of 29–44%, which showed that the high-speed mill and air classifier mill had a high pretreatment efficiency (Gu et al., 2019).

TABLE 3.1
Different Pretreatment Methods Used for Lignocellulosic Biomass Reported in Literature

Method	Process Condition	[b]Increase in Cellulose (%)	[c]Decrease in Hemicellulose (%)	[d]Decrease in Lignin (%)	Enzymatic Hydrolysis	Sugar Yield (g/ kg raw biomass)	Ethanol Yield (g/g raw biomass)	References
Physical Pretreatment	Barley straw (*Hordeum vulgare*) pretreated by nitrogen assisted wet milling at 50 Hz and 1100 RPM milling at 20% (w/v) solid loading	NA	2.0	2.0	Accellerase 1500 loading of 30 FPU/g cellulose for 24 h at 50°C	NA	23	Raud et al., 2020
	Corn stover pretreated by thermochemical disk refining treatment with 5% soda (w/w at 9% (w/v) solid loading for 30 min at 140°C	NA	NA	5.9	Accellerase® Duet enzyme 0.25 ml/g at 55°C, 200 rpm, pH 4.8 for 72 h	44.2	NA	Chen et al., 2019
Chemical Pretreatment	Paddy straw pretreated with 5% sodium hydroxide, 20% (w/v) solid loading at 90°C for 20 min	79.66	22.05	84.1	Cellulase loading 25 [a]FPU/g; 10% pretreated biomass loading; *S. cerevisae* and *Scheffersomyces stipitis* at 33.1°C for 72 h	NA	0.47	Suriyachai et al., 2013
	Corn stover pretreated with 0.45% v/v of sulfuric acid at 162.4°C for 29.5 min	10.09	99.3	11.6	CellicCTec2 loading 25 [a]FPU/g at 50°C and 300 rpm for 6 h	903.4	0.47	Suriyachai et al., 2020
	Sugarcane bagasse pretreated with 15.64% (v/v) liquid ammonia at 10.51% (w/v) solid loading and 84.9°C for 23.9 h	30.2	27.4	51.5	Cellic CTec2 loading 20 [a]FPU/g pretreated biomass at 10% (w/v) solid loading, 50°C, and 130 rpm for 72 h	545	NA	Hans et al., 2021
	Sugarcane bagasse pretreated with 2.18% (v/v) sulfuric acid at 14.35% (w/v), 121°C solid loading for 29.49 min	36.34	74.95	42.31	Cellic CTec2 loading OF 20 [a]FPU/g pretreated biomass at 10% (w/v) solid loading, and 50°C at 130 rpm for 72 h	306.14	71.45	Hans et al., 2022

(*Continued*)

TABLE 3.1 (*Continued*)
Different Pretreatment Methods Used for Lignocellulosic Biomass Reported in Literature

Method	Process Condition	[b]Increase in Cellulose (%)	[c]Decrease in Hemicellulose (%)	[d]Decrease in Lignin (%)	Enzymatic Hydrolysis	Sugar Yield (g/ kg raw biomass)	Ethanol Yield (g/g raw biomass)	References
	Paddy straw pretreated with 0.5% (v/v) sulfuric acid followed by 4% (w/v) sodium hydroxide	59.57	38.3	90.41	*Penicillium mallochii (JS17)* cellulase loading of 13 [a]FPU/g at 50 ± 2°C for 24 h	504.18	NA	Verma et al., 2022
Biological Pretreatment	Paddy straw pretreated with white-rot fungus, *Trametes hirsuta* followed by steam pretreatment at 121°C	37.6	NA	4.7	Accellerase 1500 loaded at ~10% loading (wet biomass) at 50°C and 150 rpm for 24 h	NA	0.37 g/g glucose	Arora et al., 2016
	Paddy straw pretreated with *Pleurotus sajorcaju* for 40 days incubation	14.7	27.9	35	cellulase at 5 [a]FPU/g pretreated biomass at 50°C for 2 h	119	20	Yadav & Wati, 2016
	Paddy straw was pretreated with different doses of laccase (Ganoderma lucidum MDU-7) at 10% (w/v) solid loading at 30°C and 150 rpm for 72 h	NA	NA	17.2	Cellulase of *Myceliophthora thermophila BJTLRMDU3* loading of 20 U/g at 60°C and 150 rpm for 6 h	162.82	NA	Anu et al., 2021

[a] FPU= Filter paper units

[b] $\%\text{Increase in cellulose content} = \frac{\text{Cellulose in pretreated biomass}(\%) \text{- Cellulose in raw biomass}(\%) \times 100}{\text{cellulose in raw biomass}(\%)}$

[c] $\%\ \text{Decrease in hemicellulose content} = \frac{\text{Hemicellulose in pretreated biomass}(\%) \text{-Hemicellulose in raw biomass}(\%) \times 100}{\text{Hemicellulose in raw biomass}(\%)}$

[d] $\%\ \text{Decrease in lignin content} = \frac{\text{Lignin in pretreated biomass}(\%) \text{-Lignin in raw biomass}(\%) \times 100}{\text{Lignin in raw biomass}(\%)}$

3.3.2 Chemical Pretreatment

In this pretreatment, chemicals are used to treat the robust nature of the LCB. There are some common chemical agents like alkaline, acids, ionic liquids (ILs), oxidizing agents, etc., as discussed ahead (Table 3.1).

3.3.2.1 Acid Pretreatment

Acids make the cellulose more accessible and solubilize the hemicellulose, while it does the condensation and precipitation of lignin. However, acid pretreatment makes the reaction environment corrosive and prevents the possible formation of inhibitors like hydroxymethylfurfural (HMF), furfural, and acetic acid (Hans et al., 2021). Acid pretreatment can be done in two diverse ways, as with dilute acid (0.1%) at higher temperatures (>200°C) and concentrated acid (30–70%) at relatively lower temperatures (<50°C). Dilute acid pretreatment uses lower chemical concentration but high heat energy, whereas the other one consumes less energy but higher acid concentration, which can make the reaction medium toxic and corrosive (Mankar et al., 2021). Using strong acid for pretreatment is costly because of its corrosive and toxic reaction environment, and it requires specially designed resistant and robust reactors, whereas the dilute acid pretreatment is cost-effective and has high rates of cellulose and hemicellulose hydrolysis inhibitors (Harmsen et al., 2010). Sulfuric acid and hydrochloric acid are used for strong acid pretreatment. As sulfuric acid breaks the glycosidic bonds of polymers, resulting in the removal of hemicellulose and an increase in cellulose and lignin content in biomass. Loss of potential pentose sugars and recovery of intact lignin in biomass are the major limitations in using acid for the pretreatment process, making it a less favorable choice for this purpose nowadays.

3.3.2.2 Alkaline Pretreatment

This pretreatment is replacing acid to overcome its limitations in terms of complete removal of lignin and a significant recovery of sugars in the biomass. This process uses the mild chemicals and also removes lignin and xylan efficiently from the biomass for easier access of the sugars by different enzymes and microbes. Alkaline pretreatment performs intermolecular saponification and solvation of ester bonds present in lignocellulosic biomass to remove the lignin fraction. Commonly used alkali agents are potassium hydroxide (KOH), sodium hydroxide (NaOH), calcium hydroxide (Ca(OH)2), and ammonium hydroxide, or liquid ammonia (NH4OH). The alkali pretreatment decreases the crystallinity and degree of polymerization of cellulose, which makes it easy for enzymes to approach and degrade the polymer into glucose. In a study, wheat straw was pretreated using a concentration of 4% (w/v) sodium hydroxide at 121°C for 1 h, resulting in an effective 88.8% lignin removal with a 41.8% decrease in hemicellulose content (Zheng et al., 2018). Although sodium hydroxide is a very popular alkali agent to pretreat the biomass for efficient sugar recovery, it is associated with the complete hydrolysis of xylan along with lignin, resulting in a significant loss of xylose. As xylose is the sugar of interest nowadays in 2G fuel technology due to its high potential for ethanol and other fuel production, the exploration of more efficient alkalis to recover both glucose and xylose in biomass led to another important alkali, i.e., liquid ammonia (NH_4OH), which is capable of removing lignin from biomass, leaving the xylan fraction intact in the biomass along with cellulose to be later used for the fermentation of the resulting monomeric sugars to get a high ethanol yield (Hans et al., 2021). As shown in Table 3.1, when sugarcane bagasse was pretreated with 15.64% (v/v) liquid ammonia at 10.51% (w/v) solid loading at 84.9°C for 23.9 h, it resulted in a 30.2% increase in cellulose content and a 51.5% decrease in lignin content with a 545 g/kg sugar yield after hydrolysis (Hans et al., 2021). This kind of alkali agent paves the way to integrated bio-refinery by separating lignin and sugars in solid and liquid fractions to be utilized for different valuable products.

3.3.2.3 Organosolv Pretreatment

In this method of pretreatment, organic solvent mixtures are used for the removal of lignin and hemicellulose by hydrolyzing the internal bonds as well as ether and ester interpolymer bonds between them. Depending on the type of biomass, the optimal temperature range can go up to 200°C (Aftab et al., 2019). The common organic solvents used for the process are ethanol, methanol, alcohol, triethylene glycol, ethylene glycol, and acetone, while the organic acids oxalic, salicylic, and acetyl salicylic acid are catalysts to the organosolv pretreatment process. The solvents must be removed before the fermentation process because organic solvents themselves can act as inhibitors for enzymatic hydrolysis and fermentation steps (Harmsen et al., 2010). The benefits of the organosolv pretreatment method are easy solvent recovery and the possibility of its recycling, mild reaction conditions, high efficiency, high purity of biomass fractions, and the possibility of their separation. The limitation of this treatment is the high cost of organic solvents and their recovery. In a study, wheat straw was pretreated with 60% ethanol at 200°C for 120 min, resulting in maximum lignin recovery of >90% in pretreated hydrolysate (Ramezani & Sain, 2018). In another study, sugarcane pretreated with 50% (v/v) ethanol at 190°C for 150 min resulted in 69.3% cellulose content after organosolv in comparison to 38.3% in untreated biomass, 9.1% hemicellulose (20.1% in untreated biomass), and 14.6% lignin content after pretreatments compared to 29% content in untreated biomass (Santo et al., 2018).

3.3.2.4 Ionic Liquids (IL)

Ionic liquids (IL) are the salts made up of both anionic and cationic species having a melting point below a certain temperature, i.e., 100°C. IL stays liquid at room temperature, having high viscosity, high thermal and chemical stability, low vapor pressure, and low conductivity, and is easily recyclable, and these properties make them green solvents (Paul et al., 2020). ILs are polar in nature, which makes them capable of dissolving and fractionating the wood biomass for other useful purposes (Ziaei-Rad et al., 2021). In a study, spruce sawdust and oak sawdust were pretreated with 98% (w/v) 1-ethyl-3-methylimidazolium acetate followed by hydrolysis using cellulases from *Trichoderma reesei* at 40°C for 80 h. The ethanol yields were significantly increased after 1-ethyl-3-methylimidazolium acetate pretreatment by a factor of 2.6, 2.8, and 3.9 for cellulose, spruce sawdust, and oak sawdust, respectively (Alayoubi et al., 2020). Ziaei-Rad et al. (2021) reported a good solubilization of crystalline cellulose and high saccharification efficiency of wheat straw after pretreatment with triethyl ammonium hydrogen sulfate at 20% (w/v) solid loading at 130°C for 3 h, followed by hydrolysis at 5% (w/v) solid loading using cellulase and xylanase enzymes with a loading of 28 filter paper units (FPU)/g pretreated biomass. Glucose, xylose, and ethanol yields were calculated as 87.2%, 78.2%, and 84.3%, respectively, after 84 h of fermentation. In addition to the high pretreatment efficiency, ILs further have the ability to reduce the expenditure of the process and are safe, noncorrosive, and nontoxic, unlike other chemicals used in chemical pretreatment methods.

3.3.3 Biological Pretreatment

In this method, a wide range of naturally present microorganisms are being used for the pretreatment of LCB. They degrade or alter the cell wall of the lignocellulose biomass by secreting ligninolytic enzymes, which depolymerize the lignin. This degradation is further followed by hydrolysis and fermentation processes. Biological pretreatment is influenced by several factors like temperature, moisture, incubation time, substrate size, aeration, surface area, pH, composition of culture media, source of nitrogen, source of carbon, organic and inorganic compounds, enzymes, hydrolysates, and cellulose crystallinity, which impacts the efficiency of degradation of the cell wall of LCB. Lignin degradation is well studied in fungi and much less in bacteria (Sharma et al., 2017). Cellulolytic bacteria *Cellulomonas fimi* and Thermomonospora degrade nonphenolic lignin by converting Mn^{+2} into Mn^{+3}, which makes a complex with chelators such as oxalates or malonate (Wong, 2009).

3.4 ANALYSIS OF EFFECTS OF PRETREATMENT ON BIOMASS

LCB is rebellious in nature for biodegradation due to the arrangement and composition of the cell wall. Lignin present in LCB makes it more recalcitrant and robust in nature, as it hinders the enzymatic hydrolysis and saccharification of the carbohydrates (cellulose and hemicellulose). Delignification is the most important step in the pretreatment of biomass, as it allows the enzymes to work efficiently and exposes the sugars (hexoses and pentoses) and makes them more available for the further fermentation process. From Table 3.1, it is clear that the physical pretreatment method alone is not very efficient in contrast to other pretreatment methods, but it is more successful when combined with other methods. On the other hand, alkaline pretreatment is more efficient and successful as compared to the acid pretreatment for example a study given in Table 3.1, when paddy straw was pretreated by 5% (w/v) sodium hydroxide, 20% (w/v) solid loading at 90°C for 20 min it gave 79.66% increase in cellulose content, 22.05% decrease in hemicellulose content, 84.1% decrease in lignin content, and 0.47 g/g ethanol yield by using 25 FPU cellulase/g at 10% loading and 0.31:0:69 *Saccharomyces cerevisae* and *Scheffersomyces stipitis* at 33.1°C for 72 h, whereas paddy straw pretreated using 0.5% (v/v) sulfuric acid and 4% (w/v) sodium hydroxide resulted in 59.57% increase in cellulose content, 38.3% decrease in hemicellulose content and 90.41% decrease in lignin content (Suriyachai et al., 2013). It shows that acid pretreatment followed by alkaline pretreatment (a two-step process) efficiently removes lignin from the biomass, but this process is very costly, energy-consuming, and time-consuming, whereas alkali pretreatment removes lignin selectively with partial removal of hemicellulose by saponification of ester bonds and expands the surface area of biomass, which makes the cellulases more accessible (Kim et al., 2016). The two-step process also generates solubilized lignin degradation products, which make complexes with pentose sugars, and it is not easy to separate them to make some other valuable product. Their separation requires more additional steps, making the whole process more expensive and time-consuming (Hans et al., 2019). Due to some drawbacks of thermochemical pretreatments and the hazardous nature of chemicals, which is not suitable for a greener and safer environment, biological pretreatment is considered to be the most environmentally friendly and economical, with lower energy requirements. Unlike other methods, people are more interested in this approach. White-rot fungus, e.g., *Trametes hirsute,* has an elaborate lignolytic enzyme system, which delignifies LCB, and is the most effective biological pretreatment, but its efficiency is low. Furthermore, *Pleurotus sajor-caju* is more efficient than *Phanerochaete chrysosporium* in the delignification of LCB (Yadav & Wati, 2016). Despite all other things, choosing a pretreatment method also depends upon the nature and species and also depends on the region from where we collect the LCB.

3.5 ANALYTICAL TECHNIQUES FOR PRETREATMENT EFFICIENCY OF BIOMASS

To maximize the transformation of biomass into chemicals, biofuels, and other products with added value, compositional characterization is the crucial factor. Comprehending the chemical and structural makeup of biomass is essential for developing effective processing techniques, raising yields, and cutting expenses. This paper gives a broad overview of current developments in biomass compositional analysis, emphasizing different approaches and their uses. To precisely characterize the components of biomass, comprehend how they interact during processing, and increase the overall efficiency of biomass conversion technologies, advanced analytical techniques are necessary. Some analytical techniques used for pretreatment efficiency of biomass are represented in Figure 3.2. To improve the use of biomass as a renewable resource, further research in this field is required (Deng et al., 2023).

3.5.1 Van Soest Method (Fiber Estimation)

A popular approach for estimating the amount of fiber in biomass, especially in the fields of agriculture and forage sciences, is the Van Soest method. The neutral detergent fiber (NDF) and acid detergent fiber (ADF), which are important markers of the lignocellulosic content of biomass, can only be evaluated by using this approach, which is crucial for figuring out the composition of plant materials. The idea underlying the detergent fiber analysis is that plant cells may be separated into mainly digestible cell contents (starch and sugars) and less digestible cell walls (hemicellulose, cellulose, and lignin). Van Soest used two detergents, one acidic (cetyl trimethyl ammonium bromide in 1 N H_2SO_4) and the other neutral (sodium

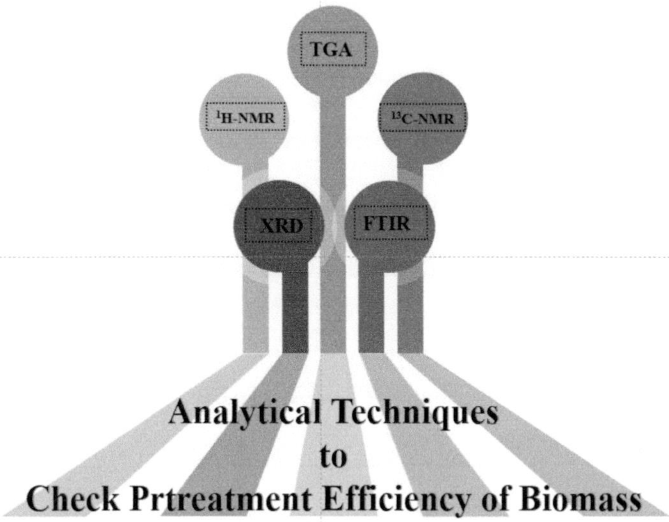

FIGURE 3.2 Techniques for compositional analysis of biomass (1HNMR: Proton Nuclear Magnetic Resonance; 13C-NMR: Nuclear Magnetic Resonance; XRD: X-ray Diffraction Crystallography; FTIR: Fourier Transform Infrared Spectroscopy).

lauryl sulfate-EDTA, pH = 7.0), to successfully separate these two components. Applications of biomass in the production of bioenergy, animal feeding, and the research of plant structural integrity all depend on an understanding of the composition of fiber. Khudyakova et al. (Khudyakova & Kosolapova, 2022) examined the Van Soest method for figuring out the amounts of lignin and structural carbohydrates in fodder. Their research shed important light on the benefits and drawbacks of the technique, especially when it came to assessing the quality of the fodder. The Van Soest approach provides a thorough understanding of fiber components, making it the recommended option for a thorough fiber analysis. Dampanaboina et al. (2021) investigated and contrasted the Van Soest technique with the thioglycolic acid (TGA) method for estimating the lignin concentration of plant biomass. Their research demonstrated the accuracy of the technique in determining total lignin concentration, which is important for comprehending plant resistance to degradation and for use in the manufacturing of biofuels. It is essential for agricultural research, animal nutrition, and bioenergy applications due to its precision in determining the composition of NDF, ADF, and lignin. A popular procedure for determining the amount of fiber in plant materials is the Van Soest technique, sometimes referred to as the detergent fiber analysis method. Neutral detergent fiber (NDF), acid detergent fiber (ADF), and acid detergent lignin (ADL) are the three primary components of plant fibers that are separated using this approach (Analytical Techniques in Aquaculture Research). NDF is the whole cell wall, which consists of lignin, hemicellulose, and cellulose. It serves as a gauge of the digestibility of feed. ADF indicates the sample's lower digestible cellulose and lignin concentration compared to hemicellulose, whereas ADL is the amount of highly indigestible lignin, which influences the forager's overall digestibility. Determination of NDF, ADF, and ADL followed by cellulose, hemicellulose, and lignin is shown in equations (i–vi) (Van Soest et al., 1991).

$$\text{NDF (\%)} = \text{weight of NDF residue / weight of original sample} \times 100 \quad \text{(i)}$$

$$\text{ADF(\%)} = \text{weight of ADF residue/weight of original sample} \times 100 \quad \text{(ii)}$$

$$\text{ADL(\%)} = \text{weight of ADL residue/weight of original sample} \times 100 \quad \text{(iii)}$$

Further,

$$\text{Cellulose (\%)} = \text{ADF-ADL} \quad \text{(iv)}$$

$$\text{Hemicellulose (\%)} = \text{NDF-ADF} \quad \text{(v)}$$

$$\text{Lignin (\%)} = \text{ADL} \quad \text{(vi)}$$

3.5.2 National Renewable Energy Laboratory (NREL) Method

The method from the National Renewable Energy Laboratory, U.S.A. (NREL), is a widely used standardized analytical approach for lignocellulosic biomass compositional characterization. The manufacture of biofuel and other bioprocessing applications depends on the precise measurement of structural carbohydrates like cellulose and hemicellulose, which is the main emphasis of this technology. The process involves drying a biomass sample, grinding it, and weighing it. The sample undergoes acid hydrolysis, which involves adding 72% sulfuric acid to the sample and reducing it to a 4% concentration. The mixture is then autoclaved at 121°C for 1 h. The hydrolysate is then filtered and washed, and the residue on the filter is analyzed for acid-insoluble lignin. The inorganic content (ash) is then subtracted from the lignin value. The hydrolysate contains sugars from cellulose and hemicellulose, which are then analyzed using high-performance liquid chromatography (HPLC) equipped with a refractive index detector (RID) for sugar content. The final calculations are based on the glucose, xylose, and arabinose content. Equations (vii–x) are used for compositional analysis (Sluiter et al., 2010).

Cellulose

$$\% = \text{Total glucose present in the sample detected by HPCL} \times \text{volume of filtrate/cellulose present} \quad \text{(vii)}$$

Hemicellulose

$$\% = 0.88 \text{ filtrate volume} \times \text{xylose concentration/sample weight} + (0.88 \times \text{filtrate volume} \times \text{arabinose concentration/sample weight}) \quad \text{(viii)}$$

Lignin soluble

$$\% = \text{Total glucose present in the sample detected by HPLC} \times \text{volume of filtrate / cellulose present} \quad \text{(ix)}$$

Lignin insoluble

$$\% = \text{recovered weight / sample weight} \quad \text{(x)}$$

The NREL method's latest developments and uses in biomass analysis are covered in the present review. Solid-state NMR spectroscopy was used by Yuan et al. (2022) to quantify the level of polymerization of hydrolyzed cellulose directly. Although this approach offers a thorough grasp of the structure of cellulose, it also enhances the overall accuracy of carbohydrate measurement by providing a better understanding of the molecular properties of biomass, which is a complement to the NREL method. Ghani et al. (2023) created a sorbent based on cellulose nanofibers that may be used in high-performance liquid chromatography to extract and quantify chlorophenols. The NREL method's adaptability in a variety of applications for biomass compositional analysis was demonstrated by its crucial role in the initial characterization of the cellulose utilized in the manufacture of the nanofibers. A mixture of graphene oxide and cellulose nanofibril for electrochemical sensors was studied by Wang et al. (2023). To maximize the material's qualities and performance in sensing applications, the cellulose content of the composite was measured using the NREL technique. Saleh et al. (2023) conducted a comparative analysis of analytical techniques for the quantification of proteins with varying glycosylation levels. The NREL technique was emphasized as a standard for carbohydrate measurement, despite the study's focus on proteins, highlighting its dependability and extensive application in analytical chemistry. In summary, the gold standard for estimating structural carbohydrates in biomass is still the NREL approach. Because of its accuracy and adaptability, environmental chemistry, materials science, and bioenergy researchers cannot do without it. In order to satisfy the increasing needs of biomass usage and bioprocessing technologies, future research should keep improving and broadening its applications.

3.5.3 X-Ray Diffraction Crystallography (XRD)

To identify the crystalline and amorphous areas of cellulose, X-ray diffraction crystallography (XRD) is a frequently utilized technique. Cellulose usually exists in two forms in biomass: crystalline (very organized) and amorphous (disordered). The enzymatic hydrolysis efficiency during the manufacture of biofuel is influenced by the degree of crystallinity, making it a crucial metric. Lower enzymatic digestibility is typically associated with higher crystallinity. Researchers may measure the decrease in crystallinity following pretreatment using X-ray reflectance (XRD), which sheds light on how successful the pretreatment procedure was (Abolore et al., 2023). The proportion of crystalline cellulose in a biomass sample is indicated by the Crystallinity Index (CrI), a metric obtained from XRD data. The ratio of the crystalline peaks' intensity to the XRD pattern's overall intensity (crystalline and amorphous peaks) is used to compute CrI. Researchers may assess how various pretreatment techniques impact the crystalline structure of cellulose by comparing the CrI before and after pretreatment. A reduction in CrI usually signifies an effective pretreatment procedure that has improved the amorphous content, raising the reactivity of the biomass during the enzymatic hydrolysis or fermentation stages that follow (Cheng et al., 2011). X-ray diffraction (XRD) is a commonly used technique for cellulose analysis, but it may also be used to monitor structural alterations in lignin and hemicellulose that occur during pretreatment. XRD can identify changes in these components' structures that may have an impact on the biomass's total crystallinity, despite the fact that they are less crystalline than cellulose. Modifications in the XRD patterns of the treated biomass may be used as an indirect indicator of the removal or alteration of lignin and hemicellulose caused by pretreatments such as organosolv or acid hydrolysis. These modifications can help determine how well these elements have been altered or eliminated, improving cellulose accessibility (Gong et al., 2018).

3.5.4 Fourier Transform Infrared Spectroscopy (FTIR)

A strong analytical method for determining a solid, liquid, or gas's infrared spectrum of absorption, emission, or photoconductivity is FTIR. For the identification and study of chemical substances, especially organic molecules, FTIR is frequently utilized. The interaction of infrared light with the sample's molecular vibrations is the basis for the FTIR technique. FTIR is used for identifying functional groups in organic molecules, quantifying component concentrations, elucidating sample molecular structure, and real-time monitoring of chemical reactions, allowing for precise analysis and understanding of sample composition (Novais et al., 2019). Molecular vibrations: Atoms bound together to form molecules are always in motion and vibrate in different ways. Stretching (a change in bond length) or bending (a change in bond angle) are two possible types of vibrations. If the energy of an infrared ray coincides with one of the molecule's vibrational modes, the energy from the radiation may be absorbed by the molecule. The molecule vibrates more violently as a result of its absorption (Zhang et al., 2020). FTIR uses polychromatic infrared radiation from a source, split into two paths by a Michelson interferometer. The interferogram, containing information about absorbed frequencies, is created when the two beams recombine. The Fourier transform converts the interferogram into the frequency domain, producing an IR spectrum showing absorption intensity at each wavelength (Jaggi et al. & Viji, 2006). FTIR is frequently employed to determine and describe the chemical structures found in biomass. When it comes to pretreatment of biomass, FTIR is especially useful for tracking chemical alterations in cellulose, hemicellulose, and lignin. These modifications are essential for enhancing the productivity of biofuel generation and other biomass conversion procedures (Hashimoto & Kameoka, 2008). The chemical makeup of biomass is frequently examined using FTIR, both before and after processing. The functional groups found in cellulose, hemicellulose, and lignin are described in depth, which aids in determining the degree of chemical changes brought on by pretreatment. For instance, following an alkaline pretreatment, FTIR is utilized to identify alterations in the functional groups connected to lignin, such as methoxy groups and aromatic rings. In order to improve the digestibility of biomass, lignin must be removed or transformed, as indicated by the decrease or alteration of these groups (Abolore et al., 2023). Further, the efficacy of pretreatment techniques in eliminating or changing lignin structures may be observed using FTIR. Through monitoring distinct absorbance bands linked to lignin, scientists may measure the degree of lignin elimination. An example of effective lignin removal is shown by the FTIR examination of biomass treated with organosolv or acid pretreatment techniques, which shows the elimination or decrease of lignin-specific bands (e.g., at 1500–1600 cm^{-1} for aromatic ring vibrations) (Mariana et al., 2021). By comparing the ratio of certain absorbance bands that represent crystalline and amorphous areas, FTIR may be used to measure the crystallinity index (CrI) of cellulose. This is especially helpful in understanding how cellulose's crystalline structure, which affects its reactivity and digestibility, is affected by pretreatment techniques. For instance, the CrI is determined by analyzing changes in the ratio of absorbance at 1420 cm^{-1} (crystalline cellulose) to 898 cm^{-1} (amorphous cellulose) in FTIR spectra. The CrI normally decreases following an efficient pretreatment (Park et al., 2010). FTIR also enables the detection of certain functional groups, including ester, carboxyl, and hydroxyl groups, in biomass. Understanding the chemical changes brought about by pretreatment and how they affect later processes of converting biomass to energy requires knowledge of this. Following pretreatment, FTIR may be used to track the chemical makeup of products obtained from biomass, such as charcoal and bio-oils. This is essential for assessing these items' quality and possible uses. FTIR offers vital information on the chemical alterations that take place in cellulose, hemicellulose, and lignin, which aids in the improvement of pretreatment procedures for increased biofuel production and other uses. Functional group transformations, lignin removal, crystallinity alterations, and hemicellulose degradation are all monitored by FTIR, which makes a substantial contribution to the development of technology for biomass conversion. The FTIR technique utilized in analyzing parameters of biomass pretreatment is summarized in Table 3.2. FTIR has been used in recent studies on biomass pretreatment, and this technique has given important new insights into the structural and chemical changes that take place throughout different pretreatment procedures. The effects of steam explosion pretreatment on rice husk were evaluated by Bhatia et al. (Bhatia Sahu, 2023) using FTIR in addition to scanning electron microscopy (SEM). They discovered substantial decreases in cellulose crystallinity and lignin re-localization, which improved the biomass's enzymatic digestibility. Rahman et al. (2021) concentrated on the use of FTIR for the compositional study of agricultural waste, discovering important functional groups like carboxyl and hydroxyl acids that are crucial for comprehending the chemical makeup of lignocellulosic biomass. Furthermore, Petrovič et al. (2021) highlighted alterations in surface functional groups that affect nutrient content and biosorption ability by using FTIR to characterize biochars produced from the pyrolysis of solid digestates. Using FTIR to detect changes in surface characteristics and elemental composition, Tsai et al. (2023) examined the effects of alkaline pretreatment on rice husk, which eventually resulted in improved fuel qualities. All of this research shows how important FTIR is for clarifying how different

TABLE 3.2
XRD, FTIR, ^{13}C-NMR, ^{1}H-NMR Analysis of Pretreated Biomass in Literature

Analytical Techniques Used	Applications	Biomass Type	Pretreatment Method	References
Measurement of cellulose crystallinity index (CrI) using XRD	Lower CrI improves enzymatic hydrolysis efficiency	Lignocellulosic biomass	Multiple pretreatment methods	Park et al., 2010
Comparison of crystallinity determination methods	XRD is reliable for assessing crystalline and amorphous phases	General cellulose samples	NA	Salem et al., 2023
Estimation of crystallinity in cellulose using XRD and FTIR	Changes in crystallinity depend on milling conditions	Banana, commercial cellulose, bacterial cellulose	Mechanical milling	Montoya-Escobar et al., 2022
Analysis of structural changes during ionic liquid pretreatment	Ionic liquids disrupt crystalline structure, improving hydrolysis	Lignocellulosic biomass	Ionic liquid pretreatment	Zhang et al., 2014
Correlation of chemical pretreatment to digestibility	Alkali pretreatment reduces crystallinity, enhancing enzymatic hydrolysis	Corn cobs	Alkali pretreatment	Sunkar et al., 2022
Correlation of chemical pretreatment to digestibility	FTIR was used to analyze chemical changes in corn cobs, showing that 2% NaOH was effective for delignification and improving digestibility	Corn cobs	Alkaline pretreatment	Sunkar et al., 2022
Analysis of functional groups using FTIR	FTIR was used to compare conventional and microwave-assisted acid pretreatment of tea waste, revealing changes in functional groups	Tea waste powder	Acid pretreatment	Hamzah et al., 2023
FTIR analysis to assess the impact of physicochemical pretreatment	FTIR was used to evaluate structural changes in rice husk after steam explosion pretreatment, showing enhanced cellulose accessibility	Rice husk	Steam explosion with NaOH and HNO3	Bhatia & Sahu, 2023

(Continued)

TABLE 3.2 (*Continued*)
XRD, FTIR, ^{13}C-NMR, ^{1}H-NMR Analysis of Pretreated Biomass in Literature

Analytical Techniques Used	Applications	Biomass Type	Pretreatment Method	References
Evaluation of surface properties and elemental composition	FTIR was utilized to observe changes in surface properties and elemental composition of rice husk after alkaline pretreatment	Rice husk	Alkaline pretreatment	Tsai et al., 2023
FTIR in cellulose nanofiber production	FTIR was used to monitor chemical changes during hydrothermal pretreatment of garlic skins for cellulose nanofiber production	Garlic skins	Hydrothermal pretreatment	Yeganeh et al., 2022
Structural analysis of lignin	^{13}C-NMR revealed significant changes in lignin structure after pretreatment, particularly in the reduction of β-O-4 linkages	Poplar wood	Organosolv pretreatment	Pu et al., 2013
Characterization of hemicellulose degradation	^{13}C-NMR identified the breakdown of hemicellulose side chains, particularly in the acetyl groups, after alkaline pretreatment	Wheat straw	Alkaline pretreatment	Li et al., 2012
NMR characterization of lignin in pretreated biomass quantitative and qualitative analysis	Degradation of lignin β-O-4 linkages during acid pretreatment and formation of new structures under alkaline conditions.	Corn Stover	Acid and alkaline pretreatment	Pu et al., 2011
In situ ^{1}H-NMR spectroscopy for monitoring biomass pretreatment: a novel approach	Real-time observation of degradation products such as furfural and acetic acid during acid	Wheat Straw	Acid pretreatment	Maitra et al., 2021
Understanding biomass dissolution in ionic liquids using ^{1}H-NMR spectroscopy	Dissolution of lignocellulosic components, with ionic liquids disrupting cellulose hydrogen bonds and solubilizing lignin.	Miscanthus	Ionic liquid pretreatment	Ben et al., 2018

(*Continued*)

TABLE 3.2 (*Continued*)
XRD, FTIR, ^{13}C-NMR, ^{1}H-NMR Analysis of Pretreated Biomass in Literature

Analytical Techniques Used	Applications	Biomass Type	Pretreatment Method	References
Quantitative ^{1}H-NMR analysis of monosaccharides in biomass hydrolysates	Rapid and accurate quantification of monosaccharides, aiding in evaluating the efficiency of pretreatment	Sugarcane Bagasse	Acid pretreatment	Mittal et al., 2009
Investigating deep eutectic solvents for biomass pretreatment using ^{1}H-NMR spectroscopy	Selective solubilization of lignin over cellulose, indicating potential for efficient biomass processing.	Eucalyptus	Deep eutectic solvent (DES)	Degam et al., 2017

pretreatment techniques affect biomass, which helps to produce biofuels more sustainably and efficiently.

3.5.5 Nuclear Magnetic Resonance (^{13}C-NMR)

An effective analytical method for examining the molecular makeup and structure of organic materials, including lignocellulosic biomass, is nuclear magnetic resonance (13C-NMR) spectroscopy. 13C-NMR offers comprehensive insights into the structural alterations and chemical transformations of cellulose, hemicellulose, and lignin in the context of biomass pretreatment. These insights are essential for improving the generation of biofuels and other biomass conversion processes. The interaction between the magnetic moments of atomic nuclei and an external magnetic field is the basis of NMR spectroscopy, in particular. 13C-NMR primarily studies carbon-13, a naturally occurring isotope of carbon with a natural abundance of around 1.1% (Ur Rehman et al., 2022). The feature of atomic nuclei known as spin may be understood as the nucleus rotating around its axis. In the same way that a spinning charged particle produces a magnetic field, its spin produces a magnetic moment. When exposed to an external magnetic field, in 13C-NMR, the carbon-13 nucleus may exist in two different spin states, aligned with the field (lower energy state) or opposed to the field (higher energy state). This is due to its nuclear spin quantum number of ½ (Levitt, 2008). 13C-NMR may be used to track the breaking of ether bonds and the removal of methoxyl groups from lignin during pretreatment procedures like organosolv or alkaline treatments. These are the crucial stages in lowering the recalcitrance of biomass. Additionally, the method can differentiate between various lignin substructures, such as p-hydroxyphenyl, guaiacyl, and syringyl units, which aids in clarifying how pretreatment affects lignin composition (Hans et al., 2023). The most prevalent polysaccharide in biomass, cellulose has both crystalline and amorphous areas, which affect how reactive it is when hydrolyzed by enzymes. 13C-NMR is a useful technique for measuring the amount of cellulose crystalline index by examining the chemical shifts connected to carbon atoms in various structural configurations. For instance, 13C-NMR can distinguish between crystalline and amorphous areas in cellulose, such as the C4 carbons in the glycosidic bond. 13C-NMR can monitor the decrease in cellulose crystallinity after pretreatment, such as steam explosion or acid hydrolysis, which usually improves its accessibility to enzymes during subsequent hydrolysis. This data is essential for maximizing the efficiency of biofuel production by adjusting pretreatment conditions (Salem et al., 2023). 13C-NMR can give precise details on the structural alterations in hemicellulose that occur during pretreatment, such as the disintegration of its side-chain sugars and backbone. The method aids in comprehending how various pretreatment settings impact hemicellulose's composition and structure, which in turn affects the process's overall efficiency of converting biomass (Jincy & Femina, 2023).

3.5.6 Proton Nuclear Magnetic Resonance (1H-NMR)

In order to better understand biomass pretreatment, one potent analytical method that is frequently utilized is proton nuclear magnetic resonance (1H-NMR). This technique uses the magnetic characteristics of protons, or hydrogen nuclei, to reveal intricate details about the interactions, dynamics, and molecular structure of a material. 1H-NMR may be used in a biomass pretreatment environment in a number of ways. The main components of biomass, such as cellulose, hemicellulose, and lignin, may be identified and measured using ^{1}H-NMR. These molecules' unique hydrogen atom chemical environments provide distinctive NMR signals that enable researchers to examine the composition and structural alterations in the biomass both before and after treatments (Sathitsuksanoh & Renneckar, 2017). The pretreatment of biomass, such as by acidic, alkaline, or enzymatic techniques, significantly alters the biomass's chemical structure. Small molecules and breakdown products, including acetic acid, furfural, and hydroxymethylfurfural (HMF), which are markers of the pretreatment process's efficacy and adverse effects, may be found using 1H-NMR (Singh et al., 2024). The use of ^{1}H-NMR in assessing different substrates during biomass pretreatment is highlighted in Table 3.2, which also contains information on the biomass substrate, pretreatment technique, key findings, and references for each study.

3.6 ADVANCEMENTS AND CHALLENGES IN ANALYSIS OF PRETREATED BIOMASS

The advancements and challenges pertaining to the analytical methods (XRD, FTIR, ^{13}C-NMR, and ^{1}H-NMR) utilized in biomass pretreatment are summarized in Table 3.3. This table provides a concise overview of the current state of each technique, highlighting both the advancements made in biomass pretreatment analysis and the ongoing challenges faced by researchers.

TABLE 3.3
Advancements and Challenges Pertaining to the Analytical Methods (XRD, FTIR, ^{13}C-NMR, and ^{1}H-NMR) Utilized in Biomass Pretreatment

Analytical Technique	Advancements	Challenges
X-ray diffraction (XRD)	Improved crystalline index analysis of pretreated biomass. • Enhanced understanding of structural changes, particularly in response to different pretreatment chemicals (e.g., ionic liquids).	Difficulty in interpreting complex and heterogeneous biomass structures. • Limited sensitivity to amorphous regions of biomass.
Fourier transform infrared spectroscopy (FTIR)	Advanced chemical characterization of lignin and hemicellulose removal. • Correlation of chemical changes with biomass digestibility.	Overlap of absorption bands makes it challenging to assign specific functional groups. • Requires careful baseline correction and normalization for quantitative analysis.
^{13}C-Nuclear Magnetic Resonance (^{13}C-NMR)	Quantitative analysis of carbon functional groups in biomass. • Detailed structural analysis of lignins and hydrochars.	Long acquisition times needed for high-resolution spectra. • Complexity in interpreting overlapping signals in heterogeneous samples.
^{1}H-Nuclear Magnetic Resonance (^{1}H-NMR)	Development of non-invasive, high-throughput methods for lipid and water content analysis in biomass. • Ability to study changes in cellulose and other biomass components during pretreatment.	Sensitivity to water and solvent content in samples, which can complicate analysis. • Interpretation challenges due to overlapping proton signals in complex biomass matrices.

3.7 CONCLUSION

Lignin present in LCB is a major bottleneck in 2G biofuels technology because it inhibits the hydrolysis of cellulose and hemicellulose, which decreases the efficiency of the whole process. It led to the evolution of various pretreatment processes based on physical, chemical, and biological principles. No one pretreatment method is 100% efficient at converting biomass into the fermentable sugars. Combining one or more pretreatment processes gives more efficient outcomes; for example, physical pretreatment alone does not give desirable results and also consumes very high energy, which makes it costlier, but when combined with chemical processes, it gives high efficiency and increases the overall yield. Out of the different pretreatment technologies, chemical pretreatment and thermo-chemical pretreatment are more efficient but must include some advances to reduce the toxic effects on the environment. Various analytical techniques like X-ray diffraction, FTIR, NMR, and compositional analyses are used to assess the effects of pretreatment on biomass structure. New efficient pretreatment processes need to be developed, or existing pretreatment methods may be upgraded to give the promising results with lesser cost, environmental friendliness, lesser biomass loss, and higher yield of products.

ACKNOWLEDGMENT

Authors gratefully acknowledge Guru Nanak Dev University, Amritsar for providing research facilities. Meenu Hans is also grateful to DST-ANRF for providing funds to carry out the necessary research work via grant no. EEQ/2023/000427. Suman Salhotra and Pragati Bhardwaj are thankful to Guru Nanak Dev University, Amritsar for full time Ph.D in Department of Biotechnology.

REFERENCES

Abolore, R. S., Jaiswal, S., Jaiswal, A. K. 2023. Green and sustainable pretreatment methods for cellulose extraction from lignocellulosic biomass and its applications: A review. Carbohydrate Polymer Technologies and Applications, 100396. https://doi.org/10.1016/j.carpta.2023.100396

Aftab, M. N., Iqbal, I., Riaz, F., Karadag, A., & Tabatabaei, M. 2019. Different pretreatment methods of lignocellulosic biomass for use in biofuel production. In Biomass for bioenergy-recent trends and future challenges. IntechOpen. http://dx.doi.org/10.5772/intechopen.84995

Alayoubi, R., Mehmood, N., Husso, E., ouzayha, A., Tabcheh, M., Chaveriat., L., Sarazin, C., Gosselin, I. 2020. Low temperature ionic liquid pretreatment of lignocellulosic biomass to enhance bioethanol yield. Renewable Energy, 145(2020), 1808–1816. https://doi.org/10.1016/j.renene.2017.07.091

Anu, Kumar, V., Singh., D., Singh, B. 2021. A green, mild, and efficient bioprocess for the pretreatment and saccharification of rice straw. Biomass Conversion and Biorefinery, 13(2023), 4121–4133. https://doi.org/10.1007/s13399-021-01450-9

Arora, A., Priya, S., Sharma, P., Sharma, S., Nain, L. 2016. Evaluating biological pretreatment as a feasible methodology for ethanol production from paddy straw. Biocatalysis and Agricultural, 8(2016), 66–72. http://dx.doi.org/10.1016/j.bcab.2016.08.006

Ben, H., Chen, X., Han., G., Shao, Y., Jiang, W., Pu, Y., Ragauskas, A. J. 2018. Characterization of whole biomasses in pyridine based ionic liquid at low temperature by 31P NMR: An approach to quantitatively measure hydroxyl groups in biomass as their original structures. Frontiers in Energy Research, 6, 13. DOI: 10.3389/fenrg.2018.00013.

Bhatia, L., Sahu., D. K. 2023. SEM & FTIR analysis of rice husk to assess the impact of physiochemical pretreatment. Journal of Agriculture and Ecology Research International, 24, 1–13. DOI: 10.9734/JAERI/2023/v24i6556.

Chen, J., Adjallé, K., Lai, T. T., Barnabé, S., Perrier, M., Paris, J. 2019. Effect of mechanical petreatment for enzymatic hydrolysis of woody residues, Corn Stover and Alfalfa. Waste and Biomass Valorization, 11(11), 5847–5856. https://doi.org/10.1007/s12649-019-00856-x

Cheng, G., Varanasi, P., Li, C., Liu, H., Melnichenko, Y. B., Simmons, B. A., Kent, M. S., Singh, S. 2011. Transition of cellulose crystalline structure and surface morphology of biomass as a function of ionic liquid pretreatment and its relation to enzymatic hydrolysis. Biomacromolecules, 12(4), 933–941. DOI: 10.1021/bm101240z.

Dampanaboina, L., Yuan, N., Mendu, V. 2021. Estimation of plant biomass lignin content using thioglycolic acid (TGA). JoVE (Journal of Visualized Experiments) (173), e62055. DOI: 10.3791/62055 (2021).

Degam, G. 2017. Deep eutectic solvents synthesis, characterization and applications in pretreatment of lignocellulosic biomass. https://openprairie.sdstate.edu/etd/1156

Deng, W., Feng, Y., Fu, J., Guo, H., Guo, Y., Han, B. et al. 2023. Catalytic conversion of lignocellulosic biomass into chemicals and fuels. Green Energy & Environment, 8(1), 10–114. https://doi.org/10.1016/j.gee.2022.07.003

Ghani., M., Khodkavandi, S., Jafar, Z., Maleki, B., Tabari, H. F. 2023. Synthesis of cellulose nanofibers-based ImSalophen@ Fe3O4 as a green sorbent for magnetic solid-phase extraction of chlorophenols followed by quantification via high-performance liquid chromatography-ultraviolet detection. Microchemical Journal, 187, 108368. https://doi.org/10.1016/j.microc.2022.108368

Gong, J., Mo, L., Li, J. 2018. A comparative study on the preparation and characterization of cellulose nanocrystals with various polymorphs. Carbohydrate Polymers, 195, 18–28. https://doi.org/10.1016/j.carbpol.2018.04.039

Gu, Y. M., Kim, S., Sung, D., Sang., B., Lee, J. H. 2019. Feasibility of continous pretreatment of Corn Stover: A comparison of three commercially available continous pulverizing devices. Bioenergies, 12, 1422. DOI: 10.3390/en12081422.

Hamzah, H. T., Sridevi, V., Surya, D. V., Palla, S., Yadav, A., Rao, P. V. 2023. Conventional and microwave-assisted acid pretreatment of tea waste powder: Analysis of functional groups using FTIR. Environmental Science and Pollution Research, 31(46), 57523–57532 1–10. https://doi.org/10.1007/s11356-023-28272-8

Hans, M., Garg, S., Pellegrini, V. O. A., Filguerias., J. G., de Azevedo, E. R., Guimares, F. E. C., Chandel, A. K., Polikarpov, I., Chadha, B. S., Kumar, S. 2021. Liquid ammonia pretreatment optimization for improved release of fermentable. Journal of Cleaner Production, 281(2021), 123922. https://doi.org/10.1016/j.jclepro.2020.123922

Hans, M., Kumar, S., Chandel, A. K., Polikarpov, I. 2019. A review on bioprocessing of paddy straw to ethanol using simultaneous T saccharification and fermentation. Process Biochemistry, 85(2019), 125–134. https://doi.org/10.1016/j.procbio.2019.06.019

Hans, M., Pellegrini, V. O., Filgueiras, J. G., de Azevedo, E. R., Guimaraes, F. E., Chandel, A. K., Kumar, S. 2023. Optimization of dilute acid pretreatment for enhanced release of fermentable sugars from sugarcane bagasse and validation by biophysical characterization. BioEnergy Research, 16(1), 416–434. https://doi.org/10.1007/s12155-022-10474-6

Hans, M., Pellegrni, V. O. A., Filgueiras, J. G., de Azevedo, E. R., Guimaraes, F. E. C., Chandel, A. K., Polikarpov, I., Chadha, B. S., Kumar, S. 2022. Optimization of dilute acid pretreatment for enhanced release of fermentable sugars from sugarcane bagasse and validation by biophysical characterization. BioEnergy Research. https://doi.org/10.1007/s12155-022-10474-6

Harmsen, P., Huijgen, W., Bermudez, L., Bakker, R. 2010. Literature review of physical and chemical pretreatment processes for lignocellulosic biomass. Wageningen UR Food & Biobased Research, 1184. https://edepot.wur.nl/150289

Hashimoto, A., Kameoka, T. 2008. Applications of infrared spectroscopy to biochemical, food, and agricultural processes. Applied Spectroscopy Reviews, 43(5), 416–451. https://doi.org/10.1080/05704920802108131

Jaggi, N., Vij, D. R. 2006. Fourier transform infrared spectroscopy. In Handbook of Applied Solid State Spectroscopy (pp. 411–450). Boston, MA: Springer US. https://doi.org/10.1007/0-387-37590-2_9

Jędrzejczyk, M., Soszka., E., Czapnik, M., Ruppert, A. M., Grams, J. 2019. Physical and chemical pretreatment of lignocellulosic biomass. The Evolution of Biofuels, 143–196. https://doi.org/10.1016/B978-0-12-815162-4.00006-9

Jincy, E. M., Femina, K. S. 2023. Heteropolymer in biomass: Hemicellulose extraction and modifications. In Handbook of Biomass (pp. 1–32). Singapore: Springer Nature Singapore. https://doi.org/10.1007/978-981-19-6772-6_26-1

Kim, J. S., Lee, Y. Y., & Kim, T. H. 2016. A review on alkaline pretreatment technology for bioconversion of lignocellulosic biomass. Bioresource Technology, 199, 42–48. https://doi.org/10.1016/j.biortech.2015.08.085

Khudyakova, K. K., Kosolapova, V. G. 2022. Determining the structural carbohydrates and lignin levels in forage using the Van Soest and Kizel methods. Russian Agricultural Sciences, 48(5), 400–404. DOI: 10.3103/S1068367422050020.

Levitt, M. H. 2008. Spin Dynamics: Basics of Nuclear Magnetic Resonance. John Wiley & Sons. https://books.google.co.in/books?id=0poQEQAAQBAJ&lpg=PP1&ots=zbwdonshQe&dq=Levitt.%20M.%20H.%202008.%C2%A0Spin%20dynamics%3A%20basics%20of%20nuclear%20magnetic%20resonance.%20John%20Wiley%20%26%20Sons.&lr&pg=PP1#v=onepage&q=Levitt.%20M.%20H.%202008.%C2%A0Spin%20dynamics:%20basics%20of%20nuclear%20magnetic%20resonance.%20John%20Wiley%20&%20Sons.&f=true

Li, H., Li, C., Liu, W., Zou, S. 2012. Optimized alkaline pretreatment of sludge before anaerobic digestion. Bioresource Technology, 123(2012), 189–194. http://dx.doi.org/10.1016/j.biortech.2012.08.017

Maitra, S., Dien, B., Long, S. P., Singh, V. 2021. Development and validation of time-domain 1H-NMR relaxometry correlation for high-throughput phenotyping method for lipid contents of lignocellulosic feedstocks. GCB Bioenergy, 13(7), 1179–1190. https://doi.org/10.1111/gcbb.12841

Mankar, A. R., Pandey, A., Modak, A., Pant, K. K. 2021. Pretreatment of lignocellulosic biomass: A review on recent advances. Bioresource Technology, 334), 125235. https://doi.org/10.1016/j.biortech.2021.125235

Mariana, M., Alfatah, T., HPS., A. K., Yahya, E. B., Olaiya, N. G., Nuryawan, A., Ismail, H. 2021. A current advancement on the role of lignin as sustainable reinforcement material in biopolymeric blends. Journal of Materials Research and Technology, 15, 2287–2316. https://doi.org/10.1016/j.jmrt.2021.08.139

Mittal, A., Scott., G. M., Amidon, T. E., Kiemle, D. J., Stipanovic, A. J. 2009. Quantitative analysis of sugars in wood hydrolyzates with 1H NMR during the autohydrolysis of hardwoods. Bioresource Technology, 100(24), 6398–6406. https://doi.org/10.1016/j.biortech.2009.06.107

Montoya-Escobar, N., Ospina-Acero, D., Fujimoto., V., Hassan, M. A., Prra, A., Gañan Rojo, P. F., Stefani, P. M. 2022. Use of fourier series in X-ray diffraction (XRD) analysis and fourier-transform infrared spectroscopy (FTIR) for estimation of crystallinity in cellulose from different sources. Polymers, 14(23), 5199. https://doi.org/10.3390/polym14235199

Novais, Â, Freitas, A. R., Rodrigues., C., Peixe, L. 2019. Fourier transform infrared spectroscopy: Unlocking fundamentals and prospects for bacterial strain typing. European Journal of Clinical Microbiology & Infectious Diseases, 38, 427–448. https://doi.org/10.1007/s10096-018-3431-3

Park, S., Baker, J. O., Himmel, M. E., Parilla, P. A., Johnson, D. K. 2010. Cellulose crystallinity index: Measurement techniques and their impact on interpreting cellulase performance. Biotechnology for Biofuels, 3, 1–10. http://www.biotechnologyforbiofuels.com/content/3/1/10

Paul, A., Muthukumar, S., Prasad, S. 2020. Electrochemical visualization of room temperature ionic liquid for the detection of functionalized 1-phenylpyridine analogue in mixed sample. Journal Electrochemical Society, 167, 137507. DOI: 10.1149/1945-7111/abba8f.

Petrovič, A., Vohl, S., CenčičPredikaka, T., Bedoić., R., Simonič, M., Ban, I., Čuček, L. 2021. Pyrolysis of solid digestate from sewage sludge and lignocellulosic biomass: Kinetic and thermodynamic analysis, characterization of biochar. Sustainability, 13(17), 9642. https://doi.org/10.3390/su13179642

Pu, Y., Cao, S., Ragauskas, A. J. 2011. Application of quantitative 31P NMR in biomass lignin and biofuel precursors characterization. Energy & Environmental Science, 4(9), 3154–3166. DOI: 10.1039/C1EE01201K.

Pu, Y., Hallac, B., Ragauskas, A. J. 2013. Plant biomass characterization: Application of solution-and solid-state NMR spectroscopy. In Aqueous Pretreatment of Plant Biomass for Biological and Chemical Conversion to Fuels and Chemicals (pp. 369–390). Springer. http://doi.org/10.1002/9780470975831.ch18

Rahman, M. S., Hasan, M. S., Nitai, A. S., Nam, S., Karmakar, A. K., Ahsan, M. S., Ahmed, M. B. 2021. Recent developments of carboxymethyl cellulose. Polymers, 13(8), 1345. https://doi.org/10.3390/polym13081345

Ramezani, N., Sain, M. 2018. Thermal and physiochemical characterization of lignin extracted from wheat straw by organosolv process. Journal of Polymers and The Environment, 26(2018), 3109–3116. https://doi.org/10.1007/s10924-018-1199-2

Raud, M., Orupold., K., Rocha-Meneses, L., Rooni, V., Trass, O., Kikas, T. 2020. Biomass pretreatment with the SzegoMill™ for bioethanol and biogas production. Processes, 8, 1327. https://doi.org/10.3390/pr8101327

Rehman, S., Xu, S., Xu, H., Ta, T., Li, Y., Yu, Z., Wang, J. 2022. The role of NMR in metal organic frameworks: Deep insights into dynamics, structure and mapping of functional groups. Materials Today Advances, 16, 100287. https://doi.org/10.1016/j.mtadv.2022.100287

Rezania, S., Oryani, B., Cho, J., Talaiekhozani., A., Sabbagh, F., Hashemi, B., Mohammadi, A. A. 2020. Different pretreatment technologies of lignocellulosic biomass for bioethanol production: An overview. Energy, 199, 117457. https://doi.org/10.1016/j.energy.2020.117457.

Saleh, O. A., Badawey, A. M., Aboul-Enein, H. Y., Fouad, M. A. 2023. Enantioseparation, quantification, molecular docking and molecular dynamics study of five β-adrenergic blockers on Lux-Cellulose-2 column. BMC Chemistry, 17(1), 22. https://doi.org/10.1186/s13065-023-00925-2

Salem, K. S., Kasera, N. K., Rahman, M. A., Jameel, H., Habibi, Y., Eichhorn, S. J., Lucia, L. A. 2023. Comparison and assessment of methods for cellulose crystallinity determination. Chemical Society Reviews. DOI: 10.1039/d2cs00569g.

Santo, E. M., Rezende, A. C., Bernardinelli, D. O. Jr., Curvelo, P. N., deAzevedo, A. S. A., Guimaraes, R. E., Polikarpov, G. E. F. 2018. Structural and compositional changes in sugarcane bagasse subjected to hydrothermal and organosolv pretreatments and their impacts on enzymatic hydrolysis. Industrial Crops & Products, 113(2018), 64–74. https://doi.org/10.1016/j.indcrop.2018.01.014

Sathitsuksanoh, N., Renneckar, S. 2017. Characterization methods and techniques. In Introduction to Renewable Biomaterials: First Principles and Concepts (pp. 107–140). https://doi.org/10.1002/9781118698600.ch4

Sharma, H. K., Xu, C., Qin, W. 2017. Biological pretreatment of lignocellulosic biomass for biofuels and bioproducts: An overview. Waste Biomass Valor, 10(2019), 235–251. DOI: 10.1007/s12649-017-0059-y.

Shukla, A., Kumar, D., Girdhar, M., Kumar, A., Goyal, A., Malik, T., Mohan, A. 2023. Strategies of pretreatment of feedstocks for optimized bioethanol production: Distinct and integrated approaches. Biotechnology for Biofuels and Bioproducts, 16(2023), 44. https://doi.org/10.1186/s13068-023-02295-2

Singh, S., Kaur, G., Singh, D. P., Arya, S., Krishania, M. 2024. Exploring rice straw's potential from a sustainable biorefinery standpoint: Towards valorization and diverse product production. Process Safety and Environmental Protection. https://doi.org/10.1016/j.psep.2024.01.105

Sluiter, J. B., Ruiz, R. O., Scarlata, C. J., Sluiter, A. D., Templeton, D. W. 2010. Compositional analysis of lignocellulosic feedstocks. 1. Review and description of methods. Journal of Agricultural and Food Chemistry, 58(16), 9043–9053. DOI: 10.1021/jf1008023.

Sunkar, B., Bhukya, B. 2022. An approach to correlate chemical pretreatment to digestibility through biomass characterization by SEM, FTIR and XRD. Frontiers in Energy Research, 10, 802522. https://doi.org/10.3389/fenrg.2022.802522

Suriyachai, N., Weersai, K., Laosiripojana, N., Champreda, V., Unrean, P. 2013. Optimized simultaneous saccharification and co-fermentation of rice straw for ethanol production by *Saccharomyces cerevisiae* and *Scheffersomyces stipitis* co-culture using design of experiments. Bioresource Technology, 142(2013), 171–178. http://dx.doi.org/10.1016/j.biortech.2013.05.003

Suriyachai, N., Weersai, K., Upajak, S., Khongchamnan, P., Wanbmolee, W., Laosiripojana, N., Champreda, V., Suwannahong, K., Imman, S. 2020. Efficiency of catalytic liquid hot water pretreatment for conversion of corn stover to bioethanol. ACS Omega, 5, 29872–29881. https://doi.org/10.1021/acsomega.0c04054

Tsai, C. H., Shen, Y. H., Tsai, W. T. 2023. Effect of alkaline pretreatment on the fuel properties of torrefied biomass from rice husk. Energies, 16(2), 679. https://doi.org/10.3390/en16020679

Van Soest, P. J., Robertson, J. B., Lewis, B. A. 1991. Methods for dietary fiber, neutral detergent fiber, and nonstarch polysaccharides in relation to animal nutrition. Journal of Dairy Science, 74, 3583–3597. https://doi.org/10.3168/jds.S0022-0302(91)78551-2

Verma, N., Taggar, M. S., Kalia, A., Kaur, J., Javed, M. 2022. Comparison of various delignification/desilication pre-treatments and indigenous fungal cellulase for improved hydrolysis of paddy straw. 3 Biotech, 12(2022), 150. https://doi.org/10.1007/s13205-022-03211-5

Wang, X., Karaman, C., Zhang, Y., & Xia, C. (2023). Graphene oxide/cellulose nanofibril composite: A high-performance catalyst for the fabrication of an electrochemical sensor for quantification of p-nitrophenol, a hazardous water pollutant. Chemosphere, 331, 138813. https://doi.org/10.1016/j.chemosphere.2023.138813

Wong, S. W. D. 2009. Structure and action mechanism of lignolytic enzymes. Applied Biochemistry and Biotechnology (2009). DOI: 10.1007/s12010-008-8279-z.

Yadav, D., Wati, L. 2016. Microbial delignification and hydrolysis of paddy straw for ethanol production. Agricultural Research Journal, 53(4), 528–531. DOI: 10.5958/2395-146X.2016.00105.8.

Yeganeh, F., Chiewchan, N., Chonkaew, W. 2022. Hydrothermal pretreatment of biomass-waste-garlic skins in the cellulose nanofiber production process. Cellulose, 29(4), 2333–2349. https://doi.org/10.1007/s10570-022-04452-6

Yuan, S., Tyufekchiev, M. V., Timko, M. T., Schmidt-Rohr, K. 2022. Direct quantification of the degree of polymerization of hydrolyzed cellulose by solid-state NMR spectroscopy. Cellulose, 29(4), 2131–2144. https://doi.org/10.1007/s10570-022-04433-9

Zhang, J., Wang, Y., Zhang, L., Zhang., R., Liu, G., Cheng, G. 2014. Understanding changes in cellulose crystalline structure of lignocellulosic biomass during ionic liquid pretreatment by XRD. Bioresource Technology, 151, 402–405. http://dx.doi.org/10.1016/j.biortech.2013.10.009

Zhang, Y., Guo, S. X., Zhang, X., Bond, A. M., Zhang, J. 2020. Mechanistic understanding of the electrocatalytic CO2 reduction reaction–New developments based on advanced instrumental techniques. Nano Today, 31, 100835.

Zheng, Q., Zhou, T., Wang, Y., Cao, X., Wu, S., Zhao, M., Wang, H., Xu, M., Zheng, B., Zheng, J., Guan, X. 2018. Pretreatment of wheat straw leads to structural changes and improved enzymatic hydrolysis. Scientific Reports, 8(2018), 1321. DOI: 10.1038/s41598-018-19517-5.

Ziaei-Rad, Z., Fooladi, J., Pazouki, M., Gummadi, S. N. 2021. Lignocellulosic biomass Pre-Treatment using low cost ionic liquid for bioethanol production: An economically viable method for wheat straw fractionation. Biomass and Bioenergy, 151(2021), 106140. https://doi.org/10.1016/j.biombioe.2021.106140

4 Lignocellulosic Ethanol Production

Bioreactor Designing and Analytical Approach

*Apurav Sharma and Sachin Kumar**

4.1 INTRODUCTION

An exhaustive use of fossil fuels and its environmental impact by emitting greenhouse gases after its combustion led to depletion of this finite resource and climate change (Saravanan et al., 2018). With the decline in petroleum reserves and significant rise in environmental interest, interest in alternative energy sources is rising (Toor et al., 2020). Bioethanol stands out as a promising alternative because of its sustainability and renewability. Its environmentally friendly, biodegradable, and carbon-neutral nature makes it a viable energy source (Lafuente-Rincón et al., 2022; Kaur et al., 2023). Bioethanol, an advanced and environmentally friendly liquid biofuel, is primarily produced from starch-based sugars. These sugars come from crops such as corn and wheat in the United States and China, as well as sugarcane juice and sugar beets in countries like Brazil and the European Union, respectively. Many countries have adopted these first-generation sugars for first-generation ethanol (1G) as transportation. However, the production of 1G ethanol from starchy crops presents challenges, as they compete with food production and require large land, along with soil, water, and nutrients (Raj et al., 2022). Bioethanol has emerged as a compelling and environmentally sustainable biofuel, serving as a viable alternative in place of conventional fossil fuels. In recent years, bioethanol has been produced through fermentation of renewable sugar sources like starch-based feedstock at a large scale (Kordala et al., 2024). However, the ethical issues surrounding the diversion of resources from food production to fuel generation have spurred an exploration for alternative, non-food-based ethanol feedstocks (Hossain et al., 2017). Contemporary scientific research has shifted its focus to producing ethanol from lignocellulosic biomass (LCB), which is mainly part of waste material. LCBs, distinguished by their cost-effectiveness, abundance, and non-competitive nature with food resources, are predominantly cultivated on arable lands. These biomass sources, which include non-edible parts of food crops (i.e., agricultural residues), forest biomass (including woody biomass), perennial grasses, dedicated energy crops, aquatic plants and their waste residues, biodegradable municipal solid wastes, and agro-industrial by-products, have become fundamental to current research efforts (Hossain et al., 2017; Devi et al., 2022). The bioethanol production process from

* Corresponding author

DOI: 10.1201/9781003485636-4

LCBs involves a systematic approach, encompassing key steps: starting from (i) pretreatment of the LCBs to liberate cellulose and hemicellulose from the intricate lignocellulosic structure, (ii) hydrolysis of the cellulose and hemicelluloses from pretreated biomass to produce fermentable sugars, (iii) fermentation of these sugars by yeast to produce ethanol, and (iv) the subsequent recovery of ethanol through distillation and dehydration (Devi et al., 2022; Beluhan et al., 2023). These steps typically occur in a reactor/bioreactor, a controlled reaction system providing an optimal environment for achieving the maximum possible yield (Pino et al., 2018). This chapter explores various bioreactor configurations and operation modes, focusing on how different process parameters such as substrate concentration, catalyst/biocatalyst concentration, temperature, pH, reaction volume, and mixing affect performance (De la Fuente et al., 2019). Among the most widely used systems is the stirred tank reactor (STR), which can be operated under the different modes batch, fed-batch, or continuous. Additionally, several innovative bioreactor designs have been proposed to enhance the processing of lignocellulosic materials, offering improved efficiency and scalability (Saha et al., 2019). This comprehensive approach not only aligns with the quest for sustainable energy sources but also addresses the ethical concerns associated with bioethanol production, ushering in an era of eco-friendly alternatives with significant potential for reducing reliance on fossil fuels. Considering the importance of engineering parameters in bioreactors for bioethanol production, it is crucial to evaluate their role in key process stages, including pretreatment, enzymatic hydrolysis, and fermentation. This chapter examines the principle of bioreactor design and analytical approaches for bioethanol production from LCBs, focusing on the operational conditions required for each step to optimize process performance.

4.2 BIOREACTOR DESIGN PRINCIPLES

In the realm of bioprocess engineering for bioethanol production from LCB, the design and selection of bioreactors play a pivotal role in the successful cultivation of microorganisms, cells, or enzymes for various applications, ranging from pharmaceutical production to environmental remediation. Understanding the principles of bioreactor design is crucial for optimizing the conditions under which biological processes occur, ensuring the efficiency, and maximizing the yields.

4.2.1 Types of Bioprocesses for Bioreactor Design

Among the fundamental considerations in the bioreactor design, the types of bioprocesses, namely, batch, continuous, and fed-batch stand out as primary configurations, each tailored to meet specific requirements and challenges inherent in diverse bioprocessing scenarios (Devi et al., 2022).

4.2.1.1 Batch Bioprocess

Batch bioprocess is a widely accepted and uncomplicated configuration wherein all components for the process, including microorganisms and culture medium, are introduced in the reactor simultaneously at the start of the process. During the bioprocessing, there is no further addition of medium or microorganisms or any removal of by-products. This bioprocess is widely adopted because it runs a batch within a short duration where a low amount of biomass or medium is utilized with a high amount of product. The operational procedure is simple, as the reactor is initially supplemented with substrate and microorganisms, and a constant volume is maintained throughout the batch without the addition of any nutrient, substrate, or microorganism (Devi et al., 2022). However, with the simplicity of the process, there are challenges associated, as there is restricted control of the process until the batch completion, substrate depletion, and compromise with productivity with time. Thus reactors for batch processes are designed with better control till the batch completion with regular analysis of various fermentation parameters, being product inhibition, an extended lag phase, and interruptions for sterilization, cleaning, and refilling (Ingledew & Lin, 2011).

4.2.1.2 Continuous Bioprocess

A continuous bioprocess emerged as the solution for challenges associated with batch bioprocesses, where a steady-state condition is maintained by a constant supply of fresh medium and cell culture is withdrawn concurrently. This continuous inflow and outflow create a stable environment for microbial or cellular growth. Continuous bioreactors offer advantages in prolonged fermentation processes by affording better control over essential parameters such as nutrient levels and waste removal. They are recognized for higher productivity and the potential for improved product consistency. However, their implementation demands sophisticated control strategies and is often associated with increased operational complexity. The system can be categorized into single, double, and multi-stage (Buyondo & Liu, 2012). In continuous single-stage fermentation, higher reactor productivity is achieved compared to batch processing, but at lower product concentration. However, double-stage fermentation mitigates inhibitory compound production by splitting the process into two separate containers. Multi-stage systems, involving 6–8 tanks, further optimize the process (Buyondo & Liu, 2012). Despite the general productivity advantage of continuous fermentation over batch fermentation, the free cell continuous process encounters the issue of low cell concentration, risking washout at significant dilution rates (Devi et al., 2022).

4.2.1.3 Fed-Batch Bioprocess

Fed-batch bioprocesses, integrating features from batch and continuous systems, aim to overcome the limitations associated with each. In this configuration, fresh medium is incrementally added during cultivation, allowing for extended cultivation times while addressing substrate limitations. Fed-batch processes provide enhanced control over nutrient levels, reducing the risk of inhibitory metabolite accumulation and ultimately improving overall productivity (Devi et al., 2022). Intermittent nutrient supply during the process helps alleviate substrate-associated growth inhibition, fostering uniformity within the living system and resulting in higher yields compared to the batch method. Furthermore, fed-batch bioprocess serves as the better alternative to address mixing challenges associated with high solid loading by preventing excessive substrate viscosity (Pino et al., 2018). Therefore, fed-batch bioprocessing is a particularly better approach for high-density cell cultures and the production of complex molecules if the effective design and operation of fed-batch bioreactors are carefully considered with better feeding strategies and monitoring to optimize the process during feeding and withdrawal.

4.2.2 Reactors/Bioreactors for Pretreatment Process

The pretreatment of LCB remains the most intricate step for the conversion into biofuels in advancing the circular bioeconomy, particularly from an economic perspective. LCB poses a challenge because of its recalcitrant nature, as it's resilient to breakdown through biological, physical, or chemical processes (Baruah et al., 2018). Various strategies have been explored to overcome these challenges, encompassing innovative pretreatment methods and the adoption of advanced reactor configurations. Diverse pretreatment techniques for biomass have been explored, such as mechanical, chemical, physicochemical, and biological processes (involving microbial consortia or enzymes), or combinations thereof, to break down the recalcitrant LCB (Kumar et al., 2020). Beyond the choice of pretreatment processes, achieving high yields and productivity in industrial-scale biomass conversion depends on selecting the optimal reactor configuration, process conditions, and operation mode during the pretreatment stage (Basak et al., 2023). Research has explored various bioreactor configurations to maximize glucose recovery during pretreatment (Pino et al., 2018). Different pretreatment techniques influence the reaction dynamics, each with distinct mechanistic characteristics. Among physical pretreatments, mechanical size reduction typically achieved through milling does not require a bioreactor but plays a critical role in determining the efficiency of the entire bioethanol production chain. Thus, identifying the optimal particle size for each LCB type is crucial (Yang et al., 2023). Other widely used physical pretreatment include microwave and hydrothermal methods. Microwave reactors (MWR) have gained interest for LCB pretreatment due to their ability to efficiently transfer energy. Microwaves, a form of electromagnetic radiation, propagate through space in the absence of material movement. They are generated by the separation of positive and negative charges, held at a fixed distance. A typical MWR consists of a generator, a cavity section, a waveguide, and an advanced control monitoring system (Lafuente-Rincón et al., 2022). Another physical pretreatment method, which can also be performed as a physio-chemical one, is hydrodynamic cavitation. This technique generates cavitations in a freely flowing liquid by passing it through a Venturi tube or orifice plate, inducing rapid pressure fluctuations. Based on Bernoulli's principle, cavitation occurs when the pressure drops below the vapor pressure of the liquid, leading to the formation, growth, and subsequent collapse of microbubbles. Hydrodynamic cavitation reactors (HCR) have been successfully employed for the delignification of various LCB, including reed grass, sugarcane bagasse, and dry mill corn. HCRs offer a significant advantage due to their lower energy consumption, making them a more cost-effective option. Additionally, HCR not only facilitates delignification but also reduces cellulose crystallinity, enhancing enzymatic hydrolysis efficiency and ultimately leading to higher bioethanol yield (Kumar et al., 2022; Lafuente-Rincón et al., 2022).

4.2.3 Bioreactors for Enzymatic Hydrolysis and Fermentation Process

During scaling up, the pretreatment stage is followed by hydrolysis and a fermentation step. In this section, recent advancements in bioreactor technology, particularly highlighting new configurations and agitation systems in hydrolysis and fermentation processes, are discussed. These advancements aim to improve the conversion of second-generation substrates and residual materials into sugars and ethanol through the following routes:

i. Separate Hydrolysis and Fermentation (SHF)
ii. Simultaneous Saccharification and Co-Fermentation (SSCF)
iii. Consolidated Bioprocessing (CBP)

4.2.3.1 Bioreactors for Separate Hydrolysis and Fermentation (SHF)

Separate hydrolysis and fermentation is the most widely used route for ethanol production. In this approach, a dual-reactor system is used where one reactor is dedicated to the hydrolysis process, while the other is dedicated to fermentation. In the first reactor, pretreated biomass containing cellulose and hemicellulose is broken down through the process of enzymatic hydrolysis. Following the hydrolysis process, the resulting sugars are transferred to the fermentation reactor along with the necessary fermenting microorganisms for ethanol production. SHF still poses challenges despite the usage of distinct reactors and specialized microbes for fermentation, like glucose inhibition, not achieving the optimal sugar concentrations for the fermenting microorganisms, higher initial investment costs, adaptability to contamination during transfer to other reactors, prolonged incubation periods, formation of inhibitors during the process, and the necessity for additional equipment (Devi et al., 2022). These drawbacks underscore the need for further innovation in reactor technology and refinement in the realm of ethanol production. The enzymatic saccharification (hydrolysis) of pretreated LCB is a critical step in biochemical biorefineries, as it converts structural polysaccharides into fermentable sugars. One of the primary challenges in the process is achieving high sugar concentrations to ensure the economic viability of large-scale bioethanol production. A key strategy to overcome this issue is operating at high solids loadings, 5% (w/v) biomass concentrations. However, increasing the solid content rises viscosity, leading to the formation of thick slurries that hinder mass transfer and mixing, ultimately limiting further increase in solid loading. Rodríguez-Jasso et al. (2023) investigated the use of a commercial enzyme cocktail of cellulase and hemicellulase for the hydrolysis of hydrothermally and CCO_2-H_2O pretreated agave biomass. Their study compared two reactor configurations: a continuous stirred tank bioreactor (CSTBR) and a horizontal bioreactor (HB). The saccharification rate reached 100%, with the CSTBR yielding 119.9 g/L of glucose at 20% (w/v) solid loading, while the HB yielded 134.4 g/L of glucose at 20% (w/v) solid loading. When using CCO_2-H_2O pretreated biomass, the HB produced 123.6 g/L of glucose at 20% (w/v) solid loading. The enzyme cocktail significantly improved hydrolysis efficiency, with hydrothermally pretreated biomass exhibiting the highest glucose conversion yields and an energy efficiency of 2.291 g sugar/MJ. These findings highlight the importance of pretreatment methods and reactor design in optimizing biomass saccharification for bioethanol production. Several attempts have been made to design reactor systems for saccharification of various biomasses at high solids—loading to ensure effective mixing and achieve bioconversion yields (Ludwig et al., 2014). Gupta et al. (2012) investigated the saccharification of sodium chlorite-retreated mesquite wood in a 3.0-L stirred tank reactor (STR) equipped with a Rushton impeller for agitation (150 rpm), a heating jacket, and heat exchangers for temperature control (maintained at 50°C). The enzymatic hydrolysis was performed using cellulase (22 FPU/g of dry substrate (gds), Sigma-Aldrich) and β-glucosidase (68 U/gds, Sigma-Aldrich). Substrate loading was varied among 5%, 10%, 115%, and 20% w/v, resulting in a significant increase in sugar concentration up to 15% solid loading (41.10 to 90.07 g/L). However, at 20% solid loading, sugar concentration declined to 80.78 g/L, with a 40.39% cellulose conversion, indicating mixing limitations at higher biomass concentrations. Similarly, Palmqvist and Lidén (2012) examined the saccharification of *Arundo donax* (an energy crop) in a 3-L STR (Belach Bioteknik, Stockholm, Sweden), using an anchor impeller (13 cm diameter, 2 cm blade width). The enzyme preparation Cellic CTec2 (Novozymes, Denmark) was applied at 0.1 g enzyme/g water-insoluble solid under two operational strategies: constant impeller speed (10 rpm) and constant impeller power input. The results showed similar glucose yields in both cases, suggesting that maintaining power input may be a viable alternative to constant impeller speed in optimizing biomass saccharification efficiency. Enzymatic hydrolysis is widely recognized as the bottleneck in the overall cost of lignocellulose bioconversion processes. Consequently, numerous kinetic models have been developed to streamline this step and reduce the associated costs. Kadam et al. (2004) developed a kinetic model that incorporates both cellulose-to-glucose and cellobiose-to-glucose conversion pathways. This model provides a more comprehensive understanding of the enzymatic hydrolysis process by considering intermediate product formation. Using this model, they performed hydrolysis experiments on sulfuric acid-treated corn stover with CPN commercial cellulase, 45 mg protein/g cellulose (Iogen Corp., Ottawa, Canada) under two different setups: 250-mL baffled Erlenmeyer flasks stirred at 130 rpm and STR with an impeller speed of 250 rpm. These experiments allowed for a comparison between small-scale flask experiments and the bioreactor-scale enzymatic hydrolysis, providing insights into reaction kinetics, enzyme performance, and mixing effects in different reactor configurations. In both systems, a maximum glucose concentration of approximately 50 g/kg of cellulose was achieved after 168 h. Although the model was unable to predict the effect of temperature, it provided a basis for optimizing the saccharification process through computational means.

Innovative bioreactor designs have been developed to efficiently convert lignocellulosic-rich wastes and agricultural/industrial residues into bioethanol. The newly developed bioreactors are usually equipped with new agitation systems that aimed at achieving efficient fiber homogenization in the reactor, further reducing the input energy at high-solid loading for SHF. Ludwig et al. (2014) developed the specialized bioreactors featuring a special segmented helical stirrer, Riedlberger and Weuster-Botz (2012) introduced S-shaped impellers, and Du et al. (2014) introduced double helical ribbon impellers at high solid loading that can be utilized for SHF. These advanced agitation systems could possibly contribute to better fiber homogenization mixing and reducing energy consumption during separate hydrolysis processes. The challenges associated with the SHF process could be overcome by an alternative method known as separate hydrolysis and co-fermentation (SHCF). In SHCF, both 5-carbon (pentose) and 6-carbon (hexose) sugars obtained from hydrolysis of pretreated biomass are simultaneously fermented through such microorganisms capable of utilizing both. The SHCF process allows for better usage of high solid loading by sugar utilization (Balan, 2014).

4.2.3.2 Bioreactors for Separate Saccharification and Fermentation (SSF)

The challenges associated with SHF and SHCF could be overcome by simultaneous saccharification and fermentation (SSF), where the cellulose content of pretreated biomass is hydrolyzed by enzymes and simultaneously converted into ethanol, resulting in the one pot sequential release and fermentation of sugars. SSF offers distinct advantages over SHF, particularly related to feedback inhibition by minimizing the inhibitory effects associated with sugar accumulation. Additionally, SSF utilizes only a single reactor system, simplifying operational procedures and reducing contamination issues considerably. Despite the advantages of SSSF, which stands for converting lignocellulosic biomass (LCB) into bioethanol, there are still some issues: the necessity to optimize conditions for both hydrolysis and fermentation processes simultaneously and the inefficient utilization of the released sugars (Devi et al., 2022). A more advanced variation of SSF, simultaneous saccharification and co-fermentation (SSCF), integrates enzymatic hydrolysis of cellulose and hemicellulose with simultaneous fermentation of both glucose and xylose. Compared to SHF, SSHCF, and SSF, SSCF offers several advantages, such as lower capital costs and improved ethanol yields. This is largely due to its ability to prevent the accumulation of sugars that can inhibit enzymatic activity of cellulases and β-glucosidases (Koppram et al., 2013). Olughu et al. (2023) developed a process model for an ethanol-based biorefinery utilizing switch grass, incorporating fungal pretreatment before enzymatic saccharification. This simulation, conducted using SuperPro Designer, involved enzymatic hydrolysis in a continuous reactor (P-11/FR-102) equipped with a stirrer. The pretreated switch grass, with a total solid loading of 20 wt%, was hydrolyzed using a commercial enzymatic blend of cellulases and hemicellulases added at a dosage of t of enzyme protein per gram of cellulose. The hydrolysis process achieved cellulose-to-glucose conversion within a residence time of 72 h. In the SSCF approach, glucose and xylose were co-fermented using recombinant *Zymomonas mobilis*, achieving conversion efficiencies of 90% and 80%, respectively (Olughu et al., 2023). The ethanol plant modeled in this study, with a processing capacity of 2000 tons of switch grass per day, yielded 211.90 L/t of ethanol per ton of biomass. The unit production cost was estimated at $1.44/L, yielding a positive net present value at ethanol selling price of $1.50/L. His integrated process enables efficient conversion of LCB into ethanol, capitalizing on the synergistic action of enzymatic hydrolysis and microbial fermentation.

4.2.3.3 Bioreactors for Consolidated Bioprocessing (CBP)

The conversion of LCB into bioethanol is indeed a multi-stage complex process, with each stage contributing significantly to the cost. The above-described processes are indeed the improved version of each other; however, the different reactor usages in SHF or SHCF and enzyme production followed by simultaneous hydrolysis and fermentation in SSF or SSSCF till incur substantial cost in bioethanol production. The consolidated bioprocessing (CBP) approach has been explored to streamline the bioethanol production process by developing structures or process modes capable of directly converting biomass into ethanol in a single step. CBP involves the hydrolysis of pretreated LCB for the release of sugars by the action of a single microorganism or consortium capable of hydrolyzing the LCB through enzyme production and their fermentation, all within the same step, reducing the cost and overall energy input. CBP is indeed in its developmental stages, but it holds immense potential for revolutionizing bioethanol production. With the integration of multiple functions into a single step, CBP can significantly lower production cost, energy input, and overall cost reduction (Sharma et al., 2024). Further research and development are essential to fully explore and optimize the CBP approach for large-scale commercial bioethanol production. CBP integrates four critical biological steps within a single reactor: the production of hydrolytic enzymes by microorganisms or CA Consortium, the hydrolysis of polysaccharides into monomeric sugars from pretreated CB, and the fermentation of both hexose and pentose sugars. This streamlined process offers a cost-effective alternative to the conventional biomass conversion process, making CBP a promising advancement in bioethanol production. However, no single natural microorganism possesses all the necessary traits for an efficient CBP system. As a result, research has focused on identifying and engineering bacterial and fungal strains that exhibit desirable characteristics (Van Zyl et al., 2007). Liu et al. (2020) investigated the potential of *Clostridium thermocellum* DSM 1237 for CBP using sugarcane bagasse (SCB) as the substrate. Initially ethanol yield reached 0.68 g/L, corresponding to 65.8% of theoretical yield. Furthermore, optimization in a 3-L STR utilizing enzyme-rich cultures with both the cellulase and xylanase activities improved ethanol production to 0.86 g/L, achieving 83.3% of the theoretical yield. Zhang et al. (2023) successfully engineered the thermophilic cellulosic fungus *Myceliophthora thermophila* to directly convert cellulose into ethanol by mimicking the Crabtree effect observed in yeast. This was achieved by optimizing the synthetic pathway, increasing the glycolytic rate, inhibiting mitochondrial NADH shuttles, and disabling the ethanol consumption pathway. Fermentations were conducted in a 5-L fermenter with a working volume of 33 L. Batch fermentation used Avicel as the carbon source, while fed-batch fermentation was performed with corncob. The final engineered strain yielded 52.8 g/L ethanol directly from cellulose and 39.8 g/L from corncob remarkably without requiring cellulase addition. This streamlined process significantly improved ethanol yields, highlighting the potential of fungal CBP as a one-step, one-strain approach to sustainable carbon-neutral bioethanol production.

4.3 SELECTION CRITERIA FOR BIOREACTOR DESIGN

Choosing the appropriate bioreactor design involves careful consideration of multiple factors. This includes the nature of the biological system, the specific process requirements, and the desired outcomes. Key criteria for bioreactor design selection encompass the following.

4.3.1 Type of Microorganism or Cells

The selection of microorganisms or cell cultures in bioreactor design is critical, taking into account their specific needs such as oxygen requirements, sensitivity to shear forces, and growth traits. The bioreactor design must be flexible to accommodate these variations, ensuring ideal conditions for both cell growth and the production of desired products. In second-generation (2G) biorefineries, the integration of thermophilic microorganisms and enzymes presents the opportunity to conduct the entire production process at higher temperatures (Di Donato et al., 2019). This approach offers numerous advantages, including enhanced substrate and product solubility, accelerated reaction rates, reduced enzyme usage, easier mixing due to decreased medium viscosity, lowered risk of contamination, simplified product recovery, and decreased cooling-related costs. However, directly replacing conventional enzymes and microorganisms with thermophilic counterparts in 2G-biorefineries would necessitate a complete redesign of the production process. This transition would require additional investments to modify existing facilities for high-temperature operations and is likely to involve a different process setup (Di Donato et al., 2019). After pretreatments, the temperature of the slurry is typically too high for saccharification to begin using commercially available enzyme cocktails, necessitating a cooling period between pretreatment and enzymatic saccharification. Instead of simply waiting for the slurry to cool, this time could be utilized for pre-saccharification with thermophilic or hyper thermophilic enzymes capable of functioning at elevated temperatures. Such an approach would enhance enzyme penetration into LCB and improve substrate solubility, leading to a more efficient sugar release. Ultimately, integrating a pre-saccharification step could increase the availability of fermentable sugars and enhance overall ethanol yield (Zuliani et al., 2021).

4.3.2 Process Scale

The process scale of the bioprocess is a crucial factor in selecting the suitable bioreactor. Bioreactors can range from laboratory-scale to industrial-scale units, where the selected design should align with the intended scale of production. Ring scale-up important factors that are taken into consideration are mixing efficiency, mass transfer, and heat transfer at larger scales (Edeh, 2021). During the scale-up phase, the primary goals include the effectiveness of utilizing constant Reynolds number and the constant impeller tip speed methodologies in laboratory settings for stepwise scale-up processes. This evaluation aims to determine whether enhancing the bioethanol yield is attainable, as the objective includes the validation of kinetic models utilized to more accurately depict the behaviors of microorganisms or cells at different stages of bioethanol production (Imamoglu & Sukan, 2013).

4.3.3 Operating Conditions

Operating conditions, including temperature, pH, and dissolved oxygen levels, play crucial roles in influencing cell growth and product formation in bioreactor systems. Temperature directly impacts the growth rate of microorganisms in any system.

High or low temperature creates stress conditions and lowers the microbial or enzymatic activity. Enzymes, which are also susceptible to temperature conditions, are denatured, leading to their inactivation at un-optimized temperature conditions. Therefore, temperature optimization before bioprocessing and regulation during bioprocessing have to be meticulously managed for efficient processing. Similarly, the pH of the fermentation broth significantly affects ethanol production, as it directly influences the growth, fermentation rate, bacterial contamination, and the formation of by-products. Thus, maintaining and regulating an optimal pH level is crucial for favorable conditions for microbial growth and obtaining maximum ethanol yield, while minimizing the formation of undesirable by-products or contamination (Azhar et al., 2017). Bioreactor designs must allow for precise control of these parameters to optimize overall process performance.

4.3.4 Sterilization and Material of Construction

Sterilization is a crucial factor impacting smooth operation of the process as well as product quality by preventing contamination. Key factors considered for sterilization of bioreactors include ease of cleaning, ability to withstand autoclaving or chemical treatment, and the choice of material of construction. The first-generation sugar-based ethanol industries face challenges associated with excessive bacterial contamination. However, achieving complete sterility at an industrial scale is expensive, labor-intensive, and hard to maintain in the long run. Despite efforts, a completely contamination-free process is difficult to achieve. To address contamination, conventional antibacterial agents like virginiamycin, sodium fluoride, and hydrogen peroxide are used in various stages of the fermentation process at the industrial level. Acids are also used to decontaminate the fermentation vessels when the vessels are made of good-quality material. Alternatively, membrane bioreactor (MBR) systems, which are physical means of decontamination, are used, as antibiotics and acids disturb the physicochemical and biological conditions of the fermentation medium. BR systems are based on semi-selective filtration membranes to segregate compounds and cells based on size-exclusion and affinity mechanisms. This method reduces contamination risk by selectively filtering components (Mahboubi et al., 2018). The choice of construction materials also affects bioreactor performance. Stainless steel is often preferred over carbon steel because it's more cost-effective and possesses a lower risk of becoming a contamination source. Furthermore, carbon steel may require thicker designs for corrosion allowance, which is more prone to contamination (Wallin, 2019). In summary, efficient sterilization, cleaning, and material of construction are essential for designing bioreactors for industrial fermentation processes along with preventing contamination, maintaining product quality, and optimizing overall process efficiency. Top of form choosing between batch, continuous, or fed-batch bioreactors depends upon the specific needs of the bioprocess, taking into consideration the nature of the product, fermentation duration, and desired productivity. Bioreactor design principles go beyond just the physical structure, as it also requires standing the microbial physiology, nutrient availability, and process kinetics. This holistic approach is crucial for the success of LCB-based bioethanol endeavors.

4.3.5 Scale-Up Considerations in Bioethanol Production

Scaling from laboratory to industrial-scale bioethanol production presents unique challenges that require careful attention. This section explores the key parameters for scaling-up bioprocesses for bioethanol production:

4.3.5.1 Mass Transfer

In a bioreactor scale-up, efficient mass transfer becomes a crucial consideration. During scaling up, the vessel's capacity is considered carefully, which is important to deliver oxygen to support microbial growth as well as ethanol output while maintaining efficiency. The volumetric mass transfer coefficient for gas-liquid characterizes the oxygenation level in stirred tank bioreactors (STBR). A major challenge in bioethanol production involves enhanced mass transfer by improving the volumetric mass transfer coefficient of liquid and gas, which can be achieved by controlling process variables within aerobic STBR using agitation. Alternatively, the reactor's design and the incorporation option of specific materials could be considered to enhance the rate of volumetric mass transfer. These strategies enable better control over mass transfer and oxygen levels, optimizing conditions for microbial growth and ethanol production (Al Awsi et al., 2023).

4.3.5.2 Heat Transfer

As bioreactors increase in size, the requirement for an efficient heat transfer mechanism to maintain optimal operating conditions also necessitates. Cooling systems and heat exchangers are usually used as heat managing systems within the biorefineries. Additionally, with the advancements in computer technology, computational fluid dynamics (CFD) and numerical heat transfer methodologies are also being employed. These methods have gained popularity for analyzing detailed flow patterns and heat transfer characteristics in different regions of the bioreactor. Using these computational tools, researchers can better understand the thermal dynamics of bioreactor systems, helping to optimize the heat management strategies (Wen et al., 2020).

4.3.5.3 Mixing and Agitation

Another challenge during upscaling in bioreactors is to achieve homogeneous mixing in larger vessels/reactors. Impellers, or spargers, are being employed to ensure uniform distribution of nutrients and prevent concentration gradients within the bioreactor for proper agitation. The agitation rate in bioreactors is a critical factor influencing nutrient transport and ethanol production. Effective mixing ensures efficient nutrient diffusion from the fermentation broth into microbial cells while facilitating ethanol removal from cells into the surrounding medium. Higher agitation rates generally enhance ethanol production by improving nutrient availability, optimizing substrate utilization, and mitigating ethanol inhibition on microbial growth. Additionally, increased agitation accelerates sugar consumption, further boosting ethanol production (Azhar et al., 2017). Temperature during different stages of bioprocessing also plays a crucial role in designing bioreactors. Temperature is considerably an important parameter during chemical and physicochemical pretreatments, where significant shifts took place depending upon the adopted process (Pino et al., 2018; Olivieri et al., 2021). Stirred tank bioreactors (STBRs) are commonly used for fermentation processes requiring moderate temperatures ranging from 25°C to 70°C. For operations at higher temperatures exceeding 100°C, bioreactors must be designed to withstand elevated pressures. Substrate concentrations play a crucial role in determining impeller speed, which can range from moderate levels (5% w/v) to high levels (exceeding 12% w/v) (Kartawiria et al., 2019; Afedzi et al., 2022). Higher concentration necessitates increasing mechanical mixing speeds, leading to greater energy demands. To optimize processing efficiency and reduce energy consumption, high-substrate-loading systems in chemical and physiochemical pretreatments often employ fed-batch, semi-batch, or continuous operation modes (Pino et al., 2018; Olivieri et al., 2021). The geometry of the impeller is a critical factor in achieving effective mixing of high-solid LCB slurries. Figure 4.1 presents various types of impellers used in STBR for high-solids lignocellulose fermentation. Impellers are generally classified based on their mixing regimes, either laminar or turbulent. In some cases, multiple impellers are incorporated to enhance mixing efficiency, particularly in tall bioreactors. A wide variety of impeller designs are available, each tailored to specific mixing requirements and applications, as summarized in Table 4.1. The choice of specific impeller directly influences the mixing pattern,

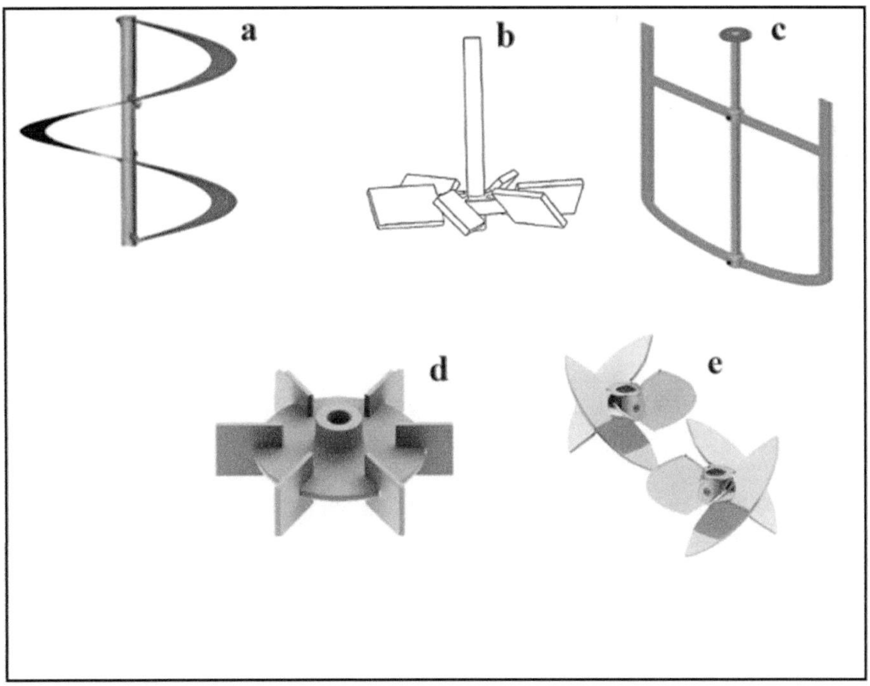

FIGURE 4.1 Various impeller designs commonly used in high-solid-loading applications (present study); (a) single helical ribbon impeller (b) rotating paddle (c) anchor impeller (d) Rushton impeller (e) three-blade impeller at a 45° angle.

TABLE 4.1
Different Types of Impeller Used for Bioethanol Production and Their Modifications

Biomass	Feeding Strategy	Impeller	Ethanol Concentration (g/L), Yield (%)	Reference
E. globulus bark kraft pulp	SHF	Rushton	55.4 g/L, 87.5 %	Amândio et al., 2024
E. globulus bark kraft pulp	Fed-Batch SSF	Rushton	70.6 g/L, 78 %	Amândio et al., 2024
Rice straw	Continuous pyro-syngas	Rushton	29.450 g/L	Manna et al., 2024
Dried oil palm	Fed-batch SSF	Marine	41.33 ± 2.02 g/L	Afedzi et al., 2023
Dried oil palm	Fed-batch SSF	Rushton	39.08 ± 1.47 g/L	Afedzi et al., 2023
Pulp and paper sludge	Fed-batch S-SSCF	Rushton	42.34 g/L, 0.53 (g/g)	Dey et al., 2021
Oil palm trunk fiber	Fed-batch PSSF	Rushton	51.66%, 0.72 (g/g)	Tareen et al., 2021
SCB	Fed-batch PSSF	Rushton	57.2 g/L, 94.65%	Raj & Krishnan, 2020
SCB	PSSF	Helical	26.8 g/L, 52.0%	de Oliveira et al., 2018
Corn stover	Fed-batch PSSF	Rushton	43.9 g/L	Zhang et al., 2010
Corn stover	Fed-batch PSSF	Helical	51.0 g/L	Zhang et al., 2010
Solka Floc	Fed-batch PSSF	Rushton	19.7 g/L, 68.6%	Um & Hanley, 2008

SHF: separate hydrolysis and fermentation; SSF: simultaneous saccharification and fermentation; PSSF: pre-saccharification and simultaneous saccharification and fermentation; Fed-Batch SSCF: fed-batch simultaneous saccharification and co-fermentation

which may follow a radial flow, axial flow, or a combination of both (Kartawiria et al., 2019).

Bioreactors must be designed to withstand extreme pH conditions, ranging from highly acidic (<1.0, e.g., sulfuric acid) to highly alkaline (>11.0, sodium hydroxide) (Kumar et al., 2022; Lafuente-Rincón et al., 2022). To enhance LCB pretreatment efficiency, reactor configuration and operation modes should be optimized with a focus on scaling up from lab scale to industrial applications. The development of new technologies and innovative methods is crucial to overcoming these challenges. Special attention should be given to reducing energy consumption and improving sugar-to-ethanol conversion yields to enhance the overall sustainability and economic feasibility of the process.

4.3.5.4 Instrumentation and Economic Considerations

Scaling up necessitates more sophisticated instrumentation and control systems to monitor and regulate key parameters accurately. Automation becomes crucial for maintaining process stability and achieving consistent results (Gomes et al., 2018).

During bioreactor designing and process scale-up of optimized conditions, cost-effectiveness becomes a crucial factor before implementation. Therefore, evaluation of the overall economics of scale-up, including capital and operating costs, helps in the determination of the feasibility and profitability of the bioethanol production process. A comprehensive understanding of the selection criteria for bioreactor design, factors affecting the design, process implementation, and the complexities associated with designing a reactor are essential for the development of efficient and economically viable bioprocesses. This knowledge is crucial for successful implementation in both laboratory and industrial contexts. Using process simulation software, the process could be improved in the near future with significantly diminished environmental impacts. However, achieving this positive outcome requires a comprehensive evaluation that considers all conversion steps and long-term regeneration of the bioresource (Sharma et al., 2020).

4.4 ANALYTICAL AND CONTROL STRATEGIES

Analytical and control strategies of operating bioreactors effectively is the key to providing an optimal environment to microorganisms or cellular components for their growth and productivity. The design and management of a bioreactor are fundamental to the fermentation process, which requires precise decisions about its design, size, control systems, and operational procedures (Pörtner et al., 2017). The success of a bioreactor is evaluated not only on the basis of output but also by the quality of the process, which can change with varying operational conditions. These changes can directly impact product quality, making the management of these varying conditions a critical focus of quality control. Maintaining the high-quality fermentation/bioprocess requires the continuous monitoring and adjustments in a bioreactor, including variables such as temperature, pH, nutrient supply, and oxygen levels to ensure they stay within optimal ranges for the organisms being cultured. Effective control and supervision of these parameters can greatly enhance the efficiency and consistency of fermentation processes, including bioethanol production. Applying these advances in process control and monitoring to bioethanol production can result in achieving consistent product quality, higher yields, and more efficient resource use. This highlights the broader importance of precision in bioreactor operation and management across various biotechnological applications. The techniques and strategies developed for controlling industrial fermentations are readily applicable to bioethanol production, offering significant opportunities for improvement in this field (Simutis & Lübbert, 2015). Moving beyond the initial phase of biorefineries, new experimental methods have significantly advanced for ethanol production. However, challenges remain due to the complexity of raw materials and difficulties in real-time fermentation/bioprocess monitoring. The implementation of online measurement of crucial process variables could greatly enhance bioethanol production for the process control, suggesting a promising avenue for improving bioethanol production methods (Gomes et al., 2018). Researchers are developing techniques and tools to characterize biomolecules for online and at-line process monitoring. These innovations enable better control strategies, integrated with quality by design (QbD) and process analytic technology (PAT). The process variables pose significant challenges in the designing of bioreactor controllers; however, recent advancements in measurements and sensor technologies offer solutions to these issues. One effective strategy to address these challenges is the integration of the QbD approach in biorefineries to minimize process variability (Singh, 2015). This has led to a paradigm shift from traditional biorefinery operations toward more controlled and predictable processes. A key component of QbD is PAT, which enables real-time monitoring and control, ensuring consistent product quality. PAT plays a crucial role in biorefineries that process raw materials from diverse sources, allowing for dynamic control strategies to maintain process efficiency and reliability (Rathore, 2009; Rathore, 2014).

4.4.1 Analytical Approaches in Bioethanol Production

In large-scale biorefineries, effective process monitoring is essential for ensuring smooth and efficient operations. However, despite their critical role in process control, the development of robust online sensors specifically designed for biorefinery applications has not received adequate attention. Accurate and reliable analytic measurement is vital for real-time monitoring and assessment of biofuel production, enabling early detection of potential issues before they become irreversible. Figure 4.2 illustrates various PAT tools utilized in bioethanol production and other bioprocessing applications. By integrating PAT tools, real-time measurements can

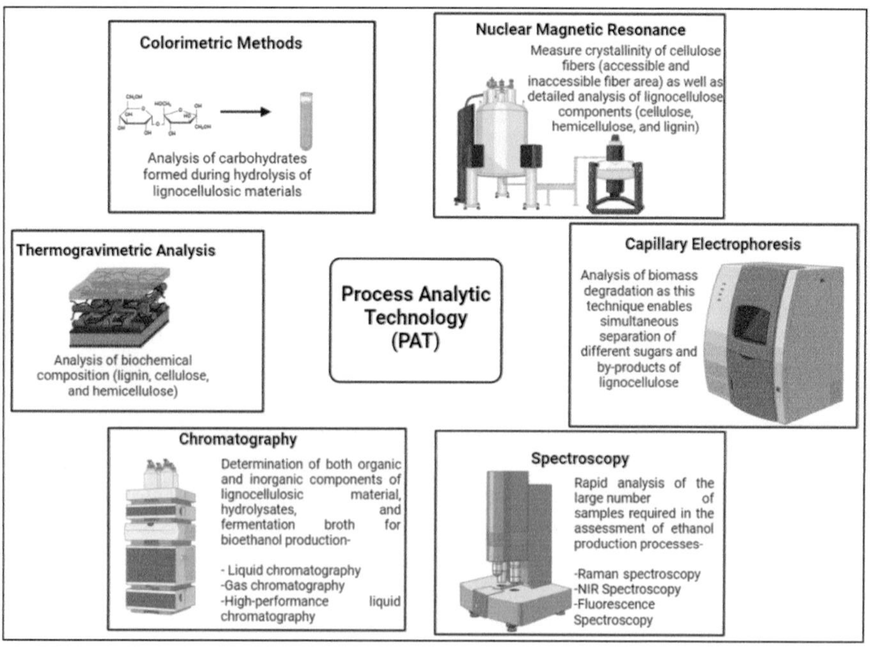

FIGURE 4.2 Process Analytic Technology (PAT) tools employed in bioethanol production (present study).

be performed, allowing the identification of process variability and enabling proactive decision-making to prevent system failures (Rathore et al., 2016).

Colorimetric assays offer straightforward analytical techniques useful for tracking carbohydrates produced in the hydrolysis of lignocellulosic substances, which directly link the absorbance (intensity of color) of a solution that is a direct indication of carbohydrates to the actual carbohydrate concentration. Colorimetric methods are gradually being phased out and replaced by more sophisticated analytical PAT tools that provide real-time and instantaneous data (Gomes et al., 2018). Thermogravimetric analysis (TGA) has proven to be a valuable technique for determining the composition of LCBs. This method involves continuously measuring the mass of a sample as its temperature changes over time, providing insights into thermal stability and decomposition characteristics (Lupoi et al., 2015). The TGA primarily measures mass, temperature, and time, from which additional measures may be derived. Thus, offering several advantages, including being less expensive, faster, and easier to use compared to traditional methods. Saldarriaga et al. (2015) proposed a methodology utilizing TGA to calculate moisture content, perform proximate analysis, determine higher heating value (HHV), and conduct elemental analysis (C, H, and O), which are essential for conditioning thermal balance in any combustion process.

The qualitative multi-component analysis of the range of chemical components present in LCB through chromatographic methods offers high levels of precision, accuracy, specificity, and sensitivity. These essential tools support both researchers and industry professionals in characterizing biomass components and monitoring process parameters. Chromatography is widely used to analyze organic and inorganic compounds in lignocellulosic materials, their hydrolysates, and the fermentation broths, ensuring composition assessment throughout processing. Gas chromatography (GC) is particularly a valuable tool for the identification of volatile and semi-volatile compounds and is widely used in biorefineries to measure sugar, ethanol, and lignin levels (Agblevor & Pereira, 2013). High-performance liquid chromatography (HPLC) is another chromatographic tool utilized, which uses a liquid mobile component of lignocellulosic biomass processing. Additionally, High-performance anion-exchange chromatography with pulsed amperometric detection (HPAEC-PAD) is an advanced tool employed for its capability to detect saccharides in lignocellulosic hydrolysates. Additionally, when coupled with mass spectrometry (MS), this method becomes a powerful analytical tool for evaluating improvements in pretreatment processes and the selective use of substrates by microbes (Coulier et al., 2013). Nuclear magnetic resonance (NMR) has emerged as a rapid and efficient at-line technique for evaluating biomass sources. Compared to traditional HPLC, which requires approximately an hour for analysis, NMR significantly reduces the processing time to just 20 h. This predictive tool is particularly valuable for analyzing monomeric sugars such as glucose, xylose, arabinose, and mannose in biomass hydrolysates. Additionally, NMR precisely determines the crystallinity of cellulose fibers, distinguishing between accessible and inaccessible regions. It also enables a comprehensive analysis, offering detailed structural insights (Pu et al., 2013).

Capillary electrophoresis (CE) is another alternative technique where traditional chromatographic techniques like GC or HPLC may not be suitable. It is commonly used for analyzing sugar samples with a strong alkaline nature or containing high salt concentrations. CE has the potential advantage of analyzing samples without the requirement of pre-derivatization or specific purification, making it valuable for examining biomass degradation. Additionally, CE facilitates the simultaneous separation of range of sugars and lignocellulose by-products (Aid et al., 2016). Spectroscopy is another advancement that has significant potential for the rapid analysis of the numerous samples needed in assessing ethanol production processes. Spectroscopy is a more time-efficient conventional method, which potentially reduces the costs associated with bioethanol production components. Techniques such as near-infrared (NIR), mid-infrared (MIR), and Fourier transform infrared (FT-IR) spectroscopy are quick, accurate, and provide detailed analyses of biomass materials, contrasting with the more labor-intensive and time-consuming wet chemical methods (Bryant et al., 2013). Recent research efforts have been directed toward lowering the operational cost of bioethanol production by utilizing low-cost feedstocks, improving microorganism efficiency and resilience, refining bioprocessing integrations, and reducing waste by-products. These collective efforts aim to make bioethanol production more cost-effective and sustainable, contributing to the advancement of renewable energy technologies. However, the areas of process control and optimization have received comparatively less attention. Furthermore, the complexity associated with the feedstock utilized in biomass-fed biorefineries underscores the urgent need for improved process efficiency and productivity, highlighting the importance of implementing online sensors for the long-term economic viability of bioethanol production processes (Gomes et al., 2018).

4.4.2 Control Strategies of Bioreactor in Bioethanol Production

Biorefinery operations have primarily relied on unsteady batch or fed-batch modes, with only limited examples of integrated and continuous operations at full-scale industrial operations. However, the intricate nature of biorefinery processes presents a significant challenge in scaling up models and control systems from laboratory and pilot scales to full-scale industrial operations (Rathore et al., 2016). Zentou et al. (2023) proposed an integrated model to optimize fermentation-pervaporation systems for bioethanol recovery, supporting the transition toward continuous fermentation in industrial settings regardless of the operational mode, analyzing the time profile of various variables remains a challenge. Key difficulties in modeling such processes include the following (Jiménez-González & Woodley, 2010; Zentou et al., 2023):

i. Predicting the rate of fermentable substrate generation in successive biochemical reactions,
ii. Understanding the microbial growth and substrate preferences in complex fermentation broth

iii. Estimating model parameters in dynamic, real-world environments,
iv. Addressing by-product inhibition and its impact on fermentation efficiency

The primary objective of any process control strategy is to maximize process efficiency while maintaining process stability. A well-defined theoretical model, combined with real-time monitoring tools or parameter estimation techniques, enables the development of a reliable control framework that guides the process along an optimal trajectory. As the number of process objectives and system variability increase, the complexity of the control algorithm also rises. Various process control strategies have been implemented, ranging from basic on/off and proportional-integral-derivative (PID) control to more intricate model-adaptive control, fuzzy logic, artificial neural network schemes, and combinatorial controls (Gomes et al., 2018). These control strategies offer more flexibility in addressing different process dynamics and can adapt to changing conditions, thereby optimizing biorefinery operations and enhancing overall productivity. PID controllers are widely used control strategy adopted in various industrial processes, including the bioethanol production process. However, controlling variable parameters like pH and dissolved oxygen (DO) in any bioreactor system is challenging due to the dynamic nature of these systems. Traditional PID controllers may struggle to accurately control these variables in reactor system environments. Researchers have been optimizing PID controllers by simulation under dynamic conditions with incorporated to-tuning features, resulting in better control performance. These auto-tuned PID controllers have improved characteristics in terms of parameters like rise time, overshoot, and residual error compared to traditional PID controllers (Imtiaz et al., 2014). Such optimization in the PID control system improves the better regulation of crucial parameters in bioreactors, ultimately improving the efficiency and obtaining stabilized bioethanol production processes. Furthermore, for specialized bioprocesses requiring unique characteristic input, the standard PID settings don't offer smooth control performance. Therefore, additional tuning of the PID parameters by adjusting the proportional, integral, and derivative settings is required to improve the system's process control for better suitability with the dynamics of these specific bioprocesses. Thus, the control system can be further optimized to respond more effectively and efficiently to enhance the stability, precision, and overall performance (Rathore et al., 2021).

Model predictive control (MPC) is an advanced process control technique that enables real-time optimization of industrial processes. By leveraging a mathematical model of the system, MPC predicts how the process will evolve over a defined time horizon based on past and present states, as well as expected inputs. It then formulates an optimization problem to identify the most suitable control actions that either minimize a given cost function or meet a predefined objective. These computed actions are applied to the system in a closed-loop manner, with the process repeating at each time step. This iterative process allows MPC to continuously adjust the control inputs based on real-time measurements and predictions, thereby improving the control performance and system operation (Rashedi et al., 2022).

The primary distinction between model predictive control (MPC) and traditional control methods such as PID lies in the potential of MPC to compute the best input for each sampling moment. Unlike conventional approaches, MPC skillfully manages both linear and nonlinear constraints, whether they are equalities or inequalities. MPC can handle multiple inputs and outputs and work with models that are either linear or nonlinear. López-Pérez et al. (2015) employed advanced MPC with nonlinear model-based control strategies, which led to a significant enhancement in bioethanol production and achieved a 33% increase over the performance of traditional controllers.

Fuzzy control offers a practical solution for complex control scenarios, providing an accessible way to develop nonlinear controllers by utilizing heuristic information. This heuristic information often comes from an operator who manually controls a process, acting as a "human-in-the-loop" controller. The design methodology for fuzzy control involves consulting with the operator to document their insights and strategies on process control in the form of rules. The design methodology for fuzzy control involves the integration of the rules obtained from the consultation with the operator, effectively replicating the human's decision-making process. By leveraging the operator's practical experience and intuition, the fuzzy controller translates human reasoning into a systematic, automated control strategy that can handle complexity with the flexibility and adaptability characteristic of human operators (Passino et al., 1998; Gomes et al., 2018). Traditional control systems rely solely on mathematical models, whereas knowledge-based control systems integrate process data with model information to make decisions. Knowledge-based systems use the integration of empirical knowledge, expert rules, and prior data to analyze the current state of the process and determine appropriate control actions. A fuzzy logic controller (FLC) is a knowledge-based system that employs a set of rules to manage "subjective" input variables through a process known as fuzzification, which enables the system to infer "logical" decisions via an inference mechanism and then translates these decisions into specific outputs through defuzzification (Prasath et al., 2023). In the fed-batch enzymatic hydrolysis process, where FLCs have been used as online feeding control tools, FLCs were particularly successful in demonstrating superior performance with a 15% increase in cellulosic conversion compared to traditional batch hydrolysis processes (Tai et al., 2016).

Neural networks (NNs) are advanced control algorithms that are gaining attraction because of their ability to learn from historical or prior data and identify process characteristics. Essentially, the primary goal of a neural network is to establish a significant mapping function that is based on the data it has been trained on, thereby predicting expected outcomes. The effectiveness of a neural network depends on its training process and its ability to adapt to new datasets (Moldes et al., 2017). Hybrid control systems combine the strengths of empirical and mechanistic models to optimize overall system performance. For instance, integrating a traditional PID controller with an advanced control module like fuzzy logic or a neural network can significantly enhance its efficiency. This hybrid approach utilizes the advantages of both empirical and mechanistic models, allowing for improved adaptability and robustness in controlling complex processes involved in biorefineries and bioethanol production. Thus, the integrative approach leverages the unique trait of each control strategy, leading to more robust and adaptable control systems capable of tackling complex and dynamic process environments (Gomes et al., 2018). Table 4.2 depicts the comparison of various control systems utilized in biorefinery process operations. The seamless integration of process monitoring with advanced process control (APC) will play a pivotal role in shaping the future of the biorefinery industry. Advancements in spectroscopic technologies have significantly allowed for real-time monitoring of bioconversion reactions and provided valuable feedback data for APC systems. The integration of these technologies enables operators to closely monitor key process parameters and make timely adjustments to optimize performance. Additionally, the development of first-principle models based on fundamental principles and kinetics has shown potential promise in the prediction of the behavior of complex fermentation processes in bioethanol production. These models capture the dynamic interactions between various process variables in the process and offer insights into process optimization. The synergizing modeling and monitoring with advanced process control could significantly enhance the process efficiency of biorefineries and reduce overall production costs. This integration allows for continuous optimization of process parameters, maximizing yields with minimum resource usage, thus improving the competitiveness of lignocellulosic-based bioethanol compared to fossil-based bioethanol. Ultimately, these technological advancements contribute to the sustainability and environmental friendliness of the bioenergy industry.

4.4.3 Ethanol Yield Calculations in Biorefineries

The interpretation of ethanol yield from sugar during alcoholic fermentation varies across academia and industry. The ethanol concentration calculation in a batch cultivation can be evaluated through distillation and a colorimetric method, chromatographic methods (either GC or HPLC or spectroscopy) (described in section 4.1, analytical approaches in bioethanol production), or indirect estimation using a densitometer, leveraging the well-established correlation between ethanol-water composition and density within industrially relevant concentration ranges (Bermejo et al., 2021).

A straightforward equation can be used to determine the ethanol yield ($Y_{ETH/S}$) from experimental data, as shown in Equation 1:

$$Y_{ETH/S} = \left(ETH_{final} - ETH_{initial}\right) / \left(TRS_{initial} - TRS_{final}\right) \tag{1}$$

where ETH and TRS represent the total mass/moles of ethanol and total reducing sugars, respectively.

For continuous process, typically conducted in a bioreactor, the $Y_{ETH/S}$ can be determined using the same analytical approach as in batch process. In the case of steady-state operation, the $Y_{ETH/S}$ is calculated from experimental data, as shown in Equation 2:

$$Y_{ETH/S} = \left(F_{out} \times [ETH]_{out} - F_{in} \times [ETH]_{in}\right) / \left(F_{in} \times [TRS]_{in} - F_{out} \times [TRS]_{out}\right) \tag{2}$$

TABLE 4.2
Different Control Strategies Used in Bioethanol Production Operations

Control Strategies	Principle	Methodological Uniqueness	Challenges Encountered during Bioprocessing	Solution
PID	PID controllers measure the current state of the system using a sensor and compare this measurement to the desired set point, identifying any discrepancy as an error. This error is then processed by the PID algorithm, which consists of three key components: Proportional, Integral, and Derivative. Each of these components adjusts the controller's output in a way that aims to minimize the error. As a result, the controller affects the system by altering physical inputs, gradually correcting any deviation from the setpoint and stabilizing the system parameter at the desired level.	PID controllers work by continuously adjusting system controls based on feedback to maintain a specific target or setpoint.	In the realm of unique bioprocesses, such as those involving microbial cultures, the pre-set PID parameters often fall short in providing optimal process control.	In The composition of the microbial culture serves as the input variable for feedback PID control loops. Actuators in PID control directly impact the growth of organism and, by adjusting the system's outcome and enhance the reproducibility of bioprocesses. Thus, actuators can include variable parameters like pH, DO, temperature, addition of inducer components, as these parameters are crucial in influencing the population dynamics within the culture.
MPC	MPC enables real-time optimization of processes by utilizing a plant model to forecast the system's future behavior over a defined horizon. MPC is employed to determine the optimal input for the closed-loop system. Consequently, at each sample instance, the MPC algorithm repeats to compute the optimal control action for the immediate future, which is then executed in the closed-loop system. This repetitive predictive approach distinguishes MPC from traditional control methods.	Unlike conventional control techniques such as PID, where control actions are determined based on current measurements alone, MPC calculates the optimal input at each instance, taking into account future predictions.	A significant obstacle encountered in the implementation of advanced control strategies within bioprocesses has been the absence of real-time monitoring tools and the challenges associated with developing comprehensive plant models.	The integration of statistical tools, chemometric methods, and spectroscopic technology in biorefinery operations holds promise for enhancing process efficiency, reducing costs, and advancing the sustainability of bio-based industries.

FLC	FLC consists of a set of control rules designed to handle "subjective" input variables and make a "logical" decision using an inference mechanism. By incorporating fuzzy logic principles, FLCs can effectively handle imprecise or vague information, allowing for more flexible and adaptive control in various applications. Top of Form	FLCs are knowledge-based control systems that leverage both process data and model information. They excel at incorporating expert knowledge into an automatic control strategy.	The effectiveness of FLCs indeed depends on the quality and quantity of process information available.	Using heuristic inference improves performance in different metabolic conditions.
NNs	NN establishes a meaningful mapping function from the trained data to generate an expected response. NNs serve as powerful tools for learning and approximating complex mappings between input and output data.	The challenge of developing a detailed mechanistic model for a biorefinery is compounded by the absence of dependable online monitoring tools and the inherent complexity of the process. However, rather than relying solely on mechanistic models, the data collected can be efficiently utilized to train neural networks. These neural networks can subsequently be deployed for effective process control.	In NN model merely mapping the input and output data are generally useful for predictive purposes. However, these predictions often lack the depth needed to fully understand the internal workings of a system which applies to both mechanistic and black box models.	Hybrid control approaches are being increasingly considered for biorefinery operations.
Hybrid control systems	The integration of empirical and mechanistic models in hybrid modeling approaches allows for the combination of their individual advantages to improve overall performance.	Hybrid control systems leverage the strengths of both black-box systems, which are data-driven and can capture complex relationships in the data, and mechanistic features, which provide a deeper understanding of the underlying processes.	The architecture of a hybrid control system plays a crucial role in determining its performance.	Integrating modeling, monitoring, and advanced process control in biorefinery operations holds significant promise for improving process efficiency and reducing production costs. By leveraging mathematical models to simulate and optimize biorefinery processes, operators can gain valuable insights into system behavior and identify opportunities for improvement.

where F represents the volumetric flow rate (volume per unit time), and [ETH] and [TRS] are the mass or molar concentrations (mass/volume or moles/volume) of ethanol and total reducing sugars, respectively. The subscripts 'in' and 'out' denote the outlet and the inlet streams of the bioreactor, respectively.

For fed-batch processes, ethanol yield can be determined using proper mass balance equations. Given that all relevant variables and parameters are known, the $Y_{ETH/S}$ can be calculated as following Equations 3 and 4:

$$d\left(V \times [TRS]\right)/dt = F_{in} \times [TRS]_{in} - F_{out} \times [TRS]_{out} - \left(\mu/YX/S\right) \times [X] \times V \quad (3)$$

$$\begin{aligned} d\left(V \times [ETH]\right)/dt = F_{in} \times [ETH]_{in} - F_{out} \times [ETH]_{out} \\ + \left(YETH/S \times \mu/YX/S\right) \times [X] \times V \end{aligned} \quad (4)$$

where V represents the cultivation volume, [TRS] is the total reducing sugars (mass/volume), F is the volumetric flow rate of the medium, μ is the specific growth rate ($time^{-1}$), [X] is the cell concentration, [ETH] is the ethanol concentration (in mass/volume), and YX/S is the biomass yield on sugar, respectively. The subscripts, 'in' corresponds for inlet and 'out' corresponds for the outlet stream of the bioreactor.

Pereira et al. (2018) introduced a refined approach for determining the ETHANOL YIELD based on mass balances, which offers greater accuracy and feasibility for widespread implementation. A key issue with traditional methodologies is their tendency to overestimate ethanol yield by failing to account for the volume occupied by yeast cells during fermentation. Additionally, many conventional approaches rely on raw material measurements, making them susceptible to variation in feedstock composition. To address these limitations, Pereira et al. (2018) proposed the exact mass of ethanol in the fermentation vat by multiplying the ethanol concentration by the volume of the cell-free broth. This method explicitly considers the volume occupied by the yeast cells, subtracting it from both the inoculum and wine volumes in the ethanol yield determination. The mass of ethanol produced during fermentation can be calculated using the following Equation 5:

$$ETH_{produced} = ETH_{final} - ETH_{initial} = V_{L\text{-}W} \times [ETH]_{wine} - V_{L\text{-}I} \times [ETH]_{inoculums} \quad (5)$$

where V_{L-W} and V_{L-I} represent the cell-free volumes of the fermented wine and of the inoculum, respectively. $[ETH]_{wine}$ and $[ETH]_{inoculum}$ are the concentrations in the fermented wine and the inoculums. While ethanol concentrations can be measured directly, the cell-free volumes are derived using Equations 6 and 7:

$$V_{L\text{-}W} = V_W \times \left[1 - \left(C_{Y\text{-}M\text{-}W \times \rho W}\right) / \left(_{\rho Y \times \sigma}\right)\right] \quad (6)$$

$$V_{L\text{-}I} = V_I \times \left[1 - \left(C_{Y\text{-}M\text{-}I \times \rho I}\right) / \left(_{\rho Y \times \sigma}\right)\right] \quad (7)$$

where V_W and V_I are the total volumes of the wine and of the inoculums, respectively. C_{Y-M-W} and C_{Y-M-I} represent the dry cell mass concentrations in the wine and in the inoculum, respectively. ρY is the yeast cell density, σ denotes the yeast dry mass fraction, and ρW and ρI correspond to the densities of the wine and of the inoculum, respectively. For fermentation industries to improve ethanol accuracy, investments are needed in better sampling designs, enhanced analytical capabilities, and additional personnel dedicated to conducting these measurements (Bermejo et al., 2021).

4.5 CONCLUSION

This chapter provides a comprehensive exploration of the complexity of the lignocellulosic ethanol production process, with a primary focus on bioreactor design intricacies and analytical techniques. Furthermore, it emphasizes the crucial role of bioreactors in facilitating key stages involving pretreatment, enzymatic saccharification, and different fermentation processes for converting biomass into ethanol. The discussion on bioreactor design principles, operational modes, and the importance of analytical approaches lays a solid foundation for understanding the complexities of bioethanol production. Highlighting the applications of bioreactors for specific processes underscores the importance of selecting the appropriate operational model to enhance efficiency. Moreover, the emphasis on analytical and control strategies for monitoring bioprocessing during different stages of bioethanol production with real-time sampling techniques is crucial for optimizing production and ensuring quality. In conclusion, advancements in bioreactor design, control strategies, and analytical methods are driving progress in lignocellulosic ethanol production. Despite the challenges, ongoing research and technological innovations offer promising prospects for achieving enhanced efficiency, sustainability, and economic viability in bioethanol production. The promise of lignocellulosic ethanol in reducing reliance on fossil fuels and lowering environmental impacts highlights the need for sustained investment in research and development. Advancing technological developments in this field will enable the bioethanol industry to pave the way for a sustainable energy future while addressing pressing global energy and environmental challenges.

REFERENCES

Afedzi, Alfred Elikem Kwami, Kittipong Rattanaporn, and Pramuk Parakulsuksatid. (2022), "Impeller selection for mixing high-solids lignocellulosic biomass in stirred tank bioreactor for ethanol production." Bioresource Technology Reports, 17, no. 2022, 100935.

Afedzi, Alfred Elikem Kwami, Kittipong Rattanaporn, Imrana Niaz Sultan, Muhammad Waseem Khan, Afrasiab Khan Tareen, and Pramuk Parakulsuksatid. (2023), "Comparison of marine and Rushton impeller in high solid loading of two-step pretreated oil palm trunk using simultaneous saccharification and fermentation for ethanol production." Results in Engineering, 19, 101273.

Agblevor, Foster A., and Junia Pereira. (2013), "Progress in the summative analysis of biomass feedstocks for biofuels production." In Aqueous Pretreatment of Plant Biomass for Biological and Chemical Conversion to Fuels and Chemicals, pp. 335–354. Wiley.

Aid, Tiina, Loore Paist, Margus Lopp, Mihkel Kaljurand, and Merike Vaher. (2016), "An optimized capillary electrophoresis method for the simultaneous analysis of biomass degradation products in ionic liquid containing samples." Journal of Chromatography A, 1447, 141–147.

Al-Awsi, Jasim, Salih A. Fatma, Salih Rushdi, and Zainab T. Al-Sharify. (2023), "A review of the effects of a variety of factors on the mass transfer coefficient in a bioreactor." In AIP Conference Proceedings, vol. 2787, no. 1. AIP Publishing.

Amândio, Mariana ST, Jorge MS Rocha, and Ana MRB Xavier. (2024), "Fed-batch SSF with pre-saccharification as a strategy to reduce enzyme dosage in cellulosic ethanol production." Fuel, 357, 129842.

Azhar, Siti Hajar, Rahmath Abdulla Mohd, Siti Azmah Jambo, Hartinie Marbawi, Jualang Azlan Gansau, Ainol Azifa Mohd Faik, and Kenneth Francis Rodrigues. (2017), "Yeasts in sustainable bioethanol production: A review." Biochemistry and Biophysics Reports, 10, 52–61.

Balan, Venkatesh. (2014), "Current challenges in commercially producing biofuels from lignocellulosic biomass." International Scholarly Research Notices, 2014, no. 1, 463074.

Baruah, Julie, Bikash Kar Nath, Ritika Sharma, Sachin Kumar, Ramesh Chandra Deka, Deben Chandra Baruah, and Eeshan Kalita. (2018), "Recent trends in the pretreatment of lignocellulosic biomass for value-added products." Frontiers in Energy Research, 6, 141.

Basak, Bikram, Ramesh Kumar, AVSL Sai Bharadwaj, Tae Hyun Kim, Jung Rae Kim, Min Jang, Sang-Eun Oh, Hyun-Seog Roh, and Byong-Hun Jeon. (2023), "Advances in physicochemical pretreatment strategies for lignocellulose biomass and their effectiveness in bioconversion for biofuel production." Bioresource Technology, 369, 128413.

Beluhan, Sunčica, Katarina Mihajlovski, Božidar Šantek, and Mirela Ivančić Šantek. (2023), "The production of bioethanol from lignocellulosic biomass: Pretreatment methods, fermentation, and downstream processing." Energies, 16, no. 19, 7003.

Bermejo, Pamela Magalí, Alberto Badino, Luciano Zamberlan, Vijayendran Raghavendran, ThiagoOlitta Basso, and Andreas Karoly Gombert. (2021), "Ethanol yield calculations in biorefineries." FEMS Yeast Research, 21, no. 8, foab065.

Bryant, David N., Elliot Firth, Naheed Kaderbhai, Stephen Taylor, Stephen M. Morris, Daniel Logan, Naroa Garcia, Andrew Ellis, Steven M. Martin, and Joe A. Gallagher. (2013), "Monitoring real-time enzymatic hydrolysis of distillers dried grains with solubles (DDGS) by dielectric spectroscopy following hydrothermal pre-treatment by steam explosion." Bioresource Technology, 128, 765–768.

Buyondo, John Paul, and Shijie Liu. (2012), "Processes and bioreactor designs for butanol production from lignocellulosic biomass." Journal of Bioprocess Engineering and Biorefinery, 1, no. 1, 53–68.

Coulier, Leon, Ying Zha, Richard Bas, and Peter J. Punt. (2013), "Analysis of oligosaccharides in lignocellulosic biomass hydrolysates by high-performance anion-exchange chromatography coupled with mass spectrometry (HPAEC–MS)." Bioresource Technology, 133, 221–231.

De la Fuente, Salcido, Norma Margarita, and Axel Misael Alejo Andrade. (2019), "Challenges of fermentation engineering." In Advances in Food Bioproducts and Bioprocessing Technologies, pp. 233–264. CRC Press.

de Oliveira, Raul Alves, Leda Maria Fortes Gottschalk, Suely Pereira Freitas, and Elba Pinto da Silva Bon. (2018), "One-vessel saccharification and fermentation of pretreated sugarcane bagasse using a helical impeller bioreactor." Biomass Conversion and Biorefinery, 8, 1–10.

Devi, Arti, Somvir Bajar, Havleen Kour, Richa Kothari, Deepak Pant, and Anita Singh. (2022), "Lignocellulosic biomass valorization for bioethanol production: A circular bioeconomy approach." Bioenergy Research, 15, no. 4, 1820–1841.

Dey, Pinaki, Vivek Rangarajan, Jayato Nayak, Diganta Bhusan Das, and Steeve Branden Wood. (2021), "An improved enzymatic pre-hydrolysis strategy for efficient bioconversion of industrial pulp and paper sludge waste to bioethanol using a semi-simultaneous saccharification and fermentation process." Fuel, 294, 120581.

Di Donato, Paola, Ilaria Finore, Annarita Poli, Barbara Nicolaus, and Licia Lama. (2019), "The production of second generation bioethanol: The biotechnology potential of thermophilic bacteria." Journal of Cleaner Production, 233, 1410–1417.

Du, Jian, Fazhan Zhang, Yuanyuan Li, Hongman Zhang, Jingrui Liang, Hongbo Zheng, and He Huang. (2014), "Enzymatic liquefaction and saccharification of pretreated corn stover at high-solids concentrations in a horizontal rotating bioreactor." Bioprocess and Biosystems Engineering, 37, 173–181.

Edeh, Ifeanyichukwu. (2021), "Bioethanol production: An overview. Bioethanol technologies." In Waste Biorefinery, vol. 1, pp. 727–749. Intechopen.

Gomes, James, Jyoti Batra, Viki R. Chopda, P. Kathiresan, and Anurag S. Rathore. (2018), "Monitoring and control of bioethanol production from lignocellulosic biomass." In Waste Biorefinery, pp. 727–749. Elsevier.

Gupta, Rishi, Sanjay Kumar, James Gomes, and Ramesh Chander Kuhad. (2012), "Kinetic study of batch and fed-batch enzymatic saccharification of pretreated substrate and subsequent fermentation to ethanol." Biotechnology for Biofuels, 5, 1–10.

Hossain, Zabed, Jitendra Nath Sahu, and Akter Suely. (2017), "Bioethanol production from lignocellulosic biomass: An overview of pretreatment, hydrolysis, and fermentation." In Sustainable Utilization of Natural Resources, pp. 145–186. CRC Press

Imamoglu, Esra, and Fazilet Vardar Sukan. (2013), "Scale-up and kinetic modeling for bioethanol production." Bioresource Technology, 144, 311–320.

Imtiaz, Ubaid, Sudhanshu S. Jamuar, Jitendra N. Sahu, and P. B. Ganesan. (2014), "Bioreactor profile control by a nonlinear auto regressive moving average neuro and two degree of freedom PID controllers." Journal of Process Control, 24, no. 11, 1761–1777.

Ingledew, William M. M., and Yih-Huei Lin. (2011), "Ethanol from starch-based feedstocks." In Comprehensive Biotechnology, pp. 37–49. Elsevier.

Jiménez-González, Concepción, and John M. Woodley. (2010), "Bioprocesses: Modeling needs for process evaluation and sustainability assessment." Computers & Chemical Engineering, 34, no. 7, 1009–1017.

Kadam, Kiran L., C. Eric Rydholm, David James McMillan. (2004), "Development and validation of a kinetic model for enzymatic saccharification of lignocellulosic biomass." Biotechnology Progress, 20, no. 3, 698–705.

Kartawiria, Irvan Setiadi, Livia Serafin, and Haznan Abimanyu. (2019), "Effect of the substrate concentration and the stirring rate on the enzymatic hydrolysis of cellulose from pre-treated corn cob. Derivation of a kinetic model." Journal of Chemical Technology & Metallurgy, 54, no. 4.

Kaur, Sandeep, Loveleen Kaur Sarao Ankita, and Harmeet Singh. (2023), "Bioenergy: Challenges ahead and future." In Agroindustrial Waste for Green Fuel Application, pp. 281–311. Springer Nature Singapore.

Koppram, Rakesh, Fredrik Nielsen, Eva Albers, Annika Lambert, Sune Wännström, Lars Welin, Guido Zacchi, and Lisbeth Olsson. (2013), "Simultaneous saccharification and co-fermentation for bioethanol production using corncobs at lab, PDU and demo scales." Biotechnology for Biofuels, 6, 1–10.

Kordala, Natalia, Marcin Walter, Bartosz Brzozowski, and Małgorzata Lewandowska. (2024), "2G-biofuel ethanol: An overview of crucial operations, advances and limitations." Biomass Conversion and Biorefinery, 14, no. 3, 2983–3006.

Kumar, Bikash, Nisha Bhardwaj, Komal Agrawal, Venkatesh Chaturvedi, and Pradeep Verma. (2020), "Current perspective on pretreatment technologies using lignocellulosic biomass: An emerging biorefinery concept." Fuel Processing Technology, 199, 106244.

Kumar, Vanka Sravan, Birupakshya Mishra, Thogaru Satish, Alka Kumari, and Siddhartha Moulik. (2022), "Hydrodynamic cavitation as a sustainable and promising technology for pre-treatment of biomass." Chemical Industry Digest, 40.

Lafuente-Rincón, David Francisco, Inty Omar Hernández De Lira, Héctor Hernández Escoto, María Alejandra Sánchez-Muñoz, Héctor Hugo Molina Correa, Cristian Emanuel Gámez-Alvarado, Perla Araceli Meléndez-Hernández, and Javier Ulises Hernández-Beltrán. (2022), "Design and engineering parameters of bioreactors for production of bioethanol." In Bioethanol, pp. 233–269. Apple Academic Press.

Liu, Yunyun, Xianli Xie, Wenhuan Liu, Huijuan Xu, and Yunqi Cao. (2020), "Consolidated bioprocess for bioethanol production from lignocellulosic biomass using *Clostridium thermocellum* DSM 1237." BioResources, 15, no. 4, 8355.

López-Pérez, Pablo A., Fernando A. Cuevas-Ortiz, Rigel V. Gómez-Acata, and Ricardo Aguilar-López. (2015), "Improving bioethanol production via nonlinear controller with noisy measurements." Chemical Engineering Communications, 202, no. 11, 1438–1445.

Ludwig, Daniel, Buchmann Michael, Thomas Hirth, Steffen Rupp, and Susanne Zibek. (2014), "High solids enzymatic hydrolysis of pretreated lignocellulosic materials with a powerful stirrer concept." Applied Biochemistry and Biotechnology, 172, 1699–1713.

Lupoi, Jason S., Seema Singh, Ramakrishnan Parthasarathi, Blake A. Simmons, and Robert J. Henry. (2015), "Recent innovations in analytical methods for the qualitative and quantitative assessment of lignin." Renewable and Sustainable Energy Reviews, 49, 871–906.

Mahboubi, Amir, Beray Cayli, Gülru Bulkan, Wim Doyen, Heleen De Wever, and Mohammad J. Taherzadeh. (2018), "Removal of bacterial contamination from bioethanol fermentation system using membrane bioreactor." Fermentation, 4, no. 4, 88.

Manna, Dinabandhu, Ranjana Chowdhury, Rajnish K. Calay, and Mohamad Y. Mustafa. (2024), "Experimental insights into the fermentation of pyro-syngas to ethanol in a semi-batch and continuous stirred bioreactor with mathematical modelling and optimization." Energies, 17, no. 3, 562.

Moldes, Olga A., Rosa Juan, Cristina Mejuto, Rosa Rial-Otero, and Jesus Simal-Gandara. (2017), "A critical review on the applications of artificial neural networks in winemaking technology." Critical Reviews in Food Science and Nutrition, 57, no. 13, 2896–2908.

Olivieri, Giuseppe, René H. Wijffels, Antonio Marzocchella, and Maria Elena Russo. (2021), "Bioreactor and bioprocess design issues in enzymatic hydrolysis of lignocellulosic biomass." Catalysts, 11, no. 6, 680.

Olughu, Onu, Lope G. Tabil, Tim Dumonceaux, Edmund Mupondwa, Duncan Cree, and Xue Li. (2023), "Technoeconomic analysis of a fungal pretreatment-based cellulosic ethanol production." Results in Engineering, 19, 101259.

Palmqvist, Benny, and Gunnar Lidén. (2012), "Torque measurements reveal large process differences between materials during high solid enzymatic hydrolysis of pretreated lignocellulose." Biotechnology for Biofuels, 5, 1–9.

Passino, Kevin M., Michael Stephen Yurkovich, and Fritz Reinfrank. (1998), Fuzzy Control, vol. 42. Addison-Wesley.

Pereira, Rauber D., Kaio CS Rodrigues, Jorge LS Sonego, Antonio JG Cruz, and Alberto C. Badino. (2018), "A new methodology to calculate the ethanol fermentation efficiency at bench and industrial scales." Industrial & Engineering Chemistry Research, 57, no. 48, 16182–16191.

Pino, Marcela Sofía, Rosa M. Rodríguez-Jasso, Michele Michelin, Adriana C. Flores-Gallegos, Ricardo Morales-Rodriguez, José A. Teixeira, and Héctor A. Ruiz. (2018), "Bioreactor design for enzymatic hydrolysis of biomass under the biorefinery concept." Chemical Engineering Journal, 347, 119–136.

Pörtner, Ralf, O. Platas Barradas, Björn Frahm, and Volker C. Hass. (2017), "Advanced process and control strategies for bioreactors." In Current Developments in Biotechnology and Bioengineering, pp. 463–493. Elsevier.

Prasath, N., A. Arun, B. Saravanan, and Kanagaraj Kamaraj. (2023), "Intelligent fuzzy edge computing for real-time decision making in IoT-based digital twin environments." Journal of Intelligent & Fuzzy Systems, Preprint, 1–12.

Pu, Yunqiao, Bassem Hallac, and Arthur J. Ragauskas. (2013), "Plant biomass characterization: application of solution-and solid-state NMR spectroscopy." In Aqueous Pretreatment of Plant Biomass for Biological and Chemical Conversion to Fuels and Chemicals, pp. 369–390. Wiley.

Raj, Kanak, and Chandraraj Krishnan. (2020), "Improved co-production of ethanol and xylitol from low-temperature aqueous ammonia pretreated sugarcane bagasse using two-stage high solids enzymatic hydrolysis and Candida tropicalis." Renewable Energy, 153, 392–403.

Raj, Tirath, K. Chandrasekhar, A. Naresh Kumar, J. Rajesh Banu, Jeong-Jun Yoon, Shashi Kant Bhatia, Yung-Hun Yang, Sunita Varjani, and Sang-Hyoun Kim. (2022), "Recent advances in commercial biorefineries for lignocellulosic ethanol production: Current status, challenges and future perspectives." Bioresource Technology, 344, 126292.

Rashedi, Mohammad, Hamid Khodabandehlou, Matthew Demers, Tony Wang, and Christopher Garvin. (2022), "Model predictive controller design for bioprocesses based on machine learning algorithms." IFAC-PapersOnLine, 55, no. 7, 45–50.

Rathore, Anurag S. (2009), "Roadmap for implementation of quality by design (QbD) for biotechnology products." Trends in Biotechnology, 27, no. 9, 546–553.

Rathore, Anurag S. (2014), "QbD/PAT for bioprocessing: Moving from theory to implementation." Current Opinion in Chemical Engineering, 6, 1–8.

Rathore, Anurag S., Viki R. Chopda, and James Gomes. (2016), "Knowledge management in a waste based biorefinery in the QbD paradigm." Bioresource Technology, 215, 63–75.

Rathore, Anurag S., Somesh Mishra, Saxena Nikita, and Priyanka Priyanka. (2021), "Bioprocess control: Current progress and future perspectives." Life, 11, no. 6, 557.

Riedlberger, Peter, and Dirk Weuster-Botz. (2012), "New miniature stirred-tank bioreactors for parallel study of enzymatic biomass hydrolysis." Bioresource Technology, 106, 138–146.

Rodríguez-Jasso, Rosa M., Iosvany López-Sandin, Miguel A. Aguilar, Claudia M. López-Badillo, and Héctor A. Ruiz. (2023), "Intensification of enzymatic saccharification at high solid loading of pretreated agave bagasse at bioreactor scale." Journal of Environmental Chemical Engineering, 11, no. 1, 109257.

Saha, Koel, Abhishek Maharana, Jaya Sikder, Sudip Chakraborty, Stefano Curcio, and Enrico Drioli. (2019), "Continuous production of bioethanol from sugarcane bagasse and downstream purification using membrane integrated bioreactor." Catalysis Today, 331, 68–77.

Saldarriaga, Juan F., Roberto Aguado, Aitor Pablos, Maider Amutio, Martin Olazar, and Javier Bilbao. (2015), "Fast characterization of biomass fuels by thermogravimetric analysis (TGA)." Fuel, 140, 744–751.

Saravanan, Azhaham Perumal, Thangavel Mathimani, Garlapati Deviram, Karthik Rajendran, and Arivalagan Pugazhendhi. (2018), "Biofuel policy in India: A review of policy barriers in sustainable marketing of biofuel." Journal of Cleaner Production, 193, 734–747.

Sharma, Apurav, Raman Soni, and Sanjeev Kumar Soni. (2024), "From waste to wealth: Exploring modern composting innovations and compost valorization." Journal of Material Cycles and Waste Management, 26, no. 1, 20–48.

Sharma, Bhawna, Christian Larroche, and Claude-Gilles Dussap. (2020), "Comprehensive assessment of 2G bioethanol production." Bioresource Technology, 313, 123630.

Simutis, Rimvydas, and Andreas Lübbert. (2015), "Bioreactor control improves bioprocess performance." Biotechnology Journal, 10, no. 8, 1115–1130.

Singh, Jatinder. (2015), "International conference on harmonization of technical requirements for registration of pharmaceuticals for human use." Journal of Pharmacology and Pharmacotherapeutics, 6, no. 3, 185–187.

Tai, Chao, Diego S. Voltan, Deepak R. Keshwani, George E. Meyer, and Pankaj S. Kuhar. (2016), "Fuzzy logic feedback control for fed-batch enzymatic hydrolysis of lignocellulosic biomass." Bioprocess and Biosystems Engineering, 39, 937–944.

Tareen, Afrasiab Khan, Vittaya Punsuvon, Imrana Niaz Sultan, Muhammad Waseem Khan, and Pramuk Parakulsuksatid. (2021), "Cellulase addition and pre-hydrolysis effect of high solid fed-batch simultaneous saccharification and ethanol fermentation from a combined pretreated oil palm trunk." ACS Omega, 6, no. 40, 26119–26129.

Toor, Manju, Smita S. Kumar, Sandeep K. Malyan, Narsi R. Bishnoi, Thangavel Mathimani, Karthik Rajendran, and Arivalagan Pugazhendhi. (2020), "An overview on bioethanol production from lignocellulosic feedstocks." Chemosphere, 242, 125080.

Um, Byung-Hwan, and Thomas R. Hanley. (2008), "High-solid enzymatic hydrolysis and fermentation of solka floc into ethanol." Journal of Microbiology and Biotechnology, 18, no. 7, 1257–1265.

Van Zyl, Willem H., Lee R. Lynd, Riaan den Haan, and John E. McBride. (2007), "Consolidated bioprocessing for bioethanol production using *Saccharomyces cerevisiae*." Biofuels, 205–235.

Wallin, Elin. (2019), "Design and Scheduling of a Bioreactor System Model for Production of 2G Ethanol." Master's Thesis in Innovative and Sustainable Chemical Engineering. University of Technology Gothenburg, Sweden.

Wen, Tao, Lin Lu, Weifeng He, and Yunran Min. (2020), "Fundamentals and applications of CFD technology on analyzing falling film heat and mass exchangers: A comprehensive review." Applied Energy, 261, 114473.

Yang, Yang, Meng Zhang, Jikai Zhao, and Donghai Wang. (2023), "Effects of particle size on biomass pretreatment and hydrolysis performances in bioethanol conversion." Biomass Conversion and Biorefinery, 13, no. 14, 13023–13036.

Zentou, Hamid, Zurina Z. Abidin, Mohammed Abdullah Issa, Robiah Yunus, and Dayang Radiah Awang Biak. (2023), "Developing an integrated mathematical model of fermentation-pervaporation system for bioethanol production." Chemical Engineering Journal, 473, 145229.

Zhang, Jian, Deqiang Chu, Juan Huang, Zhanchun Yu, Gance Dai, and Jie Bao. (2010), "Simultaneous saccharification and ethanol fermentation at high corn stover solids loading in a helical stirring bioreactor." Biotechnology and Bioengineering, 105, no. 4, 718–728.

Zhang, Yongli, Tao Sun, Taju Wu, Jinyang Li, Die Hu, Defei Liu, Jingen Li, and Tian. (2023), "Consolidated bioprocessing for bioethanol production by metabolically engineered cellulolytic fungus *Myceliophthora thermophila*." Metabolic Engineering, 78, 192–199.

Zuliani, Luca, Annabel Serpico, Mario De Simone, Nicola Frison, and Salvatore Fusco. (2021), "Biorefinery gets hot: Thermophilic enzymes and microorganisms for second-generation bioethanol production." Processes, 9, no. 9, 1583.

5 Enzymatic Hydrolysis
A Key Process

*Yashika Raheja, Tayyab Islam, Vivek Kumar Gaur, and Bhupinder Singh Chadha**

5.1 INTRODUCTION

Several biorefineries are working toward reducing the reliance on petroleum-based industrial products and are exploring lignocellulosic-based organic residues for developing sustainable and greener technologies (Mujtaba et al., 2023). The large pool of agricultural waste represents a major renewable source of lignocellulosic biomass that can be used to produce a wide portfolio of biofuels and other value-added products (Rabelo et al., 2023). More than 50% of the biomass by weight is composed of cellulose, which is a homopolymer of glucose subunits linked together by β-(1,4) glycosidic bonds and is intertwined with heterogenic polymer hemicellulose (10–40%) and a complex polymer of lignin (15–30%) (Mankar et al., 2021). However, one of the major challenges in the valorization of lignocellulosic is the recalcitrant nature and the presence of strong interconnected bonding between the matrix components of lignocellulosic (Velvizhi et al., 2022). To disintegrate and fractionate these individual components, an efficient pretreatment process is required that can disentangle cellulose fibers and improve their accessibility to the lignocellulolytic enzymes (Dahmen et al., 2019). In the recent past, many demonstration/pilot and commercial-scale plants have been established that employ thermochemical methods (dilute acid/alkali, steam explosion, liquid hot water) for the biomass pretreatment but involve high capital expenditure (CAPEX) and operational expenditure (OPEX) (Chandel et al., 2024). The steam/acid pretreatment also results in the formation of three major classes of inhibitors, i.e., organic acids (formic acid, acetic acid, and levulinic acid), furan derivatives (2-furaldehyde and 5-hydroxymethyl-2-furaldehyde (HMF)), and phenols (4-hydroxybenzaldehyde, syringaldehyde, vanillin, cinnamic acid, and ferulic acid). Owing to the process economics and to avoid losses of released sugars during substrate washing, the biorefineries prefer to use unwashed slurry for subsequent enzymatic hydrolysis (Chandel et al., 2024). However, the addition of a lignocellulolytic enzyme mixture to the pretreated biomass slurry laden with inhibitors creates a hostile environment that negatively impacts the rate of enzyme catalysis as a result, associated with enzyme denaturation (Chandel et al., 2019; Devi et al., 2021; Neis & da Silva Pinto, 2021; Reis et al., 2023). To overcome this effect, bioprospecting of catalytically efficient, robust lignocellulolytic enzymes that are resistant to inhibitors and can reduce the overall cost of the enzymatic hydrolysis is highly desirable. This chapter discusses the bioprospecting of lignocellulolytic enzymes and the

* Corresponding author

DOI: 10.1201/9781003485636-5

recent advancement in the development of the customized enzyme cocktails that are efficient for the hydrolysis of different pretreated biomass.

5.2 LIGNOCELLULOSIC BIOMASS

Cellulose is the most prominent and the most uniform component present in all lignocellulosic biomass, accounting for up to 50% by weight. Cellulose is a syndiotactic homopolymer comprised of anhydro glucopyranose subunits (up to 15000) linked together by β-(1,4) glycosidic bonds (Figure 5.1). Every anhydro glucose subunit is rotated by 180° in the plane to accommodate the acetal oxygen bridges between the C4 and C1 hydroxyl groups, resulting in the formation of a dimer known as cellobiose (Coseri, 2017). The network of intra- and inter-hydrogen bonds between the individual cellulose chains results in a crystalline region (high crystallinity) and another amorphous region (low crystallinity) that imparts strength, durability, and stiffness to the cellulose (Zhang et al., 2017). The X-ray diffraction (XRD) and nuclear magnetic resonance (NMR) studies have shown that cellulose exists in four different allomorphs (I, II, III, and IV). Of this CI form, it is considered the most important with potential application in bioenergy generation. However, highly crystalline form CI is

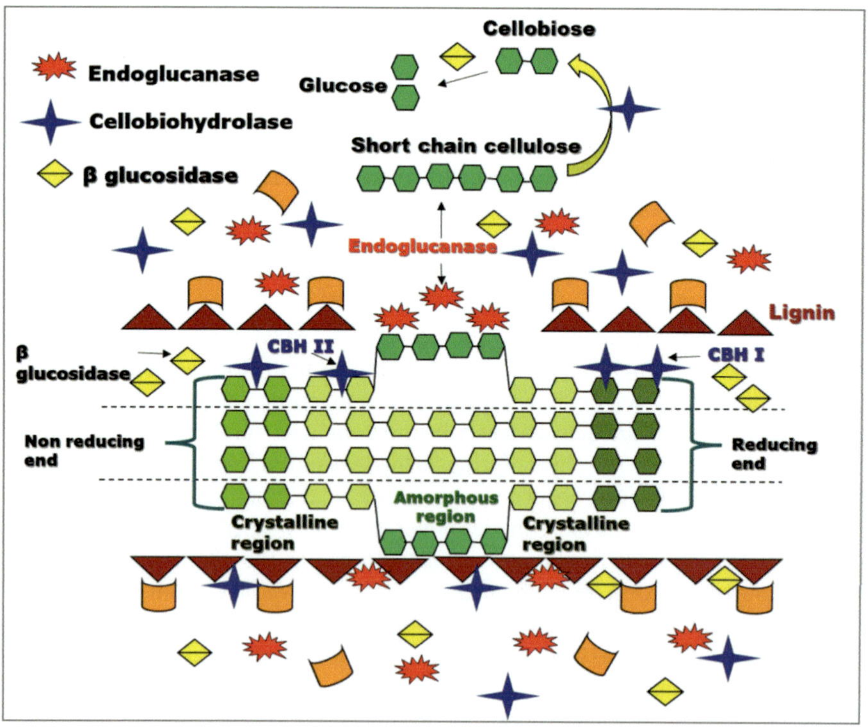

FIGURE 5.1 Conversion of cellulose into monomeric sugars by the synergistic action of cellulase enzyme (Endoglucanase, Cellobiohydrolase, and β-glucosidase) (Jain et al., 2018; Singh et al., 2021).

a mixture of two forms, i.e., triclinic (cellulose Iα) and monoclinic (cellulose Iβ), the latter being more recalcitrant to enzymatic attack (Gao et al., 2013). Unlike cellulose, the hemicelluloses are heterogeneous, branched polymers made up of different types of monomers that engulf the cellulose microfibrils. Hemicellulose is bound to lignin by aromatic esters and to cellulose by hydrogen bonds. The backbone of hemicellulosic biomass is comprised of pentose and hexose sugars, namely, xylose, galactose, arabinose, glucose, mannose, and sugar acids, such as galacturonic acid and glucuronic acid (Dodd & Cann, 2009). Xylan is the most abundant hemicellulose in cereal-derived agro-residues and is composed of β-1,4-linked *D*-xylose units, which are decorated with different side groups, such as *L*-arabinosyl, 4-*O*-methyl glucuronyl acid, feruloyl, acetyl, and *p*-coumaroyl moieties. Other important hemicelluloses are glucomannan, made of β-1,4-linked *D*-glucose and *D*-mannose with *D*-galactose side chains, and xyloglucans, which are *D*-glucose chains substituted with *D*-xylose. The two xylan chains are linked together by a di-ester bridge formed by ferulic acid, which can adhere the chains to the lignin (Rakotoarivonina et al., 2014). Lignin is a complex polymer of monolignols such as sinapyl, coumaryl, and coniferyl alcohols that polymerize to form guaiacyl (G), hydroxyphenyl (H), and syringyl (S) subunits. Lignin rich in guaiacyl alcohol exhibits stronger π–π interactions and recalcitrant β-5′ phenylcoumaran, β–β′ resinol, and 5-5′ dibenzodioxocin substructures. Sinapyl alcohol-rich lignins have higher β-*O*-4′ linkages and show less cross-linkage, and those having hydroxyphenyl features are made of low molecular weight and can be easily extracted and deconstructed (Zhao et al., 2017). The energy-dense recalcitrant form of lignin hampers the extraction of cellulosic and hemicellulosic fractions, which results in the overall rise in the cost of the bioconversion process. The research on devising methods to effectively valorize lignin to different value-added products (fuels and chemicals) is now being looked up to that can untap the economic value for the cleaner and sustainable energy production (Bilal et al., 2021). The concept of bioconversion of lignocellulosic biomass, which involves different steps: (a) reduction in the size of biomass, (b) pretreatment of LCB to make it more amenable to enzymatic attack, (c) saccharification of pretreated biomass by the lignocellulolytic enzymes into monomeric sugars, (d) microbial fermentation of sugars into bioethanol, and (e) recovery and purification of the product.

5.2.1 Carbohydrate-Active Enzymes (CAZymes) for Efficient Biomass Hydrolysis

CAZymes are categorized into cellulase (endoglucanases, cellobiohydrolases, and β-glucosidase), hemicellulases (xylanase, β-xylosidase, arabinofuranosidases, and acetyl/feruloyl esterases), and auxiliary enzymes (lytic polysaccharide monooxygenases (LPMOs), swollenin, and expansins) (Payne et al., 2015).

5.2.2 Cellulase Enzymes

The most notable cellulases include endo-1,4-β-*D*-glucanase (EG; EC 3.2.1.4), which randomly breaks the β-glucosidic bonds in the amorphous regions of cellulose, creating various glucan chain ends. In contrast, cellobiohydrolases (CBHI and CBHII; EC

3.2.1.91/176), which have a higher affinity for crystalline cellulose, act processively from the reducing or non-reducing ends of the polymer chains, releasing cellobiose as the primary end product (Igarashi et al., 2011). β-glucosidases/cellobiases (β-G; EC3.2.1.21) preferentially recognize cellobiose and soluble cello-oligosaccharides as substrates and catalyze the release of monomeric glucose units, relieving the cellobiohydrolase from the end product inhibition by cellobiose (Barbosa et al., 2020; Nguyen et al., 2018) (Figure 5.1).

5.2.3 Random Cutting Endoglucanases

The CAZy database suggests that the endoglucanases are classified into 17 different GH families: 5, 6, 7, 8, 12, 16, 44, 45, 48, 51, 64, 71, 74, 81, 87, 124, and 128 (Akram et al., 2018; Barbosa et al., 2020). Detailed analysis of the commercial cocktails shows that the cellulose disintegration is initiated by Cel5A, Cel7B, and Cel12A endoglucanases. EG II Cel5A, from the most extensively studied industrial model organism, *T. reesei*, has a shallow substrate binding pocket (2.5 Å) with an α1-β1 loop protruding toward the active site that slides toward the catalytic cleft, having two catalytic glutamates surrounded by an intensive network of hydrogen bonds. Interestingly, the XRD studies of GH5 EGs subclassified as GH5_5 have revealed the presence of only one catalytic module with the presence/absence of CBM, despite which they show high catalytic efficiency (Lee et al., 2011). GH5_5 endoglucanases specifically degrade β-1,4-glycosidic polysaccharides such as carboxymethyl cellulose (CMC) and phosphoric acid swollen cellulose (PASC); however, the presence of cellulose-binding module (CBM) attached to TrCel5A provides the ability to degrade Avicel (crystalline cellulose) as well (Vlasenko et al., 2010). Even though the TrCel5A formulates the important part of the commercial cellulase cocktail, the most widely studied class of endoglucanase is Cel7B. Most of the fungal Cel7B contains a cellulose-binding module (CBM) connecting the catalytic domain (CD) by a flexible, *O*-glycosylated linker. However, the unusual characteristic of Cel7B from *Humicola insolens* is that it possesses a single catalytic domain due to which it is catalytically active against the amorphous form of cellulose rather than the crystalline form. Strikingly, the secretome of *T. reesei* accounts for 10–15% of Cel7B EG that maintains aromatic stacking interaction naturally at the –4 subsite, mediated by tyrosine; however, Cel7B from *Humicola insolens and Fusarium oxysporum* has isoleucine at the –4 subsite and lacks aromatic interaction. Since the EGs are often able to cleave the amorphous regions rather than crystalline regions of cellulose (Figure 5.1), endo activity of Cel7B can overcome the hindrances created by the disordered region in cellulose, thereby activating the Cel7A (exo-acting) to work synergistically and increase the rate of hydrolysis. All cellulases in *T. reesei* are modular except for Cel12A, which lacks CBM and therefore poorly adsorbs on PASC and Avicel surfaces (Karlsson et al., 2002). However, an anomalous trend was observed in TrCel12A, where the activity of the enzyme was increased when incubated in the presence of cellobiose; the kinetics behind this are still not clear (Payne et al., 2015). Interestingly, Cel45A is known to be closely related to expansin-like proteins called swollenin that are the critical enzymes responsible for the depolymerization of crystalline regions by promoting swelling in the lignocellulosic biomass (Amengual et al., 2022; Yao et al., 2008). The 3D structure analysis has revealed the presence of a double psi β-barrel fold (six strands), and the substrate-binding pocket is made up of protruding loops with Asp 121 and Asp 10 in the catalytic cleft (Payne et al., 2015). Furthermore, the more detailed analysis of the structural conformity, specific activity, and enzyme kinetics of these endoglucanases can mediate in designing the efficient enzymatic cocktails for complete hydrolysis of lignocellulose.

5.2.4 Processive Cellobiohydrolases (CBHs)

CBHs are a group of enzymes that directly convert the crystalline form of cellulose into β-1,4-glucosidic dimer (cellobiose) as the major end product (Takeda et al., 2022). The efficient hydrolysis of crystalline cellulose is due to the unique 3D structure of the catalytic domain (450 residues), consisting of seven distorted strands of α/β barrel and a tunnel-shaped substrate binding site with the location of the active site inside a tunnel (Yang et al., 2018). The catalytic domain is connected by a carbohydrate-binding module with 30 residues of an intrinsically disordered *O*-glycosylated linker. Majorly, cellobiohydrolases are ascribed to six different glycosyl hydrolase families: GH 5, GH 6, GH 7, GH 9, GH 48, and GH 74. The cellobiohydrolases, Cel6A (CBH II) and Cel7A (CBH I), are the most widely studied and are considered industrially important (Nakamura et al., 2020). Both the enzymes act differently on cellulose; Cel6A and Cel7A attack by inverting and retaining mechanisms, respectively, initiating the catalysis from non-reducing and reducing ends of cellulose (Figure 5.1). Notably, the modular CBHI (Cel7A) accounts for ~80% of the total secretory protein in *T. reesei* and is one of the major ingredients in the commercial cellulase cocktail preparation (Ogunyewo et al., 2020; Taylor & Vaisman, 2010).

5.2.5 Rate Limiting β-glucosidases (BGLs)

β-glucosidases cleave the glycosidic bonds of amino-, alkyl-, and aryl-β-*D*-glucosides into two glucose molecules that are considered the end product of hydrolysis (Figure 5.1). Moreover, β-glucosidases eliminate the inhibition of EGs and CBHs by the cellobiose released in the penultimate step of hydrolysis, increasing the efficiency of cellulase machinery (Zhang et al., 2017). However, the rate-limiting step in the lignocellulosic biomass conversion to sugar production route is the accumulation of reaction product glucose that affects the kinetics of the whole biomass saccharification and therefore is of critical importance at an industrial scale (Manna & Ghosh, 2021). Owing to this obstacle recently, β-glucosidases are being engineered to increase the glucose tolerance, which is due to the presence of allosteric interaction between the catalytic domain and the secondary protein binding cleft or increased resistance to particular residues in the narrow and deep active site tunnels (Konar et al., 2019; Souza et al., 2013). β-glucosidases also catalyze the transglycosylation by assembling oligo- and disaccharide molecules and are exacerbated by the increasing glucose concentrations (Singhania et al., 2017). Transglycosylation-linked β-glucosidase stimulation by modulating the enzyme turnover kinetics has depicted that in the presence of high glucose, the synthesis of sophorose and cellobiose induces the cellulase activity (Yan et al., 2021). Based on substrate specificity, β-glucosidases are classified in three subgroups: (i) aryl-β-glucosidases that preferably cleave the aryl-glucosides; (ii) cellobiases, which convert cellobiose to glucose; and (iii) glucosidases that have wide substrate specificity (Srivastava et al., 2019). The β-glucosidases have been classified under glycosyl hydrolase families GH1, GH3, GH5, GH9, GH30, and GH116. In contrast, the classification based on amino acid sequence and conserved residues lists the β-glucosidases in clan A (GH A), which is represented by (β/α) eight-barrel structures, including families belonging to GH1, GH5, and GH30. On the other hand, GH3 has two domains, with GH9 as an exception, having (α/α) six-barrel structures. Till now, GH1 has the largest number of characterized β-glucosidases (Singhania et al., 2017). The structural constraints have revealed that GH3 β-glucosidases are less glucose tolerant, showing resistance up to 400 mM as compared to the GH1 that can resist 1 M glucose (Ramani et al., 2015). In GH1, due to the presence of conserved amino acids Trp168 and Leu173, the steric hindrance is imposed at the +2 subsite that blocks the access to glucose at the one subsite (de Giuseppe et al., 2014).

5.2.6 Xylan Backbone-Degrading Hemicellulases

A wide array of hemicellulolytic enzymes is required to degrade the robust network of hemicellulosic polymers, which are hydrogen bonded with cellulose. Since a large set of these polymers are insoluble or are attached to the insoluble cellulose microfibrils, most of the hemicellulases are modular with carbohydrate-binding modules attached to the catalytic domain that selectively target the enzymes on the insoluble polymers, and the binding of the catalytic domain is mediated by dockerin–cohesin interactions. Catalytic domains based on the sequence and cleavage pattern can be grouped into glycosyl hydrolase (hydrolytic bond cleavage) and carbohydrate esterase (cleaving ester bonds of ferulic acid and acetate side chains). For the complete degradation of biomass, the concerted action of β-1,4-endoxylanase (EC 3.2.1.8), β-xylosidase (EC 3.2.1.37), α-glucuronidase (EC 3.2.1.139), α-*L*-arabinofuranosidases (EC 3.2.1.55), acetylxylan esterase (EC 3.1.1.72), β-mannanase (EC 3.2.1.78), and feruloyl esterase (EC 3.1.1.73) (Dodd & Cann, 2009) (Figure 5.2).

The hydrolytic enzyme xylanase randomly cleaves the β-(1,4) xylosidic linkages in the xylan backbone, releasing the xylooligosaccharides (XOS). Although a diverse category of β-xylanase enzymes belonging to the GH 5, 7, 8, and 43 families *has been characterized, it is* to a lesser extent. Based on the biochemical properties, mode of action, and substrate specificity hitherto, GH10 and GH11 are the dominant class of xylanases. The structural difference in GH10 and GH11 xylanases is the major determining factor for the specific cleavage pattern of substrates. GH10 xylanases have increased activity on XOS with a low degree of polymerization (DP) and can hydrolyze the xylosidic linkages in heteroxylans and can also accommodate lower levels of glucose in the active site. However, GH11 xylanase prefers the substrate with a high degree of polymerization and can degrade the insoluble xylan in comparison to GH10 (Baker et al., 2021).

β-xylosidases belong to GH3, 30, 39, 43, 52, 54, 116, and 120 and hydrolyze the xylobiose units and the mere amount of XOS produced by xylanases that causes the end product inhibition; hence, β-xylosidases are considered as rate-limiting enzymes

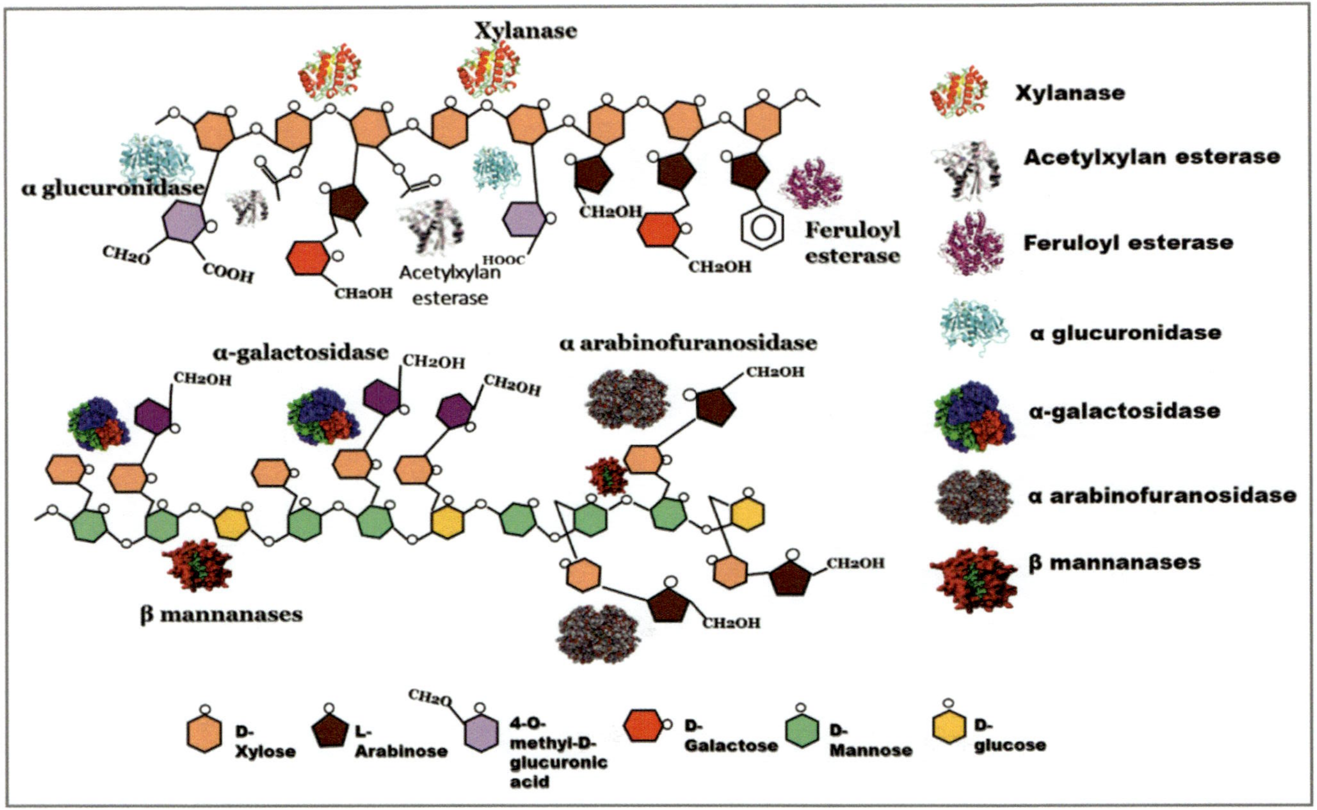

FIGURE 5.2 Mode of action of hemicellulolytic enzymes (endoxylanase, β-xylosidase, α-glucuronidase, α-*L*-arabinofuranosidases and acetylxylan esterase, β-mannanase, and feruloyl esterase) on the heteropolymeric backbone of hemicellulose polysaccharides.

in xylan hydrolysis (Bosetto et al., 2016). The commercial preparations derived from *Trichoderma reesei* (Celluclast 1.5 L) and *Humicola insolens* (Ultraflo L) were capable of complete degradation of soluble and insoluble wheat arabinoxylan due to the cooperative action of β-xylanase, β-*xylosidase*, and α-L-arabinofuranosidase. α-L-Arabinofuranosidases jointly cleave the glycosidic linkages between the arabinofuranosyl side chain and xylopyranose backbone residues of arabinoxylan (Juturu & Wu, 2013). The arabinose substitution in the xylan backbone hampers the action of xylanase and β-xylosidase enzymes; hence, the arabinofuranosidases are required to play the significant role in the hydrolysis of hemicellulose. Under the CAZy classification, the arabinofuranosidases are represented by the GH 3, 43, 51, 54, and 62 families. The studies have shown that the addition of arabinofuranosidases to other hemicellulolytic enzymes results in the release of higher sugar content after hydrolysis of differently pretreated feedstocks, such as steam explosion-treated wheat straw and delignified steam-exploded sugarcane bagasse (Delabona et al., 2013). Moreover, at a commercial scale, the arabinofuranosidases are derived from *A. niger* (GH51), *A. nidulans* (GH62), and *Bifidobacterium adolescentis* (GH43), which form an integral part of the Megazyme and Creative Enzyme (Poria et al., 2020). β-mannanases are the endo-acting enzymes that hydrolyze internal glycosidic linkages of mannan in the hemicellulose backbone of galactomannans, releasing shorter manno-oligosaccharides, whereas β-mannosidases are exo-acting hydrolases that act on the mannan polymer, releasing mannose residues. They are described under GH 5, 26, and 113 with a common crystal structure made up of a (βα)8-barrel. Supplementation of β-mannanase (MtMan26A) from *Myceliophthora thermophila* into the Celluclast 1.5 L and Novozyme 188 has been shown to hydrolyze the pretreated beechwood sawdust, resulting in the enhanced release of total reducing sugars and glucose by 13% and 12%, respectively (Dawood & Ma, 2020; Katsimpouras et al., 2016).

5.2.7 LPMOs

The discovery of LPMOs during the last decade has shifted the paradigm of lignocellulosic biomass hydrolysis. Nowadays, LPMOs are considered the major component in the commercial cellulase cocktail preparation, as they work synergistically by creating new access points for other cellulases to efficiently adsorb on the surface of lignocellulosic biomass. Unlike the other hydrolytic enzymes, LPMOs are the redox enzymes that entail the oxidative cleavage of the β-(1,4)-glycosidic bond by hydroxylation at the C1 or C4 position of glucan by using the internal electrons and the copper metal ion, resulting in the formation of aldonic acid or keto sugars (Wang et al., 2019). The LPMOs catalyze the reaction by activating the oxygen using electrons either generated by polysaccharide cleavage or artificial/enzyme donors such as ascorbate, gallic acid, cellobiose dehydrogenase, and AA3_2 (Garajova et al., 2016). The catalytic cycle starts with the priming reduction of the Cu(II) resting state to Cu(I); subsequently, by using H_2O_2 as a co-substrate, Cu(II)–O_2 intermediates are generated (Bissaro et al., 2017). The superoxo species is transformed into a Cu(II)–OOH (copper(II)–hydroperoxo) species through hydrogen atom abstraction, followed by a second electron transfer. This leads to the homolytic cleavage of the O–O bond, producing a Cu-oxyl reactive species that oxidizes the polysaccharide at the C1, C4, or both C1 and C4 positions. The regioselectivity of the LPMOs has further subcategorized them into (a) Type I: C1 specific, (b) Type II: C4 specific, and (c) Type III: both C1 and C4 specific (Hedegard & Ryde, 2017). LPMOs are classified under auxiliary activity (AA) belonging to eight different families: AA9 (known as GH61), AA10 (earlier CBM33), AA11, AA13, AA14, AA15, AA16, and AA17 (Várnai et al., 2018). The members of different families vary in their amino acid sequence and the substrate specificity. Fungal AA9 exhibits broad substrate specificity against both cellulosic and hemicellulosic substrates such as xyloglucan, xylan, glucomannan, and mixed-linkage glucan, and examples include McAA9H and PMO9A_MALCI from *Malbranchea cinnamomea* (Basotra et al., 2019; Hüttner et al., 2019) and LsAA9A from *Lentinus similis* (Simmons et al., 2017).

5.3 ESTIMATION OF CELLULASE AND HEMICELLULASE ENZYME ACTIVITY

Cellulase activity, mainly endoglucanases and total cellulase activity (FPase), is typically measured by quantifying the release of reducing sugars during the enzymatic breakdown of cellulose. Various forms of cellulose, such as filter paper (Whatman No. 1; 1 × 6 cm), carboxymethyl cellulose (CMC; 2%), or crystalline cellulose, serve as substrates for assessing enzymatic activity. A commonly employed method for this purpose is the 3,5-dinitrosalicylic acid (DNS) assay, which detects the reducing sugars produced during hydrolysis (Eveleigh et al., 2009). The assay involves incubating enzyme (500 µl) with the substrate (500 µl) prepared in sodium acetate buffer (pH 5) and is usually incubated at 50°C, followed by the addition of DNS reagent (3 ml) to halt the reaction, and the released total reducing sugars are quantified using a spectrophotometer at 540 nm. Different cellulase enzymes, including exoglucanases (CBH) and β-glucosidases, are estimated using a 96-well microtiter plate-based assay using specific substrates such as *p*-nitrophenyl-β-*D*-lactopyranoside and *p*-nitrophenyl-β-*D*-glucopyranoside, respectively (Mansour et al., 2016). A reaction mixture comprises 100 µl where enzyme (25 µl) is incubated with the respective substrate (25 µl) in the presence of sodium acetate buffer (50 mM; 50 µl) for 30–60 min (depending on substrate) at 50°C. Reaction is thereafter stopped using glycine–NaOH buffer (250 mM; pH 10.8; 100 µl) (Basotra et al., 2023). For the estimation of hemicellulolytic enzymes such as xylanase, the DNS-based method as mentioned above is most employed where birch wood xylan (1%) is used as the substrate. Similarly, other hemicellulolytic enzymes, β-xylopyranosidase and α-*L*-arabinofuranosidase activities, are also measured using a plate-based assay with the respective substrates *p*NP β-xylopyranoside (*p*NPX) and *p*NP α-*L*-arabinofuranoside (*p*NPA) (Brar et al., 2021). The amount of released para-nitro-phenol (pNP) can be quantified by measuring the intensity of the developed yellow color at 405 nm using a plate reader (Model 680 XR, Bio-Rad). The enzyme activity is defined as the amount of enzyme required to release 1 µmol of the pNP under the standard assay conditions. For the calculation of total cellulase activity (FPase), xylanase, glucose, and xylose are used as the respective standards, and the mathematical calculations have been reported previously by several studies (Al Talebi et al., 2022; Juhasz et al., 2005; Sharma et al., 2024; Wood & Bhat, 1988).

5.4 SYSTEM BIOLOGICAL APPROACHES FOR BIOPROSPECTING DIVERSE LIGNOCELLULOLYTIC ENZYMES

In recent times, remarkable advancements in the system biological and bioinformatic approaches involving genomics, transcriptomics, and proteomics tools have provided in-depth insights into the identification and discovery of novel genes encoding proteins that are critical for lignocellulosic degradation (Suravajhala et al., 2016). To bioprospect the diversity of lignocellulolytic enzymes, the genomes of 1000 biomass-degrading fungi, such as *Aspergillus flavus* (Cleveland et al., 2009), *Myceliophthora thermophila* (Berka et al., 2011), *Fistulina hepatica* and *Cylindrobasidium torrendii* (Floudas et al., 2015), and *Aspergillus niger* (Vesth et al., 2018), from various ecological niches have been sequenced and annotated. Genome-wide diversity of biomass-degrading CAZymes can now be accessed easily through increased power and efficacy of high-throughput next-generation sequencing (NGS). Whole genome assembly of the microbial cells at the DNA sequence level allows us to capture the variations that are prevalent at the genetic level, including transcriptional regulators, and helps us to understand the carbon flow through the pathways involved in the production of novel metabolites (Grigoriev et al., 2014). MycoCosm, a freely accessible online fungal genome portal (http://jgi.doe.gov/fungi), provides access to the researcher community for annotated genomes (over 2000 fungal genomes, http://jgi.doe.gov/fungi, June 2022), along with tools for multidimensional comparative genomics. Genome-wide sequencing of the thermophilic fungus *Thielavia terrestris* has been crucial in understanding the role of GH61 family proteins, now identified as LPMOs (AA9 family proteins). These enzymes are involved in enhancing the redox-mediated cleavage of cellulose (Harris et al., 2010; Levasseur et al., 2013; Westereng et al., 2011). The genome-wide annotation reveals functional genes encoding various lignocellulolytic enzymes, highlighting the fungus's potential for biomass degradation. Lignocellulolytic enzyme components expressed in the genome of thermophilic and thermotolerant fungi are of interest in biofuel biorefineries, as they can perform better under hostile conditions such as high temperature and presence of toxic inhibitors generated during pre-treatment (Atalah et al., 2019; Patel et al., 2019).

Furthermore, the discovery of thermostable enzymes derived from the microbial strains exhibiting high stability with a higher rate of catalysis, thereby increasing saccharification efficiency and reducing the process operational costs of biofuel production, is being researched (Viikari et al., 2007). Table 5.1 shows the prevalence of genes encoding biomass-degrading CAZyme proteins in the genomes of diverse fungal species. The induction of various genes involved in the degradation of biomass is substrate-dependent. Higher induction of cellulolytic and hemicellulolytic genes, especially xylanolytic, followed by mannanase and arabinan-degrading genes, was reported in the genomes of *Thielavia terrestris* and *M. thermophila* cultivated on barley (monocot plant) and alfalfa (dicot plant) (Berka et al., 2011). Furthermore, the literature also reported an upregulation of the functional genes encoding pectin lyase proteins in *M. thermophila* in comparison to *T. terrestris* (Berka et al., 2011; Kolbusz et al., 2014; Singh, 2016).

Lignocellulolytic enzymes (CAZymes) secreted by the cell and involved in biomass degradation are known as the cell's exoproteome or secretome. From 2012 to 2017,

TABLE 5.1
Genome-Wide Profiling of Different CAZymes Present in Different Fungal Strains

		CAZymes						
Microorganisms	**Genome size (Mbp)**	**GHs**	**GTs**	**PLs**	**CEs**	**AAs**	**CBMs**	**Reference**
Aspergillus sydowii	32.27	255	74	21	82	103	8	(Tulsani et al., 2022)
Thielavia terrestris	36.6	210	88	5	25	60	77	(Berka et al., 2011)
Penicillium oxalicum	37.6	165	10	10	14	17	-	(Christopher et al., 2021)
Myceliophthora verrucosa	40	160	66	11	53	76	-	(Sharma et al., 2024)
Myceliophthora thermophila	38.7	199	84	9	25	54	51	(Berka et al., 2011)
Penicillium decumbens	30.1	223	96	7	19	27	77	(Liu et al., 2013)
Penicillium chrysogenum	31.3	236	110	10	18	49	56	(de Vries & Mäkelä, 2020)
Aspergillus niger	35.2	256	119	10	21	61	72	(Vesth et al., 2018)
Aspergillus allahabadii	28.7	301	106	17	35	64	99	(http://jgi.doe.gov/fungi)
Trametes versicolor	44.7	231	83	11	17	93	48	(Floudas et al., 2015)
Chaetomium thermophilus	44.7	231	83	11	17	93	48	(Floudas et al., 2015)

proteomic research predominantly used gel-based methods for protein separation and mass spectrometry for protein identification (Guo et al., 2018). Recently, traditional proteomic techniques for identifying, profiling, and purifying lignocellulolytic proteins have advanced with high-throughput liquid chromatography coupled with mass spectrometry (LC-MS/MS). This method utilizes trypsin-based protein digestion for peptide analysis and is supported by genome-wide databases (such as JGI, MycoCosm, and FungiDB) for identifying functional proteins based on their fragmentation characteristics (Mahajan et al., 2016; Raheja et al., 2020; Rogowska-Wrzesinska et al., 2013; Sharma et al., 2022). The secretome-based approach has enabled the examination of a variety of hydrolytic enzymes expressed by *Aspergillus fumigatus* Z5 in response to complex substrates such as rice straw, crystalline cellulose, and glucose (Liu et al., 2013). Additionally, Ogunmolu et al. (2018) investigated the dynamic changes in lignocellulolytic enzyme production induced by various polymeric substrates (crystalline cellulose, untreated wheat bran, glucose, ammonia-treated wheat straw, and crystalline cellulose) added to minimal medium. This study utilized multiplex secretome profiling to analyze enzyme expression during the early and late growth phases of *Penicillium funiculosum*. Basotra et al. (2016) reported a quadrupole time-of-flight liquid chromatography and mass spectrometry (QTOF-LC/MS) proteomic strategy for the secretome-based profiling of an arsenal of cellulolytic and hemicellulolytic proteins produced by *Mycothermus thermophilus*. Similarly, LC-MS/MS orbitrap-based secretome analysis of *Penicillium* sp. Dal5 profiled the biomass-degrading CAZymes (Rai et al., 2016), while a wide array of cellulase and hemicellulase belonging to different glycoside hydrolase families, along with a high proportion of metal-dependent glycosyl hydrolases and auxiliary enzymes, was unveiled in the proteome of *Malbranchea cinnamomea* (Mahajan et al., 2016). A proteomic approach was employed by Arntzen et al. (2020) for comparing the lignocellulolytic enzyme diversity produced by five different fungi, namely, *A. terreus, Neurospora crassa, Phanerochaete chrysosporium, Trichoderma reesei*, and *M. thermophila*, in response to different substrates, i.e., birchwood, glucose, cellulose, and sugarcane bagasse. Higher abundance of xylanases (GH11) and novel β-glucanase, along with CAZymes in the exo-proteome of *Thermomyces lanuginosus* that promotes efficient hydrolysis of pre-treated lignocellulosics, was identified by Brar et al. (2021). LPMOs belonging to families AA14 and AA16 were also identified using a secretome-based approach in *Pycnoporus coccineus* and *Aspergillus* (Filiatrault-Chastel et al., 2019). Comparative secretome profiling of *T. reesei* strains (CL847 and RUT-C30) on medium supplemented with different carbon sources (lactose and lactose-xylose) has been reported by Herpoël-Gimbert et al. (2008). The secretome of *T. reesei* CL847 cultivated on lactose–xylose-containing medium was represented by Cel6A and Cel7A as the major cellulolytic proteins along with endoxylanase, β-xylosidases, and arabinofuranosidases, whereas strain RUT-C30 was unable to produce the cellulases on a similar medium. However, a 10% higher production of cellobiohydrolases in the secretome of RUT-30 was observed in the medium containing lactose in comparison to the CL847 strain (Herpoël-Gimbert et al., 2008). Similarly, QTOF-LC/MS analysis showed that the developed mutant strain of *Rasamsonia emersonii* exhibited significantly higher expression levels of auxiliary activity protein (AA9 LPMO), swollenin, xylanases, β-glucosidases, and endoglucanases compared to the industrial mutant strain TEC 210 developed by DSM-POET (Perkins et al., 2012). The integration of genomic and proteomic approaches, combined with advanced computational bioinformatics tools, has facilitated the discovery of novel proteins. One of the most extensively studied fungi being developed over 30 years of rigorous strain development is *T. reesei*, with its secretome represented by 70% of the cellobiohydrolase, highlighting it as an important CAZyme for the degradation of crystalline cellulose. Another important fungus, *Penicillium oxalicum*, a potential producer of lignocellulolytic enzymes developed in China since 1996, expresses a highly balanced composition of enzymes in comparison to *T. reesei*. Likewise, *T. reesei*, the secretome of *P. oxalicum*, has the highest abundance of cellobiohydrolase (37%) but is also rich in endo-β-1,4-xylanases. In addition, the *P. oxalicum* secretome showed the presence of the enzyme feruloyl esterase, which was absent in *T. reesei* (Liu et al., 2013). Profiling of the CAZymes secreted by the different fungal strains is represented in Figure 5.3. Surprisingly, the secretome profiling of *Rasamsonia emersonii* defied the conventional axiom that the dominant class of cellobiohydrolases are responsible for the degradation of biomass but instead points toward a distinct paradigm with the secretome representing the highest abundance of endoglucanase, β-glucosidase, LPMOs, and swollenin and low expression of cellobiohydrolases (Raheja et al., 2022).

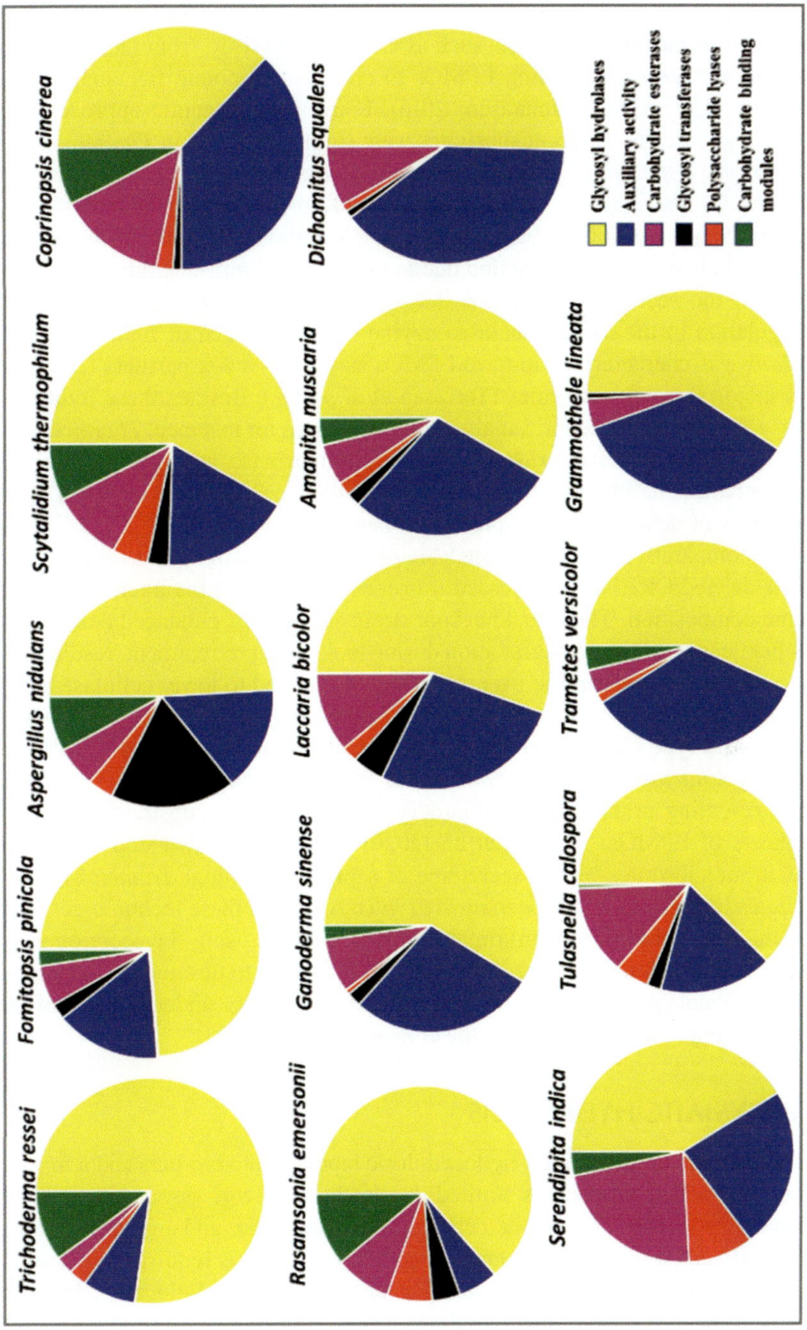

FIGURE 5.3 Comparative secretome profiling of CAZymes secreted by different fungi (Arntzen et al., 2020; Basotra et al., 2018).

The newly discovered enzyme LPMO is known for its ability to break down crystalline cellulose. When combined with benchmark cellulase enzymes, it can significantly enhance biomass hydrolysis. Consequently, Novozymes has improved their commercial enzyme cocktail series, such as Celluclast, derived from the secretome of *Trichoderma reesei*, by adding LPMOs to create more potent lignocellulolytic cocktails like CellicCTec2 (Johansen, 2016). Using the proteomic approach, the enzymatic responses to different substrates were studied in depth in *Phanerochaete chrysosporium, Aspergillus terreus, Neurospora crassa, Trichoderma reesei*, and *Myceliophthora thermophila.* Arntzen et al. (2020) conducted the quantitative analysis by NanoLC-Orbitrap-based MS/MS, highlighting the role of ascomycetes in the carbohydrate deconstruction due to the lack of lignin-degrading enzymes expressed in the secretome; however, the white rot basidiomycete fungi showed the upregulation in the oxidoreductase enzyme in the presence of lignin, whereas *A. nidulans* was characterized to have LPMOs as well as redox partners CDH and AA3 as major secretory enzymes (Terrasan et al., 2022). Besides these industrial workhorse strains, the proteomic analysis of other fungi, for instance, *Thermomyces lanuginosus* with a secretome rich in xylanase, laminarinase, and lichenanase, was found to enhance the efficacy of commercial CellicCtec2 (Brar et al., 2021) during hydrolysis of acid- and alkali-pretreated bagasse and rice straw. The LC–MS/MS-based comparative secretome analysis of *Pleurotus ostreatus* wild-type and the CRE1-deleted/CRE1-overexpressed transformants showed the alteration in the secretome composition. The *cre1* knockout strain showed the enhanced cellulolytic activity but decreased lignin degradation during biological pretreatment, resulting in less glucose yield. However, the overexpression of *cre1* led to lower cellulase activity and marginally increased lignin degradation, but the glucose yield was far below that of the wild-type strain (Yoav et al., 2018). Comparative analysis of CAZymes produced by *Chaetomium thermophilum* was studied in the presence of different inducers, revealing arabinose and microcrystalline cellulose substrates as effective inducers of LPMOs. Arntzen et al. (2020) demonstrated two-step hydrolysis of recalcitrant substrates by the secretome of saprotrophic fungi *Trametes ljubarskyi, Artolenzites elegans*, and *Laetisaria arvalis*. Although these technologies have enabled the efficient design of minimal enzyme cocktails using the secretomes of diverse filamentous fungi, further customization is needed to develop cost-effective and more efficient cocktails. Achieving this requires a deeper understanding of the secretion and regulatory pathways of the enzymatic machinery.

5.5 ENZYMATIC HYDROLYSIS

The bioconversion of recalcitrant lignocellulosic biomass into biofuels and a myriad of other value-added products is limited due to the high cost associated with the pretreatment process to fractionate cellulose, hemicellulose, and lignin. However, after pretreatment, a significant amount of cellulase enzyme is required to establish the "sugar platform" by enzymatic hydrolysis of the fractionated glucan, xylan, and solubilized oligomers (Zhu & Pan, 2022). To achieve high enzymatic conversions of glucan, 40–60 FPU/g substrate (Pielhop et al., 2015) is required after liquid hot water pretreatment of spruce, which is much higher than the 15–20 FPU/g required after chemical pretreatment (Zhao et al., 2017). Since lignocellulosic biomass is a heterogeneous polymer consisting of a complex interconnected network of hemicellulose intermingled with cellulose, the addition of hemicellulases in the cellulase cocktail can enhance the efficacy of hydrolysis. Therefore, this aspect is nowadays being intensively studied to reduce the protein loading or FPU/g substrate to achieve a high catalysis rate (Luo et al., 2019). To overcome the paucity of efficient cellulolytic cocktails, several commercial cellulase manufacturers, such as Novozymes, Genencor, Dyadic, Amano Enzyme, Danisco, and Advanced Biochemicals, are the leading companies to comply with the increasing demands of cellulase by the upcoming biorefineries (Chandel et al., 2012). Presently, a series of cocktails developed by Novozymes, i.e., CellicCTec3 and Cellic HTec3, are being endorsed by Raizen/Iotech (Brazil), Beta Renewables, Italy (shuttered now), PRAJ, and IOCL (India) (Østby et al., 2020). Moreover, several industrial companies, such as Dupont Genencor, have developed advanced enzyme formulations like ACCELLERASE® TRIO™, which contain a balanced proportion of xylanase and cellulase enzymes. These formulations are designed to mitigate the inhibitory effects of xylo-oligomers released during the deconstruction of hemicellulose. Particularly, Accelerase®1500 supplied by Genencor has been stated to completely convert the cellobiose into glucose due to the high amount of β-glucosidase in comparison to that of other commercial cocktails (Singh et al., 2022). Enzymatic preparation from the thermophilic fungal strain *Rasamsonia emersonii* developed by DSM can hydrolyze the biomass at 25% substrate loading and low enzyme dosage to achieve 40% hydrolysis at high temperatures of 65°C. Another cellulase mix is AlternaFuel®CMAX™ by Dyadic International, consisting of a C1-based in-house enzyme preparation from *Myceliophthora thermophila* that can work on a broad range of pH and high temperatures to hydrolyze a variety of feedstocks (Adsul et al., 2020). Secretomes of diverse fungi produce different sets of CAZymes on various substrates, which is an important aspect in designing the cellulolytic cocktails to hydrolyze specific substrates. Thus, designing a customized cellulase cocktail that can outperform the enzymatic degradation of different pretreated biomass at high solid concentrations is of utmost importance. Addition of *Penicillium chrysogenum* P33 secretome to the commercial cellulase resulted in a 78% increase in the release of reducing sugar from delignified corn stover (Yang et al., 2018). A cocktail comprising *Talaromyces emersonii* secretome added along with the *Achaetomium strumarium* in a 9:1 ratio showed a 20% and 28.99% increase in glucose release from acid- and alkali-pretreated rice straw, respectively (Raheja et al., 2020). Although technological advancement has reduced the cost associated with the pretreatment process and the stepwise strain improvement program has lowered the enzymatic cost, the efficient conversion of biomass remains one of the major hurdles in the commercialization route. Therefore, custom designing catalytically efficient minimal enzyme mixtures, comprising an appropriate stoichiometric ratio of enzymes utilizing minimal enzyme dosage, is important to make the process viable at an industrial scale. Several studies on designing minimal enzyme cocktails for efficient hydrolysis have been reported in the recent past. Notably, Barbosa et al. (2020) improved the production of cello-oligosaccharides (COS) from hydrothermally pretreated sugarcane straw using an optimized mixture of enzymes: endoglucanase Cel9R from *Clostridium thermocellum* and *Thermobifida fusca*, Cel9M; *Clostridium cellulolyticum*, LPMO Cel61A; *T. reesei* and CDHIIa; and *Neurospora crassa*, yielding 60.49 mg of COS/g substrate, which was higher when compared to Cellic® Ctec2 and Celluclast® 1.5 L. Acquittance of higher glucose (83% cellulose to glucose conversion efficiency) from sugarcane bagasse (SCB) with lower enzyme loading has been reported by Nwamba et al. (2021). A positive boosting effect of 2.4 mg/g xylanase and 0.4 mg/g AA9 was obtained on the hydrolysis of SCB at 30% solid loading with 3 FPU/g CellicCtec2 (Mukasekuru et al., 2020). To enhance the hydrolysis of steam-exploded wheat straw, a synthetic mix was designed by Agrawal et al. (2018) consisting of β-glucosidase, Celluclast, and xylanase that increased the hydrolysis by 75% within 6 h. Another scientific study reports that the blending of the β-glucosidase enzyme from *Aspergillus tubingensis and Pyrococcus furiosus* in the C-Mix (in-house cellulase cocktail) resulted in the release of >900 mg/g glucan from different biomasses (Fernandes et al., 2022). It was observed that the maximum yield of 96.8 mg/g and 543 mg/g of xylose and glucose, respectively, from steam-pretreated SCB using Cel7A (*Talaromyces cellulolyticus*), Cel7B (*Thielavia terrestris*), BGL, and XYN (*Aspergillus aculeatus*) in a ratio of 0.34:0.27:0.14:0.25 was obtained, which was greater than the commercial Accellerase 1500 (Bunterngsook et al., 2018). In a statistical model comprising recombinant EG (GH 5, 7), CBH (GH 6), BGL (GH 1), LPMO, and feruloyl esterase from *Myceliophthora thermophila*, a 27% enhanced reducing sugar yield from hydrothermally pretreated wheat straw with a 20 mg/g enzyme-loading rate resulted (Karnaouri et al., 2016). The most prominent of the cellulases are the endoglucanases that act randomly on amorphous regions of cellulose and cleave the internal β-glucosidic linkages, thereby generating several glucan chain ends. Cellobiohydrolases act on the cellulose chain from either the reducing end (CBH I) in a processive manner or the nonreducing end (CBH II), thereby generating cellobiose as the major end product. Subsequently, the rate-limiting enzyme β-glucosidase converts the cellobiose into glucose units and relieves the enzyme from end-product inhibition, whereas hemicellulases aid in the conversion of the redeposited hemicellulose fraction left over after the pretreatment (Dodd & Cann, 2009). The pragmatic view of biomass saccharification entails a major role of cellobiohydrolases (CBHs), which act as a primary catalyst working in coordination with endoglucanases and β-glucosidases (Van der Zwan et al., 2020). The secretome of industrial cellulolytic strains of *Penicillium oxalicum and Trichoderma reesei* is rich in CBHs ranging from 37% to 70% of the total excreted proteins, whereas endoglucanases comprise a relatively small portion of the secretome. However, studies have shown a shift in paradigm where some of the strains produce higher amounts of endoglucanases that are reported to play the dominant role in the liquefaction and in reducing the viscosity of pre-treated biomass, thereby overcoming the mass transfer issues in the reactor with cellobiohydrolase enzyme working in synergy to promote the cellulose fragmentation (Karnaouri et al., 2017; Muraleedharan et al., 2021; Van der Zwan et al., 2020). Until the last decade, the degradation of lignocellulosics was solely attributed to the hydrolytic action of glycosyl hydrolases, but with the discovery of oxidative LPMOs previously thought to be endoglucanase (GH61) and deep insight into the role of accessory enzymes (xylanases, feruloyl-esterases, and β-xylosidases) in saccharification, improved commercial cocktails have been constituted to boost the hydrolytic performance of the core cellulase enzymes (Gao et al., 2011). Today, LPMOs are considered the central component in the newly launched commercial cellulase cocktail, such as Novozymes CellicCTec3. These LPMOs can dock onto the flat crystalline surface of cellulose and oxidatively cleave the polymer for hastening the hydrolysis process. Unlike the other hydrolytic enzymes, LPMOs are the redox enzymes that entail the oxidative cleavage of the β-(1,4)-glycosidic bond by hydroxylation at the C1 or C4 position of glucan by using the internal electrons and the copper metal ion, resulting in the formation of aldonic acid or keto sugars (Wang et al., 2019). The secretome analysis of different fungi has revealed that the LPMOs are most abundantly present in *Phanerochaete chrysosporium, Myceliophthora, Penicillium sp., Neurospora, Aspergillus*, etc. (Rani Singhania et al., 2021), and systematic characterization of the fungal secretome can further help in the identification of novel enzymes required for biomass saccharification. Detailed analysis of the fungal secretome is one of the most valuable tools to unveil the complex synergistic network of enzymes required for the biomass hydrolysis. The dynamic changes in the secretome composition in fungi are regulated by signaling in response to the external environment that controls protein trafficking.

5.6 CONCLUSION

In conclusion, enzymatic hydrolysis plays a crucial role in converting lignocellulosic biomass to bioethanol, but the process faces significant challenges due to the structural complexity of the biomass and the presence of toxic inhibitors and insoluble solids after pretreatment. Achieving high saccharification efficiency with minimal enzyme loading at high solid concentrations remains a bottleneck for biorefineries. To address these issues, the development of robust and catalytically active CAZymes is essential. This chapter highlights the ongoing efforts to enhance enzymatic hydrolysis by exploring novel inhibitor-resistant enzyme cocktails that can operate efficiently at high solid loadings, ultimately aiming to reduce overall process costs.

REFERENCES

Adsul, M., Sandhu, S. K., Singhania, R. R., Gupta, R., Puri, S. K., & Mathur, A. (2020). Designing a cellulolytic enzyme cocktail for the efficient and economical conversion of lignocellulosic biomass to biofuels. *Enzyme and Microbial Technology*, *133*, 109442.

Agrawal, R., Semwal, S., Kumar, R., Mathur, A., Gupta, R. P., Tuli, D. K., & Satlewal, A. (2018). Synergistic enzyme cocktail to enhance hydrolysis of steam exploded wheat straw at pilot scale. *Frontiers in Energy Research*, *6*, 122.

Akram, F., ul Haq, I., Imran, W., & Mukhtar, H. (2018). Insight perspectives of thermostable endoglucanases for bioethanol production: A review. *Renewable Energy*, *122*, 225–238.

Al Talebi, Z. A., Al-Kawaz, H. S., Mahdi, R. K., Al-Hassnawi, A. T., Alta'ee, A. H., Hadwan, A. M., & Hadwan, M. H. (2022). An optimized protocol for estimating cellulase activity in biological samples. *Analytical Biochemistry*, *655*, 114860.

Amengual, N. G., Csarman, F., Wohlschlager, L., & Ludwig, R. (2022). Expression and characterization of a family 45 glycosyl hydrolase from Fomitopsis pinicola and comparison to Phanerochaete chrysosporium Cel45A. *Enzyme and Microbial Technology*, *156*, 110000.

Arntzen, M. Ø., Bengtsson, O., Várnai, A., Delogu, F., Mathiesen, G., & Eijsink, V. G. H. (2020). Quantitative comparison of the biomass-degrading enzyme repertoires of five filamentous fungi. *Scientific Reports, 10*(1), 20267.

Atalah, J., Cáceres-Moreno, P., Espina, G., & Blamey, J. M. (2019). Thermophiles and the applications of their enzymes as new biocatalysts. *Bioresource Technology, 280*, 478–488.

Baker, J. T., Duarte, M. E., Holanda, D. M., & Kim, S. W. (2021). Friend or foe? Impacts of dietary xylans, xylooligosaccharides, and xylanases on intestinal health and growth performance of monogastric animals. *Animals, 11*(3), 609.

Barbosa, K. L., dos Santos Malta, V. R., Machado, S. S., Junior, G. A. L., da Silva, A. P. V., Almeida, R. M. R. G., & da Luz, J. M. R. (2020). Bacterial cellulase from the intestinal tract of the sugarcane borer. *International Journal of Biological Macromolecules, 161*, 441–448.

Basotra, N., Dhiman, S. S., Agrawal, D., Sani, R. K., Tsang, A., & Chadha, B. S. (2019). Characterization of a novel lytic polysaccharide monooxygenase from Malbranchea cinnamomea exhibiting dual catalytic behavior. *Carbohydrate Research, 478*, 46–53.

Basotra, N., Joshi, S., Satyanarayana, T., Pati, P. K., Tsang, A., & Chadha, B. S. (2018). Expression of catalytically efficient xylanases from thermophilic fungus *Malbranchea cinnamomea* for synergistically enhancing hydrolysis of lignocellulosics. *International Journal of Biological Macromolecules, 108*, 185–192. https://doi.org/10.1016/j.ijbiomac.2017.11.131

Basotra, N., Kaur, B., Di Falco, M., Tsang, A., & Chadha, B. S. (2016). Mycothermus thermophilus (Syn. Scytalidium thermophilum): Repertoire of a diverse array of efficient cellulases and hemicellulases in the secretome revealed. *Bioresource Technology, 222*, 413–421.

Basotra, N., Kaur, B., Raheja, Y., Agrawal, D., Sharma, G., & Chadha, B. S. (2023). Developing and evaluating lignocellulolytic hyper producing deregulated strains of Mycothermus thermophilus for hydrolysis of lignocellulosics. *Biomass Conversion and Biorefinery, 13*(6), 5059–5071.

Berka, R. M., Grigoriev, I. V, Otillar, R., Salamov, A., Grimwood, J., Reid, I., … Moisan, M.-C. (2011). Comparative genomic analysis of the thermophilic biomass-degrading fungi Myceliophthora thermophila and Thielavia terrestris. *Nature Biotechnology, 29*(10), 922–927.

Bilal, M., Qamar, S. A., Yadav, V., Cheng, H., Khan, M., Adil, S. F., … Iqbal, H. M. N. (2021). Exploring the potential of ligninolytic armory for lignin valorization–A way forward for sustainable and cleaner production. *Journal of Cleaner Production, 326*, 129420.

Bissaro, B., Røhr, Å. K., Müller, G., Chylenski, P., Skaugen, M., Forsberg, Z., … Eijsink, V. G. H. (2017). Oxidative cleavage of polysaccharides by monocopper enzymes depends on H2O2. *Nature Chemical Biology, 13*(10), 1123–1128.

Bosetto, A., Justo, P. I., Zanardi, B., Venzon, S. S., Graciano, L., Dos Santos, E. L., & de Cássia Garcia Simão, R. (2016). Research progress concerning fungal and bacterial β-xylosidases. *Applied Biochemistry and Biotechnology, 178*, 766–795.

Brar, K. K., Raheja, Y., di Falco, M., Tsang, A., & Chadha, B. S. (2021). Novel β-glucanases along with xylanase identified in Thermomyces lanuginosus secretome for enhanced saccharification of different lignocellulosics. *Biomass Conversion and Biorefinery*, 1–14.

Bunterngsook, B., Laothanachareon, T., Chotirotsukon, C., Inoue, H., Fujii, T., Hoshino, T., … Techanan, W. (2018). Development of tailor-made synergistic cellulolytic enzyme system for saccharification of steam exploded sugarcane bagasse. *Journal of Bioscience and Bioengineering, 125*(4), 390–396.

Chandel, A. K., Albarelli, J. Q., Santos, D. T., Chundawat, S. P., Puri, M., & Meireles, M. A. A. (2019). Comparative analysis of key technologies for cellulosic ethanol production from Brazilian sugarcane bagasse at a commercial scale. *Biofuels, Bioproducts and Biorefining, 13*(4), 994–1014.

Chandel, A. K., Arora, R., Gupta, R., Narisetty, V., & Kumar, V. (2024). Global biofuel alliance and roadmap for ethanol blending program in India: Analysis and perspective. *Sugar Tech, 26*(4), 1005–1015.

Chandel, A. K., Chandrasekhar, G., Silva, M. B., & Silvério da Silva, S. (2012). The realm of cellulases in biorefinery development. *Critical Reviews in Biotechnology, 32*(3), 187–202.

Christopher, M., Sreeja-Raju, A., Kooloth-Valappil, P., Abraham, A., Gokhale, D. V., & Sukumaran, R. K. (2021). Whole Genome Sequence and CAZyme distribution of the cellulase hyper producing filamentous fungus Penicillium janthinellum NCIM 1366. *BioRxiv*, 2021–2026.

Cleveland, T. E., Yu, J., Fedorova, N., Bhatnagar, D., Payne, G. A., Nierman, W. C., & Bennett, J. W. (2009). Potential of Aspergillus flavus genomics for applications in biotechnology. *Trends in Biotechnology, 27*(3), 151–157.

Coseri, S. (2017). Cellulose: To depolymerize… or not to? *Biotechnology Advances, 35*(2), 251–266.

Dahmen, N., Lewandowski, I., Zibek, S., & Weidtmann, A. (2019). Integrated lignocellulosic value chains in a growing bioeconomy: Status quo and perspectives. *Gcb Bioenergy, 11*(1), 107–117.

Dawood, A., & Ma, K. (2020). Applications of microbial β-mannanases. *Frontiers in Bioengineering and Biotechnology, 8*, 598630.

Delabona, P. da S., Cota, J., Hoffmam, Z.B., Paixão, D.A.A., Farinas, C.S., Cairo, J.P.L.F., Lima, D.J., Squina, F.M., Ruller, R., & Pradella, J.G. da C. (2013). *'Understanding the cellulolytic system of Trichoderma harzianum* P49P11 and enhancing saccharification of pretreated sugarcane bagasse by supplementation with pectinase and α-l-arabinofuranosidase', *Bioresource Technology, 131*, 500–507. https://doi.org/10.1016/j.biortech.2012.12.105.

de Giuseppe, P. O., Souza, T. de A. C. B., Souza, F. H. M., Zanphorlin, L. M., Machado, C. B., Ward, R. J., … Murakami, M. T. (2014). Structural basis for glucose tolerance in GH1 β-glucosidases. *Acta Crystallographica Section D: Biological Crystallography, 70*(6), 1631–1639.

Devi, A., Singh, A., Bajar, S., Pant, D., & Din, Z. U. (2021). Ethanol from lignocellulosic biomass: An in-depth analysis of pre-treatment methods, fermentation approaches and detoxification processes. *Journal of Environmental Chemical Engineering, 9*(5), 105798.

de Vries, R. P., & Mäkelä, M. R. (2020). Genomic and postgenomic diversity of fungal plant biomass degradation approaches. *Trends in Microbiology, 28*(6), 487–499.

Dodd, D., & Cann, I. K. O. (2009). Enzymatic deconstruction of xylan for biofuel production. *Gcb Bioenergy, 1*(1), 2–17.

Eveleigh, D. E., Mandels, M., Andreotti, R., & Roche, C. (2009). Measurement of saccharifying cellulase. *Biotechnology for Biofuels, 2*, 1–8.

Fernandes, C. G., Sawant, S. C., Mule, T. A., Khadye, V. S., Lali, A. M., & Odaneth, A. A. (2022). Enhancing cellulases through synergistic β-glucosidases for intensifying cellulose hydrolysis. *Process Biochemistry, 120*, 202–212.

Filiatrault-Chastel, C., Navarro, D., Haon, M., Grisel, S., Herpoël-Gimbert, I., Chevret, D., … Margeot, A. (2019). AA16, a new lytic polysaccharide monooxygenase family identified in fungal secretomes. *Biotechnology for Biofuels, 12*(1), 15.

Floudas, D., Held, B. W., Riley, R., Nagy, L. G., Koehler, G., Ransdell, A. S., … Lipzen, A. (2015). Evolution of novel wood decay mechanisms in Agaricales revealed by the genome sequences of Fistulina hepatica and Cylindrobasidium torrendii. *Fungal Genetics and Biology, 76*, 78–92.

Gao, D., Chundawat, S. P. S., Sethi, A., Balan, V., Gnanakaran, S., & Dale, B. E. (2013). Increased enzyme binding to substrate is not necessary for more efficient cellulose hydrolysis. *Proceedings of the National Academy of Sciences*, *110*(27), 10922–10927.

Gao, D., Uppugundla, N., Chundawat, S. P. S., Yu, X., Hermanson, S., Gowda, K., … Dale, B. E. (2011). Hemicellulases and auxiliary enzymes for improved conversion of lignocellulosic biomass to monosaccharides. *Biotechnology for Biofuels*, *4*(1), 1–11. http://doi: 10.1186/1754-6834-4-5

Garajova, S., Mathieu, Y., Beccia, M. R., Bennati-Granier, C., Biaso, F., Fanuel, M., … Rogniaux, H. (2016). Single-domain flavoenzymes trigger lytic polysaccharide monooxygenases for oxidative degradation of cellulose. *Scientific Reports*, *6*(1), 28276.

Grigoriev, I. V, Nikitin, R., Haridas, S., Kuo, A., Ohm, R., Otillar, R., … Korzeniewski, F. (2014). MycoCosm portal: Gearing up for 1000 fungal genomes. *Nucleic Acids Research*, *42*(D1), D699–D704.

Guo, H., Wang, X.-D., & Lee, D.-J. (2018). Proteomic researches for lignocellulose-degrading enzymes: A mini-review. *Bioresource Technology*, *265*, 532–541.

Harris, P. V, Welner, D., McFarland, K. C., Re, E., Navarro Poulsen, J.-C., Brown, K., … Merino, S. (2010). Stimulation of lignocellulosic biomass hydrolysis by proteins of glycoside hydrolase family 61: Structure and function of a large, enigmatic family. *Biochemistry*, *49*(15), 3305–3316.

Hedegard, E. D., & Ryde, U. (2017). Multiscale modelling of lytic polysaccharide monooxygenases. *ACS Omega*, *2*(2), 536–545.

Herpoël-Gimbert, I., Margeot, A., Dolla, A., Jan, G., Mollé, D., Lignon, S., … Asther, M. (2008). Comparative secretome analyses of two Trichoderma reesei RUT-C30 and CL847 hypersecretory strains. *Biotechnology for Biofuels*, *1*(1), 12.

Hüttner, S., Várnai, A., Petrović, D. M., Bach, C. X., Kim Anh, D. T., Thanh, V. N., … Olsson, L. (2019). Specific xylan activity revealed for AA9 lytic polysaccharide monooxygenases of the thermophilic fungus Malbranchea cinnamomea by functional characterization. *Applied and Environmental Microbiology*, *85*(23), e01408–19.

Igarashi, K., Uchihashi, T., Koivula, A., Wada, M., Kimura, S., Okamoto, T., … Samejima, M. (2011). Traffic jams reduce hydrolytic efficiency of cellulase on cellulose surface. *Science*, *333*(6047), 1279–1282.

Jain, L., Kurmi, A. K., & Agrawal, D. (2018). Feasibility studies with lignin blocking additives in enhancing saccharification and cellulase recovery: Mutant UV-8 of T. verruculosus IIPC 324 a case study. *Enzyme and Microbial Technology*, *118*, 44–49.

Johansen, K. S. (2016). Discovery and industrial applications of lytic polysaccharide monooxygenases. *Biochemical Society Transactions*, *44*(1), 143–149.

Juhasz, T., Szengyel, Z., Reczey, K., Siika-Aho, M., & Viikari, L. (2005). Characterization of cellulases and hemicellulases produced by Trichoderma reesei on various carbon sources. *Process Biochemistry*, *40*(11), 3519–3525.

Juturu, V., & Wu, J. C. (2013). Insight into microbial hemicellulases other than xylanases: A review. *Journal of Chemical Technology & Biotechnology*, *88*(3), 353–363.

Karlsson, J., Siika-aho, M., Tenkanen, M., & Tjerneld, F. (2002). Enzymatic properties of the low molecular mass endoglucanases Cel12A (EG III) and Cel45A (EG V) of Trichoderma reesei. *Journal of Biotechnology*, *99*(1), 63–78.

Karnaouri, A., Matsakas, L., Topakas, E., Rova, U., & Christakopoulos, P. (2016). Development of thermophilic tailor-made enzyme mixtures for the bioconversion of agricultural and forest residues. *Frontiers in Microbiology*, *7*, 177.

Karnaouri, A., Muraleedharan, M. N., Dimarogona, M., Topakas, E., Rova, U., Sandgren, M., & Christakopoulos, P. (2017). Recombinant expression of thermostable processive Mt EG5 endoglucanase and its synergism with Mt LPMO from Myceliophthora thermophila during the hydrolysis of lignocellulosic substrates. *Biotechnology for Biofuels*, *10*(1), 17.

Katsimpouras, C., Dimarogona, M., Petropoulos, P., Christakopoulos, P., & Topakas, E. (2016). A thermostable GH26 endo-β-mannanase from Myceliophthora thermophila capable of enhancing lignocellulose degradation. *Applied Microbiology and Biotechnology*, *100*, 8385–8397.

Kolbusz, M. A., Di Falco, M., Ishmael, N., Marqueteau, S., Moisan, M.-C., da Silva Baptista, C., … Tsang, A. (2014). Transcriptome and exoproteome analysis of utilization of plant-derived biomass by Myceliophthora thermophila. *Fungal Genetics and Biology*, *72*, 10–20.

Konar, S., Sinha, S. K., Datta, S., & Ghorai, P. K. (2019). Probing the effect of glucose on the activity and stability of β-glucosidase: An all-atom molecular dynamics simulation investigation. *ACS Omega*, *4*(6), 11189–11196.

Lee, T. M., Farrow, M. F., Arnold, F. H., & Mayo, S. L. (2011). A structural study of Hypocrea jecorina Cel5A. *Protein Science*, *20*(11), 1935–1940.

Levasseur, A., Drula, E., Lombard, V., Coutinho, P. M., & Henrissat, B. (2013). Expansion of the enzymatic repertoire of the CAZy database to integrate auxiliary redox enzymes. *Biotechnology for Biofuels*, *6*, 1–14.

Liu, G., Zhang, L., Wei, X., Zou, G., Qin, Y., Ma, L., … Wang, C. (2013). Genomic and secretomic analyses reveal unique features of the lignocellulolytic enzyme system of Penicillium decumbens. *PloS One*, *8*(2), e55185.

Luo, X., Liu, J., Zheng, P., Li, M., Zhou, Y., Huang, L., … Shuai, L. (2019). Promoting enzymatic hydrolysis of lignocellulosic biomass by inexpensive soy protein. *Biotechnology for Biofuels*, *12*, 1–13.

Mahajan, C., Basotra, N., Singh, S., Di Falco, M., Tsang, A., & Chadha, B. S. (2016). Malbranchea cinnamomea: A thermophilic fungal source of catalytically efficient lignocellulolytic glycosyl hydrolases and metal dependent enzymes. *Bioresource Technology*, *200*, 55–63.

Mankar, A. R., Pandey, A., Modak, A., & Pant, K. K. (2021). Pretreatment of lignocellulosic biomass: A review on recent advances. *Bioresource Technology*, *334*, 125235.

Manna, B., & Ghosh, A. (2021). Molecular insight into glucose-induced conformational change to investigate uncompetitive inhibition of GH1 β-glucosidase. *ACS Sustainable Chemistry & Engineering*, *9*(4), 1613–1624.

Mansour, A. A., Da Costa, A., Arnaud, T., Lu-Chau, T. A., Fdz-Polanco, M., Moreira, M. T., & Rivero, J. A. C. (2016). Review of lignocellulolytic enzyme activity analyses and scale-down to microplate-based assays. *Talanta*, *150*, 629–637.

Mujtaba, M., Fraceto, L. F., Fazeli, M., Mukherjee, S., Savassa, S. M., de Medeiros, G. A., … Vilaplana, F. (2023). Lignocellulosic biomass from agricultural waste to the circular economy: A review with focus on biofuels, biocomposites and bioplastics. *Journal of Cleaner Production*, *402*, 136815.

Mukasekuru, M. R., Kaneza, P., Sun, H., Sun, F. F., He, J., & Zheng, P. (2020). Fed-batch high-solids enzymatic saccharification of lignocellulosic substrates with a combination of additives and accessory enzymes. *Industrial Crops and Products*, *146*, 112156.

Muraleedharan, M. N., Karnaouri, A., Piatkova, M., Ruiz-Caldas, M.-X., Matsakas, L., Liu, B., … Mathew, A. P. (2021). Isolation and modification of nano-scale cellulose from organosolv-treated birch through the synergistic activity of LPMO and endoglucanases. *International Journal of Biological Macromolecules*, *183*, 101–109.

Nakamura, A., Ishiwata, D., Visootsat, A., Uchiyama, T., Mizutani, K., Kaneko, S., … Iino, R. (2020). Domain architecture divergence leads to functional divergence in binding and catalytic domains of bacterial and fungal cellobiohydrolases. *Journal of Biological Chemistry*, *295*(43), 14606–14617.

Neis, A., & da Silva Pinto, L. (2021). Glycosyl hydrolases family 5, subfamily 5: Relevance and structural insights for designing improved biomass degrading cocktails. *International Journal of Biological Macromolecules*, *193*, 980–995.

Nguyen, S. T. C., Freund, H. L., Kasanjian, J., & Berlemont, R. (2018). Function, distribution, and annotation of characterized cellulases, xylanases, and chitinases from CAZy. *Applied Microbiology and Biotechnology*, *102*, 1629–1637.

Nwamba, M. C., Song, G., Sun, F., Mukasekuru, M. R., Ren, H., Zhang, Q., … Hong, J. (2021). Efficiency enhancement of a new cellulase cocktail at low enzyme loading for high solid digestion of alkali catalyzed atmospheric glycerol organosolvent pre-treated sugarcane bagasse. *Bioresource Technology*, *338*, 125505.

Ogunmolu, F. E., Kaur, I., Pasari, N., Gupta, M., & Yazdani, S. S. (2018). Quantitative multiplexed profiling of Penicillium funiculosum secretome grown on polymeric cellulase inducers and glucose. *Journal of Proteomics*, *179*, 150–160. https://doi.org/10.1016/J.JPROT.2018.03.025

Ogunyewo, O. A., Randhawa, A., Gupta, M., Kaladhar, V. C., Verma, P. K., & Yazdani, S. S. (2020). Synergistic action of a lytic polysaccharide monooxygenase and a cellobiohydrolase from Penicillium funiculosum in cellulose saccharification under high-level substrate loading. *Applied and Environmental Microbiology*, *86*(23), e01769–20.

Østby, H., Hansen, L. D., Horn, S. J., Eijsink, V. G. H., & Várnai, A. (2020). Enzymatic processing of lignocellulosic biomass: Principles, recent advances and perspectives. *Journal of Industrial Microbiology & Biotechnology: Official Journal of the Society for Industrial Microbiology and Biotechnology*, *47*(9–10), 623–657.

Patel, A. K., Singhania, R. R., Sim, S. J., & Pandey, A. (2019). Thermostable cellulases: Current status and perspectives. *Bioresource Technology*, *279*, 385–392.

Payne, C. M., Knott, B. C., Mayes, H. B., Hansson, H., Himmel, M. E., Sandgren, M., … Beckham, G. T. (2015). Fungal cellulases. *Chemical Reviews*, *115*(3), 1308–1448.

Perkins, J. B., Kumar, M., Van Zijl, M. F. K., Vollebregt, A. W. H., & Sarantinopoulos, P. (2012, May 10). *Talaromyces strains and enzyme compositions*. Google Patents.

Pielhop, T., Larrazábal, G. O., Studer, M. H., Brethauer, S., Seidel, C.-M., & von Rohr, P. R. (2015). Lignin repolymerisation in spruce autohydrolysis pretreatment increases cellulase deactivation. *Green Chemistry*, *17*(6), 3521–3532.

Poria, V., Saini, J. K., Singh, S., Nain, L., & Kuhad, R. C. (2020). Arabinofuranosidases: Characteristics, microbial production, and potential in waste valorization and industrial applications. *Bioresource Technology*, *304*, 123019.

Rabelo, S. C., Brenelli, L. B., Pin, T. C., Scopel, E., & da Costa, A. C. (2023). Pretreatments as a key for enzymatic hydrolysis of lignocellulosic biomass. In *Polysaccharide-Degrading Biocatalysts* (pp. 109–137). Elsevier.

Raheja, Y., Kaur, B., Falco, M., Tsang, A., & Chadha, B. S. (2020). Secretome analysis of Talaromyces emersonii reveals distinct CAZymes profile and enhanced cellulase production through response surface methodology. *Industrial Crops and Products*, *152*, 112554.

Raheja, Y., Singh, V., Kaur, B., Basotra, N., Di Falco, M., Tsang, A., & Chadha, B. S. (2022). Combination of system biology and classical approaches for developing biorefinery relevant lignocellulolytic Rasamsonia emersonii strain. *Bioresource Technology*, *351*, 127039.

Rai, R., Kaur, B., Singh, S., Di Falco, M., Tsang, A., & Chadha, B. S. (2016). Evaluation of secretome of highly efficient lignocellulolytic Penicillium sp. Dal 5 isolated from rhizosphere of conifers. *Bioresource Technology*, *216*, 958–967.

Rakotoarivonina, H., Hermant, B., Aubry, N., Rabenoelina, F., Baillieul, F., & Rémond, C. (2014). Dynamic study of how the bacterial breakdown of plant cell walls allows the reconstitution of efficient hemicellulasic cocktails. *Bioresource Technology*, *170*, 331–341.

Ramani, G., Meera, B., Vanitha, C., Rajendhran, J., & Gunasekaran, P. (2015). Molecular cloning and expression of thermostable glucose-tolerant β-glucosidase of Penicillium funiculosum NCL1 in Pichia pastoris and its characterization. *Journal of Industrial Microbiology and Biotechnology*, *42*(4), 553–565.

Rani Singhania, R., Dixit, P., Kumar Patel, A., Shekher Giri, B., Kuo, C. H., Chen, C. W., & Di Dong, C. (2021). Role and significance of lytic polysaccharide monooxygenases (LPMOs) in lignocellulose deconstruction. *Bioresource Technology, 335*, 125261. https://doi.org/10.1016/J.BIORTECH.2021.125261

Reis, C. E. R., Junior, N. L., Bento, H. B. S., de Carvalho, A. K. F., de Souza Vandenberghe, L. P., Soccol, C. R., ... Chandel, A. K. (2023). Process strategies to reduce cellulase enzyme loading for renewable sugar production in biorefineries. *Chemical Engineering Journal, 451*, 138690.

Rogowska-Wrzesinska, A., Le Bihan, M.-C., Thaysen-Andersen, M., & Roepstorff, P. (2013). 2D gels still have a niche in proteomics. *Journal of Proteomics, 88*, 4–13.

Sharma, G., Kaur, B., Raheja, Y., Agrawal, D., Basotra, N., Di Falco, M., ... Chadha, B. S. (2022). Lignocellulolytic enzymes from Aspergillus allahabadii for efficient bioconversion of rice straw into fermentable sugars and biogas. *Bioresource Technology, 360*, 127507.

Sharma, G., Kaur, B., Singh, V., Raheja, Y., Falco, M. Di, Tsang, A., & Chadha, B. S. (2024). Genome and secretome insights: Unravelling the lignocellulolytic potential of Myceliophthora verrucosa for enhanced hydrolysis of lignocellulosic biomass. *Archives of Microbiology, 206*(5), 1–18.

Sharma, G., Singh, V., Raheja, Y., & Chadha, B. S. (2024). Unlocking the potential of feruloyl esterase from Myceliophthora verrucosa: A key player in efficient conversion of biorefinery-relevant pretreated rice straw. *3 Biotech, 14*(6), 168.

Simmons, T. J., Frandsen, K. E. H., Ciano, L., Tryfona, T., Lenfant, N., Poulsen, J. C., ... Schnorr, K. (2017). Structural and electronic determinants of lytic polysaccharide monooxygenase reactivity on polysaccharide substrates. *Nature Communications, 8*(1), 1064.

Singh, A., Bajar, S., Devi, A., & Pant, D. (2021). An overview on the recent developments in fungal cellulase production and their industrial applications. *Bioresource Technology Reports, 14*, 100652.

Singh, B. (2016). Myceliophthora thermophila syn. Sporotrichum thermophile: A thermophilic mould of biotechnological potential. *Critical Reviews in Biotechnology, 36*(1), 59–69.

Singh, N., Singhania, R. R., Nigam, P. S., Dong, C.-D., Patel, A. K., & Puri, M. (2022). Global status of lignocellulosic biorefinery: Challenges and perspectives. *Bioresource Technology, 344*, 126415.

Singhania, R. R., Patel, A. K., Pandey, A., & Ganansounou, E. (2017). Genetic modification: A tool for enhancing beta-glucosidase production for biofuel application. *Bioresource Technology, 245*, 1352–1361.

Souza, F. H. M., Inocentes, R. F., Ward, R. J., Jorge, J. A., & Furriel, R. P. M. (2013). Glucose and xylose stimulation of a β-glucosidase from the thermophilic fungus Humicola insolens: A kinetic and biophysical study. *Journal of Molecular Catalysis B: Enzymatic, 94*, 119–128.

Srivastava, N., Rathour, R., Jha, S., Pandey, K., Srivastava, M., Thakur, V. K., ... Khan, A. F. (2019). Microbial beta glucosidase enzymes: Recent advances in biomass conversation for biofuels application. *Biomolecules, 9*(6), 220.

Suravajhala, P., Kogelman, L. J. A., & Kadarmideen, H. N. (2016). Multi-omic data integration and analysis using systems genomics approaches: Methods and applications in animal production, health and welfare. *Genetics Selection Evolution, 48*, 1–14.

Takeda, M., Baba, S., Okuma, J., Hirose, Y., Nishimura, A., Takata, M., ... Kondo, Y. (2022). Metagenomic mining and structure-function studies of a hyper-thermostable cellobiohydrolase from hot spring sediment. *Communications Biology, 5*(1), 247.

Taylor, T. J., & Vaisman, I. I (2010). Discrimination of thermophilic and mesophilic proteins. *BMC Structural Biology*, *10*, 1–10.

Terrasan, C. R. F., Rubio, M. V., Gerhardt, J. A., Cairo, J. P. F., Contesini, F. J., Zubieta, M. P., … Murakami, M. T. (2022). Deletion of AA9 lytic polysaccharide monooxygenases impacts A. nidulans secretome and growth on lignocellulose. *Microbiology Spectrum*, *10*(3), e0212521.

Tulsani, N. J., Jakhesara, S. J., Hinsu, A. T., Jyotsana, B., Dafale, N. A., Patil, N. V, … Joshi, C. G. (2022). Genome analysis and CAZy repertoire of a novel fungus Aspergillus sydowii C6d with lignocellulolytic ability isolated from camel rumen. *Electronic Journal of Biotechnology*, *59*, 36–45.

Van der Zwan, T., Sigg, A., Hu, J., Chandra, R. P., & Saddler, J. N. (2020). Enzyme-mediated lignocellulose liquefaction is highly substrate-specific and influenced by the substrate concentration or rheological regime. *Frontiers in Bioengineering and Biotechnology*, *8*, 917.

Várnai, A., Umezawa, K., Yoshida, M., & Eijsink, V. G. H. (2018). The pyrroloquinoline-quinone-dependent pyranose dehydrogenase from Coprinopsis cinerea drives lytic polysaccharide monooxygenase action. *Applied and Environmental Microbiology*, *84*(11), e00156–18.

Velvizhi, G., Goswami, C., Shetti, N. P., Ahmad, E., Pant, K. K., & Aminabhavi, T. M. (2022). Valorisation of lignocellulosic biomass to value-added products: Paving the pathway towards low-carbon footprint. *Fuel*, *313*, 122678.

Vesth, T. C., Nybo, J. L., Theobald, S., Frisvad, J. C., Larsen, T. O., Nielsen, K. F., … Riley, R. (2018). Investigation of inter-and intraspecies variation through genome sequencing of Aspergillus section Nigri. *Nature Genetics*, *50*(12), 1688–1695.

Viikari, L., Alapuranen, M., Puranen, T., Vehmaanperä, J., & Siika-Aho, M. (2007). Thermostable enzymes in lignocellulose hydrolysis. *Biofuels*, 121–145.

Vlasenko, E., Schülein, M., Cherry, J., & Xu, F. (2010). Substrate specificity of family 5, 6, 7, 9, 12, and 45 endoglucanases. *Bioresource Technology*, *101*(7), 2405–2411.

Wang, B., Walton, P. H., & Rovira, C. (2019). Molecular mechanisms of oxygen activation and hydrogen peroxide formation in lytic polysaccharide monooxygenases. *ACS Catalysis*, *9*(6), 4958–4969.

Westereng, B., Ishida, T., Vaaje-Kolstad, G., Wu, M., Eijsink, V. G. H., Igarashi, K., … Sandgren, M. (2011). The putative endoglucanase PcGH61D from Phanerochaete chrysosporium is a metal-dependent oxidative enzyme that cleaves cellulose. *PloS One*, *6*(11), e27807.

Wood, T. M., & Bhat, K. M. (1988). Methods for measuring cellulase activities. In *Methods in Enzymology* (Vol. 160, pp. 87–112). Elsevier.

Yan, S., Xu, Y., & Yu, X.-W. (2021). From induction to secretion: A complicated route for cellulase production in Trichoderma reesei. *Bioresources and Bioprocessing*, *8*, 1–15.

Yang, Y., Yang, J., Liu, J., Wang, R., Liu, L., Wang, F., & Yuan, H. (2018). The composition of accessory enzymes of Penicillium chrysogenum P33 revealed by secretome and synergistic effects with commercial cellulase on lignocellulose hydrolysis. *Bioresource Technology*, *257*, 54–61.

Yao, Q., Sun, T.-T., Liu, W.-F., & Chen, G.-J. (2008). Gene cloning and heterologous expression of a novel endoglucanase, swollenin, from Trichoderma pseudokoningii S38. *Bioscience, Biotechnology, and Biochemistry*, *72*(11), 2799–2805.

Yoav, S., Salame, T. M., Feldman, D., Levinson, D., Ioelovich, M., Morag, E., … Hadar, Y. (2018). Effects of cre1 modification in the white-rot fungus Pleurotus ostreatus PC9: Altering substrate preference during biological pretreatment. *Biotechnology for Biofuels*, *11*, 1–16.

Zhang, L., Fu, Q., Li, W., Wang, B., Yin, X., Liu, S., … Niu, Q. (2017). Identification and characterization of a novel β-glucosidase via metagenomic analysis of Bursaphelenchus xylophilus and its microbial flora. *Scientific Reports*, *7*(1), 14850.

Zhao, X., Li, S., Wu, R., & Liu, D. (2017). Organosolv fractionating pre-treatment of lignocellulosic biomass for efficient enzymatic saccharification: Chemistry, kinetics, and substrate structures. *Biofuels, Bioproducts and Biorefining*, *11*(3), 567–590.

Zhu, J. Y., & Pan, X. (2022). Efficient sugar production from plant biomass: Current status, challenges, and future directions. *Renewable and Sustainable Energy Reviews*, *164*, 112583.

6 Lignin Biorefinery
Value Addition to the Fuel Industry

*Kapil Mamtani, Rawel Singh, and Sandeep Kumar**

6.1 INTRODUCTION

The biggest self-inflicted adversity threatening the very existence of mankind in the 21st century is the Climate Change. The precursor to climatic change is an increase in greenhouse gas (GHG) concentration, a consequence of the rapid industrial expansion adopted globally in the last two centuries. Further, there is a rising increase in energy and materials demands globally, especially in developing nations, given the post-industrial technology revolution. As such, a global push is being made to replace energy and materials derived from the conventional petro-based industries with more sustainable and carbon-neutral (possibly carbon-negative footprint) industries, in their entirety (Kharissova et al., 2024). In the energy domain, especially in liquid fuels, biofuels have been well invested in to propel the necessary shift toward a greener and sustainable future. Recently, in the last decade, a major concern and hurdle for biofuels has been the uncompetitive nature of biofuels. In countries where biofuel usage is on the rise, it is indispensable to acknowledge the government-backed subsidies as one of the major driving forces behind the swift adaptation (I.E.A., 2023). Another issue with the biofuel industry is the lack of a wide product inventory, which makes any industry resistant to uncontrollable market forces. In short, biofuels tend to costlier and production facility lacks wide inventory, to sustain market competition with petro-based industries. Here lies an opportunity for the biofuel industry to expand its product inventory so that its adoption is not reliant on government subsidies and favorable policies to compete with petro-based fuels. Lignocellulose (LC) material, an abundant, renewable, and valuable biomass fraction, carries the potential to integrate within the biofuel industry, offering value addition via multiple pathways. A phenylpropanoid-based organic polymer, lignin is ubiquitously present in the plants. It is there to provide rigidity to the plant structure by acting as connecting adhesive (covalent linkages) between cellulose and hemicellulose. The structural configuration of a plant cell comprising cellulose, hemicellulose, and lignin is depicted in Figure 6.1.

Lignin sources are widely available and are agro and forestry residues from plant wastes. Table 6.1 lists a number of available lignin sources. Recently, owing to a high number of research investments and a growing global consciousness regarding environmental health, it has led to increased efforts in maximum feedstock utility,

* Corresponding author

 DOI: 10.1201/9781003485636-6

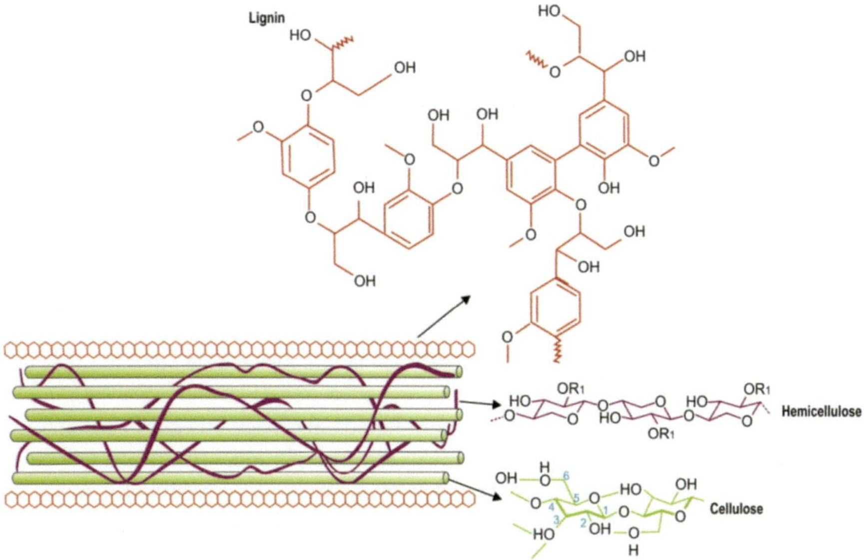

FIGURE 6.1 Generic LC biomass structural composition. (Reproduced under the CC 3.0 (Tran et al., 2019). Copyright © 2020 IntechOpen.)

with a focus on reducing waste as much as possible. This effort has resulted in more and more lignin being recovered from biomass processing (Arni, 2018; Ojo, 2023).

Valorization of the LC biomass fraction within the plant biomass refers to the concept of lignin biorefinery. Lignocellulosic (LC) material can either be sourced directly from the trees/plants or as the waste residues of the bioenergy and pulp industries. Processes or stages such as the separation of biomass into cellulose, hemicellulose, and lignin and the upgrading of individual components via various chemical, enzymatic, and physical pathways are inherent parts of a lignin biorefinery (Becker & Wittmann, 2019). LC biomass fractionation is the process of breaking down plant biomass into its primary components: cellulose, hemicellulose, and lignin. This separation is crucial for efficient biomass utilization in biorefineries. The fractionation typically involves

TABLE 6.1
Various Lignin Sources with Examples

Sources	Examples
Agro-based residues	Cotton stalk, sugarcane bagasse, corn straw, wheat straw, rice straw, corn cobs and leaves, etc.
Forestry residues	Spruce, willow, poplar, eucalyptus and cedars, among others.
Industrial wastes/rejects	The paper and pulp industry rejects brewery-based grains (waste) Kraft lignin, lignosulfonates, among others

Source: Zhao et al. (2012).

pretreatment methods such as chemical, mechanical, or enzymatic processes to disrupt the complex lignin–carbohydrate matrix. Once fractionated, cellulose and hemicellulose can be converted into biofuels or chemicals, while lignin can be upcycled into valuable products such as phenolic compounds, resins, and biofuels (Schutyser et al., 2018). Efficient fractionation is key to maximizing the value derived from LC biomass. The prominent pretreatment processes are alkali treatment, NH_3 (anhydrous) treatment, ionic liquids, and organosolv pretreatment. Owing to the differences in pretreatment processes, the obtained lignin also differs in structural configuration and chemical selectivity, which further dominates the subsequent lignin utilization capability. There has been a score of recent reviews reporting and analyzing lignin extraction or separation or pretreatment methods and their effect on the lignin obtained (Raheja et al., 2024). Typically, the pretreatment techniques may be divided into delignification (alkali and acidic), reductive catalytic fractionation, and other delignification methods based on green solvents (ionic liquids and deep eutectics), mechanical processing, ammonia fiber explosion (AFEX), steam explosion, organosolv, and many others. Post-fractionation, three conversion pathways exist for converting LC materials into value-added products (solid, liquid, and gas): thermochemical methods and chemical and biological depolymerization pathways. The thermochemical method mainly consists of gasification, pyrolysis, and hydrothermal carbonization pathways. While depolymerization technology works at much milder conditions than thermochemical counterparts, it is challenging to deconstruct lignin owing to the compositional and structural complexity of the LC biomass. However, the LC biomass pretreatment/fractionation and successive lignin valorization are not in the scope of this chapter. Hence, shifting the focus toward the product potential of any lignin-based biorefinery, at present, there exists a limited market for pulping and fractionated-derived lignin, which are mainly outsourced by lignosulfonates. The earliest (1960s) lignin-derived product is synthetic vanillin (Basotra et al., 2022), produced by the oxidation of lignosulfonates in an alkaline medium in the presence of a copper catalyst. However, an LC biorefinery presents an opportunity to exploit the full feedstock potential, especially when derived lignin is of superior quality, such as the organosolv-based lignin. The product inventory for any LC biorefinery is very diverse and high, ranging from fine chemicals to mechanical materials and fuels, as depicted in Figure 6.2. In business as usual, post-cellulose/hemicellulose extraction, the lignin fraction is rarely converted to high-value chemicals, materials, or fuels but is used as a low-quality fuel supplement for producing carbon-neutral process heat or other low-end use. The present chapter explores the scope of lignin valorization within biomass, aimed at adding value to any existing biofuel industry type.

6.2 LIGNIN BIOREFINERY: VALUE-ADDED PRODUCTS

Whenever a raw material, lignin in the present case, instead of being used as-is, is transformed from its original form, a value-added product is obtained. As discussed earlier, lignin has been underutilized industrially owing to its structural complexity and heterogeneity, which is source dependent. In the succeeding sections, the various value-added products obtained from lignin with a biorefinery concept are explored and discussed.

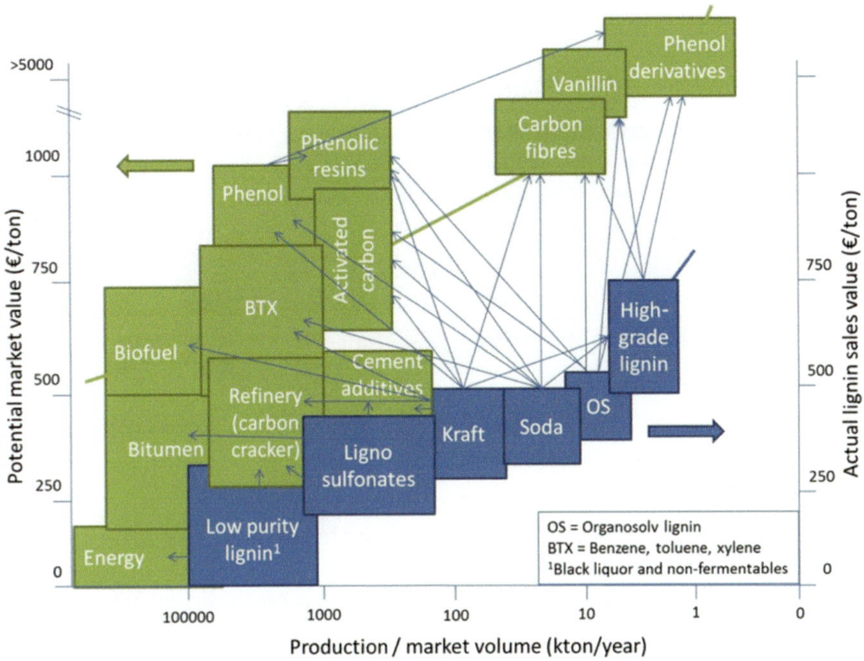

FIGURE 6.2 Market potential of lignin from a variety of sources. (Reproduced under the CC licensing (Gosselink, 2011). Copyright © 2024 Elsevier B.V.)

6.2.1 Chemicals and Materials from Lignin

Given the large availability of lignin as a high-yield side stream in the pulping industry as well as in the biorefinery industry every year, lignin carries the potential to become an excellent feedstock for the production of chemicals and polymeric materials. However, there is a pressing need for more efficient and green processing pathways for upcycling lignin. Various lignin-based products, such as chemicals, polymeric materials, carbon materials, and fuels, have been well demonstrated in research, and it has been proven that if the industrial feasibility is good, then lignin can be solely devoted toward the production of these high-value propositions (Bajwa et al., 2019; Hocking, 1997; Puziy et al., 2018). As discussed, the lignin is structurally very different owing to their respective origins and processing techniques employed. Hence, the suitable valorization techniques for the technical and biorefinery-derived lignin are a function of the structural characteristics, such as the contents of OH groups and β-*O*-4 linkages. Although it is evidently easy and cheaper to apply the lignin streams as filler or additives in the concerned industries to lower heating costs, lignin is by far more valuable in the production of a diverse range of target chemicals and materials, inclusive of fuels, as they have abundant functional characters (chemical species, e.g., phenolic and aliphatic OH groups) (Cao et al., 2019). As an example, when the hydroxyl group is plentiful, lignin can be chemically changed to serve as the base material for polymer synthesis or lignin-based

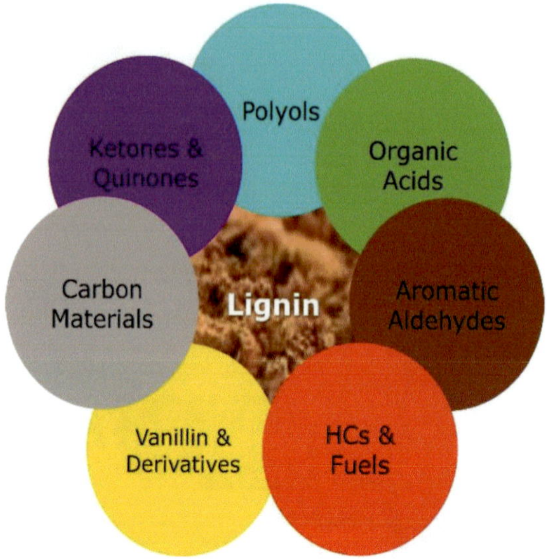

FIGURE 6.3 Various classes of value-added chemicals that can be derived from lignin.

biomaterials; when the β-*O*-4 linkages prevail, lignin can be converted into chemicals and fuels. It is worth reiterating that the structural characteristics of all extracted lignin are principal in deciding the end applications of lignin. Thus, a wide variety of chemical classes, including phenolic compounds, aromatic aldehydes, organic acids, monomeric aromatic alcohols, hydrocarbons and fuels, carbon materials, polyols, aromatic ketones, and vanillin derivatives, can be derived from various sources of agro-based lignin, as depicted in Figure 6.3; the outer periphery shows various lignocellulosic crops/plants.

Chemical changes such as hydroxy alkylation, amination, nitration, and sulfonation may introduce new chemical sites into lignin. Alkylation/dealkylation, phenolation, esterification, and etherification are a few methods of lignin modification that may functionalize the hydroxyl groups; this will help in developing the lignin's polyol derivatives. Post-chemical alterations, the lignin-derived polymers acquire lowered glass transition temperatures, increased thermal performance, and improved miscibility. Furthermore, the majority of lignin-based polymeric materials that have been reported are crosslinked networks based on the versatility of lignin (Laurichesse & Avérous, 2014). The industrial waste disposal problems related to pulping mills and biorefinery industries may be addressed in a cost-effective manner by transforming the lignin streams into value-added thermoplastic materials (Calvo-Flores & Dobado, 2010). For example, based on plasticized lignin, lignin/polymer miscible blends, and chemically modified lignin derivatives, Wang et al. (1992) had previously examined the growing interest in thermoplastic materials. The possibility of achieving economically favorable depolymerization and catalytic transformation of lignin has been gaining momentum recently. Developing an efficient and selective catalyst for targeted lignin conversions has been a core effort in biorefinery research.

Typically, the catalytic methods can be categorized as chemical oxidation, liquid-phase reforming, gasification, pyrolysis, acid/base-catalyzed depolymerization, and hydrogen reduction (Li et al., 2015).

The comparatively simple process of alkali-catalyzed depolymerization of lignin yields aromatic compounds, and catalytic reagents, including sodium and potassium hydroxides, are readily available and reasonably priced. The predominant reaction during the alkaline-catalyzed delignification process, such as the pulping process, is the cleavage of ether linkages (β-*O*-4 bonds). Similarly, the acid-catalyzed lignin depolymerization mainly proceeds with cleaving at α-aryl and β-aryl ether linkages (Wang et al., 1992). Moving on, bio-oil is obtained via pyrolytic (major thermochemical method) processing of lignin (Behling et al., 2016), with catalytic pyrolysis happening at temperatures of 450–700°C (Mohan et al., 2006). A recently proposed lignin liquefaction technique boosted the carbon yield and greatly reduced the oxygen/carbon ratio by combining pyrolysis with the addition of formic acid (Goyal et al., 2008). Table 6.2 summarizes the various processes for lignin chemical degradation pathways. When the lignin is transformed using hydrogenation reduction, valuable chemicals, including phenols and hydrocarbon fuels, are obtained. Hydrogenolysis, hydroxylation, hydro deoxygenation, hydrogenation, and other integrated hydrogen-related reactions are various types of hydro-processing lignin that can be subjected to conversion. Hydrogenolysis is one of these processes that typically occurs in alkaline conditions, and it is often aided by metal catalysts like nickel (Mukkamala et al., 2012). Multifunctional aromatic compounds ranging from aromatic aldehydes to carboxylic acids are obtained mainly from oxidative depolymerization of lignin. Additionally, there are six different types of catalytic systems

TABLE 6.2
Various Catalytic Degradation Pathways for Lignin

Method	Catalyst	Temp. (°C)	Products
Acid-catalyzed	Mineral acids, Lewis acids, ionic liquids, etc.	0–200	Monomeric phenols
Base-catalyzed	KOH, NaOH, LiOH and other organic bases	100–300	Monomeric phenols
Pyrolysis	None, zeolites	450–700	Bio-oil, phenolic compounds
Oxidation	Organometallics, metal-free organocatalytics, acids, bases, and metal salts	0–250	Aromatic alcohols, aldehydes and acids
Reduction	Platinum group catalysts, iron group catalysts, molybdenum, tungsten-based catalysts	100–350	Simple bulk aromatic compounds
Others	Liquid-phase reforming: Group VIII metals and their alloys	250–400	Hydrogen and light gases
	Gasification: noble metal salts	500–1500	Syngas

Source: *Saidi et al. (2014).*

for lignin oxidation: photocatalytic oxidation, electrocatalytic lignin cleavage, acid or base catalysis, metal salt catalysis, organic metal catalysis, and metal-free organic catalysis (Wang et al., 1992).

Synthesis gas (syngas) is typically produced by using alternative lignin catalytic techniques, such as liquid-phase reforming and gasification of lignin at varying liquid or thermal temperatures (Wang et al., 1992). Moreover, it is anticipated that combining two or more methods will result in a more sustainable output of lignin-based compounds after lignin depolymerization. At last, in order to convert lignin into valuable products, a thorough investigation of several elements is necessary. These parameters include lignin structure, catalysts, solvents, and reaction conditions.

6.2.2 Biobased Electronics and Electrochemical Applications

As iterated previously, the inexhaustible supply potential of lignin as a feedstock is a driving force, projecting it as a second-generation feedstock. The electronic electrochemical industry is a fairly young industry and thus, possesses far more flexibility in switching bio-derived and sustainable manufacturing. Lignin, similar to the holocellulose fraction of woody biomasses, offers resource renewability, biodegradability, and reduced carbon footprint, and its properties can be tailored as function specific. A major problem with inorganic petro-based electronics materials is their recalcitrant nature, being relatively impervious to microbial and fungal attacks, a necessary vantage for the biodegradability of the vast electronics dumped in landfills. This is not the case when electronics are biobased, as these should be relatively more susceptible to microbial/fungal attacks. Resins that work with the current production technology and pricing structures would be necessary if any part of the computer industry intends to switch to biopolymer-based resin systems. Other than these considerations, petroleum-based electronics materials manufacturing is an established technology, and thus, any alternative would have to come up with better economic feasibility. A few of the numerous possible applications of lignin in the electronics and electrochemical industry are discussed below.

6.2.2.1 Printed Circuit Boards (PCBs)

Most PCBs are made of epoxidized phenols, including diglycidyl ether of bisphenol A (DGEBA), which combine with amines and polyfunctional epoxies to generate a cross-linked matrix with the required performance properties (Bhaskar & Pandey, 2021). Phenolic epoxies are favored for their natural ability to withstand high temperatures and resist moisture. To be considered as a starting material for PCBs manufacturing, the biobased resins must meet the necessary performance standard. Lignin, is best suited for PCBs manufacturing of all biopolymers available, since it has phenolic and shows necessary thermal stability and moisture resistance (Figure 6.4).

Kosbar and Gelorme (1997) investigated formulating lignin-based resins, and showed that a copolymer with at least 50% lignin content (with di- and polyfunctional phenol-based epoxies) exhibits good thermal and electric properties, required PCBPCBs manufacturing. Di-functional epoxies were able to reach glass transition temperatures of up to 145°C, whereas polyfunctional epoxies could reach

FIGURE 6.4 Lignin-derived PCBs and its major advantages.

temperatures as high as 176°C. Direct mixing of rice husk (as a lignin source) with Araldite LY556 epoxy (mixed with K6 hardener in a 1:10 ratio) has been investigated by Bharath et al. (2020), reporting mediocre material performance in the majority of physical tests. Copy paper, cellulose nanofibers (CNFs), and graphitic carbon produced from lignin acting as a base substrate, binder, and conductive pigment, respectively, have been reported to create the multifunctional carbon-coated substrate for manufacturing printed electronics. Aimed for capacitive applications, the resistance, impedance, and dielectric values indicate that the carbon-coated paper was a suitable PCBs as substrate (Patermann & Altstädt, 2016). Altay et al. (2021) investigated that thermochemical lignin as a substrate in PCBs within a thickness range of 5–8 mm, employing the custom film extrusion method. Both the lignin-based substrate (5–8 mm) showed good peel strength (1.1 N/mm), a comparable permittivity (<4.54 aGHz), better loss factor values, a higher coefficient of thermal expansion than the reference material FR-4. The substrates also showed good hydrophobicity with only 0.5% swelling during 24-hour testing at room temperature. Thus, it is evident that lignin-based copolymers are well suited for PCBs manufacturing, achieving standard values as that of FR 4 substrate.

6.2.2.2 Flexible Electronics and Sensors

Polymers derived from lignin can undergo processing to form pliable substrates suitable for incorporation into flexible electronic devices, including wearable sensors

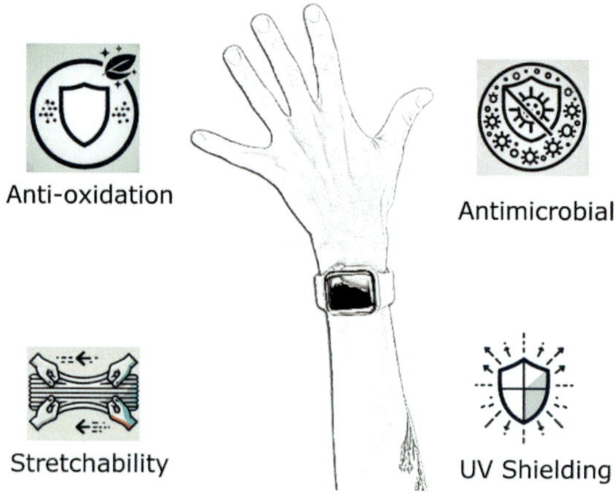

FIGURE 6.5 Lignin-based flexible wearable electronics with inherent advantages.

and displays capable of bending and stretching. Similarly, lignin-derived material can also be used in sensors to find different things like heavy metals and pollutants in different environmental settings, thus aiding bio-based sustainable evolution of environmental monitoring and medical applications. Regarding conductive hydrogel-derived electronics, there has been a lot of interest in sports science, wearable technology, artificial prostheses, and customized medicine (Lei & Wu, 2019; Yang & Suo, 2018). Responses toward external stimuli, either mechanical or electrical, in addition to good stretchability, are indispensable requirements of any material to be considered for hydrogel-based electronics. Furthermore, when considered for human usage, these wearable devices also need to be antimicrobial, anti-oxidation, and UV-shielding. An illustration of an ideal flexible wearable lignin-derived device is shown in Figure 6.5.

In this light, even though cellulose-based hydrogel has been well invested, lignin-based hydrogel for wearable, flexible electronics possesses great potential, as lignin inherently fulfils the requirements, as discussed earlier (Deng et al., 2020). Lignin and its derivatives have been applied as conductive fillers, adhesive agents, strength modifiers, and specific functional fillers in hydrogel production (Wang et al., 2021; Zhang et al., 2020). For example, to produce a conductive hydrogel, Li et al. (2021) added carbon particles produced from lignin to freezing–thawing polyvinyl alcohol, carboxymethyl chitosan, and cellulose nanofibril-based hydrogel. Adding to it, owing to the ionic nature of sulfonated lignin, it has been found to be better suited as conductive fillers in hydrogels. Regarding the role of adhesive agents, in pectin–polyacrylic acid hydrogels, silver and lignin-based composite particles have been shown to deliver lasting adhesion by working as dynamic redox systems continuously generating adhesion by providing phenolic OH groups (Gan et al., 2020). Additionally, the Ag-Lignin (alkali) composites showed excellent skin and cell biocompatibility and were effective in eliminating bacteria from both gram-negative and gram-positive strains (Gan et al., 2019). The mere addition of lignin or its derivatives naturally increases the strength of hydrogels, owing to a large number of non-covalent bonds between lignin groups and corresponding polymeric material such as polyacrylic acid. However, the non-uniform lignin distribution sometimes negatively influences the Young's modulus of elasticity for hydrogels (Thakur & Thakur, 2015); thus, this area requires further investigation.

Lignin-derived environmental monitoring sensors exhibit significant potential due to their eco-friendly nature and efficiency. For example, lignin nanoparticles (LNPs) have been used for detecting mercury (Hg^{2+}) in concentrations as low as 10^{-4} M, showing a clear decrease in UV absorption with increasing mercury levels (Moreno & Sipponen, 2020; Wanjari et al., 2023). In another study, lignin-derived laser-induced graphene (LIG) sensors demonstrated a high sensitivity in detecting pollutants like nitrogen in soil, with strain sensors showing a working range up to 14% strain and a gauge factor of 960 (Yang et al., 2022). These lignin-based sensors are not only effective in detecting environmental contaminants but also exhibit durability, with certain sensors enduring over 10,000 cycles of testing (Yang et al., 2022).

6.2.2.3 Super-capacitors and Batteries

Being a biomass derivative, lignin-based carbon materials have shown great potential as electrodes in super-capacitors and batteries, offering biodegradability and low cost when compared to petro-based carbon derivatives. As shown in Figure 6.6, lignin demonstrates significant potential in advanced energy storage and material applications. The black mesh structure illustrates how lignin-derived carbon materials, with their high surface area and conductivity, can serve as efficient electrodes in super-capacitors and batteries, enhancing ion diffusion and storage capacity. Additionally, the brown outer mesh represents lignin's natural polymeric matrix, which can be utilized as an environmentally resistant outer packaging material or

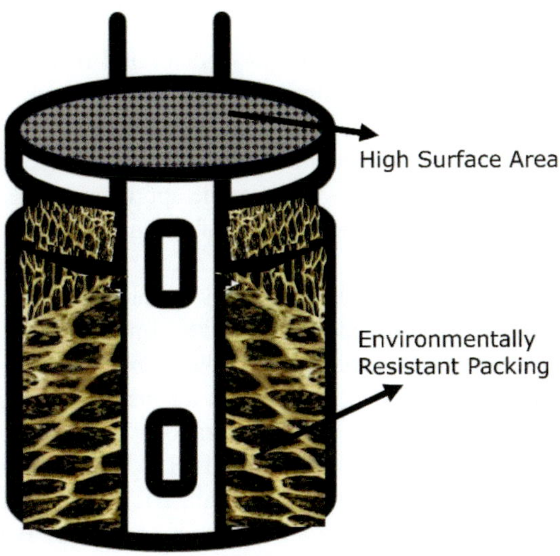

FIGURE 6.6 Lignin-derived electrodes and materials with potential applications in super-capacitors and batteries.

protective coating, providing mechanical strength and moisture resistance for electronics, batteries, and other products. These applications align with ongoing efforts to develop sustainable, bio-based alternatives to fossil-derived materials for both the energy and packaging industries. For sustainable energy storage solutions, lignin-based carbons are being explored as electrodes for Li-ion and Na-ion batteries, among other energy storage devices. In this direction, the quinine group present in lignin derived from brown liquor was reported in 2012 (Milczarek & Inganäs, 2012). Henceforth, there was heightened interest in investigating the production, characterization, and application of lignin-based electrodes with some conductive materials (Kim et al., 2014; Leguizamon et al., 2015; Milczarek & Nowicki, 2013; Nilsson et al., 2015). Quinone, as cathodes, was formed via oxidation and was further doped with polypyrrole via electrochemical polymerization to enhance its conductivity; during cyclic voltammetry measurements, redox peaks were observed in a 0.1 M perchloric acid electrolyte, and from charge/discharge experiments, the obtained capacitance values ranged from 350 to 1000 specific capacitances per weight ($F.g^{-1}$). Similar results were obtained when polypyrrole-doped lignosulfonate electrodes were tested with cotton fabrics as substrate (Zhu et al., 2014), showcasing the great potential of developed electrodes in wearable electronics. When polyaniline was used instead of polypyrrole in lignosulfonate electrodes, the current density per weight increased 100-fold (0.1 to 10 $A.g^{-1}$) (Xu et al., 2015). To further increase the specific capacitance of lignin-based electrodes, inorganic metal acid has been positively tested by Admassie et al. (2014) as an effective alternate route.

All things considered, the use of lignin in electrochemical and bio-based electronics applications presents viable paths for sustainable technological advancement, making use of plentiful biomass resources to solve environmental issues and advance a circular economy.

6.2.2.4 Biomedical Applications

Lignin's complete depolymerization into monomers and their chemical synthesis into synthetic pharmaceuticals or natural products have enormous industry potential. Owing to the aromatic ring and functional groups present in lignin monomers, compounds with substantial biological activity can be produced through chemical transformation. This section explores the possible approaches to lignin valorization and obtains high-value products related to lignin's biomedical uses in an integrated biorefinery.

Syringaldehyde and other aromatic aldehydes are produced by the catalytic oxidation of lignin and can be utilized to synthesize trimethoxybenzaldehyde, trimethoprim, and other medications (Erofeev et al., 1990). Oxidative/reductive conversion or its hybrid approaches have been investigated for medical applications to reduce the generation of by-products and repolymerization. The most desirable oxidant under mild pressure and temperature conditions is H_2O_2, which introduces carboxylate groups into lignin (He et al., 2017; Koo et al., 2017). The initial raw materials used can affect the H_2O_2-mediated oxidative conversion process. Following the treatment of raw materials with ethanol/water or ionic liquid to eliminate hemicellulose fractions, cellulose fractions from solubilized lignin can be extracted using H_2O_2. Additionally, oxidative conversion by H_2O_2 treatment can be helpful in increasing the practicality of water-insoluble kraft lignin in the industry. At 30°C for 8 hours, 2% H_2O_2 was able to solubilize about 50% of the lignin in cane bagasse (Azzam, 1989), while an increase in temperature to 90°C increased the depolymerization/oxidation efficiency. The solubilized lignin was modified into porous (nano) materials such as polyphepan (an enterosorbent) for medical applications (Kuznetsov et al., 2018). BP-Cx-1 (a promising ligand for the making of other drugs with antineoplastic activity) was obtained when polyphepan was further hydrolyzed with O_2 under basic conditions at high temperatures and pressures (Fedoros et al., 2018).

6.3 ECONOMICS

The production of high-value chemicals from lignin in a lignocellulosic biomass-based biorefinery offers significant economic potential by transforming an underutilized byproduct into profitable products. Lignin, often discarded or burned for low-efficiency energy, can be converted into aromatic chemicals such as phenols, vanillin, and bio-based polymers, which are in demand across industries like adhesives, pharmaceuticals, and fragrances. This value-added utilization not only increases the economic feasibility of biorefineries but also reduces reliance on fossil-based resources. With advancements in catalytic depolymerization and selective conversion processes, lignin valorization can substantially enhance revenue streams, offsetting operational costs and promoting sustainable industry practices. Furthermore, the global adaptation of any industry is subjected to its economics, as is the case with the lignin biorefinery. An economic analysis exploring major profitability indicators (capital and operating costs, net present value, payback time, and internal rate of return) is required to assess the industrial feasibility of lignin biorefinery. In an integrated forestry-based LC biorefinery, Huang et al. (2010) conceptualized two pathways for producing bioethanol. One pathway is based on hemicellulose extraction for ethanol production, and the other on combined hemicellulose and short fibers. Their economic analysis revealed that in both cases, the minimum selling price for produced ethanol was 901 USD/m^3 and 491 USD/m^3, respectively, for 2000 tonnes of dry wood per day of feedstock throughput. This integration of modern-day pulp mills into an integrated forest-based biorefinery boasts exclusive product inventory opportunities: cellulose fiber, ethanol (fuel grade), and additional energy, thus showcasing an excellent industrial configuration for a lignin-based biorefinery with diverse revenue streams and a lower carbon footprint. Similarly, Fornell et al. (2013) gasified the lignin-concentrated liquor (reject stream) to produce dimethyl ether (DME), where the main line produced cellulose-based ethanol. The economic analysis showed net annual earnings of 25 M€, with a payback time of seven years. The product-marked selling prices and capital investment costs dictated the economic analysis. It is plausible that a lack of product diversity or upgrading the lignin stream to a valuable chemical stream may present a more favorable economic outcome. On the same note, when a biorefinery was proposed to produce a wide range of products, including lignin, furfural, and fuel-grade ethanol, favorable economic feasibility was witnessed (Zang et al., 2020). The process design applied choline chloride (aqueous), a biphasic solvent, for simultaneous pretreatment and hemicellulose conversion. The economic sensitivity analysis showed that for furfural, the minimum selling price could go up to 37% lower than the market and still be feasible, showcasing the competitiveness of the proposed biorefinery. Thus, it is exceedingly evident that a lignin biorefinery having a diverse product range integrated into it presents a far better economic scenario with lowered carbon emissions.

6.4 CHALLENGES AND FUTURE PERSPECTIVES

With the ever-increasing population load, there is an increasing demand for energy, chemicals, and materials. Thus, there is a pressing need for sustainable yet economically feasible lignin processing technology. The lignin stream, especially low-revenue-generating streams such as in the pulping mills, needs to find applications of lignin with added value in the current biorefinery scenario. Approximately 70 million tons of industrial lignin is currently produced worldwide each year through the pulping and biorefinery processes. Of this, approximately 95% of industrial lignin is burned, while the rest is being exploited for prospective uses as surfactants, binders, additives, and dispersants. Different pretreatment and isolation techniques produce lignin with varying chemical structures and reactivity, which ultimately dictate how the particular lignin is ultimately used to produce different chemicals or energy. For example, the lignin fraction with a high quantity of hydroxyl groups is preferred if the ultimate use of lignin is to create polymeric material, as the majority of chemical alterations during polymerization depend on the quantity of hydroxyl (OH) groups.

When it comes to lignin separation, the following factors determine the separated lignin's structures and reactivity: (i) the kind of biomass, (ii) the extraction or separation technique, and (iii) the intensity of the extraction or separation technique. The latter directly affects the purity and yield of the extracted lignin. In any case, the biorefinery's ability to successfully apply these three aforementioned concepts suggests that lignin value-adding should be viewed as the main goal of the process rather than a secondary activity. In this regard, because of their low compatibility and structural variability, the majority of lignin fractions derived from pulping and pretreatment operations cannot be employed directly in the manufacture of lignin-derived polymers. One strategy to counter these limitations consists of chemical modifications of lignin; for example, various OH groups present in the lignin may be esterified or alkylated so that lignin polymer compatibility with the non-polar polymer matrix is enhanced. Furthermore, even though the typical structural configuration of lignin has been well established now, the supramolecular structure in various mediums is not very clear. Investigating the supramolecular structure and variables affecting it will ensure the right direction for the subsequent lignin-derived polymers. To summarize, lignin biorefinery involves methodical tasks that must be synchronized through many stages, including optimizing separation, characterizing lignin's structure, identifying functional groups, and, eventually, application.

6.5 CONCLUSION

The present chapter has explored the immense potential of lignin as a valuable resource within the biorefinery framework, emphasizing its applications in a wide range of industries. Beginning with an analysis of lignin sources from diverse LC biomasses, we briefly touched upon the structure and composition of lignin and different pretreatment and conversion processes to unlock its value. As discussed, lignin has proven potential as a feedstock for producing high-value chemicals and polymeric materials, offering unique opportunities for industrial applications, particularly in bio-based electronics, electrochemical devices, and biomedical fields. Lignin's incorporation into printed circuit boards, flexible electronics, sensors, super-capacitors, and batteries demonstrates its relevance in advancing sustainable, bio-based materials in electronics. Additionally, the use of lignin-derived materials in biomedical applications opens avenues for new innovations in drug delivery, wound healing, and antimicrobial technologies. Despite these promising applications, economic considerations and technical challenges such as lignin's structural variability and the need for advanced catalytic and depolymerization processes remain critical. In terms of economics, this chapter has outlined how a lignin-oriented biorefinery could significantly enhance the profitability of bio-based industries, reducing dependency on fossil-based feedstocks while aligning with circular economy principles. However, challenges like the complexity of lignin's structure, reactivity, and high processing costs necessitate further research into more efficient and scalable processing techniques. Moving forward, it is clear that the development of a comprehensive lignin-based biorefinery model will be essential for realizing its full industrial potential. Such a model would involve innovations in pretreatment techniques, catalytic processes, and techno-economic assessments to establish lignin as a cornerstone for producing chemicals, materials, and fuels. Addressing these challenges will allow lignin to not only contribute to the bioeconomy but also play a pivotal role in advancing green technology and sustainable industrial practices in the future.

REFERENCES

Admassie, S., Elfwing, A., Jager, E. W. H., Bao, Q., & Inganäs, O. (2014). A renewable biopolymer cathode with multivalent metal ions for enhanced charge storage. Journal of Materials Chemistry A, 2, 1974–1979.

Altay, B. N., Aksoy, B., Banerjee, D., Maddipatla, D., Fleming, P. D., Bolduc, M., … Demir, M. (2021). Lignin-derived carbon-coated functional paper for printed electronics. ACS Applied Electronic Materials, 3, 3904–3914.

Arni, S. A. (2018). Extraction and isolation methods for lignin separation from sugarcane bagasse: A review. Industrial Crops and Products, 115, 330–339.

Azzam, A. M. (1989). Pretreatment of cane bagasse with alkaline hydrogen peroxide for enzymatic hydrolysis of cellulose and ethanol fermentation. Journal of Environmental Science and Health, 24, 421–433.

Bajwa, D., Pourhashem, G., Ullah, A., & Bajwa, S. (2019). A concise review of current lignin production, applications, products and their environmental impact. Industrial Crops and Products, 139, 111526.

Basotra, N., Raheja, Y., Kaur, B., & Chadha, B. S. (2021). Thermophilic Fungal Lignocellulolytic Enzymes in Biorefineries. In T. Satyanarayana, S. K. Deshmukh, & M. V. Deshpande (Eds.), Progress in Mycology. Singapore: Springer. https://doi.org/10.1007/978-981-16-3307-2_2

Becker, J., & Wittmann, C. (2019). A field of dreams: Lignin valorization into chemicals, materials, fuels, and health-care products. Biotechnology Advances, 37, 107360.

Behling, R., Valange, S., & Chatel, G. (2016). Heterogeneous catalytic oxidation for lignin valorization into valuable chemicals: What results? What limitations? What trends? Green Chemistry, 18, 1839–1854.

Bharath, K. N., Madhu, P., Gowda, T. G. Y., Verma, A., Sanjay, M. R., & Siengchin, S. (2020). A novel approach for development of printed circuit board from biofiber based composites. Polymer Composites, 41, 4550–4558.

Bhaskar, T., & Pandey, A. (2021). Biomass, Biofuels, Biochemicals. Amsterdam: Elsevier.

Calvo-Flores, F. G., & Dobado, J. A. (2010). Lignin as renewable raw material. ChemSusChem, 3, 1227–1235.

Cao, Y., Chen, S. S., Zhang, S., Ok, Y. S., Matsagar, B. M., Wu, K. C.-W., & Tsang, D. C. (2019). Advances in lignin valorization towards bio-based chemicals and fuels: Lignin biorefinery. Bioresource Technology, 291, 121878.

Deng, J., Yuk, H., Wu, J., Varela, C. E., Chen, X., Roche, E. T., … Zhao, X. (2020). Electrical bioadhesive interface for bioelectronics. Nature Materials, 20, 229–236.

Erofeev, Y. V., Afanas'eva, V. L., & Glushkov, R. G. (1990). Synthetic routes to 3,4,5-trimethoxybenzaldehyde (review). Pharmaceutical Chemistry Journal, 24, 501–510.

Fedoros, E. I., Orlov, A. A., Zherebker, A., Gubareva, E. A., Maydin, M. A., Konstantinov, A. I., … Anisimov, V. N. (2018). Novel water-soluble lignin derivative BP-Cx-1: Identification of components and screening of potential targets in silico and in vitro. Oncotarget, 9, 18578–18593.

Fornell, R., Berntsson, T., & Åsblad, A. (2013). Techno-economic analysis of a kraft pulp-mill-based biorefinery producing both ethanol and dimethyl ether. Energy, 50, 83–92.

Gan, D., Shuai, T., Wang, X., Huang, Z., Ren, F., Fang, L., … Lu, X. (2020). Mussel-inspired redox-active and hydrophilic conductive polymer nanoparticles for adhesive hydrogel bioelectronics. Nano-Micro Letters, 12. https://doi.org/10.1007/s40820-020-00507-0

Gan, D., Xing, W., Jiang, L., Fang, J., Zhao, C., Ren, F., … Lu, X. (2019). Plant-inspired adhesive and tough hydrogel based on Ag-Lignin nanoparticles-triggered dynamic redox catechol chemistry. Nature Communications, 10.

Gosselink, R. J. A. (2011). Lignin as a renewable aromatic resource for the chemical industry (PhD thesis, Wageningen University). Netherlands: Wageningen University. Retrieved from https://edepot.wur.nl/186285

Goyal, H., Seal, D., & Saxena, R. (2008). Bio-fuels from thermochemical conversion of renewable resources: A review. Renewable and Sustainable Energy Reviews, 12, 504–517.

He, W., Gao, W., & Fatehi, P. (2017). Oxidation of kraft lignin with hydrogen peroxide and its application as a dispersant for kaolin suspensions. ACS Sustainable Chemistry & Engineering, 5, 10597–10605.

Hocking, M. B. (1997). Vanillin: Synthetic flavoring from spent sulfite liquor. Journal of Chemical Education, 74, 1055.

Huang, H.-J., Ramaswamy, S., Al-Dajani, W. W., & Tschirner, U. (2010). Process modeling and analysis of pulp mill-based integrated biorefinery with hemicellulose pre-extraction for ethanol production: A comparative study. Bioresource Technology, 101, 624–631.

International Energy Agency (I.E.A). (2023). Biofuel Policy in Brazil, India and the United States, Paris: IEA. https://www.iea.org/reports/biofuel-policy-in-brazil-india-and-the-united-states, Licence: CC BY.

Kharissova, A. B., Kharissova, O. V., Kharisov, B. I., & Méndez, Y. P. (2024). Carbon negative footprint materials: A review. Nano-Structures & Nano-Objects, 37, 101100.

Kim, S., Kim, Y. K., Lee, H., Lee, S. B., & Park, H. S. (2014). Superior pseudocapacitive behavior of confined lignin nanocrystals for renewable energy-storage materials. ChemSusChem, 7, 1094–1101.

Koo, H., Salunke, B. K., Iskandarani, B., Oh, W.-G., & Kim, B. S. (2017). Improved degradation of lignocellulosic biomass pretreated by Fenton-like reaction using Fe3O4 magnetic nanoparticles. Biotechnology and Bioprocess Engineering, 22, 597–603.

Kosbar, L. L., & Gelorme, J. (1997). Biobased epoxy resins for computer components and printed wiring boards. Proceedings of the 1997 IEEE International Symposium on Electronics and the Environment. ISEE-1997, 28–32. San Francisco, CA, USA. https://doi.org/10.1109/ISEE.1997.605221

Kuznetsov, B. N., Sudakova, I. G., Garyntseva, N. V., Levdansky, V. A., Ivanchenko, N. M., Pestunov, A. V., … Pinel, C. (2018). Green biorefinery of larch wood biomass to obtain the bioactive compounds, functional polymers and nanoporous materials. Wood Science and Technology, 52, 1377–1394.

Laurichesse, S., & Avérous, L. (2014). Chemical modification of lignins: Towards biobased polymers. Progress in Polymer Science, 39, 1266–1290.

Leguizamon, S., Díaz-Orellana, K. P., Velez, J., Thies, M. C., & Roberts, M. E. (2015). High charge-capacity polymer electrodes comprising alkali lignin from the kraft process. Journal of Materials Chemistry A, 3, 11330–11339.

Lei, Z., & Wu, P. (2019). A highly transparent and ultra-stretchable conductor with stable conductivity during large deformation. Nature Communications, 10(1). https://doi.org/10.1038/s41467-019-11364-w

Li, C., Zhao, X., Wang, A., Huber, G. W., & Zhang, T. (2015). Catalytic transformation of lignin for the production of chemicals and fuels. Chemical Reviews, 115, 11559–11624.

Li, M., Tu, Q., Long, X., Zhang, Q., Jiang, H., Chen, C., … Min, D. (2021). Flexible conductive hydrogel fabricated with polyvinyl alcohol, carboxymethyl chitosan, cellulose nanofibrils, and lignin-based carbon applied as strain and pressure sensor. International Journal of Biological Macromolecules, 166, 1526–1534.

Milczarek, G., & Inganäs, O. (2012). Renewable cathode materials from biopolymer/conjugated polymer interpenetrating networks. Science, 335, 1468–1471.

Milczarek, G., & Nowicki, M. (2013). Carbon nanotubes/kraft lignin composite: Characterization and charge storage properties. Materials Research Bulletin, 48, 4032–4038.

Mohan, D., Pittman, C. U., & Steele, P. H. (2006). Pyrolysis of wood/biomass for bio-oil: A critical review. Energy & Fuels, 20, 848–889. https://doi.org/10.1021/ef0502397

Moreno, A., & Sipponen, M. H. (2020). Lignin-based smart materials: A roadmap to processing and synthesis for current and future applications. Materials Horizons, 7(9), 2237–2257. https://doi.org/10.1039/D0MH00798F

Mukkamala, S., Wheeler, M. C., Heiningen, A. R. P., & DeSisto, W. J. (2012). Formate-assisted fast pyrolysis of lignin. Energy & Fuels, 26, 1380–1384.

Nilsson, T. Y., Wagner, M., & Inganäs, O. (2015). Lignin modification for biopolymer/conjugated polymer hybrids as renewable energy storage materials. ChemSusChem, 8, 4081–4085.

Ojo, A. (2023). An overview of lignocellulose and its biotechnological importance in high-value product production. Fermentation, 9, 990.

Patermann, S., & Altstädt, V. (2016). Lignin as renewable raw material for applications in electronics. AIP Conference Proceedings, Authors. https://doi.org/10.1063/1.4965524

Puziy, A. M., Poddubnaya, O. I., & Sevastyanova, O. (2018). Carbon materials from technical lignins: Recent advances. Topics in Current Chemistry, 376. https://doi.org/10.1007/s41061-018-0210-7

Raheja, Y., Sharma, G., Singh, V., Agrawal, D., & Chadha, B. S. (2024). Dependence of Enzymatic Saccharification on the Type of Biomass and Pretreatment. In V. Bisaria (Ed.), Handbook of Biorefinery Research and Technology. Dordrecht: Springer. https://doi.org/10.1007/978-94-007-6724-9_72-1

Saidi, M., Samimi, F., Karimipourfard, D., Nimmanwudipong, T., Gates, B. C., & Rahimpour, M. R. (2014). Upgrading of lignin-derived bio-oils by catalytic hydrodeoxygenation. Energy & Environmental Science, 7, 103–129.

Schutyser, W., Renders, T., Bosch, S., Koelewijn, S.-F., Beckham, G. T., & Sels, B. F. (2018). Chemicals from lignin: An interplay of lignocellulose fractionation, depolymerization, and upgrading. Chemical Society Reviews, 47, 852–908.

Thakur, V. K., & Thakur, M. K. (2015). Recent advances in green hydrogels from lignin: A review. International Journal of Biological Macromolecules, 72, 834–847.

Tran, T., Le, P., Mai, P., & Nguyen, Q. (2019). Ethanol Production from Lignocellulosic Biomass. Rijeka, Croatia: IntechOpen.

Wang, J., Manley, R., & Feldman, D. (1992). Synthetic polymer-lignin copolymers and blends. Progress in Polymer Science, 17, 611–646. https://doi.org/10.1016/0079-6700(92)90003-H

Wang, Q., Guo, J., Lu, X., Ma, X., Cao, S., Pan, X., & Ni, Y. (2021). Wearable lignin-based hydrogel electronics: A mini-review. International Journal of Biological Macromolecules, 181, 45–50.

Wanjari, V. P., Reddy, A. S., & Duttagupta, S. P. (2023). Laser-induced graphene-based electrochemical biosensors for environmental applications: A perspective. Environmental Science and Pollution Research, 30, 42643–42657. https://doi.org/10.1007/s11356-022-21035-x

Xu, H., Jiang, H., Li, X., & Wang, G. (2015). Synthesis and electrochemical capacitance performance of polyaniline doped with lignosulfonate. RSC Advances, 5, 76116–76121. https://doi.org/10.1039/C5RA12292A

Yang, C., & Suo, Z. (2018). Hydrogel ionotronics. Nature Reviews Materials, 3, 125–142.

Yang, S., Ling, Y., Wu, Q., Zhang, H., Yan, Z., Huang, G., … Wan, C. (2022). Lignin-derived porous graphene for wearable and ultrasensitive strain sensors. Journal of Materials Chemistry C, 10(32), 11730–11738. https://doi.org/10.1039/D2TC00953F

Zang, G., Shah, A., & Wan, C. (2020). Techno-economic analysis of an integrated biorefinery strategy based on one-pot biomass fractionation and furfural production. Journal of Cleaner Production, 260, 120837.

Zhang, X., Sun, D., Cai, J., Liu, W., Yan, N., & Qiu, X. (2020). Robust conductive hydrogel with antibacterial activity and UV-shielding performance. Industrial & Engineering Chemistry Research, 59, 17867–17875.

Zhao, X., Zhang, L., & Liu, D. (2012). Biomass recalcitrance. Part II: Fundamentals of different pretreatments to increase the enzymatic digestibility of lignocellulose. Biofuels, Bioproducts and Biorefining, 6, 561–579.

Zhu, L., Wu, L., Sun, Y., Li, M., Xu, J., Bai, Z., … Xu, W. (2014). Cotton fabrics coated with lignosulfonate-doped polypyrrole for flexible supercapacitor electrodes. RSC Advances, 4, 6261.

7 Biogas Production
Anaerobic Digester Design and Biogas Analysis

*Arvind Keprate, Damilola Olonisola, Aliyu Aliyu, and Ravi Kant Bhatia**

7.1 INTRODUCTION

Anaerobic digestion (AD) is a biological process that breaks down organic matter in the absence of oxygen to produce biogas, a renewable energy source, and a nutrient-rich digestate. AD can utilize a variety of feedstocks, including agricultural residues, livestock manure, food waste, and wastewater sludge. This biochemical process involves four key stages: hydrolysis, acidogenesis, acetogenesis, and methanogenesis, with various factors such as temperature, pH, and substrate composition influencing overall efficiency and output (Archana et al., 2023). Biogas primarily consists of methane and carbon dioxide, making it a valuable alternative energy source that can contribute to reducing reliance on fossil fuels and mitigating greenhouse gas emissions (Banerjee et al., 2021).

Biogas usage is expected to double in the next few years, from 14.5 GW in 2012 to 29.5 GW in 2022, according to estimates (Piechota, 2021). The components present in biogas comprise methane (55–75%), carbon dioxide (25–45%), nitrogen (0–5%), hydrogen (0–1%), hydrogen sulfide (0–1%), and oxygen (0–2%) (Sivamani et al., 2020), as shown in Figure 7.1.

The environmental impact of biogas production through AD is significant, particularly in the context of greenhouse gas (GHG) emissions and waste management. Traditional waste disposal methods, such as landfilling and incineration, are major contributors to GHG emissions, producing approximately 109 Mt CO_2e per year in the European Union alone, which constitutes about 37% of its waste-related emissions (*US EPA*, 2025). These emissions primarily include carbon dioxide (CO_2), nitrous oxide (N_2O), and methane (CH_4), the latter being a potent greenhouse gas that contributes to global warming (*US EPA*, 2025). As an alternative to landfilling, AD digesters are widely adopted for commercial biogas production and used as a fuel or source of energy. Biogas is increasingly being utilized in various industrial processes. It can serve as a feedstock in fuel cells, enabling efficient energy conversion with low emissions (*US EPA*, 2025). One of the primary applications of biogas is in energy generation. It can be utilized as fuel in internal combustion engines, turbines, and boilers, facilitating the production of electricity, heat, and steam (*US EPA*, 2025). In particular, biogas can replace natural gas in domestic and industrial applications, offering a renewable alternative that contributes to energy security and reduced greenhouse gas

* Corresponding author

DOI: 10.1201/9781003485636-7

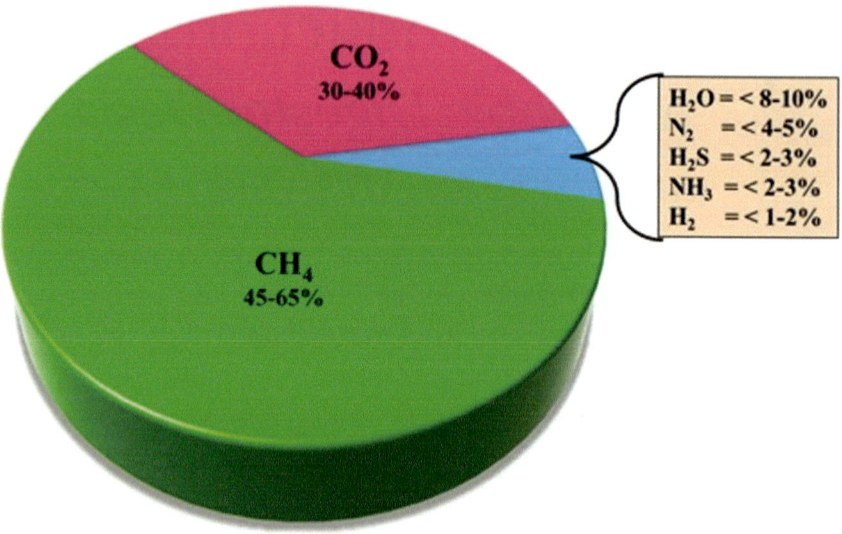

FIGURE 7.1 Typical composition of biogas.

emissions. In agricultural settings, biogas systems help farmers lower their reliance on purchased electricity by generating power on-site, thus mitigating energy costs. In many developing countries, biogas is predominantly used for cooking and heating. It is a vital resource in rural areas where traditional fuels may be scarce or expensive. Biogas stoves are commonly employed for cooking, while biogas lamps provide lighting, improving the quality of life for many households (Banerjee et al., 2021). This application enhances energy access and also contributes to environmental sustainability by reducing reliance on firewood and fossil fuels (Sivamani et al., 2020). Biogas can be upgraded to biomethane, a cleaner alternative to conventional natural gas, and used as a fuel for vehicles. The utilization of biogas in transportation helps to reduce fossil fuel consumption and lowers carbon emissions. As cities and countries seek to transition to more sustainable transport options, biogas offers an attractive solution. Additionally, biogas can be integrated into industrial heating systems, providing a renewable energy source that supports sustainable manufacturing practices (*US EPA*, 2025). However, prior to putting the biogas produced to use, it is vital to examine the composition and quality of the gas through analytical techniques such as gas chromatography, as illustrated in Figure 7.2. This method involves collecting a gas sample from the digester, regulating the flow, and injecting it into a chromatography column, where the individual gas components are separated, detected, and analyzed to determine the concentrations of methane, carbon dioxide, hydrogen sulfide, and other trace gases. Such analysis ensures that the biogas meets the required standards for specific applications, whether for energy generation, upgrading to biomethane, or use in transportation systems.

The efficiency and stability of biogas production depend heavily on reactor design, feedstock composition, operational parameters, and microbial dynamics. Anaerobic

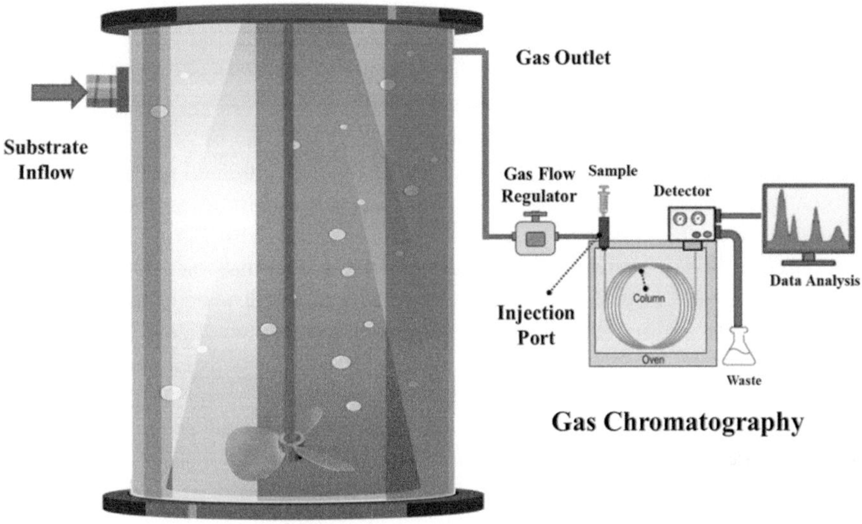

FIGURE 7.2 Biogas reactor integrated with gas chromatography for its analytics.

digesters are engineered systems that facilitate the digestion process under controlled conditions to maximize gas yield and ensure process stability. The anaerobic production of biogas can be made cheaper by designing a high-throughput reactor and operating procedures. The parameters such as substrate type, particle size, temperature, pH, carbon/nitrogen (C/N) ratio, and inoculum concentration play a major role in the design of reactors to produce biogas. Multistage systems, batch systems, continuous one-stage systems, and continuous two-stage systems are the types of digesters used in the industry for biogas production. This chapter begins by outlining the fundamentals of the biogas production process, followed by a discussion of various reactor types. It then explores key design considerations and provides an overview of biogas analysis, which is essential for the optimal operation of anaerobic digesters. Finally, the chapter presents a case study detailing the design, instrumentation, and components of an experimental biogas plant.

7.2 FUNDAMENTALS IN BIOGAS PRODUCTION PROCESS

The effectiveness and stability of the AD process heavily depend on the characteristics of the feedstock, operational parameters, and the reactor design. A deep understanding of the types of feedstocks and their properties is fundamental to ensuring optimal gas yield and maintaining digester health. The most common feedstock type used in AD digesters is discussed below:

7.2.1 Feedstocks Used in Biogas Production

Biogas production relies on a diverse array of feedstocks, each contributing unique characteristics and energy production potential. The primary feedstock types can be broadly categorized into four groups: crop residues, animal manure, organic municipal solid waste (MSW), and wastewater sludge. Animal manure is a commonly used feedstock for biogas plants. It includes waste from livestock such as cattle, pigs, and poultry. While manure typically has lower energy content due to being predigested in animals' gastrointestinal tracts, it is favored for anaerobic digestion because of its neutral pH, high buffering capacity, and natural microbial content that facilitates digestion (Esteves et al., 2019). Moreover, manure can be managed sustainably; however, improper storage and handling can lead to significant environmental issues, including greenhouse gas emissions and nutrient pollution (Pence et al., 2024). The composition of manure varies considerably based on factors like animal species, diet, and management practices. For instance, livestock manure is rich in nitrogen and phosphorus, which can be harmful if they enter water bodies unchecked (Pence et al., 2024). The specific nutrient profiles of different manures can affect their efficiency in biogas production and the resulting energy yield (Einarsson & Persson, 2017). Besides energy production, the use of animal manure in AD helps reduce greenhouse gas emissions from open manure storage and improves nutrient management through the production of stabilized digestate. Crop residues such as straw, corn stover, sugarcane bagasse, and other lignocellulosic materials are abundant agricultural by-products. These residues represent a valuable source of organic material for biogas production. However, the energy production potential from crop residues can be limited due to their lignocellulosic structure, which makes them more challenging to digest anaerobically compared to other feedstocks (IEA, 2020). Pre-treatment methods such as mechanical shredding, thermal processing, or alkaline hydrolysis are often required to improve their digestibility. Co-digestion with more easily degradable substrates can also enhance overall reactor performance. The organic fraction of municipal solid waste (MSW) includes food waste and other biodegradable materials generated by households and industries. This category of feedstock has the potential to significantly contribute to biogas generation due to the high energy content of food scraps (IEA, 2020). However, food waste also poses challenges due to its variability, potential for rapid acidification, and the need for pretreatment to remove contaminants like plastics and metals. Co-digestion of food waste with other substrates, such as manure, can enhance biogas yield by optimizing the nutrient balance in the digester (Al-Wahaibi et al., 2020). Sewage sludge, a byproduct of wastewater treatment processes, is another vital feedstock for biogas production. It contains a mix of organic materials and nutrients, which can be broken down anaerobically to produce biogas (IEA, 2020). However, the treatment and handling of sewage sludge require careful management to minimize potential contamination and health risks. AD of sewage sludge is a well-established practice that not only reduces sludge volume and pathogens but also recovers energy in the form of biogas. Sludge digestion is typically integrated into the treatment process and is optimized for mesophilic conditions. When selecting feedstocks for biogas production, several factors come into play, including biogas efficiency, availability, economic feasibility, and environmental impact (Czekała et al., 2023). The right mix of feedstocks can optimize energy output while ensuring sustainable waste management practices are upheld. However, potential investors should be aware of the economic risks associated with fluctuating prices, regulations, and the inherent uncertainties in feedstock supply and quality (Einarsson & Persson, 2017).

7.2.2 Collection and Preparation of Feedstocks

The collection and preparation of feedstocks are critical for the success of biogas production systems. Effective collection strategies ensure a reliable and continuous supply of organic material, while pretreatment enhances substrate digestibility and improves gas yields. Logistically, feedstock collection can be challenging, particularly when sources are geographically dispersed or seasonally available. The timing and frequency of collection influence storage requirements and transport costs. Transporting feedstocks over long distances not only increases operational costs but may also negate the energy benefits of biogas if the fuel used for transportation outweighs the energy produced (Einarsson & Persson, 2017). Generally, the economically viable transport distance for feedstocks is significantly lower than the theoretical limit of approximately 200 km (Einarsson & Persson, 2017).

Before biogas production, feedstocks must undergo various pretreatment processes to enhance their digestibility and biogas yield. The characteristics of each type of feedstock, such as moisture content and nutrient composition, play a significant role in determining the most effective preparation methods (Czekała et al., 2023). High-moisture feedstocks like fruits and vegetables generally yield more biogas compared to drier materials (Czekała et al., 2023). Pre-treatment techniques can be broadly categorized into three types: mechanical, chemical, and biological. Each method targets specific barriers to effective biodegradation and can significantly influence the rate and efficiency of biogas production. Mechanical pretreatment involves physical processes such as shredding, grinding, milling, or pressing to reduce the particle size of feedstocks and increase their surface area. This enhances microbial accessibility and accelerates hydrolysis, the first step in anaerobic digestion. Technologies like high-pressure homogenization and ultrasonic disintegration are often used to rupture cell walls and disintegrate fibrous materials, making them more amenable to microbial attack (Hernández-Beltrán et al., 2019). While mechanical methods are generally energy-intensive, they are particularly useful for treating lignocellulosic biomass and thick slurries. However, these techniques typically do not eliminate pathogens or modify the chemical structure of the substrate (Hernández-Beltrán et al., 2019). Chemical pretreatment employs reagents such as acids (e.g., sulfuric or hydrochloric acid), alkalis (e.g., sodium hydroxide), or oxidizing agents (e.g., ozone or hydrogen peroxide) to break down complex polymers like lignin, cellulose, and hemicellulose into simpler, more digestible compounds (Hernández-Beltrán et al., 2019). This solubilization enhances the hydrolysis and acidogenesis phases of digestion. While chemical methods are effective in increasing methane yield, they must be carefully managed due to their potential toxicity, high operational costs, and the generation of inhibitory by-products. Post-treatment neutralization may also be necessary to balance pH levels before anaerobic digestion begins (Hernández-Beltrán et al., 2019). Biological pretreatment utilizes selected microbial cultures or enzymes to initiate partial degradation of organic matter before it enters the main digester. Enzymatic hydrolysis, for instance, employs cellulases and hemicellulases to break down plant-based feedstocks. This method operates under mild conditions and is considered environmentally friendly and energy-efficient (Hernández-Beltrán et al., 2019). However, it is generally slower than mechanical or chemical alternatives and may require controlled environments to maintain microbial activity. Biological pretreatment is particularly beneficial for feedstocks with high fiber or protein content, such as crop residues and food processing waste (Hernández-Beltrán et al., 2019).

7.2.3 Microorganisms and Anaerobic Digestion Process

The AD process relies on a complex microbial consortium that works synergistically to break down organic matter and convert it into biogas. This multistep process is typically divided into four key biological stages: hydrolysis, acidogenesis, acetogenesis, and methanogenesis. Each of these stages is facilitated by specific groups of microorganisms and biochemical reactions, as illustrated in Figure 7.3.

- **Hydrolysis** is the initial step in which complex organic polymers such as carbohydrates, fats, and proteins are broken down into simpler monomers like sugars, amino acids, and fatty acids. This stage is driven by hydrolytic bacteria such as *Saccharomyces sp.* (Bhatia et al., 2020). These microorganisms secrete extracellular enzymes that help solubilize particulate organic matter, making it accessible for further microbial processing.

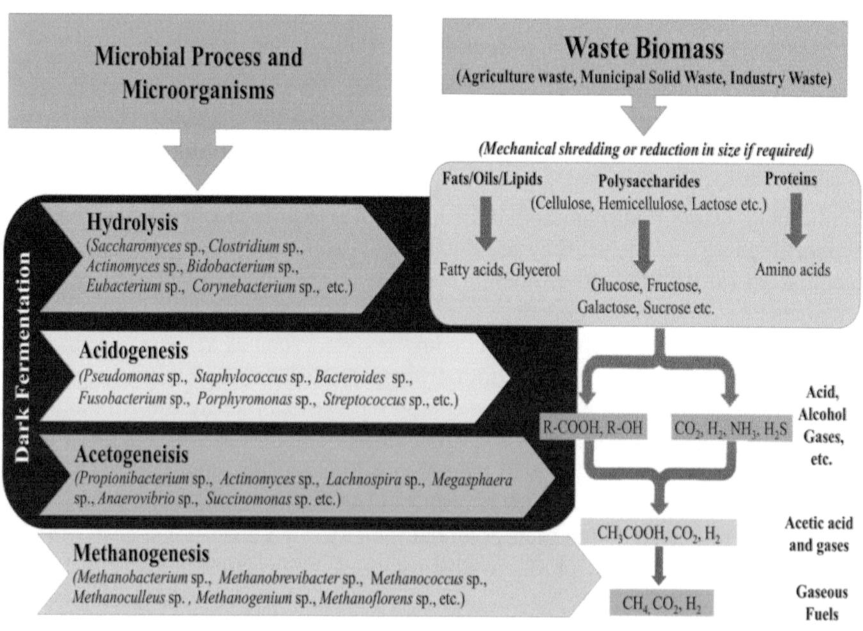

FIGURE 7.3 Microorganisms and anaerobic digestion process involved in the biogas production (Bhatia et al., 2020).

- **Acidogenesis** follows hydrolysis and involves the conversion of the monomers into short-chain volatile fatty acids (VFAs), alcohols, hydrogen (H_2), carbon dioxide (CO_2), ammonia (NH_3), and hydrogen sulfide (H_2S). This phase is facilitated by acidogenic bacteria, including *Pseudomonas sp.* (Bhatia et al., 2020). This stage can lower the pH of the digester and requires careful monitoring to avoid acidification.
- **Acetogenesis** is the third stage in which VFAs and alcohols are further metabolized into acetic acid, CO_2, and H_2 by acetogenic bacteria such as *Propionibacterium sp.* and *Actinomyces sp.* (Bhatia et al., 2020). These intermediates serve as the primary substrates for methanogens in the final stage of digestion.
- **Methanogenesis**, the final stage, involves the conversion of acetic acid, hydrogen, and CO_2 into methane (CH_4) and additional CO_2 by a specialized group of archaea known as methanogens. Dominant species include *Methanobacterium sp.* and *Methanobrevibacter sp.* (Bhatia et al., 2020). This phase is highly sensitive to changes in pH, temperature, and toxic substances and is the most critical for maximizing biogas yield.

Throughout the entire anaerobic digestion process, these microorganisms operate under strict anaerobic conditions and in a balanced ecological network. Disruption in any of the microbial populations due to changes in temperature, feedstock composition, or accumulation of inhibitors can reduce system efficiency or lead to digester failure. Therefore, maintaining optimal environmental conditions and understanding the role of microbial communities is essential for the stability and productivity of biogas plants.

7.3 TYPES OF ANAEROBIC DIGESTERS USED IN BIOGAS PRODUCTION

There exists a wide array of anaerobic reactor designs tailored to various operational needs and treatment objectives. These reactor systems serve diverse purposes; some are specifically engineered for energy recovery through biogas production, while others focus on treating wastewater by removing organic and inorganic contaminants (Orlando & Borja, 2020). The selection of a suitable reactor type is influenced by several critical factors, including the mode of operation (batch vs. continuous), the moisture content of the feedstock (wet vs. dry digestion), and the physical configuration of the reactor (e.g., stirred tank, plug flow) (Demirbas, 2008). Each reactor type presents unique advantages and limitations in terms of operational complexity, biogas yield, start-up time, and maintenance requirements. For instance, high-rate reactors like Upflow Anaerobic Sludge Blanket (UASB) systems are commonly used in industrial wastewater treatment due to their compact design and rapid start-up (Orlando & Borja, 2020), whereas Complete Mix Reactors (CSTRs) and Plug Flow Reactors (PFRs) are more commonly applied in agricultural or municipal settings for biogas generation (Senadheera et al., 2025). Understanding the distinctions among these designs is essential for achieving optimal system performance,

TABLE 7.1
Batch vs. Continuous Systems

System Type	Description	Advantages	Limitations
Batch	Digesters are filled, sealed, and allowed to digest over a set period. No input/output during digestion.	Simple operation, low cost, suitable for small-scale use	Fluctuating gas production, downtime between batches
Continuous	Feedstock is added regularly, and digestate is simultaneously removed. Maintains stable operation.	Consistent biogas output, suitable for industrial applications	Requires automation, higher operational complexity

improving energy efficiency, and ensuring long-term economic and environmental sustainability (Bachmann, 2013). Tables 7.1, 7.2, and 7.3 summarize these classifications, offering a comparative framework that can guide system designers and operators in selecting the most appropriate reactor type for a given application. The most widely adopted types of biogas digesters are described below:

7.3.1 Anaerobic Plug-flow Reactor (APFR)

APFRs are typically configured as long, narrow channels or horizontally oriented tubular digesters through which organic material flows in a single direction—from the inlet at one end to the outlet at the other. This linear, unidirectional flow minimizes internal turbulence and stratification.

In allowing the substrate to pass through sequential zones of microbial activity, APFRs are usually installed above ground, which simplifies maintenance and operational oversight. Depending on the feedstock characteristics and project goals, APFRs can be operated under mesophilic (30–40°C) or thermophilic (50–60°C) conditions (Bachmann, 2013). Thanks to their plug-flow design, these systems are recognized as high-rate digesters capable of treating a broad range of organic waste streams, including livestock manure, distillery residues, and the biodegradable fraction of

TABLE 7.2
Wet vs. Dry Digesters

Digester Type	Total Solids (TS)	Feedstock Examples	Advantages	Limitations
Wet Digesters	<15%	Liquid manure, sewage sludge	Easier mixing and pumping, compatible with CSTRs	Requires more water and larger volume
Dry Digesters	>20%	Municipal solid waste, crop residues	Compact design, water-saving, high methane yield	More challenging material handling and mixing

TABLE 7.3
Reactor Configurations

Reactor Type	Description	Suitable For	Key Features
CSTR	Continuously stirred tank reactor maintaining homogeneous conditions	Liquid substrates	High flexibility, easy mixing
Plug Flow Digester	Long, narrow tanks with unidirectional flow of substrate	High-solids feedstock	Gradient microbial zones, efficient digestion
UASB	Wastewater flows upward through a bed of granulated sludge	Industrial wastewater	High-rate system, compact design
Fixed Film Reactor	Biomass adheres to support media while substrate flows past	Low-strength wastewater	High stability, rapid digestion

municipal solid waste. The sequential flow enhances substrate retention and microbial contact, which often results in more complete digestion and increased methane yields. Furthermore, APFRs are typically more resilient under variable loading conditions and more effective at processing fibrous or dry substrates compared to single-stage CSTRs (Bachmann, 2013). In real-world applications, both APFRs and CSTRs are among the most commonly adopted technologies for anaerobic digestion. While CSTRs rely on vigorous mixing via mechanical agitators, biogas recirculation, or hydraulic means to maintain a homogeneous reactor environment—ideal for wet feedstocks with low solids content—APFRs are particularly advantageous in dry digestion setups, which typically involve feedstocks with total solids exceeding 20% (Bachmann, 2013). The choice between wet and dry digestion systems is largely influenced by the moisture content of the input material. In advanced agricultural and industrial regions, APFRs and CSTRs are both extensively deployed for biogas generation from organic wastes. Additionally, technologies such as covered lagoon digesters, hybrid configurations, and multi-phase systems are increasingly being implemented, depending on specific environmental regulations, economic feasibility, and local infrastructure (Bachmann, 2013).

7.3.2 Biofilm Reactor

Biofilm reactors represent a specialized class of anaerobic digesters where microbial communities are immobilized on inert support surfaces to facilitate the breakdown of organic matter under anaerobic conditions. These support matrices can be fixed in place or suspended and are typically composed of materials such as plastic, glass, ceramics, or other chemically stable substrates (Murshid et al., 2023). The formation of biofilms, structured microbial aggregates, on these surfaces enables efficient treatment of high-strength organic waste streams while stabilizing microbial populations. One of the primary advantages of biofilm reactors is their ability to retain a high density of active biomass within the system, even at elevated organic loading rates. The immobilized nature of the microbial consortia provides resilience to fluctuations in hydraulic retention time (HRT) and organic loading rate (OLR), allowing for more stable biogas production compared to traditional suspended-growth systems (Sivamani et al., 2021). The liquid flow around the biofilm enhances substrate diffusion and mass transfer, supporting efficient metabolic activity. Moreover, once a stable biofilm is established, biofilm reactors exhibit relatively short start-up periods, offering a practical advantage for rapid commissioning and recovery after system disturbances. However, the effectiveness of the biofilm process depends heavily on the nature of the support media, including its surface roughness, porosity, and hydrophobicity, which directly influence microbial adhesion and biofilm strength (Murshid et al., 2023). These systems are commonly applied in treating wastewater with fluctuating loads or intermittent flows, as well as in the digestion of complex industrial effluents. Their compact design, high process efficiency, and adaptability make them suitable for decentralized or space-constrained biogas facilities (Murshid et al., 2023).

7.3.3 Continuous Flow Stirred-Tank Reactor (CSTR)

CSTRs are among the most widely adopted reactor types for anaerobic digestion, particularly in large-scale biogas facilities. Their widespread use is attributed to their relatively straightforward design, which involves continuous mixing of the reactor contents to maintain a homogeneous slurry of microorganisms and organic substrate (Vincent et al., 2018). This design facilitates uniform microbial activity and stable gas production. CSTRs are best suited for the treatment of liquid and semi-liquid substrates, particularly those with total solids (TS) content ranging between 5% and 10%. They are commonly applied for processing animal manure, food-processing effluents, and other organic industrial slurries (Vincent et al., 2018). The consistent mixing helps prevent sedimentation of solids and ensures better contact between microbes and organic matter, thereby enhancing biodegradation. Despite these advantages, CSTRs are often associated with longer hydraulic retention times (HRT) and higher energy consumption due to continuous mechanical agitation or gas recirculation. These characteristics may limit their efficiency compared to other reactor configurations, such as high-rate biofilm or plug-flow systems (Bachmann, 2013). However, advancements in reactor operation, such as the recycling of microbial biomass or the incorporation of sludge retention mechanisms, have shown potential to increase microbial density and reactor productivity. Enhancing the retention of active biomass improves the degradation efficiency and reduces the required reactor volume for a given organic load (Bachmann, 2013). CSTRs remain a preferred choice for uniform waste streams and centralized biogas installations due to their operational flexibility, compatibility with automated systems, and well-established design principles.

7.3.4 Anaerobic Contact Reactor (ACR)

ACRs are a type of high-rate anaerobic system characterized by complete mixing and integrated sludge recycling. The reactor typically consists of a mechanically stirred tank where microbial biomass and organic substrate are homogeneously blended, promoting efficient degradation of high-strength organic waste. A key component of ACRs is the external solid-liquid separation unit—such as a gravity clarifier, flotation device, or inclined plate settler—which enables the separation and recycling of active biomass back into the reactor (Banerjee et al., 2021). This sludge recirculation process enhances the concentration of active microorganisms within the system, supporting high volumetric loading rates and stable biogas production. ACRs are particularly effective for treating effluents with high concentrations of readily biodegradable solids, such as distillery slops, agro-industrial wastewaters, and food processing residues. These systems exhibit excellent tolerance to fluctuations in organic loading and short hydraulic retention times, which allows for compact reactor volumes and reduced space requirements (Banerjee et al., 2021). Additionally, ACRs tend to be more resistant to common process disturbances such as acidification (souring) or transient inhibitory substances, making them a robust option for variable waste streams. Recent developments have integrated membrane-based biomass retention systems into stirred-tank designs, significantly improving the retention of slow-growing methanogenic populations and enhancing overall reactor performance (Abbasi & Abbasi, 2012). Their adaptability, operational stability, and enhanced microbial retention capabilities make ACRs a promising technology for industrial-scale anaerobic digestion where rapid stabilization and high biogas productivity are required.

7.3.5 Batch Reactor

Batch anaerobic reactors are among the most straightforward systems in terms of operation and design. These reactors are loaded with feedstock at the beginning of the process and sealed for the entire digestion cycle, which typically lasts from several days to weeks, depending on the substrate characteristics and operating temperature. Once the digestion process is complete, the reactor is emptied, cleaned, and refilled for the next cycle. This mode of operation eliminates the need for continuous feeding, mixing, or complex mechanical components, significantly reducing both capital and maintenance costs (Nkemka & Murto, 2012). Despite their simplicity, batch reactors face several operational challenges. Due to stagnant conditions, they are prone to issues such as stratification, channeling, and clogging, especially when dealing with fibrous or high-solids substrates. These limitations can reduce the efficiency of substrate conversion and lower overall biogas yield. The biogas production profile in batch systems is also non-uniform—methane generation is typically highest during the initial days of digestion when readily degradable substrates are consumed and gradually tapers off as microbial activity declines toward the end of the cycle (Lisowyj & Wright, 2018). Batch systems usually require a larger reactor volume per unit of biogas produced compared to continuous systems. However, they are still preferred in rural, small-scale, or seasonal applications where simplicity, low investment, and ease of use outweigh the drawbacks of lower productivity. Advances in passive mixing techniques and improved substrate pretreatment methods are being explored to enhance the performance of batch digesters, particularly in decentralized or off-grid energy systems.

7.3.6 Upflow Anaerobic Baffled Reactor (UASBR)

UASBRs are an advanced type of high-rate anaerobic system that builds upon the principles of the Anaerobic Baffled Reactor (ABR). These reactors utilize the upward flow of influent through a dense blanket of anaerobic sludge, enabling prolonged contact between the incoming organic material and active microbial biomass. This configuration promotes rapid hydrolysis, acidogenesis, and methanogenesis within a compact reactor volume (Sivamani et al., 2021). UASBRs are especially effective for treating the organic fraction of municipal solid waste (OFMSW), industrial effluents, and other high-strength substrates. As the influent flows upward, the microbial granules in the sludge blanket facilitate efficient degradation and biogas production. The suspended granules exhibit excellent settling characteristics, which helps in maintaining a stable biomass concentration and reactor performance even under variable loading conditions (Abbasi & Abbasi, 2012). These systems are particularly attractive for decentralized waste treatment in developing regions due to their capacity to handle high-solid waste streams with limited energy input. Their modular design, rapid start-up, and low sludge production further enhance their applicability in areas lacking sophisticated infrastructure. However, the effluent from UASBRs often contains residual organic matter and suspended solids, necessitating secondary treatment before discharge or reuse (Abbasi & Abbasi, 2012). Recent developments in UASBR technology focus on hybridizing these systems with filtration media or integrating them into multi-stage treatment processes to enhance effluent quality and methane recovery.

7.3.7 Hybrid Bioreactor

The hybrid anaerobic bioreactors are innovative systems that integrate the strengths of suspended-growth and attached-growth (biofilm) technologies within a single configuration. These reactors combine features from Upflow Anaerobic Sludge Blanket (UASB) systems and anaerobic filter reactors, offering a dual mechanism of waste degradation through both freely suspended microbial granules and biofilms attached to inert support media (Xu et al., 2011). This dual-phase architecture enhances treatment efficiency by maximizing contact time between the substrate and microbial biomass, which improves the breakdown of complex and partially soluble organic compounds. As a result, hybrid reactors are particularly effective for treating soluble and semi-soluble industrial wastewaters that are difficult to manage with traditional reactor types alone (Xu et al., 2011). The upper section of the hybrid reactor is typically packed with structured media to facilitate biofilm growth, while the lower zone behaves like a conventional UASB reactor, supporting granular sludge formation. This configuration not only stabilizes reactor performance under shock loads and variable hydraulic conditions but also extends the operational lifespan of the system by reducing sludge washout (Xu et al., 2011). Hybrid bioreactors are considered versatile and robust solutions for wastewater treatment, especially in decentralized and semi-urban settings. Their flexibility makes them suitable for retrofitting existing systems or deploying in modular treatment trains. Some advanced designs also incorporate baffled flow paths or integrate multi-stage anaerobic zones, further enhancing their efficiency and adaptability (Xu et al., 2011).

7.4 DIGESTER DESIGN CONSIDERATIONS

Designing an efficient and reliable anaerobic digester involves optimizing a combination of physical, chemical, and operational parameters that influence microbial performance and biogas yield. The effectiveness of AD is highly dependent on maintaining a stable environment that supports the activity of hydrolytic, acidogenic, acetogenic, and methanogenic bacteria. The following factors, as shown in Table 7.4 and described below, must be carefully considered:

- **Organic Loading Rate (OLR):** The OLR represents the amount of organic matter (typically measured in volatile solids) fed into the digester per unit volume per day. An appropriate OLR maximizes biogas production without overwhelming the microbial community. Overloading can lead to the accumulation of volatile fatty acids (VFAs), lowering the pH and causing digester acidification, which disrupts methanogenic activity and decreases methane yield (Yao et al., 2019).
- **Temperature:** Temperature plays a pivotal role in microbial metabolism. Digesters typically operate under mesophilic (30–40°C) or thermophilic (50–60°C) conditions. Mesophilic systems are more stable and require less energy, while thermophilic systems offer faster digestion and higher pathogen removal but require tighter control and higher operational costs. Maintaining consistent temperatures is essential, and can be achieved through insulation, in-ground construction, or supplementary heating systems (Bhatia et al., 2020).
- **pH Levels:** Maintaining pH between 6.5 and 8.0 is critical for the proper functioning of methanogenic archaea. Deviations from this range can

TABLE 7.4
Key Design Parameters in AD Design

Parameter	Description	Importance
Hydraulic Retention Time (HRT)	Average time feedstock remains in digester (typically 15–30 days)	Affects reactor sizing and digestion efficiency
Organic Loading Rate (OLR)	Organic matter input per unit volume per day (e.g., kg VS/m3/day)	Overloading can cause process imbalance
Mixing	Prevents stratification, enhances microbial contact	Improves process efficiency and stability
Temperature and pH Control	Maintains optimal microbial activity; mesophilic (30–40°C), thermophilic (50–60°C); pH 6.8–7.4	Vital for methanogen survival and activity
Gas Collection System	Captures and stores biogas safely	Minimizes methane loss and ensures safe handling

inhibit microbial activity and lead to reduced gas production. Buffering capacity, often regulated by alkalinity and ammonia levels, helps resist sudden pH changes (Yao et al., 2019).

- **C/N Ratio:** The carbon-to-nitrogen (C/N) ratio in feedstocks significantly impacts biogas production during anaerobic digestion. An optimal C/N ratio, typically between 20:1 and 30:1, is crucial for efficient methane production and process stability. Deviations from this range can lead to imbalances, resulting in reduced biogas yield and potential process inhibition (Bhatia et al., 2020).
- **Hydraulic Retention Time (HRT):** HRT is the average time that feedstock remains in the digester. It directly influences the degradation efficiency of organic material. Short HRTs risk washing out slow-growing methanogens, while excessively long HRTs can lead to under-utilization of reactor capacity. Typically, HRTs range between 15 and 30 days depending on feedstock biodegradability (Yao et al., 2019).
- **Material Selection:** The choice of construction materials affects the durability and chemical resistance of the digester. Plastics such as high-density polyethylene (HDPE) and reinforced concrete are preferred over metals, which may corrode when exposed to hydrogen sulfide and the acidic slurry environment. Material selection should also account for local availability, cost, and climatic conditions (Pal et al., 2024).

7.5 BIOGAS QUALITY ANALYSIS

Biogas produced from anaerobic digestion is primarily composed of methane (CH_4) and carbon dioxide (CO_2), with varying amounts of trace gases such as hydrogen sulfide (H_2S), ammonia (NH_3), oxygen (O_2), nitrogen (N_2), and water vapor (H_2O). These minor constituents, although present in smaller concentrations, can significantly influence the usability and safety of biogas, particularly due to their corrosive and toxic properties. For example, H_2S and water vapor can cause corrosion in pipelines and energy conversion systems, while siloxanes and VOCs can damage internal combustion engines and gas upgrading units (Yang & Corsolini, 2019). Because the calorific value and overall energy potential of biogas depend heavily on its CH_4 content, precise assessment of gas composition is essential. Contaminants reduce the efficiency of biogas as a fuel and can lead to costly damage in systems such as Combined Heat and Power (CHP) units or gas turbines. Therefore, biogas purification and upgrading processes often prioritize the removal or reduction of CO_2, H_2S, and other impurities to ensure safe and efficient energy recovery (Jameel et al., 2024). To maintain process efficiency and equipment integrity, regular analysis of biogas composition is required. Monitoring key parameters allows operators to optimize reactor conditions, adjust feedstock inputs, and select appropriate gas cleaning technologies. Analytical methods ranging from gas chromatography and IR spectroscopy to electrochemical sensors and dew point analyzers provide detailed insights into gas quality (Ohemeng-Ntiamoah & Datta, 2017). These techniques are summarized in the accompanying Table 7.5 and serve as foundational tools in both

TABLE 7.5
Biogas Analysis Methods

Method	Description	Typical Use
Gas Chromatography (GC)	Separates and quantifies gases such as CH_4, CO_2, H_2, H_2S, VOCs.	High precision; used in labs and industrial process verification.
Infrared (IR) Spectroscopy	Uses NDIR sensors to detect CH_4 and CO_2 by measuring IR absorption.	Real-time monitoring in field biogas plants.
Electrochemical Sensors	Detects low levels of toxic gases like H_2S and NH_3.	Portable; ideal for safety monitoring.
Thermal Conductivity Detectors (TCD)	Measures thermal conductivity to determine CH_4 and CO_2 levels.	Used for continuous monitoring; less selective than GC.
Water Vapor and Dew Point Measurement	Assesses moisture content to prevent condensation damage.	Important for equipment protection and gas quality.
Gas Flow and Pressure Monitoring	Monitors total biogas production and system pressure.	Used for operational diagnostics and safety.
Calorific Value Determination	Calculates energy content based on methane concentration.	Essential for energy recovery estimation.
Advanced Sensor Systems	Includes TDLAS, MEMS, photoacoustic sensors for high sensitivity.	Used in automation and remote diagnostics.

Source: Ohemeng-Ntiamoah and Datta (2017).

routine monitoring and advanced diagnostic applications. In addition, modern biogas systems increasingly employ integrated sensor networks and automation platforms to ensure real-time monitoring and data-driven decision-making.

7.6 BIOGAS PLANT CASE STUDY

In this section, we present as a case study, the design, sensor/instrumentation and components of an experimental biogas plant. This plant is the GUNT CE 642 equipment currently in use at the University of Lincoln's School of Engineering and is a model popular in various other academic and research institutions. It comprises of the supply unit, production unit, and secondary treatment unit. The supply unit has a substrate tank installed functional in continuous operation mode. Also, the production unit is made up of reactor tanks for two-stage digestion processes together with gas analysis and the programmable logic controller (PLC). The secondary treatment unit contains the digestate tank. More detailed description of the plant is given in the manufacturer's documentation (Gunt Technology Limited, 2025a, Gunt Technology Limited, 2025b).

7.6.1 Production Unit

7.6.1.1 Gas Analysis and the Programmable Logic Controller (PLC)

In the experimental plant, the produced biogas via the anaerobic degradation in the reactors is fed via the opening of valves into the gas analysis section before passing through the exhaust air system of the plant. The gas analysis section comprises sensors for analyzing the gas and respective contents (Figure 7.4). Each sensor has its specific measuring range and uncertainty; and Table 7.6 presents the magnitudes of these quantities.

The gas analysis section also contains a drying column with silica gel to protect the produced biogas being analyzed (Figure 7.4 shows the schematic diagram of the plant). The unit at the center of the schematic is the drying column. It contains the desiccant silica gel which is of a golden brown color at unsaturated conditions which begins to darken as it gets saturated with moisture or humidity. Individual pellets darken as saturation approaches. Instructions from the manufacturer state that once the humidity sensor shows a moisture content of at least 60%, the desiccant must be removed and replaced with the saturated one regenerated by heating at 130°C in an oven for 4 hours approximately.

7.6.1.2 Programmable Logic Controller (PLC)

The operation and control of the biogas plant is integrated with the programmable logic controller (PLC) or touch panel (see Figure 7.5). Therefore, the measurement of the readings from the gas analysis is made possible through this PLC's touchscreen

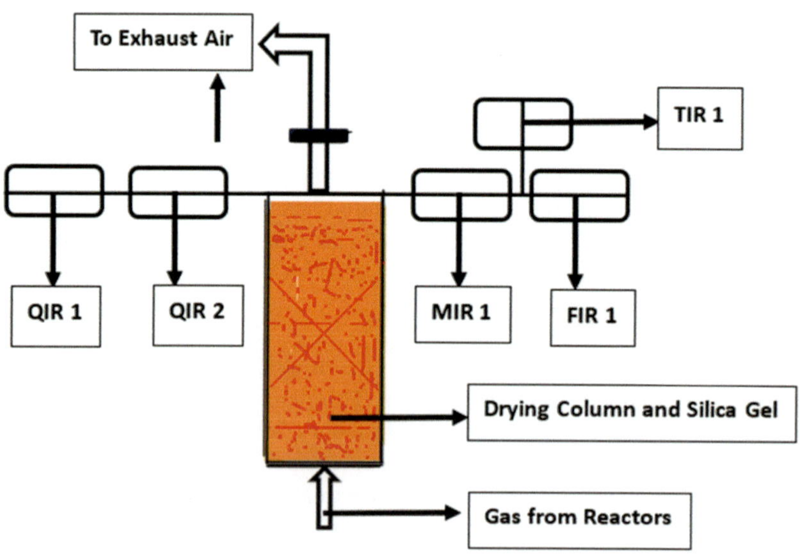

FIGURE 7.4 Schematic representation of the gas analysis measuring points on the biogas plant. QIR 1 and 2 represent the methane and CO_2 sensors, MIR 1 the moisture sensor, FIR 1 the volumetric flow meter, and TIR 1 the temperature sensor, respectively Gunt Technology Limited, 2025b.

TABLE 7.6
PLC's Gas Analysis Sensors and Measuring Range

S/N	Parameter	Measuring Range	Units	Uncertainty
1	Methane (CH_4) Content (NDIR)	0–100	volume %	±1%
2	Carbon Dioxide (CO_2) Content (NDIR)	0–100	volume %	±1%
3	Humidity or Moisture (H_2O) Content (NDIR)	0–100	% relative humidity	±1%
4	Flow Rate (turbine flow meter)	0–30	l/h	±0.25 l/h
5	Temperature (k-type thermocouple)	0–100	°C	±2.2°C

Source: Gunt Technology Limited (2025b).

control panel, and all instructions for the operation of the plant are inputted via this panel. The gas analysis of the biogas plant involves the measurement of the readings of methane for the biochemical methane potential (BMP), carbon dioxide, humidity, flow rates of the produced gas by the reactors, and temperature sensors. The non-dispersive infrared (NDIR) sensors are incorporated into the measurement section of the plant (Figure 7.4 for a schematic diagram) to detect and quantify primarily methane and carbon dioxide contents measured in volume % within a concentration measuring range of 0–100%.

7.6.1.3 Working Principle and Measurement Range of Sensors

According to Xu et al. (2023), non-dispersive infrared sensors are very sensitive in low-temperature and high-moisture environments. The deviation of the sensor

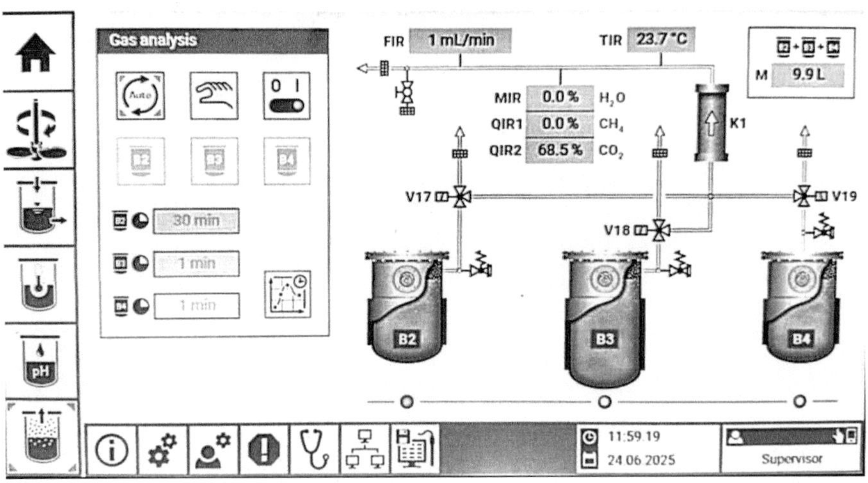

FIGURE 7.5 Gas analysis screen of the biogas plant's programmable logic controller (PLC).

readings is 1% from the final value. The sensors are integrated to work within a temperature range and have a high response time compared to that reported by Xu et al. (2023). The temperature and humidity sensors work together to measure the temperature and moisture of the produced biogas after drying with silica gel via the drying column. The sensors work with the flow meter to measure the flow rate of the produced biogas as well as detect the volume flow measurements of each reactor and digester tank (Figure 7.6).

Other sensors incorporated with the biogas plant are pH sensors in each reactor tank, temperature sensors in each reactor tank, and level measurements in each tank

FIGURE 7.6 (a) Supply unit of the biogas plant (b) Digestate tank (c) Production unit of the biogas plant with integrated programmable logic controller (PLC).

TABLE 7.7
PLC Measurement Data of Chicken Manure Pellets for 19 Days Retention Time

Days	Methane Yield (vol %)	Carbon Dioxide Yield (vol %)	Cumulative Methane Yield (ml/h)	Temperature (°C)	Humidity or Moisture (%)
1	0	99	0	27.9	0
2	0	98	0	29.9	0
3	0	97	0	25.3	0
4	0	96	0	23.8	0
5	1.2	98.8	0.72	25.3	0
6	8.8	91.2	6.00	25.0	0
7	21.1	78.9	18.66	25.8	0.1
8	25.6	74.4	34.02	22.8	0
9	31.1	68.9	52.68	23.0	0
10	35.5	64.5	73.98	24.0	0
11	32.7	67.3	93.60	25.7	0.5
12	33.4	66.6	113.64	26.7	0.4
13	31.4	68.6	132.48	26.8	0.6
14	30.4	69.6	150.72	25.4	0.9
15	28.4	71.6	167.76	25.2	0.5
16	23.2	76.8	181.68	24.7	0.6
17	23.6	76.4	195.84	23.2	0.5
18	23.2	76.8	209.76	23.4	0.3
19	23.8	76.2	224.04	26.0	0.6

(reactors 1 and 2). Table 7.7 shows measured experimental data of chicken manure pellets anaerobically digested during a hydraulic retention period of 19 days. It is seen that initially the BMP, or methane yield, is zero for the first four days when hydrolysis is taking place. Carbon dioxide yield during that period is more than 99% due to CO_2 being used to initially inert and create an anaerobic atmosphere. The methane yield then rapidly increases to a maximum of 35.5% after 12 days but starts to decrease and remains constant at around 23%, meaning that the volatiles within the substrate have started to be exhausted.

Temperature control is central to maintaining conditions for methanogenic bacteria. The temperature in the plant is set at 25°C with fluctuations around this set point value, which was decided because of previous trials with temperature ranges of 25–35°C. Under the conditions, moisture content was very low, ranging from zero initially to 0.9% maximum after two weeks, indicating that a dry gas mixture is being produced. For this unit, trace gas removal such as hydrogen sulfide (H_2S) is not measured but is estimated by subtracting all other measurements from 100%. From the methane yield percentage measurements, the actual flow rate was calculated from the total biogas flow rate. This allowed the cumulative biogas yield in ml/h to be calculated. Ultimately, further calculations from this cumulative value can be carried out, such as fitting to empirical, semi-empirical, and theoretical models such as the modified Gompertz equation (Ofon et al., 2025) or the Buswell formula (Drosg et al., 2013) for maximum biogas production.

7.7 CONCLUSION

Anaerobic digestion stands out as a robust and eco-friendly technology for converting organic waste into valuable biogas and digestate. The stability and effectiveness of the biogas production process are significantly influenced by the layout and functioning of anaerobic digesters. Engineers and practitioners can customize biogas systems to meet particular feedstocks, climate conditions, and energy requirements by carefully choosing digester types, which range from straightforward batch systems to sophisticated continuous stirred tank reactors. To maximize microbial activity and guarantee a steady methane generation, critical operating factors like temperature, pH, organic loading rate, and hydraulic retention duration must be carefully controlled. Furthermore, biogas analysis is a crucial method for process management and system protection, which are necessary for environmental compliance, energy value, and quality control. Operators can maintain ideal gas quality and identify impurities with the use of techniques like gas chromatography, infrared spectroscopy, and sophisticated sensor systems. Real-time gas sensors also allow for precise monitoring, which aids in process optimization and system troubleshooting. Additionally, by lowering greenhouse gas emissions, producing clean energy, and creating nutrient-rich biofertilizers, the incorporation of anaerobic digestion into waste management and renewable energy frameworks supports a circular economy. Continuous advancements in digester design, microbial augmentation, and process automation continue to increase practicality and adoption despite certain technical and financial obstacles, such as feedstock variability, digester scalability, and initial investment costs. Anaerobic digestion-based biogas generation provides a viable route ahead for localized energy solutions, environmental conservation, and rural and urban growth. It continues to play a crucial role in the global shift to a future with reduced carbon emissions and more effective use of resources.

REFERENCES

Abbasi, T., & Abbasi, S. (2012). Formation and impact of granules in fostering clean energy production and wastewater treatment in upflow anaerobic sludge blanket (UASB) reactors. *Renewable and Sustainable Energy Reviews*, *16*(3), 1696–1708. https://doi.org/10.1016/j.rser.2011.11.017

Al-Wahaibi, A., Osman, A. I., Al-Muhtaseb, A. H., Alqaisi, O., Baawain, M., Fawzy, S., & Rooney, D. W. (2020). Techno-economic evaluation of biogas production from food waste via anaerobic digestion. *Scientific Reports*, *10*(1). https://doi.org/10.1038/s41598-020-72897-5

Archana, K., Visckram, A., Kumar, P. S., Manikandan, S., Saravanan, A., & Natrayan, L. (2023). A review on recent technological breakthroughs in anaerobic digestion of organic biowaste for biogas generation: Challenges towards sustainable development goals. *Fuel, 358*, 130298. https://doi.org/10.1016/j.fuel.2023.130298

Bachmann, E. S. N. (2013). Design and engineering of biogas plants. In *The Biogas Handbook: Science, Production and Applications* (pp. 191–211). Elsevier eBooks. Woodhead Publishing Series in Energy https://doi.org/10.1533/9780857097415.2.191

Banerjee, S., Prasad, N., & Selvaraju, S. (2021). Reactor design for biogas production-a short review. *Journal of Energy and Power Technology, 4*(1), 1. https://doi.org/10.21926/jept.2201004

Bhatia, R. K., Ramadoss, G., Jain, A. K., Dhiman, R. K., Bhatia, S. K., & Bhatt, A. K. (2020). Conversion of waste biomass into gaseous fuel: Present status and challenges in India. *BioEnergy Research, 13*(4), 1046–1068. https://doi.org/10.1007/s12155-020-10137-4

Czekała, W., Nowak, M., & Bojarski, W. (2023). Characteristics of substrates used for biogas production in terms of water content. *Fermentation, 9*(5), 449. https://doi.org/10.3390/fermentation9050449

Demirbas, A. (2008). Biofuels sources, biofuel policy, biofuel economy and global biofuel projections. *Energy Conversion and Management, 49*(8), 2106–2116. https://doi.org/10.1016/j.enconman.2008.02.020

Drosg, B., Braun, R., Bochmann, G., & Saedi, T. A. (2013). Analysis and characterisation of biogas feedstocks. In *The Biogas Handbook: Science, Production and Applications* (pp. 52–84). Elsevier eBooks. Woodhead Publishing Series in Energy https://doi.org/10.1533/9780857097415.1.52

Einarsson, R., & Persson, U. M. (2017). Analyzing key constraints to biogas production from crop residues and manure in the EU – A spatially explicit model. *PLoS ONE, 12*(1), e0171001. https://doi.org/10.1371/journal.pone.0171001

Esteves, E. M. M., Herrera, A. M. N., Esteves, V. P. P., & Morgado, C. D. R. V. (2019). Life cycle assessment of manure biogas production: A review. *Journal of Cleaner Production, 219*, 411–423. https://doi.org/10.1016/j.jclepro.2019.02.091

Frequent Questions about Livestock Biogas Projects | US EPA. (2025, March 19). US EPA. https://www.epa.gov/agstar/frequent-questions-about-livestock-biogas-projects

Gunt Technology Limited. (2025a) *CE 642 Biogas Plant*. Hamburg Germany: Gunt Technology Limited. Available from https://www.gunt.de/images/datasheet/70/CE-642-Biogas-plant-gunt-70-pdf_1_en-GB.pdf.

Gunt Technology Limited. (2025b) *Equipment for Engineering Education*. Hamburg Germany: Available from https://www.gunt.de/en/products/process-engineering/biological-process-engineering/anaerobic-processes/biogas-plant/083.64200/ce642/glct-1:pa-148:ca-251:pr-70 [accessed 20 July].

Hernández-Beltrán, J. U., Lira, I. O. H., Cruz-Santos, M. M., Saucedo-Luevanos, A., Hernández-Terán, F., & Balagurusamy, N. (2019). Insight into pretreatment methods of lignocellulosic biomass to increase biogas yield: Current state, challenges, and opportunities. *Applied Sciences, 9*(18), 3721. https://doi.org/10.3390/app9183721

IEA. (2020). Outlook for biogas and prospects for organic growth. In *World Energy Outlook Special Report*. https://iea.blob.core.windows.net/assets/03aeb10c-c38c-4d10-bcec-de92e9ab815f/Outlook_for_biogas_and_biomethane.pdf

Jameel, M. K., Mustafa, M. A., Ahmed, H. S., Mohammed, A. J., Ghazy, H., Shakir, M. N., Lawas, A. M., Mohammed, S. K., Idan, A. H., Mahmoud, Z. H., Sayadi, H., & Kianfar, E. (2024). Biogas: Production, properties, applications, economics and challenges: A review. *Results in Chemistry, 7*, 101549. https://doi.org/10.1016/j.rechem.2024.101549

Lisowyj, M., & Wright, M. M. (2018). A review of biogas and an assessment of its economic impact and future role as a renewable energy source. *Reviews in Chemical Engineering, 36*(3), 401–421. https://doi.org/10.1515/revce-2017-0103

Murshid, S., Antonysamy, A., Dhakshinamoorthy, G., Jayaseelan, A., & Pugazhendhi, A. (2023). A review on biofilm-based reactors for wastewater treatment: Recent advancements in biofilm carriers, kinetics, reactors, economics, and future perspectives. *The Science of the Total Environment*, *892*, 164796. https://doi.org/10.1016/j.scitotenv.2023.164796

Nkemka, V. N., & Murto, M. (2012). Biogas production from wheat straw in batch and UASB reactors: The roles of pretreatment and seaweed hydrolysate as a co-substrate. *Bioresource Technology*, *128*, 164–172. https://doi.org/10.1016/j.biortech.2012.10.117

Ofon, U. A., Ndubuisi-Nnaji, U. U., Udofia, G. E., Adegoke, A. A., Orji, E., Ekaette, M. I., Ukot, C. A., Offiong, N. O., Fapojuwo, D. P., & Shaibu, S. E. (2025). Optimization of biogas production with rice straw-derived biochar: Characterization, hormetic effects, and kinetics modelling. *Cleaner Waste Systems*, 100288. https://doi.org/10.1016/j.clwas.2025.100288

Ohemeng-Ntiamoah, J., & Datta, T. (2017). Evaluating analytical methods for the characterization of lipids, proteins and carbohydrates in organic substrates for anaerobic co-digestion. *Bioresource Technology*, *247*, 697–704. https://doi.org/10.1016/j.biortech.2017.09.154

Orlando, M., & Borja, V. (2020). Pretreatment of animal manure biomass to improve biogas production: A review. *Energies*, *13*(14), 3573. https://doi.org/10.3390/en13143573

Pal, S., Bello, A., Muratova, E., & Chekanov, A. (2024). A review of the materials utilized in the design and fabrication of biogas digesters. *Renewable and Sustainable Energy Reviews*, *211*, 115167. https://doi.org/10.1016/j.rser.2024.115167

Pence, I., Kumaş, K., Cesmeli, M. S., & Akyüz, A. (2024). Future prediction of biogas potential and CH4 emission with boosting algorithms: The case of cattle, small ruminant, and poultry manure from Turkey. *Environmental Science and Pollution Research*, *31*(16), 24461–24479. https://doi.org/10.1007/s11356-024-32666-7

Piechota, G. (2021). Multi-step biogas quality improving by adsorptive packed column system as application to biomethane upgrading. *Journal of Environmental Chemical Engineering*, *9*(5), 105944. https://doi.org/10.1016/j.jece.2021.105944

Senadheera, U. E., Abeykoon, A. M. W. D. C. B., Sewmini, P. M. N., Weerasekara, W. M. R. B., Darshani, N. P., Jayasanka, J., Weerasekara, N. A., Hewawasam, C., Sanjeewa, K. K. A., & Jayawardena, T. U. (2025). Up-flow anaerobic sludge bed reactors for sustainable wastewater management: Challenges, innovations, and future directions. *Water*, *17*(4), 476. https://doi.org/10.3390/w17040476

Sivamani, S., Prasad, B. S. N., Al-Sharji, Z. A. K., Al-Rawas, K. A. M., Al-Blowshi, A. S. D., Al-Yafii, A. S. B., Al-Mashani, M. A. I., & Al-Mamari, H. S. H. (2020). Stoichiometric analysis of biogas production from industrial residues. *Clean Energy Production Technologies*, 141–153. https://doi.org/10.1007/978-981-13-8637-4_6

Sivamani, S., Saikat, B., Prasad, B. S. N., Baalawy, A. A. S., & Al-Mashali, S. M. A. (2021). A comprehensive review on microbial technology for biogas production. *Clean Energy Production Technologies*, 53–78. https://doi.org/10.1007/978-981-33-4615-4_3

Vincent, N. M., Wei, Y., Zhang, J., Yu, D., & Tong, J. (2018). Characterization and dynamic shift of microbial communities during start-up, overloading and steady-state in an anaerobic membrane bioreactor. *International Journal of Environmental Research and Public Health*, *15*(7), 1399. https://doi.org/10.3390/ijerph15071399

Xu, M., Wen, X., Yu, Z., Li, Y., & Huang, X. (2011). A hybrid anaerobic membrane bioreactor coupled with online ultrasonic equipment for digestion of waste activated sludge. *Bioresource Technology*, *102*(10), 5617–5625. https://doi.org/10.1016/j.biortech.2011.02.038

Xu, M., Xu, Y., Tao, J., Wen, L., Zheng, C., Yu, Z., & He, S. (2023). Development of a compact NDIR CO2 gas sensor for harsh environments. *Infrared Physics & Technology*, *136*, 105035. https://doi.org/10.1016/j.infrared.2023.105035

Yang, L., & Corsolini, S. I. (2019). Online removal of volatile siloxanes in solid-state anaerobic digester biogas using a biofilter and an activated carbon filter. *Journal of Environmental Chemical Engineering*, *7*(5), 103284. https://doi.org/10.1016/j.jece.2019.103284

Yao, Y., Huang, G., An, C., Chen, X., Zhang, P., Xin, X., Shen, N. J., & Agnew, J. (2019). Anaerobic digestion of livestock manure in cold regions: Technological advancements and global impacts. *Renewable and Sustainable Energy Reviews*, *119*, 109494. https://doi.org/10.1016/j.rser.2019.109494

8 Microbial Fuel Cell
Setup and Analytical Approaches

*Arshad Ali#, Saloni Mohare#, Afkham Mir, and Iyman Abrar**

8.1 INTRODUCTION

The global energy demand continues to rise, prompting the search for green, carbon-free, and other forms of energy sources with minimal environmental impact. Microbial fuel cells (MFCs) have emerged as a potential technology to address this need by converting organic waste into electricity (Flimban et al., 2019). These electrochemical devices leverage microbes as biocatalysts to convert chemical energy from biowaste materials obtained from wastewater and human waste (organic and inorganic), together with other wastes containing bacteria and substrates, to produce energy or hydrogen via appropriate metabolic pathways (Oliveira et al., 2013). Many process parameters and variables, including membrane type and design, temperature, pH concentrations, organic loading rate, microorganisms, electrode materials, substrate, and electrolyte ionic strength, might affect the operating efficiency of MFC. The interaction between the microorganisms and electrodes is largely determined by electrode conductivity and surface morphology. The energy output of an MFC is highly dependent on its configuration, which includes variables like reactor size, electrode surface area, and separation distance (Mathuriya et al., 2018). A conventional MFC has an anaerobic anode chamber that produces electrons. These electrons are then transmitted to the cathode chamber via an external circuit to produce power, as shown in Figure 8.1.

The proton exchange membrane (PEM), an essential element of MFCs, moves protons from the anode to the cathode chamber in a selective manner. Based on the number of chambers, electrode arrangement, and microbe types, different types of MFC technology have been developed (Logan, 2007). MFCs are categorized into four types: single chamber, two chamber, multi-chamber, and other specialized types like floating, photosynthetic, and sediment MFCs (Flimban et al., 2019). In a single-chamber MFC (SCMFC), passive oxygen flow to the cathode is made possible by the connection between the anodic chamber and an air-exposed cathode that is divided by a gas diffusion layer, or PEM (Xin et al., 2020). Double-chamber MFCs (DCMFC) have separate anodic and cathodic chambers, with a PEM facilitating proton transfer while preventing oxygen diffusion into the anode (Opoku et al., 2022).

# Equal Contribution of the Authors
* Corresponding author

 DOI: 10.1201/9781003485636-8

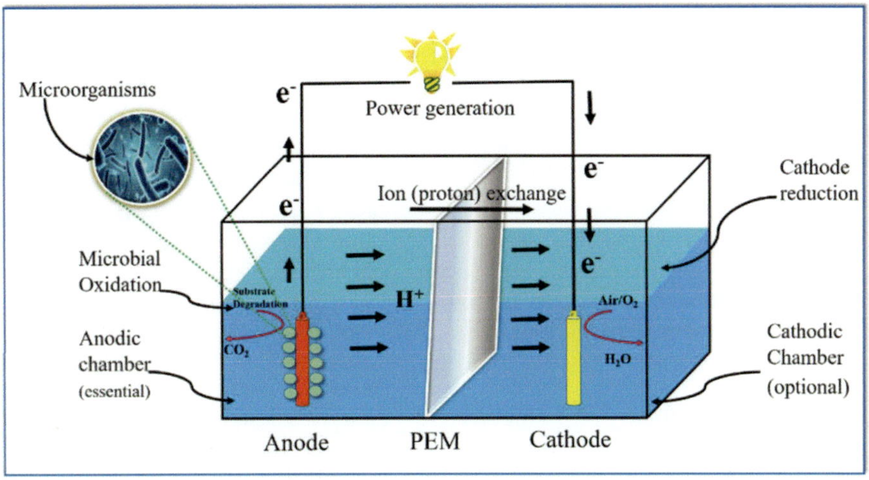

FIGURE 8.1 Representation of the basic elements of MFCs.

Multi-chamber MFCs incorporate more than two chambers, further enhancing system efficiency and output. The subsequent sections discuss these MFC technologies in detail. Table 8.1 illustrates the key contributions made by various authors to the development of MFCs, demonstrating the year-by-year technological progress and advancement in MFC technology from the past to the present.

The choice of electrode material plays a crucial role in the performance and efficiency of MFCs, as it directly influences the electron transfer processes and microbial attachment. The characterization of material properties and electrochemical behaviors is essential for enhancing the efficiency, stability, and productivity of these bioelectrochemical systems. A comprehensive analysis of the materials used in MFCs is vital for improving their performance and durability. Various advanced analytical techniques provide detailed insights into the physical and chemical properties of these materials. Scanning electron microscopy (SEM) offers high-resolution images to examine the surface morphology and microstructural properties of MFC components, such as electrodes and biofilms, helping to identify roughness and integrity that affect microbial adhesion and energy transfer (Abdullah & Mohammed, 2019). Transmission electron microscopy (TEM) provides even higher resolution than SEM, allowing observation of internal structures at the nanoscale, which is necessary for characterizing nanomaterials and understanding their interactions with cellular systems (Tang & Yang, 2017). X-ray diffraction (XRD) serves as a tool for analyzing the crystal structure and phase composition of electrode materials. This technique allows researchers to recognize the various crystalline phases and evaluate their influence on electrochemical performance (Epp, 2016). X-ray photoelectron spectroscopy (XPS) is a technique focused on the surface of materials, used to investigate their elemental composition and chemical states. It delivers insights into surface modifications, the distribution of elements, and the chemical environment surrounding electrode materials (Corcoran et al., 2010). Fourier transform infrared

TABLE 8.1
Technological Progress and Development of Microbial Fuel Cells

S. No.	Phases	Year	Contribution
1	Early discoveries	Early discoveries	M. C. Potter's work showed how microbiological activity can generate power using a plant-based MFC. Ludwig and Charles Langer used air and coal gas to make the first MFC.
2	Improvement of Performance	Late 1990s to early 2000s	Improvements in design and materials that have led to a notable boost in power production. The use of carbon-based materials as electrodes was one innovation.
3	Launch of MFCs based on mediator	Early 2000s	There was a rise in electricity production as mediators were added. Increased the range of bacteria that are energy-producing
4	Mediator less MFCs	Mid-2000s	Identifying the microbes that are electroactive and able to transmit electrons directly. This development made mediators unnecessary.
5	Scaling up and application of MFCs	2010s	Growing interest in large-scale uses, like energy generation and wastewater treatment.
6	Ongoing research	2010s to current	Ongoing research and development focusing on improving electrode materials.

spectroscopy (FTIR) identifies functional groups and molecular interactions within materials, helping to understand the chemical bonds and interactions that affect MFC performance and stability (Mohamed et al., 2017). Electrochemical characterization is crucial for evaluating the performance and efficiency of MFCs. Techniques such as electrochemical impedance spectroscopy (EIS) measure the impedance of the MFC system at different frequencies, providing insights into the resistance and capacitance of the cell components and helping to analyze charge transfer resistance, mass transport limitations, and overall electrochemical behavior (Magar et al., 2021). Cyclic voltammetry (CV) studies the redox behavior of electroactive species within MFCs, providing information on oxidation and reduction processes, peak potentials, and current responses, which indicate the electrochemical performance and stability of the electrodes (Angel et al. 2019). Chronoamperometry is a method where a constant potential is applied, and the resulting current is measured over time. This technique is utilized to investigate the kinetics of energy transfer, assess the stability of MFCs, and track current generation in response to the addition of substrates (Rossi et al., 2020). Differential pulse voltammetry (DPV) is a reliable method for examining the redox characteristics of MFC components, capable of detecting and quantifying trace amounts of electroactive species by applying potential pulses and measuring the current response, thus providing detailed information on the energy transfer processes (Choudhary et al., 2017). Together, these analytical techniques provide a deeper understanding of the materials and electrochemical behavior in

MFCs. By integrating these methods, researchers can systematically optimize MFC design and performance, paving the way for more efficient and effective bioenergy solutions.

The primary goal of this chapter is to discuss the various MFC setups, including single-chamber MFCs (such as up-flow mode MFC systems and cathode-air configurations), double-chamber MFCs (such as H-shaped, cuboid-shaped, and concentrated tubular designs), and multi-chamber MFCs (with more than two chambers). Additionally, this chapter explores the analytical methodologies employed in MFC research, such as SEM, TEM, XRD, XPS, FTIR, TGA, cyclic voltammetry, chronoamperometry, and differential pulse voltammetry, to provide a comprehensive overview of the current state of the field and support the sustainable development of the MFC techniques.

8.2 MFC ELEMENTS

A typical MFC comprises several key components, including electrodes (anode and cathode), anolytes and catholytes in anodic and cathodic chambers, PEM, substrates, and an external load circuit to complete the process to produce electricity. The anodic chamber houses an anode that interacts with microorganisms and has vital properties like biocompatibility and conductivity, which guarantee stability when substrates are present and during chemical reactions (Rashid et al., 2021). In the cathodic chamber, the cathode facilitates the reaction of electrons and protons with oxygen, forming water. Various electron-accepting compounds, such as carbon dioxide, hydrogen peroxide, perchlorate, vanadium, uranium, manganese dioxide, ferricyanide, nitrate, chromium, and triiodide, can be utilized in the cathode chamber, all functioning based on the same fundamental mechanism (Ucar et al., 2017). MFCs can be employed with a wide range of substrates, including cellulose, polysaccharides, volatile acids, and other organic materials from sources such as household biowaste and industrial and agricultural residues. These substrates provide the energy required for bacterial metabolism and subsequent electricity generation. Table 8.2 details the different components and types of materials commonly used in MFCs, highlighting the importance of material selection in optimizing MFC performance (Aryal et al., 2017; Harnisch & Schröder, 2009).

Understanding the materials and components that constitute MFCs is crucial for optimizing their design and performance. The selection of appropriate materials for each component significantly impacts the efficiency, stability, and overall functionality of the MFC, paving the way for more sustainable and effective bioenergy solutions.

8.3 MFC SETUPS

The design of MFCs has a major impact on both cost and efficiency. To more effectively harness renewable energy, a variety of MFC types and configurations have been developed. To achieve optimal outcomes, different configurations involving component types, architecture, and dimensions are carefully considered. While research and development on these designs are ongoing, numerous potential uses for MFCs are already evident, such as electricity generation, the treatment of wastewater,

TABLE 8.2
Different Components and Types of Materials Used in MFCs

S. No	Component	Type of Material Used
1	Anodic chamber	Plexiglas, polycarbonate, polyvinyl chloride, and glass
2	Cathodic chamber	Plexiglas, polycarbonate, polyvinyl chloride, and glass
3	Cathode ingredients	Carbon cloth, platinum, reticulated vitreous carbon, platinum black, carbon-based materials, soluble catholyte, metals other than platinum, Toray paper
4	Anode ingredients	Conductive polymers and nano-structures, carbon-based materials and metal and metal coating, Toray paper
5	Electrode catalyst	Platinum, manganese oxide, polyaniline, ferrous ion, electron mediator immobilized on the anode, platinum black, microbes, and electron mediators, manganese oxides
6	Types of membrane	PEM: Nafion, Ultrex, polyethylene, bipolar membranes, size-selective membranes, large porous membranes, microporous membranes, salt bridge, anion exchange membranes, porcelain septum, semipermeable membrane
7	Substrates	Acetate, arabitol, azo dye with glucose, cellulose particles, corn stover biomass, farm manure, furfural, glucose, landfill leachate, macroalgae, Ulva lactuca, malt extract, yeast extract and glucose, mannitol, microcrystalline cellulose, nitrilotriacetic acid, sodium formate, propionate, phenol, petroleum wastewater, domestic sewage

biological remediation of contaminated soil and water, recovery of valuable chemicals and metals, biosensor functions, and power supply for remote sensors. Given the diversity of these applications, multiple MFC configurations must be created, as no single type can fulfil all functions. Most MFCs have been tested in laboratories, with only a few pilot-scale and even fewer industrial-scale systems in operation (Santoro et al., 2017).

8.3.1 Single-chamber MFC Setup

SCMFCs feature a simple design with an anode and a porous or air-exposed cathode, as illustrated in Figure 8.2(a). These MFCs are cost-effective and easier to develop compared to their double-chambered counterparts. SCMFCs do not require distinct electrolytes for the anode and cathode since they lack separate anodic and cathodic chambers. Instead, the negative electrode (anode) receives the electrons produced by the microbes, which then travel through an external circuit to the cathode. The particles called protons traverse the membrane from the electrolyte, reach the cathode, and participate in the reduction of oxygen to water. Since microorganisms thrive only in anaerobic conditions, these conditions are efficiently maintained in the anode chamber. The majority of SCMFCs just have an anodic chamber; a cathodic chamber is not necessary for mixing (Sonawane et al., 2024). An early SCMFC design by Park and Zeikus included an anode inside a rectangular chamber connected to

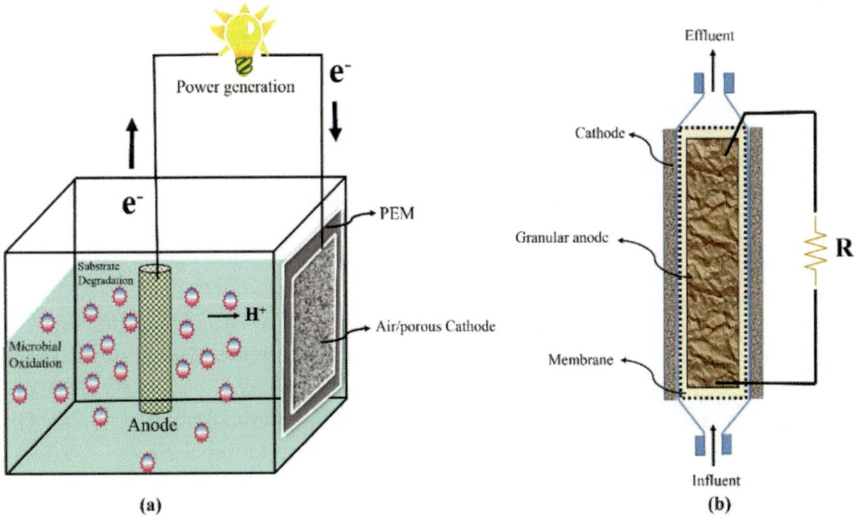

FIGURE 8.2 Various types of single-chamber microbial fuel cells (a) SCMFC with air/porous cathode (b) up-flow SCMFC.

a porous air cathode that was open to the air (Park & Zeikus, 2003). Similarly, Liu and Logan designed an SCMFC with a carbon electrode air cathode, both with and without a polymeric PEM (Liu & Logan, 2004). SCMFCs can be configured in various ways, including cathode-air and up-flow mode designs.

8.3.2 Air-cathode Single-Chamber MFC

This type of MFC features simple construction and a flexible structure, making it possible to investigate various factors affecting MFC efficiency. Air-cathode MFCs allow oxygen molecules to interact with protons and electrons produced within the cell as the cathode is fully in contact with air. This design eliminates the need for chemical oxidizing agents or aeration, which are often required in double-chambered systems, thus reducing overall costs. Reinforcing the cathode in these designs is necessary to prevent electrolyte leakage and withstand hydraulic pressure. Nimje et al. investigated the power output of mixed cultures and the degradation of various wastewaters by *Shewanella oneidensis* using a single-chamber air-cathode MFC (Nimje et al., 2012). It was observed that the highest possible power density was 493 ± 22 mW/m^2 (12.6 ± 0.4 mW/L) without PEM, compared to 261 mW/m^2 (6.5 ± 0.4 mW/L) when using glucose as the substrate.

8.3.3 Up-flow Mode Single-Chamber MFC

In up-flow SCMFCs, the substrate is injected from the bottom with high force, flowing upwards through the system. This system advantageously utilizes easily biodegradable elements in substrates, such as wastewater. Up-flow SCMFCs have an advantage over other single-chamber designs because they can operate continuously and accommodate packed bed anodes, which offer plenty of surface area for the growth of biofilms. A tubular SCMFC with continuous upflow is shown in Figure 8.2(b). Typically, wastewater and substrate-based materials that naturally degrade are used in up-flow SCMFCs. The up-flow SCMFC design, where fluid continuously flows through porous anodes toward an anodic chamber via PEM, reduces internal resistance and enhances overall MFC performance (Khan et al., 2024).

8.3.4 Double-chamber MFC Setups

DCMFCs are among the most widely used and foundational designs in MFC technology. In these systems, two independent chambers connected by a PEM—an anode and a cathode—are made possible via an external circuit. Anaerobic conditions are maintained in the anode chamber to encourage the growth of bacteria (Arun et al., 2024). The cathode chamber contains an oxygen-sparged or chemical terminal electron acceptor. Catholytes are often used to enhance MFC performance by balancing charge transfer from the anode to the cathode and controlling pH. For instance, by reducing membrane and solution resistance to 23 Ω and reducing oxygen penetration into the anode, it has been shown that employing an NaCl catholyte can raise Coulomb efficiency to 43–60% (Ahn & Logan, 2013). Similarly, Shabangu et al. conducted experiments on DCMFCs using different wastewater streams, including mixed, dairy, and biorefinery sectors, recording voltage outputs of 357 mV, 85 mV, and 145.2 mV, respectively (Shabangu et al., 2024). Dairy wastewater showed the highest voltage output, making it a practical alternative for bioelectricity production. Various practical shapes and configurations for DCMFCs are presented in Figure 8.3.

8.3.5 Simple H-shaped Double-Chamber MFC

One of the most popular double-chamber types is the H-type MFC. The key advantage of this design is its ability to control cathode performance through the addition of mediators, oxygen, and flow rate, all of which enhance overall MFC efficiency (Flimban et al., 2019). A simple H-shaped DCMFC is depicted in Figure 8.3(a). These MFCs are well-suited for lab-scale experiments using easily accessible materials and are ideal for studying bioactive organic elements, cell membranes, and microbial populations. However, scaling up H-shaped MFCs is challenging, especially the salt bridge, and the power output tends to decrease as reactor size increases. Weibel and Dodge fabricated the H-shaped DCMFC to generate energy in glass chambers that function as half-cells (Weibel & Dodge, 1975). According to experiment findings, the entire cell has a high Faradic efficiency that approaches 100%. Similarly, Bhattacharya and Chatterjee reported that H-shaped DCMFCs exhibited greater Coulombic efficiency (57%) compared to other MFC types (Bhattacharya & Chatterjee, 2024).

8.3.6 Up-flow Double-Chamber MFC

He et al. developed the first up-flow DCMFC with two cylindrical Plexiglas chambers (He et al., 2005). Up-flow MFCs have two advantages: continuous operation

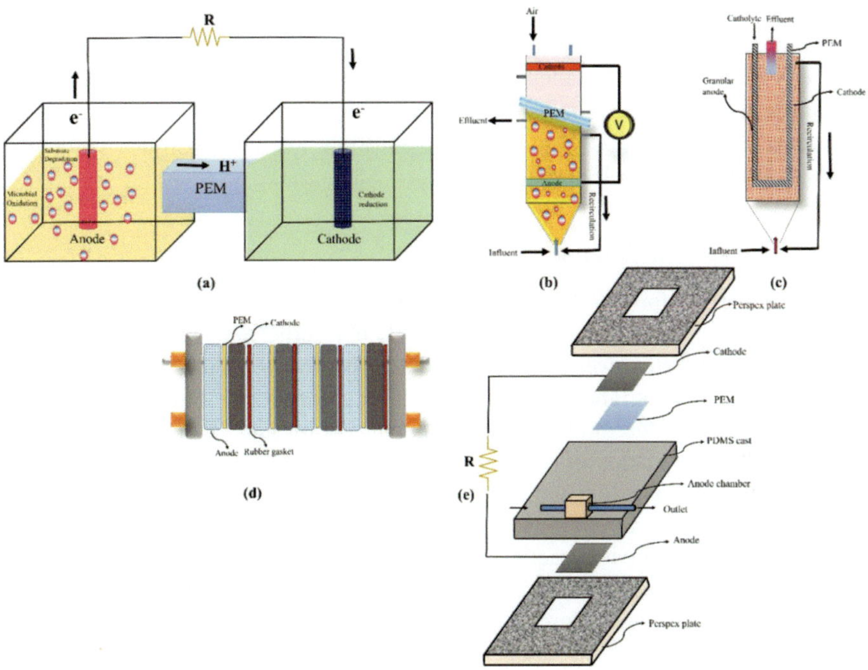

FIGURE 8.3 Visual presentation of various types of double-chamber microbial fuel cells (a) H-shape DCMFC, (b) PEM-equipped up-flow MFC, (c) up-flow MFC with U-tube cathode, (d) stacked MFC, and (e) miniature-shaped DCMFC. (Note: (b) adapted from https://journals.sagepub.com/doi/10.1177/01445987231208510. Subparts (c) and (e) adapted from https://www.sciencedirect.com/science/article/abs/pii/B9780444640529000169, further permissions related to the material excerpted should be directed to Elsevier.)

and the ability to use packed-bed cathodes and/or anodes. A PEM was positioned between the anode and cathode chambers at a 15° angle to the horizontal plane to prevent gas bubble accumulation, as shown in Figure 8.3(b). Effluent is discharged from the center of the anode chamber and forced to the bottom. In certain investigations, sand has been used as a separator to create MFCs without membranes. He et al. also developed a unique double-chambered up-flow MFC with a U-shaped cathode within the anode chamber, filled with granular activated carbon, as shown in Figure 8.3(c) (He et al., 2006). With an overall resistance inside of 17.11 Ω, this design produced a maximum power of 28.0 W/m^3 at a volumetric loading rate of 3.0 kg COD/(m^3. day).

8.3.7 Cuboid-shaped Double-Chambered MFC

Yahiro et al. developed the cuboid, or flat-plate MFC, in 1965 to examine fuel cells through biochemical substrate processes (Yahiro et al., 1964). The biochemical breakdown of substrates "D-alanine, D-glucose, and β-diphosphopyridine nucleotide via alcohol dehydrogenase, glucose oxidase, and D-amino acid oxides" produced electricity. Delaney et al. later developed a similar MFC using polymethacrylate plastic, consisting of two chambers separated by an ion-exchange membrane that allowed cation transfer (Delaney et al., 1984). The usefulness of MFC with various combinations of substrates and microbial mediators was examined. Based on the (theoretical) full oxidation of the supplied substrate to CO_2 and water, coulombic yields of 30–60% were observed.

8.3.8 Stacked MFC Setup

This configuration forms a fuel cell battery by stacking individual MFCs, as illustrated in Figure 8.3(d), preserving the Coulombic efficiency of each cell while increasing the overall power output, bringing it closer to levels comparable with traditional power sources. Theoretically, the voltage generated by a single MFC is limited to 1.20 V due to thermodynamic restrictions. For instance, the greatest potential produced by the oxidation of glucose to carbon dioxide at the anode of 0.41 V and the reduction of oxygen to water at the cathode of 0.81 V might be 1.232 V (Logan, 2007). Therefore, it is advisable to stack many MFCs to optimize performance and utilize MFC technology for practical purposes. These can be arranged either parallel or in series, both of which are useful and have high power efficiency, making them suitable options for power sources.

8.3.9 Miniature-shaped Double-Chambered MFC

Recent research has shifted toward micro-sized MFCs due to their enormous potential for power generation. Miniature MFCs, which can have capacities of several milliliters, are scaled-down versions of macro-scale MFCs (Dutta et al., 2023). Figure 8.3(e) presents a visual representation of a micro-MFC, which includes basic elements such as a Perspex plate, electrodes (cathode and anode), PEM, PDMS cast plate, and load. The primary advantage of miniaturization is the high current density produced, because smaller MFCs have more microorganisms in contact with the electrode. This makes it possible to explore the mechanisms of electron transfer from microorganisms to the electrode surface and permits higher current densities. Microsized MFCs also have unique mobility advantages, making them suitable for biological sensors and other applications in remote areas (Ali et al., 2024). It's interesting to note that because these devices use a cathode-air configuration and saliva as fuel, they don't require laboratory chemicals.

8.3.10 Multi-chamber MFC Setup

Multi-chamber MFCs represent an emerging and innovative approach in fuel cell technology, aimed at enhancing power output by incorporating more than two chambers. Although limited studies have explored these configurations, early research dates back to 1962, when Davis and Yarbrough conducted one of the first investigations using a three-chambered MFC setup (Davis & Yarbrough, 1962). This setup featured a biological chamber housing the anode and an oxidation chamber for the cathode, both of which were external chambers containing the electrodes.

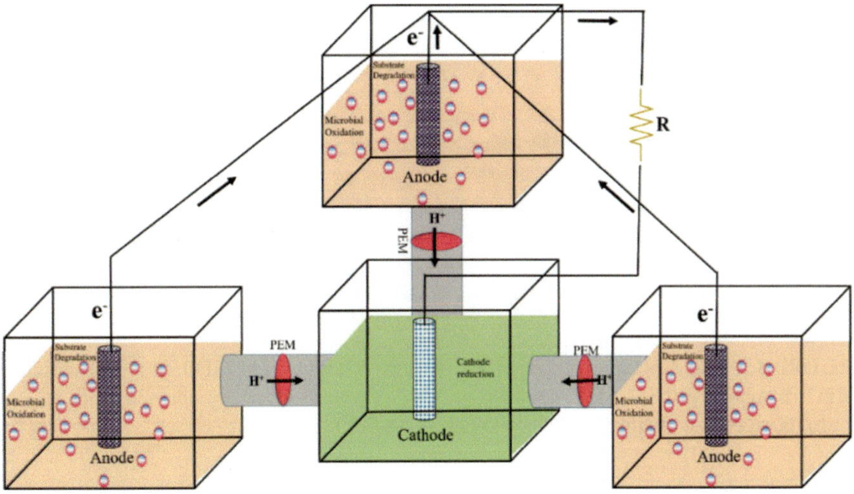

FIGURE 8.4 Multi-chamber microbial fuel cell.

An internal chamber functioned as a buffer zone, with dialysis membranes separating the chambers. Nitrogen and oxygen were continuously bubbled in the anode and cathode chambers, respectively, to maintain the necessary redox conditions. Various substrates, such as ethane-nocardia, glucose-glucose oxidase, and glucose-nocardia systems, have been utilized in multi-chamber MFCs (Naha et al., 2023).

A more recent design incorporates three anode chambers connected to a single cathode chamber, as shown in Figure 8.4. The primary goal of these triple-chambered MFCs is to mitigate the inhibitory effects observed in dual-chambered MFCs, particularly those related to leaching microorganisms, and to eliminate iron and sulfate ions from the anode solutions when present (Zheng et al., 2021). A comparative analysis of multi-chamber and single-chamber MFCs revealed that a multi-anode setup (four anode chambers) sharing a single cathode chamber exhibited the lowest internal resistance (100 Ω), nearly three times lower than that of single-chambered MFCs (Opoku et al., 2022). Additionally, the electricity output of the multi-chambered MFC was 2.7 times higher than that of the SMFC, demonstrating the potential of multi-chamber designs to significantly improve MFC performance.

8.3.11 Other Setups of MFCs

Apart from the MFC configurations already discussed, there are several MFC design configurations, including sediment, photosynthetic, and floating types of MFCs. These alternative MFC configurations expand the applicability of MFCs to diverse environments, offering tailored solutions for specific energy generation needs. Sediment or benthic MFCs utilize electroactive microorganisms that derive energy from naturally occurring organic chemicals found in soil. As illustrated in Figure 8.5(a), the use of soil as an anolyte rather than a liquid solution is the only fundamental mechanism that distinguishes sediment MFCs from other MFCs in

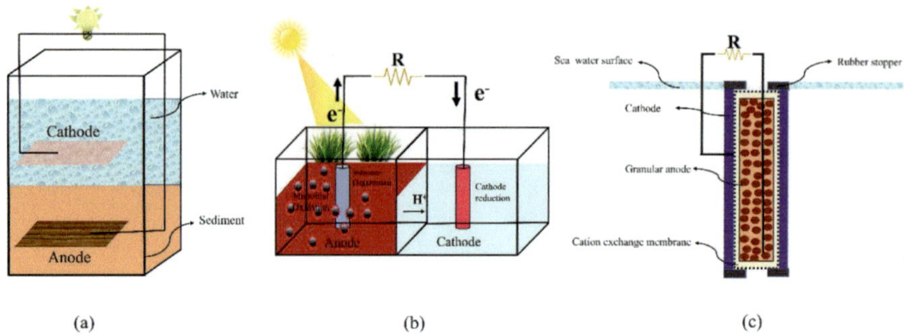

FIGURE 8.5 Schematic representation of various other types of MFC setups (a) sediment MFC (b) photosynthesis MFC (c) floating MFC. (Note: Subpart (b) adapted from https://journals.sagepub.com/doi/10.1177/01445987231208510.)

terms of operation. One significant advantage of sediment MFCs is that they do not require any membrane, simplifying the setup process. This technology can be easily adapted from laboratory settings to outdoor environments, such as lakes or seabeds. Sediment MFCs are particularly effective in soils rich in organic matter, typically found at the bottom of lakes, oceans, and seas. However, any soil containing organic matter can be used. Despite their simplicity, sediment MFCs generate less power compared to MFCs that utilize soluble organic matter (Mohd Noor et al. 2024).

Photosynthesis MFCs harness bacteria that utilize solar energy to produce electricity, as shown in Figure 8.5(b). In these MFCs, sunlight is converted into useful molecules such as oxygen, carbohydrates, and other compounds, which then participate in the energy generation process. These photosynthetic microbes function similarly to the bacteria in conventional MFCs. The efficiency of photosynthesis MFCs is directly linked to light intensity and cell density, with increased power generation observed as light exposure and microbial activity increase (Parthasarthy et al., 2024). Floating MFCs are designed for applications similar to the use of sediment MFCs, such as providing electricity to small appliances or detection devices in far-off places. The initially developed MFC that floated was fabricated by utilizing a tubular design, as depicted in Figure 8.5(c) (Huang et al., 2012). This configuration incorporates a cylindrical membrane for cation exchange and a nickel catalyst layer atop a carbon fiber cathode. Floating MFCs are equipped with buoys that allow them to remain floating slightly below the surface of the water. Before deployment in water, the necessary biofilm is cultivated in the laboratory using the appropriate inoculum. The flexibility of floating MFCs to operate at any ocean depth gives them a significant advantage over sediment MFCs. This makes them extremely adaptable, especially in tidal conditions.

8.4 MATERIAL CHARACTERIZATION TECHNIQUES

8.4.1 Morphological Characterization

The intricate world of MFCs represents a fascinating intersection of microbiology, electrochemistry, and materials science. Central to the advancement of MFC technology is the detailed understanding of the microbial communities and their interactions with electrode and membrane materials. This is where SEMs and TEMs come into play. Their images have a story to tell at scales invisible to the naked eye: structural details of biofilms and materials at micro- and nanolevels. SEM and TEM are vital techniques that deliver critical insights into the structural and morphological properties of biofilms and materials at micro- and nanoscales. SEM excels in providing detailed topographical images, revealing surface structures, while TEM offers information on internal structures and crystallography. Together, these methods enhance our understanding of biofilm formation, organization, and behavior, facilitating advancements in material science and bioengineering applications. SEM facilitates the visualization of surface topographies and biofilm formations on electrode materials with high resolution and depth of field (Abdullah & Mohammed, 2019), while TEM offers detailed internal structural information, revealing the ultrastructure of microbial cells and the fine details of membrane materials (Tang & Yang, 2017). These powerful imaging techniques enable researchers to analyze the properties and performance of various electrode materials, such as carbon-based materials, metals, and conductive polymers, as well as the durability of different membrane materials. Kramer et al. modified a carbon foam anode with polyethyleneimine (PEI) and compared its thermogravimetric analysis (TGA) data with a bare carbon foam anode. The anodes were incorporated in MFCs and used for five weeks. SEM analysis helped compare PEI-modified anodes to unmodified anodes, revealing that the modified anodes were coated with a film, as shown in Figures 8.6(a) to 8.6(f), resulting from the highly charged nature of the PEI polymer, showcasing the modification's impact. The polymer's inter-chain charge repulsion caused it to swell (Kramer et al., 2012). Quinones et al. used activated carbon and zinc-based electrodes with synthetic wastewater as a substrate. Two single-chambered MFCs made up of polypropylene were used. The anode consisted of electrodes fabricated from activated carbon, positioned at one end within the chamber. Conversely, a zinc plate served as the cathode, affixed to the chamber wall at the opposing end, with one side exposed to oxygen. The SEM analysis of the surface displayed numerous tiny pores and a smooth texture that promoted bacterial biofilm adhesion, as shown in Figures 8.6(g) and 8.6(i). A closer look at Figure 8.6(h) revealed the distribution of activated-carbon particles within these pores. The electrode structure was shown to include an aluminum mesh and copper wire embedded between layers of activated carbon, as shown in Figure 8.6 (j). This integration of materials was clearly visible in the SEM images (Agüero-Quiñones et al., 2023).

Xin et al. used nanoparticles of cuprous oxide and reduced graphene oxide as cathode catalysts of MFC. The TEM analysis confirmed that cuprous oxide nanoparticles were successfully loaded on reduced graphene oxide, as shown in Figure 8.6(l). It can be seen in Figure 8.6(m) that all cuprous oxide nanoparticles are stably attached to reduced graphene oxide, and no cuprous oxide nanoparticles are present outside the reduced graphene oxide sheet (Xin et al., 2020). Ghasemi et al. used carbon nanotubes modified with platinum nanocomposite as a cathode catalyst synthesized using chemical reduction in MFCs. TEM was used to observe dispersion of platinum nanoparticles in CNT as shown in Figure 8.6(k). The dark spots in the figure indicate the presence of platinum nanoparticles (Ghasemi et al., 2021). Both SEM and TEM

FIGURE 8.6 Scanning electron microscopy images (SEM) of (a) unmodified carbon foam anode, PEI-modified carbon foam anode, (b) unmodified carbon anode after five weeks in an MFC, (c, d) PEI-modified anode after five weeks in an MFC, (e, f) SEM micrographs of activated carbon electrodes electrode surface morphology(g), activated carbon (h), electrode structure (i), and aluminum mesh with activated carbon and carbon wire copper (j), (k) TEM image of CNT with Platinum, (l and m) TEM images of Cu_2O nanoparticles attached to Reduced graphene oxide. (Note: Subparts (a) to (f) reprinted from https://www.sciencedirect.com/science/article/pii/S0378775312006210, further permissions related to the material excerpted should be directed to Elsevier. Subparts (g) to (j) reprinted from https://www.mdpi.com/2071-1050/15/18/13767. Subparts (k) and (l) reprinted from https://www.sciencedirect.com/science/article/pii/S1385894719318492, further permissions related to the material excerpted should be directed to Elsevier. Subpart (m) reprinted from https://www.mdpi.com/2071-1050/13/14/8057.)

have proven to be invaluable tools in the detailed analysis and understanding of modifications in MFC electrodes. Together, these microscopy techniques provide a comprehensive understanding of how various modifications impact the physical and functional properties of MFC electrodes, guiding the development of more efficient and durable MFC systems.

8.4.2 Structural Characterization

XRD is an essential technique used in the study of MFCs for characterizing electrode materials and separators and understanding their structural properties. For instance, different forms of carbon, such as graphite, can be detected through their distinct diffraction patterns. This identification helps in ensuring the desired properties in electrode materials for optimal performance of MFCs. XRD operates on the principle that monochromatic X-rays interact with a crystalline sample, causing constructive interference. This scattering occurs in specific directions determined by the atomic structure of the crystal lattice. The condition for constructive interference, leading to a diffraction peak, is described by Bragg's law:

$$n\lambda = 2d \ sin\theta \tag{8.1}$$

where n is the order of reflection (an integer), λ is the wavelength of the incident X-ray, d is the distance between atomic planes in the crystal, and θ is the angle of incidence (Bragg angle). Equation 8.1 helps determine the spacing between planes in the crystal lattice (Warren, 1990). Chen et al. designed a new porous carbon anode with pores comparable in size to bacteria. Pores that match the size of bacteria provide ideal sites for colonization, leading to a stable and efficient biofilm. The specific pore size was created using silica spheres, with sucrose serving as the carbon precursor. XRD analysis confirmed the conversion of the sucrose to carbon and the removal of silica. The XRD results showed broad diffraction peaks at 22° and 44°, as shown in Figure 8.7 (a), corresponding to the (002) and (100) reflections of carbon, suggesting the existence of single-layer carbon sheets without parallel stacking (Chen et al., 2015). Lascu et al. modified the carbon network in the anode with nitrogen functional groups by carbonizing polyaniline. X-ray diffraction (XRD) analysis indicated that the two polyaniline samples, PANI-R with near-granular-shaped nanoparticles and PANI-T featuring a nanofibrillar morphology, displayed broad diffraction peaks. These broad peaks are characteristic of amorphous phases, suggesting a lack of long-range order in their structures. However, thermally treated PANI-T displayed more defined peaks, suggesting a properly ordered structure compared to the untreated PANI-R. After carbonization at 900°C, both samples displayed peaks corresponding to disordered graphite phases (Lascu et al., 2022). Jiang et al. used

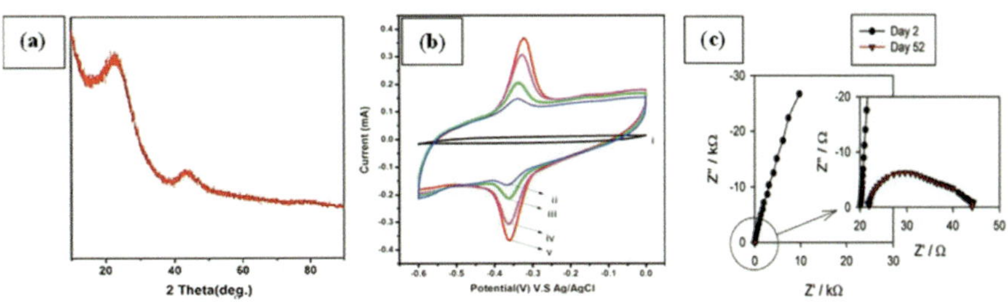

FIGURE 8.7 Chronoamperometry of the Gr (green), PPy/Gr (red), and MoO3/PPy/Gr (black) anodes on day 1 (a) XRD of single layer carbon sheets (b) CV plots after inoculation with E. coli for 0 h (curve-i), 12 h (curve-ii), 48 h (curve-iii), 75 h (curve-iv) and 104 h (curve-v), (c) Nyquist plot data on Day-2 and Day-52. Figure 8.7 (a) and (b) reprinted from https://www.sciencedirect.com/science/article/pii/S0956566315001050, and Figure 8.7 (c) reprinted from https://www.sciencedirect.com/science/article/pii/S0378775318308620. Further permissions related to the material excerpted should be directed to Elsevier.

carbon cloth modified with PANI-MXene (polyaniline-modified multilayer MXene) to improve extracellular electron transfer efficiency of dual-chambered MFC. After coating MXene with polyaniline (PANI), the XRD peak of PANI-MXene exhibited a leftward shift, indicating an increase in slice spacing and broader diffraction peaks (Jiang et al., 2022). Recent studies have leveraged XRD to confirm the successful synthesis and structural integrity of novel electrode materials, thereby advancing the field of MFC research. As the technology continues to evolve, XRD will remain a cornerstone in the characterization and optimization of materials for MFCs.

8.4.3 Degradation Studies

MFCs demonstrate significant promise for generating sustainable energy as well as treating wastewater effectively. Understanding the degradation mechanisms of MFC components is crucial for improving their longevity and efficiency. Techniques such as XPS, FTIR, and TGA are frequently used to study these degradation processes. XPS is a surface-sensitive analytical technique that provides detailed information about the elemental composition, chemical states, and electronic states of materials. It is particularly useful for studying the surface degradation of MFC components, such as electrodes and membranes. Advancements in XPS have provided insights into degradation mechanisms in MFC components, revealing significant changes in surface chemistry due to microbial activity on electrode materials. To increase power density in MFCs, Feng et al. employed carbon brush electrodes with high surface areas for bacterial growth and treated them with ammonia gas. They used single-chamber cubic MFCs with carbon fiber brush anodes and carbon cloth cathodes. XPS analysis indicated that increased power output was associated with higher N1s/C1s ratios and lower C–O composition in treated anodes compared to untreated ones (Feng et al., 2010). XPS facilitates the characterization of surface modifications and degradation mechanisms, aiding in the development of strategies to improve the stability and longevity of MFC electrodes. Thus, XPS serves as an indispensable tool for comprehensive degradation analysis in MFC research. FTIR helps identify the chemical changes occurring on the electrode surfaces by detecting functional groups and monitoring their transformations over time. By analyzing the spectral data, researchers can determine the presence and progression of degradation products, such as the formation of oxides, changes in organic functional groups, and the breakdown of polymer components. Jiang et al. used dual-chambered MFC and carbon cloth modified with PANI-MXene (polyaniline-modified multilayer MXene) as electrodes to improve extracellular electron transfer efficiency. Using FTIR, important peaks were identified for the composite modifier consisting of PANI and MXene. Peaks at 3440 cm^{-1} indicated adsorbed -OH groups, crucial for surface interactions. Vibrations at 1570 cm^{-1} and 1489 cm^{-1} corresponded to O–H bending and C = C stretching in PANI rings, respectively. Peaks at 1396 cm^{-1} and 1126 cm^{-1} signified C-N stretching and C-H vibrations in aromatic amine, revealing the composite's chemical composition and functional groups (Jiang et al., 2022). This information is crucial for understanding the mechanisms of degradation, optimizing electrode materials, and developing strategies to enhance the longevity and performance of MFCs. TGA is valuable for evaluating how surface functionalization affects material mass, offering a straightforward, rapid, and cost-effective method that requires minimal training. In the context of MFC research, where understanding bacterial biofilm adhesion to anodes is crucial, TGA can track trends in biofilm formation on electrodes. Kramer et al. modified a carbon foam anode with polyethyleneimine and compared its TGA data with a bare carbon foam anode. The anodes were incorporated in MFCs and used for five weeks. TGA data for the bare anode showed 5% degradation between 100°C and 200°C, but there was no significant degradation in anodes modified with polyethyleneimine (Kramer et al., 2012). By providing precise measurements of mass changes, TGA helps in understanding the extent of biofilm adhesion and its impact on the overall performance and longevity of the MFC system. TGA proves invaluable in degradation studies of MFC components, allowing for detailed analysis of material stability and decomposition behavior.

8.5 ELECTROCHEMICAL CHARACTERIZATION TECHNIQUES

8.5.1 Electrochemical Impedance Spectroscopy

To optimize the performance and efficiency of MFCs, it is essential to understand the electrochemical processes occurring at the electrodes. EIS provides a powerful tool to examine these processes, offering detailed insights into the charge transfer, mass transport, and biofilm dynamics that govern MFC operation. EIS entails applying a small sinusoidal voltage perturbation to an electrochemical system and recording the resulting current response. The impedance Z is then calculated as the ratio of the voltage to the current, typically over a range of frequencies. This frequency-dependent behavior reveals information about different electrochemical processes occurring at various time scales. MFC can be linked to a potentiostat using either a three-electrode or two-electrode configuration, depending on the measurement goals. The three-electrode arrangement is used to study a particular electrode, with either the anode or cathode acting as the working electrode and the other as the counter electrode. A reference electrode, such as Ag/AgCl, is placed in the respective compartment and connected to the third lead. Conversely, the two-electrode configuration is utilized to measure the internal resistance of the entire cell under cell potential. This makes the two-electrode system where one electrode functions as the working electrode and the other as the reference and counter electrode, eliminating the need for an additional reference electrode. The potentiostat can be programmed such that it can generate impedance spectra over a broad frequency range, typically from 100 kHz to 1 MHz. For MFC studies, a lower frequency limit of 1mHz or 5 mHz is generally sufficient to obtain precise information. The scan from 100 kHz to 1 MHz typically takes around 3 h. In MFCs, EIS can differentiate between the contributions of ohmic resistance, charge transfer resistance at the electrode interfaces, and diffusional resistance within the biofilm or electrolyte. EIS also helps in assessing the surface and material properties of the electrodes. Qiao et al. studied that coating a nickel electrode with carbon nanotubes doped with polyaniline increases the specific surface area of the electrode as well as improves the charge transfer capability. A conducting polymer anode doped with 20 wt.% CNT showed the maximum electrochemical performance with a maximum power of 41 mW m^{-2}

(Qiao et al., 2007). Heijne et al. employed electrochemical impedance spectroscopy (EIS) to investigate biofilm capacitance. They utilized a planar fluorinated tin oxide (FTO) electrode as an anode to analyze various contributions to anodic electrical response, including electrical resistance, biofilm diffusion resistance, and biofilm capacitance. Over time, they observed an increase in both biofilm capacitance and current density. The Nyquist plot circuit effectively differentiated the electron transfer processes occurring at the biofilm, electrolyte, and electrode interfaces. Figure 8.7(c) shows a straight line to high total internal resistance. On day 52, after growth of biofilm, a closed arc is observed (Heijne et al., 2018). Bin Wei et al. compared the effect of new and used separator electrode assembly (with biofilm) on charge transfer of single-chambered air cathode MFC. The used assemblies showed an increase in charge transfer and diffusion resistances at 0.1V. Specifically, the charge transfer resistance for the used assemblies was measured at 180.12 Ω, compared to 134.12 Ω for the new assemblies, highlighting that while the established biofilm enhances overall electron transfer, it simultaneously increases resistance within the system (Wei et al., 2013). Impedance measurement has the advantage of being not destructive and not intrusive. i.e., EIS can be conducted on an operational MFC without disrupting the system. In contrast, other characterization techniques often necessitate system disturbance and sample collection for later analysis. Consequently, EIS serves as a steady-state electrochemical method that allows for measurements without changing the current-voltage characteristics of the MFC system.

8.5.2 Cyclic Voltammetry

Cyclic voltammetry is a versatile and powerful electrochemical technique used extensively in the study of MFCs. By applying a linearly varying potential to an electrochemical system and measuring the resulting current, CV provides detailed information about the redox processes occurring at the electrodes, the kinetics of electron transfer, and the electrochemical behavior of the microbial communities involved. This technique is particularly valuable for characterizing the bio-electrochemical interactions in MFCs, which are essential for optimizing their performance and efficiency. The application of CV in MFC research allows for the investigation of various parameters such as the potential windows of microbial activity, identification of electroactive species, and the evaluation of catalytic properties of electrode materials. Additionally, CV can be used to monitor the dynamic behavior of MFCs over time, offering a comprehensive understanding of the operational stability and the impact of different operating conditions (Shoparwe et al., 2021).

Numerous studies have demonstrated the utility of cyclic voltammetry in studying the complex electrochemical processes within MFCs. Miguel et al. (2019) utilized CV to describe the electrochemical behavior of dual-chamber MFCs under different conditions, including varying analyte solutions and sweep potentials. Their work highlighted the significant impact of proton exchange membranes and potentiostat settings on the electrochemical performance of MFCs. They found that the presence of wastewater constituents and biofilm on the electrode surfaces limited electron charge transfer, suggesting that improving the conductivity of the electrolyte and enhancing catalysis could mitigate these limitations (Angel et al. 2019).

Kudier et al. utilized CV to examine the effects of electrode surface roughness on the performance of double-chambered MFC. Findings showed that the increased roughness enhanced microbial adhesion and electron transfer rates and thereby increased the current by three times that of a smooth electrode surface (Kudier et al., 2022). Another study by Elkhrachy et al. explored the use of biogenic Ag nanoparticles as cathode catalysts in MFCs, demonstrating through CV analysis that these materials significantly improve oxygen reduction reaction kinetics and overall power output as compared to platinum catalysts. CV was employed to analyze the oxygen reduction reaction (ORR) capability of AgNPs synthesized by Pseudomonas aeruginosa. The presence of distinct reduction peaks at various potentials (0.43 V, 0.42 V, 0.410 V, and 0.39 V) indicated the effective electron transfer properties of the biogenic AgNPs, essential for enhancing power production in MFC applications.

Furthermore, CV facilitated the investigation of the relationship between AgNP concentration and power output, demonstrating a steady increase in power generation with higher AgNP concentrations at the cathode. The highest power density of 4.70 W/m^3, coupled with significant chemical oxygen demand (COD) removal efficiency and Coulombic efficiency, underscores the effectiveness of biogenic AgNPs as a cost-efficient and high-performance ORR catalyst (Elkhrachy et al., 2023). Abdul et al. fabricated a double-chambered MFC using graphene and carbon-based electrodes to treat distillery effluent along with generating bioenergy. The analysis of CV indicated that the presence of a biofilm on the electrode led to identifiable oxidation and reduction peaks at 59.9 mV and −290 mV, respectively. These observations suggested that the electrochemical activities of the biofilm contribute to extracellular electron transfer. Additionally, membrane-bound redox processes facilitate electron transfer to the electrode from microbial species (Jatoi et al., 2024).

Chen et al. used the CV technique to investigate the contribution of dead biofilm to the electron transfer process. Before experiments, *E. coli* was inoculated on the electrode for 48 h to facilitate biofilm formation, after which it was kept under a UV light for 30 min. Curve-i in Figure 8.7 (b) shows capacitive behavior, confirming no formation of biofilm at the beginning, i.e., 0 h. The current of the redox reaction increases from 12 h to 108 h as shown in Figure 8.7 (b), curves ii, iii, iv, and v, detecting the growth of the biofilm (Zeng et al., 2012). The results showed that the sterilized electrodes exhibited no redox current from C-type cytochrome, confirming that active electron transfer depended solely on live biofilm, as dead bacteria lack pathways for electron transport through microbial structures (Chen et al., 2015). These studies collectively contribute to a deeper understanding of the electrochemical behavior of MFCs and demonstrate the potential of cyclic voltammetry as a tool for optimizing MFC performance. By integrating findings from various recent studies, researchers can continue to enhance the design and operation of MFCs for sustainable energy generation and environmental remediation.

8.5.3 Chronoamperometry

The chronoamperometry technique involves the application of a constant potential to an electrode and the measurement of the resulting current over time plots. This technique is particularly useful in MFC studies for monitoring the bio-electrochemical activities, evaluating the electron transfer kinetics, studying the growth of biofilms, and evaluating the diffusion coefficient. A potential step is delivered to the working electrode in chronoamperometry, and the current response which fluctuates over time is measured. The initial current response is typically high due to the charging of the electric double layer, followed by a decay as the electrochemical reaction progresses towards a steady state. The faradic current is due to the electron transfer process, and it decays as follows:

$$i = \frac{nFAD^{0.5}C_b}{(\pi t)^{0.5}}, \tag{8.2}$$

where D is the diffusion coefficient, A is the area of the electrode, Cb is the bulk concentration, and t is time.

Equation 8.2 is known as the Cottrell equation. This current-time relationship provides important information about the kinetics and mechanisms of the electrochemical processes, including the rate of electron transfer from microorganisms to the electrode. A study by Chen et al. utilized chronoamperometry to examine the performance of porous carbon anodes with defined pore sizes. The technique revealed that these anodes facilitated improved microbial adhesion and electron transfer, significantly enhancing the net efficiency of MFCs (Chen et al., 2015). Rossi et al. employed chronoamperometry to investigate the negative impact of elevated carbonate buffer concentrations on the performance of anodes in MFCs. Polarization curves were used in conjunction with anode chronoamperometry to confirm that the maximum current densities are limited to 25 Am^{-2} (Rossi et al., 2020). Hassanzadeh et al. modified a graphite anode by incorporating a molybdenum oxide and polypyrrole composite and compared it with polypyrrole/graphite and molybdenum oxide/graphite anodes in a two-chambered MFC setup. After 100 seconds of the self-charging process, the molybdenum oxide/polypyrrole/graphite anode accumulated a charge of 219 mC, whereas the polypyrrole/graphite and graphite anodes recorded charges of 160 mC and 146 mC, respectively. Chronoamperometry tests indicated that the molybdenum oxide/polypyrrole/graphite bioanode was capable of maintaining a more negative potential than the other electrodes, as illustrated in Figure 8.8 (Hassanzadeh et al., 2024).

Chronoamperometry has proven to be a powerful tool for studying and optimizing systems with slow response rates like MFCs. Its ability to provide real-time, detailed insights into the electrochemical processes makes it indispensable for advancing the field of bio-electrochemical systems. Continued research leveraging chronoamperometry will likely lead to further improvements in MFC design and operation, enhancing their viability for sustainable energy production and wastewater treatment.

8.5.4 Differential Pulse Voltammetry

In differential pulse voltammetry (DPV), small pulses of low amplitude are added to a linear potential sweep. Every pulse is preceded and followed by current measurements, which compute the difference between these values. This approach reduces

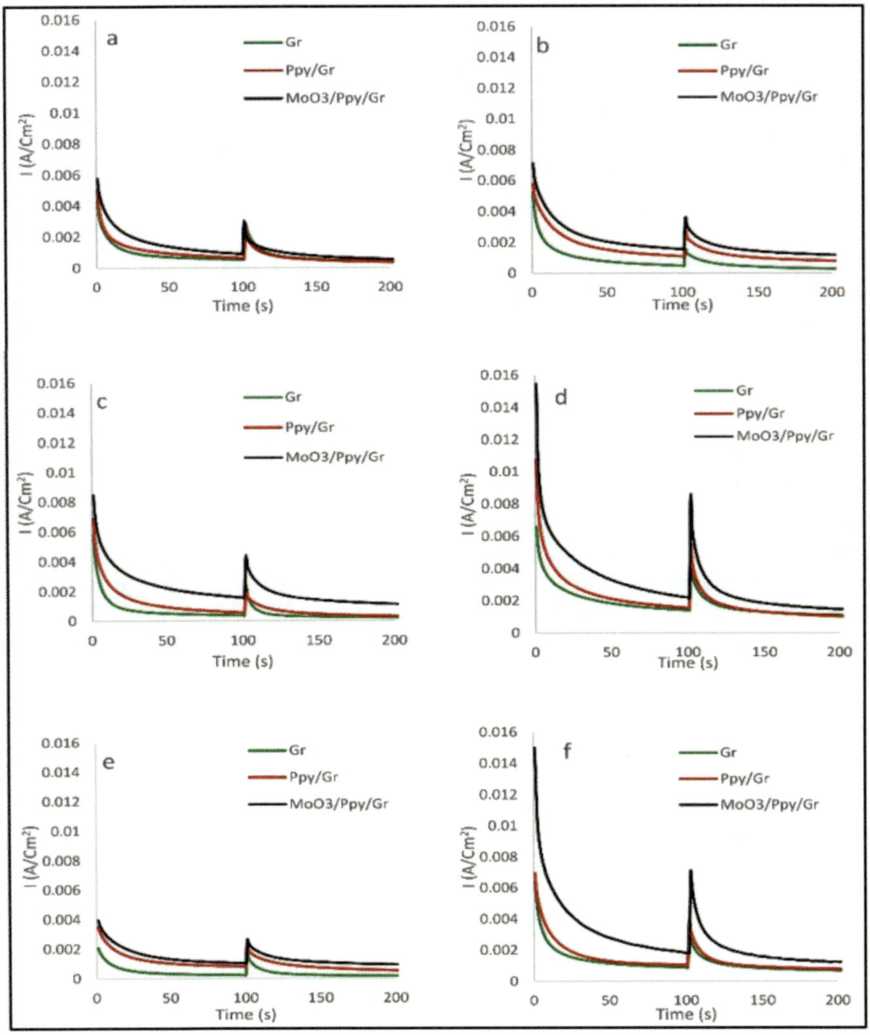

FIGURE 8.8 Chronoamperometry of the Gr (green), PPy/Gr (red), and MoO3/PPy/Gr (black) anodes on (a) day 1, (b) day 2, (c) day 3, (d) day 4, (e) day 6 of the first cycle and (f) second day of the second cycle. The second peak observed in each plot was taken after 100 seconds of anode self-recharge after the end of the first pulse. Figures reprinted from https://www.sciencedirect.com/science/article/pii/S0378775323010996, further permissions related to the material excerpted should be directed to Elsevier.

the background current of the DC ramp, yielding a Faradaic current with minimal capacitive interference. DPV has the benefit of having a low capacitive current, which raises sensitivity. The small step sizes also produce narrower voltammetric peaks, making it particularly useful for distinguishing analytes with identical oxidation potentials (Venton 2020). Recent studies have demonstrated the effectiveness of

Differential Pulse Voltammetry (DPV) in analyzing and improving the performance of electrodes in MFCs. In 2010, Chen et al. investigated the dynamics of microbial communities and their electron transfer mechanisms within a biocathode of MFC. Their analysis revealed distinct oxidation and reduction peaks in the cathodic effluent profiles obtained during differential pulse voltammetry tests. The results indicated that the presence of microorganisms significantly improved the reduction of nitrate and oxygen in the cathode compartment. Additionally, the study noted that these microbes released metabolites that aided in the electron transfer processes, benefiting both the anode and cathode compartments (Chen et al., 2010). In a study, Jiang et al. utilized DPV to investigate the extracellular electron transfer (EET) process of a carbon cloth anode, coated with polyaniline-modified multilayer MXene (PANI-MXene), in MFCs. They found that DPV was instrumental in identifying redox-active species and evaluating the electron transfer efficiency of different anode materials. The study highlighted that DPV provided distinct peaks that corresponded to the electrochemical activity of the microbial biofilms, specifically identifying the role of flavins and outer membrane cytochromes in the extracellular electron transfer (EET) processes (Jiang et al., 2022). These studies underscore the utility of DPV in providing detailed electrochemical characterizations and optimizing the efficiency of MFCs. The ability of DPV to reduce capacitive current and enhance sensitivity makes it a valuable tool for distinguishing and analyzing various redox processes within MFCs. This ongoing research continues to push the boundaries of MFC technology, promising more efficient and sustainable energy solutions.

8.6 CONCLUSION

MFCs offer a promising pathway toward sustainable energy production, leveraging bio-waste to generate green electricity. Understanding the fundamental elements and configurations of MFCs enables researchers to design more efficient bio-electrochemical systems tailored for specific applications. Authors have also underscored the significance of analytical and electrochemical characterization methods that have been instrumental in advancing the field of MFC research by providing invaluable insights about electrode materials, like their morphology, topology, and structure. The electrochemical methods described in this chapter enable readers to grasp the influence of charge transfer, mass transport, and biofilm dynamics in evaluating internal resistances and assessing material properties without disrupting operational systems. These methods have allowed researchers to evaluate microbial activity, analyze electron transfer dynamics, and identify performance-limiting factors that affect overall power output. By synthesizing information on the interactions between various microbial communities and electrode surfaces, this chapter contributed to a deeper understanding of optimizing MFC performance. Challenges in MFC characterization include the difficulty of accurately assessing biofilms. Overall, this chapter serves as a critical resource for researchers aiming to advance the efficacy of MFCs as a viable sustainable energy source. By reflecting on the insights gained from the literature, this chapter paves the way for future research directions focused on overcoming existing challenges and maximizing the potential of MFCs in the evolving energy landscape.

REFERENCES

Abdullah, A., & Mohammed, A. (2019). Scanning Electron Microscopy (SEM): A Review. *Proceedings of 2018 International Conference on Hydraulics and Pneumatics - HERVEX*, 77–85.

Agüero-Quiñones, R., Ávila-Sánchez, Z., Rojas-Flores, S., Cabanillas-Chirinos, L., De La Cruz-Noriega, M., Nazario-Naveda, R., & Rojas-Villacorta, W. (2023). Activated carbon electrodes for bioenergy production in microbial fuel cells using synthetic wastewater as substrate. *Sustainability*, *15*(18), 13767. https://doi.org/10.3390/su151813767

Ahn, Y., & Logan, B. E. (2013). Saline catholytes as alternatives to phosphate buffers in microbial fuel cells. *Bioresource Technology*, *132*, 436–439. https://doi.org/10.1016/j.biortech.2013.01.113

Ali, H. Y., Rushdi, S. A., & Alhameedi, H. A. (2024). Microbial fuel cells: Their shapes, designs, and transportation methods (mini review). *Russian Journal of Applied Chemistry*. https://doi.org/10.1134/S1070427224010142

Angel M., Mena, C. D., & Guizani, M. (2019) Use of cyclic voltammetry to describe the microbial fuel cell, *Energies*, *12*(18), 3532; https://doi.org/10.3390/en12183532

Arun, J., SundarRajan, P., Grace Pavithra, K., Priyadharsini, P., Shyam, S., Goutham, R., Hoang Le, Q., & Pugazhendhi, A. (2024). New insights into microbial electrolysis cells (MEC) and microbial fuel cells (MFC) for simultaneous wastewater treatment and green fuel (hydrogen) generation. *Fuel*, *355*, 129530. https://doi.org/10.1016/j.fuel.2023.129530

Aryal, N., Ammam, F., Patil, S. A., & Pant, D. (2017). An overview of cathode materials for microbial electrosynthesis of chemicals from carbon dioxide. *Green Chemistry*, *19*(24), 5748–5760. https://doi.org/10.1039/c7gc01801k

Bhattacharya, A., & Chatterjee, P. (2024). Impact of reactor configuration on autotrophic denitrification performance in a microbial fuel cell. *Journal of Hazardous, Toxic, and Radioactive Waste*, *28*(3). https://doi.org/10.1061/jhtrbp.hzeng-1314

Chen, G.-W., Choi, S.-J., Cha, J.-H., Lee, T.-H., & Kim, C.-W. (2010). Microbial community dynamics and electron transfer of a biocathode in microbial fuel cells. *Korean Journal of Chemical Engineering*, *27*(5), 1513–1520. https://doi.org/10.1007/s11814-010-0231-6

Chen, X., Cui, D., Wang, X., Wang, X., & Li, W. (2015). Porous carbon with defined pore size as anode of microbial fuel cell. *Biosensors and Bioelectronics*, *69*, 135–141. https://doi.org/10.1016/j.bios.2015.02.014

Choudhary, Y. S., Jothi, L., & Nageswaran, G. (2017). Electrochemical characterization. In *Spectroscopic Methods for Nanomaterials Characterization* (Vol. 2, pp. 19–54). Elsevier. https://doi.org/10.1016/B978-0-323-46140-5.00002-9

Corcoran, C. J., Tavassol, H., Rigsby, M. A., Bagus, P. S., & Wieckowski, A. (2010). Application of XPS to study electrocatalysts for fuel cells. *Journal of Power Sources*, *195*(24), 7856–7879. https://doi.org/10.1016/j.jpowsour.2010.06.018

Davis, J. B., & Yarbrough, H. F. (1962). Preliminary experiments on a microbial fuel cell. *Science*, *137*(3530), 615–616. https://doi.org/10.1126/science.137.3530.615

Delaney, G. M., Bennetto, H. P., Mason, J. R., Roller, S. D., Stirling, J. L., & Thurston, C. F. (1984). Electron-transfer coupling in microbial fuel cells. 2. Performance of fuel cells containing selected microorganism—mediator—substrate combinations. *Journal of Chemical Technology and Biotechnology. Biotechnology*, *34*(1), 13–27. https://doi.org/10.1002/jctb.280340104

Dutta, A., Mukherjee, S. K., & Hossain, S. T. (2023). Microbial fuel cell assisted wastewater treatment: A review on current trends. In *Microbial Technologies in Industrial Wastewater Treatment* (pp. 59–82). Springer Nature Singapore. https://doi.org/10.1007/978-981-99-2435-6_5

Elkhrachy, I., Singh, V., Kumar, A., Roy, A., Abbas, M., Gacem, A., Alam, M. W., Yadav, K. K., Verma, D., Jeon, B.-H., & Park, H.-K. (2023). Use of biogenic silver nanoparticles on the cathode to improve bioelectricity production in microbial fuel cells. *Frontiers in Chemistry*, *11*, 1–11. https://doi.org/10.3389/fchem.2023.1273161

Epp, J. (2016). X-ray diffraction (XRD) techniques for materials characterization. In *Materials Characterization Using Nondestructive Evaluation (NDE) Methods* (pp. 81–124). Elsevier. https://doi.org/10.1016/B978-0-08-100040-3.00004-3

Feng, Y., Yang, Q., Wang, X., & Logan, B. E. (2010). Treatment of carbon fiber brush anodes for improving power generation in air–cathode microbial fuel cells. *Journal of Power Sources*, *195*(7), 1841–1844. https://doi.org/10.1016/j.jpowsour.2009.10.030

Flimban, S. G. A., Ismail, I. M. I., Kim, T., & Oh, S.-E. (2019). Overview of recent advancements in the microbial fuel cell from fundamentals to applications: Design, major elements, and scalability. *Energies*, *12*(17), 3390. https://doi.org/10.3390/en12173390

Ghasemi, M., Sedighi, M., & Tan, Y. H. (2021). Carbon nanotube/Pt cathode nanocomposite electrode in microbial fuel cells for wastewater treatment and bioenergy production. *Sustainability*, *13*(14), 8057. https://doi.org/10.3390/su13148057

Harnisch, F., & Schröder, U. (2009). Selectivity versus mobility: Separation of anode and cathode in microbial bioelectrochemical systems. *ChemSusChem*, *2*(10), 921–926. https://doi.org/10.1002/cssc.200900111

Hassanzadeh, R., Sabzi, R. E., & Faraji, M. (2024). Polypyrrole/MoO3 composite as an appropriate anode of microbial fuel cell in both pulsed and permanent polarization with excellent performance. *Journal of Power Sources*, *589*(October 2023), 233723. https://doi.org/10.1016/j.jpowsour.2023.233723

He, Z., Minteer, S. D., & Angenent, L. T. (2005). Electricity generation from artificial wastewater using an upflow microbial fuel cell. *Environmental Science & Technology*, *39*(14), 5262–5267. https://doi.org/10.1021/es0502876

He, Z., Wagner, N., Minteer, S. D., & Angenent, L. T. (2006). An upflow microbial fuel cell with an interior cathode: Assessment of the internal resistance by impedance spectroscopy. *Environmental Science & Technology*, *40*(17), 5212–5217. https://doi.org/10.1021/es060394f

Heijne, A. ter Heijne, Liu, D., Sulonen, M., Sleutels, T., & Fabregat-Santiago, F. (2018). Quantification of bio-anode capacitance in bioelectrochemical systems using electrochemical impedance spectroscopy. *Journal of Power Sources*, *400*, 533–538. https://doi.org/10.1016/j.jpowsour.2018.08.003

Huang, Y., He, Z., Kan, J., Manohar, A. K., Nealson, K. H., & Mansfeld, F. (2012). Electricity generation from a floating microbial fuel cell. *Bioresource Technology*, *114*, 308–313. https://doi.org/10.1016/j.biortech.2012.02.142

Jatoi, A. S., Mallah, N. B., Ahmed, J., Shah, S. A., Sabzoi, N., & Ibrahim, S. M. (2024). Enhancing the performance of microbial fuel cell via use of new type of electrode for the bioenergy generation along with the treatment of distillery effluent. *Biomass Conversion and Biorefinery*. https://doi.org/10.1007/s13399-024-05286-x

Jiang, D., Zhu, C., He, Y., Xing, C., Xie, K., Xu, Y., & Wang, Y. (2022). Polyaniline-MXene-coated carbon cloth as an anode for microbial fuel cells. *Journal of Solid State Electrochemistry*, *26*(11), 2435–2443. https://doi.org/10.1007/s10008-022-05255-2

Jill Venton, B., & DiScenza J. D. (2020). *Electrochemistry for Bioanalysis* (Issue 112). Gayathri S Acquisition Editor.

Khan, S. S., Amjad, M., Shareef, H., & Larkin, S. (2024). Review of microbial fuel cell from a techno-economic perspective. *Energy Exploration & Exploitation*, *42*(1), 373–398. https://doi.org/10.1177/01445987231208510

Kramer, J., Soukiazian, S., Mahoney, S., & Hicks-Garner, J. (2012). Microbial fuel cell biofilm characterization with thermogravimetric analysis on bare and polyethyleneimine surface modified carbon foam anodes. *Journal of Power Sources*, *210*, 122–128. https://doi.org/10.1016/j.jpowsour.2012.03.022

Kudier, W., Atta, S., Majdi, H., & Hasan, B. (2022). Effect of electrode surface enhancement on the performance of microbial fuel cell under flow conditions. *Egyptian Journal of Chemistry*, *65*(13), 0–0. https://doi.org/10.21608/ejchem.2022.121700.5454

Lascu, I., Locovei, C., Bradu, C., Gheorghiu, C., Tanase, A. M., & Dumitru, A. (2022). Polyaniline-derived nitrogen-containing carbon nanostructures with different morphologies as anode modifier in microbial fuel cells. *International Journal of Molecular Sciences*, *23*(19), 11230. https://doi.org/10.3390/ijms231911230

Liu, H., & Logan, B. E. (2004). Electricity generation using an air-cathode single chamber microbial fuel cell in the presence and absence of a proton exchange membrane. *Environmental Science & Technology*, *38*(14), 4040–4046. https://doi.org/10.1021/es0499344

Logan, B. E. (2007). Microbial fuel cells. In *Microbial Fuel Cells*. Wiley. https://doi.org/10.1002/9780470258590

López Zavala, M., González Peña, O. I., Cabral Ruelas, H., Delgado Mena, C., & Guizani, M. (2019). Use of cyclic voltammetry to describe the electrochemical behavior of a dual-chamber microbial fuel cell. *Energies*, *12*(18), 3532. https://doi.org/10.3390/en12183532

Magar, H. S., Hassan, R. Y. A., & Mulchandani, A. (2021). Electrochemical impedance spectroscopy (EIS): Principles, construction, and biosensing applications. *Sensors*, *21*(19), 6578. https://doi.org/10.3390/s21196578

Mathuriya, A. S., Jadhav, D. A., & Ghangrekar, M. M. (2018). Architectural adaptations of microbial fuel cells. *Applied Microbiology and Biotechnology*, *102*(22), 9419–9432. https://doi.org/10.1007/s00253-018-9339-0

Mohamed, M. A., Jaafar, J., Ismail, A. F., Othman, M. H. D., & Rahman, M. A. (2017). Fourier Transform Infrared (FTIR) spectroscopy. In *Membrane Characterization* (pp. 3–29). Elsevier. https://doi.org/10.1016/B978-0-444-63776-5.00001-2

Mohd Noor, N. N., Oktavitri, N. I., & Kim, K. (2024). Effect of submerged and floating cathodes on sustainable bioelectricity generation and benthic nutrient removal in sediment microbial fuel cells. *Fuel*, *367*, 131438. https://doi.org/10.1016/j.fuel.2024.131438

Naha, A., Debroy, R., Sharma, D., Shah, M. P., & Nath, S. (2023). Microbial fuel cell: A state-of-the-art and revolutionizing technology for efficient energy recovery. *Cleaner and Circular Bioeconomy*, *5*, 100050. https://doi.org/10.1016/j.clcb.2023.100050

Nimje, V. R., Chen, C.-Y., Chen, H.-R., Chen, C.-C., Huang, Y. M., Tseng, M.-J., Cheng, K.-C., & Chang, Y.-F. (2012). Comparative bioelectricity production from various wastewaters in microbial fuel cells using mixed cultures and a pure strain of Shewanella oneidensis. *Bioresource Technology*, *104*, 315–323. https://doi.org/10.1016/j.biortech.2011.09.129

Oliveira, V. B., Simões, M., Melo, L. F., & Pinto, A. M. F. R. (2013). Overview on the developments of microbial fuel cells. *Biochemical Engineering Journal*, *73*, 53–64. https://doi.org/10.1016/j.bej.2013.01.012

Opoku, P. A., Guang, L., Huang, J., & Norgbey, E. (2022). Impact of wastewater volume on cathode environment of the multi-anode shared cathode and standard single anode/cathode microbial fuel cells. *Chemical Papers*, *76*(10), 6309–6321. https://doi.org/10.1007/s11696-022-02316-8

Park, D. H., & Zeikus, J. G. (2003). Improved fuel cell and electrode designs for producing electricity from microbial degradation. *Biotechnology and Bioengineering*, *81*(3), 348–355. https://doi.org/10.1002/bit.10501

Parthasarthy, V., Bhattacharya, R., Roshan, K. R., Shankar, R., Srivastava, S., & Bose, D. (2024). A systematic review of the last decade for advances in photosynthetic microbial fuel cells with bioelectricity generation. In *Clean and Renewable Energy Production* (pp. 143–173). Wiley. https://doi.org/10.1002/9781394174805.ch6

Qiao, Y., Li, C. M., Bao, S.-J., & Bao, Q.-L. (2007). Carbon nanotube/polyaniline composite as anode material for microbial fuel cells. *Journal of Power Sources*, *170*(1), 79–84. https://doi.org/10.1016/j.jpowsour.2007.03.048

Rashid, T., Sher, F., Hazafa, A., Hashmi, R. Q., Zafar, A., Rasheed, T., & Hussain, S. (2021). Design and feasibility study of novel paraboloid graphite based microbial fuel cell for bioelectrogenesis and pharmaceutical wastewater treatment. *Journal of Environmental Chemical Engineering*, *9*(1), 104502. https://doi.org/10.1016/j.jece.2020.104502

Rossi, R., Pant, D., & Logan, B. E. (2020). Chronoamperometry and linear sweep voltammetry reveals the adverse impact of high carbonate buffer concentrations on anode performance in microbial fuel cells. *Journal of Power Sources*, *476*, 228715. https://doi.org/10.1016/j.jpowsour.2020.228715

Santoro, C., Arbizzani, C., Erable, B., & Ieropoulos, I. (2017). Microbial fuel cells: From fundamentals to applications. A review. *Journal of Power Sources*, *356*, 225–244. https://doi.org/10.1016/j.jpowsour.2017.03.109

Shabangu, K. P., Mthembu, N., Chetty, M., Bwapwa, J. K., & Bakare, B. F. (2024). A comparative analysis of organic substrates from industrial wastewater streams for enhanced electricity production using a double chamber microbial fuel cell (DCMFC). *Energy Reports*, *11*, 3050–3063. https://doi.org/10.1016/j.egyr.2024.02.054

Shoparwe, N. F., Makhtar, M. M. Z., Sata, S. A., Kew, W. S., Mohamad, M., & Shukor, H. (2021). Cyclic voltammetry studies of bioanode microbial fuel fells from batch culture of Geobacter sulfurreducens. *IOP Conference Series: Earth and Environmental Science*, *765*(1), 012102. https://doi.org/10.1088/1755-1315/765/1/012102

Sonawane, A. V., Rikame, S., Sonawane, S. H., Gaikwad, M., Bhanvase, B., Sonawane, S. S., Mungray, A. K., & Gaikwad, R. (2024). A review of microbial fuel cell and its diversification in the development of green energy technology. *Chemosphere*, *350*, 141127. https://doi.org/10.1016/j.chemosphere.2024.141127

Tang, C. Y., & Yang, Z. (2017). Transmission Electron Microscopy (TEM). In *Membrane Characterization* (pp. 145–159). Elsevier. https://doi.org/10.1016/B978-0-444-63776-5.00008-5

Ucar, D., Zhang, Y., & Angelidaki, I. (2017). An overview of electron acceptors in microbial fuel cells. *Frontiers in Microbiology*, *8*. https://doi.org/10.3389/fmicb.2017.00643

Venton DJD, B. J. (2020) Electrochemistry for Bioanalysis

Warren, B. E. (1990). *X-ray Diffraction*. Courier CorporatioN.

Wei, B., Tokash, J. C., Zhang, F., Kim, Y., & Logan, B. E. (2013). Electrochemical analysis of separators used in single-chamber, air-cathode microbial fuel cells. *Electrochimica Acta*, *89*, 45–51. https://doi.org/10.1016/j.electacta.2012.11.004

Weibel, M. K., & Dodge, C. (1975). Biochemical fuel cells. *Archives of Biochemistry and Biophysics*, *169*(1), 146–151. https://doi.org/10.1016/0003-9861(75)90327-6

Xin, S., Shen, J., Liu, G., Chen, Q., Xiao, Z., Zhang, G., & Xin, Y. (2020). Electricity generation and microbial community of single-chamber microbial fuel cells in response to Cu2O nanoparticles/reduced graphene oxide as cathode catalyst. *Chemical Engineering Journal*, *380*, 122446. https://doi.org/10.1016/j.cej.2019.122446

Yahiro, A. T., Lee, S. M., & Kimble, D. O. (1964). Bioelectrochemistry. *Biochimica et Biophysica Acta (BBA) - Specialized Section on Biophysical Subjects*, *88*(2), 375–383. https://doi.org/10.1016/0926-6577(64)90192-5

Zeng, L. Z., Zhao, S. F., Wang, Y. Q., Li, H., & Li, W. S. (2012). Ni/β-Mo2C as noble-metal-free anodic electrocatalyst of microbial fuel cell based on Klebsiella pneumoniae. *International Journal of Hydrogen Energy*, *37*(5), 4590–4596. https://doi.org/10.1016/j.ijhydene.2011.05.174

Zheng, Y., Wang, L., Zhu, Y., Li, X., & Ren, Y. (2021). A triple-chamber microbial fuel cell enabled to synchronously recover iron and sulfur elements from sulfide tailings. *Journal of Hazardous Materials*, *401*, 123307. https://doi.org/10.1016/j.jhazmat.2020.123307

9 Biological Hydrogen Production

Updates and Perspectives

Sanjay K. S. Patel, Suriddhim, Vimal Topwal, Astha Ghildiyal, Anubhav Bairwan, and Kumar Sandeep*

9.1 INTRODUCTION

Government agencies encourage the development of environmentally friendly approaches toward waste management for sustainable growth. A large quantum of waste has been generated from various sectors, including agricultural, municipal, and industrial (Patel et al., 2021a; Banerjee et al., 2023). The management of biowaste has been widely demonstrated by landfill, composting, incineration, and anaerobic digestion (AD). These methods have their various advantages and disadvantages. Generating renewable energy, predominantly methane (CH_4), can produce between 50% and 80% of total biogas (Zabed et al., 2020; Budhraja et al., 2024). AD utilizes bio-organic feed via complex bioprocesses, including hydrolysis, acidogenesis, acetogenesis, and methanogenesis. Overcoming the limitations of fossil fuels extensive uses, such as finite natural reserves and the release of greenhouse gases (GHGs) on their burning, the renewable source of energy has been highly recommended due to its sustainable nature (Patel et al., 2021a; Khalekuzzaman et al., 2024). Biomass is rich in carbon sources that can produce various value-added products, including biofuels (Patel et al., 2015; Kucharska et al., 2019). Biofuels have garnered significant attention in recent years as a potential solution to the growing concerns over fossil fuel depletion and the environmental impact of GHG emissions (Narayanan, 2024). However, the sustainability of biofuel production is a complex issue that requires a comprehensive assessment of the economic, social, and environmental implications (Yang et al., 2023; Deng et al., 2024). One of the critical advantages of biofuels is their potential to reduce reliance on fossil fuels and lower the pollution associated with their use. Biofuel production can also create jobs in the production chain and contribute to reducing dependence on the volatile oil market. However, producing biofuels from energy crops and their products, such as vegetable oil for biodiesel and sugars/starch for bioethanol, faces several challenges (Kondaveeti et al., 2019; Ramzan et al., 2023). H_2 is a clean fuel because water is an end-product of its burning and is considered a suitable renewable energy source over other biofuels such as bioethanol, biomethanol, and biodiesel due to its high energy density of 140 kJ/g

* Corresponding author

DOI: 10.1201/9781003485636-9

(Patel et al., 2018a; Agrawal et al., 2024). H_2 has gained significant attention as a clean and renewable energy source, and developing efficient and sustainable H_2 production methods has become a pressing research area (Patel et al., 2012; Blay-Rogar et al., 2024). One promising approach is utilizing biological processes, which offer several advantages over traditional chemical methods (Capurso et al., 2022). Biological H_2 production (BHP) can be achieved through various pathways, including biophotolysis of water by algae, dark and photo-fermentation of organic materials by bacteria, and integrated fermentation–microbial electrolysis (Chelvam et al., 2024). Electrolysis of water using renewable electricity, such as wind and solar, is another well-established technology for producing renewable H_2 (Patel et al., 2024a). These methods have the potential to be more cost-effective and environmentally friendly compared to conventional BHP techniques, which often require high energy inputs and the use of fossil fuels (Qureshi et al., 2023; Patel et al., 2024b). Biomass electrolysis has emerged as a promising approach due to its technical ease and potential for efficient H_2 production. This method involves directly using raw biomass as a feedstock, which can lead to a smaller environmental impact than other renewable energy-based H_2 production methods (Rathi et al., 2024). Furthermore, using a catalyst-free anode in biomass electrolysis can ensure low power consumption, making it an attractive option for clean and efficient H_2 generation. In addition to biomass electrolysis, other BHP processes, such as indirect biophotolysis and biogas reforming, have also been explored and analyzed through life-cycle assessments and systems analyses (Patel et al., 2024a). These studies have highlighted these technologies' potential environmental, economic, and social impacts, providing valuable insights for further development and optimization. Recent literature reviewed suggests that BHP methods have significant advantages over traditional chemical processes, including their renewable nature, lower energy requirements, and reduced environmental impact (Patel et al., 2021; Goveas et al., 2024; Gupta et al., 2024). Due to the limitation of complete utilization of biomass in single-stage bioprocess, the various strategies through integrative approaches have been demonstrated, such as dark-fermentative H_2 followed by (i) photo-fermentative H_2, (ii) CH_4, and (iii) value-added products such as polyhydroxyalkanoate (PHA) production at the second stage. However, these methods' technical and economic feasibility still needs further improvement to enable large-scale applications. A brief overview of the BHP, primarily through dark-fermentative H_2 production, is presented in this book chapter. Further, the integration of dark fermentation into other bioprocesses has been discussed to improve the economics through a biorefinery strategy utilizing biowastes with future perspectives.

9.2 BIOHYDROGEN PRODUCTION

9.2.1 Methods of Biohydrogen Production and Analysis

Thermochemical processes, particularly gasification, partial oxidation, and steam reforming, have demonstrated the highest H_2 yields among the reviewed technologies. The various approaches to producing H_2 are presented in Figure 9.1. These methods, however, often require higher energy inputs and more complex process designs compared to biological approaches (Sahrin et al., 2022; Patel et al., 2024a,

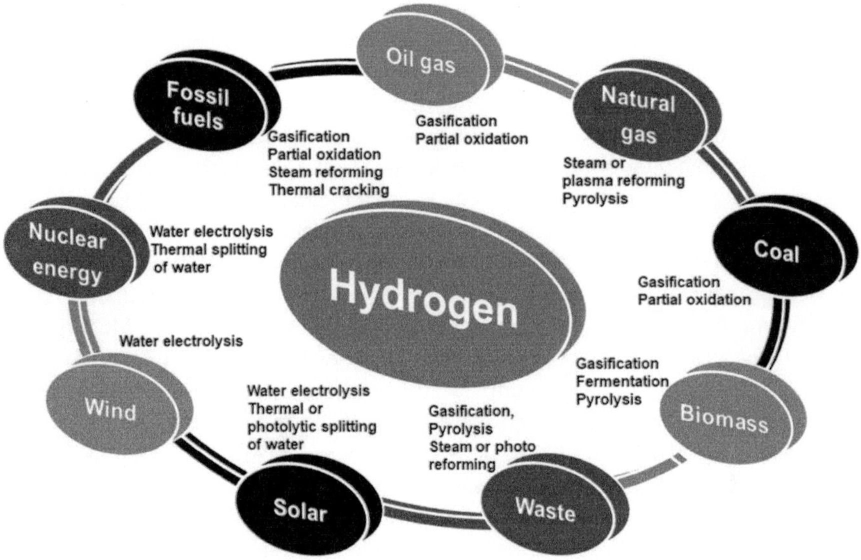

FIGURE 9.1 Hydrogen production from diverse sources: Biomass, renewable sources, fossil fuels, and nuclear energy.

2024b). BHP is a rapidly growing field in renewable energy, as it offers a promising solution to the increasing global demand for clean and sustainable energy sources. Among the various methods of BHP, key technologies have emerged as the focus of considerable research and development efforts: indirect biophotolysis, biogasification, dark and photo fermentation, and integrated fermentation–microbial electrolysis (Sarkar et al., 2022; Patel et al., 2024a). Indirect biophotolysis is when microalgae or cyanobacteria use sunlight to split water molecules, producing H_2 as a byproduct. This method does not require organic feedstocks, making it a potentially more sustainable option than other BHP techniques (Patel et al., 2024a).

Bio-gasification is a process in which biomass, such as agricultural residues or municipal solid waste, is thermochemically converted into H_2-rich syngas (a mixture of H_2 and carbon monoxide). This method leverages the energy content of the biomass feedstock to produce H_2, making it a potentially attractive option for large-scale H_2 generation (Singhvi & Kim, 2022; Singh et al., 2023). Dark fermentation involves the anaerobic conversion of organic materials, such as carbohydrates, by bacteria such as *Bacillus*, *Enterobacter*, *Klebsiella*, *Citrobacter*, *Escherichia*, and others, which can produce H_2 as a byproduct (Patel et al., 2012, 2018b). This process does not require light and can use various waste materials as feedstock. On the other hand, photo-fermentation relies on the use of photosynthetic bacteria such as *Rhodobacter* and *Rhodopseudomonas*, which can convert organic compounds into H_2 using sunlight as the energy source (Patel et al., 2012; Zhang et al., 2023). While these BHP methods offer several advantages, they are still in the development stage and face challenges in terms of efficiency and scalability. One of the critical advantages of BHP is the potential to utilize waste materials as feedstocks, which can generate a clean energy source and contribute to waste management and environmental sustainability. Several studies have explored integrating dark and photo-fermentation processes to maximize H_2 yields (Kim et al., 2006; Patel et al., 2021a; Furqon et al., 2024). On the other hand, integrated fermentation–microbial electrolysis combines dark fermentation and microbial electrolysis to produce H_2. During dark fermentation, bacteria break down organic materials, such as agricultural or food waste, to generate H_2. The H_2-rich biogas produced is then fed into a microbial electrolysis cell, where additional H_2 is generated through electrochemical reactions (Patel et al., 2024a). The research on these BHP technologies has focused on several key aspects, including improving efficiency, reducing costs, and minimizing environmental impact (Laurinavichene et al., 2018; Masihi et al., 2024). Life-cycle assessments and systems analyses have been employed to evaluate the environmental and economic viability of these technologies compared to more conventional H_2 production methods, such as hydrocarbon reforming and water electrolysis (Patel et al., 2021a; Nnabuife et al., 2023). The potential of BHP to contribute to a sustainable H_2 economy has been widely recognized. However, further technological advancements and policy support will be necessary to fully realize the benefits of these promising renewable H_2 production pathways. In BHP, such as dark fermentation, the H_2 yield mainly varies between 2 and 4 mol of H_2/mol of glucose depending on acetic and butyric acids as the primary organic end products, as follows (Eqs. 1 and 2):

$$\text{Glucose} \rightarrow 2CH_3COOH(\text{Acetate}) + 2CO_2 + 4H_2 \tag{1}$$

$$\text{Glucose} \rightarrow CH_3CH_2CH_2COOH(\text{Butyrate}) + 2CO_2 + 2H_2 \tag{2}$$

A maximum of 12 mol of H_2/mol of glucose can be generated by further integrating dark-fermentative effluents to photo-fermentative process or direct photo-fermentation of glucose (Eqs. 3–5).

$$\text{Acetate} + 2H_2O \rightarrow 4H_2 + 2CO_2 \tag{3}$$

$$\text{Butyrate} + 6H_2O \rightarrow 10H_2 + 4CO_2 \tag{4}$$

$$\text{Glucose} + 6H_2O \rightarrow 12H_2 + 6CO_2 \tag{5}$$

The amount of fermentative-evolved biohydrogen biogas (H_2 and CO_2) is primarily measured volumetrically by water displacement methods (Patel et al., 2017). The composition of H_2, CH_4, and CO_2 is analyzed using gas chromatography procedures equipped with a thermal conductivity detector and columns such as (i) the molecula sieve column for the detection of H_2, oxygen, and nitrogen gases and the Porapak Q for the detection of CH_4 and CO_2 (Patel et al., 2021b). The composition of the feed is responsible for altering the contents of produced biogas, such as nitrogen and sulfur-rich feedstock, which resulted in the production of ammonia and H_2-sulfide (Patel et al., 2023). In addition, various analytic automated instruments or sensors/

biosensors have been demonstrated to analyze the composition of biogas produced through fermentative bioprocess (Amulya et al., 2014; Chandekar et al., 2024).

9.2.2 Biohydrogen from Pure Sugars

One of the main approaches for BHP is through the fermentation of organic materials, including sugars, by specialized bacteria. Recent research has focused on using pure sugars, such as glucose and sucrose, as feedstock for BHP. These sugars can be obtained from various biomass sources, including energy crops and agricultural/municipal waste (Patel et al., 2021a; Litti et al., 2024; Chandekar et al., 2024). Using pure sugars as feedstock can lead to higher H_2 yields and more efficient conversion processes compared to using complex biomass materials. Specifically, the fermentation of glucose and sucrose by bacteria such as *Clostridium* and *Enterobacter* has shown promising H_2 production rates and yields (Patel et al., 2012; Litti et al., 2024). One of the critical advantages of BHP over pure sugars is the ease of direct utilization of raw biomass as a feedstock. This simplifies the production process and reduces the overall environmental impact compared to other renewable energy-based H_2 production methods. Furthermore, using a noble metal, catalyst-free anode in the biomass electrolysis process can ensure a significantly lower power consumption, making it an attractive option for clean and efficient H_2 generation (Singhvi & Kim, 2022; Patel et al., 2024a). In addition to the technical and environmental benefits, the economic efficiency of BHP from renewable sources is a crucial factor in driving its large-scale adoption. Several studies have examined the impact of various green technologies on the sustainable production of H_2 from biomass. These investigations have focused on optimizing the critical steps of the H_2 production process, such as reforming, shift conversion, and pressure swing adsorption, to maximize the economic recovery of H_2 (Chelvam et al., 2024; Gupta et al., 2024; Patel et al., 2024a).

Despite the potential of BHP via dark fermentation as a sustainable energy source, several challenges remain to be addressed before its large-scale implementation (Patel et al., 2021a). Various studies have investigated the use of different types of sugars, including cellobiose, glucose, fructose, galactose, mannose, lactose, sucrose, and xylose, as substrates for this process (Litti et al., 2024; Pandey et al., 2023; Chandekar et al., 2024; Jamaludin et al., 2024). The choice of substrate can significantly impact the H_2 yield, as different sugars have different metabolic pathways and conversion efficiencies. The pretreatment of biomass feedstocks is another crucial factor that can enhance the performance of dark fermentation. One of the critical advantages of dark fermentation is its potential to utilize a wide range of biomass feedstocks, including agricultural and industrial waste streams. However, the composition and properties of these feedstocks can vary significantly, requiring careful optimization of the process conditions to maximize H_2 yields (Patel et al., 2021b; Kataya et al., 2023; Singh et al., 2023). The physiological process parameters such as media composition, pH, temperature, incubation period, and mode of operation (batch, fed-batch, and continuous), along with the type of H_2-producers, highly influenced the overall productivity and yield of H_2 using pure sugars as feedstock (Patel et al., 2021a; Litti et al., 2024). The H_2 yields (mol/mol hexose) have been reported to vary under dark fermentative conditions by organisms utilizing pure sugars belonging to (i) Firmicutes: *Bacillus* spp., *B. cereus* (up to 1.98), *B. coagulans* (up to 2.6), *B. licheniformis* (up to 1.5), *B. thuringiensis* (up to 1.67), *Bacillus* sp. (up to 2.26); *Caldicellulosiruptor saccharolyticus* (up to 3.6), *Ethanoligenens harbinense* (up to 2.85), and *Clostridium* spp., *C. beijerinckii* (up to 2.8), *C. butyricum* (up to 3.26), and *C. pasteurianum* (up to 2.4). and (ii) Thermotogales: *Thermotoga neapolitana* (up to 3.8) (iii) Proteobacteria: *Enterobacter aerogenes* (up to 2.85), *Enterobacter cloacae* (up to 3.8), and *Escherichia coli* (up to 3.12) (Erikse et al., 2011; Patel et al., 2012). Under photo-fermentative conditions, H_{22} yields (mol/mol of substrate) by organisms utilizing pure sugars belonging to α-Proteobacteria: (i) *Rhodobacter* spp., *R. capsulatus* (up to 3.6), and *R. sphaeroides* (up to 7.2), and (ii) *Rhodopseudomonas* spp., R. *faecalis* (up to 3.5) and *R. palustris* (up to 6.2) (Patel et al., 2012). In recent studies, lower H_2 yields of 0.26–3.20 mol/mol of hexose from various pure sugars using phylogenetically diverse H_2-producers (Table 9.1) (Zhao et al., 2021; Litti et al., 2024; Pandey et al., 2023; Yang et al., 2023; Chandekar et al., 2024; Jamaludin et al., 2024). These H2 producers produce H2 by the action of formate dehydrogenase and hydrogenase

TABLE 9.1
Dark-fermentative Biological Hydrogen Production from Pure Sugars

Feed	Culture	Yield (mol of H_2/mol of hexose)	Reference
Cellobiose	*Clostridium butyricum* SP4	0.93	Litti et al., 2024
	Clostridium beijerinckii SP6	0.60	
Fructose	*C. butyricum* SP4	1.50	Litti et al., 2024
	C. beijerinckii SP6	1.40	
	Shigella flexneri SPD1	1.24	Chandekar et al., 2024
Galactose	*C. beijerinckii* SP6	1.30	Litti et al., 2024
	C. butyricum SP4	1.20	
	S. flexneri SPD1	0.84	Chandekar et al., 2024
Glucose	Mixed culture	0.26	Zhao et al., 2021
	C. butyricum DSM10702	1.41	Pandey et al., 2023
	C. beijerinckii SP6	1.40	Litti et al., 2024
	C. butyricum SP4	1.30	
	Clostridium sp.	2.01	Yang et al., 2023
	S. flexneri SPD1	3.20	Chandekar et al., 2024
	Thermotoga neapolitana	3.80	Erikse et al., 2011
Lactose	*C. beijerinckii* SP6	1.40	Litti et al., 2024
	C. butyricum SP4	1.30	
	S. flexneri SPD1	1.33	Chandekar et al., 2024
Mannose	*C. butyricum* SP4	1.10	Litti et al., 2024
	C. beijerinckii SP6	1.40	
Sucrose	*C. butyricum* SP4	1.20	Litti et al., 2024
	C. beijerinckii SP6	1.70	
	S. flexneri SPD1	1.65	Chandekar et al., 2024
Glucose + xylose	Mixed culture	2.94	Jamaludin et al., 2024

9.2.3 Biohydrogen from Biowastes

AD is an efficient technology that combines biofuel production with sustainable waste management. Global energy demand is high, and most energy is produced from fossil resources. Recent studies have reported that various technological trends exist in the biogas industry that enhance the production and quality of biogas (Kucharska et al., 2019; Patel et al., 2021a; Renaudie et al., 2024). Biogas production is growing in the European energy market and offers an economical alternative to bioenergy production. Additionally, biomass-to-H_2 processes, including gasification, pyrolysis, and fermentation, are less well-developed technologies that can produce H_2 from energy crops and biomass materials such as forest residue and municipal sewage (Mars et al., 2010; Patel et al., 2024a; Goren et al., 2024). In addition to using pure sugars, researchers have also explored the potential of using waste materials, such as agricultural residues and municipal solid waste, as feedstock for BHP. These waste materials are typically rich in carbohydrates and can be converted to H_2 through biological processes. One of the critical challenges in utilizing waste materials is the need for pretreatment to make the sugars and other compounds more accessible to the H_2-producing bacteria (Deng et al., 2024; Mojoodi et al., 2024). Biological routes, including dark fermentation, offer the advantage of operating under milder conditions and utilizing a more comprehensive range of feedstocks. However, the composition and properties of these feedstocks can vary significantly, requiring careful optimization of the process conditions to maximize H_2 production (Praptyana & Budiyano, 2022; Mattiello-Francisco et al., 2024). In addition to the substrate selection, the operating conditions of the dark fermentation process, such as pH, temperature, and hydraulic retention time, can also influence H_2 production. Overall, BHP from pure sugars and waste materials is a promising approach for sustainable H_2 generation (Patel et al., 2021a; Qyyum et al., 2022; Renaudie et al., 2024). Using biological processes, such as fermentation, can reduce the energy requirements and environmental impact compared to traditional H_2 production methods. Furthermore, integrating dark fermentation with other green technologies, such as pressure swing adsorption and fuel cells, can further improve the overall efficiency and sustainability of the process (Patel et al., 2024a; Sillero et al., 2024).

The various cultures showed high dark-fermentative H_2 yield (mol/mol of hexose) from biowastes such as *C. butyricum* (up to 2.29), *Caldicellulosiruptor saccharolyticus* (up to 3.80), *T. neapolitana* (up to 3.80), *Enterobacter aerogenes* (up to 3.50), *Clostridium pasteurianum* (up to 0.96), *Escherichia coli* (up to 1.37), and *Clostridium beijerinckii* to 1.02) (Ivanova et al., 2009; Patel et al., 2012, 2021a). In recent studies, pretreated biowaste-driven sugars showed H_2 yields up to 2.92 mol/mol of hexose using diverse microbes, including *Klebsiella aerogenes, S. flexneri* SPD1, *E. cloacae* WL1318, and *mirabilis, Bacillus thuringiensis, E. aerogenes* PTCC 1221, *E. aerogenes* ATCCATCC13048, *Escherichia coli,* undefined mixed cultures (Table 9.2). (Arisht et al., 2019; Kucharska et al., 2019; Zhang et al., 2020; Mechery et al., 2021; Zhu et al., 2023; Chandekar et al., 2024; Goren et al., 2024; Mojoodi et al., 2024; Renaudie et al., 2024). In contrast, defined mixed cultures showed H_2 production of up to 95.0 L/kg of total solids or up to 2.93 mol/mol of hexose from various biowaste hydrolysates (Patel et al., 2015, 2017, 2019, 2021b).

TABLE 9.2
Dark-fermentative Biological Hydrogen Production from Biowastes

Feed	Culture	Yield (mol of H_2/mol of hexose)	Reference
Apple pomace	*Enterobacter aerogenes* EGU16, *Bacillus cereus* EGU41, *Proteus mirabilis* strains (EGU21 and EGU30), *Bacillus megaterium* HPC686, and *Bacillus pumilus* HPC464 (MMC6)	83.0[a]	Patel et al., 2019
Beechwood	*Klebsiella aerogenes*	0.40	Kucharska et al., 2019
Coconut coir	*Shigella flexneri* SPD1	1.57	Chandekar et al., 2024
Coconut husk	Seed sludge	0.68	Arisht et al., 2019
Cotton	*S. flexneri* SPD1	2.92	Chandekar et al., 2024
Cotton stalk	*E. cloacae* WL1318	0.37	Zhang et al., 2020
Grape must	Mixed consortia	1.43	Renaudie et al., 2024
Groundnut shells	*S. flexneri* SPD1	1.67	Chandekar et al., 2024
Mixed vegetable wastes	MMC6	18.0–95.0[a]	Patel et al., 2017, 2019
	E. aerogenes EGU16, *P. mirabilis* EGU21, *B. cereus* EGU43, *Bacillus thuringiensis* EGU45, *B. pumilus* HPC464, and *Bacillus sp.* HPC459 (MMC4)	19.0–65.0[a]	Patel et al., 2017
Onion peels	MMC6	86.0[a]	Patel et al., 2019
Paddy straw	*P. mirabilis*	1.03	Mechery et al., 2021
Pea-shells	MMC4	54.0–70.0[a]	Patel et al., 2015
	MMC6	57.0[a]	Patel et al., 2019
Poplar leaf	Septic tank bacteria	1.09	Goren et al., 2024
Poplar biomass	*B. thuringiensis*	0.67–1.31	Patel et al., 2021b
	K. aerogenes	2.68	Patel et al., 2023
	Defined mixed culture	1.30–2.93	Patel et al., 2021
Poplar wood	*K. aerogenes*	0.39	Kucharska et al., 2019
Potato peels	MMC6	87.0[a]	Patel et al., 2019
Potato steam peels	*Thermotoga neapolitana*	3.80	Mars et al., 2010
Rice straw	*S. flexneri* SPD1	1.74	Chandekar et al., 2024
	E. aerogenes PTCC 1221	1.60	Mojoodi et al., 2024
Wheat straw	*S. flexneri* SPD1	2.66	Chandekar et al., 2024
	E. aerogenes ATCC13048	1.07	Zhu et al., 2023
Willow wood	*Escherichia coli*	0.41	Kucharska et al., 2019

[a] Yield in L of H_2/kg of total solids, volatile solids, or chemical oxygen demands

9.3 TYPES OF HYDROGEN AND THEIR ANALYSIS METHODS

H_2 is a colorless, odorless gas that plays a crucial role in various biological processes. While H_2 gas is generally considered non-toxic, it can pose a significant safety hazard due to its high flammability. Therefore, accurate and reliable detection of H_2 gas is paramount, particularly in the emerging H_2 economy (Chauhan & Bhattacharya, 2019; Li et al., 2024; Patel et al., 2024a; Saha et al., 2024). The H_2 production from different methods is categorized as brown/black H_2 (produced from coal by steam reforming or gasification with CO_2 release), gray H_2 (produced from natural gas by steam reforming with CO_2 release), blue H_2 (produced from natural gas by steam reforming with a CO_2O_2 capturing and storing system), turquoise H_2 (produced from natural gas by pyrolysis with solid carbon release), red H_2 (produced through nuclear power via catalytic splitting), pink/purple H_2 (produced through nuclear power via electrolysis), yellow H_2 (produced through solar power via electrolysis), and green H_2 (produced through use of renewable electricity via electrolysis) (Saha et al., 2024). Measuring the concentration of H_2 is one of the critical points in this study. Several methods have been developed to detect concentration, including the gas chromatography technique, the rheophore detection technique, and the oxidimetry technique (Zhai et al., 2015; Dixon & Zhang, 2016; Carrillo-Reyes et al., 2019; Mona et al., 2020). Gas chromatography is a powerful analytical technique that has been widely used for the measurement of H_2 in various applications. Its high resolution, speed, sensitivity, precision, and accuracy characterize the method, making it a preferred choice for complex mixture separation and analysis. The GC's columns separated the H_2 in the sample, and a thermal conductivity detector measured its concentration through a gas chromatogram compared with the known standard. The significant advantages of modern gas chromatography include its high resolution, speed, sensitivity, precision, and accuracy (Mona et al., 2020). Rheophore H_2 detection is a novel technique that enables the accurate and reliable measurement of H_2 levels in various industrial and environmental applications. This approach leverages alloy nanostructures' unique optical and electrical properties to detect and quantify even trace amounts of H_2 gas (Dixon & Zhang, 2016; Saha et al., 2024). One of the critical advantages of rheophore H_2 detection is its ability to overcome the limitations of traditional sensing methods. Conventional H_2 sensors often struggle with poor sensitivity, limited selectivity, and long-term stability. In contrast, rheophore-based sensors demonstrate exceptional sensitivity, with resistance variations of up to 10,000-fold in the presence of 100 ppm H_2 at room temperature. The nanoscale geometry of the rheophore nanotubes, particularly the points of tube-to-tube contact, is believed to be responsible for this outstanding performance. Beyond the impressive sensing capabilities, rheophore-based H_2 detection also offers cost, scalability, and ease of integration advantages. The computational design of alloy nanostructures can optimize optical and thermodynamic sensitivity, further enhancing the detection limits and reliability of the sensors. Rheophore H_2 detection is not limited to industrial applications; it also has the potential to contribute to the development of the H_2 economy. With the increasing focus on using H_2 as a clean energy carrier, reliable and sensitive H_2 sensors will ensure the safe handling and transportation of this flammable gas. Overall, the rheophore H_2 detection technique represents a significant advancement in gas sensing, potentially driving innovation and improving safety in a wide range of applications (Dixon & Zhang, 2016). Oxidimetry, a pivotal technique in analytical chemistry, has garnered significant attention for its ability to detect and quantify the presence of H_2 gas. This approach, typically employing high-performance liquid chromatography coupled with various detection methods, has proven invaluable in multiple applications, from forensic investigations to the advancement of the H_2 economy. One such technique, described in the literature, involves high-performance liquid chromatography with post-column derivatization and fluorescence detection (Chauhan & Bhattacharya, 2019; Saha et al., 2024).

Overall, in the qualitative measurements, the testing of H_2 is performed by a burning splint, which burns with a sound of "squeaky pop" via the combustion of H_2 in the presence of oxygen. The quantitative measurement of produced biogas through the dark/photo-fermentative process [H_2 (40–65%) and CO_2 (20–32%) with slight/trace amounts of hydrogen sulfide (<1%) or ammonia (<1%)] was measured by water displacement methods. The analytical purity and quantification of H_2 can be performed by various methods, including (i) thermal conductivity detectors (TCDs) using gas chromatography; (ii) electrochemical, catalytic combustion, or infrared sensors; (iii) inert gas fusion (involving the fusion of a sample with an inert gas followed by detection of the released H_2); (iv) thermal desorption, glow discharge optical emission, mass, laser-induced breakdown, or secondary ion mass spectrometry; (v) nuclear reaction analysis; and (vi) elastic recoil detection analysis (Zhai et al., 2015; Dixon & Zhang, 2016; Chauhan & Bhattacharya, 2019; Mona et al., 2020). The widely adapted analytical methods, such as gas chromatography and H_2-sensor setups, are presented in Figure 9.2.

It requires a well-defined protocol due to H_2 molecular mass and reactivity. Below is a typical protocol for detecting hydrogen gas using GC-MS. If a sample from the fermentation setup contains impurities or is in a gas mixture, it may be necessary to filter or dilute the sample before analysis.

9.3.1 GC-MS Setup

Choose a column appropriate for gas separation of small molecules. A common choice for H_2 detection would be a Porapak Q or Molecular Sieve column for gas mixtures.

The column length, internal diameter, and stationary phase type depend on the nature of the sample and the sensitivity requirements.

Use helium or nitrogen as the carrier gas for optimal performance. Ensure the gas flow rate is set according to the column specifications, typically around 1–2 mL/min.

Set the injector temperature to around 150–200°C to ensure proper vaporization of gases.

Ensure the injector is clean and the sample is injected accurately into the system.

Mass spectrometer (MS) should be equipped with a suitable ion source (electron impact [EI] or chemical ionization [CI] depending on sensitivity and background).

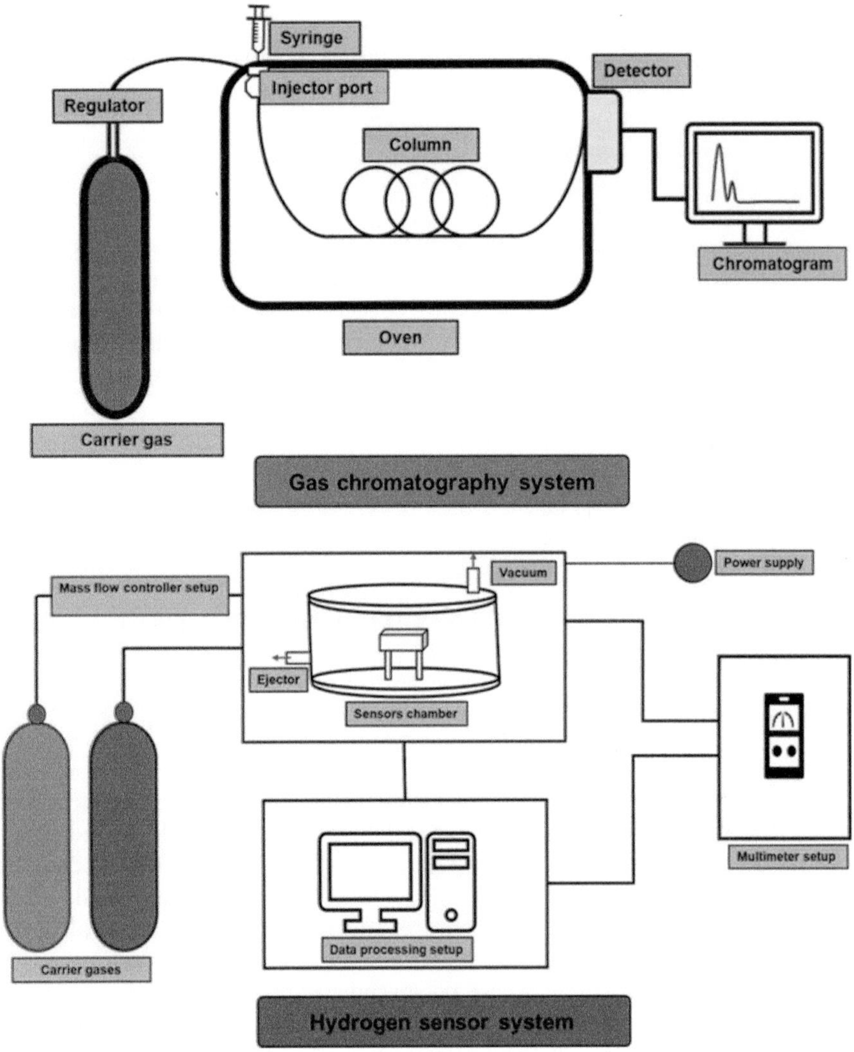

FIGURE 9.2 Biohydrogen detection by Gas Chromatography-Mass Spectrometry (GC-MS) and sensor systems setup.

Ionization mode: EI is commonly used for small molecules like H_2, but alternative modes like CI can be used for increased sensitivity.

Initial oven temperature: Set around 30–40°C, allowing light gases like H_2 to pass through the column easily.

Temperature ramping: Increase the temperature at 5–10°C/min to allow the separation of other gases in the mixture.

Final oven temperature: Typically around 150–200°C (depending on the sample matrix).

9.3.2 Analysis and Detection

Inject a small volume (usually 0.5–1 μL) of the H_2 sample into the injection port using the autosampler or manually.

The gas components will travel through the column, and H_2 will elute early due to its small molecular size.

In the MS, (H_2) should appear as an m/zn m/z2 peak (the molecular ion) in the mass spectrum.

Other compounds in the mixture may elute at different retention times, and their corresponding peaks will be visible in the mass spectrum.

Use calibration curves based on known concentrations of H_2 gas to quantify the amount present in the sample. A standard H_2 concentration range is typically used, and the peak area or height for m/z = 2 is compared to the standards.

Chromatogram analysis: Identify H_2 peaks based on retention time and m/z = 2.

Quantitative analysis: Apply standard calibration or internal standards to calculate H_2 concentrations.

9.4 INTEGRATIVE BIOPROCESSES TO BIOHYDROGEN PRODUCTION

9.4.1 Biohydrogen Production by Dark Fermentation Followed by Photo-Fermentation

H_2 has long been regarded as a promising alternative to fossil fuels due to its high energy density and clean combustion. Several methods have been developed to produce H_2 from renewable resources, including biological processes such as dark fermentation and photo-fermentation (Furqon et al., 2024; Patel et al., 2024a; Zhao et al., 2024). The dark-fermentation process utilizes anaerobic bacteria to convert organic matter, such as agricultural waste or sewage, into H_2. This process is followed by photo-fermentation, where photosynthetic bacteria further convert the remaining organic compounds into additional H_2 (Kim et al., 2006; Laurinavichene et al., 2018). The dark-fermentation process involves breaking complex organic matter by anaerobic bacteria, which produce H_2 as a byproduct (Patel et al., 2024b). This method is advantageous as it can utilize a wide range of waste materials, including agricultural residues and municipal solid waste, as feedstock. On the other hand, the photo-fermentation process employs photosynthetic bacteria that can convert the remaining organic compounds into H_2 using solar energy. Dark fermentation and photo-fermentation integration has been explored to enhance the overall H_2 yield from organic waste materials (Patel et al., 2021a; Yang et al., 2023; Masihi et al., 2024). This approach combines the strengths of both processes, where the dark-fermentation stage generates H_2 from complex organic matter, and the photo-fermentation stage further converts the remaining organic compounds into additional H_2. Literature reviews have indicated that this integrated approach can achieve higher H_2 yields than either process alone. Additionally, using renewable and waste-derived feedstocks in this process aligns with the principles of a circular economy, where waste is minimized and resources are reused or recycled (Das & Basak, 2021; Patel et al., 2021a). The environmental benefits of this approach have been examined through life-cycle assessments, which have demonstrated the potential to reduce GHG emissions and energy consumption compared to traditional H_2 production methods, such as steam reforming of natural gas (Goveas et al., 2024; Patel et al., 2024a). Furthermore, this technology's social and economic impacts have been discussed, highlighting the potential to create new job opportunities and contribute to sustainable development in rural communities. However, several challenges still need to be solved in the commercialization of this technology, including optimizing process parameters, developing efficient bioreactor designs, and scaling up production systems. Research is ongoing to address these challenges and further improve the efficiency and cost-effectiveness of integrated dark-fermentation and photo-fermentation for BHP (Patel et al., 2021a; Gupta et al., 2024). BHP's use of biowastes as a low-cost feedstock has been considered a unique strategy to improve the bioprocess economy and its management (Rathi et al., 2024). Various types of biowaste wastes originating from agricultural, municipal, and industrial sources have been demonstrated using pure or mixed cultures, especially *Clostridium*, *Bacillus*, *Klebsiella*, *Caldicellulosiruptor*, *Enterobacter*, *Escherichia coli*, *Lactobacillus,* or undefined mixed cultures for dark fermentation followed by photo-fermentative H_2 production using *Rhodopseudomonas*, *Rhodobacter*, or mixed cultures/consortia with total H_2 yields up to 8.3 mol/mol of hexose (Kim et al., 2006; Das & Basak, 2021; Patel et al., 2021a; Zhang et al., 2023; Rathi et al., 2024). Overall, these integrative approaches suggested that more than a 60-fold improvement in H_2 production can be achieved compared to the dark fermentation bioprocess alone.

9.4.2 Biohydrogen Followed by Biogas Production

H_2 and biogas production processes constitute a significant area of research within the broader field of renewable energy sources. BHP methods such as bio-photolysis, dark fermentation, and photo-fermentation have demonstrated several advantages over traditional chemical processes, including lower energy requirements and the ability to utilize various organic feedstocks, including biomass and waste materials (Patel et al., 2016; Bertasini et al., 2024; Deng et al., 2024). Recent studies have explored integrating these BHP methods with subsequent biogas generation, creating a more comprehensive and sustainable renewable energy generation approach. Indirect biophotolysis, integrated fermentation–microbial electrolysis, and biogas reforming are examples of these integrated technologies that aim to maximize the efficient utilization of biomass resources (Kim et al., 2024; Patel et al., 2024a; Pomdaeng et al., 2024). Life-cycle assessments and systems analysis have been conducted to evaluate the environmental, economic, and social impacts of these integrated H_2 and biogas production systems (Banerjee et al., 2023). The results have shown promising potential for reducing GHG emissions and improving the overall sustainability of the H_2 economy. However, further technological advancements and cost reductions are required to achieve commercial viability. The development of green technologies, such as pressure swing adsorption and fuel cells, has also played a crucial role in enhancing the efficiency and sustainability of H_2 production and utilization (Luberti & Ahn, 2022; Pérez-Barragán et al., 2024). These emerging technologies and integrating biological H_2 and biogas production present a compelling pathway toward a more sustainable energy future. In addition, biogas reforming, which involves the conversion of biogas (a mixture of CH_4 and CO_2) into H_2, is another technology that has gained attention due to its ability to utilize waste materials as feedstock (Patel et al., 2024a; Sillero et al., 2024). Integrative production of H_2 and CH_4 from biowastes has been widely demonstrated using pure (*Clostridium*, *Enterobacter*, *Caldicellulosiruptor*, and *Thermoanaerobacterium*) or defined or enriched undefined mixed cultures for dark fermentation followed by mainly naturally derived mixed cultures for methanogenesis. Using these cultures, maximum H_2 production was reported up to ~1000 L/kg of total volatile solids at the dark-fermentation stage, followed by up to 1280 L/kg of total volatile solids at the second stage of methanogenesis (Intanoo et al., 2014; Zainal et al., 2019; Patel et al., 2021a; Sillero et al., 2024). These integrative H_2 and CH_4 production approaches suggested that these combined bioprocesses are highly beneficial over their corresponding COD reduction or substrate utilization.

9.4.3 Biohydrogen Followed by Value-Added Products

BHP and PHAs are important bioproducts with significant environmental and economic implications (Patel et al., 2011; Kang et al., 2023; Righetti et al., 2024). H_2 has been identified as a crucial alternative energy source, and its production from renewable biomass can contribute to a more sustainable energy landscape. On the other hand, PHAs are biobased and biodegradable polymers with diverse applications, ranging from packaging to biomedical devices (Colombo et al., 2019; Cywar et al., 2021; Narayanasamy et al., 2024). The production of H_2 has been the subject of extensive research, with several techniques being explored (Machineni et al., 2023). Macroalgae, or seaweed, have also emerged as a potentially viable feedstock for H_2 and PHA production (Patel & Lee, 2022; de Sousa Junior et al., 2023; Narayanan, 2024). Macroalgae possess a unique biochemical composition that can be leveraged to produce a variety of high-value chemicals, including PHA, lactic acid, and ethanol, in addition to H_2 (Tan et al., 2022; Okolie et al., 2023). Integrating biorefinery-derived platform chemicals from macroalgae with the production of H_2 and PHAs can lead to a more efficient and economically viable biorefinery model (Patel et al., 2021a). Optimizing the conversion of macroalgae biomass to various bioproducts while generating renewable energy in the form of H_2 can contribute to advancing a sustainable bioeconomy (Patel et al., 2021a; Tan et al., 2022). The recent reviews have revealed several key aspects of integrated H_2 and PHA production as a biorefinery model (Amulya et al., 2014; Patel et al., 2021a). Overall, the integration of H_2 production with the synthesis of PHAs can further enhance the sustainability and economic viability of the overall process. The leftover biomass from H_2 production can be utilized as a feedstock for the synthesis of PHAs, which are biobased and biodegradable polymers with a wide range of applications in the fields of medicine, packaging, and agriculture (Mona et al., 2020; Patel et al., 2024c). Several studies have explored the potential of combining H_2 production with PHA synthesis to create a more integrated and efficient biorefinery system (Colombo et al., 2019; Patel et al., 2021a; Kang et al., 2023). The advantages of this approach include the efficient utilization of resources, the reduction of waste, and the production of multiple valuable products. It is essential to develop further and optimize these integrated processes to improve their technical and economic feasibility for large-scale implementation (Patel et al., 2021a; Kataya et al., 2023). Undefined mixed cultures have broadly demonstrated integrative bioprocesses of H_2 and PHA production from biowastes and *Bacillus* as a pure culture at the dark-fermentation stage, followed by various cultures, including *Cupriavidus*, *Bacillus*, *Pseudomonas*, *Serratia*, *Rhodobacter*, *Rhodopseudomonas,* and mixed cultures. The maximum H_2 production of up to 2.00 mol/mol of hexose or 360 L/kg of COD was achieved at the first stage and a high yield of PHA up to 3.1 g/L at the second stage (Amulya et al., 2014; Patel et al., 2015; Colombo et al., 2019; Patel et al., 2021a). The microbes can accumulat*e PHA up to 90% of their DCM with diverse polymeric compositions depending on physiological conditions, feed, precursor, and type of organism. Therefore, this strategy can be beneficial in achieving higher efficiency in the overall bioprocess. The various methods for developing integrated systems from dark-fermentation effluents are illustrated in Figure 9.3.

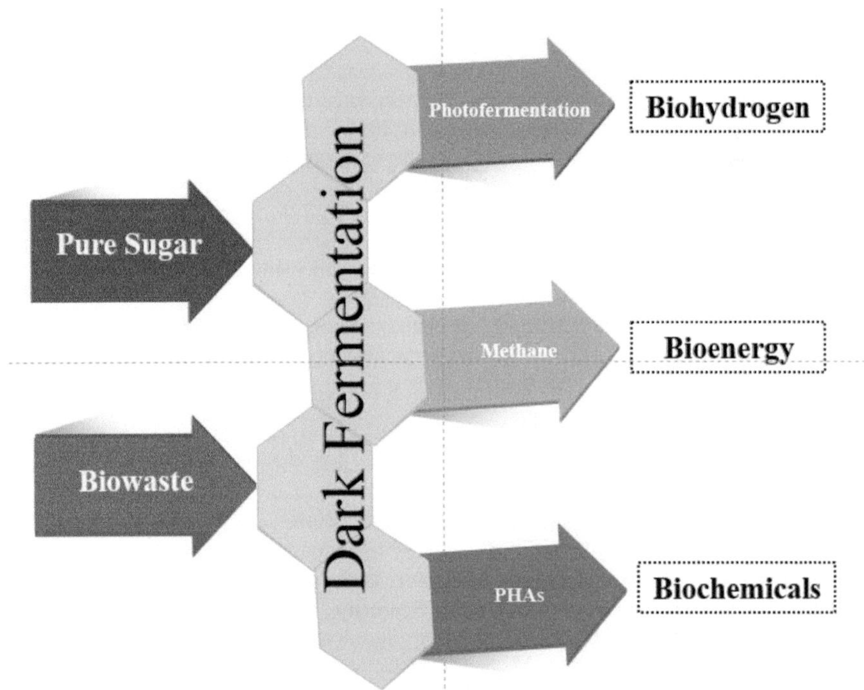

FIGURE 9.3 Integrative bioprocesses to dark-fermentative for biological hydrogen production from pure sugars and biowastes.

9.5 CONCLUSION

The push for environmentally friendly energy solutions has renewed the interest in accelerating the development of H_2 production methods. However, it is essential to note that around 96% of global H_2 production still comes from non-renewable fossil fuels, which release harmful GHGs into the atmosphere and can only produce low-purity H_2. To address these issues, the potential of green technologies, such as biomass-to-H_2 and water electrolysis using renewable electricity, to produce H_2 sustainably are explored in this chapter along with their analytical aspect. The primary biological processes utilized for H_2 gas production are bio-photolysis of water by algae and dark and photo-fermentation of organic materials, usually carbohydrates, by bacteria. These biological methods have significant advantages over chemical processes, as they are generally more cost-effective and have lower energy requirements. Life-cycle assessments and systems analyses can be conducted to evaluate the environmental, economic, and social impacts of these H_2 production technologies. Lignocellulosic biomass, such as agricultural and forestry residues, is a desirable feedstock to produce biofuels and wastewater as a co-digestion approach to enhance the nutritional requirements of H_2-producers. It appears to be an appropriate integrative tactic for efficiently valorizing residual organic wastes up to 90% present within H_2 production effluents to achieve high removal.

REFERENCES

Agrawal, D., Mahajan, N., Singh, S. P., and I. Sreedhar. 2024. Green hydrogen production pathways for a sustainable future with net zero emissions. *Fuel* 359:130131.

Amulya, K., Reddy, M. V., and S. Venkata Mohan. 2014. Acidogenic spent wash valorization through polyhydroxyalkanoate (PHA) synthesis coupled with fermentative biohydrogen production. *Bioresour. Technol.* 158:336–42.

Arisht, S. N., Abdul, P. M., Liu, C. M., Lin, S. K., Maaroff, R. M., Wu, S. Y., and J.M. Jahim. 2019. Biotoxicity assessment and lignocellulosic structural changes of phosphoric acid pre-treated young coconut husk hydrolysate for biohydrogen production. *Int. J. Hydrogen Energy* 44: 5830–43.

Banerjee, S., Pandit, C., Gundupalli, M. P., Pandit, S., Rai, N., Lahiri, D., Chaubey, K. K., and S. J. Joshi. 2023. Life cycle assessment of revalorization of lignocellulose for the development of biorefineries. *Environ. Dev. Sustain.* https://doi.org/10.1007/s10668-023-03360-4

Bertasini, D., Battisha, F., Mancini, R., Frison, N., and D. Bolzonella. 2024. Hydrogen and methane production through two-stage anaerobic digestion of straw residues. *Environ. Res.* 247:118101.

Blay-Rogar, R., Bach, W., Bobadilla, L. F., Reina, T. R., Odriozola, J. A., Amils, R., and V. Blay. 2024. Natural hydrogen in the energy transition: Fundamentals, promise, and enigmas. *Renew. Sustain. Energy. Rev.* 189:113888.

Budhraja, N., Pal, A., and R. S. Mishra. 2024. Simulation and optimization of biohydrogen production from biomass feed via anaerobic digestion. *Chem. Eng. Technol.* 47:706–15.

Capurso, T., Stefanizzi, M., Torresi, M., and S. M. Camporeale. 2022. Perspective of the role of hydrogen in the 21st century energy transition. *Energy Convers. Manage.* 251:114898.

Carrillo-Reyes, J., Tapia-Rodríguez, A., Buitron, G., Moreno-Andrade, I., Palomo-Briones, R., Razo-Flores, E., … M. Zaiat. 2019. A standardized biohydrogen potential protocol: An international round robin test approach. *Int. J. Hydrogen Energy* 44(48):26237–47.

Chandekar, B., Rohit, M. V., and S. K. S. Patel. 2024. Efficient dark-fermentation biohydrogen production by *Shigella flexneri* SPD1 from biowaste-derived sugars through green solvents approach. *Bioresour. Technol.* 410:131276.

Chauhan, P. S., and S. Bhattacharya. 2019. Hydrogen gas sensing methods, materials, and approach to achieve parts per billion level detection: A review. *Int. J. Hydrogen Energy* 44:26076–99.

Chelvam, K., Hanafiah, M. M., Woon, K. S., and K. A. Ali. 2024. A review on the environmental performance of various hydrogen production technologies: An approach towards hydrogen economy. *Energy Rep.* 11:369–83.

Colombo, B., Calvo, M. V., Sciarria, T. P., Scaglia, B., Kizito, S. S., D'Imporzano, G., and F. Adani. 2019. Biohydrogen and polyhydroxyalkanoates (PHA) as products of a two-steps bioprocess from deproteinized dairy wastes. *Waste Manag.* 95: 22–31.

Cywar, R. M., Rorrer, N. A., Hoyt, C. B., Beckham, G. T., and E. Y. X. Chen. 2021. Biobased polymers with performance advantaged properties. *Nat. Rev. Mater.* 7:83–103.

Das, S. R., and N. Basak. 2021. Molecular biohydrogen production by dark and photo fermentation from wastes containing starch: Recent advancement and future perspective. *Bioprocess Biosyst. Eng.* 44:1–25.

de Sousa Junior, R. R., Cezario, F. E. M., Antonino, L. D., dos Santos, D. J., and M. Lackner. 2023. Characterization of poly(3-hydroxybutyrate) (P3HB) from alternative, scalable (Waste) feedstocks. *Bioengineering* 10:1382.

Deng, C., Rajendran, K., Varshini, R., and R. Lin. 2024. Economic viability of two-stage biohydrogen and biomethane production from cassava stillage residue focusing on solids content and pretreatment. *Int. J. Hydrogen Energy* 52:110–21.

Dixon, B. J., and J. H. Zhang. 2016. An innovative collection in hydrogen molecular biology and medicine research. *Med. Gas Res.* 6:55–6.

Erikse, N. T., Riis, M. L., Holm, N. K., and N. Iversen. 2011. H2 synthesis from pentoses and biomass in Thermotoga spp. *Biotechnol. Lett.* 33:293–300.

Furqon, Purwanto, Y. A., Setiawan, R. P. A., and B. Susilo. 2024. Biohydrogen production from DEPOME (dark fermented effluent palm oil mill effluent) via photo-fermentation utilizing indigenous bacteria. *Int. J. Hydrogen Energy* 56:322–9.

Goren, A. Y., Kenez, M., Dincer, I., and A. Khalvati. 2024. Cleaner production of biohydrogen using poplar leaves: Experimental and optimization studies. *J. Clean. Prod.* 434:140080.

Goveas, L. C., Nayak, S., Kumar, P. S., Vinayagam, R., Selvaraj, R., and G. Rangasamy. 2024. Recent advances in fermentative biohydrogen production. *Int. J. Hydrogen Energy* 54:200–17.

Gupta, P., Toksha, B., and M. Rahaman. 2024. A critical review on hydrogen based fuel cell technology and applications. *Chem Rec.* 24:e202300295.

Intanoo, P., Rangsanvigit, P., Malakul, P., and S. Chavadej. 2014. Optimization of separate hydrogen and methane production from cassava wastewater using two-stage upflow anaerobic sludge blanket reactor (UASB) system under thermophilic operation. *Bioresour. Technol.* 173:256–65.

Ivanova, G., Rakhely, G., and K. L. Kovacs. 2009. Thermophilic biohydrogen production from energy plants by *Caldicellulosiruptor saccharolyticus* and comparison with related studies. *Int. J. Hydrogen Energy* 34:3659–70.

Jamaludin, N. F. M., Abdullah, L. C., Idrus, S., Engliman, N. S., Tan, J. P., and N. S. Jamali. 2024. Nickel-iron doped on granular activated carbon for efficient immobilization in biohydrogen production. *Bioresour. Technol.* 391:129933.

Kang, B. J., Jeon, J. M., Bhatia, S. K., Kim, D. H., Yang, Y. H., Jung, S., and J. J. Yoon. 2023. Two-stage bio-hydrogen and polyhydroxyalkanoate production: Upcycling of spent coffee grounds. *Polymers* 15:681.

Kataya, G., Cornu, D., Bechelany, M., Hijazi, A., and M. Issa. 2023. Biomass waste conversion technologies and its application for sustainable environmental development – A review. *Agronomy* 13:2833.

Khalekuzzaman, M., Jahan, N., Kabir, S. B., Hasan, M., Fayshal, M. A., and D. R. Chowdhury. 2024. Substituting microalgae with fecal sludge for biohythane production enhancement and cost saving through two-stage anaerobic digestion. *J. Clean. Prod.* 427:139352.

Kim, M. S., Baek, J. S., Yun, Y. S., Sim, S. J., Park, S., and S. C. Kim. 2006. Hydrogen production from Chlamydomonas reinhardtii biomass using a two-step conversion process: Anaerobic conversion and photosynthetic fermentation. *Int. J. Hydrogen Energy* 31:812–6.

Kim, T. H., Song, D., Jeon, Y. J., Hwang, O., Nam, J. Y., and Y. M. Yun. 2024. Enhanced production of biohydrogen and biomethane through a two-stage anaerobic fermentation of food waste mixed with conductive additives. *Chem. Eng. J.* 476:146520.

Kondaveeti, S., Kim, I. W., Otari, S., Patel, S. K. S., Pagolu, R., Losetty, V., Kalia, V. C., and J. K. Lee. 2019. Co-generation of hydrogen and electricity from biodiesel process effluents. *Int. J. Hydrogen Energy* 44:27285–96.

Kucharska, K., Cieslinski, H., Rybarczyk, P., Słupek, E., Łukajtis, R., Wychodnik, K., and M. Kaminski. 2019. Fermentative conversion of two-step pre-treated ligno cellulosic biomass to hydrogen. *Catalysts* 9:858.

Laurinavichene, T., Tekucheva, D., Laurinavichius, K., and A. Tsygankov. 2018. Utilization of distillery wastewater for hydrogen production in one-stage and two-stage processes involving photofermentation. *Enzym. Microb. Technol.* 110:1–7.

Li, F., Liu, D., Sun, K., Yang, S., Peng, F., Zhang, K., Guo, G., and Y. Si. 2024. Towards a future hydrogen supply chain: A review of technologies and challenges. *Sustainability* 16:1890.

Litti, Y. V., Khuraseva, N. D., Vishnyakova, A. V., Zhuravleva, E. A., Kovalev, A. A., Kovalev, D. A., Panchenko, V. A., and S. N. Parshina. 2024. Comparative study on biohydrogen production by newly isolated *Clostridium butyricum* SP4 and *Clostridium beijerinckii* SP6. *Int. J. Hydrogen Energy* 48:27540–56.

Luberti, M., and H. Ahn. 2022. Review of polybed pressure swing adsorption for hydrogen purification. *Int. J. Hydrogen Energy* 47:10911–33.

Machineni, L., Deepanraj, B., Chew, K. W., and A. G. Rao. 2023. Biohydrogen production from lignocellulosic feedstock: Abiotic and biotic methods. *Renew. Sustain. Energy Rev.* 182:113344.

Mars, A. E., Veuskens, T., Budde, M. A. W., van Doeveren, P. F. N. M., Lips, S. J., Bakker, R. R., de Vrije, T., and P. A. M. Claassen. 2010. Biohydrogen production from untreated and hydrolyzed potato steam peels by the extreme thermophiles *Caldicellulosiruptor saccharolyticus* and *Thermotoga neapolitana*. *Int. J. Hydrogen Energy* 35:7730–7.

Masihi, F., Rezaeitavabe, F., Karimi-Jashni, A., and G. Riefler. 2024. Optimization and enhancement of biohydrogen production in a single-stage hybrid (dark/photo) fermentation reactor using Fe3O4 and TiO2 nanoparticles. *Int. J. Hydrogen Energy* 52:295–305.

Mattiello-Francisco, L., Ferreira, F. V., Peixoto, G., Mockaitis, G., and M. Zaiat. 2024. Hydrogen production from sugarcane bagasse pentose liquor fermentation using different food/microorganism and carbon/nitrogen ratios under mesophilic and thermophilic conditions. *Fermentation* 10:432.

Mechery, J., Thomas, D. M., Kumar, C. S. P., Joseph, L., and V. P. Sylas. 2021. Biohydrogen production from acidic and alkaline hydrolysates of paddy straw using locally isolated facultative bacteria through dark fermentation. *Biomass Conv. Bioref.* 11:1263–72.

Mojoodi, S., Zilouei, H., and K. Karimi. 2024. Improvement of biohydrogen production from rice straw hydrolysate by green-self-prepared nano-silica. *Fuel* 363:130887.

Mona, S., Kumar, S. S., Kumar, V., Parveen, K., Saini, N., Deepak, B., and A. Pugazhendhi. 2020. Green technology for sustainable biohydrogen production (waste to energy): A review. *Sci. Total Environ.* 728:138481.

Narayanan, M. 2024. Marine algae biomass: A viable and renewable resource for biofuel production: A review. *Algal Res.* 82:103687.

Narayanasamy, A., Patel, S. K. S., Singh, N., Rohit, M. V., and J. K. Lee. 2024. Valorization of algal biomass to produce microbial polyhydroxyalkanoates: Recent updates and perspectives. *Polymers* 16:2227.

Nnabuife, S. G., Darko, C. K., Obiako, P. C., Kuang, B., Sun, X., and K. Jenkins. 2023. A comparative analysis of different hydrogen production methods and their environmental impact. *Clean Technol.* 5:1344–80.

Okolie, O., Kumar, A., Edwards, C., Lawton, L. A., Oke, A., McDonald, S., Thakur, V. K., and J. Njuguna. 2023. Bio-based sustainable polymers and materials: From processing to biodegradation. *J. Compos. Sci.* 7:213.

Pandey, A. K., Kim, S. M., Yoon, Y. H., Kim, Y. B., and S.-H. Kim. 2023. Influence of the permeate flux on continuous biohydrogen production, permeability, and granulation in a dynamic membrane bioreactor. *Chem. Eng. J.* 475:145684.

Patel, S. K. S., and J. K. Lee. 2022. Plastic eating enzymes: A step towards sustainability. *Indian J. Microbiol.* 62:658–61.

Patel, S. K. S., Das, D., Kim, S. C., Cho, B. K., Lee, J. K., and V. C. Kalia. 2021a. Integrating strategies for sustainable conversion of waste biomass into dark-fermentative hydrogen and value-added products. *Renew. Sustain. Energy Rev.* 150:111491.

Patel, S. K. S., Gupta, R. K, Das, D., Lee, J. K., and V. C. Kalia. 2021b. Continuous biohydrogen production from poplar biomass hydrolysate by a defined bacterial mixture immobilized on lignocellulosic materials under non-sterile conditions. *J. Clean. Prod.* 323:125037.

Patel, S. K. S., Gupta, R. K., Karthikeyan, K. K., Padhi, D. K., Ranganathan, S., Paramanantham, P., and J. K. Lee. 2024b. Nonsterile process for biohydrogen production: recent Updates, challenges and opportunities. *Indian J. Microbiol.* 64:445–56.

Patel, S. K. S., Gupta, R. K., Rohit, M. V., and J. K. Lee. 2024a. Recent developments in hydrogen production, storage, and transportation: Challenges, opportunities, and perspectives. *Fire* 7:233.

Patel, S. K. S., Kalia, V. C., and J. K. Lee. 2023. Integration of biogas derived from dark fermentation and anaerobic digestion of biowaste to enhance methanol production by methanotrophs. *Bioresour. Technol.* 367:128427.

Patel, S. K. S., Kondaveeti, S., Otari, S. V., Pagolu, R. T., Jeong, S. H., Kim, S. C., Cho, B. K., Kang, Y. C., and J. K. Lee. 2018a. Repeated batch methanol production from a simulated biogas mixture using immobilized *Methylocystis bryophila*. *Energy* 145:477–85.

Patel, S. K. S., Kumar, P., and V. C. Kalia. 2012. Enhancing biological hydrogen production through complementary microbial metabolisms. *Int. J. Hydrogen Energy* 37:10590–603.

Patel, S. K. S., Kumar, P., Singh, M., Lee, J. K., and V. C. Kalia. 2015. Integrative approach to produce hydrogen and polyhydroxybutyrate from biowaste using defined bacterial cultures. *Bioresour. Technol.* 176:136–41.

Patel, S. K. S., Lee, J. K., and V. C. Kalia. 2017. Dark-fermentative biological hydrogen production from mixed biowastes using defined mixed cultures. *Indian J. Microbiol.* 57:171–6.

Patel, S. K. S., Lee, J. K., and V. C. Kalia. 2018b. Beyond the theoretical yields of dark-fermentative biohydrogen. *Indian J. Microbiol.* 58:529–30.

Patel, S. K. S., Mardina, P., Kim, D., Kim, S. Y., Kalia, V. C., Kim, I. W., and J. K. Lee. 2016. Improvement in methanol production by regulating the composition of synthetic gas mixture and raw biogas. *Bioresour. Technol.* 218:202–8.

Patel, S. K. S., Ray, S., Prakash, J., Wee, J. H., Kim, S. Y., Lee, J. K., and V. C. Kalia. 2019. Co-digestion of biowastes to enhance biological hydrogen production by defined mixed bacterial cultures. *Indian J. Microbiol.* 59:154–60.

Patel, S. K. S., Singh, D., Pant, D., Gupta, R. K., Busi, S., Singh, R. V., and J. K. Lee. 2024c. Polyhydroxyalkanoates production by methanotrophs: Recent updates and perspectives. *Polymers* 16:2570.

Patel, S. K. S., Singh, M., and V. C. Kalia. 2011. Hydrogen and polyhydroxybutyrate producing abilities of *Bacillus* spp. from glucose in two stage system. *Indian J. Microbiol.* 51:418–23.

Pérez-Barragán, J., Martínez-Fraile, C., Muñoz, R., Quijano, G., Maya-Yescas, R., León-Becerril, E., Castro-Muñoz, R., and O. García-Depraect. 2024. Brewery spent grain valorization through fermentation: Targeting biohydrogen, carboxylic acids and methane production. *Process Safety and Environmental Protection*. 191: 206–17.

Pomdaeng, P., Kongthong, O., Tseng, C. H., Dokmaingam, P., and C. Y. Chu. 2024. An immobilized mixed microflora approach to enhancing hydrogen and methane productions from high-strength organic loading food waste hydrolysate in series batch reactors. *Int. J. Hydrogen Energy* 52:160–9.

Praptyana, I. R., and X. Budiyano. 2022. Biohydrogen production from wood dust mahogany *(Swietenia mahagony)* by dark fermentation using *Enterobacter aerogenes*: Effect of ozone pretreatment time and pH. *Mater. Today Proc.* 63:S203–9.

Qureshi, F., Yusuf, M., Khan, M. A., Ibrahim, H., Ekeoma, B. C., Kamyab, H., Rehman, M. M., Nadda, A. K., and S. Chelliapan. 2023. A state-of-the-art review on the latest trends in Hydrogen production, storage, and transportation techniques. *Fuel* 340:27574.

Qyyum, M. A., Ihsanullah, I., Ahmad, R., Ismail, S., Khan, A., Nizami, A. S., and A. Tawfik. 2022. Biohydrogen production from real industrial wastewater: Potential bioreactors, challenges in commercialization and future directions. *Int. J. Hydrogen Energy* 47:37154–70.

Ramzan, H., Usman, M., Nadeem, F., Shahzaib, M., Rahman, M. U., Singhania, R. R., Jabeen, F., Patel, A. K., Qing, C., Liu, S., Piechota, G., and N. Tahir. 2023. Depolymerization of lignin: Recent progress towards value-added chemicals and biohydrogen production. *Bioresour. Technol.* 386:129492.

Renaudie, M., Dumas, C., Vuilleumier, S., and B. Ernst. 2024. Spontaneous dark fermentation in a pre-seeded liquid-gas membrane bioreactor: Impact of wine and coffee biowaste microflora on continuous biohydrogen production. *J. Clean. Prod.* 437:140759.

Rathi, B. S., Kumar, P. S., Rangasamy, G., and S. Rajendran. 2024. A critical review on biohydrogen generation from biomass. *Int. J. Hydrogen Energy* 52:115–38.

Righetti, G. I. C., Faedi, F., and A. Famulari. 2024. Embracing sustainability: The World of bio-based polymers in a mini review. *Polymers* 16:950.

Saha, P., Akash, F. A., Shovon, S. M., Monir, M. U., Ahmed, M. T., Khan, M. F. H., Sarkar, S. M., Islam, M. K., Hasan, M. M., Vo, D. V. N., Aziz, A. A., Hossain, M. J., and R. Akter. 2024. Grey, blue, and green hydrogen: A comprehensive review of production methods and prospects for zero-emission energy. *Int. J. Green Energy* 21:1383–97.

Sahrin, N. T., Khoo, K. S., Lim, J. W., Shamsuddin, R., Ardo, F. M., Rawindran, H., Hassan, M., Kiatkittipong, W., Abdelfattah, E. A., Oh, W. D., and C. K. Cheng. 2022. Current perspectives, future challenges and key technologies of biohydrogen production for building a carbon–neutral future: A review. *Bioresour. Technol.* 364:128088.

Sarkar, O., Rova, U., Christakopoulos, P., and L. Matsakas. 2022. Organosolv pretreated birch sawdust for the production of green hydrogen and renewable chemicals in an integrated biorefinery approach. *Bioresour. Technol.* 344:126164.

Sillero, L., Perez, M., and R. Solera. 2024. Optimisation of anaerobic co-digestion in two-stage systems for hydrogen, methane and biofertiliser production. *Fuel* 365:131186.

Singh, R., Kumar, R., Sarangi, P. K., Kovalev, A. A., and V. Vivekanand. 2023. Effect of physical and thermal pretreatment of lignocellulosic biomass on biohydrogen production by thermochemical route: A critical review. *Bioresour. Technol.* 369:128458.

Singhvi, M., and B. S. Kim. 2022. Green hydrogen production through consolidated bioprocessing of lignocellulosic biomass using nanobiotechnology approach. *Bioresour. Technol.* 365:128108.

Tan, F. H. P., Nadir, N., and K. Sudesh. 2022. Microalgal biomass as feedstock for bacterial production of PHA: Advances and future prospects. *Front. Bioeng. Biotechnol.* 10:879476.

Yang, E., Chon, K., Kim, K. Y., Le, G. T. H., Nguyen, H. Y., Le T. T. Q., Nguyen H. T. T., Jae, M. R., Ahmad, I., Oh, S. E., and K. J. Chae. 2023. Pretreatments of lignocellulosic and algal biomasses for sustainable biohydrogen production: Recent progress, carbon neutrality, and circular economy. *Bioresour. Technol.* 369:28380.

Zabed, H. M., Akter, S., Yun, J., Zhang, G., Zhang, Y., and X. Qi. 2020. Biogas from microalgae: Technologies, challenges and opportunities. *Renew. Sustain. Energy Rev.* 117:109503.

Zainal, B. S., Akhbari, A., Zinatizadeh, A. A., Mohammadi, P., Danaee, M., Mohd, N. S., and S. Ibrahim. 2019. UASFF start-up for biohydrogen and biomethane production from treatment of palm oil mill effluent. *Int. J. Hydrogen Energy* 44:20725–37.

Zhai, X., Nakao, A., and X. Sun. 2015. Detection techniques for hydrogen. In: Sun, X., Ohta, S., and A. Nakao (eds), Hydrogen Molecular Biology and Medicine. Springer, Dordrecht, pp. 49–60.

Zhang, G., Liu, J., Pan, X., Abed, A. M., Le, B. N., Ali, H. E., and Y. Ge. 2023. Latest avenues and approaches for biohydrogen generation from algal towards sustainable energy optimization: Recent innovations, artificial intelligence, challenges, and future perspectives. *Int. J. Hydrogen Energy* 48:20988–1003.

Zhang, Q., You, S., Li, Y., Qu, X., and H. Jiang. 2020. Enhanced biohydrogen production from cotton stalk hydrolysate of *Enterobacter cloacae* WL1318 by overexpression of the formate hydrogen lyase activator gene. *Biotechnol. Biofuels* 13:94.

Zhao, C., Sharma, A., Ma, Q., Zhu, Y., Li, D., Liu, Z., and Y. Yang. 2021. A developed hybrid fixed-bed bioreactor with Fe-modified zeolite to enhance and sustain biohydrogen production. *Sci. Total Environ.* 758:143658.

Zhao, Z. T., Ding, J., Wang, B. Y., Bao, M. Y., Liu, B. F., Pang, J. W., Ren, N. Q., and S. S. Yang. 2024. Advances in the biomass valorization in dark fermentation systems: A sustainable approach for biohydrogen production. *Chem. Eng. J.* 481:148444.

Zhu, Z., Song, W., Chen, X., and S. Sun. 2023. Integrated process to produce biohydrogen from wheat straw by enzymatic saccharification and dark fermentation. *Int. J. Hydrogen Energy* 48:11153–61.

10 Biobutanol Production and Analysis

Kuber Singh Mehra and Varun Goel[*]

10.1 INTRODUCTION

The availability of reliable and affordable energy sources is a major concern in modern society, as the extraction and use of fossil fuels have significant social and environmental impacts, including contributing to the climate crisis (Awad et al., 2018; Kaushik et al., 2021; Mehra & Goel, 2025a). These challenges have motivated researchers to develop green technologies that can produce environmentally friendly fuels (Pathak et al., 2021; Mehra et al., 2023b). The transportation sector, in particular, requires liquid biofuels, leading to the emergence of products such as bioethanol, biodiesel, and bio-based hydrocarbon compounds (Zhang et al., 2018; Mehra et al., 2023a; Singh et al., 2020; Mehra et al., 2021). While lab-scale production of these fuels has seen success, their commercialization remains a significant hurdle (Lam et al., 2019; Mehra & Goel, 2025b; Lea et al., 2018). Extensive research has been conducted on biodiesel production from edible and non-edible sources, as well as algae, and there are numerous reports available on the topic (Amenaghawon et al., 2021; Mehra et al., 2025; Mehra & Pant, 2021; Singh et al., 2019; Mehra et al., 2014). Similarly, bioethanol production from cellulose via fermentation is well known and has been studied for decades (Abuhay et al., 2021; Boonchuay et al., 2021). More recently, bio-based liquid hydrocarbons, including petrol and diesel-range fuels, have garnered attention, with many studies reporting progress in this area (Ge et al., 2022; Menshhein et al., 2019). Biodiesel, in particular, is praised for its relatively simple production process and the low cost of its chemical components (Wang et al., 2018; Mehra et al., 2024; Goel et al., 2024). The transesterification of edible or non-edible oils is considered a potential method for biodiesel production, but pure biodiesel has been found to be incompatible with modern engines (Wang et al., 2022; Kumar et al., 2024; Singh et al., 2018). Despite the advantages and potential of these fuels, several challenges remain unsolved. Biobutanol, derived from waste cellulose fermentation, shows promise for the prospect of sustainable fuels (Huzir et al., 2018; Pugazhendhi et al., 2019).

Recent studies highlight biobutanol as a highly attractive option due to its potential as a sustainable fuel, superior industrial solvent, and precursor for important chemical reactions. Its versatility makes it highly sought after compared to other biofuels. Butanol is typically produced through two main methods: (a) the oxo process, a petrochemical method involving two stages where butyraldehyde is first produced by hydroformylating propylene, followed by hydrogenation to yield butanol, and (b) a biological approach, which is microbial fermentation (Gottumukkala et al., 2019; Xue et al., 2017). Nevertheless, petrochemical production expense is closely related

[*] Corresponding author

 DOI: 10.1201/9781003485636-10

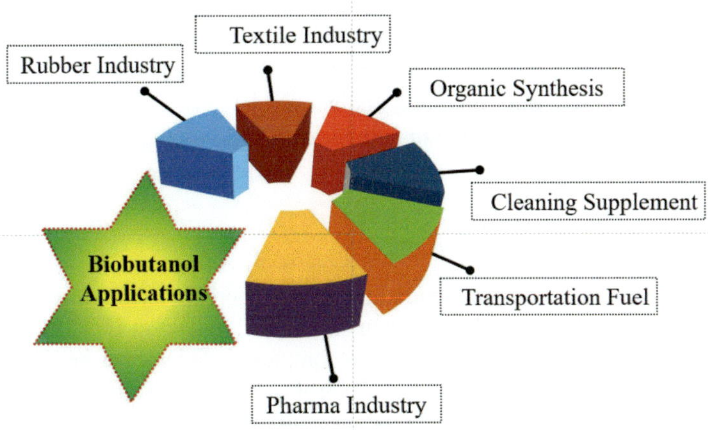

FIGURE 10.1 Biobutanol applications in different fields.

to the propylene market and fluctuates with crude oil prices (Yadav et al., 2014). Thus, microbial fermentation could be a viable alternative for butanol production, although it faces challenges such as low yields and high production costs. Various efforts have been made to overcome these obstacles, including the search for more affordable substrates (Ibrahim et al., 2017; Xin et al., 2016). Lignocellulose is considered an ideal precursor for biobutanol production due to its low cost, abundance, and availability (Dharmaraja et al., 2020; Jiang et al., 2019). Through a simple pretreatment process, lignocellulose releases sugars, which are fermented to produce alcohols and acids. After separation and purification, the resulting alcohol is known as biobutanol. Figure 10.1 outlines the main industrial applications of biobutanol.

10.2 BIOBUTANOL PROPERTIES

Butanol, a versatile four-carbon alcohol (C_4H_9OH), is a colorless and flammable substance predominantly generated by the petrochemical industry (Dürre, 2008; Antoni et al., 2007). Historically, it has been known for its use as a chemical and solvent across several industries. Today, butanol is considered a highly advanced biofuel, superior to ethanol, and is used as an additive in gasoline, diesel, and kerosene (Antoni et al., 2007). Like ethanol, which is used for ethyl-t-butyl ether production, butanol contributes to the biological production of butyl-t-butyl ether (Fortman et al., 2008). As a biofuel, butanol offers several advantages, including increased tolerance to water contamination, the ability to blend at higher concentrations without the need for vehicle modifications, and a higher energy density. It also reduces the viscosity of vegetable oils and can dissolve in them at any ratio. Butanol can be converted into jet fuel, mixed with gasoline, and serves as an effective fuel extender. As an additive in internal combustion engines, it produces only carbon dioxide, making it a more environmentally friendly option (Dürre, 2008). Compared to ethanol, butanol is preferred due to its higher calorific value (29.2 MJ/liter), lower melting point (−89.5°C), higher boiling point (117.2°C), and flash point (36°C) (Antoni et al., 2007; Wackett, 2008; Zhou & Thomson, 2009). It is less flammable, volatile, and corrosive; has a lower vapor pressure; and has a relatively low heat of vaporization, while also being highly explosive in comparison with bioethanol (Balat, 2009; Yan & Lin, 2009).

10.3 REACTION MECHANISM

The process for biobutanol production, involving anaerobic fermentation and hydrolysis, is depicted in the flowchart shown in Figure 10.2 (Ndaba et al., 2015; Zhao et al., 2015; Liu et al., 2022). In the first step, the complex carbohydrates are broken down into simple sugars like glucose through hydrolysis. These sugars are then digested by the amylase enzyme and transformed into fatty acids under anaerobic conditions and are fermented into solvents such as acetone, ethanol, and butanol by *Clostridium acetobutylicum*. The second step follows the Embden–Meyerhof pathway, where pyruvate is produced from five- and six-carbon carbohydrates, with cellulose facilitating this conversion. This process releases 2 moles of ATP and NADPH. In the

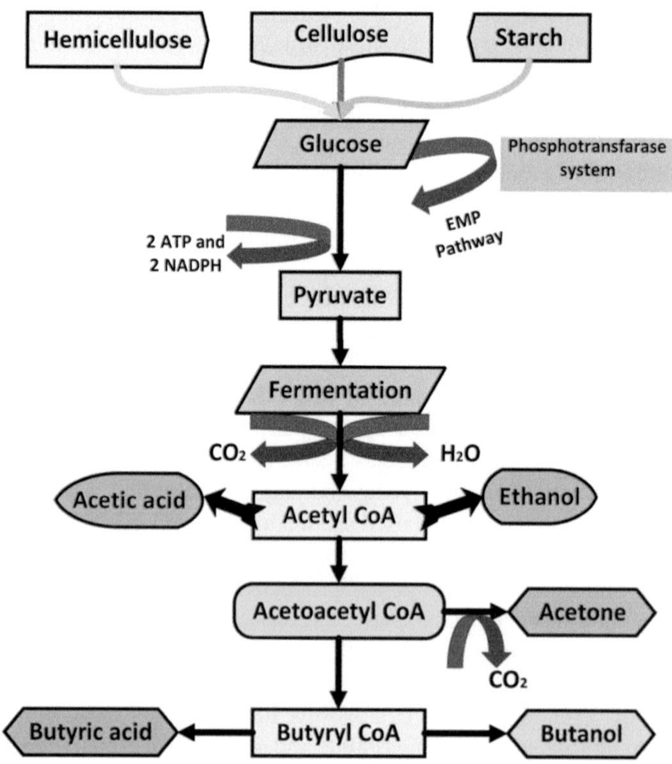

FIGURE 10.2 Reaction mechanism for biobutanol production.

third step, pyruvate undergoes fermentation to form acetyl-CoA, releasing CO_2 and H_2O as byproducts. Step four involves the production of acetone, where acetyl-CoA loses CO_2, resulting in the formation of side products acetaldehyde and acetone. The oxidation of acetaldehyde forms acetone and acetic acid. Finally, in step five, acetyl-CoA is reduced to produce butyryl-CoA, which is further reduced to generate butanol and ethanol.

10.4 BIOBUTANOL PRODUCTION

Biobutanol can be produced from sugar-containing sources, starch-rich crops, inedible or edible crops high in cellulose and hemicellulose, lignin, and whole grain. Butanol is produced through sugar fermentation, starch enzymatic hydrolysis, cellulose saccharification, and whole grain. Figure 10.3 depicts sources and routes for biobutanol production (Liu et al., 2022).

Feedstock, also known as raw materials, is an essential component of the fermentation process. In the Weizmann process for ABE (acetone–butanol–ethanol) fermentation, corn was utilized as the primary substrate, while industrial-scale ABE fermentation typically utilized molasses, with *Clostridium acetobutylicum* playing the role of fermentation agent (Liu et al., 2022). Various starchy materials have been utilized as substrates for ABE fermentation, including tapioca, rice straw, corn waste, potato, corn starch, low-grade potatoes, and byproducts of food organizations like oil palm effluent, milk fat, soya syrup, and fruit pulps. Li et al. (2020) demonstrated the production of biobutanol from the feedstock rice straw via a saccharification process followed by anaerobic fermentation, yielding approximately 10.62 g/L of biobutanol, an increase of 115.38% compared to a monoculture-type system. Additionally,

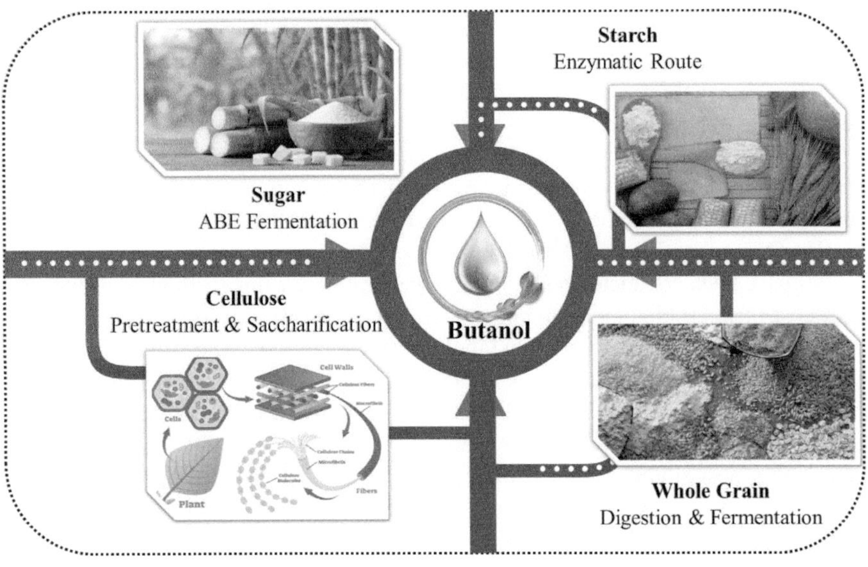

FIGURE 10.3 Biobutanol production sources and routes.

the fermentation of rice straw conducted in two stages using *Clostridium beijerinckii* NCIMB 8052 produced 230 g/kg of biobutanol, along with biohydrogen as a byproduct (Li et al., 2018). Tsai et al. (2020) observed that immobilized cells, specifically PVA-immobilized *C. acetobutylicum*, enhanced the enzymatic hydrolysis of rice straw. Valles et al. (2021) reported the recovery of 0.344 g/L of butanol from rice straw during the saccharification process and fermentation using the Cellic® CTec2 commercial enzyme blend. The use of second-generation biomass to make biobutanol appears appealing; nevertheless, the production costs reduce the method's profitability. First-generation biobutanol is produced using edible or food crops as substrates, leading to significant concerns regarding food security and rising food prices (Hönig et al., 2017). In contrast, second-generation biobutanol utilizes agricultural waste or non-edible biomass, reducing competition with food crops. As the demand for renewable energy grows rapidly, these non-edible biomasses have become increasingly important for biobutanol production. Second-generation biomass offers several advantages over first- and third-generation sources, such as not requiring fertile land and having a high lipid portion that is 10% to 50% of its dry weight. Although technological challenges currently hinder large-scale biofuel production from this type of biomass, advancements in the coming years are expected to make it a widely adopted technology. Table 10.1 presents research findings related to biobutanol synthesis from second-generation biomass. For example, *C. saccharoperbutylacetonicum* NCIMB 12,606 (N1-4) was used to convert wood hydrolysate containing 68.6 g/L of carbohydrates into 2.49 g/L of butanol (Etteh et al., 2021). *Clostridium acetobutylicum* NRRL B-591 increased biobutanol production by 49% using municipal waste containing 30.9% starch (Zhou & Thomson, 2009). Su et al. (2020) also demonstrated the use of *C. acetobutylicum* ABE 1201 to ferment sweet sorghum bagasse, producing 17.3 g/kg of butanol from the feedstock. Potato peel waste was treated using Organosolv, enzymatic hydrolysis, and ABE fermentation with *Clostridium acetobutylicum* PTCC 1492, resulting in the production of 24.8 g/L ABE (Abedini et al., 2020). In another study, Hijosa-Valsero et al. (2018) utilized *Clostridium beijerinckii* CECT 508 to obtain 7.02 g/L of butanol from the source coffee silver skin. The literature strongly supports using second-generation biomass to generate butanol.

In the late 18th century, Louis Pasteur pioneered the production of butanol using microbes. Later, Chaim Weizmann explored the anaerobic bacterial process for butanol synthesis, particularly investigating *Clostridium* bacteria for large-scale production (Anandharaj et al., 2020). The process relied on fermentation, and numerous studies have been conducted on biobutanol generation via this method. However, the high cost of microbiological processes and the availability of fossil fuels hindered its early adoption. Interest in biobutanol and bioethanol derived from *Clostridium* bacteria resurfaced after concerns about fossil fuel depletion grew. Bacteria have the capability to ferment carbohydrates into alcohols, such as bioethanol, biobutanol, and organic acids. Despite the diversity of bacteria available in nature, identifying the most suitable strains for efficient biobutanol production remains a key research focus. According to Grand View Research Inc., the global biobutanol market was estimated at USD 15.86 billion in 2023, and it is expected to rise at a 9.2% CAGR from 2024 to 2030 (Grand View Research, 2024). This expansion is being driven by an increase in demand for renewable fuels, as well as growing

TABLE 10.1
Biobutanol Production from First, Second, and Third-Generation Biomass

	Feedstock	Micro-organism Used	Conversion Method	Biobutanol Yield	Reference
First-generation biomass	Corn stover	Immobilized *Clostridium acetobutylicum*	ABE fermentation	77–136 gm/liter	Cai et al. (2017)
	Soybean meal	*Clostridium beijerinckii* TISTR 1461	Fermentation	15.33 gm/liter	Narueworanon et al. (2020)
	Chicken feather and wheat straw	*Clostridium beijerinckii* and *Clostridium saccharoperbutylacetonicum*	Fermentation and saccharification	4.6 gm/liter	Branska et al. (2020)
	Rice straw	PVA-immobilized *C. acetobutylicum*	Alkali pretreatment followed by enzymatic hydrolysis	4.32 gm/liter	Tsai et al. (2020)
Second-generation biomass	Wood hydrolysate	*C. saccharoperbutylacetonicum* NCIMB	ABE fermentation	2.49 gm/liter	Etteh et al. (2021)
	Jerusalem artichoke stalk	*Clostridium sp.*	ABE fermentation	323.4–348.7 gm/liter	Xue et al. (2017)
	Sweet sorghum bagasse	*C. acetobutylicum* ABE 1201	Integrated ABE and ethanol fermentation	17.3 gm/kg	Su et al. (2020)
	Coffee silverskin	*C. beijerinckii CECT 508*	Fermentation and autohydrolysis	7.02 ± 0.2 gm/liter	Hijosa-Valsero et al. (2018)
Third-generation biomass	Microalgal-based biodiesel residues; *C. acetobutylicum* ATCC 824;*Chlorella sorokiniana* CY1	–	ABE fermentation	3.86 gm/liter	Cheng et al. (2015)
	Nannochloropsisgaditana	–	Spectrophotometric ABE fermentation	2.9 ± 0.24 gm/liter	Onay M. (2020)

environmental concerns and rules requiring a transition away from traditional fossil fuels. Furthermore, the potential of biobutanol as a clean heat fuel and bio-based material is predicted to increase demand and drive market expansion. Algae biofuels have gained attention from researchers due to their edibility, ease of cultivation in bioreactors, and potential to replace fossil fuels (Ge et al., 2022; Nai Yuh Yek et al., 2022). Algae need nitrogen, phosphorus, carbon dioxide, and sunlight for growth, and some studies have used nitrogen- and phosphorus-rich water to promote algae cultivation. Similar large-scale methods have been employed by governments to treat areas with excessive nitrogen and pH levels, using carbon dioxide delivered through hollow fiber membranes to stimulate algae growth. The harvested algae were then dried, ground into powder, and processed to extract carbohydrates. Through a two-step fermentation process, these carbohydrates were converted into acids, with the second stage focusing on converting butyric acid into butanol. Cheng et al. (2015) investigated two methods of synthesizing biobutanol using *C. acetobutylicum* ATCC 824. They achieved a yield of 0.4 g/g glucose from algal oil fermentation, while a second method involving *Chlorella sorokiniana* CY1 biodiesel fermentation with *C. acetobutylicum* resulted in significantly higher biobutanol productivity at 3.86 g/L (Tong et al., 2020; Wang et al., 2021). This study provides a basis for using biodiesel as a precursor for biobutanol production, similar to catalytic cracking methods used to upgrade bio-oil. Onay (2020) also produced biodiesel and biobutanol from *Nannochloropsis gaditana* through ABE fermentation, yielding approximately 2.9 ± 0.24 g/L of biobutanol. Studies on *Chlorella vulgaris* JSC-6 demonstrated the potential of this algae species, with *C. acetobutylicum* ATCC 824 yielding 0.58 mol/mol sugar of biobutanol and biohydrogen (2.93 liter/liter) (Nematian et al., 2020; Wang et al., 2016). Nematian et al. (2020) also reported the production of 69.8 wt% biodiesel using *Rhizopus oryzae* lipase immobilized on magnetic Fe_3O_4 nanoparticles at 40°C and pH 7.5. These findings suggest that algae hold significant potential as feedstocks for biofuel production. Despite some variations in catalysts or reaction conditions, algae species generally produce more biodiesel compared to biobutanol, although biobutanol's fuel quality is considered superior. Studies related to first-, second-, and third-generation biobutanol production are summarized in Table 10.1.

10.5 INTENSIFICATION IN BIOBUTANOL PRODUCTION-INTEGRATED REACTOR DESIGNS

Biobutanol can be produced through three primary methods: batch, fed-batch, and continuous processes (Darkwah et al., 2018; Liu et al., 2018; Qureshi et al., 2008; Sarchami et al., 2016). Table 10.2 summarizes various reactor strategies employed for biobutanol generation. One of the main challenges in the production of biobutanol is selecting the most suitable reactor and the most economical separation technique. Each reactor type listed in Table 10.2 has distinct advantages and drawbacks. The fermentation of lignocellulosic and starch-based biomass can occur using both continuous and batch methods. Research has highlighted the benefits of continuous processes, noting their efficiency in time and yield (Pugazhendhi et al., 2019; Al-Shorgani et al., 2019). However, the ongoing nature of these operations

TABLE 10.2
Advantages and Disadvantages of Reactor Strategies

S.No.	Process	Advantages	Disadvantages
1.	Batch fermentation (Li et al., 2011; Al-Shorgani et al., 2018)	Effective and appropriate management	An unregulated pH may reduce the product yield
2.	Fibrous bed bioreactor (Bharathiraja et al., 2017; Li et al., 2019)	Minimizes cell loss brought on by immobilization	Typically limited to batch mode operation due to its user-friendly approach
3.	Fed-batch fermentation (Niglio et al., 2019)	This can be conducted when substrate inhibition and catabolite toxicity are anticipated	Compared to the continuous process, selectivity is low
4.	Continuous process (Pugazhendhi et al., 2019; Al-Shorgani et al., 2019)	Saves time and provides a favorable yield	Product recovery comes at a hefty cost

complicates the recovery of the final product. Insights into fed-batch fermentation have been provided by researchers, revealing that selectivity and low yield pose significant challenges for this approach (Li et al., 2011; Al-Shorgani et al., 2018; Niglio et al., 2019). Li et al. (2019) have examined the expected performance of a fibrous bed reactor in batch mode. Distillation remains the traditional method for separating biobutanol from the fermentation medium, and its simplicity is advantageous for this separation process (Chen et al., 2018). However, the creation of two immiscible liquid phases might prolong the recovery process, reducing economic viability (Hassankhan & Raisi, 2020). Adsorption is another basic separation method for butanol, where monitoring pressure and temperature is crucial for effective recovery (Abdehagh et al., 2017; Raganati et al., 2020). Rochón et al. (2020) and Zhu et al. (2018) have promoted gas stripping for butanol separation, with in situ methods being particularly attractive despite requiring higher biobutanol concentrations. The yield of butanol can be enhanced through liquid–liquid extraction using ionic liquids (Espinoza-Acosta et al., 2014; Zhu et al., 2020). Pervaporation is another favored separation technique, known for its simplicity and efficiency in a single-step process (Cabezas et al., 2021; Liu et al., 2014). The toxicity of biobutanol is mitigated through the membrane properties of perstractive separation, although this method generally exhibits a slower reaction rate compared to other techniques (Zheng et al., 2009). Currently, supercritical extraction is gaining attention for maximizing biobutanol recovery; however, its economic viability is challenged by the absence of standardized industrial procedures and the high costs associated with scaling up the technology (Manjare & Dhingra, 2019; Tompsett et al., 2018). Presently, integrated strategies that combine multiple separation methods are employed to enhance biobutanol extraction, with synergistic effects leading to better outcomes than single separation processes. Mariano et al. (2011) have explored the potential of flash fermentation for industrial applications, although the substantial financial investment required remains a significant drawback of the process.

10.6 BIOBUTANOL ANALYSIS

10.6.1 Spectroscopic Techniques for BioButanol Analysis

Spectroscopic methods such as FTIR, Raman, and NMR are highly effective for biobutanol analysis due to their non-destructive nature and ability to provide molecular insights. FTIR, for instance, identifies functional groups through absorption spectra, enabling the differentiation of butanol isomers. Raman spectroscopy complements this by offering high sensitivity in detecting specific bond vibrations. NMR spectroscopy further enhances analysis through chemical shift data, elucidating molecular structures even in complex mixtures. These techniques, when combined with chemometric approaches, improve quantitative and qualitative assessments significantly (Kharbach et al., 2023).

Protocol for FTIR analysis of butanol

- **Sample Preparation**: Dilute biobutanol in a suitable solvent (e.g., hexane) and place on an FTIR-compatible window.
- **Instrument Settings**: Set spectral range to 4000–400 cm^{-1} with a resolution of 4 cm^{-1}.
- **Data Acquisition**: Record background and sample spectra. Ensure baseline correction.
- **Analysis**: Identify characteristic peaks corresponding to hydroxyl and alkyl groups.

Protocol for Raman Spectroscopy analysis of butanol

- **Sample Setup**: Place the biobutanol sample in a quartz cuvette.
- **Laser Settings**: Use a 532-nm or 785-nm laser source, optimizing power to prevent thermal degradation.
- **Spectrum Collection**: Record Raman shifts from 100 cm^{-1} to 3500 cm^{-1}.
- **Data Processing**: Employ spectral smoothing and deconvolution to enhance resolution.

10.6.2 Colorimetric Assays for BioButanol Analysis

Colorimetric assays provide a cost-effective and simple approach for biobutanol quantification. By using reagents such as dichromate in acidic media, color intensity correlates with butanol concentration. This method is particularly suited for rapid field assessments. Despite its simplicity, the accuracy of colorimetric techniques can be improved with spectrophotometric calibration (Nimbalkar et al., 2019).

Protocol

- **Reagent Preparation**: Mix potassium dichromate and sulfuric acid to form the colorimetric reagent.

- **Sample Treatment**: Add a fixed volume of biobutanol to the reagent and incubate at room temperature.
- **Measurement**: Measure absorbance at 420 nm using a spectrophotometer.
- **Calibration**: Prepare standard curves with known butanol concentrations for quantitative analysis.

10.6.3 High-Performance Liquid Chromatography (HPLC) and Gas Chromatography (GC)

HPLC and GC are gold standards for biobutanol quantification and purity analysis. HPLC, equipped with a refractive index detector, is ideal for aqueous samples. GC, on the other hand, excels in separating butanol isomers using specialized capillary columns and flame ionization detection (FID). The robustness of these techniques makes them indispensable for industrial applications (Dou et al., 2021).

Protocol for HPLC analysis of butanol

- **Mobile Phase**: Use a mixture of water and acetonitrile (70:30 v/v) at a flow rate of 1 mL/min.
- **Column**: Employ a C18 column maintained at 25°C.
- **Injection**: Introduce a 10 μL sample for analysis.
- **Detection**: Monitor at 210 nm for peak identification and quantification.

Protocol for GC analysis of butanol

- **Sample Preparation**: Dilute biobutanol in ethanol or acetone.
- **Column**: Use a polar capillary column (e.g., polyethylene glycol).
- **Settings**: Set initial oven temperature to 50°C, ramping to 200°C at 10°C/min.
- **Detection**: Use FID for high sensitivity.

10.7 BIOBUTANOL AS A BIOFUEL

Biobutanol has four isomers, with isobutanol being the predominant form utilized in industrial and commercial settings (Algayyim et al., 2017). It can be blended with gasoline for engine use at concentrations of up to 12.5%. In 2005, tests were conducted using pure n-butanol as an alternative fuel in vehicles, revealing that fuel consumption was only about 10% higher than that of conventional gasoline. Despite this increase, the use of butanol significantly lowered emissions of carbon monoxide, hydrocarbons, and nitrogen oxides, demonstrating its beneficial environmental impact (Nilsson et al., 2020). During World War I, there was a pressing need for high acetone production, as it was a critical ingredient in the production of cordite, an alternative to gunpowder. After the war, large-scale production facilities for butanol were established in the United States and Canada to meet growing demand (Li et al., 2020). Following 1936, production facilities for ABE began to emerge in various countries, including China, the Soviet Union, Japan, Egypt, and South Africa. Japan started generating butanol from sugar factories in 1945, primarily to fuel military aircraft during World War II (Farmanbordar et al., 2018). The United Nations Environmental Agency permits a blend of 16% isobutanol with gasoline, which has been validated as effective (Wojcieszyk et al., 2021). Notably, a 10% ethanol blend performs similarly to a 16% isobutanol blend. Leading companies in biobutanol production include Butamax (which focuses on corn-based ethanol) and Gevo (known for its cellulosic methods, including biobutanol for commercial aircraft). In 2006, Kingston Research Ltd. was established as a joint venture by British Petroleum and DuPont, with a €25 million investment to advance biobutanol production technologies (Rome, 2019). Other companies, such as METabolic Explorer, Cobalt Tech, Green Biologics, and Tetravitae Bioscience, are also making strides in the field of biobutanol production (Schubert, 2020).

10.8 GLOBAL SCENARIO

In recent years, biofuels have gained considerable attention from both the public and the scientific community. Countries at various stages of development have shown their commitment by investing significantly to tackle energy challenges and mitigate global warming caused by environmental pollution (Bankar et al., 2013; Birgen et al., 2019; Visioli et al., 2014). Feedstocks for biofuel production include cellulose- and sugar-rich crops like sugarcane and beetroot, oil crops such as sunflower and palm, as well as starch sources like corn, all of which contain high levels of oil or sugar (Khalili & Amiri, 2020; Vivek et al., 2019; Ho et al., 2014; Rodionova et al., 2017; Alavijeh & Karimi, 2019; Hijosa-Valsero et al., 2020). Wealthy nations, particularly in the West, are focusing on the sourcing of these feedstocks and the adoption of modern, efficient conversion technologies. North America holds the largest market portion, followed by regions like Europe, Asia Pacific, and Latin America, with projections indicating a market value of USD 245.48 billion by 2027 (Raud et al., 2019). The promotion of biofuel development in North America is supported by significant incentives, legislation, and tax breaks from the government. Additionally, the infrastructure in countries like the United States and China plays a crucial role in expanding their market presence. European nations are also at the forefront of the biofuel market, having integrated biofuels into their transportation sectors and maintained relatively open regulations. Moreover, supportive legislation from developing countries such as India, China, and Indonesia is further driving the demand for biofuels. Figure 10.4 depicts the global biofuel consumption over the years and in the future (Statista, 2024). This graph shows the amount of biofuel consumed worldwide in million metric tons from 2000 to 2023 and a projection until 2030. With less than 50 million metric tons consumed, biofuel consumption was comparatively low beginning in 2000. But since 2000, consumption has been steadily rising, with a particularly significant increase in usage occurring between 2005 and 2010, when it almost doubled. The use of biofuels increased gradually once more between 2012 and 2016, after experiencing a brief plateau. Global consumption is projected to be slightly less than 200 million metric tons by 2023. According to the forecast, consumption will continue to rise and is predicted to top 225 million metric tons by 2030. The forecast period runs from 2024 to 2030.

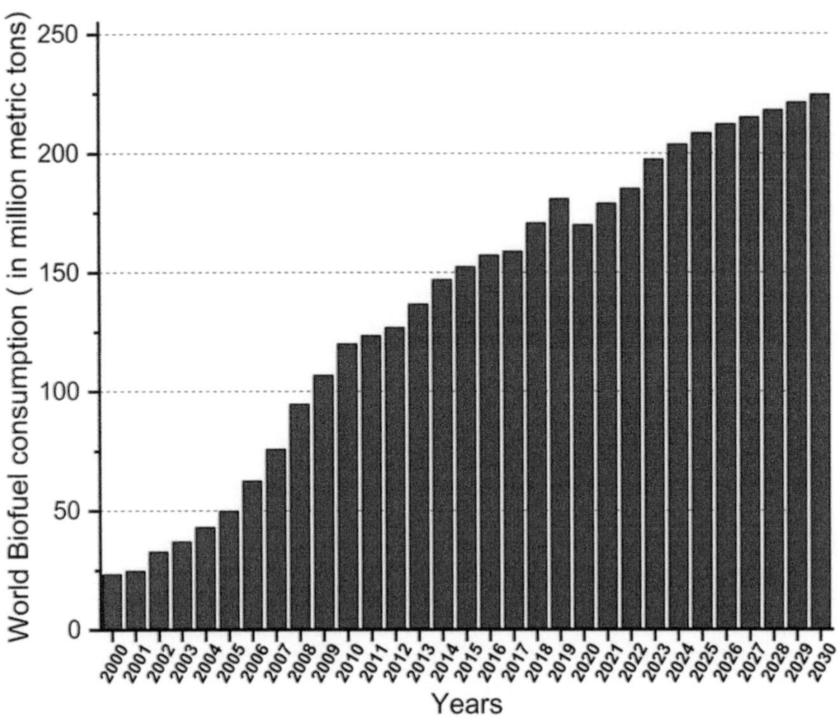

FIGURE 10.4 Projection of global biofuel consumption with years.

10.9 CONCLUSION

Biobutanol has emerged as a promising next-generation biofuel due to its superior fuel properties, such as higher energy content, reduced hygroscopicity, and better blending ability with gasoline compared to ethanol. The scope of biobutanol should extend beyond just its synthesis or use as a solvent; it has the potential to contribute to a pollution-free environment. Among various production methods, the acetone–butanol–ethanol (ABE) fermentation using *Clostridium* species remains the most studied and commercially relevant. Integrated approaches that include fermentation and separation at the same time can minimize recovery or separation step time consumption in the biobutanol manufacturing process. It is more cost-effective to focus on a single reaction for butanol production rather than generating multiple by-products, which can lead to higher production expenses. Furthermore, the valorization of lignocellulosic and industrial wastes as cost-effective feedstocks supports sustainable and circular biobutanol production. Analytical techniques such as GC, HPLC, and mass spectrometry have been critical in monitoring fermentation profiles and optimizing product recovery and purity. However, challenges such as low butanol tolerance in microbes, end-product inhibition, and substrate cost necessitate further metabolic engineering and process innovations. While significant progress has been made in the field of biobutanol production and analysis, economic feasibility and scalability remain major barriers to commercial implementation. To make biobutanol a commercial success, a collaborative effort involving biotechnologists, industry professionals, and academicians is necessary to overcoming these challenges and positioning biobutanol as a viable alternative to fossil fuels in the global bioenergy portfolio.

REFERENCES

Abdehagh, N., Dai, B., Thibault, J., & Tezel, F. H. (2017). Biobutanol separation from ABE model solutions and fermentation broths using a combined adsorption–gas stripping process. *Journal of Chemical Technology & Biotechnology*, 92, 245–251.

Abedini, A., Amiri, H., & Karimi, K. (2020). Efficient biobutanol production from potato peel wastes by separate and simultaneous inhibitors removal and pretreatment. *Renewable Energy*, 160, 269–277.

Abuhay, A., Mengie, W., Tesfaye, T., Gebino, G., Ayele, M., Haile, A., et al. (2021). Opportunities for new biorefinery products from Ethiopian ginning industry by-products: Current status and prospects. *Journal of Bioresources and Bioproducts*, 6, 195–214.

Alavijeh, M. K., & Karimi, K. (2019). Biobutanol production from corn stover in the US. *Industrial Crops and Products*, 129, 641–653.

Algayyim, S. J. M., Wandel, A. P., Yusaf, T., & Hamawand, I. (2017). The impact of n-butanol and iso-butanol as components of butanol-acetone (BA) mixture-diesel blend on spray, combustion characteristics, engine performance and emission in direct injection diesel engine. *Energy*, 140, 1074–1086.

Al-Shorgani, N. K. N., Al-Tabib, A. I., Kadier, A., Zanil, M. F., Lee, K. M., & Kalil, M. S. (2019). Continuous butanol fermentation of dilute acid-pretreated de-oiled rice bran by *Clostridium acetobutylicum* YM1. *Scientific Reports*, 9, 4622.

Al-Shorgani, N. K. N., Kalil, M. S., Yusoff, W. M. W., & Hamid, A. A. (2018). Impact of pH and butyric acid on butanol production during batch fermentation using a new local isolate of *Clostridium acetobutylicum* YM1. *Saudi Journal of Biological Sciences*, 25, 339–348.

Amenaghawon, A. N., Evbarunegbe, N. I., & Obahiagbon, K. J. C. E. (2021). Optimum biodiesel production from waste vegetable oil using functionalized cow horn catalyst: A comparative evaluation of some expert systems. *Clean Engineering and Technology*, 4, 100184.

Anandharaj, S. J., Gunasekaran, J., Udayakumar, G. P., Meganathan, Y., & Sivarajasekar, N. (2020). Biobutanol: Insight, production and challenges. In *Sustainable Development in Energy and Environment: Select Proceedings of ICSDEE 2019* (pp. 25–37).

Antoni, D., Zverlov, V. V., & Schwarz, W. H. (2007). Biofuels from microbes. *Applied Microbiology and Biotechnology*, 77, 23–35.

Awad, O. I., Mamat, R., Ali, O. M., Sidik, N., Yusaf, T., Kadirgama, K., et al. (2018). Alcohol and ether as alternative fuels in spark ignition engine: A review. *Renewable Energy*, 82, 2586–2605.

Balat, M. (2009). Recent trends in global production and utilization of bio-ethanol fuel. *Applied Energy*, 86, 2273–2282.

Bankar, S. B., Survase, S. A., Ojamo, H., & Granström, T. (2013). Biobutanol: The outlook of an academic and industrialist. *RSC Advances*, 3, 24734–24757.

Bharathiraja, B., Jayamuthunagai, J., Sudharsanaa, T., Bharghavi, A., Praveenkumar, R., Chakravarthy, M., et al. (2017). Biobutanol – An impending biofuel for future: A review on upstream and downstream processing techniques. *Renewable and Sustainable Energy Reviews*, 68, 788–807.

Birgen, C., Dürre, P., Preisig, H. A., & Wentzel, A. (2019). Butanol production from lignocellulosic biomass: Revisiting fermentation performance indicators with exploratory data analysis. *Biotechnology for Biofuels*, 12, 167.

Boonchuay, P., Techapun, C., Leksawasdi, N., Seesuriyachan, P., Hanmoungjai, P., Watanabe, M., et al. (2021). Bioethanol production from cellulose-rich corncob residue by the thermotolerant *Saccharomyces cerevisiae* TC-5. *Journal of Fungi*, 7, 547.

Branska, B., Fořtová, L., Dvořáková, M., Liu, H., Patakova, P., Zhang, J., et al. (2020). Chicken feather and wheat straw hydrolysate for direct utilization in biobutanol production. *Renewable Energy*, 145, 1941–1948.

Cabezas, R., Duran, S., Zurob, E., Plaza, A., Merlet, G., Araya-Lopez, C., et al. (2021). Development of silicone-coated hydrophobic deep eutectic solvent-based membranes for pervaporation of biobutanol. *Journal of Membrane Science*, 637, 119617.

Cai, D., Chen, C., Zhang, C., Wang, Y., Wen, H., & Qin, P. (2017). Fed-batch fermentation with intermittent gas stripping using immobilized *Clostridium acetobutylicum* for biobutanol production from corn stover bagasse hydrolysate. *Biochemical Engineering Journal*, 125, 18–22.

Chen, H., Cai, D., Chen, C., Wang, J., Qin, P., & Tan, T. (2018). Novel distillation process for effective and stable separation of high-concentration acetone–butanol–ethanol mixture from fermentation–pervaporation integration process. *Biotechnology for Biofuels*, 11, 286.

Cheng, H.-H., Whang, L.-M., Chan, K.-C., Chung, M.-C., Wu, S.-H., Liu, C.-P., et al. (2015). Biological butanol production from microalgae-based biodiesel residues by *Clostridium acetobutylicum*. *Bioresource Technology*, 184, 379–385.

Darkwah, K., Nokes, S. E., Seay, J. R., & Knutson, B. L. (2018). Mechanistic simulation of batch acetone–butanol–ethanol (ABE) fermentation with in situ gas stripping using Aspen Plus™. *Bioprocess and Biosystems Engineering*, 41, 1283–1294.

Dharmaraja, J., Shobana, S., Arvindnarayan, S., Vadivel, M., Atabani, A., Pugazhendhi, A., et al. (2020). Chapter 5 – Biobutanol from lignocellulosic biomass: Bioprocess strategies. In *Lignocellulosic Biomass to Liquid Biofuels* (pp. 169–193). Elsevier.

Dou, J., Chandgude, V., Vuorinen, T., Bankar, S., Hietala, S., & Lê, H. Q. (2021). Enhancing biobutanol production from biomass willow by pre-removal of water extracts or bark. *Journal of Cleaner Production*, 327, 129432.

Dürre, P. (2008). Fermentative butanol production: Bulk chemical and biofuel. *Annals of the New York Academy of Sciences*, 1125, 353–362.

Espinoza-Acosta, J. L., Torres-Chávez, P. I., Carvajal-Millán, E., Ramírez-Wong, B., Bello-Pérez, L. A., & Montaño-Leyva, B. (2014). Ionic liquids and organic solvents for recovering lignin from lignocellulosic biomass. *Bioresources*, 9, 3660–3681.

Etteh, C. C., Ibiyeye, A. O., Jelani, F. B., Rasheed, A. A., Ette, O. J., & Victor, I. (2021). Production of biobutanol using *Clostridia* spp. through novel ABE continuous fermentation of selected waste streams and industrial by-products. *Scientific African*, 12, e00744.

Farmanbordar, S., Amiri, H., & Karimi, K. (2018). Simultaneous organosolv pretreatment and detoxification of municipal solid waste for efficient biobutanol production. *Bioresource Technology*, 270, 236–244.

Fortman, J. L., Chhabra, S., Mukhopadhyay, A., Chou, H., Lee, T. S., Steen, E., et al. (2008). Biofuel alternatives to ethanol: Pumping the microbial well. *Trends in Biotechnology*, 26(7), 375–381.

Ge, S., Brindhadevi, K., Xia, C., Khalifa, A. S., Elfasakhany, A., Unpaprom, Y., et al. (2022). Enhancement of the combustion, performance and emission characteristics of spirulina microalgae biodiesel blends using nanoparticles. *Fuel*, 308, 121822.

Ge, S., Ganesan, R., Sekar, M., Xia, C., Shanmugam, S., Alsehli, M., et al. (2022). Blending and emission characteristics of biogasoline produced using CaO/SBA-15 catalyst by cracking used cooking oil. *Fuel*, 307, 121861.

Goel, V., Kumar, R., Nayyar, A., & Pathak, S. K. (2024). Solar drying systems for domestic/industrial purposes: A state-of-art review on topical progress and feasibility assessments. *Solar Energy*, 267, 112210.

Gottumukkala, L. D., Mathew, A. K., Abraham, A., & Sukumaran, R. K. (2019). Biobutanol production: Microbes, feedstock, and strategies. In *Biofuels: Alternative Feedstocks and Conversion Processes for the Production of Liquid and Gaseous Biofuels* (pp. 355–377). Elsevier.

Grand View Research. (2024, September 16). *Bio-butanol market size, share & trends analysis report by application*. https://www.grandviewresearch.com/industry-analysis/bio-butanol-industry

Hassankhan, B., & Raisi, A. (2020). Separation of isobutanol/water mixtures by hybrid distillation–pervaporation process: Modeling, simulation and economic comparison. *Chemical Engineering and Processing - Process Intensification*, 155, 108071.

Hönig, V., Pexa, M., Mařík, J., Linhart, Z., & Zeman, P. (2017). Biobutanol standardizing waste cooking oil as a biofuel. *Polish Journal of Environmental Studies*, 26, 69–78.

Ho, D. P., Ngo, H. H., & Guo, W. (2014). A mini review on renewable sources for biofuel. *Bioresource Technology*, 169, 742–749.

Hijosa-Valsero, M., Garita-Cambronero, J., Paniagua-García, A. I., & Díez-Antolínez, R. (2018). Biobutanol production from coffee silverskin. *Microbial Cell Factories*, 17, 154.

Hijosa-Valsero, M., Garita-Cambronero, J., Paniagua-García, A. I., & Díez-Antolínez, R. (2020). A global approach to obtain biobutanol from corn stover. *Renewable Energy*, 148, 223–233.

Huzir, N. M., Aziz, M. M. A., Ismail, S., Abdullah, B., Mahmood, N. A. N., Umor, N. A., et al. (2018). Agro-industrial waste to biobutanol production: Eco-friendly biofuels for next generation. *Renewable and Sustainable Energy Reviews*, 94, 476–485.

Ibrahim, M. F., Ramli, N., Bahrin, E. K., & Abd-Aziz, S. (2017). Cellulosic biobutanol by *Clostridia*: Challenges and improvements. *Renewable and Sustainable Energy Reviews*, 79, 1241–1254.

Jiang, Y., Lv, Y., Wu, R., Sui, Y., Chen, C., Xin, F., et al. (2019). Current status and perspectives on biobutanol production using lignocellulosic feedstocks. *Bioresource Technology Reports*, 7, 100245.

Kaushik, S., Pathak, S. K., Nayyar, A., & Goel, V. (2021). Biodiesel as a substitution for conventional diesel fuel: A comprehensive review. In *Advances in Mechanical Engineering: Select Proceedings of CAMSE 2020* (pp. 113–122). Springer.

Khalili, F., & Amiri, H. (2020). Integrated processes for production of cellulosic and hemicellulosic biobutanol from sweet sorghum bagasse using autohydrolysis. *Industrial Crops and Products*, 145, 111918.

Kharbach, M., Alaoui Mansouri, M., Taabouz, M., & Yu, H. (2023). Current application of advancing spectroscopy techniques in food analysis: Data handling with chemometric approaches. *Foods*, 12(14), 2753.

Kumar, R., Mehra, K. S., & Goel, V. (2024). A comprehensive assessment of working fluids selection in heat pipe-assisted battery thermal management systems: An integrated multi-criteria decision-making approach. *Proceedings of the Institution of Mechanical Engineers, Part A: Journal of Power and Energy*, 09576509241287668. https://doi.org/10.1177/09576509241287668

Lam, M. K., Khoo, C. G., & Lee, K. T. (2019). Chapter 19 – Scale-up and commercialization of algal cultivation and biofuels production. In *Biofuels from Algae (Second Edition)* (pp. 475–506). Elsevier.

Lea, B.-R., Yu, W.-B., & Min, H. (2018). Data visualization for assessing the biofuel commercialization potential within the business intelligence framework. *Journal of Cleaner Production*, 188, 921–941.

Li, J., Chi, X., Zhang, Y., & Wang, X. (2018). Enhanced coproduction of hydrogen and butanol from rice straw by a novel two-stage fermentation process. *International Biodeterioration & Biodegradation*, 127, 62–68.

Li, J., Du, Y., Bao, T., Dong, J., Lin, M., Shim, H., et al. (2019). *n*-Butanol production from lignocellulosic biomass hydrolysates without detoxification by *Clostridium tyrobutyricum* Δack-adhE2 in a fibrous-bed bioreactor. *Bioresource Technology*, 289, 121749.

Li, Q., Wu, M., Wen, Z., Jiang, Y., Wang, X., Zhao, Y., et al. (2020). Optimization of *n*-butanol synthesis in *Lactobacillus brevis* via the functional expression of *thl, hbd, crt,* and *ter. Journal of Industrial Microbiology & Biotechnology*, 47, 1099–1108.

Li, S., Huang, L., Ke, C., Pang, Z., & Liu, L. (2020). Pathway dissection, regulation, engineering and application: Lessons learned from biobutanol production by solventogenic *Clostridia. Biotechnology for Biofuels*, 13, Article 1.

Li, S.-Y., Srivastava, R., Suib, S. L., Li, Y., & Parnas, P. S. (2011). Performance of batch, fed-batch, and continuous A–B–E fermentation with pH-control. *Bioresource Technology*, 102(5), 4241–4250.

Liu, G., Wei, W., & Jin, W. (2014). Pervaporation membranes for biobutanol production. *ACS Sustainable Chemistry & Engineering*, 2, 546–560.

Liu, J., Liu, Z., & Guo, T. (2018). Repeated-batch fermentation by immobilization of *Clostridium beijerinckii* NCIMB 8052 in a fibrous bed bioreactor for ABE (acetone–butanol–ethanol) production. *Journal of Renewable and Sustainable Energy*, 10, 013101.

Liu, Y., Yuan, Y., Ramya, G., Singh, S. M., Chi, N. T. L., Pugazhendhi, A., … & Mathimani, T. (2022). A review on the promising fuel of the future – Biobutanol; the hindrances and future perspectives. *Fuel*, 327, 125166.

Manjare, S. D., & Dhingra, K. (2019). Supercritical fluids in separation and purification: A review. *Materials Science for Energy Technologies*, 2, 463–484.

Mariano, A. P., Keshtkar, M. J., Atala, D. I., Maugeri Filho, F., Wolf Maciel, M. R., & Maciel Filho, R. (2011). Energy requirements for butanol recovery using the flash fermentation technology. *Energy & Fuels*, 25, 2347–2355.

Mehra, K. S., & Goel, V. (2025a). Energy, exergy, environmental and sustainability assessment of renewable *n*-octanol biofuel on a compression ignition engine. *Biomass and Bioenergy*, 201, 108116.

Mehra, K. S., & Goel, V. (2025b). Unveiling next-generation biodiesel technologies: Techno-economic and energy breakthroughs for sustainable production. *Biomass and Bioenergy*, 199, 107910.

Mehra, K. S., Abrar, I., Bhatia, R. K., & Goel, V. (2025). A comprehensive review of algae consortium for wastewater bioremediation and biodiesel production. *Energy Conversion and Management*, 325, 119428.

Mehra, K. S., Goel, V., & Kumar, R. (2024). An integrated multi-attribute decision framework for sustainability assessment of renewable diesel fuel production pathways. *Energy Conversion and Management*, 309, 118461.

Mehra, K. S., Pal, J., & Goel, V. (2023b). A comprehensive review on the atomization and spray characteristics of renewable biofuels. *Sustainable Energy Technologies and Assessments*, 56, 103106.

Mehra, K. S., & Pant, G. (2021). Production of biofuel from sesame oil and its characterization as an alternative fuel for diesel engine. In *IOP Conference Series: Materials Science and Engineering* (Vol. 1020, p. 012076). IOP Publishing.

Mehra, K. S., Shah, M. P., & Singh, S. (2014). Use of non-edible vegetable oils as fuel for diesel engine: A review. *International Journal of Research in Applied Science and Engineering Technology*, 2(3), 6–11.

Mehra, K. S., et al. (2021). Performance, energy, emission and cost analysis of *Jatropha curcas* oil as a biofuel for compression ignition engine. *Materials Today: Proceedings*, 43, 348–354.

Mehra, K. S., et al. (2023a). Experimental investigation of emission characteristics of CI engine using biodiesel-diesel blends and best fuel selection: An AHP-TOPSIS approach. *Materials Today: Proceedings*.

Menshhein, G., Costa, V., Chiarello, L. M., Scharf, D. R., Simionato, E. L., Botton, V., et al. (2019). Concentration of renewable products of crude bio-oil from thermal cracking of the methyl esters in castor oil. *Renewable Energy*, 142, 561–568.

Nai Yuh Yek, P., Adibah Wan Mahari, W., Huat Kong, S., Ying Foong, S., Peng, W., Ting, H., et al. (2022). Pilot-scale co-processing of lignocellulosic biomass, algae, shellfish waste via thermochemical approach: Recent progress and future directions. *Bioresource Technology*, 347, 126687.

Narueworanon, P., Laopaiboon, L., Phukoetphim, N., & Laopaiboon, P. (2020). Impacts of initial sugar, nitrogen and calcium carbonate on butanol fermentation from sugarcane molasses by *Clostridium beijerinckii*. *Energies*, 13, 694.

Ndaba, B., Chiyanzu, I., & Marx, S. (2015). *n*-Butanol derived from biochemical and chemical routes: A review. *Biotechnology Reports*, 8, 1–9.

Nematian, T., Salehi, Z., & Shakeri, A. (2020). Conversion of bio-oil extracted from *Chlorella vulgaris* microalgae to biodiesel via modified superparamagnetic nanobiocatalyst. *Renewable Energy*, 146, 1796–1804.

Niglio, S., Marzocchella, A., & Rehmann, L. (2019). Clostridial conversion of corn syrup to acetone–butanol–ethanol (ABE) via batch and fed-batch fermentation. *Heliyon*, 5, e01401.

Nilsson, A., Shabestary, K., Brandão, M., & Hudson, E. P. (2020). Environmental impacts and limitations of third-generation biobutanol: Life cycle assessment of *n*-butanol produced by genetically engineered cyanobacteria. *Journal of Industrial Ecology*, 24, 205–216.

Nimbalkar, P. R., Khedkar, M. A., Chavan, P. V., & Bankar, S. B. (2019). Enhanced biobutanol production in folic acid-induced medium by using *Clostridium acetobutylicum* NRRL B-527. *ACS Omega*, 4(7), 12978–12982.

Onay, M. (2020). Enhancing carbohydrate productivity from *Nannochloropsis gaditana* for bio-butanol production. *Energy Reports*, 6, 63–67.

Pathak, S. K., Nayyar, A., & Goel, V. (2021). Optimization of EGR effects on performance and emission parameters of a dual fuel (Diesel + CNG) CI engine: An experimental investigation. *Fuel*, 291, 120183.

Pugazhendhi, A., Mathimani, T., Varjani, S., Rene, E. R., Kumar, G., Kim, S.-H., et al. (2019). Biobutanol as a promising liquid fuel for the future – Recent updates and perspectives. *Fuel*, 253, 637–646.

Qureshi, N., Saha, B. C., Hector, R. E., Hughes, S. R., & Cotta, M. A. (2008). Butanol production from wheat straw by simultaneous saccharification and fermentation using *Clostridium beijerinckii*: Part I—Batch fermentation. *Biomass and Bioenergy*, 32, 168–175.

Raganati, F., Procentese, A., Olivieri, G., Russo, M. E., Salatino, P., Marzocchella, A., et al. (2020). Bio-butanol recovery by adsorption/desorption processes. *Separation and Purification Technology*, 235, 116145.

Raud, M., Kikas, T., Sippula, O., & Shurpali, N. J. (2019). Potentials and challenges in lignocellulosic biofuel production technology. *Renewable and Sustainable Energy Reviews*, 111, 44–56.

Rochón, E., Cortizo, G., Cabot, M. I., Cubero, M. T. G., Coca, M., Ferrari, M. D., et al. (2020). Bioprocess intensification for isopropanol, butanol and ethanol (IBE) production by fermentation from sugarcane and sweet sorghum juices through a gas stripping–pervaporation recovery process. *Fuel*, 281, 118593.

Rodionova, M. V., Poudyal, R. S., Tiwari, I., Voloshin, R. A., Zharmukhamedov, S. K., Nam, H. G., et al. (2017). Biofuel production: Challenges and opportunities. *International Journal of Hydrogen Energy*, 42, 8450–8461.

Rome, A. (2019). DuPont and the limits of corporate environmentalism. *Business History Review*, 93, 75–99.

Sarchami, T., Munch, G., Johnson, E., Kießlich, S., & Rehmann, L. (2016). A review of process-design challenges for industrial fermentation of butanol from crude glycerol by non-biphasic *Clostridium pasteurianum*. *Fermentation*, 2(3), 13.

Schubert, T. (2020). Production routes of advanced renewable C1 to C4 alcohols as biofuel components – A review. *Biofuels, Bioproducts and Biorefining*, 14(4), 845–878.

Singh, P., Chauhan, S. R., & Goel, V. (2018). Assessment of diesel engine combustion, performance and emission characteristics fuelled with dual fuel blends. *Renewable Energy*, 125, 501–510.

Singh, P., Goel, V., & Nayyar, A. (2020). Enhancing diesel engine performance and reducing emissions using binary biodiesel fuel blend. *Journal of Energy Resources Technology*, 142(1), 012201.

Singh, P., Nayyar, A., & Goel, V. (2019). Binary biodiesel blend endurance characteristics in a compression ignition engine. *Journal of Energy Resources Technology*, 141(3), 032204.

Statista. (2024). Global biofuel consumption volume 2000–2030. *Statista*. https://www.statista.com

Su, C., Qi, L., Cai, D., Chen, B., Chen, H., Zhang, C., et al. (2020). Integrated ethanol fermentation and acetone–butanol–ethanol fermentation using sweet sorghum bagasse. *Renewable Energy*, 162, 1125–1131.

Tompsett, G. A., Boock, J. T., DiSpirito, C., Stolz, E., Knutson, D. R., Rivard, A. G., et al. (2018). Extraction rate and energy efficiency of supercritical carbon dioxide recovery of higher alcohols from dilute aqueous solution. *Energy Technology*, 6(4), 683–693.

Tong, X., Li, Q., Chen, W., & Zhao, L. (2020). Alkali extraction of xylan from poplar sawdust and preparation of xylooligosaccharide by enzymatic hydrolysis. *Journal of Forestry Engineering*, 5(3), 61–68.

Tsai, T.-Y., Lo, Y.-C., Dong, C.-D., Nagarajan, D., & Chang, J.-S. (2020). Biobutanol production from lignocellulosic biomass using immobilized *Clostridium acetobutylicum*. *Applied Energy*, 277, 115531.

Valles, A., Álvarez-Hornos, J., Capilla, M., San-Valero, P., & Gabaldón, C. (2021). Gas stripping and pervaporation hybrid system for biobutanol separation. *Bioresource Technology*, 342, 126020.

Visioli, L. J., Enzweiler, H., Kuhn, R. C., Schwaab, M., & Mazutti, M. A. (2014). Recent advances on biobutanol production. *Sustainable Chemical Processes*, 2, 15.

Vivek, N., Nair, L. M., Mohan, B., Nair, S. C., Sindhu, R., Pandey, A., et al. (2019). Bio-butanol production from rice straw – Recent trends, possibilities and challenges. *Bioresource Technology Reports*, 7, 100224.

Wackett, L. P. (2008). Biomass to fuels via microbial transformations. *Current Opinion in Chemical Biology*, 12(2), 187–193.

Wang, F., Xu, J., Jiang, J., Liu, P., Li, F., Ye, J., et al. (2018). Hydrotreatment of vegetable oil for green diesel over activated carbon supported molybdenum carbide catalyst. *Fuel*, 216, 738–746.

Wang, R., Wang, J., Lai, C., Huang, C., Lin, Z., & Yong, Q. (2021). Enzymatic preparation of glucomannan oligosaccharides and their effects on intestinal microbial proliferation. *Journal of Forestry Engineering*, 6(3), 96–103.

Wang, X., Zhang, Y., Karthikeyan, P., Boomadevic, P., Maroušek, J., Nasif, O., et al. (2022). Role of injection pressure on fuel atomization and spray penetration on the *Thevetia peruviana* and *Jatropha curcas* biodiesel blends with nanoparticle. *Fuel*, 324, 124527.

Wang, Y., Guo, W., Cheng, C. L., Ho, S.-H., Chang, J.-S., & Ren, N. (2016). Enhancing biobutanol production from biomass of *Chlorella vulgaris* JSC-6 with sequential alkali pretreatment and acid hydrolysis. *Bioresource Technology*, 200, 557–564.

Wojcieszyk, M., Knuutila, L., Kroyan, Y., de Pinto, B. M., Tripathi, R., Keskivali, J., et al. (2021). Performance of anisole and isobutanol as gasoline bio-blendstocks for spark ignition engines. *Sustainability*, 13(15), 8729.

Xin, F., Wang, C., Dong, W., Zhang, W., Wu, H., Ma, J., et al. (2016). Comprehensive investigations of biobutanol production by a non-acetone and 1,3-propanediol generating *Clostridium* strain from glycerol and polysaccharides. *Biotechnology for Biofuels*, 9, 220.

Xue, C., Zhao, J., Chen, L., Yang, S.-T., & Bai, F. (2017). Recent advances and state-of-the-art strategies in strain and process engineering for biobutanol production by *Clostridium acetobutylicum*. *Biotechnology Advances*, 35, 310–322.

Yadav, S., Rawat, G., Tripathi, P., & Saxena, R. K. (2014). A novel approach for biobutanol production by *Clostridium acetobutylicum* using glycerol: A low-cost substrate. *Renewable Energy*, 71, 37–42.

Yan, J., & Lin, T. (2009). Biofuels in Asia. *Applied Energy*, 86, S1–S10.

Zhang, Y., Xia, C., Lu, M., & Tu, M. (2018). Effect of overliming and activated carbon detoxification on inhibitors removal and butanol fermentation of poplar prehydrolysates. *Biotechnology for Biofuels*, 11, 178.

Zhao, Z., Wu, H., Wang, M., Lee, C.-F., Liu, J., & Fu, J. (2015). Computational investigation of oxygen concentration effects on a soot mechanism with a phenomenological soot model of acetone–butanol–ethanol (ABE). *Energy & Fuels*, 29(3), 1710–1721.

Zheng, Y.-N., Li, L.-Z., Xian, M., Ma, Y.-J., Yang, J.-M., Xu, X., et al. (2009). Problems with the microbial production of butanol. *Journal of Industrial Microbiology & Biotechnology*, 36(9), 1127–1138.

Zhou, A., & Thomson, E. (2009). The development of biofuels in Asia. *Applied Energy*, 86, S11–S20.

Zhu, C., Chen, L., Xue, C., & Bai, F. (2018). A novel close-circulating vapor stripping–vapor permeation technique for boosting biobutanol production and recovery. *Biotechnology for Biofuels*, 11, 128.

Zhu, Z., Li, H., Xu, Y., Zhang, W., Shen, Y., & Gao, J. (2020). Quantum chemical calculation, molecular dynamics simulation and process design for separation of heptane–butanol using ionic liquids extraction. *Journal of Molecular Liquids*, 316, 113851.

11 Biodiesel
Analytical Insights and Perspectives

*Naimi Sirjohn#, Ishita Chanana#, Duni Chand, and Pradeep Kumar**

11.1 INTRODUCTION

Biodiesel is an alternative diesel fuel derived from renewable biological materials like vegetable oils and animal fats. It is non-toxic, biodegradable, and produces fewer emissions, making it an environmentally friendly option (Knothe, 2010). With the rise of inexpensive petroleum, suitable crude oil fractions were refined to be used as fuel, leading to the co-evolution of diesel fuels and diesel engines. During the 1930s and 1940s, vegetable oils occasionally served as diesel fuels, but this was typically reserved for emergency situations. Vegetable oils and animal fats were explored as potential diesel fuels long before the energy crises of the 1970s (Murugesan et al., 2009) and early 1980s reignited interest in alternative fuel sources. The significant rise in the number of automobiles in recent years has led to a high demand for petroleum products (Jaichandar & Annamalai, 2011). With crude oil reserves expected to last only a few more decades, there is an urgent search for alternative fuels (Huang et al., 2012). The dwindling supply of crude oil could severely affect the transportation sector. Among the various alternative fuels being considered, biodiesel, which is produced from vegetable oils, stands out as the most promising substitute for conventional diesel fuel. The depletion of petroleum-based fuels and growing environmental concerns have led to the exploration of biofuels as alternative energy sources (Hassan & Kalam, 2013). However, a complete replacement of petroleum fuels with biofuels is impractical due to limitations in production capacity and engine compatibility. Nevertheless, partially substituting diesel with biofuel can help slow down the exhaustion of petroleum (Aniza et al., 2024) resources and mitigate the severe climate impacts caused by automotive emissions. Biodiesel has emerged as a promising alternative to fossil fuels for diesel engines, gaining importance due to the environmental impact of petroleum-based diesel (Rodionova et al., 2017) engines and the dwindling petroleum reserves. The two most widely used types of biofuels are biodiesel and bioethanol, which are primarily derived from vegetable oils, seeds, and lignocellulosic materials. Biodiesel serves as a substitute for diesel fuel, while bioethanol can replace petrol. Common feedstocks for biodiesel

# Equal Contribution of the Authors
* Corresponding author

DOI: 10.1201/9781003485636-11

production include plant oils like rapeseed, soybean, sunflower, and palm, as well as non-edible oils such as Mahua, Neem, and Jatropha (Hassan & Kalam, 2013). Additionally, animal fats like beef tallow and used cooking oil can be refined into biodiesel. Emerging sources like algae are considered the third generation of biofuels. Biodiesel itself does not contain petroleum but can be blended with No. 2 diesel fuel in any ratio to be used in diesel engines with minimal or no modifications (Powar et al., 2022). In contrast, bioethanol is primarily produced through the fermentation of sugar-rich crops like sugarcane, corn, wheat, maize, and potatoes. Both developing and developed countries view biofuels as essential technologies for several compelling reasons (Demirbas, 2008). Bioethanol, a substitute or additive for petrol, can potentially be produced economically from various sources, including wood, straw, or even household waste. On the other hand, biodiesel offers an eco-friendly substitute to traditionally used diesel fuel, which is compatible with any diesel engine without any modification (Ismail et al., 2024). Compared to conventional diesel, biodiesel has several advantages, including lower sulfur content, a higher flash point, reduced aromatic content, and greater biodegradability. Efficient utilization of biodiesel for energy purposes can provide substantial benefits to the environment and local communities, such as job creation (Renzaho et al., 2017), access to modern energy sources in rural areas, mitigation of urban migration, and reduction of CO_2 and sulfur emissions in the atmosphere.

While biofuels offer several advantages, they also come with notable drawbacks. One significant issue is their lower energy output compared to traditional fossil fuels, which means that more biofuel is needed to produce the same amount of energy (Deora et al., 2022). Additionally, the production of biofuels is not entirely carbon neutral. Although the combustion of biofuels tends to be cleaner, studies have shown that the overall carbon footprint can be substantial when considering the entire production process (Selvarajan et al., 2021). This includes the emissions generated by the machinery used in cultivating the crops and plants required for biofuel production. While biofuels offer several advantages, they also come with notable drawbacks. This includes the emissions generated by the machinery used in cultivating the crops and plants required for biofuel production (Hanaki & Portugal-Pereira, 2018). Furthermore, the financial aspect of biofuel poses a challenge. Substantial initial investments are necessary to improve the energy efficiency of biofuels and establish the infrastructure required for large-scale production (Torkashvand et al., 2022). This high cost can be a barrier to expanding biofuel usage and refining processes to enhance their efficiency.

11.2 RAW MATERIAL FOR BIODIESEL PRODUCTION

Biofuels are categorized into four generations first, second, third, and fourth based on their sources and methods of production, as illustrated in Figure 11.1. Extensive research has been conducted on all four generations, focusing primarily on the resources used and the production processes of various biomaterials. Numerous studies have explored different types of biofuels, aiming to improve efficiency and refine extraction methods.

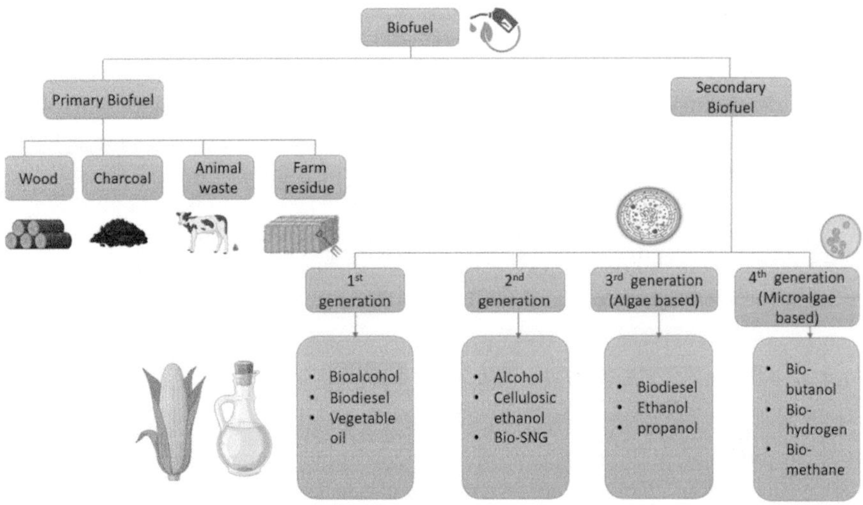

FIGURE 11.1 The four generations of biofuels can be categorized on the basis of their origin of production which includes algae or microalgae.

11.2.1 Types of Feedstocks: Edible and Non-Edible Oils

Over the past few decades, fossil fuels, including petroleum, natural gas, and coal, has been the primary. However, these resources are nonrenewable and are expected to be depleted in the foreseeable future. There is an immediate need to identify a new energy source that is renewable, environmentally friendly, reliable, and cost-effective to replace fossil fuels (Al Mubarak et al., 2024). In this context, biodiesel derived from vegetable oils has recently emerged as a viable alternative to petroleum-based diesel fuel for use in diesel engines. Biodiesel, a mono-alkyl ester made from plant or animal oils, has properties similar to those of petroleum-derived diesel fuel. Currently, approximately 84% of global biodiesel production comes from rapeseed oil, with sunflower oil contributing 13%, palm oil 1%, and soybean oil and others making up the remaining 2% (Ershov et al., 2023). Since over 95% of biodiesel production relies on edible oils, concerns have been raised about potential problems. Converting edible oils into biodiesel essentially transforms food resources into fuel for vehicles. It is believed that large-scale biodiesel production from edible oils could disrupt the global balance between food supply and demand (Shabbir et al., 2023). It has been argued that the large-scale expansion of oil crop plantations for biodiesel production has led to deforestation in countries like Malaysia, Indonesia, and Brazil, as more forests are cleared to make way for these plantations. Additionally, the distinction between food and fuel economies has become increasingly blurred, with both sectors competing for the same oil resources (Stergiou, 2023). To prevent such challenges, non-edible oil seeds are recommended for the commercial production of biodiesel. Numerous researchers have begun investigating the use of low-cost, non-edible oils as alternative feedstock for biodiesel. Among these, castor and Jatropha seeds, as well as microalgae oil, have shown great promise as reliable sources due to their high oil content (Gashaw & Lakachew, 2014). Algae or microalgae, in particular, offer an economically viable option for biodiesel production because of their abundance and low cost. Algal blooms can develop in freshwater, salt water sediment (Gaurav et al., 2023), and even within the atmosphere. Microalgae-based biodiesel production on the other hand, presents a cutting-edge renewable energy solution that harnesses the potential of microscopic, single-celled organisms known as microalgae. These organisms convert sunlight, carbon dioxide, and nutrients into organic compounds like lipids (Singha et al., 2023), which can subsequently be processed into biodiesel. A major advantage of microalgae is their ability to be cultivated on non-arable land, eliminating competition with food crops for resources.

11.2.2 Biodiesel from Waste Oil and Animal Fats

Efforts toward achieving carbon neutrality in response to global warming driven by environmental pollution are prompting governments to implement stricter air quality regulations. Notably, toxic emissions from engine combustion in automobiles are identified as major contributors to global environmental degradation among various pollutants (Ogunkunle & Ahmed, 2021). The choice of oil and fat-based raw materials is particularly critical, as they account for approximately 60–80% of the overall biodiesel production cost (Alsultan et al., 2021). Consequently, selecting the most suitable feedstocks is essential, as these are influenced by factors such as geographical location, climate, and agricultural conditions. Fatty acid composition varies across animal species but typically includes common types, most of which have carbon chains ranging from 16 to 18 carbons (Burdge & Calder, 2015). The most notable fatty acids are palmitic (16:0) and stearic (18:0) acids, classified as saturated fatty acids (SFAs); oleic acid (18:1), a monounsaturated fatty acid (MUFA); and linoleic (18:2) and arachidonic (20:4) acids, which are polyunsaturated fatty acids (PUFA) as shown in Figure 11.2. Due to the high saturated fatty acid content (approximately 40% SFA), fats from ruminants and pigs are predominantly solid, whereas chicken fat, containing about 30–33% SFA, tends to be semi-solid (Matsungo & Siziba, 2020) or liquid. Therefore, it is important to note that raw animal fats are generally solid at ambient temperatures and require preheating to 45°C for use as fuel in diesel engines. Additionally, the high saturated fatty acid content typically yields biodiesel with greater stability and higher cetane numbers, often exceeding 50 in the case of lard and tallow (Cernat et al., 2020). Pretreatment is necessary because feedstocks such as animal fats typically have a high content of free fatty acids (FFA) and water, which can lower biodiesel yield and increase production costs due to the challenges in separation and purification (Pydimalla et al., 2023). Biodiesel is synthesized via the transesterification of fats with short-chain alcohols in the presence of a catalyst. Various catalysts are employed in biodiesel production, with the most common being alkalis (Mukherjee et al., 2024) (e.g., sodium hydroxide, sodium methoxide, potassium hydroxide, potassium methoxide, sodium amide, sodium hydride, potassium amide, and potassium hydride), acids (e.g., sulfuric, phosphoric, and hydrochloric acids, or organic sulfonic acids), heterogeneous catalysts such as enzymes (lipases), and more complex catalysts including silicates, zirconias, and nanocatalysts (Raising Rathod et al., 2024). Additionally, the rate of

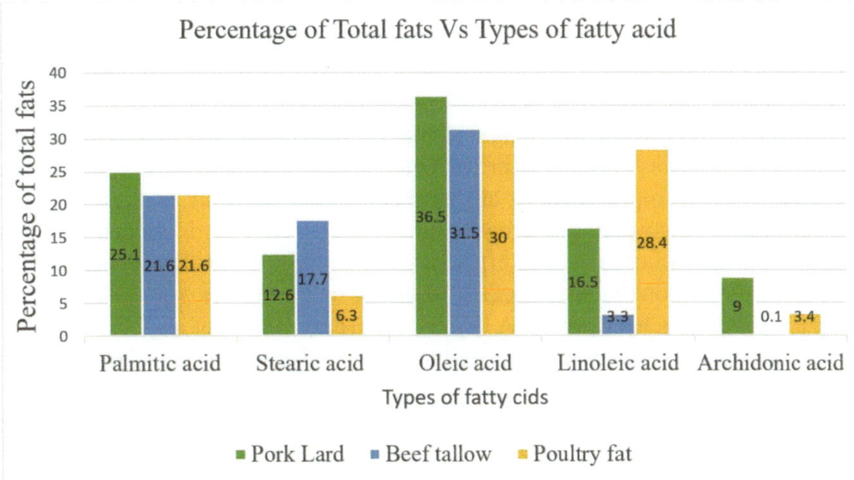

FIGURE 11.2 Graph of percentage of total fats in comparison of types of fatty acids.

transesterification reactions in methanol was found to be similar to those achieved with the conventional potassium methoxide catalyst under identical reaction conditions. Methanol is the most commonly used short-chain alcohol in commercial transesterification reactions due to its low cost, though other alcohols such as ethanol, propanol, and butanol can be utilized (Tanwar et al., 2023). Transesterification involves the stepwise conversion of triacylglycerols to diacylglycerols, releasing a fatty acid in each step. Diacylglycerols are then converted to monoacylglycerols, followed by the conversion of monoacylglycerols to glycerol, each step releasing an additional fatty acid (Mardani et al., 2024). Generally, transesterification is characterized by high conversion efficiency and low cost. However, after pretreatment, the efficiency of the reaction is influenced by several factors, including reaction time, temperature, the type and molar ratio of alcohol, the catalyst type and quantity, water content in the reaction medium, fatty acid composition, and the release of free fatty acids into the reaction medium (Kibar et al., 2023).

11.2.3 Algae as a Source of Biodiesel

Algae-based technologies present a valuable approach to mitigating greenhouse gas emissions from coal-fired plants and other carbon-intensive industries. Microalgae, which grow quickly and absorb large amounts of carbon dioxide, can produce more oil per acre than any other biodiesel feedstock (Demirbas & Fatih Demirbas, 2011), and they can be grown on non-arable land. Algae, which are simple microscopic organisms, primarily inhabit aquatic environments. Microalgae are single-celled photosynthetic organisms found in freshwater or saline environments, where they utilize sunlight, water, and carbon dioxide to generate biomass (Jaiswal et al., 2021). Algae are divided into four main categories: diatoms, green algae, blue-green algae, and golden algae. These organisms typically exist as either filamentous or phytoplankton algae, with phytoplankton known for their ability to rapidly multiply and cause algae blooms. The production of biodiesel from microalgae (Saber et al., 2022) demands substantial amounts of algal biomass. Currently, both macroalgae and microalgae are primarily utilized in food products, animal feed, aquaculture feed, and as biofertilizers. Microalgal biomass is often dried and sold in the human health food market as powders or pressed into tablets. The majority of algal species are obligate phototrophs (Patil et al., 2005), meaning they rely on light for their growth. These phototrophic microalgae are typically cultivated in open ponds or photobioreactors, where they can efficiently utilize light for photosynthesis. While the exact composition (Narayanan, 2024) depends on the species, some algae consist of up to 40% fatty acids by mass. A key distinguishing feature of algal oil is its high yield, making it especially efficient for biodiesel production. Algae utilized in biodiesel production are typically unicellular aquatic green algae. These photosynthetic eukaryotes are known for their rapid growth and ability to achieve high population densities, such as *Chlorophyceae* sp. (Mathushika & Gomes, 2022). Under optimal conditions, green algae can double their biomass within 24 hrs. Moreover, they can accumulate significant lipid content, often exceeding 50%. This combination of high biomass yield and lipid concentration makes them well-suited for intensive cultivation and a promising resource for biodiesel production (Osman et al., 2023). The annual oil yield and productivity of algae far surpass that of traditional seed crops. For instance, soybeans produce approximately 450 L of oil/hectare, while canola yields about 1,200 liters, and palm oil generates 6,000 liters per hectare. In contrast, algae can produce an impressive 90,000 liters of oil per hectare, making them a significantly more efficient source of oil (Demirbas & Fatih Demirbas, 2011).

11.3 PRODUCTION METHODS OF BIODIESEL

Producing microalgal biodiesel necessitates substantial amounts of algal biomass. At present, both macroalgae and microalgae are predominantly employed in food production, animal and aquaculture feed, and biofertilizers (Yaamini et al., 2023). Dried microalgal biomass is commonly sold in the health food industry in powder form or as pressed tablets. Furthermore, aquatic biomass holds promise as a raw material for electricity generation through co-firing, liquid fuel (bio-oil) production via pyrolysis, and biomethane generation (Pandey & Sharma 2024) through fermentation. Algae can thrive in nearly any environment with sufficient sunlight, and some species can even grow in saline water. They contain varying proportions of proteins, carbohydrates, lipids, and nucleic acids (Sarkar et al., 2024). Microalgae, being the fastest-growing photosynthetic organisms, can complete a growth cycle in just a few days. Diatom algae, for example, can produce approximately 46 tons of oil per hectare per year. Oil yields vary across algae species, with some producing up to 50% (Ismail et al., 2024) of their biomass as oil. As highly efficient converters of solar energy, microalgae can generate a diverse range of metabolites. Algal cultures typically yield 30–50% oil, with theoretical projections suggesting that oil production from algae could range between 47,000 and 308,000 liters per hectare annually (Demirbas & Fatih Demirbas, 2011). Biodiesel is a type of biofuel consisting of mono-alkyl esters derived from vegetable oils or animal fats. It is produced through a chemical process called transesterification, which converts the parent oil or fat into fatty esters, reducing its viscosity to levels similar to petro-diesel. This process typically results in methyl esters, which form the primary component of biodiesel (Tesfaye et al., 2016) and can be sourced from both plant and animal oils. The reaction for the same can be given as

$$\text{Triglycerides} + 3\ \text{methanol Glycerine} + 3\ \text{methyl esters (Biodiesel)}$$

11.3.1 Transesterification

Transesterification, also known as alcoholysis, is a chemical reaction where an ester undergoes transformation by replacing its original alcohol component with a different alcohol, akin to hydrolysis but involving alcohol instead of water (Kayode & Hart, 2019). This reaction is commonly employed to decrease the viscosity of triglycerides. While transesterification is a reversible process that can occur simply by mixing the reactants, the use of catalyst-typically strong acids (Tan et al., 2019) or bases can expedite the reaction. The process yields fatty acid alkyl esters and glycerol, with the latter settling at the bottom of the reaction vessel (Thakur & Kundu, 2016). Intermediate compounds such as diglycerides and monoglycerides are formed during the reaction. The transesterification process takes place in multiple (Ejikeme et al., 2010) reversible steps. To push the equilibrium toward ester production, a small surplus of alcohol is used. When alcohol is in excess, the forward reaction behaves as pseudo-first-order, whereas the reverse reaction follows second-order kinetics (Trejo-Zárraga et al., 2018). It has also been observed that using an alkali catalyst speeds up the transesterification reaction significantly. Initially, the alkoxide ion attacks the carbonyl carbon of the triglyceride molecule, forming a tetrahedral intermediate. In the next step, this intermediate reacts with alcohol, regenerating the alkoxide ion (Jothiramalingam & Wang, 2009). Finally, the rearrangement of the tetrahedral intermediate leads to the production of an ester and a diglyceride. Alkali catalyzed the transesterification mechanism as shown in the illustration below:

$$\underset{\text{Ester}}{RCOOR^1} + \underset{\text{Alcohol}}{R^2OH} \overset{\text{Catalyst}}{\leftrightarrow} \underset{\text{Ester}}{RCOOR^2} + \underset{\text{Alcohol}}{R^1OH}$$

Transesterification can also be catalyzed by Bronsted acids, with sulfonic and sulfuric acids being the preferred choices (Marchetti et al., 2007). While these catalysts result in high yields of alkyl esters, the reactions proceed slowly, typically requiring temperatures above 100°C and more than 3 hrs to achieve full conversion. Protocol for the production of biodiesel from microalgae using the transesterification process is given below. This protocol describes the transesterification of 1 g of microalgal oil to produce fatty acid methyl esters (FAME), followed by purification and production of biodiesel:

- **Microalgae used:** the species used for the extraction are *Clamydomonas sp., Spirulina sp., Chlorella sp., Anabaena cylindrica, Nostoc muscorum, and Spirulina maxima.*

- **Setup:** The reaction was conducted in 50 cc airtight glass vials at 60°C, stirred for 2 min every 30 min using a cyclomixer.
- **Variables Studied:** The methanol-to-oil ratios tested were 30:1, 45:1, 60:1, 75:1, and 90:1, with catalyst types including HCl and H_2SO_4 at concentrations ranging from 2.5 M to 5.5 M. The reactions were carried out at temperatures of 50°C, 60°C, 70°C, and 80°C, while the reaction times were varied to achieve optimization.
- **Procedure:** Acid catalyst was dissolved in methanol to prevent oil scorching before adding it to the oil.
- **Optimization:** Using response surface methodology (RSM) with Design Expert software, a central composite rotary design (CCRD) can be applied to optimize methanol ratio, catalyst concentration, temperature, and time for maximum FAME yield.
- **Post-Transesterification & Purification:** After cooling, 5 ml of deionized water and 5 ml of hexane were added to the reaction mixture. After vortexing and phase separation, the top FAME-rich organic layer was extracted, while the aqueous layer (containing residual methanol, glycerol, and catalyst) was discarded. The FAME layer was washed with warm deionized water (50°C) until neutral, and the hexane and residual water were evaporated at 105°C to obtain purified biodiesel (Mandal et al., 2013).

11.3.2 Lipase Enzyme Catalysis for Biodiesel Production

Lipase-based or lipase-supported catalyst systems are emerging as a promising area of research in biodiesel production. Enzyme-catalyzed transesterification of renewable oils offers an environmentally friendly and safe process, free of harmful by-products (Moazeni et al., 2019). Optimizing biodiesel synthesis using enzyme catalysts involves fine-tuning various factors, including the substrate molar ratio, reaction temperature, choice of organic solvent, and control of water activity (Quayson et al., 2020). Al-Zhuair et al. investigated biodiesel production from simulated waste cooking oil (SWCO) using immobilized lipase from various sources, both in the presence and absence of n-hexane (Jothiramalingam & Wang, 2009). The study found that immobilizing lipase on ceramic beads was effective for catalyzing biodiesel synthesis, even with high water content in the SWCO. Additionally, immobilized lipase from bacterial sources outperformed that from yeast. However, the presence of n-hexane negatively impacted the production rate by diluting the reactants and increasing mass transfer resistance (Dhawane et al., 2021). Observing the kinetic reaction, it was observed to be slightly more inhibited by methanol than by the ester bonds within the triglycerides. This was evident in the inhibition constant for alcohol, which was slightly higher compared to that of the substrate. Zheng et al. (2023) explored a lipase-catalyzed transesterification process without the use of solvents, synthesizing feruloylated diacylglycerol (FDAG) through a selective reaction between ethyl ferulate and triolein. The process achieved a conversion rate of 73.9% and a selectivity of 92.3% toward FDAG under optimal conditions of 338 K, a reaction time of 5.3 days, an enzyme concentration of 30.4 mg/mL, a water activity of 0.08, and a substrate molar ratio of 3.7 (Islam & Ravindra, 2017). However, the main limitation of this approach is that it is more time-consuming compared to other lipase-catalyzed transesterification processes.

11.3.3 Solid Acid Catalysis (Heterogeneous) for Biodiesel Production

Recent advancements in heterogeneous catalysis research have primarily focused on developing recyclable solid acid catalysts to simplify biodiesel production. These catalysts have emerged as promising alternatives to commercially used catalysts (Ganesan et al., 2021) in biodiesel synthesis. Solid acid catalysts, including Bronsted-type (e.g., sulfonic acid-containing materials) and Lewis-type (e.g., mixed sulfated oxides), offer the combined benefits of heterogeneous base catalysts and mineral acids. One of the key advantages of solid acid catalysts is their ability to (Su & Guo, 2014) facilitate both esterification and transesterification reactions simultaneously. Hence, zeolites or zeotype materials occur in their natural crystalline aluminosilicate form with oxygen bonds. They showcase a three-dimensional structure that has equal bond length, and this imparts pores (Warrender et al., 2007) and channels in their structures. They absorb molecules that fit into their sizes perfectly and exclude the ones that are bigger in size, thus acting as molecular sieves. An interesting property of zeolites is their ability to enhance adsorption, which is attributed to the strong electric fields generated by loosely bound cations (Chai et al., 2021) within the catalyst framework. This feature, along with their capacity to introduce active sites within the pores and channels, contributes to improved catalyst activity and selectivity. However, in transesterification reactions (Kamer et al., 2017), the activity increase with most microporous zeolites is relatively modest, typically ranging between 1% and 4%, when compared to non-catalyzed reactions. The efficiency of these catalysts is largely dependent on the interplay between their acid strength and hydrophobicity (Cavuoto, 2023), with higher SiO_2/Al_2O_3 ratios increasing the hydrophobic nature of the catalyst. To enhance the catalytic performance, it is important to develop zeolites and zeotypes with larger pores and channels (Al-Ani et al., 2020). These modifications improve access to active sites, ultimately boosting the catalyst's effectiveness.

11.4 CHEMICAL PROPERTIES OF BIODIESEL

11.4.1 Cetane Number and Composition of Biodiesel

Cetane number (CN) serves as an important indicator of a fuel's auto-ignition quality. Biodiesel generally exhibits a higher CN than petroleum diesel, primarily due to its composition, which is predominantly made up of long-chain hydrocarbon groups that possess minimal branching and aromatic structures (Giakoumis & Sarakatsanis, 2019). As the biodiesel content in fuel blends increases, the CN of the blend correspondingly rises. However, there are notable exceptions to this trend. When low-CN biodiesel is mixed with high-CN petroleum diesel, the addition of more biodiesel can lead to a reduction in the blend's overall CN (Imdadul et al., 2016). The composition of biodiesel is influenced by the length and degree of unsaturation of the fatty acid alkyl chains. Most fatty acids, with the exception of certain types like fish oil, are linear molecules containing carbon chains that range from 8 atoms to 20 atoms (Zamora & Hidalgo, 2004). These fatty acids can be classified as either saturated containing only single bonds and incapable of incorporating additional hydrogen or unsaturated, which possess at least one double bond that allows for hydrogen addition (Lunn & Theobald, 2006). For each double bond present in a fatty acid, one hydrogen molecule can be added. In examining various feedstocks, it is observed that most primarily consist of the unsaturated fatty acids oleic, linoleic, and linolenic. However, significant exceptions exist. For instance, castor oil is rich in ricinoleic acid, a monounsaturated fatty acid characterized by a hydroxyl group, which contributes to its notably high viscosity and density (Talwan et al., 2024). On the other hand, fish oils are abundant in polyunsaturated acids that feature carbon chains longer than 18 atoms and can contain up to six double bonds. These variations in composition not only affect the physical and chemical properties of biodiesel but also influence its performance characteristics (Ghazali et al., 2015), including ignition quality and viscosity, making the choice of feedstock a critical factor in biodiesel production.

11.4.2 The Qualitative Attributes of Biodiesel

- **Kinematic viscosity:** Both biodiesel and petro-diesel standards include kinematic viscosity as a crucial metric that needs to be followed (Ateeq, 2015). Because of hydrogen bonding, ricinoleic acid's hydroxyl group greatly raises oil's viscosity. Thus, while processing plant oils to produce biodiesel, lowering viscosity is a key goal. Apart from transesterification, dilution, microemulsion, pyrolysis, and catalytic cracking are further techniques to reduce viscosity.
- **Presence of fatty acids:** Free fatty acids (FFAs) are harmful to engines and are unsuitable for fuel tanks. The ASTM D 6751 standard specifies that biodiesel should contain no more than 0.5% FFAs, as higher levels can lead to engine deposits and the corrosion of rubber components (Etim et al., 2022). While sources allow for FFA levels of up to 5%, Murugesan et al. recommend keeping FFAs below 3% in the feedstock for biodiesel synthesis. In base-catalyzed transesterification, high FFA levels deplete the catalyst and promote saponification, which lowers yields and increases emulsion formation (Tran-Nguyen et al., 2020), making glycerin separation from biodiesel more difficult.
- **Flash point:** The term "flash point" refers to the temperature at which biodiesel becomes volatile enough for its alcohol content to pose a safety risk during storage and transport (Ateeq 2015). It is also a key quality indicator related to the structure of fatty acids in biodiesel. The flash point can be adjusted by blending petro-diesel and biodiesel in appropriate ratios. Several researchers have recommended blends such as B20, which contains 20% biodiesel and 80% petro-diesel, as a viable option (Chauhan et al., 2016).
- **Density:** Various guidelines establish density requirements to ensure that only appropriate materials are used as feedstock for biodiesel production (Moser, 2009). Density is also used to help determine the viscosity of biodiesel.

- **Heat of combustion:** Plant oils commonly used as raw materials for biodiesel production have a calorific value, or heat of combustion, ranging from 5,443 kJ/kg to 14,654 kJ/kg (Tutunea et al., 2013). This falls within the range of hexadecane, also known as cetane (10,714 kJ/mole), which serves as a reference material for evaluating the ignition quality of petro-diesel. Hence, the composition and physicochemical properties of oils have been shown to differ depending on the plant's location and the agricultural methods employed in cultivating the raw materials (Aremu et al., 2015).

Due to its relatively high oxygen content (around 11%), biodiesel has lower carbon and hydrogen levels compared to regular diesel, which leads to about a 10% reduction in its mass energy content. However, because biodiesel has a higher fuel density, its volumetric energy content is only about 5–6% less than that of petroleum diesel (Ajala et al., 2015). Biodiesel generally has a higher molecular weight than petroleum diesel, which results in slightly elevated distillation temperatures (as indicated by T90). Since biodiesel primarily consists of straight-chain esters, it typically has excellent cetane numbers, often higher than those of No. 2 diesel (Knothe 2010). Additionally, biodiesel's viscosity is significantly greater than that of petroleum diesel, usually by a factor of two.

11.5 QUALITY STANDARDS AND SPECIFICATIONS

Biodiesel standards are essential for ensuring quality, safety, and performance in fuel applications (Atadashi et al., 2010). Various organizations have established guidelines and specifications that govern the production, testing, and use of biodiesel. Compliance with these standards is crucial for maintaining engine performance by enabling biodiesel to function effectively in different diesel engines without causing damage or reducing efficiency. It also addresses environmental impact by setting guidelines that promote sustainable practices and reduce harmful effects on the environment. Additionally, it aids in market acceptance by facilitating the integration of biodiesel into the broader fuel market, fostering consumer confidence, and ensuring regulatory compliance. In conclusion, adherence to international and national standards for biodiesel not only enhances its viability as a renewable fuel source but also contributes to the overall sustainability and efficiency of the energy sector (Anuar & Abdullah, 2016). Among the most recognized standards are those set by the American Society for Testing and Materials (ASTM) and international bodies (Nadkarni, 2007).

11.5.1 ASTM Standards

ASTM International has developed several standards for biodiesel, primarily focused on the fuel's properties and performance (Hoekman et al., 2012). The most relevant ASTM standard for biodiesel is

- **ASTM D6751:** This standard specifies the requirements for biodiesel (B100) used as a blend component with diesel fuel (Jamaluddin et al., 2019). It covers a range of parameters, including physical properties such as density, viscosity, and flash point; chemical properties like fatty acid methyl ester (FAME) content, acid value, and free fatty acid (FFA) limits; and performance requirements, including cetane number, oxidative stability, and filtration properties. In addition to D6751, ASTM has developed standards for biodiesel blends:

- **ASTM D7467:** This standard specifies the requirements for diesel fuel oil and biodiesel blend (B6 to B20), focusing on the performance and quality (Howell, 2016) characteristics when biodiesel is blended with conventional diesel fuel.

11.5.2 International Standards

Internationally, several organizations have set standards for biodiesel, including

- **EN 14214:** Developed by the European Committee for Standardization (CEN), this standard specifies the requirements for FAME used as a biodiesel fuel. It addresses similar parameters as ASTM D6751, emphasizing the importance of quality for performance in diesel engines (Noor et al., 2018).
- **ISO 14040 / ISO 14044:** This standard provides guidelines for the sustainable use of biodiesel, focusing on lifecycle assessment and sustainability criteria (Akpan and Olanrewaju, 2022).
- **ASTM D6751 vs. EN 14214:** While both standards aim to ensure biodiesel quality, they have different testing methods and acceptable limits for various properties (Carrero and Pérez, 2012). Manufacturers and suppliers often need to ensure compliance with both standards, depending on their market.

11.6 METHODS FOR QUALITY CONTROL AND ASSURANCE

Another approach to ensuring high-quality biodiesel production involves the implementation and enforcement of quality control and quality assurance (QC/QA) programs. In Germany, the Association for Quality Management of Biodiesel (AGQM) was established as early as 1999 to address the quality of biodiesel in use (Körbitz et al., 2004). In the United States, the National Biodiesel Board tackled QC/QA through the creation of the National Biodiesel Accreditation Commission, which manages the BQ-9000 Quality Management System. This Commission has set forth three types of requirements: one for biodiesel producers, another for biodiesel marketers, and a third for biodiesel testing laboratories (Barabás & Todoruţ, 2011). The BQ-9000 Program combines ASTM biodiesel quality requirements with its quality management system, ensuring proper storage, sampling, testing, blending, shipping, distribution, and fuel management practices. Accreditation is granted to the company, not to individual fuel batches once these requirements are met.

11.6.1 Analytical Techniques for Biodiesel Analysis

Analytical techniques are used in biodiesel analysis to quantify the final product (mono-alkyl esters), co-products (glycerol),unreacted and unprocessed reactants and intermediates (alcohol, monoglycerides, diglycerides, triglycerides, catalyst, free/bonded fatty acids, free/bonded glycerol, and trace metals), and any other inhibitory compounds and contaminants that might cause operational and environmental problems and affect the quality control for market production (Jamil, 2023). Chromatography and spectroscopy are the analytical methods used for characterizing biodiesel by determining the carbon chain length and its mixtures, structure and size for molecular formula, fatty-ester content, inhibitory compounds in the transesterification process, and additional products in the crude sample (Vinoth Kumar et al., 2020).

i. **Gas Chromatography**
The GC-MS analysis is used to determine derived glycerol (mono-, di-, tri-), glycerol esters, free and total glycerol, acylglycerols, methyl linolenate, sterol and sterol esters glycerol, methyl esters, and un-transesterified components in a biodiesel sample w.r.t. the standard ASTM D6584 method through retention time. Bis(trimethylsilyl)trifluoroacetamide (BSTFA), N-methyl-N-trimethylsilyl-trifluoroacetamide (*MSTFA),* and *trimethylchlorosilane* (TMCS) are used as derivatization agents, releasing trimethylsilyl (TMS) derivatives of the hydroxyl groups (Vendruscolo et al., 2019). The analysis follows the separation of the sample by a capillary column with a constant gas (hydrogen, helium) flow rate. The split ratio of the sample is set before sample injection at a set temperature for FAME analysis and quantification with respect to a standard. For FAME analysis, 70% cyanopropyl-modified polydimethylsiloxane was identified as the most efficient option, better than polyethylene glycol, nitroterephthalic acid-modified polyethylene glycol, and 35% phenyl-modified polydimethylsiloxane, which are set between 60°C and 270°C at a 1.5 ml/min flow rate. This states the role of column type and temperature for quantitative analysis of biodiesel (Goding et al., 2013).

ii. **High-Performance Liquid Chromatography (HPLC)**
HPLC analysis of biodiesel is carried out for detection of glycerol (mono-, di-, tri-), bound glycerol, alkyl and benzyl esters, and complex mixtures, including blends, by taking in ASTM D6591 and IP391 as reference methods. The first step after the preparation of dilutions for samples and standards to be analyzed is followed by setting up the column pump and temperature (40°C), vacuum degasser, autosampler at the desired injection volume (between the set injection capacity of 0.5–500 µl), and mobile phase with a desired flow rate (usually 1 ml/min) (Lau et al., 2024). As the samples are injected, they are detected by the RI detector and UV detector set between 200 nm, and the data is analyzed by the software installed. As the chromatogram is obtained, the retention time tends to fluctuate with respect to the molecular weight of the compound and unsaturated C-C bonds. The quantitative determination of the compound depends largely on the sensitivity of the UV detection.

iii. **Fourier Transform Infrared Spectroscopy (FTIR)**
As FTIR is used to assess adulteration of biodiesel with unreacted oil, which is detected by the peaks acquired in between 5,000 cm^{-1} and 1,000 cm^{-1}. It is used for feedstock analysis, rapid free fatty acid determination, functional

group identification, and monitoring transesterification kinetics (Bukkarapu & Krishnasamy, 2021). As the instrument is set by cleaning the accessory and purged with zero air, the temperature is set at room temperature with a maintained spectral resolution. The samples are scanned for obtaining spectra for quantitative analysis; the software identifies the IR region where the absorbance varies with concentration.

iv. **Nuclear Magnetic Resonance (NMR) Spectroscopy**
NMR provides detailed structural information, without any derivatizing agents, by working on the principle of interaction of nuclei with magnetic fields. NMR spectroscopy identifies characteristic peaks for triacylglycerols and methyl esters, unsaturated fatty acids, and reaction kinetics of the process, usually confirming the results received from other analytical methods applied (Doudin, 2021). A suitable solvent such as deuterated chloroform ($CDCl_3$) is used to determine the signal area ratio for peak determination in an NMR spectrometer operating at 200–500 MHz. For determination of glycerol conversion to esters, it can be obtained by integration of the areas of the selected signals. As the spectra denote the ester content, this is further given by the addition of the characteristic peaks and high esters in the mixture. While observing the correlation between carbon and hydrogen signals, the key signals signify the presence of different functional groups.

v. **Other Emerging Techniques in Biodiesel Analysis**
Other different techniques such as the iodine method (total unsaturation), kinetic viscosity measurement (viscosity), acid value (measurement of free fatty acid), iodine value (total unsaturation), UV-V is spectroscopy (polyunsaturated esters), sulfated ash test (residual metals from catalysts), carbon residue test (coking tendency), sodium and potassium test (trace metals), inductively coupled plasma mass spectroscopy (various elements in biodiesel), and many more also exist for quality assessment of biodiesel (Knothe, 2010). techniques such). Others include NIR (distinguishing between triacylglycerol and biodiesel), PLS and PCA (identification of natural abundance radiocarbon analysis (14C levels in biodiesel)), electrothermal atomic absorption spectrometry (ETAAS), and flame atomic absorption spectrometry (FAAS).

11.7 LIFE CYCLE ANALYSIS OF BIODIESEL PRODUCTION

The life cycle assessment (LCA) of biodiesel production is carried out systematically by the evaluation of the environmental impact of the processes involved in the flow of materials used for biodiesel development, its formation, its applicability to its surroundings, and its disposal as depicted in Figure 11.3. Every process, from cultivation, transportation, extraction, production, and combustion, is evaluated differently because of the utilization of different raw materials and process methodology for biodiesel production and thus cannot be compared to each other (Pikula et al., 2020).

Uppalapati et al. (2022) studied a comparison of LCA in the biodiesel production process between soybean, Jatropha, *Calophyllum inophyllum,* and microalgae. Soybean had 32.53 g CO_2 Eq/MJ GHG emissions, while microalgae showed 49.4 g CO_2 Eq/MJ. Jatropha and *Calophyllum inophyllum* were observed to be better

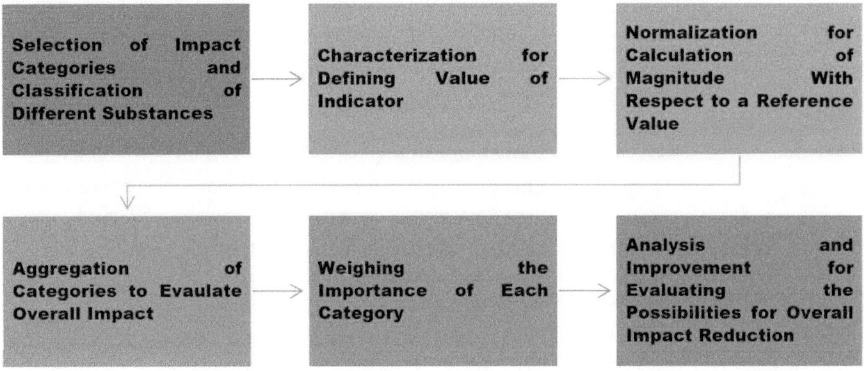

FIGURE 11.3 Process steps followed for life cycle assessment of biodiesel production.

sustainable solutions because of lower GHG emissions and balanced energy output. Some of the recent LCA studies on biodiesel production from different feedstocks have been mentioned in Table 11.1.

11.7.1 Greenhouse Gas Emissions and Carbon Footprint

The environmental impact of agricultural feedstocks is greatly influenced by the land footprint area used for operation, geographical location, deforestation and forest degradation that occurred for increasing croplands, and increased NOX generation due to the increase in the use of fertilizers and fuel sources in biorefineries. This brings about variability in data due to the use of varied sources, location, allocation, applied methods, and interaction ability with the surrounding environment (Jeswani et al., 2020). Jeswani and his co-workers reviewed the Global Warming Potential (GWP) of biodiesel from first-generation feedstocks (3 to 111 g CO_2 eq. MJ^{-1}), with only palm oil meeting the Renewable Energy Directive (RED) requirement with a 60% reduction in global warming without any land use change. While biofuel processing and the transesterification process greatly affect the GWP value of second-generation biodiesel (−88 to 150 g CO_2 eq. MJ^{-1}), only biodiesel from Jatropha can yield up to 30% GWP reduction. Algal biodiesel has a high variability of −2,400 to 2,880 g CO_2 eq. MJ^{-1} GWP as compared to fossil biodiesel due to high energy requirements in cultivation, harvesting, drying, and processing. This was observed with meta-regression analysis (MRA), which was performed by Garcia et al., (2020) on 36 LCA, reviewing 308 individual result studies to evaluate the impact of different parameters on GHG emissions from the microalgae biodiesel production process. This includes species type, nutrient media components used for cultivation, cultivation system type used, CO_2 source, pretreatment, extraction, and transesterification. Methodology and system applied for conversion, geographical location, surrounding environment, machinery, and investment variation. This concluded with a statement that no process can outperform the other. For each and every type of feedstock source and plant species, methods can be varied to find the most suitable process for the individual with respect to environmental and social conditions.

TABLE 11.1
Life Cycle Assessment Analysis in Different Research Studies on Biodiesel Production

Title	LCA Method	Environment Impact	Conclusion	Reference
Life cycle assessment of biodiesel produced from sunflower oil from soil phytoremediation	All stages from cultivation to final product	Climate change: control soil: 1.16×10^2; industrial soil: 1.56×10^2 freshwater ecotoxicity; water usage	A sustainable option promoting land reuse and biomass valorization	Caetano et al., 2023
Life cycle analysis of biodiesel derived from freshwater microalgae and Karanja	All stages from cultivation to final product	GHG emissions: Karanja 63.81 g CO_2 Eq/MJ; microalgae: 85.3 g CO_2 Eq/MJ; Energy ratio: Karanja (3.57) and microalgae (0.45)	Karanja might be preferred over microalgae but microalgae offer more sustainability for a longer time without affecting food supply	Kumar and Pal, 2023
Used-cooking-oil biodiesel: Life cycle assessment and comparison with first- and third-generation biofuel	All stages from cultivation to final product	Carbon footprint: 14 g CO_2 Eq/MJ; environment footprint: 58.3 Pt; mazut (23.6 Pt/t) Valorization concerns over cooking oil dumping, Glycerol and potassium sulfate add to burden. Three times reduced impact. Human health 24.71 Pt Resources 24.06 Pt, Ecosystem: 9.360 Pt	Environment footprint: Three times lower than petro-diesel; 40% lower than first generation. Thus, advantageous over other fuel sources	Foteinis et al., 2020
Life cycle assessment of biodiesel from estuarine microalgae	All stages from cultivation to final product	Energy requirement with fertilizer and acid catalyst: 11,738 MJ/ha; transesterification energy: 1.81–2.13 MJ/kg	economically and environmentally favorable alternative to fossil fuels	Saranya & Ramachandra, 2020
Life cycle assessment of biodiesel production from rapeseed oil: Influence of process parameters and scale	All stages from cultivation to final product	Global Warming Potential: Large scale (LS): 2.63t CO_2 Eq/t biodiesel; small scale (SS): 2.88t CO_2 Eq/t; CO_2 emissions:.65% both; GWP reduction: LS: 14.1;SS: 33.6%; energy demand: LS: 32.15 GJ/t; SS: 30.82 GJ/T	Glycerol disposal, rape straw/meals for energy, rapeseed yield fluctuation, oil extraction capacity are key influencing factors	Gupta et al., 2022

11.7.2 Role of Biodiesel against Pollution

Biodiesel has been observed to show a significant decrease in pollution by reducing CO_2 emissions, which has been observed in the case of B100 (78% w.r.t conventional) and B20 (2–4%). While the CO_2 can be recycled, i.e., reabsorbed by the plants for cultivation purposes. Methane emissions were observed to be reduced by B20 (0.51%) and B100 (2.57%), and NOX emissions were reduced by 16.54 gms/MJ energy w.r.t conventional. 20% less CO, 40% less particulate matter emissions, sulfur content reduced to be less than 0.0024 ppm, with reduced hydrocarbon and polycyclic aromatic hydrocarbons (PAHs) (Siddeg et al., 2022). Comparatively, biodiesel reduces net GHG emissions and waste materials, captures CO2, prevents land conversions, and promotes biodiversity and resource utilization (Suhara et al., 2024).

11.7.3 Techno-economic Assessment of Biodiesel Production

The global biodiesel market was valued at $35.5 billion. It is expected to increase by $17 billion in the next four years with a 7.38 CAGR (Malik et al., 2024). Techno-economic analysis (TEA) of biodiesel production evaluates the total production cost, profit earned, and total economic viability of the process, i.e., the total investment vs. the return value of the product. The economic analysis of biodiesel production involves evaluating the total cost of raw materials, utilities, process development, processing of raw materials, product formation, tools and machinery, purification, packaging, manpower, etc. TEA of biodiesel production from different feedstock sources or through different conversion processes applied has been mentioned in Table 11.2.

11.8 FUTURE TRENDS AND INNOVATIONS IN BIODIESEL RESEARCH

Targeting the production process by introducing modifications and upgrades in the extraction and transesterification process can bring about an increase in the biodiesel yield. Extraction techniques such as accelerated solvent extraction, ultrasound, microwave, supercritical fluid extraction, and enzymatic have been identified as the advanced techniques that, upon being introduced in the biodiesel formation process, can increase the yield (Akhihiero, 2022; Neag et al., 2023; Pikula et al., 2022). The significance of an extraction or conversion technique applied for biodiesel production is to obtain maximum fatty acid and free fatty acid content from the feedstock. Application of different techniques targets the cell differently, thus leading to obtaining fatty acid in high content, which further leads to production of biodiesel in high yield. The pretreatment and the extraction process applied to obtain the fatty acid content from biomass play an important role in the biodiesel content obtained. Steam distillation, mechanical extraction, chemical leaching, Soxhlet extraction, enzyme-assisted extraction, supercritical fluid extraction, microwave-assisted extraction, and ultrasound-assisted extraction are the commonly used extraction method options for biodiesel extraction from feedstock (Athar & Zaidi, 2020; Karmakar & Halder, 2019). Each method is applied to different feedstock types depending upon their

TABLE 11.2
Techno-Economic Assessment of Biodiesel Production Reported in Several Studies

Title	Process	TEA	Conclusion	Reference
Net energy and techno-economic assessment of biodiesel production from waste cooking oil using a semi-industrial plant: A Malaysia perspective	From feedstock acquisition to biodiesel production	Production capacity: 15 t/ batch/day; capital investment: USD 490.3 thousand; operating costs: USD 1.78 million; gross profit: USD 219.38 thousand; net present value: USD 1.47 million Annual revenue: USD 2 million	Overall process is profitable while feedstock acquisition, methanol, and KOH contribute vastly to total investment costs	Farid et al., 2020
Techno-economic assessment of biodiesel production from canola oil through ultrasonic cavitation	Overall process from acquisition to production, highlighting mainly the transesterification process	Total investment: Ultrasonication cultivation: $4,030,000, Mechanical stirring cultivation $5,090,000; Production Costs: Ultrasonication: $53,710,000, Mechanical Stirring: $56,673,000; IRR: 18.3%	Ultrasonic cultivation is a better alternative to conventional methods for biodiesel production	Gholami et al., 2021
Techno-economic assessment of coconut biodiesel as a potential alternative fuel for compression ignition engines	Feedstock preparation to Biodiesel production	Production costs: $1,134 million; Operating costs $5.068 million. Cost per liter: 0.859 $/L	Production costs of coconut diesel are higher than conventional, but represents a better renewable source of energy option for Indian Railways	Thangaraja et al., 2019
Optimization and techno-economic analysis of biodiesel production from Calophyllum inophyllum oil using heterogeneous nanocatalyst	Feedstock preparation to production	Capital costs: $ 2.753 thousand Operating costs: $14.831 million; net unit production costs: $0.68 /kg; gross margin: 17.92%; IRR: 66.90% NPV: $11.673 million	The optimized factors bring about favorable conditions to the overall revenue.	Naveen Kumar et al., 2020

characteristics, fatty acid content and profile, presence of free fatty acids, cellulosic crystallinity, reduced formation of inhibitory products, and many other factors depending upon the feedstock source. Modifications introduced in the extraction methods can change the total biodiesel produced at the end of the process. Also, it's important to select the feedstock type based on oil content, cost, feedstock type, its fatty acid profile, and catalyst selection (Ambat et al., 2020; Mohiddin et al., 2021). Some of the optimization studies have been mentioned in Table 11.3. It is important

TABLE 11.3
New Advanced and Optimized Conditions Applied for Biodiesel Production

S. No	Title	Methodology	Process Modified	Yield	Reference
1.	The optimization of tetrahydrofuran as a co-solvent on biodiesel production from rubber seeds using response surface methodology	Response surface methodology (RSM) applied by optimizing tetrahydrofuran effects reaction time and temperature.	Transesterification	97% yield was achieved at 52°C, 30 min and mass ratio 1.4:1	Ridwan et al., 2023
2.	Microwave energy enhanced oil extraction from Millettia ferruginea (Birbira) seed for biodiesel application: effects of microwave energy on oil extraction yield	Effect of microwave energy, its power, irradiation time, particle size, solvent-to-seed ratio and extraction duration	Microwave-based extraction	20.65% increase in yield was observed to be more than untreated seeds in microwave energy treated seeds (450 W for 150 s)	Woldie et al., 2024
3	Simultaneous oil sono-extraction and sono-transesterification (in situ) of soybean and sunflower seeds for the production of biodiesel	Ultrasound-assisted integrated extraction and transesterification process carried out at 20kHz frequency and 106W power on sunflower and soyabean seeds	Extraction and transesterification	At 1:5 seed to solvent ratio and 1 min sonication in sunflower seeds: 94% yield; At 1:10 soyabean: n-hexane ratio 1 min sonication, methanol to oil ratio, reaction time and varying catalyst concentration: 59% yield with 99% biodiesel conversion rate	Dolores et al., 2020

(Continued)

TABLE 11.3 (*Continued*)
New Advanced and Optimized Conditions Applied for Biodiesel Production

S. No	Title	Methodology	Process Modified	Yield	Reference
4	Combination of oscillatory and hydrodynamic system for biodiesel production from sunflower oil	RSM-based optimization of biodiesel production in oscillatory-hydrodynamic system	Transesterification	Increasing residence time from 30 s to 90 s, catalyst concentration, molar ratio from 6:1 to 9:1, increased yield by 6%. Also by increasing temperature 40°C to 60°C, increased methyl ester content by 4.16%	Taki et al., 2023
5	Biodiesel production from spent coffee grounds by using ethanolic extraction and supercritical transesterification	Ethanol: used as a solvent and reactant; extraction temperature: 50–70 C and the reactor conditions was varied in both batch and continuous (275–325°C) with the reaction time (5–40 min)		Batch: high ester content (88.37 ± 3.00 wt% at 275°C 40 min; Continuous: ester content 83.38 ± 5.86 wt% at 325°C and 29.4 min.	Supang et al., 2024

to optimize and develop a process for biodiesel production that yields more product with fewer impurities and unreacted products, as well as to turn the overall process into an efficient and slightly less costly one. For optimization, the use of AI tools, software, machinery, and reactants is applied during the pretreatment, conversion, and purification processes. Use of nanomaterials in the production process can not only increase yield by a significant percentage, achieving more than 90% production yield, but they can also be recovered and reused in the process (Pandit et al., 2023). Combining the advanced tools, techniques, and optimization processes developed with the use of a feedstock source from waste material can turn the overall process into a sustainable one. Biodiesel production from spent coffee beans yielded 88.37±3.00 –±3.00 wt% at 40 min in a batch reactor, while 83.wt% at 29.4 min in a continuous reactor (Supang et al., 2024). The NaOH/Chitosan-Fe_3O_4 catalyst, which was first synthesized, gave a fatty acid methyl ester yield of 92% from waste cooking oil when an RSM study was used to optimize the production process factors: methanol-to-oil ratio (6:1 to 12:1), catalyst weight (0.25–1%), and reaction time (4–8 hrs). The optimal conditions were observed with 0.5% catalyst weight, a 6:1 methanol-to-oil ratio, a reaction time of 4.5 hrs, and temperature conditions of 25°C (Helmi & Hemmati, 2021).

11.9 CONCLUSION

Biodiesel is widely recognized as a viable alternative fuel capable of replacing conventional diesel to a significant extent. Its environmental impacts, production costs, and overall economic returns remain subjects of considerable debate that depend mainly on the feedstock source type, conversion process, and technology or machinery used. This also includes the geographical location and market requirements and size of the location (country). It took millions of years for the formation of the non-renewable fuel we use in our machinery and engines. The long formation time of fossil fuels translates into high economic costs when seeking sustainable alternatives, posing significant investment challenges since investment is preferred towards what is easier and cheaper, whether it compromises earth's environment. Therefore, increasing investment in sustainable, low-carbon fuel technologies is essential for conserving natural resources and mitigating environmental degradation.

REFERENCES

Ajala, Olawale E., Folorunsho Aberuagba, Temitope E. Odetoye, and Adejoke M. Ajala. "Biodiesel: Sustainable energy replacement to petroleum-based diesel fuel–a review." *ChemBioEng Reviews* 2, no. 3 (2015): 145–156.

Akhihiero, Ejiroghene T. *Recent Advances in Biodiesel from Plants*. IntechOpen, 2022. https://doi.org/10.5772/intechopen.106924

Akpan, Joseph, and Oludolapo A. Olanrewaju. "Asset management models brief review and framework development for energy sustainability & sustainable development." In *Proceedings of the International Conference on Industrial Engineering and Operations Management*, pp. 7–10, 2022.

Al-Ani, Aqeel, Cátia Freitas, and Vladimir Zholobenko. "Nanostructured large-pore zeolite: The enhanced accessibility of active sites and its effect on the catalytic performance." *Microporous and Mesoporous Materials* 293 (2020): 109805.

Al Mubarak, Faisal, Reza Rezaee, and David A. Wood. "Economic, societal, and environmental impacts of available energy sources: A review." *Engineering* 5, no. 3 (2024): 1232–1265. https://doi.org/10.3390/eng5030067

Alsultan, Abdulkareem G., Nurul Asikin-Mijan, Zueriani Ibrahim, Robiah Yunus, Siti Zulaika Razali, Nasar Mansir, Aminul Islam, Sivasangar Seenivasagam, and Yun H. Taufiq-Yap. "A short review on catalyst, feedstock, modernised process, current state and challenges on biodiesel production." *Catalysts* 11, no. 11 (2021): 1261.

Ambat, Indu, Varsha Srivastava, Sidra Iftekhar, Esa Haapaniemi, and Mika Sillanpää. "Effect of different co-solvents on biodiesel production from various low-cost feedstocks using Sr–Al double oxides." *Renewable Energy* 146 (2020): 2158–2169. https://doi.org/10.1016/j.renene.2019.08.061

Aniza, Ria, Wei-Hsin Chen, Eilhann E. Kwon, Quang-Vu Bach, and Anh T. Hoang. "Lignocellulosic biofuel properties and reactivity analyzed by thermogravimetric analysis (TGA) toward zero carbon scheme: A critical review." *Energy Conversion and Management: X* 22 (2024): 100538.

Anuar, Mohd R., and Ahmad Z. Abdullah. "Challenges in biodiesel industry with regards to feedstock, environmental, social and sustainability issues: A critical review." *Renewable and Sustainable Energy Reviews* 58 (2016): 208–223. https://doi.org/10.1016/j.rser.2015.12.296

Aremu, Michael O., H. Ibrahim, and Timothy O. Bamidele. "Physicochemical characteristics of the oils extracted from some Nigerian plant foods – A review." *Chemical and Process Engineering Research* 32 (2015): 36–52.

Atadashi, Issac M., Mohamed K. Aroua, and Abdul A. Aziz. "High quality biodiesel and its diesel engine application: A review." *Renewable and Sustainable Energy Reviews* 14, no. 7 (2010): 1999–2008. https://doi.org/10.1016/j.rser.2010.03.020

Ateeq, Eman. "Biodiesel viscosity and flash point determination." PhD diss., 2015.

Athar, Moina, and Sadaf Zaidi. "A review of the feedstocks, catalysts, and intensification techniques for sustainable biodiesel production." *Journal of Environmental Chemical Engineering* 8, no. 6 (2020): 104523. https://doi.org/10.1016/j.jece.2020.104523

Barabás, István, and Ioan-Adrian Todoruţ. *Biodiesel Quality, Standards and Properties*, vol. 2. InTech, 2011.

Bukkarapu, Kiran R., and Anand Krishnasamy. "Fourier-transform-infrared-spectroscopy-based approach to predict engine fuel properties of biodiesel." *Energy & Fuels* 35, no. 9 (2021): 7993–8005. https://pubs.acs.org/doi/10.1021/acs.energyfuels.0c03927.

Burdge, Graham C., and Philip C. Calder. "Introduction to fatty acids and lipids." *Intravenous Lipid Emulsions* 112 (2015): 1–16.

Caetano, Nídia de Sá, António A. Martins, and Cristiana Ribeiro. "Life cycle assessment of biodiesel produced from sunflower oil from soil phytoremediation." In *CICITEC 2023-4th International Congress of Engineering Sciences and Technology*. Autonomous University of Baja California, 2023.

Carrero, A., and Á. Pérez. "Advances in biodiesel quality control, characterisation and standards development." In *Advances in Biodiesel Production*, pp. 91–130. Woodhead Publishing, 2012. https://doi.org/10.1533/9780857095862.1.91

Cavuoto, D. "Green catalytic strategies for the production of biobased building-blocks." (2023).

Cernat, Alexandru, Constantin Pana, Niculae Negurescu, Gheorghe Lazaroiu, Cristian Nutu, Dinu Fuiorescu, Marius Toma, and Adrian Nicolici. "Combustion of preheated raw animal fats-diesel fuel blends at diesel engine." *Journal of Thermal Analysis and Calorimetry* 140 (2020): 2369–2375.

Chai, Yuchao, Weili Dai, Guangjun Wu, Naijia Guan, and Landong Li. "Confinement in a zeolite and zeolite catalysis." *Accounts of Chemical Research* 54, no. 13 (2021): 2894–2904.

Chauhan, Bhupendra S., Ram K. Singh, Hyun M. Cho, and Hee C. Lim. "Practice of diesel fuel blends using alternative fuels: A review." *Renewable and Sustainable Energy Reviews* 59 (2016): 1358–1368. https://doi.org/10.1016/j.rser.2016.01.062

Demirbas, A. "The importance of bioethanol and biodiesel from biomass." *Energy Sources, Part B* 3, no. 2 (2008): 177–185.

Demirbas, Ayhan, and M. Fatih Demirbas. "Importance of algae oil as a source of biodiesel." *Energy Conversion and Management* 52, no. 1 (2011): 163–170.

Deora, Prayagraj S., Yash Verma, Ram A. Muhal, Chandramani Goswami, and Tej Singh. "Biofuels: An alternative to conventional fuel and energy source." *Materials Today: Proceedings* 48 (2022): 1178–1184. https://doi.org/10.1016/j.rser.2017.04.072

Dhawane, Sumit H., Eman G. Al-Sakkari, Tarkeshwar Kumar, and Gopinath Halder. "Comprehensive elucidation of the apparent kinetics and mass transfer resistances for biodiesel production via in-house developed carbonaceous catalyst." *Chemical Engineering Research and Design* 165 (2021): 192–206.

Dolores, Reyman D., and Quiñones G. Almudena. "Simultaneous oil sono-extraction and sono-transesterification (in situ) of soybean and sunflower seeds for the production of biodiesel." (2020).

Doudin, Khalid I. "Quantitative and qualitative analysis of biodiesel by NMR spectroscopic methods." *Fuel* 284 (2021): 119114. https://doi.org/10.1016/j.fuel.2020.119114

Ejikeme, Patrick M., Ifeanyi D. Anyaogu, Chika L. Ejikeme, Nnaemeka P. Nwafor, Chinwe A. C. Egbuonu, K. Ukogu, and Joseph A. Ibemesi. "Catalysis in biodiesel production by transesterification processes-an insight." *Journal of Chemistry* 7, no. 4 (2010): 1120–1132.

Ershov, Mikhail A., Vsevolod D. Savelenko, Ulyana A. Makhova, Alisa E. Makhmudova, Aleksandr V. Zuikov, Vladimir M. Kapustin, and Tamer M. M. Abdellatief, et al. "Current challenge and innovative progress for producing HVO and FAME biodiesel fuels and their applications." *Waste and Biomass Valorization* 14, no. 2 (2023): 505–521.

Etim, Anietie O., Chiazor Faustina Jisieike, Taiwo Hassan Ibrahim, and Eriola Betiku. "Biodiesel and its properties." In *Production of Biodiesel from Non-Edible Sources*, pp. 39–79. Elsevier, 2022.

Farid, Mohammed A. A., Ahmad M. Roslan, Mohd A. Hassan, Muhamad Y. Hasan, Mohd R. Othman, and Yoshihito Shirai. "Net energy and techno-economic assessment of biodiesel production from waste cooking oil using a semi-industrial plant: A Malaysia perspective." *Sustainable Energy Technologies and Assessments* 39 (2020): 100700. https://doi.org/10.1016/j.seta.2020.100700

Foteinis, Spyros, Efthalia Chatzisymeon, Alexandros Litinas, and Theocharis Tsoutsos. "Used-cooking-oil biodiesel: Life cycle assessment and comparison with first-and third-generation biofuel." *Renewable Energy* 153 (2020): 588–600. https://doi.org/10.1016/j.renene.2020.02.022

Ganesan, Ramya, S. Manigandan, Sabarathinam Shanmugam, V. P. Chandramohan, Raveendran Sindhu, Sang-Hyoun Kim, Kathirvel Brindhadevi, and Arivalagan Pugazhendhi. "A detailed scrutinize on panorama of catalysts in biodiesel synthesis." *Science of the Total Environment* 777 (2021): 145683. https://doi.org/10.1016/j.scitotenv.2021.145683

Garcia, Rita, Filipa Figueiredo, Miguel Brandão, Marika Hegg, Érica Castanheira, João Malça, Astrid Nilsson, and Fausto Freire. "A meta-analysis of the life cycle greenhouse gas balances of microalgae biodiesel." *The International Journal of Life Cycle Assessment* 25 (2020): 1737–1748.

Gashaw, Alemayehu, and Amanu Lakachew. "Production of biodiesel from non edible oil and its properties." *International Journal of Science, Environment and Technology* 3, no. 4 (2014): 1544–1562.

Gaurav, Kumar, Krishna Neeti, and Reena Singh. "Microalgae-based biodiesel production and its challenges and future opportunities: A review." *Green Technologies and Sustainability* (2023): 100060.

Ghazali, Wan N. M. W., Rizalman Mamat, Haji Hassan Masjuki, and Gholamhassan Najafi. "Effects of biodiesel from different feedstocks on engine performance and emissions: A review." *Renewable and Sustainable Energy Reviews* 51 (2015): 585–602. https://doi.org/10.1016/j.rser.2015.06.031

Gholami, Ali, Fathollah Pourfayaz, and Akbar Maleki. "Techno-economic assessment of biodiesel production from canola oil through ultrasonic cavitation." *Energy Reports* 7 (2021): 266–277. https://doi.org/10.1016/j.egyr.2020.12.022

Giakoumis, Evangelos G., and Christos K. Sarakatsanis. "A comparative assessment of biodiesel cetane number predictive correlations based on fatty acid composition." *Energies* 12, no. 3 (2019): 422.

Goding, Julian C., Dorisanne Y. Ragon, Jack B. O'Connor, Sarah J. Boehm, and Amber M. Hupp. "Comparison of GC stationary phases for the separation of fatty acid methyl esters in biodiesel fuels." *Analytical and Bioanalytical Chemistry* 405 (2013): 6087–6094. https://doi.org/10.1007/s00216-013-7042-7

Gupta, Rohit, Ruairidh McRoberts, Zhibin Yu, Cindy Smith, William Sloan, and Siming You. "Life cycle assessment of biodiesel production from rapeseed oil: Influence of process parameters and scale." *Bioresource Technology* 360 (2022): 127532. https://doi.org/10.1016/j.biortech.2022.127532

Hanaki, Keisuke, and Joana Portugal-Pereira. 2018. "The effect of biofuel production on greenhouse gas emission reductions." In *Biofuels and Sustainability: Holistic Perspectives for Policy-Making*, pp. 53–71. Springer,

Hassan, Masjuki H., and Md A. Kalam. "An overview of biofuel as a renewable energy source: Development and challenges." *Procedia Engineering* 56 (2013): 39–53.

Helmi, Maryam, and Alireza Hemmati. "Synthesis of magnetically solid base catalyst of NaOH/Chitosan-Fe3O4 for biodiesel production from waste cooking oil: Optimization, kinetics and thermodynamic studies." *Energy Conversion and Management* 248 (2021): 114807. https://doi.org/10.1016/j.enconman.2021.114807

Hoekman, S. Kent, Amber Broch, Curtis Robbins, Eric Ceniceros, and Mani Natarajan. "Review of biodiesel composition, properties, and specifications." *Renewable and Sustainable Energy Reviews* 16, no. 1 (2012): 143–169. https://doi.org/10.1016/j.rser.2011.07.143

Howell, Steven A. "Discussion on Uses of the Specifications for Biodiesel Fuel Blend Stock B100 (ASTM D6751) and Biodiesel Blends B6 to B20 (ASTM D7467)." (2016). https://doi.org/10.1520/MNL6920150006

Huang, Daming, Haining Zhou, and Lin Lin. "Biodiesel: An alternative to conventional fuel." *Energy Procedia* 16 (2012): 1874–1885. https://doi.org/10.1016/j.egypro.2012.01.287

Imdadul, Hossain K., Hassan H. Masjuki, Mohd A. Kalam, Nor W. M. Zulkifli, Abdullah Alabdulkarem, Md. M. Rashed, Yee H. Teoh, and Hock G. How. "Higher alcohol–biodiesel–diesel blends: An approach for improving the performance, emission, and combustion of a light-duty diesel engine." *Energy Conversion and Management* 111 (2016): 174–185. https://doi.org/10.1016/j.enconman.2015.12.066

Islam, Aminul, and Pogaku Ravindra. *Biodiesel Production With Green Technologies*. Springer International Publishing, 2017.

Ismail, Mona M., Gehan A. Ismail, and Mostafa M. El-Sheekh. "Potential assessment of some micro-and macroalgal species for bioethanol and biodiesel production." *Energy Sources, Part a: Recovery, Utilization, and Environmental Effects* 46, no. 1 (2024): 7683–7699. https://doi.org/10.1080/15567036.2020.1758853

Jaichandar, Sundar, and Kumar Annamalai. "The status of biodiesel as an alternative fuel for diesel engine–an overview." *Journal of Sustainable Energy & Environment* 2, no. 2 (2011): 71–75.

Jaiswal, Krishna K., Swapnamoy Dutta, Ishita Banerjee, Cheryl B. Pohrmen, and Vinod Kumar. "Photosynthetic microalgae–based carbon sequestration and generation of biomass in biorefinery approach for renewable biofuels for a cleaner environment." *Biomass Conversion and Biorefinery* (2021): 1–19.

Jamaluddin, Nor A. M., Teuku M. I. Riayatsyah, Arridina S. Silitonga, Md. Mofijur, Abd H. Shamsuddin, Hwai C. Ong, Teuku M. I. Mahlia, and SM Ashrafur Rahman. "Techno-economic analysis and physicochemical properties of Ceiba pentandra as second-generation biodiesel based on ASTM D6751 and EN 14214." *Processes* 7, no. 9 (2019): 636. https://doi.org/10.3390/pr7090636

Jamil, Farrukh. "A review on biodiesel production, analysis, and emission characteristics from non-edible feedstocks." *ChemistrySelect* 8, no. 31 (2023): e202300800. https://doi.org/10.1002/slct.202300800

Jeswani, Harish K., Andrew Chilvers, and Adisa Azapagic. "Environmental sustainability of biofuels: A review." *Proceedings of the Royal Society A* 476, no. 2243 (2020): 20200351.

Jothiramalingam, Rajabathar, and Ming K. Wang. "Review of recent developments in solid acid, base, and enzyme catalysts (heterogeneous) for biodiesel production via transesterification." *Industrial & Engineering Chemistry Research* 48, no. 13 (2009): 6162–6172.

Kamer, Paul C. J., Vogt Dieter, and Joris W. Thybaut, eds. *Contemporary Catalysis: Science, Technology and Applications*. Royal Society of Chemistry, 2017 https://doi.org/10.1016/j.scitotenv.2021.145683

Karmakar, Bisheswar, and Gopinath Halder. "Progress and future of biodiesel synthesis: Advancements in oil extraction and conversion technologies." *Energy Conversion and Management* 182 (2019): 307–339. https://doi.org/10.1016/j.enconman.2018.12.066

Kayode, Bayonle, and Abarasi Hart. "An overview of transesterification methods for producing biodiesel from waste vegetable oils." *Biofuels* 10, no. 3 (2019): 419–437.

Kibar, Murat E., Lina Hilal, Begüm T. Çapa, Buse Bahçıvanlar, and Brahim B. Abdeljelil. "Assessment of homogeneous and heterogeneous catalysts in transesterification reaction." *ChemBioEng Reviews* 10, no. 4 (2023): 412–422. https://doi.org/10.1002/cben.202200021.

Knothe, Gerhard. "History of vegetable oil-based diesel fuels." In *The Biodiesel Handbook*, pp. 5–19. AOCS Press, 2010. https://doi.org/10.1016/B978-1-893997-62-2.50007-3

Körbitz, Werner, Christian Berger, Erich Waginger, and Manfred Wörgetter. "Best case studies on biodiesel production plants in Europe." *International Energy Agency Bioenergy Task* 39 (2004). https://task39.sites.olt.ubc.ca/files/2013/05/Best-case-studies-on-biodiesel-production-plants-in-Europe.pdf

Kumar, Sanjeev, and Amit Pal. "Life cycle analysis of biodiesel derived from fresh water microalgae and karanja." *Sustainable Energy Technologies and Assessments* 56 (2023): 103082.

Lau, Chi H., Harrison L. N. Lau, Hoon K. Ng, Suchithra Thangalazhy-Gopakumar, Lai Y. Lee, and Suyin Gan. "Evaluation of oxidation stability for palm biodiesel with additives using Fourier transform infrared spectroscopy, gas chromatography-mass spectrometry and high-performance liquid chromatography." *Results in Engineering* 22 (2024): 102165. https://doi.org/10.1016/j.rineng.2024.102165

Lunn, Joanne, and Heather E. Theobald. "The health effects of dietary unsaturated fatty acids." *Nutrition Bulletin* 31, no. 3 (2006): 178–224.

Malik, Muhammad A. I., Sadaf Zeeshan, Muhammad Khubaib, Adeel Ikram, Fayaz Hussain, Hayati Yassin, and Atika Qazi. "A review of major trends, opportunities, and technical challenges in biodiesel production from waste sources." *Energy Conversion and Management: X* (2024): 100675. https://doi.org/10.1016/j.ecmx.2024.100675

Mandal, Sudip, Rashmi Patnaik, Akhilesh K. Singh, and Nirupama Mallick. "Comparative assessment of various lipid extraction protocols and optimization of transesterification process for microalgal biodiesel production." *Environmental Technology* 34, no. 13–14 (2013), 2018–2024. https://doi.org/10.1080/09593330.2013.827730

Marchetti, Jorge M., Vera U. Miguel, and Alberto F. Errazu. "Possible methods for biodiesel production." *Renewable and Sustainable Energy Reviews* 11, no. 6 (2007): 1300–1311. https://doi.org/10.1016/j.rser.2005.08.006

Mardani, Mohsen, Katalin Badak-Kerti, Judit Tormási, and Jamshid Farmani. "Selective partial hydrolysis as a novel strategy to produce specialty structured fats from vegetable oils: Optimization of monoacylglycerol formation and assessment of the final product." *Food and Bioprocess Technology* (2024): 1–21.

Mathushika, Judy, and Chandima Gomes. "Development of microalgae-based biofuels as a viable green energy source: Challenges and future perspectives." *Biointerface Research in Applied Chemistry* 12 (2022): 3849–3882.

Matsungo, Tonderayi M., and Linda P. Siziba. "Lipids and nutrition security." In *Lipids and Edible Oils*, pp. 1–43. Academic Press, 2020. https://doi.org/10.1016/B978-0-12-817105-9.00001-X

Moazeni, Faegheh, Yen-Chih Chen, and Gaosen Zhang. "Enzymatic transesterification for biodiesel production from used cooking oil, a review." *Journal of Cleaner Production* 216 (2019): 117–128.

Mohiddin, Mohd N. B., Yie H. Tan, Yee X. Seow, Jibrail Kansedo, Nor M. Mubarak, Mohammad O. Abdullah, Yen S. Chan, and Mohammad Khalid. "Evaluation on feedstock, technologies, catalyst and reactor for sustainable biodiesel production: A review." *Journal of Industrial and Engineering Chemistry* 98 (2021): 60–81. https://doi.org/10.1016/j.jiec.2021.03.036

Moser, Bryan R. "Biodiesel production, properties, and feedstocks." *In Vitro Cellular & Developmental Biology-Plant* 45 (2009): 229–266.

Mukherjee, Sohini, Debasish Sarkar, Tanima Bhattacharya, Tanmoy Das, Avery Sengupta, Subham Preetam, Vaidehi Ulaganathan, and Sarvesh Rustagi. "Reaction kinetics involved in esterification between the fatty acids in castor oil and furfuryl alcohol." *Industrial Crops and Products* 213 (2024): 118393.

Murugesan, Arumugam, Chandrasekaran Umarani, Ramasamy Subramanian, and Natarajan Nedunchezhian. "Bio-diesel as an alternative fuel for diesel engines – A review." *Renewable and Sustainable Energy Reviews* 13, no. 3 (2009): 653–662. https://doi.org/10.1016/j.rser.2007.10.007

Nadkarni, Ramesh A. *Guide to ASTM Test Methods for the Analysis of Petroleum Products and Lubricants*, vol. 44. ASTM International, 2007.

Narayanan, Mathiyazhagan. "Promising biorefinery products from marine macro and microalgal biomass: A review." *Renewable and Sustainable Energy Reviews* 190 (2024): 114081.

Naveenkumar, Rajendiran, and Gurunathan Baskar. "Optimization and techno-economic analysis of biodiesel production from Calophyllum inophyllum oil using heterogeneous nanocatalyst." *Bioresource Technology* 315 (2020): 123852. https://doi.org/10.1016/j.biortech.2020.123852

Neag, Emilia, Zamfira Stupar, S. Andrada Maicaneanu, and Cecilia Roman. "Advances in biodiesel production from microalgae." *Energies* 16, no. 3 (2023): 1129.

Noor, Chow W. M., Mohd M. Noor, and Rozlan Mamat. "Biodiesel as alternative fuel for marine diesel engine applications: A review." *Renewable and Sustainable Energy Reviews* 94 (2018): 127–142.

Ogunkunle, Oyetola, and Noor A. Ahmed. "Overview of biodiesel combustion in mitigating the adverse impacts of engine emissions on the sustainable human–environment scenario." *Sustainability* 13, no. 10 (2021): 5465.

Osman, Mohamed E. H., Atef M. Abo-Shady, Saly F. Gheda, Samy M. Desoki, and Mostafa E. Elshobary. "Unlocking the potential of microalgae cultivated on wastewater combined with salinity stress to improve biodiesel production." *Environmental Science and Pollution Research* 30, no. 53 (2023): 114610–114624. https://doi.org/10.1016/B978-0-12-823764-9.00024-8

Pandey, Archana, and Yogesh C. Sharma.. "Advancements in biomass valorization in integrated biorefinery systems." *Biofuels, Bioproducts and Biorefining*. 18, no. 6 (2024): 2078–2090.

Pandit, Chetan, Srijoni Banerjee, Soumya Pandit, Dibyajit Lahiri, Vinod Kumar, Kundan Kumar Chaubey, Rayyan Al-Balushi, Saif Al-Bahry, and Sanket J. Joshi. "Recent advances and challenges in the utilization of nanomaterials in transesterification for biodiesel production." *Heliyon* 9, no. 4 (2023). https://doi.org/10.1016/j.heliyon.2023.e15475

Patil, Vishwanath, Kjell I. Reitan, Gjert Knutsen, Leiv M. Mortensen, Torsten Källqvist, Elisabeth Olsen, Gjermund Vogt, and Hans R. Gislerød. "Microalgae as source of polyunsaturated fatty acids for aquaculture." *Plant Biol* 6, no. 6 (2005): 57–65.

Pikula, Konstantin, Alexander Zakharenko, Antonios Stratidakis, Mayya Razgonova, Alexander Nosyrev, Yaroslav Mezhuev, Aristidis Tsatsakis, and Kirill Golokhvast. "The advances and limitations in biodiesel production: Feedstocks, oil extraction methods, production, and environmental life cycle assessment." *Green Chemistry Letters and Reviews* 13, no. 4 (2020): 275–294. https://doi.org/10.1080/17518253.2020.1829099

Powar, Ranveer S., Anil Singh Yadav, Ch Siva Ramakrishna, Swastika Patel, Man Mohan, Sanjay G. Sakharwade, Mayank Choubey, Alok Kumar Ansu, and Abhishek Sharma. "Algae: A potential feedstock for third generation biofuel." *Materials Today: Proceedings* 63 (2022): A27–A33. https://doi.org/10.1016/j.matpr.2022.07.161

Pydimalla, Madhuri, Sadia Husaini, Akshara Kadire, and Raj Kumar Verma. "Sustainable biodiesel: A comprehensive review on feedstock, production methods, applications, challenges and opportunities." *Materials Today: Proceedings* 92 (2023): 458–464.

Quayson, Emmanuel, Jerome Amoah, Shinji Hama, Akihiko Kondo, and Chiaki Ogino. "Immobilized lipases for biodiesel production: Current and future greening opportunities." *Renewable and Sustainable Energy Reviews* 134 (2020): 110355.

Raising Rathod, Dhnyaneshwar, Sandesh S. Karkal, Akil S. Jamadar, Aliaa M. A. Hashem, Prashanth V. Suresh, Sowmya S. Mamatha, and Tanaji G. Kudre. "Prospects of novel heterogeneous base catalysts and nanocatalysts in achieving sustainable biodiesel production." *International Journal of Green Energy* 21, no. 5 (2024): 1017–1042 https://doi.org/10.1080/15435075.2023.2228886

Renzaho, Andre M. N., Joseph K. Kamara, and Michael Toole. "Biofuel production and its impact on food security in low- and middle-income countries: Implications for the post-2015 sustainable development goals." *Renewable and Sustainable Energy Reviews* 78 (2017): 503–516. https://doi.org/10.1016/j.rser.2017.04.072

Ridwan, Iwan, Herawati Budiastuti, Retno Indarti, Ninik L. E. Wahyuni, Hasna Mutiara Safitri, and Rama Luthfi Ramadhan. "The optimization of tetrahydrofuran as a co-solvent on biodiesel production from rubber seeds using response surface methodology." *Materials Science for Energy Technologies* 6 (2023): 15–20. https://doi.org/10.1016/j.mset.2022.11.002

Rodionova, Margarita V., Roshan Sharma Poudyal, Indira Tiwari, Roman A. Voloshin, Sergey K. Zharmukhamedov, Hong G. Nam, Bolatkhan K. Zayadan, Barry D. Bruce, John M. Harvey, Suleyman I. Hou, and Sergi I. Allakhverdiev. "Biofuel production: Challenges and opportunities." *International Journal of Hydrogen Energy* 42, no. 12 (2017): 8450–8461 https://doi.org/10.1016/j.ijhydene.2016.11.125

Saber, Abdullah A., Ahmed A. El-Refaey, Hani Saber, Prashant Singh, Sanet J. V. Vuuren, and Marco Cantonati. "Cyanoprokaryotes and algae: Classification and habitats." In *Handbook of Algal Biofuels*, pp. 1–38. Elsevier, 2022. https://doi.org/10.1016/B978-0-12-823764-9.00024-8

Saranya, G., and T. V. Ramachandra. "Life cycle assessment of biodiesel from estuarine microalgae." *Energy Conversion and Management: X* 8 (2020): 100065. https://doi.org/10.1016/j.ecmx.2020.100065

Sarkar, Pradip, Tarun Kanti Bandyopadhyay, Konga Gopikrishna, Onkar Nath Tiwari, Biswanath Bhunia, and Muthusivaramapandian Muthuraj. "Algal carbohydrates: Sources, biosynthetic pathway, production, and applications." *Bioresource Technology* (2024): 131489.

Selvarajan, Ramganesh, Cecilia Oluseyi Osunmakinde, Timothy Sibanda, Anofi Ashafa, and Sanniyasi Elumalai. "Biofuels from microalgae: Future bio-energies for sustainable development." In *Phycobiotechnology*, pp. 109–140. Apple Academic Press, 2021.

Shabbir, Muneeba, Muhammad N. Anwar, Hira Saif, Eza Tahir, Ajwa Tahir, Mohammad Rehan, Rameesha Tanveer, Mortaza Aghbashlo, Meisam Tabatabaei, and Abdul S. Nizami. "Policy and regulatory constraints in the biodiesel production and commercialization." In *Sustainable Biodiesel*, pp. 357–372. Academic Press, 2023. https://doi.org/10.1016/B978-0-12-820361-3.00007-3

Siddeg, Abu B., Umar Adam, Norrizam Jaat, Amir Khalid, Azwan Sapit, Sofian Mohd, and Mzahar A. Jalal. "Analysis of the influences of biodiesel on performance and emissions of a diesel engine." *Journal of Automotive Powertrain and Transportation Technology* 2, no. 2 (2022): 1–8.

Singha, Golap C., M. Vijayakumar, R. Rajkumar, K. R. Jayappriyan, and S. Pujithaa. "Algae materials for advanced biofuel production through the cost-effective process and integration of nanocatalysts." In *Algae Materials*, pp. 29–62. Academic Press, 2023. https://doi.org/10.1016/B978-0-443-18816-9.00025-3

Stergiou, Andreas. "Eastern Mediterranean energy geopolitics revisited: Green economy instead of conflict." *Journal of Balkan and Near Eastern Studies* 25, no. 4 (2023): 604–625.

Su, Fang, and Yihang Guo. "Advancements in solid acid catalysts for biodiesel production." *Green Chemistry* 16, no. 6 (2014): 2934–2957.

Suhara, Ade, Karyadi, Safarudin Gazali Herawan, Andy Tirta, Muhammad Idris, Muhammad Faizullizam Roslan, Nicky Rahmana Putra, April Lia Hananto, and Ibham Veza. "Biodiesel sustainability: Review of progress and challenges of biodiesel as sustainable biofuel." *Clean Technologies* 6, no. 3 (2024): 886–906. https://doi.org/10.3390/cleantechnol6030045

Supang, Wirasinee, Somkiat Ngamprasertsith, Winatta Sakdasri, and Ruengwit Sawangkeaw. "Biodiesel production from spent coffee grounds by using ethanolic extraction and supercritical transesterification." *BioEnergy Research* (2024): 1–11. https://doi.org/10.1007/s12155-024-10782-z

Taki, Kimia, Bahram Hosseinzadeh Samani, Sajad Rostami, and Shahin Besharati. "Combination of oscillatory and hydrodynamic system for biodiesel production from sunflower oil." *Biofuels* 14, no. 3 (2023): 235–242. https://doi.org/10.1080/17597269.2022.2130510

Talwan, Poonam, Darsh Gautam, Rishi Kumar, Savita Sharma, Sahil Dhiman, Rahul Gill, Aparna Thakur, Deexa Sharma, Saksham Sharma, and Akhil Kumar. "A review of the chemical composition, modification, and biomedical application of Ricinus communis." *Indian Journal of Natural Products and Resources (IJNPR)[Formerly Natural Product Radiance (NPR)]* 15, no. 1 (2024): 21–42.

Tan, Shiou X., Steven Lim, Hwai C. Ong, and Yean L. Pang. "State of the art review on development of ultrasound-assisted catalytic transesterification process for biodiesel production." *Fuel* 235 (2019): 886–907. https://doi.org/10.1016/j.fuel.2018.08.021

Tanwar, Manju D., Felipe A. Torres, Ali M. Alqahtani, Pankaj K. Tanwar, Yashas Bhand, and Omid Doustdar. "Promising bioalcohols for low-emission vehicles." *Energies* 16, no. 2 (2023): 597.

Tesfaye, Feleke. "Performance evaluation of castor biodiesel and petro diesel blends with diethyleether additive." (2016).

Thakur, Deepak S., and Arunabha Kundu. "Catalysts for fatty alcohol production from renewable resources." *Journal of the American Oil Chemists' Society* 93, no. 12 (2016): 1575–1593.

Thangaraja, Jeyaseelan, and Vignesh Srinivasan. "Techno-economic assessment of coconut biodiesel as a potential alternative fuel for compression ignition engines." *Environmental Science and Pollution Research* 26 (2019): 8650–8664.

Torkashvand, Mohammad, Atefe Hasan-Zadeh, and A. Torkashvand. "Mini review on importance, application, advantages and disadvantages of biofuels." *Journal of Materials and Environmental Science* 13, no. 6 (2022): 612–630.

Tran-Nguyen, Phuong L., Lu K. Ong, Alchris Woo Go, Yi-H. Ju, and Artik E. Angkawijaya. "Non-catalytic and heterogeneous acid/base-catalyzed biodiesel production: Recent and future developments." *Asia-Pacific Journal of Chemical Engineering* 15, no. 3 (2020): e2490.

Trejo-Zárraga, Fernando, Felipe J. Hernández-Loyo, Juan C. Chavarría-Hernández, and Rogelio Sotelo-Boyás. "Kinetics of transesterification processes for biodiesel production." *Biofuels-State of Development* (2018): 149–179.

Tutunea, Dragos, Bica Marin, and Cernaianu Corina, and Alexandru Dima. "Correlation between physico-chemical properties of biodiesel blends and fossil diesel." (2013): 67–70

Uppalapati, Sudhakar, Stanly P. Jani, Mahaboob B. A. Khan, Manjunathan Alagarsamy, Mathankumar Manoharan, Hitesh Panchal, and Kishor K. Sadasivuni. "A comparative assessment on life cycle analysis of the biodiesel fuels produced from soybean, Jatropha, Calophyllum inophyllum, and microalgae." *Energy Sources, Part A: Recovery, Utilization, and Environmental Effects* 44, no. 2 (2022): 3253–3272.

Vendruscolo, Raquel G., Mariane Bittencourt Fagundes, Eduardo Jacob-Lopes, and Roger Wagner. "Analytical strategies for using gas chromatography to control and optimize microalgae bioprocessing." *Current Opinion in Food Science* 25 (2019): 73–81. https://doi.org/10.1016/j.cofs.2019.02.008

Vinoth Kumar, R., I. Ganesh Moorthy, Lalit Goswami, G. Pugazhenthi, Kannan Pakshirajan, Adrián M. T. Silva, and Sergio Morales-Torres. "Analytical methods in biodiesel production." *Biomass Valorization to Bioenergy* (2020): 197–219. https://doi.org/10.1007/978-981-15-0410-5_13

Warrender, Stewart J. "Structure direction in the formation of zeolitic materials." PhD diss., University of St Andrews, 2007.

Woldie, Aklilu A., and Sintayehu M. Hailegiorgis. "Microwave energy enhanced oil extraction from Millettia ferruginea (Birbira) seed for biodiesel application: Effects of microwave energy on oil extraction yield." *Biomass Conversion and Biorefinery* (2024): 1–14. https://doi.org/10.1007/s13399-024-05530-4

Yaamini, Matheswaran, Chidambaram K. Venil, and Laurent Dufossé. "Valuable bio-products from algae and their commercial applications." In *Algal Metabolites*, pp. 25–45. Apple Academic Press, 2023.

Zamora, Rosario, and Francisco J. Hidalgo. "Fatty acids." *Food Science and Technology-New York-Marcel Dekker* 138, no. 1 (2004): 221.

Zheng, Jiawei, Yudong Liang, Jiaxi Li, Shuping Lin, Qiangyue Zhang, Kanghua Zuo, and Nanjing Zhong, and Xuebing Xu. "Enzymatic preparation of mono- and diacylglycerols: A review." *Grain & Oil Science and Technology*, 6, no. 4 (2023): 185–205.

12 Analytical Advancements in Biofuels Industry

*Sougata Rajak, Hassna Jaber, Saanya Chaturvedi, Vivek Raj, and Deboja Sharma**

12.1 INTRODUCTION

Biofuels are renewable energy sources derived from biological materials like plants, agricultural by-products, and wastes, which can be transformed into fuel using biochemical, thermochemical, or physical-chemical methods (Ahorsu et al., 2018). Familiar examples are bioethanol and biodiesel. In contrast to fossil fuels, biofuels are within the short-term carbon cycle, offering potential carbon dioxide reduction if produced sustainably (Martens et al., 2017; Saleh & Hassan, 2023). The urgent need to control climate change, increase global energy supply, and decrease the use of fossil fuels has shifted focus to biofuels as part of renewable energy solutions, particularly in bioenergy for aviation, shipping, and heavy-duty transportation.

Biofuels were first addressed in the 19th century, but interest declined with the rise of inexpensive petroleum until the oil crises in the 1970s shifted focus to alternative energy (Dahlgren, 2022; Nemett & Adams, 2022). Biofuels have developed in three generations over time; first generation biofuels raise concerns of land use and food security due to being produced from food crops; second-generation biofuels are produced from non-food biomass (Asase et al., 2024; Trivedi & Joshi, 2025) and seek to mitigate these problems but still face technological obstacles (Singla et al., 2023); and third-generation biofuels, derived from microalgae, hold promise but are hindered by high production costs and scalability issues.

The motivation for the development of biofuels stems from environmental and geopolitical considerations. Biofuels are an appealing substitute for fossil fuels due to climate change and the need to decarbonize the world's energy systems, especially in areas where electrification is difficult (Ali et al., 2025; Mignogna et al., 2024). In addition, biofuels enhance energy security by diversifying access to oil and imported petroleum, while aiding rural economies through the local biomass supply chains. Promotion of decentralized biomass-based energy generation in India's National Bioenergy Programme (NBP) aims at valorizing local agricultural residues and reducing transportation costs and open-field burning (Thaha et al., 2025; Verma et al., 2024). Achieving 20% ethanol blending in petrol (E20) by 2025 is one of the central goals to significantly reduce dependence on imported oil (Chandel et al., 2024; Kubendran, 2024). The policy also emphasizes the non-food feed stocks crop residues and dedicated energy crops to avert clashes with food production (Bhuvaneshwari et al., 2019; Rathour et al., 2023). The policy addresses environmental concerns while promoting inclusive economic development by integrating

* Corresponding author

 DOI: 10.1201/9781003485636-12

rural labor markets, supporting second- and third-generation biorefineries, and subsidizing rural biogeographical dispersal regions with decentralized biorefinery infrastructures with second- and third-generation biofuel technologies (Makepa & Chihobo, 2024).

Agencies such as the International Energy Agency (IEA) focus on the role of biofuels in attaining international climate and energy targets (Junginger et al., 2019). The IEA stresses other reports like Net Zero by 2050 and annual Renewables outlooks, stating biofuels, particularly advanced and sustainably derived ones, are paramount in decarbonizing hard-to-abate sectors such as aviation, maritime transportation, and heavy road freight (Anika et al., 2022). The agency calls for immediate action in the form of research and development investments, targeted policy frameworks, and cross-border collaborations to increase biofuel production. Besides, the IEA insists that without substantial deployment of bioenergy, including liquid biofuels, achieving global net-zero emissions would not be possible (Friedmann et al., 2020). This strategic framework highlights the need to integrate biofuels into overarching energy transition strategies within countries and beyond their borders.

Even with their ever-increasing importance as substitutes for fossil fuels, biofuels still have challenges that relate to the availability of feedstocks, conversion efficiency, and economic feasibility. According to the Arnoult and Brancourt-Hulmel (2015) and Williams et al. (2016) study, it is discussed that biomass resources vary significantly by region and season, giving inconsistencies in the composition and supply, which impact process performance and product quality. Microalgae and lignocellulosic residues are advanced biofuels that require sophisticated procedures for production, such as pretreatment and the use of specialized catalysts. Recent studies (John et al., 2025; Tran et al., 2024) explain how these processes are costly and use intensively more energy. Infrastructure limitations are also an issue; since there is a lack of decentralized biorefineries and compatibility with already used fuel systems, scalability is hindered (Singh et al., 2024). High precision and control are necessary to manage all elements of the biofuel value chain, such as raw material characterization, real-time monitoring, and final product validation. Process optimization is made possible with advanced analytical techniques that ensure quality standards as well as enhance yield and decrease variability. The aim of this work is to illustrate the contribution these techniques have toward the sustainable industrial-scale deployment of biofuels.

12.2 THE ROLE OF ANALYTICAL TECHNIQUES IN BIOFUEL PRODUCTION

Modern biofuel production involves a series of complex biochemical and thermochemical transformations that require precise monitoring and control to ensure efficiency, high yield, and environmental sustainability (Adeoye et al., 2023; Malode et al., 2022). From the beginning of the process, with the preparation of the raw materials, to the end of the production of biofuels, each step of the process comes with its very own analytical challenges, which have a very big influence on the economic and ecological rationale of the fuel (Asghar et al., 2022). The usage of corresponding analytical instruments helps to monitor and control the goods flow in time to protect the highest goods quality and to produce the products underlying a certain legal regulation (Da Silva & Queiroz, 2017). On the other hand, the sustainability trend and their contribution to the new energy concept in energy generation and mobility have certainly led to extensive research (Nisansala, 2025).

12.2.1 Analytical Requirements Across Production Stages

12.2.1.1 Feedstock Characterization

In terms of biochemical or thermochemical processes, biomass feedstocks should be preceded with a detailed and comprehensive characterization. As noted by Williams et al. (2016), as well as Mishra et al. (2022), the physicochemical properties of the biomass feedstock impact its conversion efficiency, how the process is designed, and the resultant yield. Esteves et al. (2023) and Shadangi et al. (2023) suggest that the compositional parameters encompass moisture and ash content, structural polysaccharides (cellulose and hemicellulose), lignin, extractives, such as lipids and phenolics, and the elemental composition of carbon, hydrogen, nitrogen, sulfur, oxygen, and certain trace metals. In analyzing these attributes, both classical and advanced analytical methods are utilized (Ajorloo et al., 2022; Bernier et al., 2018; Bojkovic et al., 2021; Liu et al., 2022).

- **Thermogravimetric Analysis (TGA):** TGA measures the amount and rate of change in the weight of a material as a function of temperature or time.
- **Near-Infrared Spectroscopy (NIR):** NIR is a fast and non-destructive method for predicting the major chemical constituents, and in particular for high-throughput applications.
- **High-Performance Liquid Chromatography (HPLC):** HPLC quantifies the monosaccharides, oligosaccharides, and degradation products upon biomass hydrolysis.
- **Gas Chromatography (GC):** GC is used to characterize volatile extractives, especially lipids and degradation products in lignocellulose feedstocks.
- **Fourier Transform Infrared Spectroscopy (FTIR):** FTIR is used to identify the functional groups in the biomass and in lignocellulosic materials and hence helps to distinguish between cellulose, hemicellulose, and lignin.
- **X-ray Fluorescence (XRF):** XRF provides quick and accurate multi-elemental qualitative and quantitative information, in particular for inorganic parameters such as the detection of alkali metals, silicon, and metal impurities without further sample preparation.
- **Elemental Analysis (CHNS-O):** EA determines carbon, hydrogen, nitrogen, sulfur, and oxygen to assess the energy and combustion properties of the biomass.
- **Pyrolysis-Gas Chromatography/Mass Spectrometry (Py-GC/MS):** Py-GC/MS is an analytical technology used to acquire evolution profiles of the products (gases, liquids, or solids) resulting from the thermally induced decomposition of materials in an inert environment (Liu et al., 2022)

- **Laser-Induced Breakdown Spectroscopy (LIBS):** LIBS is seen as a potential alternative to many existing analysis techniques for trace element analysis in mineral ores.
- **Hyperspectral Imaging (HSI):** It utilizes imaging and spectroscopy for the purpose of visualizing chemical differences across diverse biomass samples, thus delivering localized compositional information.

These approaches are essential not only to establish a precise profile of the feedstocks but also to generate biomass conversion processes that are designed specifically for the regional context and to manufacture sustainable biofuels. Multimodal analytical workflows make it possible today to bring together the most relevant information in terms of structural organization and chemical composition within the same sample.

12.2.1.2 Monitoring the Conversion Process

Transforming biomass into biofuels is achieved through a range of intricate biochemical and thermochemical processes, including fermentation, pyrolysis, gasification, and transesterification (Kumar et al., 2017; Okolie et al., 2022; Poornima et al., 2024). These highly complex processes are very sensitive to the slightest change in operating conditions, thus necessitating precise and real-time monitoring for optimizing reaction kinetics, maximizing yields, and ensuring product purity. The process parameters such as temperature, pressure, pH, dissolved oxygen, reactant concentration, and catalyst activity are closely controlled and monitored (Hossain & Mahlia, 2019; Lee et al., 2021). A range of analytical techniques is utilized to follow the time-dependent changes of reactions and to gain insight into the formation of intermediates and end products (Ong et al., 2020):

- **Gas Chromatography (GC):** Most commonly used for qualitative and quantitative analyses of volatile compounds, including alcohols, esters, and fatty acid methyl esters (FAMEs) in biodiesel production. GC provides reliable analytical information about the degree of reaction completion and side-reaction product formation.
- **Nuclear Magnetic Resonance (NMR) Spectroscopy:** Enables real-time, non-invasive structural elucidation of intermediates and products. In transesterification or fermentation, NMR provides molecular-level insights into chemical transformations and kinetics.
- **UV-Visible Spectroscopy (UV-Vis):** Often applied in fermentation monitoring, UV-Vis tracks changes in absorbance related to enzymatic activity, microbial growth, and chromophoric metabolite production. It is also used to monitor cofactors or pigments indicative of reaction progress.
- **High-Performance Liquid Chromatography (HPLC):** Essential in fermentation and enzymatic hydrolysis processes for monitoring sugars, alcohols, organic acids, and inhibitory compounds. HPLC ensures that substrate utilization and product formation are accurately quantified.
- **Fourier-Transform Infrared Spectroscopy (FTIR):** For in situ studies or for real-time monitoring of chemical transformations such as the

degradation of lignin to form bio-oil and biodiesel during the continuous upgrading of the crude bio-oil to biodiesel.

- **Online Sensors and Process Analyzers:** Advanced instrumentation such as pH probes, dissolved CO2/O2 sensors, and Raman spectrometers are increasingly integrated into bioreactors and continuous flow systems for automated control and feedback regulation.

Introducing such advanced process analytical techniques enhances the transparency, reproducibility, and flexibility of processes. Moreover, in a manufacturing location where the Process Analytical Technology (PAT) framework, which is supported by authorities, is implemented, it is feasible to implement continuous quality control in real-time and stay consistent in terms of quality while scaling up from the laboratory to the full-scale industrial processing.

12.2.1.3 Product Analysis and Quality Control

The final biofuel product must meet strict quality requirements and fulfil international norms that are safe and eco-friendly to ensure the quality, safety, and sustainability of the product (Ahmad, 2018; Nisansala, 2025). Detailed examination of biofuels is warranted to assess such properties as fitness for purpose, energy content, thermal- and oxidation stability, and the absence of (harmful) impurities (e.g., methanol, glycerol, soap, free fatty acids, by-products of oxidation, heavy metals) (Borecki et al., 2024; Wongsurakul et al., 2022). Such measures are necessary to ensure the quality of biofuels intended for use in the transport sector as well as to maintain the compatibility with vehicle and fuel distribution systems. Several different cutting-edge analytical tools can be a source of information concerning the quality of the final product.

- **Gas Chromatography–Mass Spectrometry (GC-MS):** Provides both qualitative and quantitative identification of volatile and semi-volatile organic compounds. It is particularly effective for detecting trace contaminants such as residual methanol or oxidation by-products in biodiesel and ethanol (Muhammed Raji et al., 2025).
- **Fourier-Transform Infrared Spectroscopy (FTIR):** Offers rapid, non-destructive analysis of functional groups present in biofuels. FTIR is routinely used to confirm the presence of ester bonds, alcohols, or residual feedstock components, and to assess chemical conformity to fuel standards (Liu et al., 2022).
- **Differential Scanning Calorimetry (DSC):** Assesses thermal properties, such as heat of combustion, melting point, and oxidative stability, which are essential for evaluating the fuel's energy content and performance under various operating conditions (Uğuz et al., 2019).
- **Nuclear Magnetic Resonance (NMR):** Complements FTIR by offering more detailed structural insights and quantification of specific functional groups, especially useful in biodiesel and advanced bio-oil analysis (Nazir et al., 2025). As seen in Table 12.1 below, indicate the functions and some representative applications of those analytical techniques in the process of biofuel production.

TABLE 12.1
Common Analytical Techniques in Biofuel Production

Technique	Function	Application in Biofuel Production
Gas Chromatography (GC)	Separates and quantifies volatile compounds	Analysis of fatty acid methyl esters (FAMEs) in biodiesel
High-Performance Liquid Chromatography (HPLC)	Separates non-volatile compounds	Monitoring sugars, alcohols, and inhibitors during fermentation
Mass Spectrometry (MS)	Molecular identification and quantification	Detection of trace contaminants and profiling of bio-oils
Thermogravimetric Analysis (TGA)	Measures thermal stability and composition	Determination of moisture, volatile content, and ash in biomass
Differential Scanning Calorimetry (DSC)	Heat flow and thermal transition analysis	Evaluation of fuel energy content and thermal stability
UV-Vis/FTIR/NIR/NMR	Spectroscopic structural characterization	Reaction monitoring, purity assessment, and compositional analysis

Analytical techniques are indispensable to the advancement of biofuel production technologies. By enabling accurate, real-time process control and comprehensive product evaluation, they ensure high-quality outputs, operational safety, and regulatory compliance. These methodologies are vital for bridging the gap between laboratory-scale experimentation and the commercialization of sustainable, efficient, and scalable bioenergy solutions.

12.2.2 Spectroscopic Techniques in Biofuel Analysis

Spectroscopy represents a range of techniques that are sensitive to the interaction of electromagnetic radiation with matter to enable detailed molecular information to be obtained (Beć et al., 2020). They play an indispensable role in the biofuel value chain, offering quick, non-invasive, and often real-time analysis of complex chemical systems (Da Silva & Queiroz, 2017). Their significance lies in the ease of sample preparation and the availability for high-throughput analysis (Ahmed et al., 2024), which leads to a wide range of laboratory and industrial applications with respect to the support of quality issues, process optimization, and environmental legislation in modern bioenergy systems. A variety of spectroscopic tools have been used throughout the different steps of biofuel production, from the evaluation of feedstock to the analysis of the final product. Each of these tools is based on specific physical principles that, altogether, allow the detection of molecular structures, functional groups, and the evolution of chemical makeup over time (Table 12.2).

12.2.2.1 Advantages and Limitations of Spectroscopic Methods

While spectroscopic methods are incredibly useful, every type of spectroscopy has its own strengths and limitations, which means the choice of the best method of analysis will depend on the questions that are being asked and the situations (Eberhardt et al., 2015). Things like resolution, cost (Table 12.3), portability of the instrumentation,

TABLE 12.2
Spectroscopic Techniques and Their Applications in Biofuel Analysis

Technique	Analytical Principle	Typical Applications in Biofuel Production
FTIR (Fourier Transform Infrared)	Absorption of IR radiation due to molecular vibrations of functional groups	Identification of ester bonds in biodiesel; tracking transesterification progress
NIR (Near-Infrared Spectroscopy)	Absorption of overtone and combination bands of molecular vibrations	Determination of biomass composition (e.g., lignocellulose, moisture); inline process monitoring
UV-Vis (Ultraviolet–Visible Spectroscopy)	Absorbance of UV/visible light by chromophores	Monitoring of enzymatic activity and pigment changes during fermentation; ethanol quantification
Raman Spectroscopy	Inelastic scattering of monochromatic light revealing vibrational modes	Structural characterization of bio-oils; complementary to FTIR for complex matrices
NMR (Nuclear Magnetic Resonance)	Analysis of magnetic properties of atomic nuclei under a magnetic field	Structural elucidation and purity assessment of biodiesel, bioethanol, and biopolymers

and ease of use help to decide if a particular method of spectroscopy is best carried out in a laboratory or in an industrial environment (Emwas, 2015).

Eventually, the decision on which spectroscopic technique to use should be made based on the precise target molecules, the required sample throughput, and the ease of inline integration into the production process.

TABLE 12.3
Comparison of Spectroscopic Techniques: Advantages and Limitations

Technique	Resolution	Cost	Portability	Complexity
FTIR	High (for functional group detection)	Moderate to High	Low (bench-top systems)	Moderate (requires spectral interpretation)
NIR	Moderate (mostly qualitative)	Moderate	High (portable instruments available)	Low (minimal sample prep)
UV-Vis	Moderate (depends on analyte chromophores)	Low to Moderate	High (compact devices common)	Low (user-friendly interface)
Raman	High (rich molecular vibrational data)	High	Low (limited portable options)	High (interpretation of complex spectra)
NMR	Very High (detailed structural insight)	Very High	Very Low (requires large, specialized equipment)	Very High (sophisticated operation and data analysis)

12.2.3 Chromatographic and Mass Spectrometric Methods

Chromatography is a physical method of separation in which the components to be separated are distributed between two phases: a stationary phase and a mobile phase. The diversity of chromatographic techniques is based on differences in the physical states of the mobile and stationary phases and the mechanisms of separation involved. The core principle of chromatography involves the differential partitioning of analytes between a stationary and a mobile phase. As the mobile phase traverses the stationary phase, analytes interact with these phases to varying degrees, resulting in their temporal and spatial separation. The most prevalent types of chromatography include gas chromatography (GC), liquid chromatography (LC), high-performance liquid chromatography (HPLC), and ultra-performance liquid chromatography (UPLC) (Niessen, 2006). A comprehensive understanding of chromatographic parameters is critical for optimizing separation efficiency and resolution:

i. **Retention Factor (k′):** Describes the relative retention of an analyte:

$$K' = \frac{tR - tO}{to}$$

t_R = retention time of analyte
t_o = column dead time (time for unretained compound)

ii. **Selectivity (α):** Indicates the ability to distinguish between two analytes:

$$\alpha = \frac{K2'}{K1'} \text{ where k2}' > \text{k1}'$$

iii. **Resolution (Rs): Quantifies the separation between two analyte peaks:**

$$Rs = \frac{2(tR2 - tR1)}{W1 + W2}$$

tR1, tR2 = retention times of peak 1 and 2
w1, w2 = baseline widths of the peaks

iv. **Theoretical Plates (N) and Height Equivalent to a Theoretical Plate (HETP): Indicators of column efficiency:**

$$N = 16\left(\frac{tR2}{w}\right)^2$$

w = peak width at base

$$HETP = \frac{L}{N}$$

L = column length
N = number of theoretical plates

12.2.4 Mass Spectrometry (MS)

Mass spectrometry is an analytical technique that measures the mass-to-charge ratio (m/z) of ions. By converting analytes into gas-phase ions and analyzing their mass-to-charge characteristics, MS provides highly specific molecular information. The essential components of a mass spectrometer include an ion source, a mass analyzer, and a detector (Watson & Sparkman, 2020). Ionization is a critical process in MS, wherein analytes are converted into charged species. The choice of ionization method significantly influences the sensitivity and applicability of the technique:

- **Electrospray Ionization (ESI):** A soft ionization technique ideal for polar and high-molecular-weight biomolecules, commonly used in LC-MS.
- **Atmospheric Pressure Chemical Ionization (APCI):** Suitable for moderately polar compounds, also used in LC-MS.
- **Matrix-Assisted Laser Desorption/Ionization (MALDI):** Predominantly employed in proteomics and polymer analysis.
- **Electron Ionization (EI):** A hard ionization method primarily used in GC-MS, generating extensive fragmentation useful for structural elucidation.

12.2.5 Mass Analyzers

Mass analyzers separate ions based on their mass-to-charge ratio. Various analyzers offer distinct performance characteristics:

- **Quadrupole Analyzer:** Offers unit mass resolution and is widely used in routine quantitative analyses.
- **Time-of-Flight (TOF):** High-resolution analyzer that measures ion travel time across a fixed distance.
- **Ion Trap and Orbit rap:** Provide high sensitivity and accurate mass measurement; Orbitraps are particularly valuable in complex mixture analysis.
- **Mass-to-Charge Ratio:** The following equation fundamental for ion separation in MS,

$$\frac{m}{z} = \frac{mass\ of\ ion}{charge\ on\ ion}$$

- **Time-of-Flight Equation**:
  Used in TOF mass analyzers.

$$t = \frac{d}{\sqrt{2v}} \cdot \sqrt{m / z}$$

t = flight time
d = length of flight tube
V = accelerating voltage
m/z = mass-to-charge ratio

- **Kinetic Energy Relationship**: In mass spectrometry, the kinetic energy (KE) of ions is related to their mass (m) and velocity (v) by the equation:

$$KE = \frac{1}{2}mv^2 - zV$$

m = mass of ion
ν = velocity
z = ion charge
V = voltage applied

This relationship is crucial for understanding how ions are accelerated and separated based on their mass-to-charge ratio. These equations form the mathematical basis for ion acceleration and detection in mass spectrometry.

12.2.6 Hyphenated Techniques: LC-MS and GC-MS

The coupling of chromatographic separation with mass spectrometric detection, known as hyphenated techniques, significantly enhances analytical capabilities. LC-MS and GC-MS are the most widely employed techniques in both targeted and untargeted analysis. The LC-MS system integrates the separation capability of LC with the detection power of MS. Figure 12.1 shows the workflow of the LC-MS hyphenated technique and what the basic principle of it is.

Liquid chromatography–mass spectrometry (LC-MS) and gas chromatography–mass spectrometry (GC-MS) serve as powerful analytical tools in various scientific and industrial fields, each tailored to specific types of analytes and applications. LC-MS is extensively employed in biofuel studies, where it facilitates precise measurement of biomolecules concentrations, thereby supporting biofuel development and regulatory approval processes (Zhang et al., 2021). It is also integral to metabolomics and proteomic profiling, enabling the comprehensive analysis of metabolites and proteins within complex biological systems. Additionally, food contaminant analysis benefits greatly from LC-MS due to its sensitivity and ability to detect trace levels of harmful substances such as inhibitors and other impurities in biofuel.

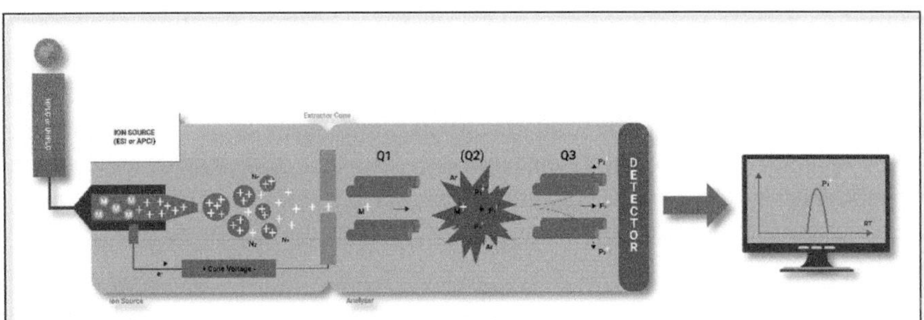

FIGURE 12.1 Illustrates this simplified LC-MS workflow, highlighting the stepwise progression from sample introduction to data output (Zhou et al., 2022).

TABLE 12.4
Table of the Comparison of LC-MS and GC-MS

Feature	LC-MS	GC-MS
Analyte Type	Polar, thermally labile	Volatile, thermally stable
Ionization Method	ESI, APCI	EI, CI
Sample Preparation	Often minimal	Frequently requires derivatization
Sensitivity	High for large, polar molecules	High for small, volatile molecules
Common Applications	Biofuel, pharmaceuticals, clinical analysis	Environmental, food, forensic

On the other hand, GC-MS is especially effective for volatile and thermally stable compounds. It plays a critical role in the detection of volatile organic compounds (VOCs), which are important markers in environmental monitoring and indoor air quality assessment. GC-MS is also widely utilized in pesticide and herbicide residue testing, ensuring compliance with safety standards in biofuel sectors. A comparison between LC-MS and GC-MS has been summarized in Table 12.4.

12.3 ADVANTAGES AND LIMITATIONS

Chromatographic and mass spectrometric techniques offer several notable advantages that make them indispensable in analytical science. One of the most prominent strengths is their high sensitivity and specificity, allowing the detection and identification of analytes even at trace levels. This is crucial in fields such as environmental monitoring, clinical diagnostics, and forensic analysis. Additionally, these techniques boast wide applicability, proving effective across a diverse spectrum of molecular classes, from small organic compounds to large biomolecules (Scigelova & Makarov, 2020). Another major benefit is their combined quantitative and qualitative capabilities, enabling researchers to obtain both structural information and precise concentration data from a single analysis.

However, despite these strengths, there are certain limitations associated with these advanced methods. The most significant drawback is the high cost of instrumentation, which includes not only the initial purchase but also ongoing maintenance expenses. Moreover, the operation of chromatographic-MS systems and interpretation of data require technical expertise, often necessitating trained personnel and extensive experience. Another challenge is matrix interference, where complex sample backgrounds can either suppress or enhance ionization, potentially leading to inaccurate quantification or misidentification of analytes. These limitations must be considered when implementing these technologies in routine or resource-constrained settings (Cappiello et al., 2023).

12.4 ADVANCED SEPARATION TECHNIQUES

The underlying goal of separation science is to exploit differences in the physicochemical properties of substances to achieve spatial or temporal resolution. Key properties utilized include molecular size, charge, polarity, volatility, and affinity toward specific phases. Advanced techniques build upon traditional principles, integrating enhanced detection systems, automation, and computational analysis for improved performance (Poole, 2023).

12.4.1 High-Performance Liquid Chromatography (HPLC)

HPLC remains a foundation in modern analytical labs. Recent advances include:

- **Ultra-High-Performance Liquid Chromatography (UHPLC):** Operates at higher pressures (>15,000 psi) using sub-2 μm particles, leading to faster separations and improved resolution.
- **Multidimensional LC (2D-LC):** Enhances peak capacity by coupling two different separation mechanisms (e.g., reverse-phase × ion-exchange) (Snyder et al., 2022).
- **Chiral Chromatography:** Employs chiral stationary phases for the enantiomeric separation of optically active compounds (Zhang et al., 2022).
- **Supercritical Fluid Chromatography (SFC):** SFC utilizes supercritical CO_2 as the mobile phase, combining the advantages of GC (speed) and LC (versatility). It is especially useful for thermally labile or non-volatile compounds (Berger & Deye, 2021).
- **Hydrophilic Interaction Chromatography (HILIC):** HILIC is optimized for polar analytes poorly retained in reverse-phase systems. It uses a polar stationary phase and an aqueous-organic mobile phase to retain hydrophilic molecules (Alpert, 2023).

12.4.2 Electrophoretic and Electro Kinetic Techniques

- **Capillary Electrophoresis (CE):** Capillary electrophoresis separates analytes based on their charge-to-size ratio in a narrow capillary under a high-voltage electric field. It offers high efficiency, minimal sample requirements, and rapid analysis. Key variants include:
- **Capillary Zone Electrophoresis (CZE):** Capillary zone electrophoresis (CZE) is a separation technique that utilizes an electric field to separate charged molecules based on their electrophoretic mobility within a narrow capillary filled with a buffer solution. It is a type of capillary electrophoresis (CE) and is known for its high resolution, speed, and minimal sample volume requirements.
- **Micellar Electrokinetic Chromatography (MEKC):** Micellar electrokinetic chromatography (MEKC), or micellar electrokinetic capillary chromatography (MECC), is a separation technique used in analytical chemistry. It combines principles of both capillary electrophoresis and chromatography to separate a wide range of analytes, including both charged and neutral molecules. This is achieved by adding surfactants to the buffer solution at concentrations above their critical micelle concentration, forming micelles that interact with the analytes.

- **Capillary Isoelectric Focusing (CIEF):** Capillary isoelectric focusing (cIEF) is a powerful analytical technique used to separate proteins and other amphoteric molecules based on their isoelectric point (pI). It's essentially a miniaturized version of slab gel isoelectric focusing, performed in a capillary tube. This method is particularly useful for analyzing charge variants of biomolecules (Breadmore et al., 2021)
- **Microchip Electrophoresis:** An advancement of capillary electrophoresis, microchip electrophoresis integrates microfluidic platforms to conduct high-throughput, miniaturized separations ideal for point-of-care testing (Whitesides, 2023).

12.4.3 Membrane and Filtration-Based Techniques

- **Ultrafiltration and Nanofiltration**: Membrane technologies allow size-based separations of macromolecules. Applications span protein purification, wastewater treatment, and food processing (Singh & Kim, 2022).
- **Solid-Phase Extraction (SPE)**: SPE is a pre-concentration and purification technique where analytes are retained on a sorbent and selectively eluted. Advances include automated platforms and novel sorbent chemistries (Hu et al., 2023).

12.4.4 Hybrid and Emerging Techniques

- **Hyphenated Techniques (e.g., LC-MS, GC-MS)**: Combining separation with detection enhances analytical specificity and sensitivity. Examples include:
- **LC-MS/MS:** Quantifies low-abundance biomarkers in biological matrices.
- **GC-MS:** Ideal for volatile and semi-volatile organic compounds.
- **Field-Flow Fractionation (FFF)**: FFF separates analytes based on their differential migration under an applied external field (e.g., flow, temperature, or magnetic). It's particularly suitable for nanoparticles and macromolecules (Giddings & Myers, 2021).
- **Affinity-Based Methods**: Techniques like immunoaffinity chromatography or DNA aptameric-based separation leverage specific bimolecular interactions for highly selective isolation (Song et al., 2022).

12.4.5 Miniaturization and Lab-on-a-Chip Technologies

Recent advances focus on reducing the footprint and resource consumption of analytical systems:

- **Microfluidics:** Enables automated, high-throughput processing with reduced reagent volumes.
- **Lab-on-a-Chip (LOC):** Combines multiple analytical steps—sample prep, separation, detection-into a single micro device (Sackmann et al., 2023).

12.5 REAL-TIME MONITORING TECHNOLOGIES

Real-time monitoring refers to the continuous or frequent measurement of parameters without significant delay between data acquisition and interpretation, as shown in Figure 12.2. The core principle relies on the integration of sensors, data acquisition systems, and computational tools to deliver instant or near-instant information. These systems typically include:

- **Sensors and Detectors:** Devices that convert physical or chemical stimuli into measurable signals.
- **Data Acquisition Systems (DAQs):** Hardware and software that collect, digitize, and store sensor data.
- **Communication Interfaces:** Wireless or wired transmission modules that relay data to control centers.
- **Data Analysis Software:** Algorithms and artificial intelligence tools that interpret raw data and generate actionable insights.

Several technologies underpin real-time monitoring systems. These include but are not limited to:

- **Spectroscopic Techniques:** Such as near-infrared (NIR), Raman, and UV-Vis spectroscopy, which are widely used for chemical identification and quantification in real time.
- **Chromatography Coupled with Mass Spectrometry:** Techniques like LC-MS and GC-MS are increasingly being optimized for real-time or at-line analysis.
- **Electrochemical Sensors:** These are particularly useful in environmental and clinical settings for detecting ions and biomolecules.

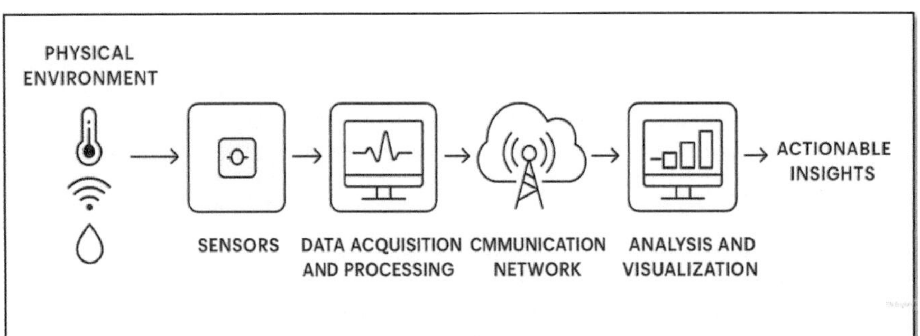

FIGURE 12.2 A schematic representation of a real-time monitoring system showing the interconnection between sensors, data acquisition units, communication modules, and cloud-based analytical platforms. The diagram illustrates how raw data from the physical environment is converted into actionable insights.

TABLE 12.5
Comparison of Selected Real-Time Monitoring Technologies

Technology	Detection Principle	Response Time	Typical Applications	Advantages
Near-Infrared Spectroscopy	Absorption of NIR light	Seconds	Pharmaceutical, Food, Agriculture	Non-destructive, fast
Electrochemical Sensors	Redox Reactions	Milliseconds	Medical diagnostics, Water quality	High sensitivity, compact
Raman Spectroscopy	Scattering of light	Seconds	Chemical identification, Bioanalysis	Specific, requires minimal prep
Microfluidic Devices	Fluid flow analysis	Seconds	Lab-on-a-chip, Bioreactors	Portable, low sample volume
Wireless Sensor Networks	Variable (depends on sensor)	Variable	Smart agriculture, Industrial monitoring	Remote sensing, scalable

- **Microfluidics and Lab-on-a-Chip Systems:** These integrate miniaturized fluidic channels with sensors for real-time monitoring of biochemical reactions.
- **Wireless Sensor Networks (WSNs):** Used in industrial and environmental monitoring to collect and transmit data over large areas (Zhang et al., 2021).

Real-time monitoring technologies are extensively used in biofuel sectors: The FDA's process analytical technology (PAT) initiative encourages the use of real-time monitoring to ensure product quality and consistency of the products (FDA, 2004). A comparison of various real-time technologies has been summarized in Table 12.5. Real-time sensors measure various parameters related to biofuel and its emission like air quality, water contaminants, and soil nutrients. This information is vital for pollution control and ecosystem management (Zhang et al., 2021). Glucose monitors, wearable biosensors, and implantable devices allow continuous monitoring of the overall biofuel process (Heikenfeld et al., 2022). Real-time systems detect spoilage, contamination, and composition changes, ensuring quality (Dey et al., 2021). In sectors like petrochemicals and materials manufacturing, real-time monitoring improves yield, reduces downtime, and ensures safety.

12.6 RECENT ADVANCEMENTS

Technological progress continues to expand the scope and precision of real-time monitoring systems: Recent advancements in real-time monitoring technologies are reshaping the landscape of analytical science by leveraging smart, interconnected, and efficient systems. One of the leading innovations is the use of artificial intelligence (AI) and machine learning (ML). These tools enable predictive analytics and adaptive control within monitoring setups, allowing systems to anticipate changes and optimize response actions in real time (Singh et al., 2023). AI-driven platforms enhance both the speed and accuracy of data interpretation, making real-time monitoring more autonomous and intelligent. Another significant trend is miniaturization and portability. Technological progress has led to the development of smaller, lighter, and more energy-efficient devices that are ideally suited for deployment in remote or mobile environments. This has expanded the scope of monitoring applications to include field-based studies and on-site diagnostics, where traditional lab setups would be impractical. Analytical chemistry is rapidly advancing, driven by the need for faster, more precise, and automated techniques. High-resolution mass spectrometry (HRMS) and ambient ionization methods like desorption electrospray ionization (DESI) and direct analysis in real time (DART) have improved compound identification and streamlined sample preparation. Portable mass spectrometers now enable on-site analysis, while artificial intelligence enhances data interpretation and peak detection. Integration with omics technologies such as metabolomics and proteomics offers deeper biological insights. Separation science is evolving through the convergence of chemistry, engineering, and informatics. Artificial intelligence (AI) and machine learning are transforming method development and real-time analysis, improving efficiency and accuracy. Green chemistry is gaining traction, promoting solvent-free techniques for sustainable lab practices. Innovations like 3D-printed devices are making analytical tools more accessible and customizable. The future of real-time monitoring is equally dynamic. Self-calibrating and quantum sensors promise greater accuracy and sensitivity, while deep learning is enabling predictive diagnostics and smarter decision-making. On a global scale, Internet of Things (IoT) powered monitoring systems are being developed to support environmental health, disaster management, and public safety ushering in a new era of intelligent, connected analytics.

12.7 CONCLUSION

Real-time monitoring technologies are transforming the landscape of analytical and operational systems across a multitude of disciplines. By delivering timely, accurate, and actionable data, these systems improve safety, efficiency, and quality control. With ongoing innovations in sensor technology, data analytics, and wireless communication, the reach and impact of real-time monitoring are expected to grow significantly in the coming years. Chromatographic and mass spectrometric methods are indispensable in analytical science, offering unmatched capabilities in separation and detection. Their integration in LC-MS and GC-MS formats has set new standards for analytical precision and reliability. As technological advancements continue to emerge, these methods will become increasingly powerful, accessible, and integral to scientific discovery and regulatory compliance. Advanced separation techniques have revolutionized the analytical capabilities of modern laboratories. By offering greater speed, resolution, and sensitivity, they play a pivotal role in driving innovation across multiple scientific and industrial domains. Continuous technological integration and interdisciplinary collaboration promise to further refine these methods, making them more accessible, robust, and adaptable to emerging analytical challenges.

REFERENCES

Adeoye, A. O., Lawal, O. S., Quadri, R. O., Malomo, D., Aliyu, M. T., Dang, G. E., Emojevu, E. O., Maikato, M. J., Yahaya, M. G., Omonije, O. O., Edidem, V. G., & Hikon, B. N. (2023). Sustainable energy via thermochemical and biochemical conversion of biomass wastes for biofuel production. In *Transportation Energy and Dynamics* (pp. 245–306). Springer Nature Singapore.

Ahmad, N. B. (2018). Responsive regulation and resiliency: The renewable fuel standard and advanced biofuels. *Virginia Environmental Law Journal*, *36*, 40.

Ahmed, M. W., Esquerre, C. A., Eilts, K., Allen, D. P., McCoy, S. M., Varela, S., Singh, V., Andrew Leakey, A. D. B., & Kamruzzaman, M. (2024). Rapid and high-throughput determination of sorghum (Sorghum bicolor) biomass composition using near infrared spectroscopy and chemometrics. *Biomass and Bioenergy*, *186*, 107276.

Ahorsu, R., Medina, F., & Constantí, M. (2018). Significance and challenges of biomass as a suitable feedstock for bioenergy and biochemical production: A review. *Energies*, *11*(12), 3366.

Ajorloo, M., Ghodrat, M., Scott, J., & Strezov, V. (2022). Recent advances in thermodynamic analysis of biomass gasification: A review on numerical modelling and simulation. *Journal of the Energy Institute*, *102*, 395–419.

Ali, S. A., Bangash, I. A., Sajjad, H., Karim, M. A., Ahmad, F., Ahmad, M., & Qudoos, A. (2025). Review on the role of electrofuels in decarbonizing hard-to-abate transportation sectors: Advances, challenges, and future directions. *Energy & Fuels*, *39*(11), 5051–5098.

Alpert, A. J. (2023). Hydrophilic interaction chromatography: Fundamentals and recent developments. *Journal of Chromatography A*, *1682*, 463471.

Anika, O. C., Nnabuife, S. G., Bello, A., Okoroafor, E. R., Kuang, B., & Villa, R. (2022). Prospects of low and zero-carbon renewable fuels in 1.5-degree net zero emission actualisation by 2050: A critical review. *Carbon Capture Science & Technology*, *5*, 100072.

Arnoult, S., & Brancourt-Hulmel, M. (2015). A review on miscanthus biomass production and composition for bioenergy use: Genotypic and environmental variability and implications for breeding. *BioEnergy Research*, *8*(2), 502–526. https://doi.org/10.1007/s12155-014-9524-7.

Asase, R. V., Okechukwu, Q. N., & Ivantsova, M. N. (2024). Biofuels: Present and future. *Environment, Development and Sustainability*, 1–29.

Asghar, A., Sairash, S., Hussain, N., Baqar, Z., Sumrin, A., & Bilal, M. (2022). Current challenges of biomass refinery and prospects of emerging technologies for sustainable bioproducts and bioeconomy. *Biofuels, Bioproducts and Biorefining*, *16*(6), 1478–1494.

Beć, K. B., Grabska, J., & Huck, C. W. (2020). Biomolecular and bioanalytical applications of infrared spectroscopy–A review. *Analytica Chimica Acta*, *1133*, 150–177.

Berger, T. A., & Deye, J. F. (2021). Supercritical fluid chromatography in modern analysis. *Analytical Chemistry*, *93*(14), 5803–5811.

Bernier, F., Tahara, R., & Gendron, M. (2018). Additive manufacturing powder feedstock characterization using x-ray tomography. *Metal Powder Report*, *73*(3), 158–162.

Bhuvaneshwari, S., Hettiarachchi, H., & Meegoda, J. N. (2019). Crop residue burning in India: Policy challenges and potential solutions. *International Journal of Environmental Research and Public Health*, *16*(5), 832.

Bojkovic, A., Vermeire, F. H., Kuzmanovic, M., Dao Thi, H., & Van Geem, K. M. (2021). Analytics driving kinetics: Advanced mass spectrometric characterization of petroleum products. *Energy & Fuels*, *36*(1), 6–59.

Borecki, M., Geca, M., Zan, L., Prus, P., & Korwin-Pawlowski, M. L. (2024). Multiparametric methods for rapid classification of diesel fuel quality used in automotive engine systems. *Energies*, *17*(16), 4189.

Breadmore, M. C. et al. (2021). Capillary electrophoresis: Trends and recent advances. *Electrophoresis*, *42*(5), 567–588.

Cappiello, A., Famiglini, G., Palma, P., & Termopoli, V. (2023). Advancements in miniaturized and field-deployable MS systems. *Journal of Analytical Science and Technology*, *14*(1), 18.

Chandel, A. K., Arora, R., Gupta, R., Narisetty, V., & Kumar, V. (2024). Global biofuel Alliance and roadmap for ethanol blending program in India: Analysis and perspective. *Sugar Tech*, *26*(4), 1005–1015.

Dahlgren, S. (2022). Biogas-based fuels as renewable energy in the transport sector: An overview of the potential of using CBG, LBG and other vehicle fuels produced from biogas. *Biofuels*, *13*(5), 587–599.

Da Silva, J. C., & Queiroz, A. (2017). Advances in the application of spectroscopic techniques in the biofuel area over the last few. *Frontiers in Bioenergy and Biofuels*, 25.

Dey, D., Bhattacharya, S., & Chakravorty, A. (2021). Advances in real-time food safety monitoring: A review. *Critical Reviews in Food Science and Nutrition*, *61*(17), 2840–2855.

Eberhardt, K., Stiebing, C., Matthäus, C., Schmitt, M., & Popp, J. (2015). Advantages and limitations of Raman spectroscopy for molecular diagnostics: An update. *Expert Review of Molecular Diagnostics*, *15*(6), 773–787.

Emwas, A. H. M. (2015). The strengths and weaknesses of NMR spectroscopy and mass spectrometry with particular focus on metabolomics research. *Metabonomics: Methods and Protocols*, 161–193.

Esteves, B., Sen, U., & Pereira, H. (2023). Influence of chemical composition on heating value of biomass: A review and bibliometric analysis. *Energies*, *16*(10), 4226.

Friedmann, J., Zapantis, A., Page, B., Consoli, C., Fan, Z., Havercroft, I., Liu, H., Ochu, E., Raji, N., Rassool, D., Sheerazi, H., & Townsend, A. (2020). *Net-Zero and Geospheric Return: Actions Today for 2030 and Beyond*. Columbia University, Center on Global Energy Policy.

Giddings, J. C., & Myers, M. N. (2021). Field-flow fractionation: Theory, instrumentation, and applications. *Analytical Chemistry Reviews*, *93*(7), 1249–1261.

Heikenfeld, J., Jajack, A., Rogers, J., Gutruf, P., Tian, L., Pan, T., & Kim, J. (2022). Wearable sensors: Modalities, challenges, and prospects. *Lab on a Chip*, *22*(4), 758–784.

Hossain, N., & Mahlia, T. M. I. (2019). Progress in physicochemical parameters of microalgae cultivation for biofuel production. *Critical Reviews in Biotechnology*, *39*(6), 835–859. https://doi.org/10.1080/07388551.2019.1624945.

Hu, X. et al. (2023). Solid-phase extraction: Developments in sorbent chemistry and automation. *TrAC Trends in Analytical Chemistry*, *157*, 116777.

John, S. U., Onu, C. E., Ezechukwu, C. M. J., Nwokedi, I. C., & Onyenanu, C. N. (2025). Multi-product biorefineries for biofuels and value-added products: Advances and future perspectives. *Academia Green Energy*, *2*(1).

Junginger, H. M., Mai-Moulin, T., Daioglou, V., Fritsche, U., Guisson, R., Hennig, C., Thrän, D., Heinimö, J., Hess, J. R., Lamers, P., Li, C., Kwant, K., Olsson, O., Proskurina, S., Ranta, T., Schipfer, F., & Wild, M. (2019). The future of biomass and bioenergy deployment and trade: A synthesis of 15 years IEA bioenergy task 40 on sustainable bioenergy trade. *Biofuels, Bioproducts and Biorefining*, *13*(2), 247–266.

Kubendran, N. (2024). Does ethanol blended petroleum program modulate current account deficit in India. *Library of Progress-Library Science, Information Technology & Computer*, *44*(3), 20496–20508.

Kumar, G., Shobana, S., Chen, W. H., Bach, Q. V., Kim, S. H., Atabani, A. E., & Chang, J. S. (2017). A review of thermochemical conversion of microalgal biomass for biofuels: Chemistry and processes. *Green Chemistry*, *19*(1), 44–67.

Lee, C. S., Conradie, A. V., & Lester, E. (2021). Review of supercritical water gasification with lignocellulosic real biomass as the feedstocks: Process parameters, biomass composition, catalyst development, reactor design and its challenges. *Chemical Engineering Journal, 415*, 128837.

Liu, J., Li, X., Li, M., & Zheng, Y. (2022). Lignin biorefinery: Lignin source, isolation, characterization, and bioconversion. In *Advances in Bioenergy* (Vol. 7, pp. 211–270). Elsevier.

Liu, P., Zhuang, H., Qian, Y., Yang, J., Pan, Y., Zhou, Z., & Qi, F. (2022). Recent advances in mass spectrometric studies on the reaction process of biomass pyrolysis. *Fuel Processing Technology, 237*, 107473.

Makepa, D. C., & Chihobo, C. H. (2024). Barriers to commercial deployment of biorefineries: A multi-faceted review of obstacles across the innovation chain. *Heliyon, 10*(12).

Malode, S. J., Gaddi, S. A. M., Kamble, P. J., Nalwad, A. A., Muddapur, U. M., & Shetti, N. P. (2022). Recent evolutionary trends in the production of biofuels. *Materials Science for Energy Technologies, 5*, 262–277.

Martens, J. A., Bogaerts, A., De Kimpe, N., Jacobs, P. A., Marin, G. B., Rabaey, K., & Verhelst, S. (2017). The chemical route to a carbon dioxide neutral world. *ChemSusChem, 10*(6), 1039–1055.

Mignogna, D., Szabó, M., Ceci, P., & Avino, P. (2024). Biomass energy and biofuels: Perspective, potentials, and challenges in the energy transition. *Sustainability, 16*(16), 7036.

Mishra, R. K., Kumar, P., & Mohanty, K. (2022). Hydrothermal liquefaction of biomass for bio-crude production: A review on feedstocks, chemical compositions, operating parameters, reaction kinetics, techno-economic study, and life cycle assessment. *Fuel, 316*, 123377.

Raji, A. M., Manescau, B., Chetehouna, K., & Ogabi, R. (2025). Overview of thermal and analytical characterization techniques for biofuels and their blends. *Journal of Thermal Analysis and Calorimetry*. https://doi.org/10.1007/s10973-024-13907-w.

Nazir, Z., Rehman, K. U., Hussain, I., Mir, I. M., Aziz, R., & Aslam, M. (2025). Analytical techniques for bio-crude analysis: Fuel property and international standards. In *Biocrude Oil Biorefinery: An Emerging Biorefining Approach* (pp. 285–312). Springer Nature Switzerland.

Nemett, A., & Adams, K. B. (2022). *National Renewable Energy Laboratory History: 1977-2016 (No. NREL/BK-6A42-84180)*. National Renewable Energy Lab.

Niessen, W. M. A. (2006). *Liquid chromatography–mass spectrometry* (3rd ed.). CRC Press. https://doi.org/10.1201/9781420014549.

Nisansala, H. D. (2025). Sustainability certifications and standards for algae biofuel production. In *Microalgal Biofuels* (pp. 439–458). Woodhead Publishing.

Okolie, J. A., Epelle, E. I., Tabat, M. E., Orivri, U., Amenaghawon, A. N., Okoye, P. U., & Gunes, B. (2022). Waste biomass valorization for the production of biofuels and value-added products: A comprehensive review of thermochemical, biological and integrated processes. *Process Safety and Environmental Protection, 159*, 323–344.

Ong, H. C., Chen, W. H., Singh, Y., Gan, Y. Y., Chen, C. Y., & Show, P. L. (2020). A state-of-the-art review on thermochemical conversion of biomass for biofuel production: A TG-FTIR approach. *Energy Conversion and Management, 209*, 112634.

Poole, C. F. (2023). *The Essence of Chromatography* (2nd ed.). Elsevier.

Poornima, S., Manikandan, S., Prakash, R., Deena, S. R., Subbaiya, R., Karmegam, N., & Govarthanan, M. (2024). Biofuel and biochemical production through biomass transformation using advanced thermochemical and biochemical processes–a review. *Fuel, 372*, 132204.

Rathour, R. K., Behl, M., Dhashmana, K., Sakhuja, D., Ghai, H., Sharma, N., Meena, K. R., Bhatt, A. K., & Bhatia, R. K. (2023). Non-food crops derived lignocellulose biorefinery for sustainable production of biomaterials, biochemicals and bioenergy: A review on trends and techniques. *Industrial Crops and Products, 204*, 117220.

Sackmann, E. K. et al. (2023). Lab-on-a-chip: Applications and challenges. *Annual Review of Biomedical Engineering, 25*, 327–355.

Saleh, H. M., & Hassan, A. I. (2023). Green conversion of carbon dioxide and sustainable fuel synthesis. *Fire, 6*(3), 128.

Scigelova, M., & Makarov, A. (2020). Orbitrap mass spectrometry. *Mass Spectrometry Reviews, 39*(5), 493–523.

Shadangi, K. P., Sarangi, P. K., & Behera, A. K. (2023). Characterization techniques of biomass: Physico-chemical, elemental, and biological. In *Bioenergy Engineering* (pp. 51–66). Woodhead Publishing.

Singh, D., Tembhare, M., Dikshit, A. K., Dangi, M. B., & Kumar, S. (2024). Technical and operational challenges in setting up a decentralized biogas plant: Opportunities and future perspective toward sustainable nation. *Process Safety and Environmental Protection.*

Singh, R., & Kim, S. J. (2022). Membrane-based separation in food and pharmaceutical industries. *Separation and Purification Technology, 291*, 120948.

Singh, R., Sharma, M., & Kumar, A. (2023). AI-powered real-time monitoring systems: Trends and future directions. *Journal of Real-Time Image Processing, 20*(1), 1–15.

Singla, A., Singh, Y., Abd Rahim, E., Singh, N. K., & Sharma, A. (2023). Challenges and future prospects of biofuel generations: An overview. In Y. Singh, V. Strezov, & P. Negi (Eds.), *Biowaste and biomass in biofuel applications* (pp. 49–61). CRC Press. https://doi.org/10.1201/9781003265597-4

Snyder, L. R. et al. (2022). *Practical HPLC Method Development* (3rd ed.). Wiley.

Song, S. et al. (2022). Aptamer-based separation and sensing technologies. *Biosensors and Bioelectronics, 202*, 113988.

Thaha, A. N., Ghamari, M., Jothiprakash, G., Velusamy, S., Karthikeyan, S., Ramesh, D., & Sundaram, S. (2025). High impact biomass valorization for second generation biorefineries in India: Recent developments and future strategies for sustainable circular economy. *Biomass, 5*(1), 16.

Nguyen, V. G., Tran, M. H., Paramasivam, P., Le, H. C., & Nguyen, D. T. (2024). Biomass: A versatile resource for biofuel, industrial, and environmental solution. *International Journal on Advanced Science, Engineering and Information Technology, 14*(1), 268–286. https://doi.org/10.18517/ijaseit.14.1.17489.

Trivedi, A., & Joshi, S. (2025). Biofuels as energy for tomorrow. *Optimizing Biofuel Production With Artificial Intelligence*, 53–81.

U.S. Food and Drug Administration. (2004). *PAT—A framework for innovative pharmaceutical development, manufacturing, and quality assurance* (Guidance for industry). https://www.fda.gov/media/71012/download

Uğuz, G., Atabani, A. E., Mohammed, M. N., Shobana, S., Uğuz, S., Kumar, G., & Al-Muhtaseb, A. A. H. (2019). Fuel stability of biodiesel from waste cooking oil: A comparative evaluation with various antioxidants using FT-IR and DSC techniques. *Biocatalysis and Agricultural Biotechnology, 21*, 101283.

Verma, M., Singh, P., & Dhanorkar, M. (2024). Sustainability in residue management: A review with special reference to Indian agriculture. *Paddy and Water Environment, 22*(1), 1–15.

Watson, J. T., & Sparkman, O. D. (2020). *Introduction to Mass Spectrometry.* Wiley.

Whitesides, G. M. (2023). Microfluidics: A roadmap for emerging applications. *Nature Reviews Materials, 8*(1), 36–48.

Williams, C. L., Westover, T. L., Emerson, R. M., Tumuluru, J. S., & Li, C. (2016). Sources of biomass feedstock variability and the potential impact on biofuels production. *BioEnergy Research*, *9*, 1–14.

Wongsurakul, P., Termtanun, M., Kiatkittipong, W., Lim, J. W., Kiatkittipong, K., Pavasant, P., Kumakiri, I., & Assabumrungrat, S. (2022). Comprehensive review on potential contamination in fuel ethanol production with proposed specific guideline criteria. *Energies*, *15*(9), 2986.

Zhang, H. et al. (2022). Recent advances in chiral separation techniques. *Journal of Chromatography A*, *1662*, 462588.

Zhang, X. et al. (2021). Advances in high-resolution mass spectrometry for biomedical applications. *TrAC Trends in Analytical Chemistry*, *135*, 116161.

Zhang, X., Li, Y., Wang, Y., & Liu, Z. (2021). Environmental monitoring technologies and their application in smart cities. *Sensors*, *21*(5), 1736.

Zhou, Y. et al. (2022). Recent advances in LC-MS techniques for bioanalysis. *TrAC Trends in Analytical Chemistry*, *153*, 116628.

13 Current Status and Challenges of Biofuel Sector

*Dolly Kumari, Gulab Singh Maurya, and Yogita Jain**

13.1 INTRODUCTION

The drastic changes in lifestyle result in a continuous need for energy in modern civilization. However, because of the limited supply, significant price fluctuations, and detrimental effects on the environment, fossil fuels are unable to completely meet the current energy needs (Singh et al., 2021). Fossil fuels, including coal, oil, and natural gas, are responsible for the majority of the world's primary energy source, which was about 81% in 2016 and is estimated to decrease to 73% by 2030 (IEA, 2023a; Soares et al., 2020). In recent years, overuse of fossil fuels has contributed to an increase in greenhouse gas emissions, which results in global warming and various other climatic changes, leading to impinged health and well-being (Mood et al., 2013; Watts et al., 2018a). Fossil fuels' shortcomings and rapid depletion have reinforced the notion of alternate, renewable, affordable, and sustainable energy sources (Kumari & Singh, 2018). To address the current environmental issues as well as the expanding global energy demand, there is a need to shift the paradigm from fossil fuels to renewable energy sources. The term "green economy," which refers to an environmentally conscious approach to the market, suggests that research efforts should concentrate on novel sources and methods of producing renewable energy (Chatellard et al., 2017). The majority of countries aim to implement renewable energy generation to meet their energy needs, which also fulfills the 7th and 13th UN SDGs focused on affordable energy and climate action (D'Silva et al., 2023). According to reports, a shift has been witnessed toward renewable energy sources because of their many benefits, like cheap cost, longevity, dependability, and minimal environmental effect (Tusher et al., 2022; Efroymson et al., 2021; Voca & Ribic, 2020). However, excessive research and technology interventions are needed, as at present only 14% of total energy demand is fulfilled by renewable sources (Watts et al., 2018b). Since non-carbonaceous fuels, or biofuels, manufactured from renewable feedstock, emit fewer greenhouse gases during fuel production and combustion, substantial efforts are being made globally to employ them (Cheng et al., 2011). The energy-rich compounds produced by biological processes or obtained from the biomass of some biotic components, including microalgae, plants, and bacteria, are termed "biofuels," such as bio-hydrogen, bio-ethanol, bio-methanol, bio-butanol, and

* Corresponding author

DOI: 10.1201/9781003485636-13

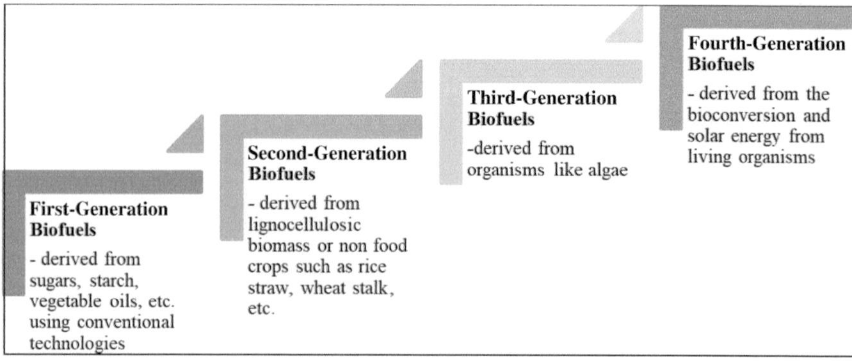

FIGURE 13.1 Categories of biofuel (Kumari & Singh, 2018; Gopinathan & Sudhakaran, 2009).

biodiesel (Razzak et al., 2013; Voloshin et al., 2015). Biofuels may now be extracted in solid, liquid, and gaseous forms from an array of biomass sources, including wood crops and biodegradable agricultural and livestock wastes, which are the result of sophisticated and effective conversion technologies (Gopinathan & Sudhakaran, 2009). Based on the source and processing types, biofuels can be classified into first-, second-, third-, and fourth-generation biofuels, as shown in Figure 13.1.

From 2001 to 2006, the production of biofuel surged about sixfold, and by 2030, it will account for over 6% of global energy supply (IEA, 2023b). The effective use of biofuels, however, is still heavily dependent on the policies surrounding these technologies in terms of investment, trade, and subsidies. Biofuel can be effectively produced from lignocellulosic biomass, which includes agricultural and forest left-overs, microalgae, and even municipal solid waste (Cheah et al., 2020). A wide range of pretreatment technologies can be used to enhance the production of biofuel. These methods and technologies have their pros and cons, which should also be economically suitable and environmentally benign (Rathour et al., 2023). Examining the technological, financial, and environmental difficulties involved in the production and use of biofuels as well as their possible influence on lowering greenhouse gas emissions worldwide, the present chapter aims to determine how biofuels can support a greener and more sustainable energy revolution. It takes an integrated approach to provide a comprehensive understanding of the role of biofuels, focusing on the challenges and opportunities of various feedstocks and biofuels synthesized from them. Later on, the current scenario and future aspects of biofuel will be discussed concerning leading countries and their role in the renewable energy sector.

13.2 ANALYSIS OF RAW MATERIALS USED FOR BIOFUEL PRODUCTION

13.2.1 Biomass from Food Crops

In the context of biofuel production, "biomass" refers to the natural resources derived from either living or non-living organisms. Three main sources provide the majority of the biomass used today: forestry, agriculture, and waste (Malode et al., 2021).

Energy crops have great promise as a source of future energy, and for the subsequent biofuel conversion process, the choice of biofuel crop is crucial to guarantee low production costs (Kour et al., 2019). There are various food crops that have been grown to extract biofuel and to fulfill the future energy demands, like switchgrass, miscanthus, sweet sorghum, etc. (Hirani et al., 2018). During the first generation of biofuels, the contentious questions of "food vs fuel" surfaced, and a hike in food prices was observed. Biodiesel and bioethanol made from corn, sugarcane, seeds, grains, palm oil, or entire plants had a significant impact on food security, instability in food commodity prices globally, and competition for arable lands, all of which were viewed as barriers to reaching the necessary targets (Hafid et al., 2022). There is a possibility that food production might be jeopardized in favor of biofuels (HLPE, 2013) as a large quantity of land has to be devoted to raising crops for energy fuel instead of food because these biofuels necessitate a lot of harvest as raw material (Pragya et al., 2013). Figure 13.2 shows the feedstock demand in India for biofuel production, indicating that in the coming five years, there will be an increase in the demand for sugar and starch food crops. There is always a running international debate about the extent to which biofuel production can affect the price of food crops (Ajanovic, 2011).

In the US, the cost of biofuel has increased by 1%, while fertilizer cost climbed by 8%, which results in the hike in the food prices. Climatic conditions for the growth of food crops are also a very important factor, which are characterized by various factors and indicators (Ahmad et al., 2023). Several signs of climate change have been seen, including increased levels of carbon dioxide in the atmosphere, loss of the ozone layer, and melting of glaciers, which lead to higher sea levels and altered weather patterns (Abbas et al., 2021). Extensive research aimed at ascertaining the ecological prerequisites for biofuel-cum-oilseed crops might be employed to identify appropriate regions for current and future production. There are various other factors also that are studied using SWOT analysis for the production of biofuel from

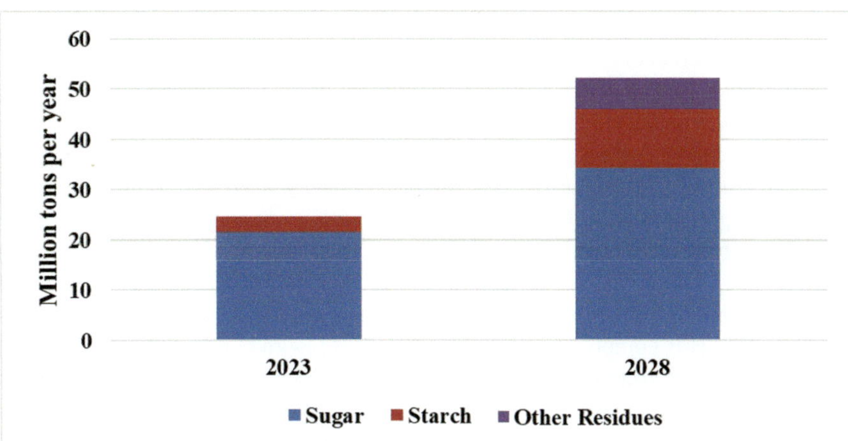

FIGURE 13.2 Feedstock demand in India (IEA, 2023a).

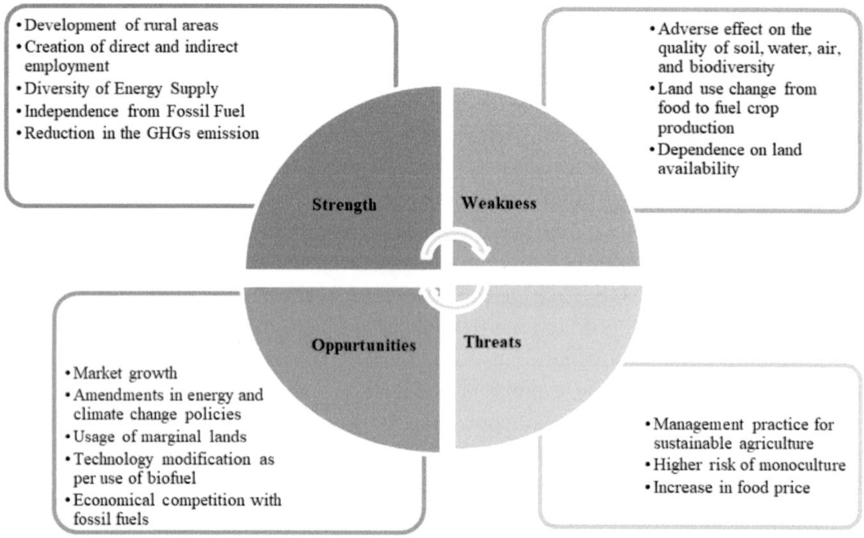

FIGURE 13.3 SWOT analysis of energy crops for biofuel production (Paschalidou et al., 2016).

food crops (Figure 13.3). All these factors have put researchers on utilizing second-generation biofuels for energy needs, as they mostly use agricultural leftovers and non-crop plant species, minimizing conflict with food crops (Farooq et al., 2022).

13.2.2 Lignocellulosic Biomass

Lignocellulosic biomass (LCB) such as agricultural leftovers, woody forest residues, algae, and even municipal waste is a more advantageous source for the synthesis of biofuel as compared to food crops (Kumari et al., 2021). Over 220 billion tons of lignocellulose biomass leftovers are estimated to be produced annually globally, the majority of which is either disposed of, burned, piled, or utilized as mulch for subsequent crops (Alawad & Ibrahim, 2024; Ludfiani et al., 2024). Burning LCB is an inexpensive way to get rid of waste, but it contaminates and degrades the air quality, resulting in adverse impacts on health. Roughly 730 Mt of biomass, comprising that from both natural and human sources, is burned every year in Asia, with 18% of that amount being generated in India (Semwal et al., 2023). Second-generation biofuels are mainly produced by the LCB; opportunities and inadequacies of these are listed in Table 13.1. The primary barrier preventing lignocellulose biomass from being commercialized as biofuel is LCB structural resistance (Mupondwa et al., 2017). The majority of LCB is composed of cellulose (40–50%), with hemicellulose making up 25–35% and lignin roughly 15–20%. Lignin, being an amorphous, nonsugar structure, acts as a resistant layer and may slow down the hydrolysis process and prevent cellulose from breaking down (Banerjee, 2023). For an efficient conversion of LCB to biofuel, a wide range of different pretreatment technologies were used from a decade ago till now, and optimizations are under study (Kumari & Jain, 2020).

TABLE 13.1
Opportunities and Inadequacies of Lignocellulosic Biomass to Biofuel Conversion

Opportunities	Inadequacies
No adverse impact on food security	Biofuel production costs a lot of money, and biofuel itself is pricey.
Using leftover biomass for bioenergy promotes waste management.	Pretreatment is costly and needs optimization
50–70% decrease in greenhouse gases	Energy-intensive process
No additional requirement of land and water	Marketing is a challenge

Source: Bhatia et al. (2020); Banerjee (2023).

The pretreatment method for the breakdown of lignin and the release of cellulose and hemicellulose depends on the composition of the LCB. According to a new research report, the pretreatment procedure of LCB is anticipated to cost between 70 and 150 US dollars per ton of feedstock (Preethi & Gunasekaran, 2023). An efficient and less energy-intensive pretreatment method should be used to remove lignin and increase the porosity of the material for the conversion process (Kumari & Singh, 2022). These pretreatments are mainly classified as physical, chemico-biological, and physico-chemical methods (Kumari & Singh, 2020a). Unfortunately, the majority of the pretreatment methods now in use have a number of drawbacks, including the limited separation of lignin and cellulose, energy-intensive and catalyst usage, and the release of by-products that obstruct further convergence steps (Chaudhary et al., 2024). It is discovered that the physico-chemical and chemical methods are more efficient in terms of cost and conversion. Analyzing each pretreatment technique, it can be concluded that, based on the characteristics and features of the biomass being used, each technique has its own advantages and disadvantages (Kumari & Singh, 2020b).

13.2.3 Algal Biomass

Algae are marine creatures that come in a variety of types and sizes, from microalgae to macroalgae. Microalgae, particularly solitary or clustered cells, have garnered attention for producing high-value products, and many have significant lipid yields, which make them ideal for biodiesel production. Macroalgae, however, often contain a high amount of carbohydrates and a low amount of lipids and are occasionally call seaweeds. They are mostly utilized as food supplies and cultivated on a huge scale, mainly in the Far East, encompassing China and Indonesia (Griffiths et al., 2021). Microalgae have gained recognition in recent decades as a highly potential source of fuel for the generation of bioenergy. The concept of harnessing energy from algae has been proposed for more than six decades, with the aim of utilizing the lipids found in algae to produce liquid fuels (Signoretto et al., 2019). Their particular significance is in the business due to the ease with which they may be cultivated in a laboratory, photobiological, or terrestrial pond setting within a very brief timeframe (Branco-Vieira et al., 2020). *Chlamydomonas sp., Chlorella sp., Spirulina sp., Spirogyra sp.*, and *Dunaliella sp.* are microalgae species that are rich in carbohydrates and might be used as feedstock since they contain up to 64% starch per dry cell weight (Cheah et al., 2020). On top of that, microalgae are incredibly efficient at photosynthetic processes, have a high capacity to bioseques CO_2 and also do not have a lignin cross-linking structure. This allows them to thrive in saltwater and wastewater, where they can also grow and float while creating more sugar substrates for fermentation. An alga possesses desirable characteristics such as increased productivity, a short life cycle, and the ability to be available throughout the year, making it an excellent choice for biofuel production. The minimal amount of pretreatment needed to use algae for biofuel generation is an immense benefit (Mandotra et al., 2021). Extracting algae from aquatic environments to produce biomass for the generation of biofuel and biogas not only purifies the water but also safeguards it against eutrophication (Chia et al., 2022). In addition, it diminishes the emission of gas pollutants resulting from the utilization of fossil fuels. Extraction of energy in the form of biofuel from algae biomass needs to optimized and some challenges should also be address properly, the conversion process, scale-up cultivation, and harvesting (Halder & Azad, 2019; Saad et al., 2019).

13.2.4 Fruit and Vegetable Wastes

Food waste, often known FW comprising fruit and vegetable waste, is a type of garbage that includes biodegradable materials and is produced by a wide range of sources, including households, food processing enterprises, eateries, and other hospitality facilities. It has been estimated that the annual generation of food waste around the world amounts to 1.6 billion tons (Bt), which in turn liberates 3.3 billion tons of carbon dioxide (Maalouf & Agamuthu, 2023). The FAO reports that in 2019, over 14% of the world's food supply was lost or squandered after harvest (FAO, 2019). Approximately 1.3 billion tons of food is wasted or discarded annually during production and processing, with one-third of the loss occurring during consumption. FSSAI (2023) statistics reveal India as the second-largest food producer, contributing 10.1% to global food output and producing 68.7 million metric tons of food waste. According to reports, over 42% and 30% of fruits/vegetables and grains, respectively, produced in the Asia and Pacific region are wasted or lost before they reach the customer (Suhartini et al., 2022). There are various factors that contribute to a high amount of food waste generated, such as unsustainable food consumption practices, poor public participation and awareness, and lack of conversion technologies. The ISO guidelines recommend anaerobic digestion, composting, incineration, sewerage, and landfilling as the main methods for handling food waste (Ahmed et al., 2022). Landfills are often the primary technique for disposing of these wastes, leading to significant land and air pollution, and degradation of these wastes releases dangerous methane gas, a greenhouse gas with a higher potential than carbon dioxide, resulting in global warming. However, the low cost of composting and incineration makes them more popular waste treatment options than anaerobic digestion. The dumping of food waste is responsible for around 7% of the greenhouse gas

TABLE 13.2
Comparison of AD and DF

Conversion Technology	Anaerobic Digestion	Dark Fermentation
Merits	• High moisture food waste can be treated • High methane yield • Mitigating climate change by reducing GHGs • Employed at large scale • Produce biofertilizer • No pretreatment is required	• Environment benign process • High carbohydrate and TCOD results in better hydrogen yield • Cost-effective • 80% reduction in GHG emission
Challenges	• High operation and maintenance cost • Time-consuming • VFA accumulation	• Time-consuming • Pretreatment is required to release fermentable sugar • Costly • Scale-up issues

Source: Suhartini et al. (2022); Nayak and Bhusan (2019).

emissions that are produced all over the world (Tang et al., 2023). Anaerobic digestion (AD) and dark fermentation (DF) processes have been used for several decades to treat food waste and for the production of biogas and biohydrogen, respectively (Kumari & Singh, 2019). A detailed comparison of both processes is given in Table 13.2. The dark fermentation process, as compared to the anaerobic digestion process, has a low technology readiness level (Neuling & Kaltschmitt, 2017).

Incorporating FW into the bioeconomy necessitates the implementation of productive and economical manufacturing techniques. However, it is important to focus on resolving bottlenecks, such as improving efficiency, creating cost-effective valorization techniques, and implementing post-treatment processes for slurry (Sahota et al., 2024).

13.3 ANALYSIS OF CURRENT STATUS AND CHALLENGES IN THE PRODUCTION OF DIFFERENT BIOFUELS

13.3.1 Biohydrogen Production

The efficiency and adaptability of hydrogen have made it a promising candidate for use as a fuel in the future. Combustion of this fuel results only in the release of water vapor, making it an ideal and efficient transportation fuel (Elbeshbishy et al., 2017). It is not easy to extract hydrogen from biomass since biomass contains only around 6% hydrogen compared to 25% methane (Samrot et al., 2023). Microbial-mediated biochemical processes produce biohydrogen, an environmentally friendly, non-degradable fuel with high energy density and low emissions (Singh et al., 2017). At present, the methods for producing biohydrogen from organic waste that have received the greatest attention, discussion, and development include dark fermentation, photo-fermentation, bio-photolysis, microbial electrolysis, and gasification with a percentage conversion efficiency of 4.3, 5.1, 4.0, 11.3, and 88.1, respectively (Samrot et al., 2023). Presently, biohydrogen production has been explored from various types of carbohydrate-rich sources, including food crops, agricultural waste, municipal solid waste, and microalgae (Chatellard et al., 2017; Singhal & Singh, 2014). There are various internal and external factors that play a vital role during biohydrogen production; feedstock availability, pretreatment of lignocellulosic waste, pretreatment of microbial source, and storage of biohydrogen are some of them. Nevertheless, the current techniques for sustainable biohydrogen generation are hindered by their poor yields and efficiency. Currently, there is no operational biohydrogen plant of significant size due to its lack of economic viability and the typically subpar performance of the process (Osman et al., 2020).

13.3.2 Ethanol Production

In the past, ethanol was commonly utilized in the manufacturing of alcoholic beverages. Nevertheless, due to advancements in purification processes, the utilization of ethanol has significantly expanded to encompass several different applications (Tse et al., 2021). As an illustration, the substitution of gasoline with bioethanol can effectively decrease carbon dioxide emissions (Rodionova et al., 2017). Bioethanol can be produced from a wide range of food crops and waste, presently focusing on corn and sugarcane, responsible for nearly 85% of production in the USA and Brazil (Broda et al., 2022). There are various steps involved during the production of bioethanol from feedstock collection and transportation, pretreatment, hydrolysis, and fermentation. To achieve the most efficient fermentation rate and extent, it is necessary to consider both biotic variables, such as microbial contamination, and abiotic ones, such as shortages in nutrients, trace metals, and vitamins. Currently, employing just lignocellulosic biomass as a sustainable source for producing bioethanol does not provide a smooth shift from petroleum-based to green bioenergy. By combining first- and second-generation biofuels, it is possible to optimize ethanol production and profits. Additionally, this approach would necessitate a lower initial investment to include lignocellulosic processes into an industrial design (Rajendran et al., 2016). The use of cost-effective manufacturing procedures built with green chemistry to fully utilize all components of the lignocellulosic complex appears to have a higher chance of success. The current knowledge base is extensive, but further multidisciplinary research is also necessary to fully optimize bioethanol production for widespread commercial viability.

13.3.3 Methanol Production

The chemical process by which methanol is produced involves the conversion of natural gas (CH4, methane) into methanol through three stages: steam reforming, synthesis, and distillation. Due to the costly purification process involved, methanol recovery is significantly simpler than ethanol (Rodionova et al., 2017). An estimated

110 million metric tons of methanol are produced annually by 90 global methanol production facilities. Bio-methanol possesses several advantageous characteristics that make it a viable candidate for energy utilization, such as a high octane number (87–110), low combustibility, excellent efficiency, and low emissions, making it suitable for use in automobiles (Subudhi et al., 2023). In the biomass pathway, organic substrates obtained from fermentative processes are subjected to high temperatures without oxygen to produce syngas. This syngas is then subjected to additional catalysis, during which the methane present in the gas is transformed into methanol. A major issue with the biochemical approach is the low accessibility of gases like hydrogen, CO, and methane to methanogens due to their lack of solubility in the culture medium, resulting in reduced methanol synthesis. While standard biochemical processes are viable in labs, scaling up requires metabolic engineering (Gautam et al., 2020). To address technical, logistical, and economic challenges, a bio-methanol production plant from biomass can be established, considering the location and feedstock distribution network.

13.3.4 Butanol Production

Butanol, a valuable biofuel and significant precursor chemical, has garnered considerable attention owing to its exceptional fuel characteristics, including high energy density, excellent miscibility, low volatility, and low corrosiveness (Guo et al., 2022). The ABE fermentation process is frequently employed for the production of acetone, butanol, and ethanol. However, this method has significant drawbacks, including low butanol productivity, expensive pretreatment methods for hydrolyzing cellulosic feedstocks, etc. (Sahu & Singh, 2023). Butanol from different lignocellulosic biomass can be synthesized using appropriate microbial species, but pretreatment of LCB makes it a costly process (Dhanya, 2022). To address this financial limitation, one potential solution is to develop and build resilient cellulolytic and bio-alcohol-generating microorganisms and utilize them inside a consolidated bio-processing (CBP) system (Kumari & Singh, 2018). Integration of the pretreatment methods, hydrolysis and intensification, can successfully lead to the induction of the efficient production of biobutanol. Detailed and deep research is currently needed in the field of butanol production from biomass.

13.3.5 Biodiesel Production

Biodiesel has emerged as a viable alternative to traditional fossil fuels in recent years and is gaining global recognition and interest. Biodiesel can be derived from sustainable organic materials and can serve as a replacement for petroleum-based diesel fuels. Direct oil use/blending, pyrolysis (thermal cracking), micro-emulsion, and transesterification are the four basic procedures documented for biodiesel generation. Among these methods, transesterification is considered the most cost-effective because it achieves the highest product yield under the most favorable circumstances of temperature and pressure (Mathew et al., 2021). It can be derived from several seed oils, with rapeseed and soybean oils being the most prevalent. *Jatropha curcas*, a plant with significant economic importance, has been recognized as a promising candidate for the manufacture of biodiesel (Rodionova et al., 2017). Currently, it has been reported that microalgae can produce biodiesel more efficiently as compared to food crops because of its easy and fast harvesting. The major challenge in the utilization of biodiesel is its high cost, which needs to be addressed to make it economically viable.

13.4 BIOFUEL SECTOR AND FUTURE PROSPECTS IN INDIA

India has surpassed Russia and Japan to become the third-largest oil user in the world. Research on biofuels is now being done all around the world due to two key factors: renewability and sustainability. Due to the depletion of oil supplies, concerns about energy security, and the growing environmental problems brought on by greenhouse gas emissions and climate change, biofuels have become very important. Heat conversion is the technique that is typically used to manufacture biofuels from densified materials (Shweta et al., 2024). In India, biofuels like biohydrogen, biogas, bioethanol, bio-methanol, etc., have been used for energy generation and transportation. Many new technologies developed today for biofuel production are still limited to the lab-scale production of biofuel, and there is a need to develop such techniques that are feasible, cost-effective, and can be commercialized easily (Kumari et al., 2024; Kumari & Singh, 2018).

13.4.1 Biohydrogen Sector

In order to address climate change, reduce greenhouse gas emissions, and reduce the excessive use of fossil fuels, hydrogen production is essential. The methods for producing hydrogen that are currently in use rely more on steam methane reforming (SMR) or electrolysis-dependent technologies. In order to address climate change and provide energy security, affluent nations, particularly those in the West, have started producing hydrogen on their own (Kovalenko et al., 2021). In an effort to establish India as a global leader in hydrogen generation, the Indian government has introduced the National Hydrogen Mission. The purpose is to produce "green hydrogen" with an emphasis on clean energy transition, energy self-sufficiency, and self-reliance (D'Silva et al., 2023). With its advantages, renewable hydrogen production via photolysis, photo-fermentation, and the biological process of DF should likewise receive the attention it merits on the "Green hydrogen" platform. This boosts the national hydrogen mission's goal of creating jobs and green growth while enabling self-sufficient biohydrogen production. Additionally, in order to engage solid waste management, clean energy, and transportation from a wider range of perspectives, the National Hydrogen Mission can be combined with other missions like Swachh Bharat Abhiyaan, a solid waste management scheme, and Sustainable Alternative towards Sustainable Transportation (SATAT), a clean vehicular energy scheme based on compressed biomethane (D'Silva et al., 2023). Das et al. (2022) used a SWOT (strengths, weaknesses, opportunities, and threats) analysis to ascertain whether biological biogas upgrading systems were feasible. Table 13.3 displays the findings of the SWOT analysis for the two-stage AD-based biorefinery idea.

TABLE 13.3
SWOT Analysis of the Two-Stage Anaerobic Digestion-Based Biorefinery Concept

Strength	Weaknesses	Opportunities	Threats
• Available of Plenty of waste biomass • Eco-friendly and sustainable technology in comparison to alternative methods • Easy-to-use, flexible technology that requires less complexity	• Gathering, moving, and classifying the leftover biomass resources • Adopting the technology without taking into account downstream technologies is not economically possible • Rigid optimization techniques are necessary for both startup and long-term steady operation • Insufficient expertise with pilot-scale operations • Low energy recovery productivity	• More research and developments (collaborations and partnerships) inside the countries between academic institutions and industries and between the countries in order to meet the combined demand for hydrogen and methane fuels • Valorization of bio-slurry, biohydrogen, bio-carbon dioxide, bio-methane, VFAs, and bio-slurry in place of chemical fertilizers and conventional energy fuels • The sustainable management of trash is facilitated by the appropriate handling of waste biomass • Achieving the hydrogen and methane fuel demand in single process	• Inadequate policy framework supporting the production of "biohydrogen" • Inconsistent performance depending on the kind and composition of the substrate, the inoculum, and the microbes involved • Absence of cost-effective methods for the purification, storage, and transportation of hydrogen

Source: D'Silva et al. (2023); Das et al. (2022).

13.4.2 Methane and Biogas Sector

Given that 535.78 million animals reside in India, of which 302 million are bovines (cattle, buffalo, mithun, and yak), there is considerable potential for the establishment of biogas facilities in the country. The cattle industry will continue to grow and contribute substantially to India's GDP. The widespread adoption of biogas technology offers both direct and indirect advantages to Indian farmers (Ministry of New and Renewable Energy, 2022). Methane makes up between 40% and 70% of biogas, carbon dioxide 30% to 40%, and trace amounts of other gases like nitrogen and ammonia (1–2%) and hydrogen sulfide (1–2%) (Singh et al., 2022). In its unprocessed state, biogas can be utilized for energy production, illumination, motive power, and clean cooking fuel similar to LPG. It may be utilized in diesel engines to replace diesel up to 80% of the time, and when using 100% biogas engines, it can replace diesel 100% of the time. Additionally, biogas can be refined to a methane concentration of up to 98% purity, which qualifies it as compressed bio-gas (CBG), a clean and environmentally friendly fuel that can be utilized for transportation or filled into cylinders at pressures as high as 250 bar (Ministry of New and Renewable Energy, 2022). The quickest way to advance the biomass-based energy (BBE) industry in India is through the establishment of CBG facilities. The sustainable options toward the SATAT program intend to produce 15 MT of CBG from 5000 facilities spread throughout India by 2023. Agricultural leftovers (150 MT/year), animal waste (190 MT/year), press mud (20 MT/year), municipal solid waste (MSW) (62 MT/year), and sewage sludge (50 MT/year) are some of the most widely used feedstocks (Arias et al., 2021) for CBG plants. On November 2, 2022, the Indian government's Ministry of New and Renewable Energy (MNRE) announced the National Bioenergy Programme. The National Bioenergy Programme has been extended by MNRE to run from financial year 2021–2022 to financial year 2025–2026. It has been suggested that the program be implemented in two stages. The first phase of the program, which has a budget of Rs. 858 crores, includes Rs. 100 crores for the Biogas Program, which supports the establishment of small- and medium-sized biogas plants (those that generate between 1 m^3 and 25 m^3 of biogas per day) and above 25 m^3 to 2500 m^3 of biogas per day for equivalent power generation capacities ranging from 3 kW to 250 kW from biogas or unprocessed biogas for thermal power/cooling applications. As of March 2023, 4090 potential financiers have received letters of intent (LOI) for the establishment of CBG facilities, and 46 CBG plants have been put into operation in India, according to the SATAT portal (Singh et al., 2023). At the moment, these CBG plants have been established in the states of Tamil Nadu, Punjab, Maharashtra, Gujarat, Haryana, and Uttar Pradesh (Singh et al., 2023).

13.4.3 Bioethanol Sector

India's petroleum consumption increased to a record high of 222.3 million tons in the fiscal year 2022–2023. Nonetheless, the nation's production of crude oil stayed around 29.2 million tons over that time, which resulted in a significant reliance on imports. In actuality, India bought 232.4 million tons of crude oil in 2022–2023 at a cost of almost USD 158.3 billion (Madhurjya et al., 2024). India's energy demand could quadruple by 2040 if current trends continue, which would be extremely dangerous for the nation's energy security (IEA, 2021). For a number of years, the Indian government has pushed ethanol as a green energy substitute in an effort to reduce its reliance on imports. Even though ethanol-blended program policies have been in place since 2003, they failed to achieve their goals until the national biofuel policy was introduced in 2018. The policy addressed the inadequacies of earlier attempts and included a number of revisions (Madhurjya et al., 2024). In the Oct. 2021–2022, India produced 5000 lakh metric tons (LMT) of sugarcane, according to the Press Information Bureau of the Government of India (2021). 394 metric tons of sugar was produced by sugar mills from 3574 metric tons of sugarcane. Then, 35 LMT of sugar will be utilized in 2021–2022 to create 113 crore liters of ethanol. The production of ethanol is predicted to rise by 12.68% in the upcoming years. It is possible for ethanol to reduce oil imports. They are also far less harmful to the environment than fossil fuels. There have been certain feedstock-related issues with the ethanol blending program and the manufacturing of ethanol, the delivery of ethanol to final users, and automotive technology for the acceptance of the novel fuel. The challenges may include feedstock, vehicle technology barriers, and issues with oil marketing companies (Madhurjya et al., 2024).

13.4.4 Methanol Sector

India has a 2 MT annual installed capacity for producing methanol. By 2025, 20 MT of methanol can be produced yearly utilizing Indian high ash coal, trapped gas, and biomass, according to the NITI Aayog plan. With 500 million tons of annual biomass production, 125 billion tons of confirmed coal reserves, and enormous amounts of flared and stranded gases, India has enormous potential for guaranteeing energy security through the use of alternative feedstocks and fuels. A roadmap has been developed by NITI Aayog to replace 10% of crude imports by methanol alone by 2030, and around 30 MT of methanol is needed for this. India might aim to cut its fuel expense by 30% by 2030 since methanol and dimethyl ether (DME) are significantly less expensive than gasoline and diesel. With very minor changes made to the current fuel distribution system and engines (vehicles), methanol and DME can be produced from 15% of all vehicle fuels. India will soon introduce a 15% methanol blend program with gasoline, which is expected to immediately reduce the price of gasoline by 10%. Additionally, an M100 plan for trucks and buses will soon be launched (Press Information Bureau, Gov. of India, 2017).

13.4.5 Butanol Sector

From the 2018–2019 fiscal years, Andhra Petrochemicals Limited (the sole manufacturer of N-butanol in India) will share its output market with the BPCL Kochi refinery. Due to its increasing popularity as a gasoline substitute, butanol usage inside the country is expected to rise sharply. Between fiscal years 2008–2009 and 2016–2017, domestic demand for n-Butanol increased by a healthy double digit, and this trend is expected to continue. Because of the low margin of imports and the way that international markets use India as an export hub for n-butanol, the rate of utilization of the capacity of n-butanol is fragile. The expanding butyl acrylate manufacturing sector, robust Chinese demand, and significant expenditures in Saudi Arabia's oil and gas downstream sector are the main factors propelling the n-butanol market. The market for n-butanol is expected to increase significantly due to the rising use of the product in butyl acetate, butyl acrylate, plasticizing agents, and glycol ethers due to its popularity in engineering plastics, automotive coatings, marine coatings, adhesives, and super absorbent polymers (Indian Petro Chem, 2024).

13.4.6 Biodiesel Sector

Without requiring significant changes, biodiesel has the potential to be a workable replacement for fossil fuels in a number of industries, including internal combustion engines and transportation. The utilization of alternative energy sources, such as biodiesel, is imperative in order to adequately address the increasing demand primarily noted in the transportation industry (Damian et al., 2024). In 2022, 185 million gallons of biodiesel fuel were produced in India. In 2023, this is expected to rise to 200 million liters. Vegetable oil, fats from animals, and wasted oils can all be used to make biodiesel (Figure 13.4) (https://www.statista.com/statistics/1051902/india-biodiesel-production-volume/). In 2022, India's nameplate capacity (the total quantity of biodiesel that can be generated over the course of a year under standard operating conditions is known as its nameplate capacity) for producing biodiesel was 577 million liters. An estimate of 600 million liters was made for this in 2023 (https://www.statista.com/statistics/1052762/india-nameplate-capcity-biodiesel-production/). Indian Oil Corporation and Indian Railways have planned to cultivate crops over a one million square kilometer region in order to promote the use of biodiesel in India. In the past, Madhya Pradesh and Jharkhand have also partnered with Indian Oil to cultivate Jatropha over sizable areas in order to promote biodiesel (Dewangan et al., 2018). Numerous public and commercial organizations are involved in the development and marketing of biofuels in India. The infrastructure for biodiesel harvesting, extraction, processing, production, storage, and distribution is supported by a number of India's top financial institutions, including the Small Industries Development Bank of India (SIDBI), National Bank for Agriculture and Rural Development (NABARD), and Indian Renewable Energy Development Agency (IREDA) (Reddy et al., 2008). India's 20% blending mandate will be difficult to implement in a mere five years, but even at 11% blending, the country would rank third in the world's ethanol market, behind the US and Brazil (IEA, Renewables, 2021).

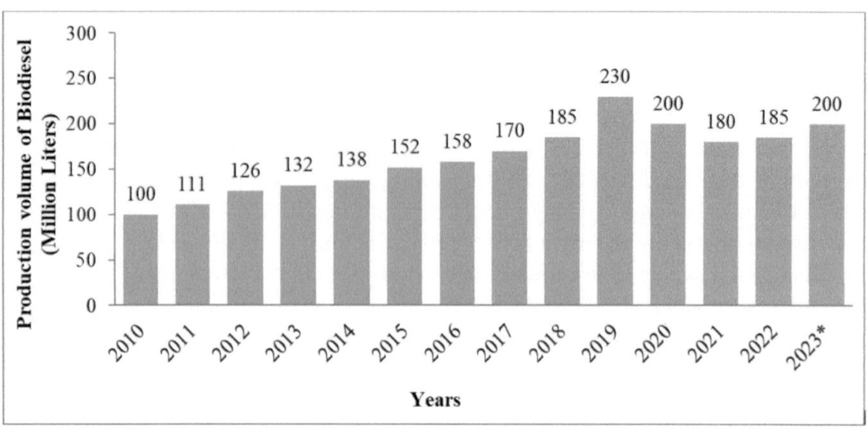

FIGURE 13.4 Production yield of biodiesel in India (2010 to 2022), with a forecast for 2023 (*in million liters*) (https://www.statista.com/statistics/1051902/india-biodiesel-production-volume/).

13.5 BIOFUEL SECTOR OF OTHER COUNTRIES

In 2022, the market for biofuels was estimated to be worth approximately 117 billion US dollars. Notwithstanding the decline in 2021, statistics are expected to rise steadily until 2030, when the market for biofuels is anticipated to reach 201.2 billion US dollars. The biofuel market is anticipated to grow at the fastest rate across the Middle East and Africa (Figure 13.5) (Biofuels, 2025). Many Latin American nations have looked for biofuel substitutes for gasoline in recent years. Brazil has established itself as the benchmark in the area and globally in the production of renewable energy by integrating biofuels into the transportation sector and enacting a biofuels strategy. The main sources of energy in Argentina have historically been fossil fuels like oil and gas. But as fossil fuel supplies have steadily declined recently, this nation has opted to turn to alternative energy sources, such as renewable energy (RE) (Vega et al., 2024). According to IRENA (2023), hydropower accounted for the largest percentage of RE in 2022, followed by wind and solar. The proportion of RE for the year 2023 was 15076 MW. Because it is made from oil from soybean farming, biodiesel produced in this nation is first generation (Saharrea, 2017). Argentina generated 2,147,270 tons of biodiesel in 2019; 1,157,363 tons in 2020; 1,723,668 tons in 2021; and 1,909,618 tons in 2022 (IRENA, 2023). This biodiesel was used domestically and exported, mostly to the Netherlands. There are presently seventeen bioethanol producers in Argentina, six of which use corn and eleven of which use sugarcane (Ciani, 2020). Argentina produced 1,072,000 m^3 of bioethanol in 2019; of that amount, 553,000 m^3 came from corn and 519,000 m^3 from sugarcane; in 2020, the production was 808,725 m^3; in 2021, it was 1,008,057 m^3; and in 2022, it was 1,154,856 m^3 (IRENA, 2023). As a result, at least 12% of the bioethanol is blended with gasoline for use in automobiles (Boletín Oficial de la República Argentina, 2016).

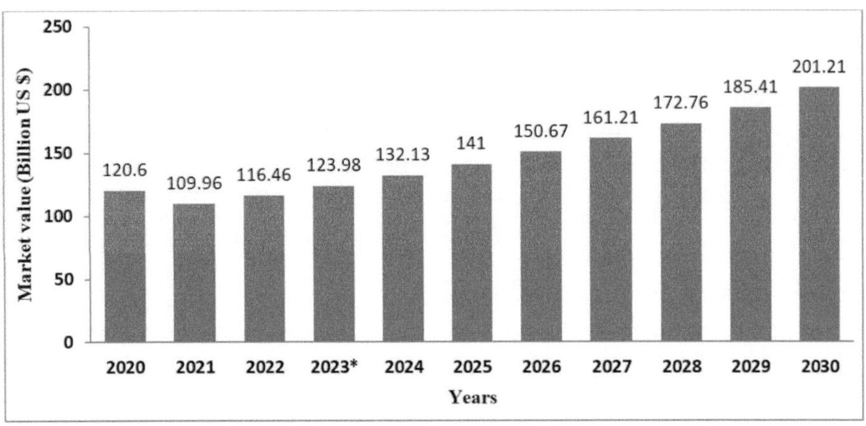

FIGURE 13.5 Market size of biofuel sector worldwide (2020–2022) with a forecast (2023–2030) (*in billion U.S. $*) (Biofuels, 2025).

13.5.1 Renewable Energy Sector of Indonesia

The government of Indonesia has joined the Clean Energy Demand Initiative (CEDI), according to the Ministry of Energy and Mineral Resources in Indonesia. By doing this, the global community can enhance the green economy and undertake mitigation strategies for climate change. Consequently, in order to meet the targets of 2030 and 2060 for the nationally determined contribution (NDC) and net-zero emissions (NZE), respectively, the president's directions regarding the CEDI are required to expedite the essential actions. In addition, its goal is to achieve a 23% share of renewable sources of energy in the primary energy mix by 2025, accompanied by a 41% reduction in emissions (Santika et al., 2020). As per the PRRI (2021), Indonesia accomplished a 25.93% decrease in its greenhouse gas emissions in 2020. Nevertheless, this decrease was lowered to 23.55% in 2021. Therefore, it is essential to successfully execute low-carbon development strategies in the upcoming years by increasing government initiatives and funding (Tiawon & Miar, 2023). Reforestation, stopping deforestation, increasing the capacity of renewable energy sources, and improving energy efficiency are some possible approaches to address environmental issues (Raihan, 2023). With 34 provinces, Indonesia is a group of islands with abundant energy potential. To accomplish Indonesia's goal of being emission-free by 2050, this must also facilitate the country's shift to using only renewable energy sources (Presiden Republik Indonesia, 2017). Given the current geographic circumstances, renewable energy production is feasible; government-developed power plants have been erected in a number of locations (Pambudi et al., 2023). Regarding the Nationally Determined Contribution (NDC), every nation on the planet, including Indonesia, pledged to keep the rise in global temperature within the 1.5°C to 2°C ranges during the first period. With individual efforts, this aims to cut emissions by 29%, and with international collaboration, it aims to minimize emissions by 41%. The government bill ratifying the United Nations Framework Convention on Climate Change (UNFCC) in Paris has enhanced this commitment. In order to do this, Indonesia established a target of not less than 23% and 31%, respectively, for renewable energy in the country's energy mix by 2025 and 2050. Furthermore, it possesses all the potential for renewable energy sources, including geothermal, hydro, sun, wind, and biomass or bioenergy (Pusat Studi Lingkungan Hidup, 2022). As Table 13.4 demonstrates, despite Indonesia's 419 GW of renewable energy potential, very little of it is actually used. Moreover, problems getting low-interest financing and a lack of homegrown industrial support are two other factors that tend to impede the development of renewable energy (Pambudi et al., 2023).

13.5.2 Biofuel Segment and Economic Sustainability in USA

The primary feedstocks used in the production of ethanol and biodiesel in the US are maize and soybean oil. With approximately 15.36 billion gallons produced in 2022 alone, the US is the worldwide leading producer of ethanol. 85% of the world's ethanol was produced in the United States and Brazil combined, totaling 27.05 billion gallons (Renewable Fuels Association, 2024). It is significant to remember that, since 2017, the biofuel sector has evolved. Significant development has been observed in

TABLE 13.4
Energy Potential in Indonesia

S. No.	Energy Type	Potential (GW)
1.	Solar	207.8
2.	Hydro energy	75
3.	Wind	60.6
4.	Bioenergy	32.6
5.	Geothermal	23.7
6.	Micro-hydro	19.3
	Total	**419**

Source: Sekretariat Jenderal Dewan Energi Nasional (2021).

the previous five years for renewable diesel (a biofuel that is chemically identical to conventional diesel); since 2017, production volumes have increased by more than four times. In 2022, the production of renewable diesel was almost equal to that of biodiesel. But throughout the last ten years, maize ethanol output has stayed mostly stable, and it is still by far the most common type of biofuel (U.S. Energy Information Administration, 2023). The findings obtained from the study carried out by Smith et al. (2023) indicated that, when indirect land use change emissions are taken into account, these two biofuels (ethanol and diesel) produced 99.5 million Mg CO2-eq in total greenhouse gas emissions volume in 2017 (94.8 million Mg from maize ethanol and 4.7 million Mg from soybean biodiesel). In comparison to gasoline derived from petroleum, this volume translates to a cumulative carbon footprint of 65.3 gCO2-eq/MJ, a decrease in greenhouse gas emissions (conventional gasoline has a carbon intensity of 93.0 gCO2-eq/MJ). But the cost of producing these biofuels is higher ($6.2 billion) than that of replacing gas and diesel. Furthermore, in comparison to petroleum-based fuels, these biofuels have negative effects on other environmental categories, especially those related to soil, water, and nitrogen intensity. By 2030, the airline industry in the US is expected to use 11 billion liters of sustainable aviation fuel (SAF), or 15% of the current jet fuel demand, according to the Sustainable Aviation Challenge. SAFs are jet fuels produced from non-fossil fuel source materials, such as trash and agricultural residues, and approved for use in business jet aircraft. They also meet GHG and other criteria for sustainability (IEA, Renewables, 2021).

13.5.3 Current Status and Future Prospects of Biofuel Sector in North America

According to the Energy Information Administration (EIA), North America consumed 38,240 PJ of oil products, 18,739 PJ of natural gas, 16,703 PJ of electricity, 4464 PJ of biofuels and waste, 857 PJ of coal, 356 PJ of crude oil, 277 PJ of heat, and 143 PJ of renewable energy in 2019 (Wang et al., 2020). According to estimates, coal may be used as a source of energy for 150 years at the current rate of use. Gas and oil supplies are expected to last for an additional 52 years, while uranium will run out in a few decades (Azarpour et al., 2022). These non-RE sources have a significantly shorter lifespan due to the fast rate of globalization. By 2040, the amount of energy consumed worldwide is expected to rise by 28%, according to the US Energy Information Administration (US EIA) (Evans et al., 2009). IRENA published information on the amount of electricity generated in North America in 2019 from renewable energy resources, including 109 TWh solar, 348 TWh wind, 425 TWh biofuel, 693 TWh hydropower, and 24 TWh geothermal. 105.8 GW of solar, 154.7 GW of wind, 17.05 GW of biofuel, 175.3 GW of hydropower, and 4.8 GW of geothermal energy were among the installed capacities of various renewable energy sources in North America in 2021 (Azarpour et al., 2022).

13.5.4 The Energy Transition in Colombia

Colombia, as a signatory to the 2015 Paris Agreement and afterwards the Conferences of the Parties (COP), committed itself to decarbonization and, consequently, to an energy transition by the middle of this century at the latest (Escobar, 2022). Globally, the energy sector (buildings, transportation, and energy) accounted for 79% of GHGs in 2019, while agriculture, forestry, and land use change (AFOLU) accounted for 22% (IPCC, 2023, p. 4). This report's analyzed countries deviate significantly from this distribution (Thema & Roa García, 2023). Columbia is highly dependent on fossil fuels, wind power, and thermal power; there is an urgent need to switch to biofuel (ethanol, hydrogen, butanol, etc.) production to replace the conventional fuels. These days, novel methods are being used to produce energy, one of which is energy derived from biomass. Colombia presently contributes 0.8% of the global biomass energy generated in this field. Similarly, the government has actively supported the growth and productivity of biofuels; in 2001, laws pertaining to biofuels were approved (IRENA, 2023). Regulations for the utilization of liquid fuels were created by this statute, which also significantly advanced the manufacture of biodiesel and bioethanol (Ramirez & Rodriguez, 2019). African oil palm is the most commonly utilized oil crop in Colombia for the manufacturing of biodiesel, with a potential planting area of 150,000 ha. According to research conducted internationally, Colombia produces some of the greatest biodiesel in the world. Transesterification is the technology used most frequently in Colombia to produce biodiesel since it is a quick process with high yields. In this procedure, a catalyst, sodium or potassium hydroxide, primarily is used to react vegetable oil with an alcohol.

In Colombia, twelve biodiesel factories are situated in the country's northeast and northwest. Plants producing Biodiesel (200,000 tons annually) and Ecodiesel Colombia (115,000 tons annually) have the largest capacity (Ramirez & Rodriguez, 2019). In Colombia, the sources having the most potential for producing bioethanol include beans, rice, sugarcane, and coffee pulp. Colombia produces 0.4% of the 26.3 million liters of bioethanol globally (Asociación Colombiana del Petróleo, 2019). According to Lombana et al. (2015), there are currently seven bioethanol production facilities, the biggest of which are Bioenergy (500,000 L/day) and Ingenio Riopaila Castilla (400,000 L/day). In Colombia, the installed capacity for biogas generation

TABLE 13.5
Biogas Projects of Columbia with Their Biogas Production Capacity and Technology Used

S. No.	Project Name	Substrate Used	Technology Applied	Biogas/Energy Production Capacity	References
1.	Huevos Kikes	Chicken and water	Anaerobic digestion in a biodigester	4600 m^3 of biogas	LEDS LAC and LEDS Global Partnership, 2021
2.	WWTP Cali	Sludge from the city wastewater treatment plant	By turbines	0.8 MW of electricity	LEDS LAC and LEDS Global Partnership, 2021
3.	Bavaria WWTP	Wastewater from beer production	UASB Reactor	-	Vel´asquez & Rinc´on, 2018
4.	Aliar S.A Fazenda	Pig manure	Anaerobic digestion using four continuous flow biodigesters	136,528 m^3/day of biogas	Vel´asquez & Rinc´on, 2018
5.	Manuelita	Sludge generated by the oil palm production process	Anaerobic digestion using two continuous flow biodigesters	317,732 m^3 of biogas per month	Vel´asquez & Rinc´on, 2018

was 9 MW in 2022. Manure and municipal solid waste are the primary sources of its production (IRENA, 2023) using anaerobic digestion and gasification technologies. If all of Colombia's biomass, which is produced from agricultural and agro-industrial residues was put to this use, it is projected that 204.8–235.3 PJ of energy might be produced (Rodríguez et al., 2023). Since 2017, five initiatives for the production of biogas have been launched (Table 13.5). Given the scope of the examined nations, Columbia has great potential for implementing environmentally friendly alternatives centered on the generation of energy from biomass because the sectors dealing with agriculture, livestock, and/or urbanization allow for the utilization of the raw materials they produce to bet on environmental change (Vega et al., 2024).

13.5.5 Energy Segment and Economic Sustainability in Nigeria

Nigeria's biomass resources include wood, livestock manure, fodder grasses, bushes, etc. (Adeoti et al., 2014; Mohammed et al., 2013). Animal excrement is another resource for bioenergy in Nigeria. Nigeria may theoretically produce over 6.8 million m^3 of biogas per day through the process of accumulated animal waste alone, since the country is anticipated to generate about 227,500 t of newly collected animal waste daily, with each kg of fresh animal waste producing roughly 0.03 m^3 of biogas. More than 80% of Nigerians rely on fuel wood for heating and cooking purposes, yet the country's present energy strategy devalues fuel wood consumption. Nigeria's bioenergy resources hold great promise for producing bioelectric power and for reducing the nation's current electricity issue (Mohammed et al., 2014). In Nigeria, the concept of bioenergy might not be well received, particularly when it comes to food crops. The creation of biogas trash, which eliminates the need for irrigation and land use and has the potential to clean up the environment, is the idea behind biofuel that would immediately benefit Nigeria (Ishola et al., 2013). Nigeria's reliance on firewood for energy would decline as a result of the creation of biogas from trash. The Nigerian National Petroleum Corporation (NNPC) drafted the country's biofuels strategy and incentives in 2007 with the goal of combining agricultural practices with the production and exploration of oil and gas. The policy aims to address the main government initiatives concerning the nation's production of biodiesel and ethanol. Galadima et al. (2011) provide a thorough explanation of the goal, expected benefits, and investment motivations. Nigeria's biodiesel industry has a lot of room to grow even if it remains in its infancy. Nigeria's capacity to produce biodiesel is currently limited, but interest and investment in the industry have recently surged. The main crops used in Nigeria for biodiesel production are jatropha, palm oil, and other locally grown crops. Due to its wide availability and affordable price, palm oil is the most widely used feedstock. Jatropha has garnered attention as a promising biodiesel feedstock due to its significant oil content as well as its capacity to thrive in challenging environments (Adewuyi, 2020). One of the difficulties facing Nigeria's biodiesel industry is feedstock quality. Both the quantity and the quality of biodiesel have been discovered to be influenced by the feedstock's composition, free fatty acid content, and water concentration. The extent to which this influence is felt, however, depends on the production techniques used (Monirul et al., 2015). One more obstacle impeding the growth of Nigeria's biodiesel business is the absence of sufficient personnel and experience in the field. This may have an adverse effect on feedstock, processing, and output quality as well as impede the industry's expansion (Igwebuike, 2023). It is crucial to make investments in the development of human capital, to train and educate employees, and to promote the growth of local experts in the industry in order to conquer these obstacles. In addition, cooperation with foreign institutions and organizations could aid in bringing in the technical know-how and experience required to assist the sector in overcoming these obstacles and achieving long-term, sustainable growth (Hassan & Ayodeji, 2019).

13.5.6 Energy Sector of Gulf Countries

The economies of the Gulf nations Saudi Arabia, Bahrain, the United Arab Emirates, Kuwait, Oman, and Qatar are booming. Oil and gas prices are kept high by the discipline of the cartel of oil-producing nations and global geopolitical tensions, which also support the external and fiscal accounts of those nations that still rely heavily on oil earnings (Economic Research, 2023). To fulfill the demands of its growing energy demand over the next 20 years, the region will need to make significant energy investments (Bhutto et al., 2014). Therefore, the Gulf Cooperation Council (GCC) countries' true opportunity is in creating and implementing innovative technologies as well as establishing a robust management framework for carbon sequestration, energy efficiency, and alternative energy mix. By enabling the diversification of the energy mix and lowering reliance on finite fossil fuel supplies, investments in renewable energy would enhance their long-term energy security and open up prospects for capital investment in the area (Alharbi & Csala, 2020). Presently, the GCC countries' common future concerns for bottom-line value generation (both ecological and economic) include security of energy and renewable energy development. One problem with sustainability projects, according to Ramady (2012), is that banks and other private investors are especially hesitant to fund projects in arid and desert regions. The GCC is among the most alluring places globally for the establishment of large-scale renewable energy initiatives since the area has an abundance of solar and wind energy resources found in Saudi Arabia, Oman, Kuwait, and the coastal regions of the Red and Arab seas, together with a long history of leadership in the world energy market. By means of overhead lines and submarine wires, the six GCC nations are already electrically connected to a large, contemporary regional energy interconnection grid that is currently significantly overcapacity. This grid may facilitate the implementation of significant amounts of renewable energy in the region and aid in balancing the regional differences in energy supply and demand. The robust socioeconomic circumstances in the GCC countries lend assistance to the availability of money and the flexibility of competitive financing resources, enabling them to advance and succeed in their environmentally friendly energy projects (Alharbi & Csala, 2020).

13.6 CONCLUSION

The chapter concludes that the major cause of global warming is GHG emission due to the use of fossil fuels for energy production and transport. Although it is now impossible to avoid using energy produced from nuclear and fossil fuels, eliminating GHG emissions by the middle of the twenty-first century a goal established at the Conference of the Parties (COP21) in Paris clearly points the way toward sustainability. Almost all the countries are suffering from the harmful effects of GHGs, and the production of low-carbon-containing fuel (biofuel) is the best option to reduce these effects. Production of second-generation biofuels is a doubly beneficial process, as it reduces waste and produces energy. But switching totally to biofuels for energy requirements is not reliable; instead, blending second-generation biofuels with fossil fuels is most trustworthy. Many countries, including India, are working in this direction for the betterment of the environment. Afro-Asian nations must accelerate the establishment of national and international policies and partnerships in order to embrace hydrogen as an environmentally friendly fuel. More fast efforts are needed to retain the environmental conditions by reducing conventional fuel uses.

REFERENCES

Abbas, R. N., A. Iqbal, M. A. Iqbal, O. M. Ali, R. Ahmed, R. Ijaz, et al. 2021. Weed-free durations and fertilization regimes boost nutrient uptake and paddy yield of direct-seeded fine rice (*Oryza sativa L.*). *Agronomy* 11:2448.

Adeoti, O., T. A. Ayelegun, S. O. Osho. 2014. Nigeria biogas potential from livestock manure and its estimated climate value. *Renew Sustain Energy Rev* 37:243–8.

Adewuyi, A. 2020. Challenges and prospects of renewable energy in Nigeria: A case of bioethanol and biodiesel production. *Energy Rep* 6:77–88. https://doi.org/10.1016/j.egyr.2019.12.002

Ahmad, M., E. A. Waraich, M. B. Hafeez, U. Zulfiqar, Z. Ahmad, M. A. Iqbal, et al. 2023. Changing climate scenario: Perspectives of Camelina sativa as low-input biofuel and oilseed crop. In *Global Agricultural Production: Resilience to Climate Change* (pp. 197–236). Ed. Ahmed M. Springer, Cham. https://doi.org/10.1007/978-3-031-14973-3_7

Ahmed, A., W. Li, S. Varjani, S. You. 2022. Chapter 23 – Waste-to-energy technologies for sustainability: life- cycle assessment and economic analysis. In S. Varjani, A. Pandey, T. Bhaskar, S. V. Mohan, D. C. W. Tsang (eds.), *Biochemicals, Biofuels, and Bioplastics from Biomass* (pp. 599–612). Elsevier.

Ajanovic, A. 2011. Biofuels versus food production: Does biofuels production increase food prices? *Energy* 36(4):2070–6.

Alawad, I. and H. Ibrahim. 2024. Pretreatment of agricultural lignocellulosic biomass for fermentable sugar: Opportunities, challenges, and future trends. *Biomass Convers Biorefin* 14(5):6155–83.

Alharbi, F. R. and D. Csala. 2020. GCC countries' renewable energy penetration and the progress of their energy sector projects. *IEEE Power Energy Soc Sect* 8:211986–2002.

Arias, D. E., C. Veluchamy, M. B. Habash, B. H. Gilroyed. 2021. Biogas production, waste stabilization efficiency, and hygienization potential of a mesophilic anaerobic plug flow reactor processing swine manure and corn stover. *J Environ Manage* 284:112027. https://doi.org/10.1016/j.jenvman.2021.112027

Asociaci´on Colombiana del Petr´oleo. 2019. Importaciones de etanol: respaldo a la producci´on nacional y beneficio para los consumidores. https://acp.com.co/portal/importaciones-de-etanol-respaldo-a-la-produccion-nacional-y-beneficio-para-los-consumidores/. (Accessed 19 March, 2024)

Azarpour, A., O. Mohammadzadeh, N. Rezaei, S. Zendehboudi. 2022. Current status and future prospects of renewable and sustainable energy in North America: Progress and challenges. *Energy Convers Manag* 269:115945.

Banerjee, N. (2023). Biomass to energy—an analysis of current technologies, prospects, and challenges. *BioEnergy Res* 16(2):683–716.

Bhatia, S. K., S. S. Jagtap, A. A. Bedekar, et al. 2020. Recent developments in pretreatment technologies on lignocellulosic biomass: Effect of key parameters, technological improvements, and challenges. *Biores Technol* 300:1227.

Bhutto, A. W., A. A. Bazmi, G. Zahedi, J. J. Kleme. 2014. A review of progress in renewable energy implementation in the gulf cooperation council countries. *J Cleaner Prod* 71:168–80. DOI: 10.1016/j.jclepro.2013.12.073.

Boletín Oficial de la República Argentina, Ministerio de Energía y Minería. 2016. https://www.boletinoficial.gob.ar/pdf/linkQR/cVUvOEdIcEZTdGMrdTVReEh2ZkU0dz09%0AQue. (Accessed 19 March, 2024)

Branco-Vieira, M., D. Costa, T. M. Mata, A. A. Martins, M. A. V. Freitas, N. S. Caetano. 2020. A life cycle inventory of microalgae-based biofuels production in an industrial plant concept. *Energy Rep* 6:397–402.

Broda, M., D. J. Yelle, K. Serwańska. 2022. Bioethanol production from lignocellulosic biomass—challenges and solutions. *Molecules* 27(24):8717.

Chatellard, L., A. Marone, H. Carrère, E. Trably. 2017. Trends and challenges in biohydrogen production from agricultural waste. In *Biohydrogen Production: Sustainability of Current Technology and Future Perspective* (pp. 69–95). ed. Singh A. Rathore D., Springer New Delhi. https://doi.org/10.1007/978-81-322-3577-4_4

Chaudhary, G., N. Chaudhary, S. Saini, Y. Gupta, V. Vivekanand, A. Panghal. 2024. Assessment of pretreatment strategies for valorization of lignocellulosic biomass: Path forwarding towards lignocellulosic biorefinery. *Waste Biomass Valor* 15(1):1–36.

Cheah, W. Y., R. Sankaran, P. L. Show, T. Ibrahim, T. N. Baizura, K. W. Chew, et al. 2020. Pretreatment methods for lignocellulosic biofuels production: Current advances, challenges and future prospects. *Biofuel Res J* 7(1):1115–27.

Cheng, C. L., Y. C. Lo, K. S. Lee, D. J. Lee, C. Y. Lin, J. S. Chang. 2011. Biohydrogen production from lignocellulosic feedstock. *Bioresour Technol* 102(18):8514–23.

Chia, S. R., S. B. H. M. Nomanbhay, K. W. Chew, H. S. H. Munawaroh, P. L. Show. 2022. Algae as potential feedstock for various bioenergy production. *Chemosphere* 287:131944.

Ciani, M. 2020. Informe Biocombustibles Enero 2020 Direcci´on de Bioenergía, 2020. *CIBIOG´AS* 5(3):248–53.

D'Silva, T. C., S. A. Khan, S. Kumar, D. Kumar, A. Isha, S. Deb, et al. 2023. Biohydrogen production through dark fermentation from waste biomass: Current status and future perspectives on biorefinery development. *Fuel* 350:128842.

Damian, C. S., Y. Devarajan, R. Jayabal. 2024. Biodiesel production in India: Prospects, challenges, and sustainable directions. *Biotechnol Bioeng* 121(3):894–902. https://doi.org/10.1002/bit.28643

Dewangan, A., A. K. Yadav, A. Mallick. 2018. Current scenario of biodiesel development in India: prospects and challenges. *Energy Sources A: Recover Util Environ Eff.* DOI: 10.1080/15567036.2018.1502849.

Dhanya, M. S. (2022). Production of biomass-based butanol: Strategies and challenges. In *Sustainable Microbial Technologies for Valorization of Agro-Industrial Wastes* (pp. 169–198). CRC Press.

Economic Research. 2023. Eco Conjunction: Gulf countries and the challenge of the energy transition. Https://economic-research.bnpparibas.com/html/en-us/gulf-countries-challenge-energy-transition-3/9/2023,48341. (Accessed 5 April, 2024)

Efroymson, R. A., H. I. Jager, S. Mandal, E. S. Parish, T. J. Mathews. 2021. Better management practices for environmentally sustainable production of microalgae and algal biofuels. *J Clean Prod* 289:125150.

Elbeshbishy, E., B. R. Dhar, G. Nakhla, H. S. Lee. 2017. A critical review on inhibition of dark biohydrogen fermentation. *Renew Sustain. Energy Rev* 79:656–68.

Escobar, A. 2022. Reinterpretando las civilizaciones: De la crítica a las transiciones. *ARQ (Santiago)* 111:24–41. https://doi.org/10.4067/S0717-69962022000200024

Evans A., V. Strezov, T. J. Evans. 2009. Assessment of sustainability indicators for renewable energy technologies. *Renew Sustain Energy Rev* 13(5):1082–8.

FAO. 2019. *The State of Food and Agriculture 2019. Moving Forward on Food Loss and Waste Reduction*. FAO.

Farooq, M. S., M. Uzair, A. Raza, M. Habib, Y. Xu, M. Yousuf, et al. 2022. Uncovering the research gaps to alleviate the negative impacts of climate change on food security: A review. *Front Plant Sci* 13:927535.

FSSAI, 2023. The Food Safety Standards Authority of India (FSSAI): guiding the optimisation of surplus food donation. https://nutritionconnect.org/food-safety-standards-authority-india-fssai-guiding-optimisation-surplus-food-donation (Accessed 18 October, 2025)

Galadima, A., Z. N. Garba, B. M. Ibrahim, M. N. Almustapha, L. Leke, I. K. Adam. 2011. Biofuels production in Nigeria: The policy and public opinions. *J Sustain Dev* 4:22–31.

Gautam, P., S. N. Upadhyay, S. K. Dubey. 2020. Bio-methanol as a renewable fuel from waste biomass: Current trends and future perspective. *Fuel* 273:117783.

Gopinathan, M. C., R. Sudhakaran. 2009. Biofuels: Opportunities and challenges in India. *Vitro Cell Dev Biol-Plant* 45:350–71.

Griffiths, G., A. K. Hossain, V. Sharma, G. Duraisamy. 2021. Key targets for improving algal biofuel production. *Clean Technol* 3(4):711–42.

Guo, Y., Y. Liu, M. Guan, H. Tang, Z. Wang, L. Lin. 2022. Production of butanol from lignocellulosic biomass: Recent advances, challenges, and prospects. *RSC Adv* 12(29):18848–63.

Halder, P. and A. K. Azad. 2019. Recent trends and challenges of algal biofuel conversion technologies. *In Advanced Biofuels: Applications, Technologies and Environmental Sustainability*, ed Azad A. K. and M. Rasul. Woodhead Publishing Series in Energy. 167–179. https://doi.org/10.1016/B978-0-08-102791-2.00007-6

Hafid, H. S., F. N. Omar, N. Abdul Rahman, M. Wakisaka. 2022. Innovative conversion of food waste into biofuel in integrated waste management system. *Crit Rev Environ Sci Technol* 52(19):3453–92.

Hassan, A. B. and O. V. Ayodeji. 2019. Benefits and challenges of biodiesel production in West Africa. *Niger J Technol* 38:621. https://doi.org/10.4314/njt.v38i3.12

Hirani, A. H., N. Javed, M. Asif, S. K. Basu, A. Kumar. 2018. A review on first-and second-generation biofuel productions. In *Biofuels: Greenhouse Gas Mitigation and Global Warming: Next Generation Biofuels and Role of Biotechnology* (pp. 141–154). (ed.) Kumar A., Ogita S., Yau, Y. Y., Springer, New Delhi. https://doi.org/10.1007/978-81-322-3763-1_8

HLPE. 2013. Biofuels and Food Security, a Report by the High Level Panel of Experts on Food Security and Nutrition of the Committee on World Food Security, 13–14, Rome, 2013. (Accessed 2 February, 2024)

https://www.statista.com/statistics/1051902/india-biodiesel-production-volume/. (Accessed 16 March, 2024)

https://www.statista.com/statistics/1052762/india-nameplate-capcity-biodiesel-production/. (Accessed 16 March, 2024)

Biofuels: Market Size, 2025. https://www.statista.com/statistics/217179/global-biofuels-market-size/. (Accessed 18 October, 2025)

IEA. 2021. India. Energy Outlook 2021. https://www.iea.org/reports/indiaenergy-outlook-2021. (Accessed 15 March, 2024)

IEA. 2023a. Feedstock Demand in India, 2015 to 2028, IEA, Paris. https://www.iea.org/data-and-statistics/charts/feedstock-demand-in-india-2015-to-2028, IEA. Licence: CC BY 4.0.

IEA. 2023b. World Energy Outlook. www.iea.org/reports/world-energy-outlook-2023

IEA, Renewables. 2021. Analysis and Forecast to 2026, Biofuels. https://iea.blob.core.windows.net/assets/5ae32253-7409-4f9a-a91d-1493ffb9777a/Renewables2021-Analysisandforecastto2026.pdf. (Assessed April, 2024)

Igwebuike, C. M. 2023. Biodiesel: Analysis of production, efficiency, economics and sustainability in Nigeria. *Clean Technol Recycl* 3(2):92–106. DOI: 10.3934/ctr.2023006.

Indian Petro Chem, Report on N-Butanol in India. 2024. https://indianpetrochem.com/report/n-butanolreport. (Accessed 15 April, 2024)

IPCC. 2023. Synthesis Report of the IPCC Sixth Assessment Report (AR6). Summary for Policymakers. https://report.ipcc.ch/ar6syr/pdf/IPCC_AR6_SYR_SPM.pdf. (Accessed 15 March, 2024)

IRENA, Renewable Energy Statistics 2023, International Renewable Energy Agency. 2023. [Online]. https://www.irena.org/Publications/2023/Jul/Renewable-energy-statistics-2023. (Accessed 19 March, 2024)

Ishola, M. M., T. Brandberg, S. A. Sanni, M. J. Taherzadeh. 2013. Biofuels in Nigeria: A critical and strategic evaluation. *Renew Energy* 55:554–60.

Kour, D., K. L. Rana, N. Yadav, A. N. Yadav, A. A. Rastegari, C. Singh, et al. 2019. Technologies for biofuel production: Current development, challenges, and future prospects. In *Prospects of Renewable Bioprocessing in Future Energy Systems* (pp. 1–50). Ed. Rastegari, A., Yadav, A., Gupta, A., vol 10. Springer, Cham. https://doi.org/10.1007/978-3-030-14463-0_1

Kovalenko, N., T. Hutsol, V. Kovalenko, S. Glowacki, S. Kokovikhin, V. Dubik, et al. 2021. Hydrogen production analysis: prospects for Ukraine. *Agric Eng 2021* 25:99–114. https://doi.org/10.2478/agriceng-2021-0008

Das, J., Ravishankar, H., P. N. L. Lens. 2022. Biological biogas purification: Recent developments, challenges and future prospects. *J Environ Manage* 304:114198. https://doi.org/10.1016/j.jenvman.2021.114198

Kumari, D., R. K. Bhatia, R. Singh. 2021. Technologies and challenges to recover energy products from carbohydrate rich effluents: A mini review. *Int J Plant Anim Environ Sci* 11(3):423–42. DOI: 10.26502/ijpaes.202113.

Kumari, D. and Y. Jain. 2020. Enhanced biofuel production from lignocellulosic biomass: an overview of advanced physico-chemical and biological technologies. In *Biorefineries: A Step Towards Renewable and Clean Energy*, ed Verma, P. Springer. 151-172. 978-981-15-9592-9. ISBN: 978-981-15-9593-6 (Online), 978-981-15-9592-9 (Print). https://doi.org/10.1007/978-981-15-9593-6_7

Kumari, D., G. S. Maurya, R. Singh. 2024. Coupled pretreatment of rice straw with dairy wastewater and ultrasonication for biofuel production using mixed microbial source. *Bioresour Technol Reports* 25:101738.

Kumari, D. and R. Singh. 2019. Recent advances in bioremediation of wastewater for sustainable energy products. In *Zero Waste: Management Practices for Sustainability* (pp. 247–76), ed Rathoure, A. K. CRC Press, Taylor & Francis Group. ISBN 978-0-367-18039-3.

Kumari, D. and R. Singh. 2020a. Coupled green pretreatment of petha wastewater and rice straw. *Envirn Sustain Indicators* 5:100013. https://doi.org/10.1016/j.indic.2019.100013

Kumari, D. and R. Singh. 2020b. Ultrasonic assisted petha waste water pretreatment of rice straw for optimum production of methane and ethanol using mixed microbial culture. *Renew Energy* 145:682–90.

Kumari, D., and R. Singh. 2018. Pretreatment of lignocellulosic wastes for biofuel production: A critical review. *Renew Sustain. Energy Rev* 90:877–91.

Kumari, D. and R. Singh. 2022. Rice straw structure changes following green pretreatment with petha wastewater for economically viable bioethanol production. *Sci Rep* 12:10443. https://doi.org/10.1038/s41598-022-14627-7

LEDS LAC and LEDS Global Partnership, Proyecto De la pr´actica a la política: an´alisis de las barreras a la inversi´on en biog´as en Colombia y las medidas para abordarlas, a partir de la experiencia de los desarrolladores y otros actores relevantes, 2021.

Lombana J., J. Vega Jurado, E. Britton Acevedo, S. Herrera Vel´asquez. 2015. An´alisisdel sector biodi´esel en Colombia y su cadena de suministro.

Ludfiani, D. D., F. D. Arianti, A. Prabowo, B. Haryanto, M. Megawati, N. A. Sasongko. 2024. Prediction of bioethanol from production of lignocellulosic biomass waste from agriculture and livestock using regression analysis model. *F1000Research* 13:111.

Maalouf, A. and P. Agamuthu. 2023. Waste management evolution in the last five decades in developing countries – A review. *Waste Manag Res* 41(9):1420–34.

Madhurjya, S., S. Pranjal, D. Rupam, D. Dipankar. 2024. Ethanol blending in petrol: Current status and prospects in India. *Renew Energy Res Appl* 5(1):45–9.

Malode, S. J., K. K. Prabhu, R. J. Mascarenhas, N. P. Shetti, T. M. Aminabhavi. 2021. Recent advances and viability in biofuel production. *Energy Convers Manag: X* 10:100070.

Mandotra, S. K., C. Sharma, N. Srivastava, A. S. Ahluwalia, P. W. Ramteke 2021. Current prospects and future developments in algal bio- hydrogen production: A review. *Biomass Convers Biorefin* 11:1–18.

Mathew, G. M., D. Raina, V. Narisetty, V. Kumar, S. Saran, A. Pugazhendi, et al. 2021. Recent advances in biodiesel production: Challenges and solutions. *Sci Total Environ* 794:148751.

Ministry of New and Renewable Energy (MNRE), Government of India, 2022. Biogas Programme (Phase-I) for FY 2021-22 to 2025-26. https://biogas.mnre.gov.in/about-the-programmes. (Accessed 28 March, 2024)

Mohammed, Y. S., M. W. Mustafa, N. Bashir, A. S. Mokhtar. 2013. Renewable energy resources for distributed power generation in Nigeria: A review of the potential. *Renew Sustain Energy Rev* 22:257–68.

Mohammed, Y. S., M. W. Mustafa, N. Bashir, M. A. Ogundola, U. Umar. 2014. Sustainable potential of bioenergy resources for distributed power generation development in Nigeria. *Renew Sustain Energy Rev* 34:361–70.

Monirul, I. M., H. H. Masjuki, M. A. Kalam, et al. 2015. A comprehensive review on biodiesel cold flow properties and oxidation stability along with their improvement processes. *RSC Adv* 5:86631–55. https://doi.org/10.1039/C5RA09555G

Mood, S. H., A. H. Golfeshan, M. Tabatabaei, G. S. Jouzani, G. H. Najafi, M. Gholami. 2013. Lignocellulosic biomass to bioethanol, a comprehensive review with a focus on pretreatment. *Renew Sustain. Energy Rev* 27:77–93.

Mupondwa, E, X. Li, L. Tabil, et al. 2017. Status of Canada's lignocellulosic ethanol: Part I: Pretreatment technologies. *Renew Sustain Energy Rev* 72:178–90.

Nayak, A. and B. Bhushan. 2019. An overview of the recent trends on the waste valorization techniques for food wastes. *J Environ Manage* 233:352–70.

Neuling, U. and M. Kaltschmitt. 2017. Review of biofuel production-feedstock, processes and markets. *J Oil Palm Res* 29:137–67.

Osman, A. I., T. J. Deka, D. C. Baruah, D. W. Rooney. 2020. Critical challenges in biohydrogen production processes from the organic feedstocks. *Biomass Convers Biorefin* 10:1–19.

Rodríguez -, C. P. P. L. A. Ríos, C. S. Duarte-Gonz´alez, A. Monta˜na, C. Marcía-Marroquín. 2023. Aprovechamiento de la biomasa residual como fuente de energía renovable en Colombia: escenario de gasificaci´on potencial. *Palmas* 44(1): 65–82.

Pambudi, N. A., R. A. Firdaus, R. Rizkiana, D. K. Ulfa, M. S. Salsabila, S. Suharno. 2023. Renewable energy in Indonesia: Current status, potential, and future development. *Sustainability* 15:2342. https://doi.org/10.3390/su15032342

Paschalidou, A., M. Tsatiris, K. Kitikidou. 2016. Energy crops for biofuel production or for food?-SWOT analysis (case study: Greece). *Renew Energy* 93:636–47.

Pragya, N., K. K. Pandey, P. K. Sahoo. 2013. A review on harvesting, oil extraction and biofuels production technologies from microalgae. *Renew Sustain Energy Rev* 24:159–71.

Preethi and M. Gunasekaran. 2023. Indexing energy and cost of the pretreatment for economically efficient bioenergy generation. *Front. Energy Res* 10:1060599.

Presiden Republik Indonesia. Lampiran I Perpres Nomor 22 Tahun 2017. Tentang Rencana Umum Energi Nasional; Peraturan Presiden; Sekretariat Negara Republik Indonesia: Jakarta, Indonesia, 2017.

Press Information Bureau, 2021. India Emerges as the World's Largest Producer and Consumer of Sugar and World's 2nd Largest Exporter of Sugar. https://www.pib.gov.in/PressReleseDetailm.aspx?PRID=1865320. (Accessed 15 March, 2024)

Press Information Bureau, Government of India. 2021. India's Sugar Production 2021–22. https://pib.gov.in/PressReleasePage.aspx?PRID=1777916. (Accessed 15 March, 2024)

Press Information Bureau, Government of India, Ministry of Road Transport & Highways, 2017. Methanol Economy for India: Energy Security, Make in India and Zero Carbon Foot Print. https://pib.gov.in/newsite/printrelease.aspx?relid=174919. (Accessed 19 March, 2024)

PRRI. 2021. Presidential Regulation of the Republic of Indonesia (Number 115 of 2021 Concerning Updating the 2022 Government Work Plan), State Secretariat of the Republic of Indonesia. Jakarta, Indonesia.

Pusat Studi Lingkungan Hidup (PSLH). 2022. Geliat Pemanfaatan Energi Terbarukan. https://pslh.ugm.ac.id/geliat-pemanfaatan-energi-terbarukan/. (Accessed 26 March, 2024)

Raihan, A. 2023. An overview of the energy segment of Indonesia: Present situation, prospects, and forthcoming advancements in renewable energy technology. *J Technol Innov Energy* 2(3):37–63. https://doi.org/10.56556/jtie.v2i3.599

Rajendran, K., S. Rajoli, M. J. Taherzadeh. 2016. Techno-economic analysis of integrating first and second-generation ethanol production using filamentous fungi: An industrial case study. *Energies* 9(5):359.

Ramady, M. A. 2012. *The GCC Economies: Stepping up to Future Challenges*. Springer.

Ramirez, J. and S. Rodriguez. 2019. Conveniencia de la producci´on de biodiesel en Colombia desde la perspectiva del desarrollo sostenible, 2019. https://repository.ucatolica.edu.co/server/api/core/bitstreams/3f1cd4d1-721c-4da2-bf0d-9b957a7491e4/content

Rathour, R. K., M. Devi, P. Dahiya, N. Sharma, N. Kaushik, D. Kumari, et al. 2023. Recent trends, opportunities and challenges in sustainable management of rice straw waste biomass for green biorefinery. *Energies* 16(3):1429. https://doi.org/10.3390/en16031429

Razzak, S. A., M. M. Hossain, R. A. Lucky, A. S. Bassi, H. De Lasa. 2013. Integrated CO2 capture, wastewater treatment and biofuel production by microalgae culturing—a review. *Renew Sustain. Energy Rev* 27:622–53.

Reddy, B. V. S., S. Ramesh, A. A. Kumar, S. P. Wani, R. Ortiz, H. Ceballos, et al. 2008. Biofuel crops research for energy security and rural development in developing countries. *Bioenerg Researcher* 1:248–58. DOI:10.1007/s12155-008-9022-x.

Renewable Fuels Association. 2024. Global Ethanol Production. (Accessed 15 April, 2024)

Rodionova, M. V., R. S. Poudyal, I. Tiwari, R. A. Voloshin, S. K. Zharmukhamedov, H. G. Nam, et al. 2017. Biofuel production: Challenges and opportunities. *Int J Hydrog Energy* 42(12):8450–61.

Saad, M. G., N. S. Dosoky, M. S. Zoromba, H. M. Shafik. 2019. Algal biofuels: Current status and key challenges. *Energies* 12(10):1920.

Saharrea, J. 2017. *El sector biodiesel en la República Argentina y la cuesti´on del desarrollo sustentable*. Universidad Torcuato Di Tella.

Sahota, S., S. Kumar, L. Lombardi. 2024. Biohythane, biogas, and biohydrogen production from food waste: Recent advancements, technical bottlenecks, and prospects. *Energies* 17(3):666.

Sahu, R. B. and P. Singh. 2023. Current status and challenges of biobutanol production from biomass. In *Production of Biobutanol from Biomass* (pp. 301–22).ed. A. Kuila and M. Mukhopadhyay, Wiley.

Samrot, A. V., D. Rajalakshmi, M. Sathiyasree, S. Saigeetha, K. Kasipandian, N. Valli. 2023. A review on biohydrogen sources, production routes, and its application as a fuel cell. *Sustainability* 15(16):12641.

Santika, W. G., T. Urmee, Y. Simsek, P. A. Bahri, M. Anisuzzaman. 2020. An assessment of energy policy impacts on achieving sustainable development goal 7 in Indonesia. *Energy Sustain Dev* 59:33–48.

Sekretariat Jenderal Dewan Energi Nasional. 2021. *Laporan Hasil Analisis Neraca Energi Nasional 2021*. Dewan Energi Nasional.

Semwal, P., B. Majhi, D. D. Pandey, S. Misra, S. K. Mishra, P. S. Chauhan. 2023. Agriwaste burning management through microbial intervention. In *Recent Trends in Solid Waste Management* (pp. 309–31). Elsevier.

Shweta, S. C. Capareda, B. R. Kamboj, K. Malik, K. Singh, D. K. Bhisnoi, S. Arya. 2024. Biomass resources and biofuel technologies: A focus on Indian development. *Energies* 17:382. https://doi.org/10.3390/en17020382

Signoretto, M., S. Taghavi, E. Ghedini, F. Menegazzo. 2019. Catalytic production of levulinic acid (LA) from actual biomass. *Molecules* 24(15):2760.

Singh, D., M. Tembhare, S. Kumar. 2023. Compressed biogas plants in India: Existing status, technological advances and challenges. *Biomass Bioenergy* 178:106964. https://doi.org/10.1016/j.biombioe.2023.106964

Singh, M., D. Kumari, P. Chahar, Y. Singhal, R. Singh. 2017. Effect of different microbial culture on biohydrogen production from *Benincasa hispida* waste. In *Humanitarian Technology Conference (R10-HTC), 21st-23rd December, 2016 IEEE Region 10, DEI, Dayalbagh, Agra.* Institute of Electrical and Electronics Engineers, IEEE Xplore.

Singh, D., Y. Aryan, D. Chavan, M. Tembhare, A. K. Dikshit, S. Kumar. 2022. Mask consumption and biomedical waste generation rate during Covid-19 pandemic: A case study of central India. *Environ Res* 212(C):113363. https://doi.org/10.1016/j.envres.2022.113363

Singh, T., A. Alhazmi, A. Mohammad, N. Srivastava, S. Haque, S. Sharma, V. K. Gupta. 2021. Integrated biohydrogen production via lignocellulosic waste: Opportunity, challenges & future prospects. *Bioresour Technol* 338:125511.

Singhal, Y. and R. Singh. 2014. Effect of microwave pretreatment of mixed culture on biohydrogen production from waste of sweet produced from Benincasa hispida. *Int J hydrog energy* 39(14):7534–40.

Smith, J. P., B. J. Limb, C. M. Beal, K. R. Banta, J. L. Field, S. J. Simske, et al. 2023. Evaluating the sustainability of the 2017 US biofuel industry with an integrated techno-economic analysis and life cycle assessment. *J Clean Prod* 413:137364. https://doi.org/10.1016/j.jclepro.2023.137364

Soares, J. F., T. C. Confortin, I. Todero, F. D. Mayer, M. A. Mazutti. 2020. Dark fermentative biohydrogen production from lignocellulosic biomass: Technological challenges and future prospects. *Renew Sustain Energy Rev* 117:109484.

Subudhi, S., K. Saha, D. Mudgil, P. K. Sarangi, R. K. Srivastava, M. K. Sarma. 2023. Biomethanol roduction from renewable resources: A sustainable approach. *Environ Sci Pollut Res* 30:1–17.

Suhartini, S., N. A. Rohma, I. Elviliana, Santoso, R. Paul, P. Listiningrum, L. Melville. 2022. Food waste to bioenergy: Current status and role in future circular economies in Indonesia. *Energy Ecol Environ* 7(4):297–339.

Tang, R., Y. Liu, R. Ma, L. Zhang, Y. Li, G. Li, et al. 2023. Effect of moisture content, aeration rate, and C/N on maturity and gaseous emissions during kitchen waste rapid composting. *J Environ. Manage* 326:116662.

Thema, J. and M. C. Roa García. 2023. *The Energy Transition in Colombia. Current Situation, Projections, Challenges, Narratives and Public Policies – In Relation to the Energy Transition in Germany (Wuppertal Paper No. 200).* Wuppertal Institute.

Tiawon, H. and M. Miar. 2023. The role of renewable energy production, energy efficiency and green finance in achieving sustainable economic development: Evidence from Indonesia. *Int J Energy Econ Policy* 13(1):250.

Tse, T. J., D. J., Wiens, M. J. Reaney. 2021. Production of bioethanol – A review of factors affecting ethanol yield. *Fermentation* 7(4):268.

Tusher, T. R., J. J. Chang, M. I. Saunivalu, S. Wakasa, W. H. Li, C. C. Huang, et al. 2022. Second-generation bioethanol production from phytomass after phytoremediation using recombinant bacteria-yeast co-culture. *Fuel* 326:124975.

U.S. Energy Information Administration. 2023. https://www.eia.gov/outlooks/aeo/. (Assessed 1 April, 2024)

Vega L. P., K. T. Bautista, H. Campos, S. Daza, G. Vargas. 2024. Biofuel production in Latin America: A review for Argentina, Brazil, Mexico, Chile, Costa Rica and Colombia. *Energy Rep* 11:28–38.

Vel´asquez, M. E. and J. M. Rinc´on. 2018. *Estimaci´on del potencial de conversi´on a biog´as de la biomasa en Colombia y su aprovechamiento* (pp. 1–216). Inf. Unidad De. Planeac. Min. Energ´etica UPME - Univ. Nac De Colomb.

Voca, N. and B. Ribic. 2020. Biofuel production and utilization through smart and sustainable biowaste management. *J Clean Prod* 259:120742.

Voloshin, R. A., V. D. Kreslavski, S. K. Zharmukhamedov, V. S. Bedbenov, S. Ramakrishna, S. I. Allakhverdiev. 2015. Photoelectrochemical cells based on photosynthetic systems: A review. *Biofuel Res J* 2(2):227–35.

Wang S., S. J. Mason, H. Gangammanavar. 2020. Stochastic optimization for flow-shop scheduling with on-site renewable energy generation using a case in the United States. *Comput Ind Eng* 149:106812.

Watts, N., M. Amann, N. Arnell, S. Ayeb-Karlsson, K. Belesova, H. Berry, et al. 2018b. The 2018 report of the lancet countdown on health and climate change: Shaping the health of nations for centuries to come. *Lancet* 392(10163):2479–514.

Watts, N., M. Amann, S. Ayeb-Karlsson, K. Belesova, T. Bouley, M. Boykoff, et al. 2018a. The lancet countdown on health and climate change: From 25 years of inaction to a global transformation for public health. *Lancet* 391(10120):581–630.

Index

For Product Safety Concerns and Information please contact our EU representative GPSR@taylorandfrancis.com Taylor & Francis Verlag GmbH, Kaufingerstraße 24, 80331 München, Germany